# The Physiology
## and
# Biochemistry of
# Prokaryotes

# The PHYSIOLOGY
## *and*
# BIOCHEMISTRY *of*
# PROKARYOTES

SECOND EDITION

David White
*Indiana University*

*New York   Oxford*
OXFORD UNIVERSITY PRESS
2000

Oxford University Press

Oxford New York
Athens Auckland Bangkok Bogotá Buenos Aires Calcutta
Cape Town Chennai Dar es Salaam Delhi Florence Hong Kong Istanbul
Karachi Kuala Lumpur Madrid Melbourne Mexico City Mumbai
Nairobi Paris São Paulo Singapore Taipei Tokyo Toronto Warsaw

and associated companies in
Berlin Ibadan

Copyright © 1995, 2000 by Oxford University Press, Inc.

Published by Oxford University Press, Inc.
198 Madison Avenue, New York, New York 10016
http://www.oup-usa.org

Oxford is a registered trademark of Oxford University Press

**Library of Congress Cataloging-in-Publication Data**

White, David, 1936–
    The physiology and biochemistry of prokaryotes / by David White. —
2nd ed.
      p.   cm.
    Includes bibliographical references and index.
    ISBN 0-19-512579-7 (alk. paper)
    1. Prokaryotes—Physiology.   2. Microbial metabolism.   I. Title.
QR88.W48   1999
571.2'93—dc21                                                                98-38420
                                                                                          CIP

Printing (last digit): 9 8 7 6 5 4 3 2 1

Printed in the United States of America
on acid-free paper

# CONTENTS

CONTENTS

# PREFACE

The prokaryotes are a diverse assemblage of organisms that consists of the Bacteria (eubacteria) and the Archaea (archaebacteria). This text covers major aspects of the physiology and biochemistry of these organisms at the advanced undergraduate and beginning graduate levels. Topics include the structure and function of the prokaryotic cell (Chapter 1), growth physiology (Chapter 2), membrane and cytosol bioenergetics (Chapters 3 and 7), electron transport and photosynthetic electron transport (Chapters 4 and 5), metabolic regulation (Chapter 6), intermediary metabolism (Chapters 8, 9, and 12–14), the metabolism of DNA, RNA, and protein (Chapter 10), cell wall, capsule, and polysaccharide biosynthesis (Chapter 11), homeostasis (Chapter 15), solute transport (Chapter 16), protein export (Chapter 17), cell–cell signaling and responses to environmental signals (Chapter 18), and responses to environmental stresses (Chapter 19). A summary at the end of each chapter emphasizes the major points in the chapter and is followed by a set of study questions.

It must be emphasized that the notes at the end of each chapter that are cited in the text are particularly important. Not only do they contain references that can be consulted, but they also have vital background information as well as detailed explanations of text material, including descriptions of experimental approaches. Students are therefore urged to examine the endnotes, particularly when the text directs them to do so.

The diversity of the prokaryotes is in part due to adaptations to the different habitats in which they grow. The habitats range in pH from approximately 1 to 12, temperatures from below 2°C to over 100°C, pressures that can be several thousands of atmospheres (at the bottom of the oceans), habitats without oxygen, and habitats with salt concentrations that can reach saturating levels. The physiological types among the prokaryotes mirror the diversity in the habitats and the sources of nutrient. Thus there are aerobes, anaerobes, facultative anaerobes, heterotrophs, autotrophs, phototrophs, chemolithotrophs, alkaliphiles (alkalophiles), acidophiles, and halophiles. Some of these physiological types are found only among the prokaryotes, with rare exceptions. For example, among microorganisms that have been cultured in the laboratory, the ability to live indefinitely in the absence of air (anaerobic growth) is confined almost entirely to prokaryotes. Also, growth using the energy extracted from inorganic compounds (lithotrophy) seems not to have evolved in the eukaryotes, and is restricted to certain groups of prokaryotes. The same can be said about the reduction of nitrogen gas to ammonia (nitrogen fixation), although it is more widespread among the prokaryotes than is lithotrophy. Another typically prokaryotic capability is the use of inorganic compounds such as nitrate and sulfate as electron acceptors during respiration (anaerobic respiration).

On the other hand, it is also true that most of the metabolic pathways that exist in the prokaryotes are the same as those in all living organisms, reflecting a unity in biochemistry dictated by principles of chemistry and (with respect to bioenergetics) physics. This text emphasizes the underlying principles of chemistry and physics in explaining the various physiological and metabolic features of prokaryotic cells, and thus provides the backround for further advanced studies in this area.

Chapter 1 is an overview of the parts of prokaryotic cells and the functions of these parts. The chapter also introduces the student to the two evolutionary lines of prokaryotes: Bacteria (eubacteria) and Archaea (archaebacteria).

Chapter 2 is a general treatment of bacterial growth. It includes a description of the different phases of population growth, an explanation of the growth equations and their use and applications, continuous growth, as well as selected topics such as responses to different growth conditions, to starvation conditions, the stringent response, as well as a description of the physiology of cell division.

Chapter 3 is devoted to membrane bioenergetics. It explains the principles of the chemiosmotic theory, including the basic tenets of the theory, and the different ways that prokaryotes generate the proton and sodium ion potentials that drive membrane activities such as solute transport, motility, and ATP synthesis. This chapter, as well as Chapter 7, which explains bioenergetics in the cytosol, brings together principles of physics, thermodynamics, and organic chemistry to explain how prokaryotes use and interconvert light, chemical, and electrochemical energies. The first part of the chapter explains the chemiosmotic theory and provides an overview of how the proton potential is created by respiration, photosynthesis, and ATP hydrolysis. The details of respiration and photosynthesis are covered in the two subsequent chapters. The second part of the chapter includes detailed discussions of specialized processes that generate a proton or sodium potential

in certain groups of prokaryotes. This includes light-driven proton pumping by bacteriorhodopsin in the halophiles, as well as sodium-dependent decarboxylases, and proton and sodium-coupled end-product efflux in fermenting bacteria.

Chapter 4 describes electron transport pathways. The first part is a general description of electron transport and coupling sites in both mitochondria and bacteria, and includes a discussion of the $bc_1$ complex and the Q cycle, as well as linear Q loops and proton pumps. The second part of the chapter describes electron transport pathways in certain well-studied bacteria. Electron transport pathways in particular bacteria are also discussed in Chapter 11 (the lithotrophs and sulfate reducers).

Chapter 5 covers photosynthesis. It begins with an introduction to the various photosynthetic prokaryotes, and gives an overview of electron flow in the different photosynthetic systems. Later sections of the chapter are devoted to a more detailed examination of photosynthesis, including a closer look at the reaction centers, the light-harvesting pigments, and energy transfer between pigment complexes.

Chapter 6 is an introduction to the principles of metabolic regulation and provides the necessary background for learning the metabolic regulation of the pathways discussed in Chapters 8 and 9. It includes a discussion of the patterns of metabolic regulation, the Michaelis–Menten equation, allostery, and covalent modification of enzymes.

Chapter 7 explains the principles of bioenergetics in the soluble part of the cell, the cytosol, and is the counterpart to Chapter 3 that is concerned with membrane bioenergetics. It includes a discussion of what is meant by the term *high-energy molecule* and how these molecules are synthesized, group transfer potentials, high-energy intermediates, substrate-level phosphorylation, and coupled reactions. This chapter provides the background for understanding the energetics of the pathways discussed in Chapters 8 and 9.

Chapters 8 and 9 describe the central pathways of intermediary metabolism, including the metabolism of carbohydrates, organic acids, lipids, purines, pyrimidines, amino acids, and aliphatic hydrocarbons. The chapters on intermediary metabolism are by no means inclusive with respect to the pathways that are covered. They were written to provide an overview of metabolism in both aerobically and anaerobically growing prokaryotes, as well as to prepare the student for the chapter on sugar fermentations (Chapter 13). Major pathways found in most bacteria (as well as eukaryotes) are discussed, as well as some that are seen only in certain groups of prokaryotes (e.g., the modified Entner–Doudoroff pathway in some archaea and bacteria, and the "non-phosphorylated" Entner–Doudoroff pathway in some archaea). Special emphasis is given to the relationships between the pathways (i.e., how they interconnect) and their physiological roles.

Chapter 10 reviews DNA, RNA, and protein synthesis. The chapter begins with a discussion of DNA structure and replication, including termination and partitioning of the daughter chromosomes to the daughter cells. Mechanisms of repairing errors made during replication are also described. This is followed by a description of RNA replication and its regulation, as well as protein synthesis and the folding of newly synthesized proteins.

Chapter 11 discusses the biosynthesis of peptidoglycan, lipopolysaccharide, and polysaccharides, including extracellular polysaccharides (e.g., capsules).

Chapter 12 describes the metabolism of inorganic molecules. This includes ammonia and sulfate assimilation, nitrogen fixation, nitrate and sulfate respiration, and lithotrophy.

Chapter 13 explains the pathways for the assimilation of single-carbon compounds as well as methanogenesis.

Chapter 14 describes the main classes of carbohydrate fermentations.

Chapters 15, 16, and 17 cover pH and osmotic homeostasis, solute transport, and protein export, respectively.

Chapter 18 is concerned with how bacteria respond to signals from other bacteria, as well as environmental influences on metabolism.

This includes responses to changes in nutrient and oxygen levels, chemotaxis, how bacteria sense cell density (quorum sensing), bioluminescence, cell-to-cell signaling in myxobacteria, in *Bacillus* during sporulation and development of competence, and the production of virulence factors by pathogenic bacteria. Much of Chapter 18 is devoted to two-component regulatory systems.

Chapter 19 discusses responses to environmental stress such as heat stress, oxidative stress, and damage to DNA.

As can be seen from this introduction, the organization of the text is according to topics rather than organisms, although the physiology of specific groups of prokaryotes is emphasized. This pattern of organization lends itself to the elucidation of general principles of physiology and metabolism.

Most of the carboxyl groups are drawn as nonionized and the primary amino groups as nonprotonated. However, at physiological pH these groups are ionized and protonated, respectively. The names of the organic acids indicate that they are ionized (e.g., acetate rather than acetic acid).

There are two distinct evolutionary lines of prokaryotes commonly referred to as eubacteria and archaebacteria. However, this terminology implies a specific relationship between eubacteria and archaebacteria, whereas at the molecular level, the archaebacteria are not any more related to the eubacteria than they are to the eukaryotes. In recognition of this fact, Woese et al. suggested that the eubacteria be referred to simply as bacteria and that the archaebacteria be called archaea.[1] That terminology is followed in this text.

## NOTES

1. Woese, C. R., Kandler, O., and M. L. Wheelis. 1990. Towards a natural system of organisms: Proposal for the domains Archaea, Bacteria and Eucarya. *Proc. Natl. Acad. Sci. USA* 87:4576–4579.

# ACKNOWLEDGMENTS

I would like to express gratitude to the following individuals who thoughtfully reviewed portions of this second edition: Carl Bauer, Yves Brun, Neena Din, Jim Drummond, Jeffrey Favinger, Clay Fuqua, George Hegeman, Arthur Koch, and Larry Shimkets. Special thanks also to the reviewers of the first edition: Tom Beatty, Martin Dworkin, Howard Gest, George Hegeman, Alan Hooper, Arthur Koch, Jorg Overmann, Norman Pace, Carlos Ramirez-Icaza, and Palmer Rogers. I am deeply indebted to Kirk Jensen of Oxford University Press for his editorial advice and confidence in the project. I would also like to thank John Bauco of Oxford University Press for his thoughtful help in preparing the second edition. The text was illustrated by Eric J. White. His careful work and enthusiasm for the project are greatly appreciated, and it was a personally rewarding experience working with him.

# SYMBOLS

| | |
|---|---|
| $c$ | Speed of light ($3.0 \times 10^8$ m/s) |
| C | Coulomb |
| cal | Calorie (4.184 J) |
| $E_0$ | Standard redox potential (reduction potential) at pH 0 |
| $E'_0$ | Standard redox potential at pH 7 |
| $E_m$ | Midpoint potential at a specified pH (e.g., $E_{m',7}$). For pH 7, also written as $E'_m$, which is numerically equal to $E'_0$. |
| $E_h$ | Actual redox potential at a specified pH. For pH 7, $E'_h$ or $E_{h,7}$. |
| eV | Electron volt. The work required to raise one electron through a potential difference of one volt. It is also the work required to raise a monovalent ion (e.g., a proton) through an electrochemical potential difference of one volt. One eV $= 1.6 \times 10^{-19}$ J. |
| $F$ | Faraday constant (approximately 96,500 coulombs). The charge carried by one mole of electrons or monovalent ion. It is the product of the charge carried by a single electron and Avogadro's constant. |
| $g$ | Generation time (time/generation) |
| $\Delta G_0$ | Standard Gibbs free energy change at pH 0 (J/mol or cal/mol) |
| $\Delta G'_0$ | Gibbs free energy at pH 7 |
| $\Delta G_p$ | Phosphorylation "potential" (energy required to synthesize ATP using physiological concentrations of ADP, inorganic phosphate, and ATP) |
| $h$ | Planck's constant ($6.626 \times 10^{-34}$ J s) |
| J | Joule. One coulomb volt (C V). The work required to raise one coulomb through a potential difference of one volt. |
| $k$ | Instantaneous growth rate constant (time$^{-1}$) |
| kJ | Kilojoule |

| | |
|---|---|
| $K$ | Absolute equilibrium constant |
| $K'$ | Apparent equilibrium constant at pH 7 |
| K | Degrees kelvin ($273.16 + °C$) |
| $\Delta\mu_{ion}$ | Electrochemical potential difference between two solutions of an ion separated by a membrane. Units are joules. |
| $\Delta\mu_{ion}/F$ | Electrochemical potential difference expressed as volts or millivolts |
| $\Delta\mu_{H^+}/F$ | The protonmotive force. Electrochemical potential difference in volts or millivolts of protons between two solutions separated by a membrane. Also written as $\Delta p$. |
| $N$ | Avogadro's number ($6.023 \times 10^{23}$ particles/mol) |
| $\Delta p$ | See $\Delta H_{H^+}/F$ |
| $\Delta$pH | Difference in pH between the inside and outside of the cell. Usually, $pH_{in} - pH_{out}$. |
| $R$ | The ideal gas constant ($8.3144$ J deg $K^{-1}$ mol$^{-1}$ or $1.9872$ cal deg $K^{-1}$ mole$^{-1}$) |
| V | Volt. The potential difference across an electric field. |
| $\Delta\Psi$ | Membrane potential, usually $\Psi_{in} - \Psi_{out}$. |

# CONVERSION FACTORS, EQUATIONS, AND UNITS OF ENERGY

### Electrode potential at pH 7

$E'_h = E'_0 + [RT/nF] \ln[(\text{ox})/(\text{red})]$ volts. The symbol $n$ refers to the number of electrons, and (ox) and (red) refer to the concentrations of oxidized and reduced forms, respectively. When (ox) = (red), then $E'_h = E'_0 = E'_m$.

$E'_h = E'_0 + (60/n) \log_{10}[(\text{ox})/(\text{red})]$ mV at 30°C.

### Electron, charge

$1.6023 \times 10^{-19}$ C

### Gibbs energy

For the reaction $aA + bB \longleftrightarrow cC + dD$,

$\Delta G = \Delta G_0 + RT \ln[C]^c[D]^d/[A]^a[B]^b$

### Gibbs energy and equilibrium constant

$\Delta G'_0 = -RT \ln K'_{eq}$ or $-2.303RT \log_{10} K'_{eq}$

### Gibbs energy for solute uptake and concentration gradient

$\Delta G = RT \ln[C]_{in}/[C]_{out}$

at 30°C, $\Delta G = 5.8 \log_{10}[C]_{in}/[C]_{out}$ kJ/mole

or

$\Delta G/F = 60 \log_{10}[C]_{in}/[C]_{out}$ mV

### Growth

$g(k) = 0.693$

$x = x_0 2^Y$, the equation for exponential growth. $Y$ is the number of generations. $X$ is mass or any parameter that changes linearly with mass.

$Y = t/g$

### Light, energy in a quantum

$E = \nu h = hc/\lambda$, where $\nu$ is the frequency ($c/\lambda$); $h$, Planck's constant; $\lambda$, wavelength

$E(kJ) = 1.986 \times 10^{-19}/\lambda$, where $\lambda$ is in nm

$E(eV) = 1.24 \times 10^3/\lambda$, where $\lambda$ is in nm

## Light, energy in an einstein

$E = Nh\upsilon = Nhc/\lambda = 1.197 \times 10^5$ kJ/$\lambda$, where $\lambda$ is the wavelength in nanometers, $N$ is Avogadro's number, and $c$ is the speed of light

$E(\text{kJ}) = 1.196 \times 10^5/\lambda$, where $\lambda$ is in nm

## Nernst equation

$\Delta\Psi = -(RT/nF) \ln[S]_{in}/[S]_{out}$ V, or, $\Delta\Psi = -(60/n) \log [S]_{in}/[S]_{out}$ mV at 30°C, where S is the concentration of a diffusible cation ($S_{in} > S_{out}$) of valency $n$, and $\Delta\Psi$ is the membrane potential, inside negative. The equation states that at equilibrium the chemical driving force due to the outward diffusion of S along its concentration gradient is equal to the electrical driving force drawing S into the cell. According to the equation, each ten-fold concentration difference of a permeant monovalent cation corresponds to a potential difference of 60 mV.

## Phosphorylation potential

$\Delta G_p = \Delta G'_0 + RT \ln[\text{ATP}]/[\text{ADP}][P_i]$

## Proton potential

$\Delta p = \Delta\mu_{H^+}/F = \Delta\Psi - 60 \Delta\text{pH}$ mV at 30°C

## Proton potential and $\Delta E_h$ at equilibrium

$-n \Delta E_h = y \Delta p$, where $n$ is the number of electrons transferred over a redox potential difference of $\Delta E_h$ volts, and $y$ is the number of protons translocated over a proton potential difference of $\Delta p$ V.

## 2.303$RT$

5.8 kJ/mol or 1.39 kcal/mol at 30°C

## 2.303$RT/F$

0.06 V or 60 mV at 30°C

# DEFINITIONS

**Acetogenic bacteria**　Anaerobic bacteria that synthesize acetic acid from $CO_2$ and secrete the acetic acid into the medium.

**Acidophile**　Grows between pH 1 and 4 and not at neutral pH.

**Aerobe**　Uses oxygen as an electron acceptor during respiration.

**Aerotolerant anaerobe**　Cannot use oxygen as an electron acceptor during respiration but can grow in its presence.

**Anaerobe**　Does not use oxygen as an electron acceptor during respiration.

**Alkaliphile (alkalophile)**　Grows at pH above 9, often with an optimum between 10 and 12.

**Autotroph**　Uses $CO_2$ as sole or major source of carbon.

**Chaperone protein**　Proteins that transiently bind to other proteins and assist in proper folding of the target protein and/or transport to a correct cellular site.

**Chaperonins**　Multisubunit complexes of chaperone proteins.

**Chemolithotroph**　See Lithotroph.

**Cytoplasm**　The fluid material enclosed by the cell membrane.

**Cytosol**　The liquid portion of the cytoplasm.

**Dalton**　All atomic and molecular weights refer to the carbon isotope, $^{12}C$, which is 12 D or $1.661 \times 10^{24}$ g. Daltons are numerically equal to molecular weights and can be used as units when molecular weight units of grams per mole are not appropriate (e.g., when referring to ribosomes).

**Einstein**　One "mole" of light ($6.023 \times 10^{23}$ quanta).

**Facultative anaerobes**　Can grow anaerobically in the absence of oxygen or will grow by respiration if oxygen is available.

| | |
|---|---|
| **Facultative autotroph** | Can grow on $CO_2$ as sole or major source of carbon or on organic carbon. |
| **Gel electrophoresis** | Procedure in which macromolecules such as proteins or nucleic acids are applied to a polyacrylamide or agarose gel and subjected to an electrical field. The macromolecules having a net electrical charge migrate in the electrical field and can be separated according to their size. |
| **Growth yield constant, $Y$** | The amount of dry weight of cells produced per weight of nutrient used. |
| **Halophile** | Requires high salt concentrations for growth. |
| **Heat-shock proteins** | Proteins that transiently increase in amount relative to most cell proteins when the temperature is elevated. Several are chaperone proteins. |
| **Heterotroph** | An organism that uses organic carbon as a major source of carbon. |
| **Holliday junction** | An intermediate in homologous recombination in which a cruciform-like (cross-shaped) structure is formed. |
| **Hyperthermophile** | An organism whose growth temperature optimum is 80°C or greater. |
| **Leader region** | The region of mRNA for an operon that is 5′ of the coding region for the first gene. |
| **Lithotroph** | An organism that oxidizes inorganic compounds as a source of energy for growth. |
| **Mesophile** | An organism whose growth temperature optimum is between 25 and 40°C. |
| **Molar growth yield constant ($Y_m$)** | Grams of dry weight of cells produced per mole of nutrient used. |
| **Neutrophile** | An organism that grows at a pH optimum near neutrality. |
| **Obligate anaerobes** | Will grow only in the absence of oxygen but is not necessarily killed by oxygen. |
| **Phosphorylation potential** | Energy required to phosphorylate one mole of ADP using physiological concentrations of ADP, $P_i$, and ATP. |
| **Photon** | A quantum "particle" of light. |
| **Photosynthesis** | The use of light as a source of energy for growth. |
| **Photoautotroph** | An organism that uses light as a source of energy for growth and $CO_2$ as the source of carbon. |
| **Photoheterotroph** | An organism that uses light as a source of energy for growth and organic carbon as the source of cell carbon. |
| **Phototroph** | An organism that uses light as the source of energy for growth. |

| | |
|---|---|
| **Psychrophile** | An organism that grows best at temperatures of 15°C or lower, and does not grow above 20°C. |
| **Quantum** | A particle of light (photon). |
| **Regulon** | A set of noncontiguous genes controlled by the same transcription regulator. |
| **Standard conditions** | All reactants and products are in their "standard states." This means that solutes are at a concentration of one molar (1 mol of solute per liter) and gases are at one atmosphere. Biochemists usually take the standard state of $H^+$ as $10^{-7}$ M (pH 7). By convention, if water is a reactant or product, its concentration is set at 1.0 M, even though it is 55.5 M in dilute solutions. |
| **Strict anaerobes** | Will grow only in the absence of oxygen and is also killed by traces of oxygen. |
| **Thermophile** | An organism that can grow at temperatures greater than 55°C. |

*The* Physiology
*and*
Biochemistry *of*
Prokaryotes

# 1

# Structure and Function

Although prokaryotes[1] are devoid of organelles such as nuclei, mitochondria, chloroplasts, Golgi vesicles, and so on, their cell structure is far from simple. Despite the absence of organelles comparable to those found in eukaryotic cells, metabolic activities in prokaryotic cells are nevertheless compartmentalized, which is necessary for efficient metabolism and growth. For example, compartmentalization occurs: within multienzyme granules that house enzymes for specific metabolic pathways; in intracellular membranes and the cell membrane; within the periplasm in gram-negative bacteria; and within the cell wall itself. There are also various inclusion bodies that house specific enzymes, storage products, or photosynthetic pigments. In addition, prokaryotic cells display appendages such as fimbriae and pili that are used for adhesion to other cells, and flagella that are used for swimming. There is also a suggestion that bacteria may possess something analogous to the cytoskeleton found in eukaryotic cells.[2] This chapter describes these structural cell components and their functions. As we shall see, there are actually two distinct types of prokaryotes, the Bacteria and the Archaea.

## 1.1 Phylogeny

Figure 1.1 shows a current phylogeny of life forms based upon comparing ribosomal RNA nucleotide sequences. (For a discussion, see endnote[3].) Notice that there are three lines (domains) of evolutionary descent—Bacteria (eubacteria), Eucarya (eukaryotes), and Archaea (archaebacteria)—that diverged in the distant past from a common ancestor.[4–6] (The term *archaeon* may be used to describe particular archaea.) Archaea differ from bacteria in ribosomal RNA nucleotide sequences, in cell chemistry, and in certain physiological aspects, described in Section 1.2. Table 1.1 lists examples of prokaryotes in the different subdivisions within the domains Bacteria and Archaea.

Notice that the gram-positive bacteria are a tight grouping. Although there is no single grouping of gram-negative bacteria, most of the well-known gram-negative bacteria are in the purple bacteria group.

### 1.1.1 Archaea

#### Phenotypes

The archaea commonly manifest one of three phenotypes: *methanogenic, extremely halophilic,* and *extremely thermophilic,* which also correspond to phylogenetic groups.

1. The *methanogenic* archaea (kingdom Euryarchaeota) are obligate anaerobes that grow in environments such as anaerobic ground waters, swamps, and sewage. They derive energy by reducing carbon dioxide to

**Table 1.1** Major subdivisions of Bacteria

**Bacteria and their subdivisions**

Purple bacteria
- $\alpha$ a subdivision
  Purple non-sulfur bacteria (*Rhodobacter, Rhodopseudomonas*) rhizobacteria, agrobacteria, rickettsiae, *Nitrobacter, Thiobacillus* (some), *Azospirillum, Caulobacter*
- $\beta$ subdivision
  *Rhodocyclus* (some), *Thiobacillus* (some), *Alcaligenes, Bordetella, Spirillum, Nitrosovibrio, Neisseria*
- $\gamma$ subdivision
  Enterics (*Acinetobacter, Erwinia, Escherichia, Klebsiella, Salmonella, Serratia, Shigella, Yersinia*), vibrios, fluorescent psedomonads, purple sulfur bacteria, *Legionella* (some), *Azotobacter, Beggiatoa, Thiobacillus* (some), *Photobacterium, Xanthomonas*
- $\delta$ subdivision
  Sulfur and sulfate reducers (*Desulfovibrio*), myxobacteria, bdellovibrios

Gram-positive eubacteria
- A. High (G + C) species
  *Actinomyces, Streptomyces, Actinoplanes, Arthrobacter, Micrococcus, Bifidobacterium, Frankia, Mycobacterium, Corynebacterium*
- B. Low (G + C) species
  *Clostridium, Bacillus, Staphylococcus, Streptococcus*, mycoplasmas, lactic acid bacteria
- C. Photosynthetic species
  *Heliobacterium*
- D. Species with Gram negative walls
  *Megasphaera, Sporomusa*

Cyanobacteria and chloroplasts
  *Oscillatoria, Nostoc, Synechococcus, Prochloron, Anabaena, Anacystis, Calothrix*

Spirochaetes and relatives
- A. Spirochaetes
  *Spirochaeta, Treponema, Borrelia*
- B. Leptospiras
  *Leptospira, Leptonema*

Green sulfur bacteria
  *Chlorobium, Chloroherpeton*

Bacteroides, flavobacteria and relatives
- A. Bacteroides group
  *Bacteroides, Fusobacterium*
- B. Flavobacterium group
  *Flavobacterium, Cytophaga, Saprospira, Flexibacter*

Planctomyces and relatives
- A. Planctomyces group
  *Planctomyces, Pasteuria*
- B. Thermophiles
  *Isocystis pallida*

Chlamydiae
  *Chlamydia psittaci, C. trachomatis*

Radio-resistant micrococci and relatives
- A. Deinococcus group
  *Deinococcus radiodurans*
- B. Thermophiles
  *Thermus aquaticus*

Green nonsulfur bacteria and relatives
- A. Chloroflexus group
  *Chloroflexus, Herpetosiphon*
- B. Thermomicrobium group
  *Thermomicrobium roseum*

**Archaea subdivisions**

Extreme halophiles
  *Halobacterium, Halococcus morrhuae*

Methanobacter group
  *Methanobacterium, Methanobrevibacter, Methanosphaera stadtmaniae, Methanothermus fervidus*

Methanococcus group
  *Methanococcus*

"Methanosarcina" group
  *Methanosarcina barkeri, Methanococcoides methylutens, Methanotrhix soehngenii*

Methospirillum group
  *Methanospirillum hungatei, Methanomicrobium, Methanogenium, Methanoplanus limicola*

Thermoplasma group
  *Thermolasma acidophilum*

Thermococcus group
  *Thermococcus celer*

Extreme thermophiles
  *Sulfolobus, Thermoproteus tenax, Desulfurococcus mobilis, Pryodictium occultum*

*Source*: Hodgson, D. A. 1989. Bacterial diversity: The range of interesting things that bacteria do, p. 4–22. In: *Genetics of Bacterial Diversity*. D. A. Hopwood and K. F. Chater, eds. Academic Press, London.

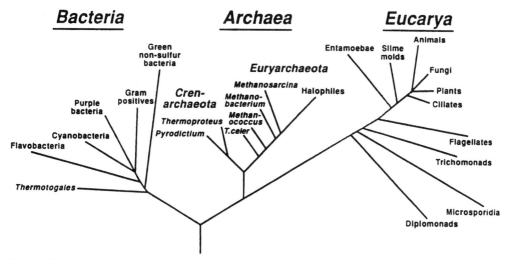

**Fig. 1.1** Phylogenetic relationships among life forms based upon rRNA sequences. The lengths of the lines are proportional to the evolutionary differences. The position of the root in the tree is approximate. *Source:* From Woese, C. R., and N. R. Pace. 1993. Probing RNA structure, function, and history by comparative analysis. *The RNA World.* Cold Spring Harbor Press.

methane or by converting acetate to carbon dioxide and methane. The metabolism of the methanogens is discussed in chapter 13.

2. The *extremely halophilic* archaea (also Euryarchaeota) require very high sodium chloride concentrations (at least 3–5 M) for growth. They grow in salt lakes and solar evaporation ponds. The halophilic archaea have unique light-driven proton and chloride pumps called *bacteriorhodopsin* and *halorhodopsin*, respectively (Sections 3.8.4 and 3.9). Most extreme halophiles are archaea. Some exceptions are the bacterium *Ectothiorhodospira* and the alga *Dunaliella*.

3. The *extremely thermophilic* archaea (kingdom Crenarchaeota) grow in thermophilic environments (generally 55–100°C).[7] Some of these have an optimal growth temperature near the boiling point of water. (Indeed, most extremely thermophilic prokaryotes are archaea, although some some, i.e., *Thermotoga* and *Aquifex,* are bacteria.) They use inorganic sulfur either as an electron donor or as an electron acceptor in energy-yielding redox reactions. (The pathways for sulfate reduction and sulfur oxidation are described in Sections 12.2.2 and 12.4.1, respectively.) For this reason they are also called *sulfur dependent.* For example, some, including *Sulfolobus* and *Acidianus,* oxidize inorganic

sulfur compounds such as elemental sulfur and sulfide using oxygen as the electron acceptor and derive ATP from the process, while others are anaerobes that oxidize hydrogen gas using elemental sulfur or thiosulfate as the electron acceptor. The latter include *Thermoproteus, Pyrobaculum, Pyrodictium,* and *Archaeoglobus*. (*Pyrobaculum* and *Pyrodictium* use $S^o$ as an electron acceptor during autotrophic growth, i.e., growth on $CO_2$ as the carbon source, whereas *Archaeglobus* uses $S_2O_3^{2-}$.) Archaea belonging to the genus *Pyrodictium* have the highest growth temperature known, being able to grow at 110°C. A few of the sulfur-oxidizing archaea are *acidophiles*, growing in hot sulfuric acid at pH values as low as 1.0. They are called *thermoacidophiles* in recognition of the fact that they grow optimally in hot acid. For example, *Sulfolobus* grows at pH values of 1–5 and at temperatures up to 90°C in hot sulfur springs, where it oxidizes $H_2S$ (hydrogen sulfide) or $S^o$ to $H_2SO_4$ (sulfuric acid). Although most of the extreme thermophiles are obligately sulfur dependent, some are facultative. For example, *Sulfolobus* can be grown heterotrophically on organic carbon and $O_2$ as well as autotrophically on $H_2S$ or $S^o$, $O_2$, and $CO_2$. Interestingly, some of the sulfur-dependent archaea have metabolic pathways for sugar degradation not found

Table 1.2 Comparison between Bacteria, Archae, and Eucarya

| Characteristic | Bacteria | Archaea | Eucarya |
|---|---|---|---|
| Peptidoglycan | Yes | No | No |
| Lipids | Ester linked | Ether linked | Ester linked |
| Ribosomes | 70S | 70S | 80S |
| Initiator tRNA | Formylmethionine | Methionine | Methionine |
| Introns in rRNA | No | Yes | Yes |
| Ribosome sensitive to diphtheria toxin | No | Yes | Yes |
| RNA polymerase | One (4 subunits) | Several (8–12 subunits each) | Three (12–14 subunits each) |
| Ribosome sensitive to chloramphenicol, streptomycin, kanamycin | Yes | No | No |

among the bacteria. These are described in Section 8.4.

Recently, a new genus of thermoacidophilic archaea was discovered.[8] *Picrophilus,* a member of the kingdom Euryarchaeota, order *Thermoplasmales,* is an obligately aerobic, heterotrophic archaeon that grows at temperatures between 45 and 65°C at a pH of 0 with an optimum pH of 0.7. It was isolated from hot geothermally heated acidic solfataras[9] in Japan. Despite the sulfur content of the habitat, *Picrophilus* is not sulfur dependent.

## Comparison of domains Archaea, Bacteria, and Eucarya

Listed below are several structural and physiological differences that distinguish bacteria from archaea. In addition, some similarities between archaeal and eukaryotic RNA polymerase, and archaeal and eukaryotic ribosomes are noted (Table 1.2).

1. Archaeal membrane lipids are long-chain hydrocarbons linked by *ether-linkage* to glycerol.[10] The lipids in the membranes of bacteria and eukaryotes are *fatty acids ester linked to glycerol* (Fig. 1.16 and 9.5), whereas the archaeal lipids are methyl-branched, isopranoid *alcohols ether linked to glycerol* (Fig. 1.17). Archaeal membranes are discussed in Section 1.2.5, and the biosynthesis of archaeal lipids is discussed in Section 9.1.3.

2. Archaea lack peptidoglycan, *a universal component of bacterial cell walls.* The cell walls of some archaea contain pseudomurein,

a component absent in bacterial cell walls (Section 1.2.3).

3. Archaea contain histones that resemble eukaryal histones and bind archaeal DNA into compact structures resembling eukaryal nucleosomes.[11, 12]

4. The archaeal RNA polymerase differs from bacterial RNA polymerase by having 8 to 10 subunits, rather than four subunits, *and* not being sensitive to the antibiotic rifampicin. The difference in sensitivity to rifampicin reflects differences in the proteins of the RNA polymerase. It is interesting that the archaeal RNA polymerase resembles the eukaryotic RNA polymerase, which also has many subunits (10–12) and is not sensitive to rifampicin.

5. Some protein components of the archaeal protein synthesis machinery differ from those found in the bacteria. Archaeal ribosomes are not sensitive to certain inhibitors of bacterial ribosomes (i.e., erythromycin, streptomycin, chloramphenicol, and tetracycline). These differences in sensitivity to antibiotics reflect differences in the ribosomal proteins. In this respect, archaeal ribosomes resemble cytosolic ribosomes from eukaryotic cells. Other resemblances to eukaryotic ribosomes are the use of methionine rather than formylmethionine to initiate protein synthesis, and the requirement for an elongation factor, EF-2, that can be ADP-ribosylated by diphtheria toxin. In contrast, bacteria use formylmethionine to initiate protein synthesis, and their EF-2 is not sensitive to diphtheria toxin. However, the archaeal ribosomes

are similar to bacterial ribosomes in being 70S.

6. The halophilic archaea have light-driven ion pumps not found among the bacteria. The pumps are bacteriorhodopsin and halorhodopsin, which pump protons and chloride ions, respectively, across the membrane. The proton pump serves to create a proton potential that can be used to drive ATP synthesis. This is discussed in Section 3.6.2.

7. The methanogenic archaea have several coenzymes that are unique to archaea. The coenzymes are used in the pathway for the reduction of carbon dioxide to methane and in the synthesis of acetyl–CoA from $H_2$ and $CO_2$. These coenzymes and their biochemical roles are described in Section 13.1.5.

## 1.2 Cell Structure

Much of the discussion of cell structure will refer to the more well-studied bacteria because structural studies of archaea are not as extant; however, there are well-known differences between archaea and bacteria with respect to cell walls and cell membranes, and these will be discussed. The structure and function of the major cell components will be described, beginning with cellular appendages (various filaments and flagella) and working our way into the interior of the cell.

### 1.2.1 Appendages

Numerous appendages, each designed for a specific task, extend from the surface of bacteria. There are three classes of appendages:

1. Flagella, used for motility.

2. Fimbriae (sometimes called pili) used for adhesion.

3. Sex pili, used for mating by some bacteria.

### Flagella

Swimming bacteria have one or several flagella, which are organelles of locomotion that protrude from the cell surface.[13, 14] (Many bacteria move without flagella on solid surfaces by gliding.[15]) The *flagellum* is a stiff (semirigid) helical filament (either left handed or right handed, depending upon the species) that rotates like a propeller. It is unrelated in composition, structure, and mechanism of action to the eukaryotic flagellum. The arrangement and number of flagella vary with the species. They can be located at one or both cell poles, or inserted laterally around the entire cell. (See Note 16 for a more complete discussion of the arrangements of flagella.) The flagella that are studied in most detail are those of *Escherichia coli* and *Salmonella typhimurium*, and most of the following discussion concerns these flagella. The flagella of other bacteria, except for the spirochaetes, are similar in general structure to those of *E. coli* and *S. tyhphimurium*. Each bacterial flagellum has a small rotating motor at its base, built of protein, embedded in the cell membrane. The motor is attached to a helical protein filament of about 20 nm in diameter that extends approximately 5–10 $\mu$m from the cell surface (Fig. 1.2). The motor is actually an electrochemical machine. The energy to drive the motor comes from a current of protons that moves down a proton potential gradient through the flagellar motor from the outside of the membrane to the inside. (Extreme alkaliphiles and some marine bacteria use a sodium ion current. This is reviewed in Ref. 17.) The passage of protons turns the motor in a way that is not understood. When the motor turns, the attached filament rotates and propels the bacterium through the medium, often quite rapidly (e.g., 20–80 $\mu$m/sec). Flagellar rotation and swimming, as well as tactic responses, are described in more detail in Chapter 18. The use of ion currents across membranes to do work is described more fully in Chapter 3.

As discussed in Chapter 18, bacteria are capable of swimming toward more favorable environments with respect to nutrient, light, and electron acceptors, and are able to avoid toxic environments. These swimming bacterial responses, called tactic responses, are due to the fact that bacteria can sense environmental signals and modify the rotation of their flagella accordingly. Another role for

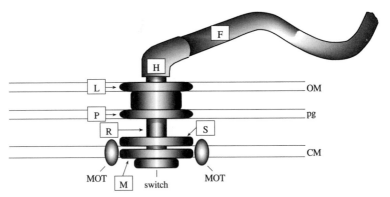

**Fig. 1.2** Bacterial flagellum in a gram-negative envelope. The basal body is 22.5 × 24 nm and is composed of four rings, L, P, S, and M, connected by a central rod. The M ring is embedded in the cell membrane, and the S ring appears to lie on the surface of the membrane. Recently it has been concluded that the S and M rings are actually one ring, the MS ring. (See Wilson and Beveridge, Ref. 14.) The P ring may be in the peptidoglycan layer, and the L ring seems to be in the outer membrane. The P and L rings may act as bushings that allow the central rod to turn. Gram-positive bacteria have similar flagella but lack the L and P rings. The MotA and MotB proteins couple the proton potential to the rotation of the motor. The switch complex consists of three peripheral membrane proteins, FliG, FliM, and FliN, which are probably closely apposed to the cytoplasmic face of the M ring. Not shown are hook accessory or adaptor proteins called HAP1 and HAP3 between the hook and filament, and a protein cap called HAP2 on the end of the filament. The flagellum is assembled from the proximal to the distal end, with the filament being assembled last. It appears that the HAP1 and HAP3 proteins are required for the proper asssembly of the filament onto the hook. Abbreviations: OM, outer membrane; pg, peptidoglycan; CM, cell membrane; R, central rod; MOT, Mot A and MotB; H, hook; F, filament.

flagella is perhaps in virulence of pathogenic bacteria.[18] For example, it has been suggested that the ability of spirochetes (spiral-shaped bacteria) such as *Treponema pallidum,* the causative agent of syphilis, to swim in a corkscrew fashion though viscous liquid aids in their dissemination (e.g., through connective tissue or the intracellular junctions between endothelial cells).

### 1. General structure

The flagellum consists of three parts: a *basal body,* a *hook,* and a *filament*—plus additional proteins required for motor function. Intact flagella have been isolated and their components subjected to analysis.[19, 20] The flagellum contains about 20 different polypeptides and requires approximately 40 different genes for its assembly and function. The motor can rotate either clockwise or counterclockwise, which determines the directionality of swimming, and responds to chemotactic signals. Section 18.4 contains a detailed description of chemotaxis. (Some bacteria, e.g., *Rhodobac-*

*ter sphaeroides* and *Rhizobium meliloti,* have flagella that rotate in only one direction. They are discussed in Section 18.4.8.)

**a. The basal body** (Fig. 1.2). At the base of the flagellum there is a *basal body* embedded in the membrane. Although the basal body has frequently been speculated to be the motor, the isolated structure does not contain several proteins described later that are required for motor function. It is therefore probably not the entire motor. The basal body consists of two stacked rings in gram-positive bacteria and four rings in gram-negative bacteria, through which a central rod passes, which is attached to the flagellum filament. The innermost ring (M) appears to be in the cell membrane. Next to the M ring is the S ring, which appears to be located on the external membrane surface. Both the M and S rings are assembled from one type of protein called FliF protein. Electron microscopic evidence suggests that the M and S rings are actually one ring, called the

MS ring. Gram-negative bacteria have an additional pair of rings (L and P rings) that may act as bushings, allowing the central rod to rotate in the peptidoglycan and outer membrane. The P ring is in the area of the peptidoglycan, and the L ring is in the outer envelope. The L ring in *S. typhimurium* has recently been shown to be a lipoprotein.[21] Presumably the lipid portion helps to anchor the protein in the lipid regions of the outer envelope.

**b. Additional proteins required for motor function.** There are proteins that are not part of purified flagella and that are probably components of the motor because mutations in the genes that code for these proteins affect motor function. For example, mutations in the *motA* and *motB* genes result in paralyzed flagella (Mot⁻ phenotype). The MotA and MotB proteins are in the membrane, and it is suggested that they surround the MS ring in a ring of 8–12 proteins. Such rings of particles were seen in electron micrographs of freeze-fractured cytoplasmic membranes, but not when either MotA or MotB were missing.[22] The MotA and MotB proteins are believed to transduce the proton potential in an unknown fashion into mechanical rotation of the motor. Membrane vesicles prepared from strains synthesizing wild-type MotA were more permeable to protons than were vesicles prepared from strains synthesizing mutant MotA, indicating that MotA is likely to be a proton channel.[23] Evidence using suppressor mutations indicates that MotA and Mot B form a complex that together might be the transmembrane proton channel.[24] In bacteria such as *E. coli* and *S. typhimurium*, the motor changes its direction of rotation periodically. Mutants that fail to change flagellar rotation map in three genes, *fliG*, *fliM*, and *fliN*, which code for the switch proteins FliG, FliM, and FliN. (For more information about these proteins, see endnote 25.) The location and workings of the switch are, of course, of great interest. It appears that the three switch proteins form a complex of peripheral membrane proteins closely associated with the cytoplasmic side of the M ring. In particular, FliG seems to be bound to the M ring, whereas FliM and FliN may be part of a cytoplasmic cylindrical attachment to the M ring called the C ring.[26] Mutant studies have suggested that FliG is directly involved in torque generation and may be part of the rotor in the flagellar motor.[27] In this model, the MotA/B complex would be the stator portion of the motor.

**c. The hook.** The central rod is attached to a curved hook that is made of multiple copies of a special protein called the hook protein (product of *flgE* gene). There are also two hook-associated proteins (HAP1, product of the *flgK* gene, and HAP3, product of the *flgL* gene), which are necessary to form the junction between the hook and the filament, and a third HAP (HAP2, product of the *fliD* gene), which caps the flagellar filament. Mutants that lack the HAPs secrete flagellin into the medium.

**d. The filament.** Attached to the hook is a semirigid helical filament that along with the hook protrudes from the cell. The protein in the filament is called *flagellin*, which is present in thousands of copies. Flagellin is not identical in different bacteria. For example, the protein can vary in size from 20 to 65 kD depending upon the species of bacterium. Furthermore, although there is homology between the C-terminal and N-terminal ends of most flagellins, the central part can vary considerably and is distinguished immunologically in different bacteria. In some cases, there is no homology at all. For example, nucleotide-derived amino acid sequences for the flagellins from *Rhizobium meliloti* showed almost no relationship to flagellins from *E. coli*, *S. typhimurium*, or *Bacillus subtilis*, but were 60% similar to the N and C termini of flagellin from *Caulobacter crescentus*.[28] The flagellin subunits are arranged so that there is a central 60 angstrom unit hole, which may be important for transporting flagellin subunits from the base to tip during growth.

*2. Mechanics of motor function*
The rotational force originates in the cell membrane, presumably in the M–S ring and its associated proteins, and is transmitted to the filament through the rod and hook. The P and L rings are assumed to be bushings through which the rod passes through the outer

envelope. The flagellar motor must have both a rotor and a stator. Because of the large mass of the filament and the significant viscous drag that it encounters, the stator must be attached to a structural element sufficiently massive so as to preclude the stator from spinning in the membrane while the filament barely moves. It is usually assumed that whatever functions as the stator is firmly anchored to the peptidoglycan. Precisely how torque is generated by the flux of protons through the MotA/B complex is not understood.

### 3. Growth of the flagellum

The filament grows at the tip as demonstreated by the use of the phenylalanine analogue fluorophenylanine or radioisotopes. Incorporation of the phenylalanine analogue fluorophenylalanine by *Salmonella* resulted in curly flagella that had only one-half the normal wavelength. When the analogue was introduced to bacteria that had partially synthesized flagella, then the completed flagella were normal at the proximal ends and curly at the distal ends, implying that the fluorophenylalanine was incorporated at the tips during flagellar growth.[29] When *Bacillus* flagella were sheared off the cells and allowed to regenerate for 40 min before adding radioactive amino acids ([³H]-leucine), radioautography showed that all the radioactivity was at the distal region of the completed flagella.[30] The flagellin monomers are possibly transported through the central hole in the filament to the tip, where they are assembled.

### 4. Differences in flagellar structure

Although the basic structure of flagella is similar in all bacteria thus far studied, there are important species-dependent differences. For example, some bacteria have *sheathed flagella*, whereas others do not. In some species (e.g., *V. cholerae*) the sheath contains lipopolysaccharide and appears to be an extension of the outer membrane.[31] In spirochaetes the sheath is made of protein. (See later.) There are also differences regarding the number of different flagellins in the filaments. Depending upon the species, there may be only one type of flagellin, or two or more different flagellins in the same filament. For example, *E. coli* has one flagellin, and *R. meliloti* and *B. pumilis* each have

two different flagellin proteins, but *Caulobacter crescentus* has three.[32–34] The data for *B. pumilis* and *C. crescentus* support the conclusion that the different flagellins reside in the same filament. It has been proposed that the filaments of *R. meliloti* are composed of heterodimers of the two different flagellins. Whereas most bacteria that have been studied have flagella that have a smooth surface as seen with the electron microscope, called *plain filaments* (e.g., *E. coli),* certain bacteria (e.g., the soil bacteria, *Pseudomonas rhodos, R. lupini,* and *R. meliloti*) have "complex" filaments that have obvious helical patterns of ridges and grooves on the surface. Flagella with plain filaments rotate either clockwise or counterclockwise, whereas flagella with complex filaments rotate only clockwise with intermittent stops. It is thought that complex filaments are more rigid (because they are brittle) than are plain filaments, and are better suited for propelling the bacteria in viscous media such as the gelatinous matrix through which *R. lupini* and *R. meliloti* must swim in order to infect root hairs of leguminous plants.[35] A discussion of the role that bacterial flagella play in pathogenicity can be found in the review by Moens.[35] Although *E. coli* and *S. typhimurium* have four rings through which the central rod passes, other bacteria may have fewer or more. For example, gram-positive bacteria lack the outer two rings (the L and P rings). Additional structural elements of unknown function (e.g., additional rings or arrays of particles surrounding the basal body) have also been observed in certain bacteria.

The spirochete flagella are called *axial filaments*. A major difference between the spirochete flagella and those found in other bacteria is that the spirochete flagella do not protrude from the cell; rather, they are wrapped around the length of the helical cell between the cell membrane and the outer membrane (i.e., in the periplasm). Spirochetes are helical cells (i.e., they are coiled) and have two or more flagella (some have 30 or 40 or more) inserted near each cell pole. The number inserted at each pole are the same. The flagella are usually more than half the length of the cell and overlap in the middle. Another difference is that the spirochete flagellum is often surrounded by a *proteinaceous sheath*. The rotation of the periplasmic filaments is thought to propel the cell by propagating a helical wave down the length of the cell, which causes the cells to "corkscrew" through viscous media. (An exception is *Borrelia burgdorferi,* the causative agent of Lyme disease. Its axial filaments are not surrounded by sheaths. Furthermore, this organism has a planar waveform type of motility rather than the corkscrew type.[36]) There are five antigenically related flagellins in the axial filaments, but it is not known whether individual filaments contain more than one type of flagellin.[37] Despite these differences, the structure of the spirochete flagella is the same as the flagella found in other bacteria in having a basal body composed of a series of rings surrounding a central rod, a hook, and a filament.

### 5. Archaeal flagella

Archaeal flagella have not been as well studied as have bacterial flagella. In general they appear to be similar, but there are some differences.[38] The primary sequences of the flagellins from several archaea show no homology to bacterial flagellins, although they do show sequence homology to each other at the N-terminal end.[39] Also, the basal body structure in *Methanospirillum hungatei* appears to be simpler than for bacterial flagella in that a simple knob is present rather than rings.[40]

## Fimbriae, pili, filaments, and fibrils

Protein fibrils have been called various names, including fimbriae, pili, filaments, and simply fibrils. They extend from the surface of most gram-negative bacteria (Fig. 1.3).[41–44] Similar fibrils are rarely seen in gram-positive bacteria, but are present in the gram-positive *Corynebacterium renale* and *Actinomyces viscosus.*[45, 46] (See endnote 47 for the diseases caused by these bacteria.) All such appendages are not alike, nor are their functions known in all instances. The size varies considerably. They can be quite short (0.2 $\mu$m) or very long (20 $\mu$m), and differ in width from 3 to 14 nm or greater. Some originate from basal bodies in the cell membrane. However, most seem not to originate in the cell membrane at all, and how they are attached to the cell is not known. Furthermore, they are not always present. Although freshly isolated strains frequently have fibrils, they are often lost during subculturing in the laboratory. They are apparently useful in the natural habitat but dispensable in laboratory cultures. What do they do?

### 1. Fimbriae

Many of the fibrils can be observed to mediate attachment of the bacteria to other cells, such as to other bacteria, animal, plant, or fungal cells. They are therefore important for colonization because they help the bacteria to stick to surfaces. In nature most bacteria grow while attached to surfaces where the concentration of nutrients is frequently highest. Attachment can be via fibrils that extend from the surface of the cells, and/or nonfibrillar material that may be part of the cell wall or glycocalyx (Section 1.2.2). It has been proposed that fibrils that mediate attachment be called *fimbriae*. That convention will be followed here, although many researchers instead refer to adhesive protein fibrils as adhesive pili. Fimbriae have *adhesins* (i.e., molecules that cause bacteria to stick to surfaces). The adhesins are proteins in the fimbriae, often minor proteins at the tip, that recognize and bind to specific receptors on the surfaces of cells.

Fimbrial proteins (and other adhesins) are of important medical significance, as the following discussion will illustrate. Adhesion is studied in two types of experimental situations: attachment of bacteria to erythrocytes causing them to clump (*hemagglutination*), and attachment of bacteria in vitro to host cells to which they normally attach in vivo. (See endnote 48.) An example of the latter is the attachment of the causative agent of gonorrhea, *Neisseria gonorrhoeae,* to epithelial cells of the urogenital tract. When studying such attachments, it has often been found that specific monosaccharides and oligosaccharides inhibit the attachment or hemagglutination when added to the suspension. The implication is that at least some, if not all, fimbriae bind to oligosaccharides in cell surface receptors, and that the added monosaccharides and oligosaccharides are inhibitory because they compete with the receptor for the adhesin. Receptors on animal cell surfaces include

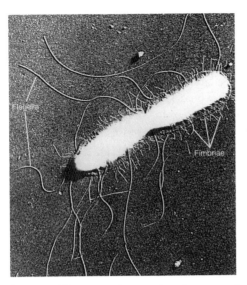

**Fig. 1.3** Electron micrograph of a metal-shadowed preparation of *Salmonella typhi* showing flagella and fimbriae. The cell is about 0.9 μm in diameter. *Source:* Reprinted with permission of J. P. Duguid.

*glycolipids* and *glycoproteins,* which are embedded in animal cell membranes via the lipid and protein portions, and present their oligosaccharide moieties to the outer surface. It has been reported that many strains of *E. coli, Salmonella,* and *Shigellae* carry fimbriae whose hemagglutinin activity is prevented by D-mannose and methyl-α-mannoside. These fimbriae, therefore, are believed to attach to mannose glycoside residues on the cell surface receptors. They are called *common fimbriae,* or *mannose-sensitive fimbriae,* or (more frequently) *type-1 fimbriae.* Other fimbriae recognize different cell surface receptors and are sometimes called mannose-resistant fimbriae. For example, although most clinical isolates of *E. coli* from the urinary and gastrointestinal tracts possess type-1 fimbriae, many also (or instead) have galactose-sensitive fimbriae, called *P-type fimbriae.*[49, 50] P fimbriae bind the α-D-galactopyranosyl-(1-4)-β-D-galactopyranoside [Galα(1-4)Gal] in the glycolipids on cells lining the upper urinary tract.[51] Other known fimbriae include the *S, K88, F17,* and *K99* fimbriae of *E. coli,* the *type-4 fimbriae* produced by *N. gonorrhoeae, Moraxella bovis, Bacteroides nodosus, P. aeruginosa,* and the *Tcp fimbriae*

produced by *Vibrio cholerae.* (See endnote 52 for the diseases caused by these bacteria.) The point being made here is that different types of fimbriae are specialized for attachment to specific receptors and can account for specificity of bacterial attachment to hosts and tissues. They can be distinguished by inhibition of binding by mono- and oligosaccharides, as well as by a variety of other methods, including morphology, antigenicity, molecular weight of the protein subunit, isoelectric point of the protein subunit, and amino acid composition and sequence. Since gram-positive bacteria generally do not have fimbriae, their adhesins are part of other cell surface components (e.g., the glycocalyx described in Section 1.2.2).

### 2. Sex pili

Bacteria are capable of attaching to each other for the purpose of transmitting DNA from a donor cell to a recipient cell. This is called mating. Some bacteria (not all) use fimbriae for mating attachments. The fimbriae that mediate attachment between mating cells are different from the other fimbriae and are called *sex pili.* The requirement of sex pili for mating is found for enteric bacteria such as *E. coli* and pseudomonads, but it is not universal among the bacteria. For example, gram-positive bacteria do not have sex pili. The sex pilus grows on "male" strains that donate DNA to recipient ("female") strains. In *E. coli,* it is coded for by a conjugative transmissible plasmid, the F plasmid, that resides in the donor strains. (See endnote 53.) An illustration of how the F pilus in *E. coli* works is shown in Fig. 1.4. The tip of the F pilus adheres to receptor molecules on the recipient cell surface. This is followed by a retraction of the pilus (depolymerization of the F-pilin subunits into the cell membrane), bringing the two cells closer together until their surfaces are in contact. DNA transfer takes place at the site of contact (not through the sex pilus). Sex pili seem to be designed for mating in cell suspension where the bacteria are not in intimate contact. The sex pilus presumably helps to stabilize the mating pairs of bacteria. They are not needed for many bacteria, perhaps because they mate in colonies

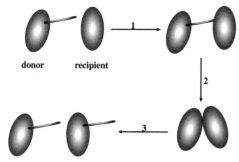

donor          recipient

**Fig. 1.4** F-pilus mediated conjugation. Transfer of a sex plasmid. The donor cell has a plasmid and an F-pilus encoded by plasmid genes. (1) The F-pilus binds to the recipient cell. (2) The pilus retracts, bringing the two cells together. This is due to a depolymerization of the pilus subunits. 3. The plasmid is transferred as it replicates so that when the cells separate, each has a copy of the plasmid and is a potential donor.

or aggregates on solid surfaces where the cells are in close contact. However, it should be pointed out that the gram-positive bacterium *Streptococcus faecalis* (now called *Enterococcus faecalis*) forms efficient mating aggregates in liquid suspension and does not have sex pili. In other words, their adhesins are located on the cell surface rather than on pili. Mating in *E. faecalis* is induced by a sex pheromone secreted into the medium by recipient cells. The sex pheromones signal the donor cells to synthesize cell surface adhesins that promote cell aggregate formation (clumping) and subsequent DNA transfer.[54] Mating interactions mediated by surface adhesins is widespread in the microbial world. Other well-studied examples include *Chlamydomonas* mating and yeast mating.[55]

## 1.2.2 The Glycocalyx

The term *glycocalyx* is often used to describe all extracellular material *external to the cell wall*.[56–59] All bacteria are probably surrounded by glycocalyces as they grow in their natural habitat, although they often lose these external layers when cultivated in the laboratory. The extracellular polymers may be in the form of S layers, capsules, or slime. The synthesis of extracellular polysaccharides is discussed

in Section 11.3. For a review of bacterial capsules and their medical significance, see Ref. 60.

## S layers

These are an array of protein or glycoprotein subunits on the cell wall. The S layers are found on the cell wall surfaces of a wide range of gram-positive and gram-negative eubacteria. They are also present in the archaea, where the S layer sometimes covers the cell membrane and serves as the cell wall itself. If the S layer is the wall itself, it is not called a glycocalyx.

## Capsules

Capsules are composed of fibrous material at the cell surface (Figs. 1.5 and 1.6). They may be rigid, flexible, integral (i.e., very closely associated with the cell surface), or peripheral (i.e., loose) material that is sometimes shed into the medium. Material that loosely adheres to the cell wall is sometimes called *slime or slime capsule*. If the loosely adhering or shed material is polysaccharide, it is often referred to as extracellular polysaccharide (EPS). The capsular polysaccharide is usually covalently bound to phospholipid or lipid-A, which is embedded in the surface of the cell. This is not the case for EPS. The synthesis of capsular material is described in Section 11.3.

## Chemical composition

The glycocalyces from several bacteria have been isolated and characterized. Although many are polysaccharide, some are protein. For example, some *Bacillus* species form a glycocalyx that is a polypeptide capsule. Also, pathogenic *Streptococcus* species have a fibrillar protein layer, the M protein, on the external face of the cell wall. The capsular polysaccharides are extremely diverse in their chemical composition and structure.[61] Some consist of one type of monosaccharides (homopolymers), whereas others are composed of more than one type of monosaccharide (heteropolymers). The monosaccharides are linked together via glycosidic linkages to form straight-chain or branched molecules.

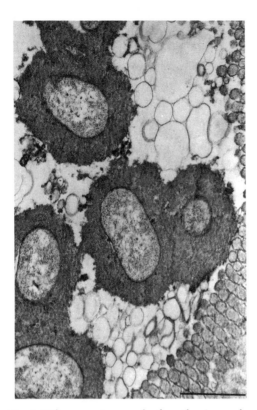

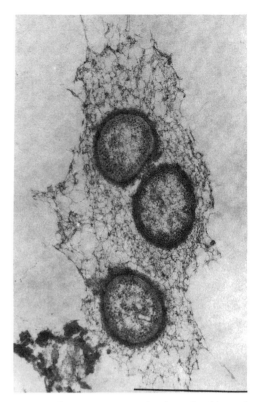

Fig. 1.5 Electron micrograph of a ruthenium red-stained thin section of *E. coli* shown adhering to the neonatal calf ileum. *Source:* From Costerton, J. W., T. J. Marrie, and K.-J. Cheng. 1985. Phenomena of bacterial adhesion, pp. 3–43. In: Bacterial Adhesion, D. C. Savage and M. Fletcher (Eds.). Plenum Press, New York and London.

Fig. 1.6 Electron micrograph of a ruthenium red-stained thin section of bacteria adhering to a rock surface in a subalpine stream showing fimbriated glycocalyx. *Source:* From Costerton, J. W., T. J. Marrie, and K.-J. Cheng. 1985. Phenomena of bacterial adhesion, pp. 3–43. In: *Bacterial Adhesion.* D. C. Savage and M. Fletcher (Eds.). Plenum Press, New York and London.

They can be substituted with organic or inorganic molecules, which further increases their diversity. For example, *Escherichia coli* strains make more than 80 different capsular polysaccharides, called K antigens. Sometimes the same capsular polysaccharide is made by different bacterial species. For example, the K1 polysaccharide of *E. coli* is identical to the group B capsular polysaccharide of *Neisseria meningitidis.*

### Role

An important role for the glycocalyx is adhesion to the surfaces of other cells or to inanimate objects to form a biofilm. Such adhesion is necessary for colonization of solid surfaces or growth in biofilms (Figs. 1.5 and 1.6). The bacteria can adhere to a nonbacterial surface or to each other via these attachments. An example is the complex ecosystem of several different bacteria maintained by intercellular adherence in human oral plaques.[62] Advantages to such adherence include the higher concentrations of nutrients that may be found on the surfaces.

Another role of the glycocalyx is protection from phagocytosis. For example, mutants of pathogenic strains of bacteria that no longer synthesize a capsule, such as unencapsulated strains of *Streptococcus pneumoniae*, are more easily phagocytized by white blood cells, making the pathogens less virulent. The glycocalyx can also prevent dehydration of the bacterial cell. This is because polyanionic polysaccharides, including polysaccharide capsules, are heavily hydrated.

### 1.2.3 Cell walls

Most bacteria are surrounded by a cell wall that lies over the external face of the cell membrane and protects the cell from bursting due to the turgor pressure that exists within the cell.[63, 64] The turgor pressure is due to the fact that bacteria generally live in environments that are more dilute than the cytoplasm. As a consequence, there is a net influx of water, resulting in a large hydrostatic pressure (turgor) of several atmospheres directed out against the cell membrane. Two different types of walls exist among the bacteria. Bacteria possessing one type of wall can be stained using the Gram stain procedure. They are called gram positive. Bacteria possessing the second wall type do not stain and are called gram negative. The Gram stain is described next, followed by a description of peptidoglycan, a cell wall polymer found in both gram-positive and gram-negative walls responsible for the strength of bacterial cell walls. This will be followed by descriptions of the gram-positive and gram-negative cell walls.

### The Gram stain

The Gram stain was invented in the nineteenth century by Christian Gram to visualize bacteria in tissues using ordinary bright-field microscopy. When appropriately stained, bacteria with cell walls can usually be divided into two groups, depending upon whether they retain a crystal violet–iodine stain complex (gram positive) or not (gram negative) (Fig. 1.7). Gram-negative bacteria are visualized with a pink counter stain called safranin. During the staining procedure the complex of crystal violet and iodine can be removed with alcohol or acetone from gram-negative cells, but not from gram-positive cells. This is thought to be related to the thickness and composition of the gram-positive wall as compared to the much thinner gram-negative wall. The crystal violet–iodine complex is trapped within gram-positive cells by the thick cell wall. The difference in gram

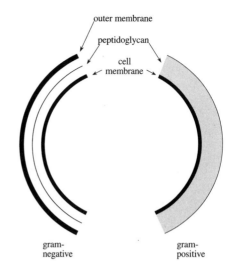

**Fig. 1.7** Bacterial cell walls. Schematic illustration of a gram-negative wall and a gram-positive wall. Note the presence of an outer membrane (also called outer envelope) in the gram-negative wall and the much thicker peptidoglycan layer in the gram-positive wall.

staining is also related to the outer wall layer of gram-negative bacteria (also called the outer envelope or outer membrane), which is rich in phospholipids and thus made leaky by the lipid solvents, alcohol, and acetone. The archaea can stain either gram positive or gram negative, but their cell wall composition is different from that of the bacteria, as will be described later. Although useful for identification, the gram stain cannot be solely relied upon to determine relationships among the bacteria. Nor can one rely on the gram stain as a probe for cell wall structure or composition.

### Peptidoglycan

The strength and rigidity of bacterial cell walls is due to a glycopeptide called peptidoglycan or murein, which consists of glycan chains crosslinked by peptides (Figs. 1.8 and 11.3). The glycan consists of alternating residues of N-acetylglucosamine (GlcNAc or G) and N-acetylmuramic acid (MurNAc or M) attached to each other via $\beta$-1,4 linkages. Attached to the MurNAc is a tetrapeptide that crosslinks the glycan chains via peptide bonds. The tetrapeptide usually consists of L-alanine, D-glutamate, a diamino acid, and D-alanine. The

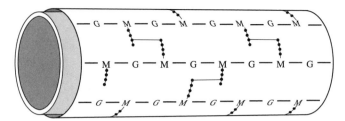

**Fig. 1.8** Schematic drawing of a peptidoglycan layer. The peptidoglycan surrounds the cell membrane and consists of glycan chains (–G–M–) cross-linked by tetrapeptides (closed circles). The thickness of the peptidoglycan varies from approximately a monomolecular layer in gram-negative bacteria to several layers thick in gram-positive bacteria. The direction in which the glycan chains are depicted to be running is not to be taken literally. Symbols: M, N-acetylmuramic acid; G, N-acetylglucosamine.

structural diversity of the peptidoglycan as well its synthesis is discussed in more detail in Section 11.1. The peptidoglycan forms a three-dimensional network surrounding the cell membrane, covalently bonded throughout by the glycosidic and peptide linkages. It is the covalent bonding that gives the peptidoglycan its strength. The shape of the peptidoglycan determines the shape of the cell as well as shape changes that occur during cellular morphogenesis, such as during the formation of round cysts from rod-shaped vegetative cells. Destruction of the peptidoglycan by the enzyme lysozyme (which hydrolyzes the glycosidic linkages) or interference in its synthesis by antibiotics such as penicillin, vancomycin, or bacitracin results in the inability of the cell wall to restrain the turgor pressures. Under these circumstances the influx of water bursts the cell in dilute media. Since the strength of the peptidoglycan is due to the fact that it is covalently bonded throughout, changes in its shape, or expansion during growth of the cell, must be accompanied by hydrolysis of some of the covalent bonds and synthesis of new bonds. However, because of the high intracellular turgor pressure, this must be done very carefully so that a lethal weakness in the peptidoglycan structure is not produced when its bonds are cleaved. This problem was discussed by Koch.[65]

There are some important differences between the peptidoglycan in gram-positive and gram-negative bacteria. The peptidoglycan of gram-negative bacteria can be isolated as a sac of pure peptidoglycan that surrounds the cell membrane in the living cell. It is called

the *murein sacculus*. The sacculus is elastic and believed to be under stress in vivo because of the expansion due to turgor pressure against the cell membrane.[58, 66] In contrast, the peptidoglycan from gram-positive bacteria is covalently bonded to various polysaccharides and teichoic acids and not isolatable as a pure murein sacculus (see the discussion later of the gram-positive wall). The peptidoglycan from gram-negative bacteria differs from the peptidoglycan from gram-positive bacteria in two other ways: (1) diaminopimelic acid is generally the diamino acid in gram-negative bacteria, whereas it is the diamino acid in only some of the gram-positive bacteria; and (2) the crosslinking is usually direct in gram-negative bacteria, whereas there is usually a peptide bridge in gram-positive bacteria (Fig. 11.3).

As described later, the peptidoglycan in gram-negative bacteria is attached noncovalently to the outer envelope via lipoprotein. Whether it is also attached to the cell membrane is a matter of controversy. The evidence for attachment to the cell membrane is that when *E. coli* is plasmolyzed (placed in hypertonic solutions so that water exits the cells), and prepared for electron microscopy by chemical fixation followed by dehydration with organic solvents, the cell membrane is seen to have adhered to the outer envelope in numerous places while generally shrinking away in other locations. It is reasonable to suggest that the zones of adhesion may be areas where the peptidoglycan bonds the cell membrane to the outer membrane because the peptidoglycan lies between the two membranes. However, this is currently

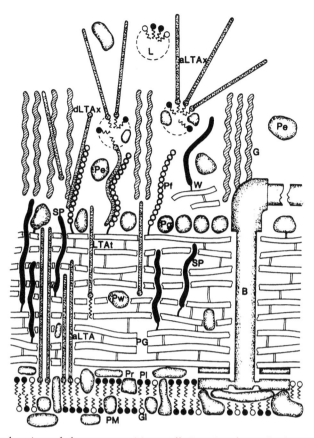

**Fig. 1.9** Schematic drawing of the gram-positive wall. Starting from the bottom up, PM, plasma cell membrane consisting of protein (Pr), phospholipid (Pl), and glycolipid (Gl). Overlaying the cell membrane is highly cross-linked peptidoglycan (PG) to which is covalently bound teichoic acids, teichuronic acids, and other polysaccharides (SP). Acylated lipoteichoic acid (aLTA) is bound to the cell membrane and extends into the peptidoglycan. Some of the LTA is in the process of being secreted (LTAt). LTA that is already secreted into the glycocalyx is symbolized as aLTAx. Some of the LTA in the glycocalyx is deacylated and symbolized as dLTAx. (Deacylated means that the fatty acids have been removed from the lipid moiety.) Within the glycocalyx can also be found lipids (L), proteins (Pe), pieces of cell wall (W), and polymers that are part of the glycocalyx proper (G). Also shown is an inserted flagellum (B). *Source:* From Wicken, A. 1985. Bacterial cell walls and surfaces, pp. 45–70. In: *Bacterial Adhesion.* D. C. Savage and M. Fletcher (Eds.). Plenum Press, New York and London.

a controversial subject. It has been argued that the zones of adhesion seen in electron micrographs of plasmolyzed, chemically fixed, and dehydrated cells are artifacts since they were not visible in cells prepared for electron microscopy by rapid cryofixation (also called freeze substitution) after plasmolysis.[67] In cryofixation the water in and around the cells is rapidly frozen by liquid helium and the frozen water is substituted with osmium tetroxide in acetone.

## Gram-positive walls

### 1. Chemical composition

The gram-positive cell wall in bacteria is a thick structure approximately 15–30 nm wide and consists of several polymers, the major one being peptidoglycan (Figs. 1.8 and 1.9). The kinds and amounts of the other polymers in the wall vary according to the genus of bacterium. The nonpeptidoglycan polymers can comprise up to 50% of the dry weight of the wall. The ones most commonly found are usually covalently bound to the glycan chain of the peptidoglycan. These polymers include (Fig. 1.10):

*Teichoic acids.* Teichoic acids are polymers of either ribitol phosphate or glycerol phosphate.

*Teichuronic acids.* Teichuronic acids are acidic polysaccharides containing uronic acids (e.g., some have N-acetylgalactosamine and D-glucuronic acid).

*Neutral polysaccharides.* Neutral polysaccharides are particularly important for the classification of streptococci and lactobacilli, where they are used to divide the bacteria into serological groups (e.g., groups A, B, and C streptococci).

*Lipoteichoic acids.* A teichoic acid found in most gram-positive walls is lipoteichoic acid (LTA), which is a linear polymer of phosphodiester-linked glycerol phosphate covalently bound to lipid. The C2 position of the glycerol-phosphate is usually glycosylated and/or D-alanylated. Because of the negatively charged backbone of glycerol phosphate and the hydrophobic lipid, the molecule is amphipathic (i.e., it has both a polar and a nonpolar end). The lipid portion is bound hydrophobically to the cell membrane, whereas the polyglycerol phosphate portion extends into the cell wall. Unlike the other polymers thus far discussed, LTA is not covalently bound to the peptidoglycan. Its biological role at this location is not understood,

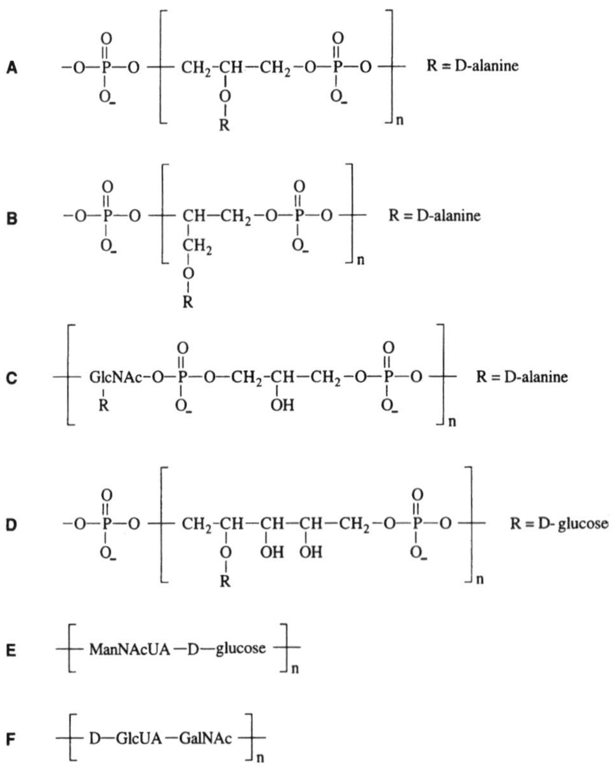

Fig. 1.10 Some teichoic and teichuronic acids found in different gram-positive bacteria. (A) Glycerol phosphate teichoic acid with D-alanine esterified to the C2 of glycerol. (B) Glycerol phosphate teichoic acid with D-alanine esterified to the C3 of glycerol. (C) Glycerol phosphate teichoic acid with glucose and N-acetylglucosamine in the backbone subunit. D-alanine is esterified to the C6 of N-acetylglucosamine. (D) Ribitol phosphate teichoic acid, with D-glucose attached in a glycosidic linkage to the C4 of ribitol. (E) Teichuronic acid with N-acetylmannuronic acid and D-glucose. (F) Teichuronic acid with glucuronic acid and N-acetylgalactosamine. It is believed that teichoic and teichuronic acids are covalently bound to the peptidoglycan through a phosphodiester bond between the polymer and a C6 hydroxyl of muramic acid in the peptidoglycan.

but in some bacteria it is secreted and can be found at the cell surface, where it is thought to act as an adhesin. For example, LTA is secreted by *S. pyogenes*, where it binds with the M protein and acts as a bridge to receptors on host tissues. (See Note 68 for a further explanation.) Under these circumstances one should consider the secreted LTA as part of the glycocalyx along with the M protein.

*Other glycolipids.* There is a growing list of gram-positive bacteria that do not contain LTA but have instead other amphiphilic glycolipids that might substitute for some of the LTA functions, whatever they might be.[69] Bacteria having these cell-surface glycolipids (also called macroamphiphiles or lipoglycans) belong to various genera including *Micrococcus, Streptococcus, Mycobacterium, Corynebacterium, Propionibacterium, Actinomyces,* and *Bifidobacterium.*

*Mycolic acid.* Still another cell wall polymer is formed by bacteria belonging to the genus *Mycobacterium*, which include the causative agents of tuberculosis and leprosy. They have cell walls rich in waxy lipids called mycolic acids. When these bacteria are appropriately stained, the dye (basic fuschin) is not removed by dilute hydrochloric acid in ethanol (acid alcohol) because of the presence of the mycolic acid. They are therefore called *acid-fast* bacteria.

In summary, it can be stated that gram-positive cell walls have diverse types of neutral and acidic polysaccharides, glycolipids, lipids, and other compounds either free in the wall or covalently bound to the peptidoglycan. The functions of most of these polymers is largely unknown, although some may act as adhesins and/or presumably affect the permeability characteristics of the cell wall.

## Gram-negative wall

The gram-negative cell wall is structurally and chemically complex. It consists of an outer membrane composed of lipopolysaccharide, phospholipid, and protein, and an underlying peptidoglycan layer (Fig. 1.11). Between the outer and inner membrane (the cell membrane) is a compartment called the *periplasm*, wherein the peptidoglycan lies.

### 1. Lipopolysaccharide structure and function

The lipopolysaccharide (LPS) consists of three regions: *lipid A, core,* and a *repeating oligosaccharide,* sometimes called *o-antigen* or *somatic antigen.* Its chemical structure and synthesis are described in Section 11.2. In the Enterobacteriaceae (e.g., *E. coli*), the lipopolysaccharide is confined to the outer leaflet of the outer membrane and is arranged so that the lipid A portion is embedded in the membrane as part of the lipid layer, and the core and oligosaccharide extend into the medium (Fig. 1.11). Mutants that lack the oligosaccharide experience loss of virulence, and it is believed that the LPS can increase pathogenicity. Loss of most of the core and oligosaccharide in *E. coli* and related bacteria is associated with increased sensitivity to hydrophobic compounds (e.g., antibiotics, bile salts, and hydrophobic dyes such as eosin and methylene blue). This is because the LPS provides a permeability barrier to hydrophobic compounds. It is advantageous for the enteric bacteria to have such a permeability barrier because they live in the presence of bile salts in the intestine. In fact, the basis for selective media for gram-negative bacteria is the resistance of these bacteria to bile salts and/or hydrophobic dyes, which are included in the media. The bile salts and dyes inhibit the growth of gram-positive bacteria, but not gram-negative bacteria, because of the LPS. An example of such a selective medium is eosin–methylene blue (EMB) agar, which is used for the isolation of gram-negative bacteria because it inhibits the growth of gram-positive bacteria.

The low permeability of the outer membrane to hydrophobic compounds is apparently due to the fact that the phospholipids are confined primarily to the inner leaflet of the outer envelope and the LPS is in the outer leaflet. The LPS presents a permeability barrier to hydrophobic substances. One suggestion as to how this might occur is that, since lipid A contains only saturated fatty acids, the LPS presents a somewhat rigid matrix, and this, plus the tendency of the large LPS molecules to engage in lateral noncovalent interactions, makes it difficult for hydrophobic molecules to penetrate between the LPS molecules to

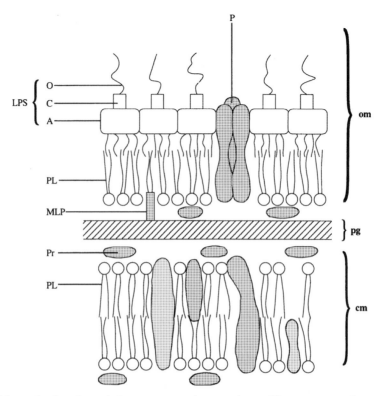

**Fig. 1.11** Schematic drawing of the gram-negative envelope. The outer membrane consists of lipopolysaccharide, phospholipid, and proteins, most of which are porins. Underneath the outer membrane is the peptidoglycan layer that is noncovalently bonded to the outer membrane via murein lipoproteins, themselves covalently attached to the peptidoglycan. The cell membrane is composed of phospholipid and protein. The area between the outer membrane and the cell membrane is called the periplasm. The wavy lines are fatty acid residues, which anchor the phospholipids and lipid A into the membrane. Abbrev. LPS, lipopolysaccharide; O, oligosaccharide; C, core; A, lipid A; P, porin; PL, phospholipid; MLP, murein lipoprotein; pg, peptidoglycan; Pr, protein; om, outer membrane; cm, cell membrane.

the phospholipid layer. The fact that mutants that lack a major region of the oligosaccharide and core are more permeable to hydrophobic compounds is due to the fact that there is more phospholipid in the outer leaflet of these mutants, and is also due to the likelihood that there is less lateral interaction between the LPS molecules. The asymmetric distribution of phospholipid to the inner leaflet of the outer envelope seems to be an adaptive evolutionary response of the enteric bacteria to hydrophobic toxic substances in the intestines of animals. Accordingly, not all gram-negative bacteria have an outer envelope with an asymmetric distribution of lipopolysaccharide and phospholipid.

## 2. Lipoproteins

In addition to lipopolysaccharide and phospholipid, a major component of the outer membrane is protein, of which there are several different kinds. One of the proteins is called the *murein lipoprotein*. This is a small protein with lipid attached to the amino terminal end (Fig. 1.12). The lipid end of the molecule extends into and binds hydrophobically with the lipids in the outer envelope. The protein end of some of the molecules is covalently bound to the peptidoglycan, thus anchoring the outer envelope to the peptidoglycan. In *E. coli*, about one-third of the murein lipoprotein is bound to the peptidoglycan. Mutants unable to synthesize the murein

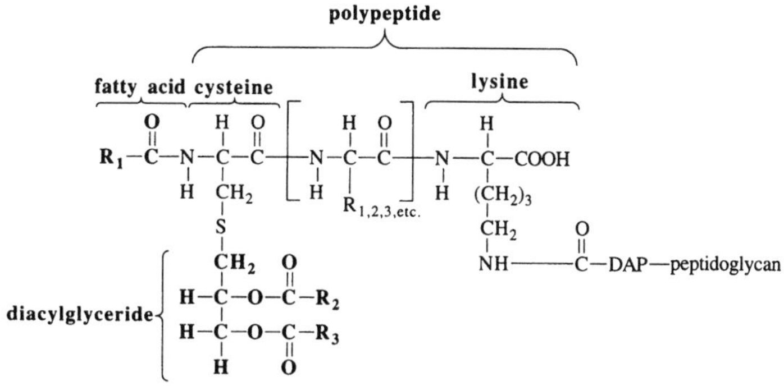

**Fig. 1.12** Murein lipoprotein. Attached to the amino terminal cysteine is a diacylglyceride in thioether linkage and a fatty acid in amide linkage. The lipid portion extends into the outer envelope and binds hydrophobically with the fatty acids in the phospholipids and lipopolysaccharide. The carboxyterminal amino acid is lysine, which can be attached via an amide bond to the carboxyl group of diaminopimelic acid (DAP) in the peptidoglycan. The murein lipoprotein therefore holds the outer envelope to the peptidoglycan.

lipoprotein have unstable outer envelopes that bleb off into the medium at the cell poles and septation sites. Therefore, the murein lipoprotein may play a structural role in keeping the outer membrane attached to the cell surface. There are a small number of other outer membrane or cell membrane lipoproteins.[70] These were discovered by chemically crosslinking the petidoglycan in whole cells to closely associated proteins with a bifunctional crosslinking reagent. (See endnote 71 for how crosslinking reagents work.) There is no evidence that these additional lipoproteins are covalently bonded to the peptidoglycan, and their functions remain to be elucidated.

*3. Porins and other proteins*

The major proteins in the outer envelope are called porins. The porins form small nonspecific hydrophilic channels through the outer envelope, allowing the diffusion of neutral and charged solutes of molecular weights less than 600 Da. The channels are necessary to allow passage of small molecules into and out of the cell. *E. coli* has three major porins: OmpF, OmpC, and PhoE. Each porin makes a separate channel. Thus there are OmpF, OmpC, and PhoE channels. The OmpC channel is approximately 7% smaller than the OmpF channel and is expected to make the outer envelope less permeable to larger molecules. OmpF and OmpC are both present under all growth conditions, although

the ratio of the smaller OmpC to the larger OmpF increases in high osmolarity media and at high temperature. The increased amounts of OmpC relative to OmpF presumably also occur in the intestine, which has a higher osmolarity and temperature than in lakes and streams, where *E. coli* is also found. This may be an advantage to the enterics because the smaller OmpC channel should present a diffusion barrier to toxic substances in the intestine, whereas the larger OmpF channel should be an advantage in more dilute environments outside the body. The protein PhoE is produced only under conditions of inorganic phosphate limitation. This is because PhoE is a channel for phosphate (and other anions) whose synthesis under limiting phosphate conditions reflects the need to bring more phosphate into the cell. Porins appear to be widespread among gram-negative bacteria, although they are not all identical. As mentioned earlier, *E. coli* regulates the amounts of the various porins according to growth conditions. This is discussed in Chapter 18.

Since the porins exclude molecules with molecular weights larger than 600 Da, one would expect to find other proteins in the outer membrane that facilitate the translocation of larger solutes across the outer membrane. This is the case. *E. coli* has an outer membrane protein called the LamB protein that forms channels for maltose and maltodextrins. Other proteins in the outer membrane of

*E. coli* facilitate the transport of vitamin $B_{12}$ (BtuB), nucleosides (Tsx), and several other solutes.

## Archaeal cell walls

Archaeal cell walls are not all alike and are very different from bacterial cell walls. For example, *no archaeal cell wall contains peptidoglycan.* Archaeal cell walls may be either pseudopeptidoglycan, polysaccharide, or protein (the S layer). Pseudopeptidoglycan (also called pseudomurein) resembles peptidoglycan in consisting of glycan chains crosslinked by peptides (Fig. 1.13). However, the resemblance stops here. In pseudopeptidoglycan, although one of the sugars is N-acetylglucosamine as in peptidoglycan, the other is N-acetyltalosaminuronic acid instead of N-acetylmuramic acid. Furthermore, the sugars are linked by a $\beta-(1 \to 3)$ glycosidic linkage rather than a $\beta-(1 \to 4)$, and the amino acids in the peptides are L-amino acids rather than D-amino acids. (The latter are present in peptidoglycan; compare Fig. 1.13 with Fig. 11.2.) Thus pseudomurein is very different from murein.

## 1.2.4 Periplasm

Gram-negative bacteria have a separate compartment between the cell membrane and the outer membrane called the *periplasm*[72] (Fig.1.14). It can be seen in electron micrographs of thin sections of cells as a space but should be considered an aqueous compartment containing protein, oligosaccharide, salts, and the peptidoglycan. It seems that the peptidoglycan and oligosaccharides may exist in a hydrated state forming a periplasmic gel.[73] There is an unresolved controversy in the literature as to whether the periplasm has zones of adhesion (Bayer's patches) between the inner and outer membranes.[66, 74] Whether or not zones of adhesion are seen depends upon how the cells are prepared for electron microscopy. One school of thought holds that the zones are artifacts of fixation, whereas the counterargument is that the zones are seen only if the proper techniques are employed to preserve the rather fragile adhesion sites. Whether or not the zones of adhesion actually exist is an important issue because they have been postulated to be sites of export of lipopolysaccharide and protein that become part of the outer membrane. The periplasm should be considered a cellular compartment with specialized activities. These activities include oxidation–reduction reactions (Chapters 4 and 12), osmotic regulation (Chapter 15), solute transport (Chapter 16), protein secretion (Chapter 17), and hydrolytic activities such as are mediated by phosphatases and nucleases. The phosphatases and nucleases

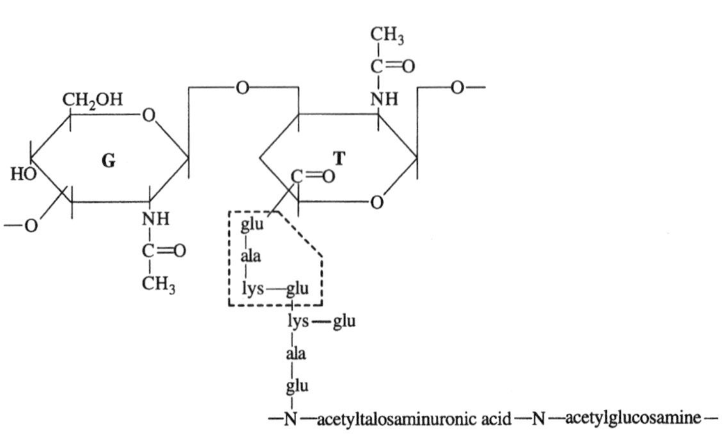

Fig. 1.13 Pseudopeptidoglycan in archaea. Pseudopeptidoglycan resembles peptidoglycan in being a cross-linked glycopeptide. It differs from peptidoglycan in the following ways: (1) N-acetyltalosaminuronic acid replaces N-acetylmuramic acid. (2) The glycosidic linkage is $\beta$1–3 instead of $\beta$1–4. (3) There are no D-amino acids. Abbreviations: G, N-acetylglucosamine; T, N-acetyltalosaminuronic acid. The peptide subunit is enclosed by dotted lines.

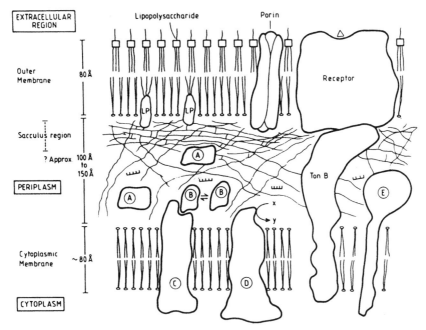

**Fig. 1.14** A model of the periplasm in *E. coli*. The outer region of the periplasm is thought to consist of cross-linked peptidoglycan attached to the outer envelope via lipoprotein (LP) covalently bound to the peptidoglycan. The inner region of the periplasm (approaching the cell membrane) is believed to consist of fewer cross-linked peptidoglycan chains and oligosaccharides that are hydrated and form a gel. The gel phase is thought to contain periplasmic proteins, e.g., A and B. For example, A might be a periplasmic enzyme, and B could be a solute binding protein that interacts with a membrane transporter (C). D and E are integral membrane proteins, perhaps enzymes. The outer envelope is depicted as consisting of lipopolysaccharide, porins, and specific solute transporters (receptor) that require a second protein (TonB) for uptake. *Source:* From, Ferguson, S. J. 1991. The periplasm. pp. 311–339. In: *Prokaryotic Structure and Function, A New Perspective.* S. Mohan, C. Dow, and J. A. Coles (eds.). Cambridge University Press.

degrade organophosphates and nucleic acids that might enter the periplasm from the medium and transport the hydrolytic products into the cell.

*Periplasmic components*

The periplasm is chemically complex and carries out diverse functions. The following list of components and their functions reflects the importance of the periplasm. The list should be considered as a partial inventory emphasizing those periplasmic functions about which most is known.

*1. Oligosaccharides*
The oligosaccharides in the periplasm are thought by some to be involved in osmotic regulation of the periplasm because their amounts decrease when the cells are grown in

media of high osmolarity. This is a complex subject and is discussed more fully in Section 15.2.

*2. Solute binding proteins*
There are also solute binding proteins in the periplasm that assist in solute transport. The solute-binding proteins bind to solutes (e.g., sugars and amino acids that have entered the periplasm through the outer envelope) and deliver the solutes to specific transporters (carriers) in the membrane. This is an important means of bringing nutrients into the cell and is discussed in Section 16.3.3.

*3. Cytochromes c*
Some of the enzymes in the periplasm are cytochromes c that oxidize carbon compounds or inorganic compounds and deliver the

electrons to the electron transport chain in the cell membrane. These oxidations are called periplasmic oxidations. There are other oxidoreductases in the periplasm as well, but the various cytochromes c are very common. Periplasmic oxidations are important for energy metabolism in many different gram-negative bacteria and are discussed in Chapters 4 and 12.

### 4. Hydrolytic enzymes

There are also hydrolytic enzymes in the periplasm that degrade nutrients to smaller molecules that can be transported across the cell membrane by specific transporters. For example, the enzyme amylase is a periplasmic enzyme that degrades oligosaccharides to simple sugars. Another example is alkaline phosphatase, which removes phosphate from simple organic phosphate monoesters. The inorganic phosphate is then carried into the cell via specific inorganic phosphate transporters.

### 5. Detoxifying agents

Some periplasmic enzymes are detoxifying agents. For example, the enzyme to degrade penicillin ($\beta$-lactamase) is a periplasmic protein.

### 6. TonB protein

Figure 1.14 also shows an interesting periplasmic protein anchored to the cell membrane in E. coli called the TonB protein, whose mechanism of action is not understood. It is known that TonB is required for the uptake of several solutes that do not diffuse through the porins; rather, they require specific transport systems (also called receptors) in the outer envelope. All these solutes have molecular weights larger than 600 Da, which is the upper limit for molecules that enter via the porins. Examples of solutes with specific outer membrane receptors that require TonB for uptake are iron siderophores[75] and vitamin $B_{12}$ (cobalamin). The interesting thing is that these solutes are brought into the periplasm against a large concentration gradient, sometimes 1,000 times higher than the concentration outside the cell. In a way that is not understood, the TonB protein couples the electrochemical energy (the proton motive force, $\Delta p$) in the cell membrane to the uptake of certain solutes through the outer envelope and into the periplasm.[76–78] One suggestion is that TonB is energized by the electrochemical potential that exists across bacterial cell membranes and that in the energized state TonB causes a conformational change in the outer membrane receptor protein that leads to binding of substances such as ferric siderophores and $B_{12}$ to the receptor and the opening of a receptor channel accompanied by translocation through the receptor channel into the periplasm.

## Is there a periplasm in gram-positive bacteria?

In the past it has been assumed that there is no space between the cell membrane and the cell wall in gram-positive bacteria equivalent to the periplasm of gram-negative bacteria. This is because thin sections of gram-positive cells do not indicate that such a space exists, and of course gram-positive bacteria do not have an outer envelope. Recent evidence suggests, however, that perhaps gram-positive bacteria do have a compartment analogous to the periplasm found in gram-negative bacteria, albeit one that is much more compact than the gram-negative periplasm.[79] The evidence in favor of the existence of a periplasm in Bacillus subtilis is the release of putative periplasmic proteins after making protoplasts of the cells with lysozyme. (Protoplasts are cells from which the cell wall has been removed. They can be stabilized by suspension in isotonic buffer, i.e., buffer that is isosmolar with the cell contents so that water does not rush in and lyse the protoplasts. Proteins solubilized by removing the cell wall during protoplast formation may include proteins resident in the area between the cell membrane and the cell wall.) The proteins released by protoplast formation include nucleases that are distinct from cytoplasmic nucleases. The exact nature of this putative periplasm, including its contents and the relationships of the proteins found therein with the cell membrane and cell wall, as well as the similarities to the gram-negative periplasm, are not known and must await further studies.

## 1.2.5 Cell Membrane

We now come to what is certainly the most functionally complex of the cell structures (i.e., the cell membrane). The cell membrane is responsible for a broad range of physiological activities, including solute transport, electron transport, photosynthetic electron transport, the establishment of electrochemical gradients, motility, ATP synthesis, biosynthesis of lipids, biosynthesis of cell wall polymers, secretion of proteins, intercellular signaling, and responses to environmental signals. To refer to the cell membrane simply as a lipoprotein bilayer does not do justice to the machinery embedded in the lipid matrix. It is clearly a complex mosaic of parts whose structure and interactions at the molecular level are not well understood. As expected, the protein composition of cell membranes is complex. There can be more than 100 different proteins. Many of the proteins are clustered in functional aggregates (e.g., the proton translocating ATPase, the flagella motor, electron transport complexes, and certain of the solute transporters). At the molecular level, the membrane is certainly a complex and busy place. What follows is a general description of the membrane, without reference to its microheterogeneity.

### Bacterial cell membranes

Bacterial cell membranes consist primarily of phospholipids and protein in a fluid mosaic structure in which the phospholipids form a bilayer (Fig. 1.15). The structure is said to be fluid because there is extensive lateral mobility of bulk proteins and phospholipids. Although this is the case, it should be realized that certain protein aggregates (e.g., complex solute transporters and electron transport aggregates) remain as aggregates within which the proteins interact to catalyze sequential reactions.

The phospholipids are fatty acids esterified to two of the hydroxyl groups of phosphoglycerides (Fig. 1.16). The structure and synthesis of phospholipids is described in detail in Section 9.1.2. The third hydroxyl group in the glycerol backbone of the phospholipid is covalently bound to a substituted phosphate

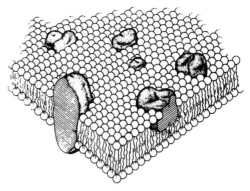

**Fig. 1.15** Model of the cell membrane showing bimolecular lipid leaflet and embedded proteins. The phospholipid molecules are shown interacting with each other via their hydrophobic (apolar) "tails." The hydrophilic (polar) "heads" of the phospholipids face the outside of the membrane, where they interact with proteins and ions. Proteins can span the membrane or be partially embedded. *Source:* From Singer, S. J., and G. L. Nicolson. 1972. The fluid mosaic model of the structure of cell membranes. *Science* **175**:720–731. Copyright 1972 by the AAS.

group, which makes one end of the molecule very polar due to a negative charge on the ionized phosphate group. Because the phospholipids are polar at one end and nonpolar at the other end (the end with the fatty acids), they are said to be *amphipathic*, and will spontaneously aggregate with their nonpolar fatty acid regions interacting with each other by hydrophobic bonding, while their polar phosphorylated regions face the aqueous phase, where ionic interactions with cations, water, and polar groups on proteins occurs. Phospholipids accomplish this by spontaneously forming lipid bilayers in water solutions or in cell membranes.

There are two classes of proteins in membranes, *integral* and *peripheral*. Integral proteins are embedded in the membrane and are bound to the fatty acids of the phospholipids via hydrophobic bonding. They can be removed only with detergents or solvents. Peripheral proteins are attached to membrane surfaces to the phospholipids by ionic interactions, and can be removed by washing the membrane with salt solutions. The lipids and proteins diffuse laterally in

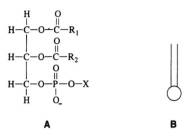

**A**   **B**

**Fig. 1.16** Phospholipids have both a polar and a nonpolar end. (A) Phospholipid with two fatty acids (R) esterified to glycerol. The phosphate is substituted by X, which determines the type of phospholipid. In bacteria, X is usually serine, ethanolamine, a derivative of glycerol, or a carbohydrate derivative. See Section 9.1.3 for a more complete description of bacterial phospholipids. (B) A schematic drawing of a phospholipid showing the polar (circle) and nonpolar (straight lines) regions.

the plane of the membrane but do not rotate across the membrane. The insertion of the proteins into the membrane during membrane synthesis is discussed in Section 17.2.

The phospholipid bilayer acts as a permeability barrier to virtually all water-soluble molecules so that most solutes diffuse or are carried across the membrane through or on special protein transporters that bridge the phospholipid bilayer. This is discussed in the context of solute transport in Chapter 16. (However, the lipid bilayer is permeable to water molecules, gases, and small hydrophobic molecules.) An important consequence of the lipid matrix is that ions do not freely diffuse across the membrane unless they are carried on or through protein transporters. Because of this, the membrane is capable of holding a charge that is due to the unequal transmembrane distribution of ions. This is discussed in Chapter 3.

*Archaeal cell membranes*

*1. The lipids*
Archaeal membrane lipids differ from those found in bacterial membranes.[80–82] The archaeal lipids consist of *isopranoid alcohols,* either 20 or 40 carbons long, *ether-linked* to one glycerol to form monoglycerol diethers or to two glycerols to form diglycerol tetraethers. These are illustrated in Fig. 1.17. Their synthesis is described in Section 9.1.3. (Recall

that bacterial glycerides are fatty acids esterified to glycerol. Refer to Fig. 1.16 for a comparison.) The $C_{20}$ alcohol is a fully saturated hydrocarbon called *phytanol*. The $C_{40}$ molecules are two phytanols linked together head to head in the diglycerol tetraether lipids. Thus the lipids are either phytanyl glycerol diethers or diphytanyl diglycerol tetraethers. The diethers and tetraethers occur in varying ratios depending upon the bacterium. For example, there may be from 5 to 25 different lipids in any one bacterium. This is really quite a diverse mix and can be contrasted to the four or five different phospholipids found in a typical bacterium. The diversity of the archaeal lipids is due to the different polar head groups that exist, as well as to the mix of core lipid to which the head groups are attached (Fig. 1.17). Although the polar head group is responsible for the polarity of most phospholipids, there is some polarity at one end of the archaeal lipids without a polar head group because of the free hydroxyl group on the glycerol. (Recall that hydroxyl groups are capable of forming hydrogen bonds with water and proteins.) It is usually stated that the ether linkages are an advantage over ester linkages in the acidic and thermophilic environments in which archaea live, because of the greater stability of the ether linkage to hydrolytic cleavage. It is clear, however, that ether-linked lipids are not necessary for growth at high temperatures. A bacterium called *Thermotoga* does not have ether-linked lipids, yet grows in geothermally heated marine sediments alongside the sulfur-dependent thermophilic archaea. This emphasizes that the correlation between ether-linked lipids and habitat is not precise.

*2. The proteins*
There is little information regarding archaeal membrane proteins. An exception is bacteriorhodopsin and halorhodopsin in *Halobacterium*, whose conformational array in the cell membrane is dependent upon interaction with polar membrane lipids.[81] The functions of these two proteins are discussed in Sections 3.8.4 and 3.9.

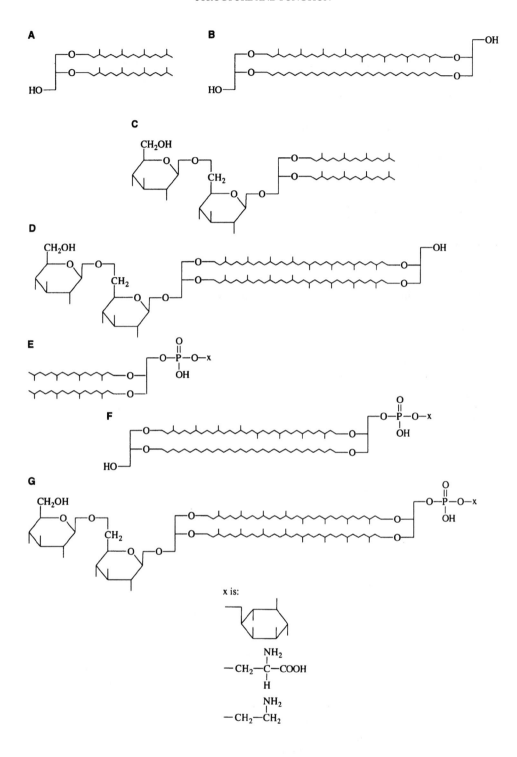

**Fig. 1.17** Major lipids of *Methanobacterium thermoautotrophicum*. (a) Glycerol diether (archaeol); (b) diglycerol tetraether (caldarchaeol); (c) a glycolipid (gentiobiosyl archaeol); (e) a phospholipid (archaetidyl-X), where X can be inositol, serine, or ethanolamine; (f) a phospholipid (caldarchaetidyl-X); (g) a phosphoglycolipid (gentiobiosyl caldarchatidyl-X). *Source:* From Nishihara, M., H. Morii, and Y. Koga. 1989. Heptads of polar ether lipids of an archaebacterium, *Methanobacterium thermoautotrophicum*: Structure and biosynthetic relationship. *Biochemistry* 28:95–102.

### 3. The membrane

The thermoacidophilic archaea and some methanogens have tetraether glycerolipids. These lipids have a polar head group at both ends and span the membrane forming a *lipid monolayer* (Fig. 1.18). This is the only known example of a membrane where there is no midplane region. Since there is no midplane region, the lipid monolayer is more resistant to levels of heat that would disrupt the hydrophobic bonds holding the two lipids in the lipid bilayer together. The increased resistance to heat of the lipid monolayer may confer an advantage to organisms living at high temperatures. However, it cannot be claimed that diether lipids or tetraether lipids are a *specific* adaptation to high temperatures, although they may be advantageous in these environments. This is because some mesophilic methanogens have tetraether lipids, whereas two extremely thermophilic archaeons, *Methanopyrus kandleri* and *Thermococcus celer*, do not have tetraether lipids.

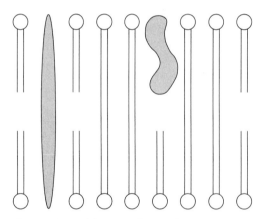

**Fig. 1.18** Lipid layer with membrane proteins in archaebacteria membranes. The glycerol diethers form a lipid bilayer, and the tetraethers form a monolayer. Some archaebacteria, e.g., the extreme halophiles, contain only the diethers. Most of the sulfur-dependent thermophiles have primarily the tetraethers, with only trace amounts of the diethers. Many methanogens have significant amounts of both the di- and tetraethers. The shaded areas represent protein.

## 1.2.6 Cytoplasm

The cytoplasm is defined as everything enclosed by the cell membrane. Cytoplasm is a viscous material containing a heavy concentration of protein (100–300 mg/ml),[83] salts, and metabolites. In addition, there are large aggregates of protein complexes designed for specific metabolic functions, various inclusions, and highly condensed DNA. Intracytoplasmic membranes are also present in many prokaryotes. The soluble part of the cytoplasm is called the *cytosol*.

### Intracytoplasmic membranes

Many prokaryotes have intracytoplasmic membranes that have specialized physiological functions.[84] Intracytoplasmic membranes are often connected to the cell membrane and are generally believed to be derived from invaginations of chemically modified areas of the cell membrane. However, in some cases connections to the cell membrane are not seen, and it is unknown whether the intracytoplasmic membranes are derived from an invagination of the cell membrane or are synthesized independently of the cell membrane (e.g., the thylakoids of cyanobacteria). The following lists examples of a few prokaryotes with intracytoplasmic membranes and their physiological roles.

### 1. Methanotrophs

Bacteria that grow on methane as their sole soure of carbon (methanotrophs) possess intracytoplasmic membranes that are suggested to function in methane oxidation. Methane oxidation is discussed in Section 13.2.1.

### 2. Nitrogen fixers

Bacteria utilizing nitrogen gas as a source of nitrogen use an oxygen-sensitive enzyme called *nitrogenase* to reduce the nitrogen to ammonia, which is subsequently incorporated into cell material. Many of these organisms have extensive intracytoplasmic membranes. One such nitrogen-fixing bacterium is *Azotobacter vinelandii*, whose intracytoplasmic membranes increase with the degree of aeration of the culture. Since respiratory activity is localized in the membranes, it is probable that an important role for *Azotobacter* intracytoplasmic membranes is to increase the cellular respiratory activity in order to provide more ATP for nitrogen fixation and to remove oxygen from the vicinity of the nitrogenase. Nitrogen fixation is discussed in Section 12.3.

### 3. Nitrifiers

Intracellular membranes are also found in nitrifying bacteria (i.e., those bacteria that oxidize ammonia and nitrite as the sole source of electrons: *Nitrosomonas, Nitrobacter, Nitrococcus*). Several of the enzymes that catalyze ammonia and nitrite oxidation are in the membranes. This is discussed in Section 12.4.

### 4. Phototrophs

In bacteria that use light as a source of energy for growth (phototrophs) the intracytoplasmic membranes are the sites for the photosynthetic apparatus. The membrane structure varies and can be flat membranes, vesicles, flat sacs (thylakoids in cyanobacteria), and tubular invaginations of the cell membrane (photosynthetic bacteria) (Fig. 5.16). Photosynthesis and photosynthetic membranes are discussed in Chapter 5.

## Inclusion bodies, multienzyme aggregates, and granules

Some bacteria contain specialized organelles in the cytoplasm. These differ from eukaryotic organelles in not being surrounded by a lipid bilayer–protein membrane, although they do have a membrane or coat. They are referred to by some researchers as inclusion bodies rather than organelles. In addition, there are numerous aggregates and multienzyme complexes of large size in all bacteria.

### 1. Gas vesicles

Aquatic bateria such as cyanobacteria, certain photosynthetic bacteria, and some nonphotosynthetic bacteria have *gas vesicles* surrounded by a protein coat. Gas vesicles are hollow spindle–shaped structures about 100 nm long filled with gas in equilibrium with the gases dissolved in the cytoplasm. The gas vesicles allow the bacteria to float in lakes and ponds at depths that support growth because of favorable light, temperature, or nutrients. Many of these bacteria and cyanobacteria are plentiful in stratified freshwater lakes, but they are not as abundant in isothermally mixed waters. Other bacteria containing gas vesicles live in hypersaline waters (*Halobacterium*), and a few marine species of cyanobacteria belonging to the genus *Trichodesmium* have gas vesicles. When gas vesicles are collapsed by experimentally subjecting cells to high hydrostatic pressure or turgor pressure, the cells are no longer buoyant and sink. Collapsed vesicles do not recover, and the cells acquire gas-filled vesicles only by de novo synthesis of new vesicles. Thus, during synthesis of the vesicles, water is excluded, presumably because of the hydrophobic nature of the inner protein surface. The green sulfur bacterium *Pelodictyon phaeoclathratiforme* forms gas vesicles only at low light intensities.[85] Perhaps this allows the bacteria to float at depths where the light is optimal for photosynthesis.

### 2. Carboxysomes

Bacteria that obligately grow on $CO_2$ as their sole or major source of carbon (strict autotrophs) sometimes have large (100 nm) polyhedral inclusions called *carboxysomes*.[86] These inclusions have been observed in nitrifying bacteria, sulfur oxidizers, and cyanobacteria. The distribution appears to be species specific (e.g., not all sulfur oxidizers have carboxysomes). Ribulose–bisphosphate carboxylase (RuBP carboxylase), the enzyme in the Calvin cycle that incorporates $CO_2$ into organic carbon, is stored in carboxysomes. The enzyme is discussed in Section 13.1.1. The physiological role for carboxysomes is not clear, since many autotrophs do not have them.

### 3. Chlorosomes

Green sulfur photosynthetic bacteria (e.g., *Chlorobium*) have ellipsoid inclusions, *chlorosomes* (formerly called chlorobium vesicles) that lie immediately underneath the cytoplasmic membrane. The chlorosomes are surrounded by a nonunit membrane of galactolipid, with perhaps some protein. At one time it was believed that such vesicles were found only in the green sulfur photosynthetic bacteria. However, similar vesicles have been found in the green photosynthetic gliding bacteria, *Chloroflexus*. The major light-harvesting photopigments are located in the chlorosomes, whereas the photosynthetic reaction centers are in the cell membrane. This means that during photosynthesis in these organisms light is absorbed by pigments in the chlorosomes and energy is transmitted to the reaction centers in the cell membrane, where photosynthesis takes place. In those photosynthetic bacteria that do not have chlorosomes, the light-harvesting pigments surround the reaction centers in the cell membrane. The structure and function of chlorosomes are discussed in Section 5.6.

### 4. Granules and globules

Bacteria often contain cytoplasmic granules whose content varies with the bacterium. Many bacteria store a lipoidal substance called *poly-β-hydroxybutyric acid* (PHB) as a carbon and energy reserve. Other bacteria may store *glycogen* for the same purpose. Other granules found in bacterial cells can include *polyphosphate*, and in some sulfur-oxidizing bacteria, *elemental sulfur globules*.

### 5. Ribosomes

Ribosomes are the sites of protein synthesis. They are small ribonucleoprotein particles, approximately 22 nm by about 30 nm or about the size of the smallest viruses, and consist of over 50 different proteins and three different types of RNA (23S, 16S, and 5S). Bacterial ribosomes sediment in a centrifugal field at a characteristic velocity of 70 Svedberg units (i.e., they are 70S), as opposed to eucaryotic cytosolic ribosomes, which are 80S. Bacterial ribosomes are very similar regardless of the bacterium. However, there are some differences from archaeal ribosomes, which are also 70S. These differences are described in Section 1.1.1.

### 6. Nucleoid

The site of DNA and RNA synthesis is the nucleoid, an amorphous mass of DNA unbounded by a membrane, lying approximately in the center of the cell. Faster-growing bacteria may contain more than one nucleoid, but each nucleoid has but one chromosome, and all the chromosomes are identical. The DNA is very tightly coiled. If the DNA from *E. coli* were stretched out, it would be 500 times longer than the cell. The DNA in the nucleoids is bound to several proteins as well as nascent chains of RNA. Of course, one of the proteins bound to the DNA is RNA polymerase molecules engaged in transcription, but several other proteins can be present as well. When nucleoids are isolated from *E. coli* under low salt conditions (which preserves the attachment of proteins to the DNA), membrane proteins are present as well as DNA-binding proteins HU, IHF, H-NS, and Fis. HU, often referred to as a histonelike protein because it has physical properties and an amino acid composition similar to eukaryotic histones, is the major DNA-binding protein present in *E. coli*.[87] It binds to DNA without any apparent sequence specificity, wrapping the DNA and causing it to bend. IHF (integration host factor) binds to specific sequences and also bends DNA. H-NS binds to curved or bent DNA. It is believed that all the DNA binding proteins facilitate bending of DNA and/or have a high affinity for bent DNA, or restraining of DNA supercoils in the nucleoid, and for the compact structure of the nucleoid.[88] The DNA-binding proteins are important not only for nucleoid structure but also for the regulation of expression of certain genes. See the discussion of the H-NS and Fis proteins in Section 2.2.2 (ribosomal RNA genes), the discussion of H-NS and IHF in Section 18.19.4 (virulence genes), and the discussion of IHF in Section 18.3 (nitrate regulation), Section 18.8.2 (Ntr regulon), and Section 18.10 (porin biosynthesis). Another protein important for DNA structure is DNA gyrase (topoisomerase II), which is responsible for negative supercoiling (i.e., the twisting of the double helix about its axis in the direction opposite to the right-handed double helix; Section 10.1). Cellular DNA is mostly in a negative supercoil, and if it were not for this, it would be much more difficult to unwind to obtain single strands for replication and transcription (Section 10.1) There is also an enzyme, topoisomerase I, that removes negative supercoils. Supercoiling of DNA and topoisomerases are discussed in Chapter 10.

### 7. Multienzyme complexes

It should not be thought that the cytoplasm is a random mixture of solutes and proteins. There are many examples of enzymes in the same pathway forming stable multienzyme complexes, reflecting strong intermolecular bonding.[89] For example, *pyruvate dehydrogenase* from *E. coli* is a complex of three different enzymes, each present in multiple copies (50 proteins total), that oxidizes pyruvic acid to acetyl-CoA and $CO_2$. The size of the pyruvate dehydrogenase complex is (4.6–4.8) $\times$ $10^6$ Da. Contrast this with the size of a 70S ribosome (another multienzyme complex), which is about 2.7 $\times$ $10^6$ Da. Other enzyme complexes that catalyze a consecutive series of biochemical reactions include the $\alpha$-*ketoglutarate dehydrogenase* complex, which consists of three different enzymes present in multiple copies (i.e., 48 proteins), 2.5 $\times$ $10^6$ daltons (in *E. coli*), and oxidizes $\alpha$-ketoglutarate to succinyl-CoA and $CO_2$. Another example is *fatty acid synthase* in yeast, which consists of seven different enzymes, 2.4 $\times$ $10^6$ Da, which synthesizes fatty acids from acetyl-SCoA. (For comparison to bacteria, see endnote[90].) One of the advantages of multienzyme complexes is that they facilitate the channeling of metabolites, thus increasing the efficiency of catalysis. For example, in these stable enzyme complexes the biochemical intermediates in the pathway are transferred directly from one enzyme to the next without entering the bulk phase. Thus there is no dilution of intermediates and no reliance on random diffusion to reach a second enzyme.

## Cytosol

The cytosol is the liquid portion of the cytoplasm. This can be isolated in diluted form as the supernatant fraction obtained after centrifuging broken cell extracts at 105,000 $\times$ *g* for 1–2 hours, which should sediment membranes, the DNA, the ribosomes, very large protein aggregates, and other intracellular inclusions. In the cytosol are found the enzymes that catalyze a major portion of the biochemical reactions in the cell, such as the enzymes of the central pathways for the metabolism of carbohydrates (glycolysis, the pentose phosphate pathway, and the Entner–Doudoroff pathway), as well as the central pathways for organic acid metabolism (citric acid cycle and glyoxylate pathways), and enzymes for other pathways such as the biosynthesis and degradation of amino acids, lipids, and nucleotides.

The concentration of proteins in the cytosol is very high, making it viscous, and it is expected that there are extensive protein-protein interactions among the enzymes, even those that do not exist in tight complexes. However, if such interactions exist, they must be weak, because most of the enzymes that catalyze metabolic pathways in the cytosol (e.g., the glycolytic enzymes that catalyze the degradation of glucose to pyruvic acid or lactic acid) cannot be isolated as complexes. It has generally been assumed that they exist either as independent proteins or in loose associations that are easily disrupted during cell breakage and accompanying dilution of the proteins. (It should be realized that the concentration of proteins in broken cell extracts is orders of magnitude lower than in the aqueous portion of the unbroken cell because of the addition of buffers during the washing and suspension of the cells prior to breakage.)

## Analogues to tubulin and actomyosin

There has been increasing evidence that prokaryotes possess analogues of eukaryotic tubulin and actomyosin that contribute towards a bacterial "cytoskeleton."[2] Two of these, FtsZ and FtsA, are cell division proteins, which are discussed in Section 2.2.5.

## 1.3 Summary

There are two evolutionary lines of prokaryotes, the Bacteria and the Archaea. Archaea are similar to bacteria, but they differ in certain fundamental aspects of structure and biochemistry. These include differences in ribosomes, cell wall chemistry, membrane lipids, and coenzymes. In addition, certain archaea have metabolic pathways (e.g., methanogenesis) not found in bacteria.

Surrounding most bacteria are fimbriae and extracellular material called glycocalyx. The fimbriae are protein fibrils. The glycocalyx encompasses all extracellular polymers, polysaccharide or protein, including capsules and slime layers. The fimbriae and glycocalyx anchor the cell to specific animal and plant cell surfaces, as well as to inanimate objects, and in several cases to each other.

Some bacteria possess a fibril called a sex pilus that attaches the cell to a mating partner and retracts to draw the cells into intimate contact. When the cells are in contact, DNA is transferred unilaterally from the donor cell (i.e., the one with the sex pilus) to the recipient.

Protruding from many bacteria are flagella filaments that aid the bacterium in movement. At the base of the flagellum is a basal body that is embedded in the cell membrane. Part of the basal body is a rotary motor that in most bacteria runs on a current of protons. Extreme halophiles and some marine bacteria use a sodium ion current instead of a proton current. Some bacteria have flagella covered by a membranous sheath (e.g., *Vibrio cholerae*, *V. parahaemolyticus*, *Helicobacter*, and *Bdellovibrio bacteriovorus*). There is a proteinaceous sheath covering the flagella (axial filaments) of most spirochaetes (an exception is *Borrelia burgdorferi*, the causative agent of Lyme disease), which is actually located in the periplasm rather than outside the cell. Spirochaetes are shaped like a coil, and when the flagellum turns, the cell moves in a corkscrewlike fashion, which apparently is suited for movement through high-viscosity media. (Again, *B. burgdorferi* is an exception and does not move in a corkscrewlike fashion.[91]) Bacterial flagella are important for the pathogenicity of certain bacteria, and this is discussed in the review by Moens.[35]

There are two types of bacterial cell walls. One kind has a thick peptidoglycan layer to which there are covalently bonded polysaccharides, teichoic acids, and teichuronic acids. Bacteria with such walls stain gram positive. A second type of wall has a thin peptidoglycan layer and an outer envelope consisting of lipopolysaccharide, phospholipid, and protein. Cells with such walls do not stain gram positive and are called gram negative. There is also a separate compartment in gram-negative bacteria called a periplasm, between the outer envelope and the cell membrane. The periplasm is the location of numerous proteins and enzymes, including proteins required for solute transport and enzymes that function in the oxidation and degradation of nutrients in the periplasm. Archaea have very different cell walls. No archaeon has peptidoglycan,

although some have a similar polymer called pseudopeptidoglycan or pseudomurein.

The cell membranes of the bacteria are all similar. They consist primarily of phosphoglycerides in a bilayer and protein. The phosphoglycerides are generally fatty acids esterified to glycerol phosphate. Archaeal cell membranes have lipids that are long-chain alcohols ether linked to glycerol. Archaeal membranes can be a bilayer or a monolayer, or perhaps a mixture of the two.

The cytoplasm of prokaryotes is a viscous solution of protein. Salts, sugars, amino acids, and other metabolites are dissolved in the proteinaceous cytoplasm. In many bacteria intracytoplasmic membranes are present. In many cases these membranes are invaginations of the cell membrane. They are sites for electron transport and specialized biochemical activities. Numerous inclusion bodies are also present. These include ribosomes, which are the sites of protein synthesis, large aggregates of enzymes that catalyze short metabolic pathways, gas vesicles, and carbon and energy reserves (e.g., glycogen particles). The DNA is also present in the cytoplasm, and is very tightly packed and referred to as a nucleoid. It is clear that although the cytoplasm of prokaryotes is not compartmentalized into membrane-bound organelles as is the eukaryotic cytoplasm, it nonetheless represents a very complex pattern.

The central metabolic pathways that are described in the ensuing chapters (i.e., glycolysis, the Entner–Doudoroff pathway, pentose phosphate pathway, and the citric acid cycle) all take place in the liquid part of the cytoplasm (i.e., the cytosol). Also found in the cytosol are most of the other pathways, including the enzymatic reactions for the synthesis and degradation of amino acids, fatty acids, purines, and pyrimidines.

The cell membrane is the site of numerous other metabolic pathways, including phospholipid biosynthesis, protein secretion, solute transport, electron transport, cell wall biosynthesis, the generation of electrochemical ion gradients, and ATP synthesis. The prokaryote's cell can therefore be considered as having three metabolic domains: cytosol, particulate (ribosomes, enzyme aggregates, etc.), and membrane.

## Study Questions

1. What chemical and structural differences distinguish the Archaea from the Bacteria?

2. What are the differences between gram-positive and gram-negative cell walls?

3. What structures enable bacteria to adhere to surfaces? What is known about the molecules that mediate the adhesion (i.e., their chemistry, location, and receptors)?

4. Contrast the functions and cellular location of peptidoglycan, phospholipid, and lipopolysaccharide. What is it about the chemical structure of these three classes of compounds that is suitable for their functions and/or cellular location?

5. What are porins, and what do they do? Are they necessary for gram-positive bacteria? Explain.

6. What are the physiological and enzymatic functions associated with the periplasm?

7. Summarize the cell compartmentalization of metabolic activities in prokaryotes.

8. What are some multienzyme complexes? What is the advantage to these?

9. Which metabolic pathways or activities are found in the cytosol and which in the membranes?

## NOTES AND REFERENCES

1. Prokaryotes are organisms without a membrane-bound nucleus. It is also spelled pro-caryote. 'Karyo' (caryo) means nucleus. It is derived from the Greek word *karuon*, which means kernel or nut.

2. Norris, V., G. Turnock, and D. Sigee. 1996. The *Escherichia coli* enzoskeleton. *Molec. Microbiol.* 19:197–204.

3. The evolutionary relationships among all living organisms have been deduced by comparing the ribosomal RNAs of modern organisms. The structures of the ribosomal RNA molecules from different living organisms are sufficiently conserved in certain regions that it is possible to align conserved sequences and secondary structures in order to compare the differences in base sequences between RNA molecules in homologous regions. What is analyzed are the number of positions that differ between pairs of sequences as well as other features (e.g., which positions vary

and the number of changes that have been made in going from one sequence to another). The number of nucleotide differences between homologous sequences is used to calculate the evolutionary distance between the organisms and to construct a phylogenetic tree. Most recently published phylogenetic trees are based upon 16S rRNA sequences.

There are two major ways in which phylogenetic trees are constructed from the nucleotide differences. One is called the *evolutionary distance* method. In this method the number of nucleotide differences is used as a measure of the evolutionary distance between the organisms. A second method is called the *maximum parsimony* method. This method is more complicated and takes into account not only the nucleotide differences but also the positions where the differences occur and the nature of the differences. Parsimony means "stingy," or extremely economical, and the method attempts to find the simplest evolutionary tree that can explain the differences in nucleotide sequences. The simplest tree is the one that requires the smallest number of nucleotide changes to evolve the collection of extant sequences from a postulated ancestral sequence.

Woese has pointed out that that it is difficult to assess the validity of the phylogenetic trees. There are several reasons for this. For example, none of the methods accounts for the possibility that multiple changes may have taken place at any single position. This results in an underestimate of the distances. Underestimation of the distances might bring two distantly related lineages much closer, and in fact might make them appear specifically related to each other when they are not. Other ambiguities in the phylogenetic trees could arise if different positions in the sequence alignments change at very different rates. Furthermore, the branching pattern itself can differ even using the same database if a different algorithm is used. The outcome is that there exist a variety of phylogenetic trees, and no single published tree has been accepted as perfectly representing the four billion years or so of bacterial evolution. Despite these problems there is consistency in the trees that are based upon 16S rRNA sequences, and it is believed that most of the phylogenetic trees derived from sequencing bacterial 16S rRNA are at least plausible. However, it has been pointed out that it is an unproven assumption that the relationships between the 16S rRNA represents the phylogenies of the organisms rather than simply the phylogeny of the 16s rRNA. Because of this, it is imperative that the relationships between the different organisms be tested using other characters (e.g., other appropriate molecules besides the 16S rRNA) and various phenotypic characteristics of the organisms. For a further discussion of the various methods to construct phylogenetic trees and the problems therein, see the articles by Woese, Stackebrandt, and Felsenstein.

Woese, C. R. 1991. Prokaryotic systematics: The evolution of a science, pp. 3–11. In: *The Prokaryotes*, Vol. I. A. Balows, H. G. Trüper, M. Dworkin, W. Harder, and K.-H. Schleifer (Eds.). Springer-Verlag, Berlin.

Stackebrandt, E. 1991. Unifying phylogeny and phenotypic diversity, pp. 19–47. In: *The Prokaryotes*, Vol. I. A. Balows, H. G. Trüper, M. Dworkin, W. Harder, and K.-H. Schleifer (Eds.). Springer-Verlag, Berlin.

Felsenstein, J. 1982. Numberical methods for inferring evolutionary trees. *Quart. Rev. Biol.* 57:379–404.

4. Woese, C. R., O. Kandler, and M. L. Wheelis. 1990. Towards a natural system of organisms: Proposal for the domains Archaea, Bacteria, and Eucarya. *Proc. Natl. Acad. Sci. USA* 87:4576–79.

5. Woese, C. R. 1987. Bacterial evolution. *Microbiol. Rev.* 51:221–271.

6. Pace, N. R., D. A. Stahl, D. J. Lane, and G. J. Olsen. 1985. Analyzing natural microbial populations by rRNA sequences. *ASM News* 51:4–12.

7. Stetter, K. O. 1989. Extremely thermophilic chemolithoautotrophic archaebacteria, pp. 167–176. In: *Autotrophic Bacteria*. H. G. Schlegel and B. Bowien (Eds.). Springer-Verlag, Berlin.

8. Schleper, C., G. Puehler, I. Holtz, A. Gambacorta, D. Janekovic, U. Santarius, H.-P. Klenk, and W. Zillig. 1995. Picrophilus gen. nov., fam. nov.: A novel aerobic, heterotrophic, thermoacidophilic genus and family compromising archaea capable of growth around pH 0. *J. Bacteriol.* 177:7050–7059.

9. Solfataras are acidic geothermally heated habitats containing inorganic sulfur compounds, including sulfuric acid. These include volcanic fissures, springs, basins, and dried soil that previously contained solfataric water. They have an acidic pH in the aerobic portions due to sulfuric acid that arises from the oxidation of hydrogen sulfide, which occurs spontaneously and also as a result of the growth of sulfur-oxidizing bacteria.

10. Koga, Y., M. Nishihara, H. Morii, and M. Akagawa-Matsushita. 1993. Ether polar lipids of methanogenic bacteria: Structures, comparative aspects, and biosynthesis. *Microbiol. Rev.* 57:164–182.

11. Grayling, R. A., Sandman, K., and J. N. Reeve. 1996. Histones and chromatin structure in hyperthermophilic *Archaea*. *FEMS Microbiol. Rev.* 18:203–213.

12. Pereira, S. L., Grayling, R. A., Lurz, R., and J. N. Reeve. 1997. Archaeal nucleosomes. *Proc. Natl. Acad. Sci. USA* 94:12633–12637.

13. Joys, T. M. 1988. The flagellar filament protein. *Can. J. Microbiol.* 34:452–458.

14. Wilson, D. R., and T. J. Beveridge. 1993. Bacterial flagellar filaments and their component flagellins. *Can. J. Microbiol.* 39:451–472.

15. A great many bacteria move by gliding motility on solid surfaces. Organelles for gliding locomotion in bacteria have not been discovered.

16. Some rod-shaped or curved cells have flagella that protrude from one or both of the cell poles. A bacterium with a single, polar flagellum is said to be *monopolar*. Bacteria with a bundle of flagella at a single pole are *lophotrichous*. Bacteria with flagella at both poles are said to be *bipolar*. They may have either single or bundles of flagella at the poles. *Amphitrichous* refers to bundles of flagella at both poles. Some bacteria (e.g., spirochaetes) have *subpolar* flagella, where the flagella are inserted near but not exactly at the cell poles, and some curved bacteria (e.g., *Vibrio*) have a single, *medial* flagellum. If the flagella are arranged laterally all around the cell, then the condition is known as *peritrichous* (e.g., *Escherichia* and *Salmonella*). Peritrichous flagella coalesce into a trailing bundle during swimming.

17. Ivey, D. M., M. Ito, R. Gilmour, J. Zemsky, A. A. Guffanti, M. G. Sturr, D. B. Hicks, and T. A. Krulwich. 1998. Alkaliphile bioenergetics, pp. 181–210. In: *Expremophiles: Microbial Life in Extreme Environments*. K. Horikoshi and W. D. Grant (Eds.). John Wiley and Sons, Inc., New York.

18. Moens, S., and J. Vanderleyden. 1996. Functions of bacterial flagella. *Crit. Rev. Microbiol.* 22:67–100.

19. DePamphilis, M. L., and J. Adler. 1971. Purification of intact flagella from *Escherichia coli* and *Bacillus subtilis*. *J. Bacteriol.* 105:384–395.

20. Aizawa, Shin-Ichi, G. E. Dean, C. J. Jones, R. M. Macnab, and S. Yamaguchi. 1985. Purification and characterization of the flagellar hook-basal body complex of *Salmonella typhimurium*. *J. Bacteriol.* 161:836–849.

21. Schoenhals, G. J., and R. M. Macnab. 1996. Physiological and biochemical analyses of FlgH, a lipoprotein forming the outer membrane L ring of the flagellar basal body of *Salmonella typhimurium*. *J. Bacteriol.* 178:4200–4207.

22. Khan, S., M. Dapice, and T. S. Reese. 1988. Effects of mot gene expression on the structure

of the flagellar motor. *J. Mol. Biol.* **202**:575–584.

23. Blair, D. F., and H. C. Berg. 1990. The MotA protein of *E. coli* is a proton-conducting component of the flagellar motor. *Cell* **60**:439–449.

24. Garza, A. G., L. W. Harris-Haller, R. A. Stoebner, and M. D. Manson. 1995. Motility protein interactions in the bacterial flagellar motor. *Proc. Natl. Acad. Sci. USA* **92**:1970–1974.

25. FliG, FliM, and FliN are probably also involved in flagellar assembly, because null mutants do not have flagella. This may be because the switch complex must be made prior to more distal portions of the flagellum. FliG, FliM, and FliN are involved in the turning of the flagellar motor (generating torque), because certain mutations result in paralysis. However, the evidence suggests that only FliG is directly involved in generating torque. (See Section 1.2.1, "Additional proteins required for motor function.") It has been concluded that FliG, FliM, and FliN interact with one another in a complex because certain mutations in *fliG*, *fliM*, and *fliN* can be suppressed by other mutations in any of the *fli* genes. CheY-P, the molecule that causes the flagellar motor to reverse its direction of rotation, binds to FliM (Section 18.4.5).

26. Francis, N. r., G. E. Sosinsky, D. Thomas, and D. J. DeRosier. 1994. Isolation, characterization and structure of bacterial flagellar motors containing the switch complex. *J. Mol. Biol.* **235**:1261–1270.

27. Lloyd, S. A., H. Tang, X. Wang, S. Billings, and D. F. Blair. 1996. Torque generation in the flagellar motor of *Escherichia coli*: Evidence of a direct role for FliG but not FliM or FliN. *J. Bacteriol.* **178**:223–231.

28. Pleier, E., and R. Schmitt. 1989. Identification and sequence analysis of two related flagellin genes in *Rhizobium meliloti. J. Bacteriol.* **171**:1467–1475.

29. Iino, T. 1969. Polarity of flagellar growth in *Salmonella. J. Gen. Microbiol.* **56**:227–239.

30. Emerson, S. U., K. Tokuyasu, and M. I. Simon. 1970. Bacterial flagella: Polarity of elongation. *Science* **169**:190–192.

31. Fuerst, J. A., and J. W. Perry. 1988. Demonstration of lipopolysaccharide on sheathed flagella of *Vibrio cholerae* 0:1 by protein A-gold immunoelectron microscopy. *J. Bacteriol.* **170**:1488–1494.

32. Weissborn, A., H. M. Steinman, and L. Shapiro. 1982. Characterization of the proteins of the *Caulobacter crescentus* flagellar filament. *J. Biol. Chem.* **257**:2066–2074.

33. Pleier, E., and R. Schmitt. 1989. Identification and sequence analysis of two related flagellin genes in *Rhizobium meliloti. J. Bacteriol.* **171**:1467–1475.

34. Driks, A., R. Bryan, L. Shapiro, and D. J. DeRosier. 1989. The organization of the *Caulobacter crescentus* flagellar filament. *J. Mol. Biol.* **206**:627–636.

35. Moens, S., and J. Vanderleyden. 1996. Functions of bacterial flagella. *Crit. Rev. Microbiol.* **22**:67–100.

36. Goldstein, S. F., N. W. Charon, and J.A. Kreiling. 1994. *Borrelia burgdorferi* swims with a planar waveform similar to that of eukaryotic flagella. *Proc. Natl. Acad. Sci. USA.* **91**:3433–3437.

37. Parales, J., Jr., and E. P. Greenberg. 1991. N-Terminal amino acid sequences and amino acid compositions of the *Spirochaeta aurantia* flagellar filament polypeptides. *J. Bacteriol.* **173**:1357–1359.

38. Jarrell, K. F., D. P. Bayley, and A. S. Kostyukova. 1996. The archaeal flagellum: A unique motility structure. *J. Bacteriol.* **178**:5057–5064.

39. Faguy, D. M., K. F. Jarrell, J. Kuzio, and M. L. Kalmokoff. 1994. Molecular analysis of archaeal flagellins: Similarity to the type IV pilin-transport superfamily widespread in bacteria. *Can. J. Microbiol.* **40**:67–71.

40. Faguy, D. M., S. F. Koval, and K. F. Jarrell. 1994. Physical characterization of the flagella and flagellins from *Methanospirillum hungatei. J. Bacteriol.* **176**:7491–7498.

41. Pearce, W. A., and T. M. Buchanan. 1980. Structure and cell membrane-binding properties of bacterial fimbriae, pp. 289–344. In: *Bacterial Adherence* E. H. Beachey (Ed.). Chapman and Hall, New York.

42. Ottow, J. C. G. 1975. Ecology, physiology, and genetics of fimbriae and pili. *Ann. Rev. Microbiol.* **29**:79–108.

43. Jones, G. W., and R. E. Isaacson. 1983. Proteinaceous bacterial adhesins and their receptors. *Crit. Rev. Microbiol.* **10**:229–260.

44. Hultgren, S. J., S. Abraham, M. Caparon, P. Falk, J. W. St. Geme III, and S. Normark. 1993. Pilus and nonpilus bacterial adhesins: Assembly and function in cell recognition. *Cell* **73**:887–901.

45. Yanagawa, R., and K. Otsuki. 1970. Some properties of the pili of *Corynebacterium renale. J. Bacteriol.* **101**:1063–1069.

46. Cisar, J. O., and A. E. Vatter. 1979. Surface fibrils (fimbriae) of *Actinomyces viscosus* T14V. *Infect. Immun.* 24:523–531.

47. *C. renale* is the causative agent of bovine pyelonephritis and cystitis, and *A. viscosus* is part of the normal flora of the human mouth, where it adheres to other bacteria in plaque and to teeth.

48. The fimbriae presumably attach to the erythrocytes because the erythrocytes carry cell surface molecules resembling the adhesin receptors found on the natural host cells.

49. Eisenstein, B. I. 1987. Fimbriae, pp. 84–90. In: Escherichia coli *and* Salmonella typhimurium: *Cellular and Molecular Biology.* Vol. 1. F. C. Neidhardt (Ed.). American Society for Microbiology, Washington, D.C.

50. Hultgren, S. J., S. Normark, and S. N. Abraham. 1991. Chaperone-assisted assembly and molecular architecture of adhesive pili. *Ann. Rev. Microbiol.* 45:383–415.

51. Hultgren, S. J., S. Abraham, M. Caparon, P. Falk, J. W. St. Geme III, and S. Normark. 1993. Pilus and nonpilus bacterial adhesins: Assembly and function in cell recognition. *Cell* 73:887–901.

52. Pathogenic strains of *Escherichia coli* cause gastrointestinal and urinary tract infections, *Neisseria gonorrhoeae* causes gonorrhoeae, *Bacteroides nodosus* causes bovine foot rot, *Moraxella bovis* causes bovine keratoconjunctivitis, *Vibrio cholerae* causes cholera, and *Pseudomonas aeruginosa* causes infections of the urinary tract and wounds. The adherence of these bacteria to their host tissue is the first stage in pathogenesis.

53. In addition to the bacterial chromosome, most bacteria carry smaller extrachromosomal DNA molecules called *plasmids*. There are many types of plasmids. They all carry genes for self-replication. Conjugative plasmids also carry genes for DNA transfer into recipient cells (transmissible plasmids). Other genes that may be carried by plasmids include genes for antibiotic resistance (resistance transfer factors or R factors), and genes for the catabolism of certain nutrients. For example, *Pseudomonas* carries plasmids for the degradation of camphor, octane, salicylate, and naphthalene. Some plasmids confer virulence because the plasmids carry genes for toxins or other virulence factors, such as fimbriae. The F plasmid in *E. coli* carries genes for self-transmission, including genes for the F pilus, and is also capable of integrating into the chromosome and promoting the transfer of chromosomal DNA.

54. Dunny, G. M. 1991. Mating interactions in Gram-positive bacteria. pp. 9–33. In: *Microbial Cell Interactions*. M. Dworkin (Ed.). American Society for Microbiology, Washington, DC.

55. An excellent reference for microbial cell–cell interactions is the book edited by Dworkin: *Microbial Cell–Cell Interactions*, 1991. M. Dworkin (Ed.). American Society for Microbiology, Washington, D.C.

56. Costerton, J. W., T. J. Marrie, and K.-J. Cheng. 1985. Phenomena of bacterial adhesion, pp. 3–43. In: *Bacterial Adhesion: Mechanisms and Physiological Significance*. D. C. Savage and M. Fletcher (Eds.). Plenum Press, New York.

57. Sleytr, U. B., and P. Messner. 1983. Crystalline surface layers on bacteria. *Ann. Rev. Microbiol.* 37:311–339.

58. Costerton, J. W., R. T. Irvin, and K.-J Cheng. 1981. The bacterial glycocalyx in nature and disease. *Ann. Rev. Microbiol.* 35:299–324.

59. Roberts, I. S. 1995. Bacterial polysaccharides in sickness and in health. *Microbiology* 141:2023–2031.

60. Bacterial capsules. 1990. Current Topics in Microbiology and Immunology, K. Jann and B. Jann (Eds.). Springer-Verlag, Berlin.

61. Roberts, I. S. 1996. The biochemistry and genetics of capsular polysaccharide production in bacteria. *Ann. Rev. Microbiol.* 50:285–315.

62. Kolenbrander, P. E., N. Ganeshkumar, F. J. Cassels, and C. V. Hughes. 1993. Coaggregation: Specific adherence among human oral plaque bacteria. *FASEB* 7:406–409.

63. Nikaido, H., and M. Vaara. 1987. Outer membrane, pp. 7–22. In: Escherichia coli *and* Salmonella typhimurium: *Cellular and Molecular Biology*, Vol. 1. F. C. Neidhardt (Ed.) ASM Press.

64. Weidel, W., and H. Pelzer. 1964. Bagshaped macromolecules—a new outlook on bacterial cell walls. *Adv. Enzymol.* 26:193–232.

65. Koch, A. L. 1983. The surface stress theory of microbial morphogenesis. *Adv. Microb. Physiol.* 24:301–366.

66. Koch, A. L., and S. Woeste. 1992. Elasticity of the sacculus of *Escherichia coli*. *J. Bacteriol.* 174:4811–4819.

67. Kellenberger, E. 1990. The "Bayer bridges" confronted with results from improved electron microscopy methods. *Mol. Microbiol.* 4:697–705.

68. *S. pyogenes* is an important pathogen that causes most streptococcal infections in humans, including "strep" throat. The M protein is a cell surface adhesin that is thought to bind to the LTA that in turn binds to host cell surfaces.

69. Sutcliffe, I. C., and N. Shaw. 1991. Atypical lipoteichoic acids of Gram-positive bacteria. *J. Bacteriol.* **173**:7065–7069.

70. Leduc, M., K. Ishidate, N. Shakibai, and L. Rothfield. 1992. Interactions of *Escherichia coli* membrane lipoproteins with the murein sacculus. *J. Bacteriol.* **174**:7982–7988.

71. Crosslinking reagents can be useful for determining which molecules are physically associated. These are bifunctional reagents (i.e., they have a chemical group at both ends of the molecule that can form a covalent bond to functional groups on proteins or other molecules such as peptidoglycan), thus crosslinking two molecules that are within the length of the crosslinking reagent. For example, the reagent dithio-bis-succinimidylpropionate (DSP) forms covalent crosslinks between amino groups that are less than 12 angstrom units (1.2 nm) apart. Therefore, DSP can crosslink proteins that are physically associated with the peptidoglycan. After treatment with DSP, the peptidoglycan and its crosslinked proteins can be isolated. DSP has two arms held together by a disulfide bond that can be cleaved with mercaptoethanol. It is called a cleavable crosslinking reagent. The crosslinked proteins that were held to the peptidoglycan by the DSP are released after mercaptoethanol treatment and can be analyzed by gel electrophoresis. Thus the proteins crosslinked to the peptidoglycan can be identified. In addition to various lipoproteins, including the murein lipoprotein, an outer membrane protein, OmpA, can also be crosslinked to the peptidoglycan, indicating close association. The crosslinking of OmpA to the peptidoglycan reflects the fact that it extends through the outer envelope.

72. Ferguson, S. J. 1992. The periplasm. In: *Procaryotic Structure and Function: A New Perspective.* S. Mohan, C. Dow, and J. A. Cole (Eds.). Society for General Microbiology Symposium 47, Cambridge University Press.

73. Hobot, J. A., E. Carleman, W. Villiger, and E. Kellenberger. 1984. Periplasmic gel: New concept resulting from the reinvestigation of bacterial cell envelope ultrastructure by new methods. *J. Bacteriol.* **160**:143–152.

74. Bayer, M. E. 1991. Zones of membrane adhesion in the cryofixed envelope of *Escherichia coli. J. Struct. Biol.* **107**:268–280.

75. Bacteria secrete specific iron chelators called siderophores for scavenging iron from the environment. Special transport systems bind the iron siderophores and bring the iron into the cell.

76. Postle, K. 1990. TonB and the Gram-negative dilemma. *Molec. Microbiol.* **4**:2019–2025.

77. Braun, V. 1995. Energy-coupled transport and signal transduction through the gram-negative outer membrane via TonB-ExbB-ExbD-dependent receptor proteins. *FEMS Microbiol. Rev.* **16**:295–307.

78. Bradbeer, C. 1993. The proton motive force drives the outer membrane transport of cobalamin in *Escherichia coli. J. Bacteriol.* **175**:3146–3150.

79. Merchante, R., H. M. Pooley, and D. Karamata. 1995. A periplasm in *Bacillus subtilis. J. Bacteriol.* **177**:6176–6183.

80. De Rosa, M., A. Gambacorta, and A. Gliozzi. 1986. Structure, biosynthesis, and physiochemical properties of archaebacterial lipids. *Microbiol. Rev.* **50**:70–80.

81. Sternberg, B., C. L'Hostis, C. A. Whiteway, and A. Watts. 1992. The essential role of specific *Halobacterium halobium* polar lipids in 2D-array formation of bacteriorhodopsin. *Biochem. Biophys. Acta.* **1108**:21–30.

82. Gambacorta, A., A. Gliozzi, and M. De Rosa. 1995. Archaeal lipids and their biotechnological applications. *World J. Microbiol.* **11**:115–131.

83. Westerhoff, H. V., and G. R. Welch. 1992. Enzyme organization and the direction of metabolic flow: Physiochemical considerations, pp. 361–390. In: *Current Topics in Cellular Regulation*, Vol. 33. E. R. Stadtman and P. B. Chock (Eds.). Academic Press, Inc., New York.

84. Reviewed by G. Drews. 1991. Intracytoplasmic membranes in bacterial cells, pp. 249–274. In: *Prokaryotic Structure and Function, A New Perspective.* S. Mohan, D. Dow, and J. A. Coles (Eds.). Cambridge Univ. Press.

85. Overmann, J., S. Lehmann, and N. Pfennig. 1991. Gas vesicle formation and buoyancy regulation in *Pelodictyon phaeoclathratiforme* (Green sulfur bacteria). *Arch. Microbiol.* **157**:29–37.

86. Codd, G. A. 1988. Carboxysomes and ribulose bisphosphate carboxylase/oxygenase, pp. 115–164. In: *Advances in Microbial Physiology*, Vol. 29. A. H. Rose and D. W. Tempest (Eds.). Academic Press, New York.

87. Rouviere-Yaniv, J., and F. Gros. 1975. Characterization of a novel, low-molecular-weight DNA-binding protein from *Escherichia coli. Proc. Natl. Acad. Sci. USA* **72**:3428–3432.

88. Pettijohn, D. E. 1996. The nucleoid, pp. 158–166. In: *Escherichia coli* and *Salmonella*: Cellular and Molecular Biology. F. C. Neidhardt (Ed.). ASM Press, Washington, D.C.

89. Molecular interactions among enzymes is reviewed by D. K. Srivastava and S. A. Bernhard. 1986. Enzyme–enzyme interactions and

the regulation of metabolic reaction pathways, pp. 1–68. In: *Current Topics in Cellular Regulation*, Vol. 28. B. L. Horecker and E. R. Stadtman (Eds.). Academic Press, Inc., New York.

90. In bacteria fatty acid synthesis is catalyzed by similar enzymatic reactions as found in yeast and mammals, but the enzymes cannot be isolated as a complex.

91. Golstein, S. F., N. W. Charon, and J. A. Kreiling. 1994. *Borrelia burgdorferi* swims with a planar waveform similar to that of eukaryotic flagella. *Proc. Natl. Acad. Sci. USA* **91**:3433–3437.

# 2

# Growth and Cell Division

The study of growth of microbial populations is at the heart of microbial cell physiology because population growth characteristics reflect underlying physiological events in the individual cells. In fact, these cellular events are frequently manifested to the investigator only by changes in the growth of populations. It is therefore important to understand the changes in growth that microbial populations undergo and to be able to measure population growth accurately. Additionally, the investigator must be able to manipulate population growth (e.g., by placing the culture in balanced growth or continuous growth) in order to investigate certain aspects of cell physiology (e.g., the interdependencies between the rates of synthesis of the individual classes of macromolecules such as DNA, RNA, and protein). This chapter describes methods to measure growth, an analysis of exponential growth kinetics, growth in batch culture and in continuous culture, some important aspects of growth physiology under certain nutritional conditions, and cell division.

## 2.1 Measurement of Growth

Growth is defined as an increase in mass, and one can measure any growth parameter provided it increases proportionally to the mass of the culture. The most commonly used measurements of growth are: turbidity, total and viable cell counts, dry weight, and protein.

### 2.1.1 Turbidity

Routinely, bacterial growth is measured using turbidity measurements because of the simplicity of the procedure. Within limits, the amount of light scattered by a bacterial cell is proportional to its mass. Therefore, light scattering is proportional to the total mass. (The relationship between total mass and light scattering deviates from linearity at very high cell densities.) If the average mass per cell remains a constant, then one can also use light scattering to measure changes in cell number. What is actually measured is the *fraction of incident light transmitted* through the culture (i.e., not the scattered light). The more dense the culture, the less light is transmitted. This relationship is given by the *Beer–Lambert law* (eq. 2.1). Thus if $I_0$ is the incident light and $I$ is the transmitted light, then the Beer–Lambert law states that in a population whose cell density is $x$, the fraction of light that is transmitted ($I/I_0$) will decrease as the logarithm of $x$ to the base 10.

$$I/I_0 = 10^{-xl} \qquad (2.1)$$

where $l$ is the light path in centimeters. The situation can be drawn schematically as shown in Fig. 2.1.

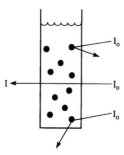

**Fig. 2.1** Illustration of light scattering. The incident light is $I_0$, and the transmitted light is $I$. The spectrophotometer usually provides the logarithm of the reciprocal of the fraction of transmitted light, i.e., $\log(I_0/I)$, which is called the optical density or turbidity.

If one takes the log to the base 10 of both sides of eq. 2.1, then $\log_{10} (I/I_0) = -xl$, where $I/I_0$ is the fraction of incident light that is transmitted. Note that the fraction of light that is transmitted decreases as a logarithmic function (base 10) (i.e., exponentially with the density of the culture). Bacteriologists, however, prefer to measure something that increases with cell density. They therefore use the reciprocal of $\log_{10} (I/I_0)$ which is $\log_{10} (I_0/I)$. The log $(I_0/I)$ is called *turbidity* or *absorbancy* or *optical density,* and has the symbol OD (for optical density) or $A$ (for absorbancy). Thus turbidity or OD is directly proportional to cell mass in the culture (or cell density if the mass/cell remains constant) and is written as follows:

$$OD = A = xl \qquad (2.2)$$

The turbidity is measured with a colorimeter or spectrophotometer. In practice, one constructs a standard curve by measuring the turbidity of several different cell suspensions where the cell number is counted independently and used as a measure of cell mass. A straight line is generated as shown in Fig. 2.2. However, the line deviates from linearity at high cell densities. The reason for that is that when the cell density is too high, then some of the light is rescattered and directed toward the phototube, thus lowering the turbidity reading.

Whenever turbidity measurements are made on an unknown sample, a standard curve must be constructed to determine the

cell density. Turbidity measurements are the simplest way to measure growth. Methods to determine the cell density by direct cell counts are described next.

### 2.1.2 Total cell counts

If the mass per cell is constant, one can measure growth by sampling the culture over a period of time and counting the cells. There are two ways to do this: total cell counts and viable cell counts. For *total cell counts* one can use a counting chamber. This is simply a glass slide divided into tiny square wells of known area and depth. A drop of culture is placed on the slide, and, when a cover slip is applied, each square well in the grid holds a known volume of liquid. The number of cells is counted microscopically, and the value is converted to cells per milliliter using a conversion factor based on the total number of square wells counted and their volume. Although this is a simple procedure, and one used widely, there are two limitations:

1. They do not distinguish between live and dead cells.

2. They cannot be performed on populations whose cell density is too low to count microscopically (less than $10^6$ cells/ml).

A more sophisticated method is to use *electronic cell counting*. For electronic counting, the bacteria, suspended in a saline solution, are placed in a chamber with an electrode separated by a second chamber filled with the same saline solution and also provided with an electrode. A microscopic pore separates the two chambers. The bacterial suspension

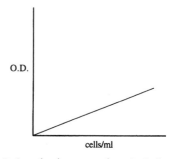

**Fig. 2.2** Standard curve of optical density vs. cells/ml. The line deviates from linearity at very high cell densities.

is pumped through the pore into the second chamber. Whenever a bacterium passes through the pore, the electrical conductivity of the circuit decreases because the electrical conductivity of a bacterial cell is less than that of the saline solution. This results in a voltage pulse that is counted electronically. The electronic counter has an advantage in that one can also measure the *size* of the bacterium. This is because the size of the pulse is proportional to the size of the bacterium. Thus one can obtain both a total cell count and a size distribution.

### 2.1.3 Viable cell counts

A *viable cell count* is one in which the cells are serially diluted and then deposited on a solid growth medium. Each viable cell grows into a colony, and the colonies are counted. Viable cell counts are routinely performed in microbiology laboratories. However, it is important to recognize that they may underestimate the number of viable cells for the following reasons:

1. A viable cell count will underestimate the number of live cells if the cells are clumped, since each clump of cells will give rise to a single colony.

2. Some bacteria plate with a poor efficiency; that is, single cells give rise to colonies with a low frequency.

### 2.1.4 Dry weight and protein

The most direct way to measure growth is to quantify the *dry weight* of cells in a culture. The cells are harvested by either centrifugation or filtration, dried to a constant weight, and carefully weighed. Since protein increases in parallel with growth, one can also measure growth by doing *protein measurements* of the cells. The cells are either harvested by centrifugation or filtration or (more usually) precipitated first with acid or alcohol, and then recovered by centrifugation or filtration. A simple colorimetric test for protein is then performed.

## 2.2 Growth Physiology

Growth physiology is a large topic. It includes the regulation of rates of synthesis of macromolecules, the regulation of timing of DNA synthesis and cell division, adaptive physiological responses to nutrient availability, including changes in gene expression, homeostasis adaptations to the external environment, and the coupling of the rates of biosynthetic pathways to the rates of utilization of products for growth (metabolic regulation). From this point of view, this entire text is concerned with the physiology of growth, and we therefore return to the topic in subsequent chapters. What follows here is an introduction to growth physiology as it pertains to the synthesis of macromolecules during steady-state growth, and very general adaptations of cells to nutrient depletion. We begin with a discussion of the sequence of growth phases through which a population of bacteria progresses when inoculated into a flask of fresh media.

### 2.2.1 Phases of population growth

When one measures the growth of populations of bacteria grown in batch culture, a progression through a series of phases can be observed (Fig. 2.3). The first phase is frequently a *lag phase*, where no net growth occurs (i.e., no increase in cell mass). This is followed by a phase of *exponential growth*, where cell mass increases exponentially with time. Following exponential growth, the culture enters a phase of no net growth called the *stationary phase*. After a stationary period, cell death occurs in a final stage called the *death phase*. Notice in Fig. 2.3 that, prior to the exponential phase of growth, the cells increase in mass (i.e., they grow larger) before cell division occurs (solid line). At the end of exponential growth, the cells continue to divide after growth has ceased and become smaller (dotted line). It would seem to be an advantage to a population of bacteria to continue to divide after growth has ceased in the population so as to produce more cells for distribution to new sites where conditions for continued growth might be better. One consequence of the uncoupling of growth from cell division during lag and

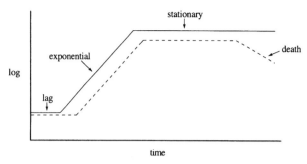

Fig. 2.3 Growth kinetics in batch culture. The solid line is mass; the dotted line is viable cell number. Note that if we define growth as an increase in mass, then only the solid line accurately reflects the growth of the culture. The dotted line reflects growth only when it is parallel to the solid line.

late log phase is that the size of the cell varies during growth in batch culture. A practical consequence for the investigator of the non-coincidence of growth and cell division during stages of batch growth is that cell counts are not always a valid measurement of growth.

## Lag phase

When cells in the stationary phase of growth are transferred to fresh media, a lag phase often occurs. Basically, the lag phase is due in this situation to the time required for the physiological adaptation of stationary phase cells in preparation for growth. Usually the longer the cells are kept in stationary phase, the longer is the lag phase when they are transferred to fresh media. Some of the physiological changes that occur in the stationary phase are described in Section 2.2.2. There are several possible reasons for the lag phase. The lag phase can be due to the time required for recovery of cells from toxic products of metabolism that may accumulate in the external medium, such as acids, bases, alcohols, or solvents. Sometimes, new enzymes or coenzymes must be synthesized before growth resumes if the fresh medium is different from the inoculum medium and requires a change in the enzyme composition of the cells. If significant cell death occurs in the stationary phase, then an apparent lag phase will be measured because the inoculum includes dead cells that contribute to the turbidity. The lag phase can be avoided if the inoculum is taken from the exponential phase of growth and transferred to fresh medium of the same composition.

## Stationary phase

The reason why cells stop growing and enter stationary phase can vary. It may be because of the exhaustion of nutrients, limitation of oxygen, or the accumulation of toxic products (e.g., alcohols, solvents, bases, or acids). The accumulation of toxic products is frequently a problem for fermenting cells because most of the nutrient is not converted to cell material but excreted as waste products. The excretion of fermentation end products is discussed in Chapter 14.

## Death

Death can result from several factors. Common causes of death include the depletion of cellular energy and the activity of autolytic (self-destructive) enzymes. Some bacteria begin to die within hours of entering the stationary phase. However, many bacteria remain viable for longer periods. For example, some bacteria sporulate or form cysts when exponential growth ceases. The spores and cysts are resting cells that remain viable and germinate in fresh media. As discussed next, even nonsporulating bacteria can adapt to nutrient depletion and remain viable for long periods in stationary phase.

## 2.2.2 Adaptive responses to nutrient limitation

In the natural environment bacteria are frequently faced with starvation conditions and enter intermittent periods of no growth or very slow growth. In some environments the generation time may be many days or even months because the nutrient levels are so dilute. When a bacterial culture faces the depletion of an essential nutrient, one of two things happens. The culture may keep growing by inducing specific uptake systems to scavenge the environment for the nutrient or a source thereof, and/or induce the synthesis of enzymes that can use an alternative source of the nutrient. For example, when bacteria are starved for inorganic phosphate, they may induce the synthesis of a high-affinity inorganic phosphate uptake system as well as the synthesis of enzymes capable of degrading organic phosphates to release inorganic phosphate (phosphatases). (See the discussion of the Ntr and PHO regulons in Sections 18.8 and 18.9.) If the bacteria cannot bring into the cells the essential nutrient, then the population will stop growing and dividing. Under these circumstances, the population enters stationary phase, or in the case of certain bacteria, the cells may sporulate or encyst. (See Section 18.15 for a description of sporulation.) Lately, increased attention is being given to physiological changes that occur in bacteria that enter the stationary phase when they are experimentally subjected to starvation.[1-4] These bacteria undergo physiological changes that result in metabolically less active cells that are more resistant to environmental hazards. This has been known for a long time for those bacteria that sporulate or form desiccation-resistant cysts upon nutrient deprivation. The spores and cysts represent metabolically inactive or less active stages of the life cycle of the organism. When nutrients become available once more, the spores and cysts germinate into vegetative cells that grow and divide. More recently, it has been discovered that even bacteria such as *E. coli*, *Salmonella*, *Vibrio*, and *Pseudomonas*, which do not form spores, undergo adaptive changes when faced with nutrient deprivation and enter stationary phase. Some of these changes also result in resistant, metabolically less active cells, and such responses may be common in most if not all bacteria. However, not all of the effects can be rationalized in terms of survivability, and their physiological role is not yet known. Some of the changes that occur in cells that

are starved are described next. For a review of starvation in bacteria, see Note 5.

### Changes in cell size

As discussed earlier, cells that enter stationary phase upon carbon exhaustion generally become smaller due to the fact that they keep dividing for 1 to 2 h after growth has ceased. This results in the production of more cells that may be advantageous for dispersing the population. In some cases there may be several cell divisions in the absence of growth so that the cells size differs radically from the growing cell. Some bacteria decrease in length from several micrometers (e.g., 5–10 $\mu$M) to approximately 1–2 $\mu$M or even less. (See Note 6.) Frequently, the decrease in size is accompanied by a change from rod-shaped cells to coccoid-shaped cells, as discussed next.

### Morphological changes

Along with a reduction in size, some bacteria (e.g., *Arthrobacter*) change from rod-shaped cells to coccoid cells in stationary phase. Similar morphological changes from rods to small coccoid shapes can occur upon starvation in several other bacteria, including *Klebsiella*, *Escherichia*, *Vibrio*, and *Pseudomonas*.

### Changes in surface properties

When certain marine bacteria are starved, the cell surface becomes hydrophobic and the cells are more adhesive.[7] *Vibrio* synthesizes surface fibrils and forms cell aggregates when starved for a long time. These changes in the surfaces of starved cells make them more adhesive. Presumably, it is advantageous for nutrient-limited cells to adhere to particles where they might feed on nutrient adsorbed to the surfaces of the particles.

### Changes in membrane phospholipids

In *E. coli* all the unsaturated fatty acids in the membrane phospholipids become converted to the cyclopropyl derivatives by the methylation of the double bonds. The advantage to the cyclopropane fatty acids is not known, since mutants unable to synthesize cyclopropane fatty acids appear not to be at a survival disadvantage when faced with environmental stresses such as starvation and high or low oxygen tension.

## Changes in metabolic activity

When bacteria enter stationary phase, their overall metabolic rate slows. In addition, for many bacteria there is a significant increase in the turnover (metabolic breakdown and resynthesis) of protein and RNA when they are starved. Presumably, in starved cells the protein and ribosomal RNA can serve as an energy source to maintain viability and crucial cell functions. The latter would include solute transport systems, an energized membrane, and ATP pool levels.

## Changes in protein composition

Bacteria may synthesize as many as 50–70 or more new proteins under conditions of carbon, nitrogen, or phosphate starvation. Many of the proteins serve specific functions related to the nutrient that is in low concentration. For example, phosphate starvation induces the synthesis of PhoE porin, which is an outer membrane channel for anions, including phosphate (Sections 1.2.3 and 18.9). This helps the bacterium to bring in more phosphate. Another example is the nitrogen fixation genes that are induced when the cells are starved for nitrogen (Sections 12.3 and 18.8). In addition to the proteins of known function, there are also many proteins made under all conditions of starvation for which there is currently no known function, although many of these are presumed to be involved in resistance to stress.

## Changes in resistance to environmental stress

Cells entering stationary phase also become more resistant to environmental stresses such as heat, osmotic stress, and certain chemicals (e.g., hydrogen peroxide). For example, *E. coli* has been reported to become more resistant in stationary phase to high temperature, hydrogen peroxide, high salt, ethanol, solvents such as acetone and toluene, and acidic or basic pH. Some of these resistant properties are due to the synthesis of a starvation sigma factor, called $\sigma^s$ (or KatF), as discussed next.

## The rpoS gene in E. coli encodes a sigma factor required for transcription of genes expressed during stationary phase and starvation

The response to starvation in *E. coli* is mediated in part by the *rpoS* gene (also called *katF*), which is required for the synthesis of at least 30 proteins induced by carbon starvation when cells enter stationary phase and for the transcription of several genes whose activities increase during phosphate and nitrogen starvation.[8–10] Proteins whose synthesis depends upon $\sigma^s$ include a catalase made during stationary phase (HPII), and which presumably accounts for resistance to $H_2O_2$, an exonuclease III that can repair damaged DNA due to $H_2O_2$ and near-UV radiation, and an acid phosphatase. The predicted amino acid sequence of the *rpoS* gene product, based upon nucleotide sequence data, suggests that it is a sigma factor; that is, it is a subunit of RNA polymerase that is necessary for the recognition of specific promoter regions in genes (Section 10.2.4). Studies with purified protein have confirmed that the protein binds to RNA polymerase and directs transcription of *rpoS*-dependent genes. Accordingly, the protein is also called $\sigma^s$ for starvation sigma factor. It is also known as $\sigma^{38}$ because it has a MW of 38 kD. The products of *rpoS*-dependent genes are necessary for survival under stressful conditions during stationary phase, since RpoS mutants do not have the resistance properties (measured as % survival) of the wild type, including stationary-phase resistance to heat, high salt (osmotic shock), near-UV, $H_2O_2$, or prolonged starvation (e.g., several days starvation for carbon or nitrogen).[8] The transcription of the *rpoS* gene (measured using a *rpoS–lacZ* fusion plasmid) increases dramatically in a nutrient-rich medium as cells approach and enter stationary phase and during starvation for specific nutrients such as phosphate.[11, 12] (See Note 13 for a description of lacZ gene fusions.) Homologues to *rpoS* exist in other bacteria, including pathogens. Mutations in *rpoS* in the intestinal pathogen *Salmonella* result in the attenuation of virulence in mice.[14, 15] One might suppose that RpoS aids pathogens to survive stress such as oxidative stress that they might encounter inside host cells, or pH stress (e.g., in the stomach).

## Rpos is a global regulator that is important for stress reponses in exponentially growing cells as well as in stationary phase

The $\sigma^s$ subunit of RNA polymerase should be viewed as a *global regulator* that also functions during responses to stress during exponential growth.[16] Increased RpoS levels can be caused by slow growth, temperature upshift from 30 to 42°C, and high osmolarity. For example, during osmotic upshifts in exponentially growing cultures RpoS levels increase. Such cultures become not only osmotolerant but also thermotolerant and $H_2O_2$ resistant.

## Regulaton of RpoS levels can be at the post-transcriptional as well as transcriptional level

Interestingly, the reason for the increased levels of RpoS during stress responses is not simply increased transcription of the *rpoS* gene. The regulation of the levels of RpoS occurs not only at the transcriptional level but also at the translational and post-translational levels.[17] For example, the rate of degradation of RpoS decreases when cells enter stationary phase. The half-life of RpoS in exponential cultures of *E. coli* growing at 37°C is less than 2 min. However, the half-life is greater than 30 min when cells enter stationary phase. Thus the regulation of RpoS levels is complex. See Note 18 for a discussion of how the levels of RpoS are adjusted, including transcriptional regulation of the *rpoS* gene, regulation of translation of the RpoS mRNA, and the role of proteases in the degradation of RpoS.

## Other regulators for stationary-phase gene expression

Although $\sigma^s$ is certainly a major transcription factor for gene expression during stationary phase, additional regulatory factors exist. (Reviewed in Ref. 4.) This conclusion follows, in part, from the fact that the timing of expression during stationary phase of $\sigma^s$-dependent genes varies, depending upon the gene in question, clearly pointing to regulatory factors in addition to $\sigma^s$. In fact, there are several $\sigma^s$-dependent genes that are also controlled by other global regulators, both positive and negative. One of these, cyclic AMP (c-AMP)–CRP complex, stimulates the expression of approximately 2/3 of the genes expressed during carbon starvation. Furthermore, not all genes whose expression increases during stationary phase require $\sigma^s$. It is clear from the available data that gene expression during stationary phase is complex and poorly understood at the present time. Nevertheless, it is extremely important because bacteria in their natural habitat are frequently in stationary phase or growing extremely slowly due to limitations of carbon, phosphate, or nitrogen sources. How they survive such nutritional stress is fundamental to an understanding of bacterial physiology in natural ecosystems.

## The control of ribosomal RNA synthesis

There is much controversy surrounding the question of what regulates the rates of synthesis of ribosomes, although it is clear that ribosome synthesis is coupled to growth rates (Section 2.2.3).[19] As shown in Fig. 2.4, faster growing cells have more ribosomes per unit cell mass. For example, the number of ribosomes in an *E. coli* cell can vary from fewer than 20,000 to about 70,000 depending upon the growth rate. In addition to growth rate control, amino acid starvation results in the inhibition of ribosome synthesis. Probably, several factors are involved in regulating rates of ribosome synthesis. For example, Fis, a DNA-binding protein whose synthesis is increased in rich media, positively regulates transcription of rRNA genes. Another DNA-binding protein, H-NS, antagonizes Fis stimulation (see later). A very important regulator is ppGpp, which increases when cells are subjected to amino acid starvation or carbon and energy limitation, and slows the synthesis of rRNA (and tRNA). The global regulator, ppGpp, is best known as the effector that inhibits rRNA and tRNA synthesis

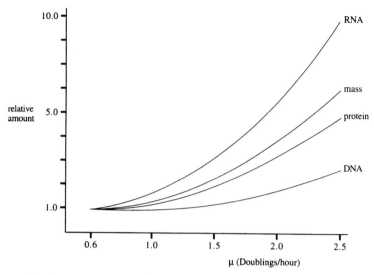

**Fig. 2.4** Macromolecular composition of *E. coli* grown at different growth rates. The values are amounts per cell and have been normalized to values at a doubling time of 0.6 doublings/h. *Source:* From Neidhardt, F. C., J. L. Ingraham, and M. Schaechter. 1990. *Physiology of the Bacterial Cell.* Sinauer Associates, Inc., Sunderland, MA.

during the *stringent response*, which was originally discovered as the inhibition of rRNA and tRNA synthesis due to amino acid starvation.[20] See Note 21 for other physiological effects of ppGpp.

### 1. The stringent response

The stringent response in *E. coli* is a temporary inhibition of the synthesis of ribosomal RNA and transfer RNA when the cells are shifted to a medium where they are starved for an amino acid,. It works in the following way: When cells are shifted from a medium rich in amino acids to a minimal medium, there occurs an accumulation of "uncharged" tRNA. This results in ribosomes that are "idling" or "stalled" because of an insufficiency of aminoacylated tRNA. An enzyme on the "stalled" ribosomes called RelA [(p)ppGpp synthetase I], which is the product of the *relA* gene, becomes activated by uncharged tRNA that binds to the A site. Activated RelA synthesizes a signaling molecule called guanosine tetraphosphate (ppGpp) and to a lesser extent the pentaphosphate (i.e., pppGpp). Together they are referred to as (p)ppGpp. The ppGpp somehow slows the transcription of ribosomal RNA (and transfer RNA). This leads to an inhibition of ribosomal protein synthesis by an interesting mechanism

called translational coupling, as explained in Note 22.

### 2. Carbon and energy limitation cause an increase in the levels of ppGpp in a pathway independent of RelA

The regulator (p)ppGpp also accumulates during carbon and energy limitation due to a pathway that is independent of RelA—it occurs in *relA* mutants—and which requires the product of the *spoT* gene. The SpoT protein is most probably a bifunctional enzyme that both synthesizes and degrades (p)ppGpp. The ratio of biosynthetic to degradative activity is somehow regulated by carbon and energy limitation. This may account at least in part for why cells growing slowly in poor medium have fewer ribosomes. It is therefore clear that the inhibition of rRNA and tRNA synthesis coincides with essentially any nutritional condition (not simply amino acid starvation) that slows the growth rate. See Note 23 for a further discussion of the synthesis of (p)ppGpp and the contributions of RelA and SpoT.

### 3. Other factors that control rRNA synthesis: Fis, H-NS

It must be emphasized that the control of ribosomal RNA synthesis is complex, and ppGpp is only one factor in the regulation. Other proteins involved in transcriptional regulation of the ribosomal RNA genes include the Fis and H-NS proteins.[24] (See Note 25 for a further discussion of Fis.) Fis binds to DNA sequences upstream of the rRNA promoter and activates the transcription of the operon. (Fis regulates the expression, both positively and negatively, of many genes besides rRNA genes.) Importantly, Fis synthesis is under growth rate control. Synthesis increases when exponentially growing cells are shifted to a nutritionally richer medium and when stationary-phase cells, which have very low Fis levels, are subcultured into a nutrient-rich medium.[26] One would expect an increase in the rate of rRNA synthesis under these conditions and Fis accounts in part for the increased synthesis.

Fis-dependent activation of ribosomal RNA genes is antagonized by another protein called H-NS (also called H1), which can therefore act as a repressor of rRNA transcription.[27] (H-NS is discussed in Section 1.2.6 as part of the nucleoid. It does not bind to specific nucleotide sequences, but it does bind to curved or bent DNA.) The cellular levels of H-NS are highest under conditions where rRNA synthesis is slow (e.g., stationary phase). H-NS proteins belong to a remarkable group of DNA-binding proteins that have other cellular functions besides the regulation of rRNA gene transcription. (See Section 1.2.6 and Note 28.) So, although ppGpp somehow slows ribosomal RNA synthesis, it appears to be only one component of a complex regulatory system(s) that must control ribosomal RNA synthesis under a variety of growth conditions.

## 2.2.3 Macromolecular composition as a function of growth rate

Much has been learned about growth physiology by observing changes in the macromolecular composition of cells whose growth rates have been altered by changing the nutrient composition of the medium, or by manipulation of the dilution rates of cells growing in continuous culture (Section 2.4).

Figure 2.4 illustrates the changes in the ratios of cell mass, RNA, protein, and DNA of *E. coli* grown at increasingly faster growth rates due to differences in the nutrient composition of the growth media. On the ordinate are plotted relative amounts of macromolecules *per cell* normalized to cells undergoing 0.6 doublings per hour. On the abscissa are plotted the growth rate in units of doublings per hour. Notice that the mass of the cell as well as the relative concentrations of the macromolecules increase as exponential functions of the growth rate. As shown in Fig. 2.4, faster-growing cells have a higher ratio of RNA to protein, a larger mass, and more DNA.

### Why faster growing cells have a larger RNA-to-protein ratio

Faster-growing cells are enriched for RNA with respect to the other cell components. This reflects a higher proportion of ribosomes in faster-growing cells and the fact that ribosomes are approximately 65% RNA and 35% protein by weight. The reason for the increase in ribosomes is that, over a wide range of rapid growth rates, a ribosome polymerizes amino acids at approximately a constant rate. When a cell increases or decreases its growth rate and therefore the number of proteins it must make per unit time, it adjusts the number of ribosomes rather than making each ribosome work substantially slower or faster.

### Why faster-growing cells have more DNA

Figure 2.4 also shows that faster-growing cells have more DNA per cell. The increased DNA per cell is rationalized in the following way. In *E. coli*, the minimum time required between the initiation of DNA replication and completion of cell division (cell separation) for rapidly growing cells (i.e., those with doubling times between 20 and 60 min) is about 60 min.[29] (See Note 30 for an explanation of the C and D periods.) If the generation time is shorter than 60 min, then DNA replication must begin in a previous cell cycle, and cells are born with partially replicated DNA, which increases the average amount of DNA per cell. This is illustrated in Fig. 2.4, which shows that cells with a generation time of less than 60 min have more DNA.

### Why faster-growing cells have more mass

As seen in Fig. 2.4, cells that are grown at a faster growth rate have a greater mass (i.e., they are larger). For example, *E. coli* when growing at a doubling rate of 2.5 doublings per hour is approximately six times larger in mass than when it is growing at 0.6 doublings per hour. Why is that? It is because the initiation of DNA replication requires a minimum cell mass per replication origin, called the critical cell mass ($M_i$), faster-growing cells have several replication origins, and replication must begin in a previous cell cycle when the doubling times are less than 60 min. (See Note 31 for a more complete explanation.) (The significance of $M_i$ was reported by Donachie several years ago.[32] See Note 33 for how this was done. There is some question whether it is an absolute constant or whether it decreases slightly with increasing growth rate.[34, 35] For this discussion, we will assume it is constant.)

### 2.2.4 Diauxic growth

Many bacteria, including *E. coli*, will grow preferentially on glucose when presented with mixtures of glucose and other carbon sources. The result is a biphasic growth curve as shown in Fig. 2.5 for glucose and lactose. Preferential growth on one carbon source before growth on a second carbon source is called *diauxic growth*. The bacteria first grow exponentially on glucose, then enter a lag period, and finally grow exponentially on the second carbon source.

### 2.2.5 Catabolite repression by glucose

It is now known that in many bacteria glucose represses the synthesis of enzymes required to grow on certain alternative carbon sources such as lactose, and this can account for the diauxic growth discussed in Section 2.2.4. Glucose also inhibits the uptake of certain other sugars into the cell. The result is that growth on the second carbon source does not proceed until the glucose is exhausted from the medium. During the lag period following the exponential growth on glucose, the bacteria synthesize the enzymes required to grow on the second carbon source. The repression of genes by glucose is called *catabolite repression* or *glucose repression*, and as mentioned is widespread among bacteria. The inhibition of uptake of alternative sugars is called *inducer exclusion*. Catabolite repression in gram-negative and gram-positive bacteria is discussed Section 18.13. A rationale for glucose repression is that glucose is one of the most common carbon sources in the environment, and bacteria frequently grow

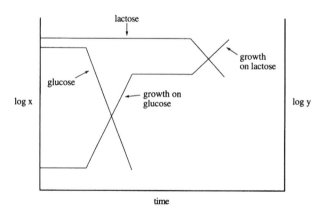

**Fig. 2.5** Diauxic growth on glucose and lactose. (x), Bacterial mass; (y), glucose or lactose concentration in the medium. There are two phases of growth. Initially the cells grow on the glucose. When the glucose is sufficiently depleted, a lag phase occurs followed by growth on lactose. During the lag phase the cells induce the synthesis of enzymes necessary for growth on lactose. Glucose prevents the induction of the genes necessary for growth on lactose. The inhibition by glucose is called catabolite repression and is discussed in Section 16.3.4.

more rapidly on glucose than on other carbon sources, especially other sugars. Therefore, bacteria in certain ecological habitats that preferentially utilize glucose may be at a competitive advantage with respect to the other cells that depend upon glucose for carbon and energy. Additionally, the enzymes required to metabolize glucose are generally constitutive (i.e., present under all growth conditions) and the cell is prepared to grow on glucose at any time.

However, not all bacteria preferentially grow on glucose, nor are the glucose catabolic enzymes necessarily constitutive. Several obligately aerobic bacteria will grow first on organic acids when given a mixture of glucose and the organic acid. This is also called catabolite repression except that it is the organic acid that is the repressor of glucose utilization. An example is the nitrogen-fixing bacteria belonging to the genus *Rhizobium* that live symbiotically in root nodules of leguminous plants. These bacteria grow most rapidly in laboratory cultures on $C_4$ carboxylic acids that are part of the citric acid cycle such as succinate, malate, and fumarate. When presented with a mixture of succinate and glucose, *Rhizobium* shows diauxic growth by growing on the succinate first and the glucose second. This is because in these bacteria succinate represses key enzymes of the Entner–Doudoroff and Embden–Meyerhof–Parnas pathways that degrade glucose.[36] The mechanism of repression is not understood but is probably not the same as glucose repression. Another example where diauxic growth occurs first on organic acids and then on carbohydrates (e.g., glucose) occurs with *Pseudomonas aeruginosa*. If presented with glucose or other carbohydrates and any one of several organic acids such as acetate, or one of the intermediates of the citric acid cycle, *P. aeruginosa* will grow on the organic acid first and repress the enzymes required to degrade the carbohydrate.[37]

## 2.2.5 Cell divison

Two major events that occur in the cell cycle are DNA replication and cell division. Cell division is the division of a mother cell into two daughter cells separated by a septum. Cell division is accompanied by, or followed by, cell separation. Most research attention has been focused on the gram-negative bacterium *E. coli* (and the closely related *Salmonella typhimurium*), *Caulobacter crescentus,* and the gram-positive bacterium, *Bacillus subtilis.*

### The period prior to septum formation

When *E. coli* reaches a critical cell mass during growth, DNA synthesis is initiated (Section 2.2.3). Initiation requires the DnaA protein, which binds to the replication origin (*oriC*) and opens up the duplex, allowing other proteins required for DNA replication to enter (Section 10.1.2). After the chromosome is replicated, the sister chromosomes separate and move toward opposite halves of the cell, so that when cell division takes place, each daughter cell receives a chromosome. The movement of the sister chromosomes to opposite cell poles is called chromosome partitioning. As discussed in Section 2.2.3, rapidly growing *E. coli* requires approximately 40 min (the C period) to replicate its chromosome, followed by a period (the D period) of about 20 min before cell division is complete.

How the sister chromosomes might be partitioned to the cell poles prior to formation of the septum is not understood. See Section 10.1.5 for a more complete discussion. In some bacteria an early stage of the partitioning process appears to involve the Par proteins that attach to the DNA and align the nucleoids along the long axis of the cell. It is speculated that the Par proteins anchor the replication origins to opposite poles and that this is a necessary step in moving the nucleoids to the poles of the cell prior to septum fomation. Genes homologous to the *par* genes have been detected in the chromosomes of *B. subtilis, C. crescentus,* and *Pseudomonas putida,* but not *E. coli* chromosomes, although they are present in *E. coli* plasmids where they function in plasmid partitioning. In *E. coli* the MukB protein has been postulated to move the nucleoids to opposite cell poles.[38–40] It is required for chromosome partitioning (at temperatures higher than 22°C) but not for the partitioning of plasmid DNA. The *mukB* gene has thus far been detected only in *E. coli* and *Hemophilus influenzae* (reviewed in

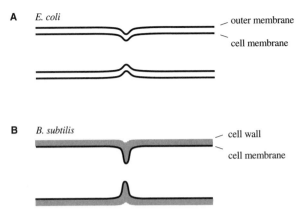

**Fig. 2.6** Cell division in gram-negative and gram-positive bacteria. (A) Many gram-negative bacteria divide by constriction as the outer membrane invaginates with the cell membrane. Thus septation and cell separation occur together. (B) In gram-positive bacteria a septum consisting of cell membrane and peptidoglycan forms, but there is generally little constriction in the early stages. Thus septation precedes cell separation.

Ref. 41). For a further discussion of the role of the MukB protein in nucleoid partitioning, see Note 42.

## Cell division

Once the sister chromosomes have moved to opposite cell poles, a septum forms in the center of the cell. The site of septum formation is regulated in some way in *E. coli* by the products of three linked genes (i.e., *minC*, *minD*, and *minE*). These genes are in the *minB* operon. When the operon is deleted, division occurs not only at midcell but also at the poles to produce a population of anucleate minicells. The first morphological sign of cell division is the centripetal synthesis of a septum, which forms by inward growth of the cell membrane and peptidoglycan layers. In *E. coli*, this begins soon after chromosome replication is complete (i.e., at the end of the C period). In many gram-negative bacteria, including *E. coli*, the outer envelope also invaginates at this time (i.e., cell separation and septation occur at approximately the same time) so that septa are usually not seen in thin sections of cells prepared for electron microscopy. Invagination is manifested as a constriction that becomes narrower as septation proceeds (Fig. 2.6 A). Initially there is a double layer of peptidoglycan in the septum, but this splits in half as constriction proceeds and the

cells separate. For those bacteria that have peptidoglycan, septation and peptidoglycan synthesis are coupled (i.e., inhibitors of septal peptidoglycan synthesis prevent septation). In *E. coli* septation occupies approximately the first 10–13 min of the D period, depending upon the growth rate (slower-growing cells have a longer period of septation), and daughter cells separate within 7 min of completion of the septum. Cell division in *B. subtilis* differs from that in *E. coli* in that septation proceeds cell separation, although cell separation may begin before septation is complete (Fig. 2.6 B).[43] The result is that thin sections of dividing cells are seen in the electron microscope to have partially completed septa, and there is no visible constriction at the beginning of septation. As in gram-negative bacteria, cell separation is effected by the splitting of the septum. Under certain conditions of rapid growth, septum formation in *B. subtilis* can be completed before cell separation begins, and chains of cells grow attached to each other via their septa.

## Cell division genes

The physiological events underlying cell division in bacteria are incompletely undersood, but there has been progress in identifying the genes and proteins involved.[44–46] Most research in the area has focused on *E. coli*, where several temperature-sensitive division mutants have been isolated, as well as *Bacillus* and *Caulobacter*. (See Note 47 for a description of temperature-sensitive mutants.) Temperature-sensitive mutations in genes required for cell division result in the growth of filamentous cells (i.e., cells that grow longer but do not divide) at the restrictive temperature. Since mutations in genes not directly involved in septation or cell separation can also result in filamentous growth at restrictive temperatures (e.g., mutations that cause a defect in DNA synthesis), one must be careful in assigning a role for a particular gene in cell division per se. (See Note 48.) Cell division genes are called *fts* genes for filamentation temperature-sensitive phenotype. At the restrictive temperature the cells do not divide but grow into long filaments that eventually lyse. Two of the *fts* gene protein products FtsZ and FtsA, are cytoplasmic, and the others (FtsI, FtsL, FtsQ, FtsN, ZipA, and FtsW) are membrane associated. FtsZ forms a ring in the cell center. The FtsZ ring is made before invagination and is believed to contract, leading to the constriction of the cell in the center. Other proteins join the FtsZ ring. The FtsZ ring with the other proteins is called a *septal ring*. ZipA (FtsZ interacting protein A) becomes associated with the FtsZ ring, perhaps as it is forming. FtsA, FtsI, FtsW, and FtsN becme associated with the septal ring after it forms and are required for its function. We shall see that it is not known exactly how the Fts proteins interact in aiding the septal ring to constrict the cell. As will be referred to later, the *envA* gene is strictly speaking not a cell division gene but codes for an enzyme required for lipid A biosynthesis. A mutation in this gene results in cells that divide but are unable to separate. Other gene products involved in cell division, such as genes involved in the positioning of the septum and in nucleoid partitioning, are discussed in Ref. 45. As will be made clear in the following discussion, the precise roles that most of the gene products play in cell division are not yet known. Figure 2.7 is a model of how the cell division proteins might be arranged at the site of septation.

The following discussion refers primarily to research done with *E. coli*. Although all bacteria and archaea thus far examined have

the *ftsZ* gene, and those with peptidoglycan have the *ftsI* gene (involved in peptidoglycan synthesis), this is not the case for the other cell division genes. Genome sequences reveal that some bacteria and archaea lack some or all of the other cell division genes identified in *E. coli*. It is not known whether prokaryotes missing some of the cell division proteins described next have evolved different proteins to take their place or that they are simply not always necessary. The occurrence of the *E. coli* cell division genes in eight other bacteria and one archaean, as well as the implications of their absence, is discussed in Ref. 49.

### 1. FtsZ

For a model of how FtsZ and the other cell division proteins might interact at the septal ring, see Fig. 2.7. The FtsZ protein is a cytoplasmic protein that aggregates on the inner surface of the cell membrane at the division site as a circumferential ring (septal ring) early in the cell cycle, and remains at the leading edge of the invaginating membrane during septation (Fig. 2.8).[50, 51] (This appears also to be true for ZipA and FtsA, as discussed later.) The ring is not present in newborn cells or at the ends of cells indicating that the septal ring is disassembled after septation and is assembled

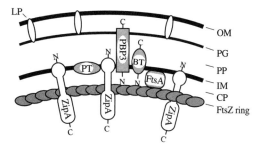

**Fig. 2.7** Septal ring model. A ring of FtsZ proteins is suggested to be connected to the cell membrane by ZipA. PT represents FtsK and FtsW division proteins; BT represents FtsL, FtsN, and FtsQ division proteins; PP, periplasm; CP, cytoplasm; LP, lipoprotein; OM, outer membrane; PG, peptidoglycan; IM, inner (cell) membrane. *Source*: From, Hale, C. A., and P. A. J. de Boer. 1997. Direct binding of FtsZ to ZipA, an essential component of the septal ring structure that mediates cell division in *E. coli*. *Cell* **88**:175–185.

**Fig. 2.8** Schematic drawing showing recruitment of FtsZ to leading edge of invaginating cell membrane during septation.

again in the daughter cells shortly after cell division (it may even form before cell separation). The cellular location of FtsZ can be seen using immunoelectron microscopy, immunofluorescence, and GFP fusions. (See the discussion of ZipA and FtsI for an explanation of GFP fusions and immunofluorescence microscopy, respectively.) Mutations in *ftsZ* lead to filaments without constrictions, and it is clear that FtsZ is necessary for septation. Precisely how FtsZ is involved in septation is not clear, but it is thought to form a contractile ring that constricts the cell during division. The analogy of FtsZ to tubulin, which forms the eukaryotic cytoskeletal microtubules, has been made, and it has been suggested that both tubulin and FtsZ may have evolved from the same protein. (See Note 52 for a discussion of tubulin and FtsZ.) It is present in all bacteria thus far examined, including mycoplasma, as well as in the plant *Arabidopsis*, where it was suggested to be a choroplast protein and is presumably involved in fission of the chloroplast, and in members of the domain Archaea.[53, 54] However, FtsZ is not sufficient for division, since mutations in the other *fts* genes will also lead to filamentous growth even though such mutants still form a ring of FtsZ in the center of the cell. Indeed, there are immunofluorescent data suggesting that FtsA and FtsN are also located in the FtsZ ring after it forms. The model is that the FtsZ ring forms, perhaps with a protein called ZipA discussed later, and then recruits FtsA and FtsN to form a septal ring that constricts. (See the discussion of FtsA next.) The way in which FtsZ finds the center of the cell is not known. Presumably, it binds to a molecule, not yet identified, that is anchored in the membrane at that site.

## 2. FtsA

Mutants in *ftsA* begin but do not complete septation. They form indented filaments. FtsA is suspected to be an ATPase because it possesses an amino acid sequence domain found in other ATPases. How it functions is not known, but it appears that FtsA interacts with FtsZ so that septation can continue. This suggestion is prompted by the observation that the ratio of FtsZ to FtsA must be within a narrow range in order for cell division to continue, and that FtsA, along with FtsN, becomes localized at the site of septum formation in a ring with FtsZ.[55, 56] It appears that the recruitment of FtsA to the septal ring requires the prior localization of FtsZ. FtsZ and FtsA have been shown to interact using the yeast two-hybrid system.[57] (See Note 58 for a description of the yeast two-hybrid system.) It has been demonstrated with *Caulobacter* that the C-terminal portion of FtsZ is important for both its localization and interaction with FtsA.[59]

## 3. ZipA

ZipA is a cell membrane protein in *E. coli* that is bound to FtsZ in the septal ring.[60] It was identified by its ability to bind to radioactively labelled FtsZ after separation of the membrane proteins by gel electrophoresis. It was also shown to be located in the septal ring using a fluorescent ZipA–GFP fusion protein. (GFP is an abbreviation for green fluorescent protein, which is a fluorescent protein found in certain jellyfishes.[61] (See Note 62 for a further description of these experiments.) Fluorescent ZipA–GFP rings are found in young cells before invagination is seen, and appear to coincide with the FtsZ rings. Mutants lacking ZipA form filaments without constrictions. Thus it appears that the recruitment of ZipA to the septal ring is an early event that may occur at around the same time that FtsZ is recruited, or shortly after. One model proposes that ZipA binds FtsZ to the cell membrane (Fig. 2.7).

## 4. FtsI and FtsW

The *ftsI* gene, also called *phpB*, codes for the FtsI protein, which is also called

PBP3 for penicillin-binding protein. This is an enzyme required for peptidoglycan synthesis specifically in the septum. (See Note 63.) Inhibition of this enzyme results in stopping cell division, but the cells maintain their rod shape. (Inhibition of PBP2, a peptidoglycan-synthesizing enzyme required for cell elongation, results in loss of rod shape.) It has transglycosylase and transpeptidase activity and may act together with FtsW in this regard. (See Section 11.1.2 for a discussion of peptidoglycan synthesis.) The reason for suggesting that FtsW acts with FtsI is that FtsW is homologous to RodA, a protein that augments the synthetic activity of PBP2. FtsA and FtsQ might interact with FtsI and FtsW in a complex that takes part in septum formation.[43] FtsI and FtsW are located in the septal ring, along with FtsZ, FtsA, FtsN, and ZipA.[64, 65] As explained in Note 66, their localization was determined using immunofluorescence microscopy.

### 5. FtsK, FtsL, FtsN, FtsQ

The exact functions of these four genes required for cell division are not understood. Their protein products are all membrane associated and at least two of them, FtsK and FtsN, are also associated with FtsZ as well as FtsA in the septal ring.[61, 67] They were recognized either because mutations resulted in an inhibition of cell division resulting in filamentation or because they were able to suppress other cell division mutants. A model for the sequence of events in the formation of the septal ring is that first FtsZ forms the ring, perhaps with the help of ZipA. Then FtsA joins the ring, followed by FtsN. The addition of FtsN to the ring requires FtsI and FtsQ, but the additon of FtsA does not. Shortly after or around the time of the addition of FtsN, constriction begins at the septal site. Mutants that lack FtsN form filaments that do not have constrictions. It is believed that FtsK is added very late because mutations in FtsK result in very highly indented filaments. (See Note 68 for a summary of evidence for the model.) As discussed later, another protein, called MinE, appears to be necessary in order for the FtsZ ring to form. However, MinE is not found in all bacteria. (See the sections: *MinE forms a ring at midcell independent of the FtsZ ring* and *A model for early events in cell division.*)

### 6. EnvA

The product of the *envA* gene is necessary for cell separation. (Mutants in *envA* form chains of cells.) As discussed in Section 11.2.2, the gene actually codes for an enzyme required for an early step in lipopolysaccharide biosynthesis (lipid A synthesis), suggesting that synthesis of the outer envelope lipopolysaccharide is necessary, perhaps in an indirect way, for cleavage of the peptidoglycan in the septum and consequent cell separation. Thus *envA* should be considered a gene involved in lipopolysaccharide biosynthesis that has pleiotropic effects.

### 7. The min genes control the site of septation

*E. coli* has an operon, called the *minB* operon, that consists of three genes, *minC*, *minD*, and *minE*, that somehow determine the site of septation.[42, 43] MinCD inhibit septation at either the center of the cell or at other sites nearer to the cell poles, and MinE interacts with MinCD to release the inhibition at the cell center. It is not known what directs MinE (or FtsZ for that matter) to the cell center. The evidence for these conclusions include:

1. Deletion of the *min* locus results in cells that divide at the poles as well as in the center to produce anucleate minicells as well as large multinucleate cells. Thus it can be concluded that the Min proteins are not required for septation per se, but rather for the localization of the septum to the center of the cell.

2. When MinE is made in excess, then inhibition by MinCD is relieved both in the center and at the poles, and minicells are produced. This is the same phenotype that is obtained with *minC* and/or *minD* mutants.

3. When MinCD is in excess, then cell division is prevented.

It is not known how the Min proteins work, except that the target of MinCD may be FtsZ, since excess FtsZ can overcome the inhibition by MinCD. In a way that is not understood,

MinCD prevents the formation of the FtsZ septal ring.

## MinE forms a ring at midcell independent of the FtsZ ring

Somehow, MinE finds its way to the cell center, where it acts to prevent the inhibition by MinCD of septum formation. Precisely how this occurs is not known. It is clear, however, that MinE forms a ring at approximately midcell. This was demonstrated with the use of MinE–GFP fusion proteins.[69] (See the previous discussion of ZipA and Note 62 for an explanation of the use of the green fluorescent protein from the jellyfish *Aequorea victoria* to localize proteins in situ.) The formation of the MinE ring does not require the presence of the FtsZ ring, which would be expected if the role of MinE is to relieve the inhibition by MinCD and thus allow the FtsZ ring to form. (This was demonstrated by inducing the *sulA* gene, whose product prevents the formation of the FtsZ ring, as described in Note 70.)

## A model for early events in cell division

In light of the data discussed above, it has been suggested that early events in cell divison are as follows[69]:

1. MinE forms a membrane-associated ring at approximately the cell center. This relieves the inhibition by MinCD of FtsZ ring assembly.

2. FtsZ assembles into a ring at the cell center and recruits other proteins to form the septal ring.

3. The septal ring constricts, and the MinE ring disassembles.

4. When constriction is completed, MinE forms a new ring at the centers of the daughter cells, and MinCD prevent FtsZ ring formation at other sites.

The *minC* and *minD* genes are found in *B. subtilis*, whose genome has been sequenced over 95%. However, it does not have *minE* linked to *minCD*, indicating that perhaps a different protein substitutes for MinE, or the *minE* gene is located elsewhere in the chromosome. There are several other bacteria whose genome sequences indicate that they are missing one or more of the *min* genes. For example, *minC*, *minD*, and *minE* are missing from *Mycoplasma genitalium*, *Mycoplasma pneumoniae*, and *Hemophilus influenziae*, whose genomes have been sequenced 100%, as well as from *Streptococcus pyogenes*, whose genome has been sequenced over 95%. (See Ref. 49 for a review of the distribution of cell division genes in prokaryotes.)

## 2.2.6 Growth yields

When a single nutrient (e.g., glucose) is the sole source of carbon and energy, and when its quantity limits the production of bacteria, it is possible to define a *growth yield constant, Y*. The growth yield constant is the amount of dry weight of cells produced per weight of nutrient used (i.e., $Y$ = wt cells made/wt nutrient used). It has no units, since the weights cancel out. [The molar growth yield constant is $Y_m$, which is the dry weight of cells produced (in grams) per *mole* of substrate used.] For example, the $Y_{glucose}$ for aerobically growing cells is about 0.5, which means that about 50% of the sugar is converted to cell material, and 50% oxidized to $CO_2$. For certain sugars and bacteria, the efficiency of conversion to cell material can be much lower (e.g., 20%). These differences are thought to relate to the amount of ATP generated from the unit weight of the carbon source. The more ATP that is made, the greater the growth yields. This makes sense if you consider that growth is, after all, an increase in dry weight, and it requires a certain number of ATP molecules to synthesize each cell component. A value of 10.5 g of cells per mol of ATP (called the $Y_{ATP}$ or molar growth yield) has been determined for fermenting bacterial cultures growing on glucose as the energy source. The experiments are performed in media in which all the precursors of the macromolecules (e.g., amino acids, purines, pyrimidines) are supplied so that the glucose serves only as the energy source, and essentially all the glucose carbon is accounted for as fermentation end products. Knowing the amount of ATP produced in the fermentation pathways, it is possible to calculate the $Y_{ATP}$ from the $Y_{glucose}$. For example, 22 g of *Streptococcus faecalis* cells are produced per mole of glucose. This organism ferments a mole of glucose to fermentation end products using the Embden–Meyerhof–Parnas pathway, which produces 2 moles of ATP (Section 8.1). Thus the $Y_{ATP}$ is 22/2, or 11. On the other hand, *Zymomonas mobilis* produces only 8.6 g of cells per mole of glucose fermented. These organisms use a different pathway for glucose fermentation, the Entner–Doudoroff pathway, which yields only 1 ATP per mol of glucose fermented (Section 8.6). Thus the $Y_{ATP}$ is 8.6. As mentioned, a comparison of several different fermentations by bacteria and yeast has produced an average $Y_{ATP}$ of 10.5. Although this value can vary, knowledge of the growth yields and the $Y_{ATP}$ has allowed some deductions as to which fermentation pathway might be operating. Also, an unexpectedly high growth yield in fermenting bacteria can point to unrecognized sources of metabolic energy (Section 3.8.3).

## 2.3 Growth Kinetics

### 2.3.1 The equation for exponential growth

During exponential growth the mass in the culture doubles each generation. The equation for exponential growth is:

$$x = x_0 2^Y \tag{2.3}$$

where $x$ is anything that doubles each generation, $x_0$ is the starting value, and $Y$ is the number of generations. $x$ can be cell number, mass, or some cell component, for example, protein or DNA. Taking the $\log_{10}$ of both sides,[71]

$$\log_{10} x = \log_{10} x_0 + 0.301Y \tag{2.4}$$

Let us define $g$ as the generation time, the time per generation. Therefore, $g = t/Y$ and thus $Y = t/g$. $t$ is the time elapsed.

Therefore, eq. 2.4 becomes

$$\log_{10} x = \log_{10} x_0 + (0.301/g)t \tag{2.5}$$

You will recognize this as an equation for a straight line. When $x$ is plotted against $t$ on semilog paper (which is to base 10), the slope is $0.301/g$ and the intercept is $x_0$. See Fig 2.9.

### The generation time, g

As stated, the generation time is the time that it takes for the population of cells to double. It is an important parameter of growth and is probably the one most widely used by microbiologists. It is important that the student learns how to find the generation time. The simplest way is graphically. In practice, one determines the generation time from inspection of a plot of $x$ versus $t$ on semilog paper (Figs. 2.3 or 2.9), and simply reads off the time it takes for $x$ to double. It can also be done using eq. 2.3 if $x$, $x_0$, and $t$ are known accurately for the exponential phase of growth.

### The growth rate constant, k

The growth rate constant, $k$, is a measure of the *instantaneous* growth rate and has the units of reciprocal time. It is also a widely used parameter of growth. It is not to be confused with the generation time, $g$, which is the *average* time it takes for the population of cells to double. The distinction is made clear below. Recall that since each bacterial cell gives rise to two, the rate at which the population is growing at any instant, that is, the instantaneous rate of growth, $dx/dt$, must be equal to the number of cells at that time $(x)$ times a growth rate constant $(k)$. Thus

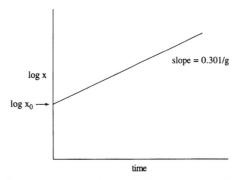

Fig. 2.9 Exponential growth plotted on a semilog scale. Symbol: g, generation time (time per generation).

$$\frac{dx}{dt} = kx, \quad \text{or} \quad \frac{dx}{x} = k \, dt \quad (2.6)$$

Integrate: $\ln x = kt + \ln x_0$, or

$$x = x_0 e^{kt} \quad (2.7)$$

Taking the $\log_{10}$ of both sides of eq. 2.7 converts the expression from the natural log ($\log_e$) to log base 10 so that $x$ can be plotted against $t$ on semilog paper:

$$\log_{10} x = 0.4342(kt) + \log_{10} x_0 \quad (2.8)$$

or

$$\log x = kt/2.303 + \log_{10} x_0$$

Note that this gives the same line as in Fig. 2.9. However, the slope is equal to $k/2.303$. Since the slope in Fig. 2.9 is equal to $0.301/g$, it follows that $k/2.303$ must be equal to $0.301/g$. Thus $k = 0.693/g$ (or $kg = 0.693$). Therefore, $k$ and $g$ vary as the reciprocal of one another with the proportionality constant of 0.693. The slope of the curve in Fig. 2.9 can give either $k$ (the instantaneous growth rate) or $g$ (the doubling time for the population).[72]

## Using the growth equations

The growth equations can be used for finding $g$ and $k$. Another, and more routine, circumstance is when one must choose the proper size inoculum when growing a culture to be used at a later time. A convenient equation to remember for this calculation is eq. 2.3 or 2.5. Suppose you have an exponentially growing culture at a density of $10^8$ cells/ml and you would like to subculture it so that 16 h later the density of the new culture will also be $10^8$ cells/ml. The generation time ($g$) is 2 h. What should $x_0$ be? One way to do this problem is to estimate the number of generations ($Y$) and then to use eq. 2.3 or 2.5. For example, since $Y = t/g$, the number of generation is 16/2, or 8. Using eq. 2.3, $10^8 = x_0 2^8$ or $x_0 = 10^8/2^8 = 3.9^5$ cells/ml. This means that the initial cell density in the growth flask should be $3.9 \times 10^5$ cells/ml. But the cell density of the inoculum is $10^8$/ml. This means that the inoculum must be diluted $10^8/(3.9 \times 10^5)$, or $2.6 \times 10^2$ times. If you wanted to grow 1 liter of cells, the inoculum size would have

to be 3.8 ml.[73] Most growth problems that one might have to do routinely can be done with the following equations:

$$x = x_0 2^Y$$

$$Y = t/g$$

$$g(k) = 0.693$$

## 2.3.2 The relationship between the growth rate (k) and the nutrient concentration (S)

In the natural environment the concentrations of nutrients is so low that the growth rates are limited by the rates of nutrient uptake or the rate at which a previously stored nutrient is used. At very low nutrient concentrations, the growth rate can be shown to be a function of the nutrient concentration (Fig. 2.10). The curve approximates saturation kinetics and is probably due to the saturation of a transporter in the membrane that brings the nutrient into the cell. The curve can be compared to the kinetics of solute uptake on a transporter described in Section 16.2. One can rationalize the kinetics by assuming that at very low concentrations of nutrient the growth rate is proportional to the percent of transporter that has bound the nutrient. At high nutrient concentrations, the rate of nutrient entry approaches its maximal value because all the

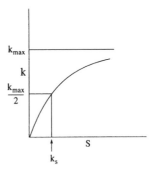

Fig. 2.10 Variation of growth rate as a function of substrate concentration. Symbols: $k_{max}$, maximal growth rate constant; $k$, specific growth rate constant; $k_s$, nutrient concentration that gives $\frac{1}{2} k_{max}$; S, nutrient concentration. The concentrations where the growth rate is proportional to the nutrient concentrations are very low, in the micromolar range.

transporter has bound the nutrient. Therefore, the growth rate becomes independent of the nutrient concentration, and becomes limited by some other factor. The curve in Fig. 2.10 can be described by the following equation, where $k$ is the specific growth rate constant, $k_{max}$ is the maximal growth rate constant, $S$ is the nutrient concentration, and $K_s$ is the nutrient concentration that gives one-half $k_{max}$

$$k = k_{max}S/(K_s + S) \qquad (2.9)$$

These are the same kinetics that describe the saturation of an enzyme by its substrate, Michealis–Menten kinetics (Section 6.2.1). Equation 2.9 can be used to calculate growth yields during continuous growth (Section 2.4).

## 2.4 Steady-State Growth and Continuous Growth

A culture is said to be in steady-state growth (balanced growth) when all of its components double at each division and maintain a constant ratio with respect to one another. Steady-state growth is usually achieved when a culture is maintained in exponential growth by subculturing; that is, it is not allowed to enter stationary phase, or during continuous growth.

### 2.4.1 The chemostat

Continuous growth takes place in a device called a chemostat (Fig. 2.11). A reservoir continuously feeds fresh medium into a growth chamber at a flow rate ($F$) set by the operator. In the chemostat the concentration of the limiting nutrient in the reservoir is kept very low so that growth is limited by the availability of the nutrient. When a drop of fresh medium enters the growth chamber, the growth-limiting nutrient is immediately used up and the cells cannot continue to grow until the next drop of media enters. Because of this, the growth rate is manipulated by adjusting the rate at which fresh medium enters the growth chamber.

### The dilution rate, D

The dilution rate is equal to $F/V$, where $F$ is the flow rate and $V$ is the volume of medium in the growth chamber. For example, if the flow

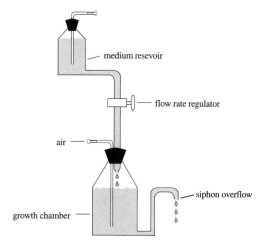

**Fig. 2.11** A chemostat. A chemostat works because the growth of the cells in the growth chamber is limited by the rate at which a particular limiting nutrient, e.g., glucose, is supplied. That is to say, the concentration of the limiting nutrient in the reservoir is sufficiently low so that there is never an excess in the growth chamber. Each moment when fresh medium comes into the growth chamber the incoming limiting nutrient is rapidly utilized, and growth cannot continue until fresh nutrient enters.

rate ($F$) is 10 ml/h and the volume ($V$) in the growth chamber is 1 liter, then the dilution rate ($D$) is 10 ml h$^{-1}$/1000 ml or 0.01 h$^{-1}$. Notice that the units of $D$ are reciprocal time. If one multiplies the dilution rate $D$ by the number of cells $x$ in the growth chamber, then the product $Dx$ represents the rate of loss of cells from the outflow, and has the units of cells/time.

### Relationship of dilution rate (D) to growth rate constant (k)

In the steady state, $k = D$. To show this, consider that the rate of growth is $dx/dt = kx$. When a steady state is reached in the growth chamber, the rate of formation of new cells ($kx$) equals the rate of loss of cells from the outflow ($Dx$). That is to say, $kx = Dx$, and therefore, $k = D$. Therefore, in a chemostat the growth rate of a culture can be changed merely by changing the flow rate $F$, which determines the dilution rate $D$.

## Dependence of cell yield, x, on concentration of limiting nutrient in reservoir, $S_R$

The actual concentration of cells in the growth chamber is manipulated by changing the concentration of limiting nutrient in the reservoir. Let $S_r$ be the concentration of limiting nutrient in the reservoir and $S$ be the concentration of the nutrient in the growth chamber. The difference, $S_r - S$, must be equal to the amount of nutrient used up by the growing bacteria. If we define the growth yield constant $Y$ as being equal to the mass of cells ($x$) produced per amount of nutrient used up ($S_r - S$), then $Y = x/(S_r - S)$, or rearranging,

$$x = Y(S_R - S) \qquad (2.10)$$

$S$ can be calculated from eq. 2.9 by substituting $D$ for $k$. $S$ is much smaller than $S_R$ and can be conveniently ignored in most calculations.

## Varying the cell density (x) and growth rate constant (k) independently of one another

Equation 2.10 predicts that if one were to increase the concentration of rate-limiting nutrient in the reservoir ($S_R$), then the cell density in the growth chamber will increase. When $S_R$ increases, more nutrient enters the growth chamber per unit time. Initially, $S$ must increase. Since the growth rate is limited by the concentration of $S$ (Fig. 2.10), the cells will respond initially by growing faster, and the cell density will increase. A new steady state will be reached in which the increased number of cells use up all the available nutrient as it enters. Thus, even though $S_R$ is increased, the dilution rate still controls the growth rate. What has been accomplished, therefore, when $S_R$ is increased is a higher steady-state value of $x$, according to eq. 2.9, but no change in the growth rate constant ($k$). There are two conclusions to be drawn:

1. The only way to change the steady-state growth rate constant is to change the dilution rate ($D$), because this changes the rate of supply of $S$.

2. The only way to change the steady-state growth yield is to change the concentration of limiting nutrient in the reservoir, $S_r$.

## 2.5 Summary

Growth is defined as an increase in mass and can be conveniently measured turbidometrically. Other methods can be used provided that they measure something that parallels mass increase. These methods include measuring the rate of increase in cell number (viable or total), or specific macromolecules, such as protein, DNA, RNA. Steady-state (continuous) growth is defined as the growth of a population of cells during which all the components of the cell double each division.

Growth in batch culture can progress through a lag and log to stationary phase, where net growth of the population (measured as mass or its equivalent) has ceased. Eventually, the viable cells may decrease in number, and this is referred to as the death phase. The availability of nutrients, the need to synthesize specific enzymes to metabolize newly encountered nutrients, and the accumulation of toxic end products in the medium can explain the different stages of the growth curve. In addition, specific physiological changes of an adaptive nature occur when cells enter the stationary phase because of nutrient depletion.

Growth during the log phase can be described by a simple exponential equation depicting a first-order autocatalytic process. That is to say, the mass doubles at each generation. From this equation one can derive a generation time and an instantaneous growth rate constant. These constants are used to characterize growth under different physiological situations, and to predict growth yields at specific times for experimental purposes.

A variety of physiological and morphological changes takes place in bacteria when they are subjected to nutrient limitation and enter stationary phase. These may include the induction of specific uptake systems to scavenge the environment for limiting nutrient or ions, sporulation, or encystment. Even bacteria that do not sporulate or encyst undergo significant physiological changes when they are starved. Cells may become metabolically less active and more resistant to environmental stresses such as heat, osmotic stress, and certain chemicals such as hydrogen peroxide. There may also occur an inhibition of ribosomal RNA synthesis when starved for a required amino acid, or for a carbon and energy source (stringent response), which is correlated with increased synthesis of guanosine tetra- and pentaphosphates. The response to starvation is mediated in part by an increase in the transcription of certain genes. The transcription of many of these genes requires $\sigma^s$, also called $\sigma^{38}$. Cells also become smaller and may change their morphological shape when they are starved. There may also occur changes in the chemistry of the cell surface.

When cells are grown at different growth rates because of nutritional alterations or chemostat growth, then the macromolecular composition changes. Faster-growing cells are larger, have proportionally more ribosomes, and have more DNA per cell. This can be rationalized in terms of an approximate constancy of ribosome efficiency in protein synthesis, and an almost constant period between the initiation of DNA replication and cell division.

Diauxic growth is characterized by two phases of population growth separated by a stationary phase when the cells are incubated with certain pairs of carbon and energy sources. For example, consider diauxic growth with glucose. The cells grow on the glucose first because it represses the expression of the genes necessary to grow on the second carbon source and because it prevents the uptake of other sugars. The repression is called catabolite repression or glucose repression. Prevention of sugar uptake is called inducer exclusion. There are at least two possible rationales for this. Glucose is the most widely used carbon and energy source, and cells are in a better situation to outgrow their neighbors if they use the glucose first. In doing so they lower the supply of glucose to other cells. Furthermore, many bacteria always express the genes to metabolize glucose (i.e., they are constitutive), but the genes to metabolize other carbon sources are often not expressed unless the carbon source is present in the medium. This lowers the energy burden of carrying genetic information. However, glucose is certainly not a universal catabolite repressor among the bacteria. For example, several obligately aerobic bacteria such as *Rhizobium* are known to grow preferentially on organic acids such as succinate, malate, and fumarate, when given a mixture of glucose and one of these acids. This makes physiological sense for these organisms because growth on the $C_4$ carboxylic acids is faster than growth on glucose.

*E. coli* begins DNA replication when it reaches a critical mass in the cell cycle. Following DNA replication the sister chromosomes are partitioned to the ends of the cells, and cell division takes place. Cell division is a complex process requiring at least 11 genes in *E. coli* to form a septum in the center of the cell. Most of the gene products, including FtsZ, are part of a septal ring that constricts the cell, forming two cells.

Cells can be grown in continuous culture using a chemostat. Two advantages to growing cells this way are: (1) They can be maintained in balanced growth; and (2) the growth rate constant can be easily changed simply by altering the flow rate. The growth yields can be separately manipulated by changing the concentration of limiting nutrient in the reservoir. Growth of continuous cultures is possible when the growth rate of the culture is limited by the supply of a nutrient that is continuously fed into the growth chamber.

## Study Questions

1. What is the generation time ($g$) of a culture with a growth rate constant ($k$) of 0.01 $min^{-1}$?

   *ans.* 69.3 min

2. Assume you wanted to grow a culture to $10^8$ cells/ml in 3 h. The generation time is 30 min. What should be the starting cell density ($x_0$)?

   *ans.* $1.6 \times 10^6$

3. Assume an inoculum whose cell density is $10^8$/ml. The generation time is 30 min. If you started with a $10^{-2}$ dilution, how many hours would you grow the culture to reach $10^8$/ml?

   *ans.* 3.3 h

4. Assume you have a stock culture at $5\times10^9$/ml and you wish to inoculate 1

liter of fresh medium so that in 15 h the cell density will be $2 \times 10^8$/ml. Assume a generation time of 3.5 h. What should be the dilution? What size inoculum should be used?

*ans.* 1/500 or $2 \times 10^{-3}$; 2 ml

5. Assume the yield coefficient for glucose ($Y_g$) is 0.5 g cells per g of glucose consumed. In a glucose-limited chemostat, what should be the concentration of glucose in the reservoir so that the mass of cells in the growth chamber ($x$) will be 0.1 mg/ml? For this problem, ignore $S$ in eq. 2.9 because it is much smaller than $S_R$.

*ans.* 0.2 mg/ml

6. In a 500 ml chemostat, what should be the flow rate ($F$) in minutes for a generation time ($g$) of 6 h?

*ans.* 0.95 ml/min

7. Suppose you were operating a chemostat where $S$ is the limiting nutrient. Assume that $D$ is 0.2 h$^{-1}$, $K_s$ is $1 \times 10^{-6}$ M, and $k_{max}$ is 0.4 h$^{-1}$. (a) What is the concentration of $S$ in the growth chamber? (b) If the cell density were 0.25 mg/ml, what would be the concentration of $S$ in the reservoir if $Y_S$ is 0.5? (c) What is the concentration of $S$ in the growth chamber when the cells are growing one-half as fast, that is, $D = 0.1$ h$^{-1}$?

*ans.* (a) $1 \times 10^{-6}$ M; (b) 0.5 mg/ml; (c) $3.3 \times 10^{-7}$ M

8. During which phases of population growth would you not use cell number as an indicator of growth?

9. What is the physiological role of RpoS? When is it made?

## NOTES AND REFERENCES

1. Siegele, D. A., and R. Kolter. 1992. Life after log. *J. Bacteriol.* **174**:345–348.

2. Matin, A. 1991. The molecular basis of carbon-starvation-induced general resistance in *Escherichia coli*. *Molec. Microbiol.* **5**:3–10.

3. Kolter, R., D. A. Siegele, and A. Tormo. 1993. The stationary phase of the bacterial life cycle. *Ann. Rev. Microbiol.* **47**:855–874.

4. Hengge-Aronis, R.. 1996. Regulation of gene expression during entry into stationary phase. pp. 1497–1512. In: *Escherichia coli* and *Salmonella*: Cellular and Molecular Biology. F. C. Neidhardt, (Ed. in chief). ASM Press, Washington, D.C.

5. Kjelleberg, S. 1993. *Starvation in Bacteria*. Plenum Press, New York and London.

6. This is especially obvious among the cytophages and flexibacteria, the latter sometimes decreasing in length from over 100 $\mu$M to approximately 10–30 $\mu$M.

7. Kjelleberg, S., M. Hermansson, and P. Marden. 1987. The transient phase between growth and nongrowth of heterotrophic bacteria, with emphasis on the marine environment. *Ann. Rev. Microbiol.* **41**:25–49.

8. McCann, M. P., J. P. Kidwell, and A. Matin. 1991. The putative $\sigma$ factor KatF has a central role in development of starvation-mediated general resistance in *Escherichia coli*. *J. Bacteriol.* **173**:4188–4194.

9. Loewen, P. C., and R. H-Aronis. 1994. The role of the sigma factor $\sigma^s$ (KatF) in bacterial global regulation. *Ann. Rev. Microbiol.* **48**:53–80.

10. O'Neal, C. R., W. M. Gabriel, A. K. Turk, S. J. Libby, F. C. Fang, and M. P. Spector. 1994. RpoS is necessary for both the positive and negative regulation of starvation survival genes during phosphate, carbon, and nitrogen starvation in *Salmonella typhimurium*. *J. Bacteriol.* **176**:4610–4616.

11. Mulvey, M. R., J. Switala, A. Borys, and P. C. Loewen. 1990. Regulation of transcription of *katE* and *katF* in *Escherichia coli*. *J. Bacteriol.* **172**:6713–6720.

12. Gentry, D. R., V. J. Hernandez, L. H. Nguyen, D. B. Jensen, and M. Cashel. 1993. Synthesis of the stationary-phase sigma factor $\sigma^s$ is positively regulated by ppGpp. *J. Bacteriol.* **175**:7982–7989.

13. Gene fusions are valuable probes to monitor the expression of genes of interest. Consider a *lacZ* fusion. The fused gene has the promoter region of the target gene but not the promoter for the *lacZ* gene. Expression of the fused gene is therefore under control of the promoter region of the target gene. The fusions produce a hybrid protein whose amino-terminal end is derived from the target gene and the carboxy-terminal end from $\beta$-galactosidase. The hybrid protein has $\beta$-galactosidase activity. Therefore, an assay for $\beta$-galactosidase is a measure of the expression of the target gene. Thus one can measure the expression of virtually any gene simply by constructing the proper gene fusion

and performing an assay for $\beta$-galactosidase. One can construct gene fusions in vitro or in vivo. In vitro construction involves using restriction endonucleases to cut out a portion of the gene with its promotor region from a plasmid containing the cloned DNA, and ligating it to a *lacZ* gene, without its promotor or ribosome binding site, in a second plasmid. The plasmid containing the fused gene is then introduced into the bacterium, and transformants are selected on the basis of being resistant to an antibiotic-resistant marker on the plasmid, and the production of $\beta$-galactosidase.

14. Fang, F. C., S. J. Libby, N. A. Buchmeier, P. C. Loewen, J. Switala, J. Harwood, and D. G. Guiney. 1992. The alternative sigma factor KatF (RpoS) regulates *Salmonella* virulence. *Proc. Natl. Acad. Sci. USA* 89:11978–11982.

15. Wilmes-Riesenberg, M. R., J. W. Foster, and R. Curtiss III. 1997. An altered *rpoS* allele contributes to the avirulence of *Salmonella tyhphimurium* LT2. Infect. Immun. 65:203–210.

16. Hengge-Aronis, R. 1996. Back to log phase: $\sigma^S$ as a global regulator in the osmotic control of gene expression in *Escherichia coli*. *Mol. Microbiol.* 21:887–893.

17. Hengge-Aronis, R. 1996. Back to log phase: $\sigma^S$ as a global regulator in the osmotic control of gene expression in *Escherichia coli*. *Mol. Microbiol.* 21:887–893.

18. The regulation of transcription of *rpoS* (*katF*) appears to be complicated and not well understood. There are probably several signaling pathways that influence the transcription of the *rpoS* gene, allowing the cells to respond to a variety of environmental challenges. The transcription of *rpoS* does not always increase markedly when cells enter stationary phase because expression during growth may be high. It depends upon how the cells are grown. Anaerobiosis has been reported to stimulate *rpoS* expression in exponential cells grown in a rich medium. Several other factors appear to influence the levels of *rpoS* expression. For example, depending upon the strain of *E. coli*, *rpoS* expression may be high during exponential growth in minimal medium and increase only slightly during stationary phase. Additionally, starvation induced by transferring exponential-phase cells to a medium lacking a specific nutrient might stimulate *rpoS* expression significantly or only marginally, depending on which component is missing from the medium. The amounts of RpoS also increase in response to an upshift in osmolarity, even in exponentially growing cells. The increase is due to an increase in translation of the RpoS mRNA as well as to a decrease in the turnover of RpoS due to ClpXP. This, then, is an example of

translational and post-translational control over the levels of a protein.

The stability of the RpoS protein increases during stationary phase. Apparently RpoS is sensitive to proteolytic digestion by the ClpXP protease during exponential growth but less so during stationary phase. (See Section 19.1.3 for a description of the Clp proteases.) The evidence for this is that RpoS is stabilized during exponential growth in *clpP* and *clpX* mutants. (Schweder, T., K-H Lee, O. Lomovskaya, and A. Matin. 1996. Regulation of *Escherichia coli* starvation sigma factor ($\sigma^S$) by ClpXP protease. *J. Bacteriol.* 178:470–476; Zhou, Y., and S. Gottesman. 1998. Regulation of proteolysis of the stationary-phase sigma factor RpoS. *J. Bacteriol.* 180:1154–1158.

For more information on the role of RpoS as well as factors that regulate its levels, see Loewen, P. C., and R. Hengge-Aronis. 1994. The role of the sigma factor $\sigma^S$ (KatF) in bacterial global regulation. *Ann. Rev. Microbiol.* 48:53–80, and Hengge-Aronis, R. 1996. Back to log phase: $\sigma^S$ as a global regulator in the osmotic control of gene expression in *Escherichia coli*. *Mol. Microbiol.* 21:887–893.)

19. Gourse, R. L., T. Gaal, M. S. Bartlett, J. A. Appleman, and W. Ross. 1996. rRNA transcription and growth rate-dependent regulation of ribosome synthesis in *Escherichia coli*. *Ann. Rev. Microbiol.* 50:645–677.

20. Cashel, M., D. R. Gentry, V. J. Hernandex, and D. Vinella. 1996. The stringent response, pp. 1458–1496. In: *Escherichia coli* and *Salmonella*: Cellular and Molecular Biology, Vol. 1. F. C. Neidhardt (Ed. in chief). ASM Press, Washington, D.C.

21. In addition to an inhibition of ribosomal RNA synthesis, ppGpp seems to be responsible for:

1. A large decrease in the rate of synthesis of protein.
2. A temporary cessation in the initiation of new rounds of DNA replication.
3. An increase in the biosynthesis of amino acids.
4. A decrease in the rates of synthesis of phospholipids, nucleotides, peptidoglycan, and carbohydrates.

One must therefore conclude that the increase in levels of ppGpp and the responses due to those increases can be viewed as a general response to conditions that limit growth, and not simply due to amino acid starvation. This is because it occurs as a consequence of any nutrient or energy limitation e.g., nitrogen limitation, or a shift to a poorer carbon or energy source. Although ppGpp or a metabolically related compound(s) is clearly important in mediating the stringent response, little is known concerning how it is responsible for all the myriad effects that appear to be correlated with increased ppGpp levels. One might think of ppGpp as a second messenger that receives a signal from the environment, i.e., nutrient or energy depletion, and transfers the signal to the genome either activating or inhibiting transcription of relevant genes, or affecting some other cellular process such as the activity of an enzyme. Of course, ppGpp need not act directly on the target and perhaps is one component in a longer signal transduction sequence. Interestingly, ppGpp also has been reported to stimulate the synthesis of sigma factor $\sigma^s$, which is induced during starvation. (See Mulvey, M. R., J. Switala, A. Borys, and P. C. Loewen. 1990. Regulation of transcription of *katE* and *katF* in *Escherichia coli. J. Bacteriol.* **172**:6713–6720.)

22. Ribosomal RNA and ribosomal proteins are synthesized independently of one another and then assembled into ribosomes. However, the synthesis of ribosomal proteins is dependent upon continued synthesis of rRNA. Hence, a decrease in the synthesis of rRNA leads to an inhibition of ribosomal protein synthesis. This is because certain ribosomal proteins in the absence of rRNA to which they normally bind will bind to ribosomal mRNA and inhibit translation. In order to undertand this, it is important to recognize that ribosomal proteins are encoded in operons in which the translation of the genes is coupled. *E. coli* has at least 20 such operons that encode the ribosomal proteins, and in certain instances also DNA primase, RNA polymerase, and elongation factors. Translational coupling means that the operon produces a polycistronic mRNA and that the translation of the gene immediately upstream is required for translation of the downstream gene. A model for translational coupling postulates that the translational start region, including the start codon, for the downstream gene is inside a hairpin loop in the mRNA so that it cannot bind to a ribosome. However, when the ribosome translating the upstream gene comes to the stop codon of the upstream gene, then the ribosome disrupts the secondary structure of the mRNA and "opens it up" so that the translational start region for the downstream gene is accessible to a second ribosome. One of the ribosomal proteins encoded by the operon is believed to be able to prevent translation of the entire operon when the protein is present in excess. It is therefore a translational repressor. The repressor ribosomal protein can bind not only to rRNA but also to the initiating region of the mRNA of one of the genes (usually the first one) in the operon. It appears that in at least some instances the sequence in the mRNA to which the repressor protein binds is similar to the sequence in the rRNA, thus accounting for why the repressor ribosomal protein can bind to both the ribosomal RNA and the mRNA. As an example of such translational autoregulation, consider translation of the *rplK–rplA* operon, which encodes the 50S ribosomal subunit proteins L11 (*rplK*) and L1 (*rplA*). The *rplA* gene is downstream of the *rplK* gene, and its translation requires prior translation of the *rplK* gene. The L1 protein regulates the translation of the entire operon. It can bind to either rRNA or the mRNA in the translational initiation region of the *rplK* gene and block translation not only of the *rplK* gene but also of the *rplA* gene that follows it. All the ribosomal protein operons are regulated in a similar manner by one of the protein products.

23. The ppGpp can be synthesized on ribosomes that are "stalled" because of a restriction in the supply of amino acylated tRNA. This occurs during aminoacid starvation, e.g., by restricting amino acids to auxotrophs, and requires the product of the *relA* gene. The *relA* gene was discovered in a search for mutants that failed to respond to amino acid starvation with the stringent response. These mutants were termed "relaxed," hence the name of the gene. The RelA protein [also called (p)ppGPP synthetase I or PSI] is a ribosome-associated protein that synthesizes either ppGpp or pppGpp by displacing AMP and transferring a pyrophosphoryl group from ATP to the 3′ OH of either GDP or GTP.

$$GDP + ATP \longrightarrow ppGpp + AMP$$

Because the pool size for GDP in *E. coli* is small relative to GTP, it is believed that the product is pppGpp, which is then dephosphorylated by a 5′-phosphohydrolyase (gpp) to ppGpp. This suggestion is in agreement with the kinetics of appearance of pppGpp and ppGpp. The synthetase is activated in starved (stalled) ribosomes when aminoacylated tRNA becomes limiting during amino acid starvation. The levels of (p)ppGpp may also rise upon carbon and energy starvation (e.g., glucose starvation) even in null *relA* mutants, indicating alternative ways to synthesize (p)ppGpp. Thus, even in a relaxed strain, RNA accumulation stops when the cells are shifted from a good carbon and energy source to a poorer medium (but not when starved for an amino acid). This constitutes the evidence for a relA-independent (p)ppGpp synthetase, called PSII. PSII is thought to be encoded by the *spoT* gene, and is believed to be a 3′-pyrophosphotransferase that uses GTP as the acceptor and synthesizes pppGpp. SpoT is also a pyrophosphohydrolase that degrades (p)ppGpp to GDP and GTP. In other words, it has been proposed that SpoT is a bifunctional enzyme that takes part in either the synthesis or degradation of ppGpp, and the ratio of biosynthetic to degradative activity is somehow controlled by energy starvation. The result is an increase during energy starvation of ppGpp. See the review article by Cashel et al. for a discussion of these points. Cashel, M., D. R. Gentry, V. J. Hernandex, and D. Vinella. 1996. The stringent response, pp. 1458–1496. In: *Escherichia coli* and *Salmonella*: Cellular and Molecular Biology, Vol. 1. F. C. Neidhardt (Ed. in chief). ASM Press, Washington, D.C.

24. Xu, J., and R. C. Johnson. 1995. Identification of genes negatively regulated by Fis: Fis and RpoS comodulate growth-dependent gene expression in *Escherichia coli*. *J. Bacteriol.* 177:938–947.

25. Fis negatively regulates the expression of the *fis* gene and positively or negatively regulates the expression of several other genes. Two-dimensional gel electrophoresis of *fis* mutants has revealed over twenty proteins whose synthesis is decreased (indicating Fis-dependent transcription) and a similar number of proteins whose synthesis has increased (indicating Fis repression). The evidence suggests that Fis binds to both the rRNA DNA as well as to the RNA polymerase and stimulates transcription by interacting with the polymerase rather than by causing an alteration in DNA structure such as bringing upstream DNA closer to the polymerase. See Gourse, R. L., T. Gaal, M. S. Bartlett, J. A. Appleman, and W. Ross. 1996. rRNA transcription and growth rate-dependent regulation of ribosome synthesis in *Escherichia coli*. *Ann. Rev. Microbiol.* 50:645–677. Xu, J., and R. C. Johnson. 1995. Identification of genes negatively regulated by Fis: Fis and RpoS comodulate growth-dependent gene expression in *Escherichia coli*. *J. Bacteriol.* 177:938–947.

26. Ball, C. A., R. Osuna, K. C. Ferguson, and R. C. Johnson. 1992. Dramatic changes in Fis levels upon nutrient upshift in *Escherichia coli*. *J. Bacteriol.* 174:8043–8056.

27. Reviewed in: Wagner, R. 1994. The regulation of ribosomal RNA synthesis and bacterial cell growth. *Arch. Microbiol.* 161:100–109.

28. In addition to being a transcriptional activator for rRNA genes, Fis stimulates Hin-, Gin- and Cin-mediated inversion and site-specific recombination of phage lambda (both excision and integration) with the bacterial chromosome. Fis also binds to the *E. coli* origin of replication (*oriC*) and possibly plays a role in the initiation of DNA replication. H-NS is involved in transcriptional regulation of several genes whose activity is modulated by cellular stress (e.g., gene regulation in stationary phase, osmotic shock). See the discussion of the nucleoid in Section 1.2.6.

29. Cooper, S., and C. R. Helmstetter. 1968. Chromosome replication and the division cycle of *Escherichia coli* B/r. *J. Mol. Biol.* 31:519–539.

30. The length of time between the onset of DNA replication and the completion of cell division is the sum of two time periods [i.e., the time it takes for the chromosome to replicate (the C period) and the interval between the end of a round of DNA replication and the completion of cell division (the D period)]. The shortest C and D periods are approximately 40 and 20 min, respectively, in rapidly growing *E. coli*. Hence 60 min is the minimum time required between the onset of a round of DNA replication and the completion of cell division. These times do change with growth rate. The C period decreases from approximately 67 min at 0.6 doublings per hour to about 42 min at 2.5 doublings per hour. The D period also decreases (but less so) as the growth rate increases, i.e., from 30 min at 0.6 doublings per hour to 23 min at 2.5 doublings per hour. See Bipatnath, M., P. P. Dennis, and H. Bremer. 1998. Initiation and velocity of chromosome replication in *Escherichia coli* B/r and K-12. *J. Bacteriol.* 180:265–273, and Bremer, H., and P. P. Dennis. 1996. Modulation of chemical composition and other parameters of the cell by growth rate. In: *Escherichia coli* and *Salmonella typhimurium*: Cellular and Molecular Biology, Second Edition, Vol. 2, pp. 1553–1569. F. C. Neidhardt (Ed. in chief). American Society for Microbiology, Washington, D.C.

31. For illustrative purposes, let us assume that the C and D periods are constants at 40 and 20 min, respectively, so that a cell divides 60 min after it initiates DNA replication. We will also assume that each cell grows exponentially and that $x = x_0 2^Y$, where $x_0$ is the mass at birth, and $x$ is the mass at a point in the cell cycle where $Y$ is the fraction of cell cycle time. For example, if the cell is halfway through the cell cycle, then $x = x_0 2^{0.5}$. Let us call a cell dividing every 60 min a 60 min cell, a cell dividing every 40 min a 40 min cell, and a cell that divides every 30 min a 30 min cell. The 60 min cell must be born with one chromosome (and one replication origin) and begins replicating that chromosome at the begining of the cell cycle so that 60 min later it divides. Let us say that the mass of the 60 min cell at the beginning of the cell cycle when it begins replicating the chromosome is $x_0$. This means that the mass ($x$) 20 min later (when, as we shall see, the 40 min cell begins DNA replication) is about $1.15x_0$ ($x = x_0 2^{0.2}$). The 40 min cell also needs 60 min between the initiation of a round of DNA replication and cell division. This means that the 40 min mother cell must initiate DNA synthesis at two replication origins (one for each of the daughter cell chromosomes) 20 min into the cell cycle. Now if you compare the 40 min

cell with the 60 min cell, you will realize that the 40 min cell must on average be larger. This is because by 20 min the 40 min cell must grow to a mass that will initiate replication at two origins of replication; i.e., it must be $2x_0$ where $x_0$ is the size of the 60 min cell at birth. Since that size is reached by the 40 min cell 0.5 into the cell cycle, the 40 min cell must be 1.41 times larger than its mass at birth ($x = x_0\ 2^{0.5} = x_0\ 1.41$). That is to say, $2x_0$ is 1.41 times larger than the mass at birth of the 40 min cell. The mass at birth must therefore be $2x_0/1.41$ or about $1.41x_0$. This can be contrasted to the 60 min cell that was born with a mass of $x_0$. Now let's look at the 30 min cell. It must begin replicating the daughter chromosomes at the beginning of the cell cycle in order that the daughter cells can divide 60 min later. Thus replication must begin at two origins (one for each of the daughter chromosomes), and the mass at the begining of the cell cycle must be $2x_0$, as opposed to $1.42x_0$ for the 40 min cell and $x_0$ for the 60 min cell.

32. Donachie, W. D. 1968. Relationship between cell size and time of initiation of DNA replication. *Nature* 219:1077–1079.

33. Donachie assumed that each cell grew exponentially and doubled in mass during the cell cycle. Knowing the size of the cell at the time of cell division, he was able to calculate the size at any time during the cell cycle. With this information and the knowledge that DNA initiation occurred 60 min before cell division, he was able to compute the mass of the cell at the time of initiation of DNA initiation. This mass, divided by the number of replication origins, was a constant.

34. Wold, S., K. Skarstad, H. B. Steen, T. Stokke, and E. Boye. 1994. The initiation mass for DNA replication in *Escherichia coli* K-12 is dependent on growth rate. *EMBO J.* 13:2097–2102.

35. Cooper, S. 1997. Does the initiation mass for DNA replication in *Escherichia coli* vary with the growth rate? *Mol. Microbiol.* 26:1138–1143.

36. Chandra, N. M., and P. K. Chakrabartty. 1993. Succinate-mediated catabolite repression of enzymes of glucose metabolism in root-nodule bacteria. *Curr. Microbiol.* 26:247–251.

37. Collier, D. N., P. W. Hager, and P. V. Phibbs, Jr. 1996. Catabolite repression control in the *Pseudomonads. Res. in Microbiol.* 147:551–561.

38. Niki, H., A. Jaffé, R. Imamura, T. Ogura, and S. Hiraga. 1991. The new gene *mukB* codes for a 177 kD protein with coiled-coil domains involved in chromosome partitioning of *E. coli. EMBO J.* 10:183–193.

39. Niki, H., R. Imamura, M. Kitaoka, K. Yamanaka, T. Ogura, and S. Hiraga. 1992.

*E. coli* MukB protein involved in chromosome partition forms a homodimer with a rod-and-hinge structure having DNA binding and ATP/GTP binding activities. *EMBO J.* 11:5101–5109.

40. Yamanaka, K., T. Ogura, H. Niki, and S. Hiraga. 1996. Identification of two new genes, *mukE* and *mukF*, involved in chromosome partitioning in *Escherichia coli. Mol. Gen. Genet.* 250:241–251.

41. Erickson, H. P. 1997. FtsZ, a tubulin homologue in prokaryote cell division. *Trends in Cell Biol.* 7:362–367.

42. A previous model that *E. coli* separates its nucleoids to the cell poles by attaching a DNA strand to the cell envelope and inserting new cell envelope material between the two chromosome attachment sites cannot be correct because (1) cell envelope growth occurs by random insertion of new material over the whole surface throughout most of the cell cycle and (2) separation of the nucleoids requires postreplication protein synthesis. In the presence of inhibitors of protein synthesis, DNA replication is completed but the nucleoids do not separate and remain in the center of the cell. If protein synthesis is allowed to resume after DNA replication, then the nucleoids separate toward the cell poles without cell elongation, suggesting some sort of machinery that moves the nucleoids. The Muk proteins are part of that machinery in *E. coli*. The *mukB* gene was discovered by isolating *E. coli* mutants that produce normal-size anucleate cells in addition to nucleated cells. Whereas the wild type produced less than 0.03% anucleate cells, the mutant produced approximately 5% anucleate cells. A phenotype used for mapping was that the mutant formed minute colonies on L agar plates at 42°C, at which temperature nucleoid separation was more aberrant. Mutations in two other genes also lead to the production of anucleate cells. These are the *mukF* and *mukE* genes. The mechanistic role of the Muk proteins in chromosome partitioning is a matter of speculation. (They are not required for plasmid partitioning.) MukB is an ATP/GTP and DNA-binding protein that has been suggested to be a molecular motor that moves the nucleoids apart. MukF might be involved in chromosome condensation, which occurs at opposite ends of the cell, leaving a gap in the center for septum formation. There are probably alternatives to the Muk proteins. This conclusion is based upon the finding that null mutants in *mukB*, *mukF*, or *mukE* show a decreased plating efficiency only at temperatures higher than 22°C, indicating that the proteins are not essential at the lower temperatures. A more complete discussion of chromosome partitioning can be

found in Section 10.1.4. See Niki, H., A. Jaffé, R. Imamura, T. Ogura, and S. Hiraga. 1991. The new gene *mukB* codes for a 177 kD protein with coiled-coil domains involved in chromosome partitioning of *E. coli*. *EMBO J.* **10**:183–193, and Niki, H., R. Imamura, M. Kitaoka, K. Yamanaka, T. Ogura, and S. Hiraga. 1992. *E. coli* MukB protein involved in chromosome partioning forms a homodimer with a rod-and-hinge structure having DNA binding and ATP/GTP binding activities. *EMBO J.* **11**:5101–5109, and Yamanaka, K., T. Ogura, H. Niki, and S. Hiraga. 1996. Identification of two new genes, *mukE* and *mukF*, involved in chromosome partitioning in *Escherichia coli*. *Molec. Gen. Genet.* **250**:241–251, for a discussion of these points.

43. Nanninga, N., L. J. H. Koppes, and F. C. deVries-Tijssen. 1979. The cell cycle of *Bacillus subtilis* as studied by electron microscopy. *Arch. Microbiol.* **123**:173–181.

44. Donachie, W. D. 1993. The cell cycle of *Escherichia coli*. *Ann. Rev. Microbiol.* **47**:199–230.

45. Vicente, M., and J. Errington. 1996. Structure, function and controls in microbial division. *Molec. Microbiol.* **20**:1–7.

46. Lutkenhaus, J., and A. Mukherhjee. 1996. Cell division. In: *Escherichia coli* and *Salmonella*: Cellular and Molecular Biology. pp. 1615–1626. Vol. 1. F. C. Neidhardt et al. (Eds.). ASM Press, Washington, D.C.

47. Temperature-sensitive mutants display the mutant phenotype at an elevated temperature (e.g., 42°C) but have a wild-type phenotype at a lower temperature (e.g., 30°C). They are especially useful to study genes that are essential for growth, since the culture can be maintained at the lower growth temperature but the mutant phenotype can be studied by raising the temperature.

48. Many filamentous mutants are known, some of which are called *fts* mutants. However, because they are defective in other cell events that can affect cell division (e.g., DNA synthesis, nucleoid segregation, protein secretion, heat-shock response), they are not considered to be specifically involved in cell division.

49. Erickson, H. P. 1997. FtsZ, a tubulin homologue in prokaryote cell division. *Trends in Cell Biol.* **7**:362–367.

50. Bie, E., and J. Lutkenhaus. 1991. FtsZ ring structure associated with division in *Escherichia coli*. *Nature* **354**:161–164.

51. Lutkenhaus, J., and S. G. Addinall. 1997. Bacterial cell division and the Z ring. *Ann. Rev. Biochem.* **66**:93–116.

52. Microtubules are found only in eukaryotic cells. They are hollow cylindrical tubes about 25 nm in diameter made from the protein tubulin. The tubulin is a dimer consisting of the subunits alpha-tubulin and beta-tubulin. The dimers exist as 13 rows (called protofilaments) surrounding the central hollow core of the tube. Microtubules are found in the cytoplasm next to plant cell walls and in the cytoplasm of animal cells, where they participate, with kinesin, in moving organelles from one cellular location to another. Microtubules are also important for shape determination of cells, and they make up the spindle apparatus that is necessary for the separation of daughter chromosomes during nuclear division. Tubulin undergoes a GTP-dependent polymerization to extend microtubules, e.g., during the construction of the spindle apparatus. Depolymerization is associated with GTPase activity. Another location of microtubules in eukaryotic cells is in the cilia and flagella. The microtubules are arranged as nine outer microtubule doublets surrounding two singlets in the center of the flagellum or cilium and are responsible for the ATP-dependent bending of these structures during their whiplike beating. FtsZ is not tubulin. However, it resembles tubulin in that (1) it has GTPase activity, and (2) it is polymerized in vitro in the presence of GTP to form protofilaments. These can be straight, curved, or shaped in "minirings." A third resemblance to tubulin is that FtsZ has a 7 amino acid glycine-rich sequence (GGGTGTG) important for GTP binding that is similar to a 7 amino acid sequence in tubulins called the "tubulin signature" (G/AGGTG(S/A)G).

53. Osteryoung, K. W., and E. Vierling. 1995. Conserved cell and organelle division. *Nature* **376**:473–474.

54. Wang, Xunde, and J. Lutkenhaus. 1996. FtsZ ring: The eubacterial division apparatus conserved in archaebacteria. *Molec. Microbiol.* **21**:313–319.

55. Addinall, S. G., and J. Lutkenhaus. 1996. FtsA is localized to the septum in an FtsZ-dependent manner. *J. Bacteriol.* **178**:7167–7172.

56. Addinall, S. G., C. Chune, and J. Lutkenhaus. 1997. FtsN, a late recruit to the septum in *Escherichia coli*. *Molec. Microbiol.* **25**:303–309.

57. Wang, X., H. Jian, Mukherjee, A., Cao, C., and J. Lutkenhaus. 1997. Analysis of the interaction of FtsZ with itself, GTP, and FtsA. *J. Bacteriol.* **179**:5551–5559.

58. The yeast two-hybrid system is based upon the fact that the yeast GAL4 protein, which is a transcriptional activator of genes encoding enzymes for galactose metabolism, has an N-terminal end that binds to the target DNA and a C-terminal end that activates transcription. The N- and C-terminal ends have been cloned

separately and they can be genetically fused to two different proteins, e.g., FtsA and FtsZ, to form hybrid proteins encoded on separate plasmids. If FtsA and FtsZ form a complex such that the N- and C-terminal ends of the GAL4 protein are in close proximity, then gene transcription is activated when a yeast cell contains both hybrids (both plasmids). Transcription can be monitored in yeast deleted for GAL4 but containing the fusion reporter gene GAL1-lacZ and assaying for β-galactosidase.

59. Din, N., E. M. Quardokus, M. J. Sackett, and Y. V. Brun. 1998. Dominant C-terminal deletions of FtsZ that affect its ability to localize in Caulobacter and its interaction with FtsA. Mol. Microbiol. 27:1051–1063.

60. Hale, C. A., and P. A. J. de Boer. 1997. Direct binding of FtsZ to ZipA, an essential component of the septal ring structure that mediates cell division in E. coli. Cell 88:175–185.

61. Helm, R., A. B. Cubitt, and R. Y. Tslen. 1995. Improved green fluorescence. Nature 373:663–664.

62. Hale and de Boer were able to demonstrate that ZipA was capable of binding to FtsZ using affinity blotting. The probe was radioactive FtsZ, which was a fusion protein in which a 54 amino acid tag sequence containing a stretch of histidine residues (HFKT) was fused to the N terminus of FtsZ. The tag sequence also had the substrate site for the catalytic portion of heart muscle kinase. The FtsZ fusion protein was purified by nickel chelation chromatography and labeled with $^{32}P$ via enzymatic phosphorylation. The membrane proteins were separated using SDS-Page electrophoresis and then electrophoretically transferred to a nitrocellulose filter. The filters were then incubated with the radioactive FtsZ fusion protein, and the radioactivity was detected using X-ray film. The radioactive FtsZ bound to ZipA. To localize the ZipA protein in the cells, a fusion gene called zipA–GfpS65t was created. This encodes a green fluorescent protein called GfpS65T fused to the C terminus of ZipA and has a promoter stimulated by IPTG. The green fluorescent protein (GFP) is found naturally in the jellyfish Aequorea victoria, and when the gene for this protein is expressed in other organisms, such as E. coli, a prominent green fluorescence results. When the gene is correctly fused to a gene specifying a protein of interest such as ZipA, then a fused protein is made that can be visualized by its green fluorescence. The fused gene was introduced into E. coli on a phage and induced with IPTG. Cells were then observed with a fluorescent microscope, and a bright ring in the septal area indicated the presence of ZipA.

63. E. coli has three other transpeptidases that are used for peptidoglycan synthesis and that bind penicillin (i.e., PBP1a, PBP1b, and PBP2). However, PBP3 preferentially binds certain β-lactams (i.e., furazlocillin, cephalexin, or benzyl penicillin at low concentration). The result is the inhibition of septum formation but not of net peptidoglycan synthesis.

64. Weiss, D. S., K. Pogliano, M. Carson, L-M. Gusman, C. Fraipont, M. Nguyen-Distèche, R. Losick, and J. Beckwith. 1997. Localization of the Escherichia coli cell division protein FtsI (PBP3) to the division site and cell pole. Molec. Microbiol. 25:671–681.

65. Wang, L., M. K. Khattar, W. D. Donachie, and J. Lutkenhaus. 1998. FtsI and FtsW are localized to the septum in Escherichia coli. J. Bacteriol. 180:2810–2816.

66. Immunofluorescence microscopy is more sensitive than immunoelectron microscopy for detecting proteins within the cell. Antibodies are raised against the protein, e.g., against FtsI. The cells are permeabilized, e.g., with lysozyme, and incubated with the specific antibody. The antibody–antigen complex is then visualized using fluorescein-conjugated anti-IgG. For example, the IgG might be from rabbit, whereas the antirabbit IgG would be from another animal.

67. Yu, X-C., A. H. Tran, Q. Sun, and W. Margolin. 1998. Localization of cell division protein FtsK to the Escherichia coli septum and identification of a potential N-terminal targeting domain. J. Bacteriol. 180:1296–1304.

68. The evidence for the FtsZ ring is from immunofluorescent staining. Antiserum to purified FtsZ is bound to the cells. Then a fluorescein-labelled antibody to the anti-FtsZ antibody is added to locate the anti-FtsZ antibody. A similar approach can be used to locate the FtsA and FtsN proteins to the FtsZ ring. It should be noted that although FtsA is a cytoplasmic protein, FtsN is membrane associated. The model is that FtsN has an N-terminal cytoplasmic region that associates with the FtsZ ring and a carboxy terminus that is periplasmic. There is a single transmembrane region. The evidence that FtsZ forms a ring independently of the other Fts proteins is that the ring forms even in mutants that do not make the other Fts proteins. The evidence that FtsZ recruits FtsA and FtsN is that if the formation of the FtsZ ring is blocked, then FtsA and FtsN do not localize to the septum. This can be done using temperature-sensitive mutants that do not form a Z ring at the nonpermissive temperature or by the overproduction of SulA in wild-type cells. When the sulA gene carried on, a plasmid is induced, then the cells stop dividing, form filaments, and do not have Z rings. (The SulA protein interacts with FtsZ.) The requirement

for FtsQ and FtsI for the recruitment of FtsN but not of FtsA to the ring is deduced from filamentous mutants of *FtsQ* and *FtsI*. See papers by Addinall and Lutkenhaus and Addinall, Cao, and Lutkenhaus for a further discussion of these experiments.

69. Raskin, D. M., and P. A. J. Boer. 1997. The MinE ring: An FtsZ-independent cell structure required for selection of the correct divison site in *E. coli*. Cell **91**:685–694.

70. FtsZ ring assembly can be prevented by inducing the *sulA* gene (*sfiA*). The *sulA* gene is normally induced as part of the SOS response to DNA damage. (See Section 19.2.) The SulA protein binds to FtsZ and prevents ring assembly. Hence the cells filament. To show that the assembly of the MinE ring is independent of the FtsZ ring, Raskin and de Boer constructed an *E. coli* strain in which *sulA* is under the control of the *lac* promoter and induced synthesis of

SulA with IPTG. Even in these filamentous cells MinE formed regularly spaced rings. See Raskin, D. M., and P. A. J. de Boer. 1997. The MinE ring: An FtsZ-independent cell structure required for selection of the correct division site in *E. coli*. Cell **91**:685–694.

71. Suppose you want to convert $\log_2$ to $\log_{10}$. First write the exponential equation, e.g., $X = X_0 2^Y$. Now take $\log_{10}$ of both sides of the equation: $\log_{10}X = \log_{10}X_0 + \log_{10}2$ $(Y)$. Since $\log_{10}2 = .301$, the equation becomes $\log X = \log X_0 + 0.301Y$. Note that $\log_{10}$ is sometimes written simply as log.

72. Some investigators use the symbol $\mu$ for the instantaneous growth rate constant and $k$ for the reciprocal of the generation time: $k = 1/g$.

73. The dilution is $2.6 \times 10^2$. Assume the volume of inoculum is $X$. Therefore, $2.6 \times 10^2(X)$ must equal the final volume, which is 1000 ml $+ X$. Solving for $X$ gives 3.8 ml.

# 3

# Membrane Bioenergetics:
# The Proton Potential

A major revolution in our conception of membrane bioenergetics[1,2] has taken place in the past 30 years as a result of the theoretical ideas of Peter Mitchell referred to as the chemiosmotic theory.[3–5] Briefly, the chemiosmotic theory states that energy-transducing membranes (i.e., bacterial cell membranes, mitochondrial, and chloroplast membranes) pump protons across the membrane, thereby generating an electrochemical gradient of protons across the membrane (the proton potential) that can be used to do useful work when the protons return across the membrane to the lower potential. In other words, bacterial, chloroplast, and mitochondrial membranes are energized by proton currents. Of course, the return of the protons across the membrane must be through special proton conductors that couple the translocation of protons to do useful cellular work. These proton conductors are transmembrane proteins. *Some membrane proton conductors are solute transporters, others synthesize ATP, and others are motors that drive flagellar rotation.* The proton potential provides the energy for other membrane activities besides ATP synthesis, solute transport, and flagellar motility (e.g., reversed electron transport and gliding motility). Because the chemiosmotic theory is central to energy metabolism, it lies at the foundation of all bacterial physiology. As explained in this chapter, the chemiosmotic theory brings together principles of physics and thermodynamics in explaining membrane bioenergetics. The student should study the principles of the chemiosmotic theory for a deeper understanding of how the bacterial cell uses ion gradients to couple energy-yielding (exergonic) reactions to energy-requiring (endergonic) reactions. This chapter explains the principles of the theory.

## 3.1 Chemiosmotic Theory

According to the chemiosmotic theory, protons are translocated out of the cell by exergonic (energy-producing) driving reactions, which are usually biochemical reactions (e.g., respiration, photosynthesis, or ATP hydrolysis). Some of the translocated protons leave behind negative counterions (e.g., hydroxyl ions), thus establishing a membrane potential, outside positive. Protons may also accumulate electroneutrally in the extracellular bulk phase, establishing a proton concentration gradient, high on the outside (outside acid). When the protons return to the inside, moving down the concentration gradient and toward the negative pole of the membrane potential, work can be done. Figure 3.1 illustrates the proton circuit in the bacterial cell membrane. The cell membrane is similar to a battery in that it maintains a potential difference between the inside and outside, except that the current that flows is one of protons

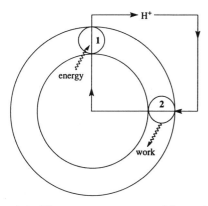

**Fig. 3.1** The proton current. There is a proton circuit traversing the bacterial cell membrane. Protons are translocated to the cell surface, driven there by either chemical or light energy through a proton pump (1) and return through special proton transporters (2) that do work. The accumulation of protons on the outside surface of the membrane establishes a membrane potential, outside positive. A pH gradient can also be established, outside acid. In several gram-negative bacteria oxidizing certain inorganic compounds (lithotrophs), or single-carbon compounds such as methanol, protons that are released into the periplasm via periplasmic oxidations contribute to the proton current (Chapter 12). In some cases, periplasmic oxidations are the sole provider of protons for inward flux.

rather than electrons. In Fig. 3.1, the potential difference is maintained by reactions 1 that translocate protons to the outside. Reactions 1 include redox reactions that occur during electron transport (Fig. 3.2, 1) and an ATP-driven proton pump (the ATP synthase) (Fig. 3.2, 7). These will be discussed later (Sections 3.7.1 and 3.7.2). The cell membrane does work via reactions 2. The work that is done by the protons that enter the cell includes the extrusion of sodium ions (Fig. 3.2, 3), solute transport (Fig. 3.2, 4), flagellar rotation (Fig. 3.2, 6), and the synthesis of ATP via the ATP synthase (Fig. 3.2, 7, and Section 3.6.2). As Mitchell emphasized, it is important that the membrane has a low permeability to protons so that the major route of proton re-entry is via the energy-transducing proton transporters rather than by general leakage. This, of course, would be expected of a lipid bilayer that is relatively

nonpermeable to protons. Some bacteria couple respiration or the decarboxylation of carboxylic acids to the extrusion of sodium ions (Fig. 3.2, 2) (see Sections 3.7.1 and 3.8.1). The re-entry of sodium ions can also be coupled to the performance of work (e.g., solute uptake Fig. 3.2, 5). Once established, the membrane potential can energize the secondary flow of other ions. For example, the influx of potassium ions can be in response to a membrane potential, inside negative, created by proton extrusion. Mitochondrial and chloroplast membranes are also energized by proton gradients. Therefore, this is a widespread phenomenon, not restricted to prokaryotes.

## 3.2 Electrochemical Energy

When bacteria translocate protons across the membrane to the outside surface, energy is conserved in the proton gradient that is established. The energy in the proton gradient is both electrical and chemical. The *electrical* energy has to do with the fact that a positive charge (i.e., the proton) has been moved to one side of the membrane, creating a charge separation, and therefore a membrane potential. When the proton moves back into the cell toward the negatively charged surface of the membrane, the membrane potential is dissipated (i.e., energy has been given up and work can be done). The energy dissipated when the proton moves to the inside of the cell is equal to the energy required to translocate the proton to the outside.

Stated more precisely, energy is required to move a charge *against* the electric field (i.e., to the side of the same charge). This energy is stored in the electric field. The energy that is stored in the electric field is called *electrical energy*. Conversely, the electric field gives up energy when a charge moves *with* the electric field (i.e., to the opposite pole) and work can be done. The amount of energy is the same, but opposite in sign.

The same description applies to chemical energy. Energy is required to move the proton against its concentration gradient. This energy is stored in the concentration gradient. The energy that is stored in a concentration gradient is called *chemical energy*. When the

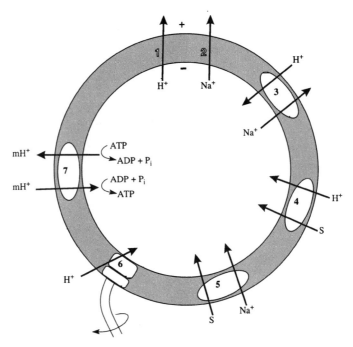

**Fig. 3.2** An overview of the proton and sodium ion currents in a generalized bacterial cell. Driving reactions (metabolic reactions) deliver energy to create proton (1) and sodium ion (2) electrochemical gradients, high on the outside. The major driving reactions encompassed by (1) are the redox reactions that occur during electron transport. The establishment of sodium ion potentials coupled to metabolic reactions such as respiration are not widespread. When the ions return to the lower electrochemical potentials on the inside, work can be done. Built into the membrane are various transporters (porters) that translocate protons and sodium ions back into the cell, completing the circuit, and in the process doing work. There are three classes of porters: (a) *antiporters* that carry two solutes in opposite directions, (b) *symporters* that carry two solutes (S) in the same direction, and (c) *uniporters* that carry only a single solute. Illustrated are: (3) a $Na^+/H^+$ antiporter, which is the major mechanism for extruding $Na^+$ in bacteria, and also functions to bring protons into the cell for pH homeostasis in alkaliliphilic bacteria. In most bacteria the $Na^+/H^+$ antiporter creates the sodium potential necessary for the $Na^+$/solute symporter because a primary $Na^+$ pump is not present. The antiporter uses the proton electrochemical potential as a source of energy; (4) a $H^+$/solute symporter that uses the proton potential to accumulate solutes; (5) a $Na^+$/solute symporter that uses the sodium electrochemical potential to accumulate solutes; (6) a flagella motor that turns at the expense of the proton electrochemical potential; (7) an ATP synthase that synthesizes ATP at the expense of the proton electrochemical potential. The ATP synthase is reversible and can create a proton electrochemical potential during ATP hydrolysis.

proton returns to the lower concentration, then the energy in the concentration gradient is dissipated and work can be done.

The sum of the changes in electrical and chemical energies is called *electrochemical energy*. The symbol for electrochemical energy is $\Delta\tilde{\mu}$, which is equal to $\tilde{\mu}_{in}-\tilde{\mu}_{out}$. For the proton, it would be $\Delta\tilde{\mu}_{H^+}$. The units of electrochemical energy are joules per mole, or calories per mole. (One calorie is equal to 4.18 joules.) The electrochemical energy

is discussed in more detail in the following sections.

### 3.2.1 The electrochemical energy of protons

#### The proton motive force

The electrochemical work that is performed when an ion crosses a membrane is a function of both the membrane potential, $\Delta\Psi$, and the difference in concentration between the solutions separated by the membrane. For example, for one mole of protons:

$$\Delta \tilde{\mu}_{H^+} = F \, \Delta \Psi$$
$$+ RT\ln([H^+]_{in}/[H^+]_{out}) \text{ J} \quad (3.1)$$

In eq. 3.1, $F \, \Delta \Psi$ represents the electrical energy when one mole of protons moves across a potential difference of $\Delta \Psi$ volts and $RT \, \ln([H^+]_{in}/[H^+]_{out})$ represents the chemical energy when one mole of protons moves across a concentration gradient of $[H^+]_{in}/[H^+]_{out}$.

To express eq. 3.1 in millivolts (mV), we simply divide by the faraday (F). Since $(RT/F) \ln([H^+]_{in}/[H^+]_{out}) = -60 \, \Delta pH$ at 30°C, where $\Delta pH = pH_{in} - pH_{out}$, eq. 3.1 in mV is:

$$\Delta p = \Delta \tilde{\mu}_{H^+}/F =$$
$$\Delta \Psi - 60 \Delta pH \quad \text{mV, at 30°C} \quad (3.2)$$

Usually, $\Delta \tilde{\mu}_{H^+}/F$ is called the proton motive force and is denoted as $\Delta p$. The $\Delta p$ is the potential energy in the electrochemical proton gradient. When protons move toward the lower electrochemical potential, the $\Delta p$ gives up energy (is dissipated) and work can be done (e.g., flagellar rotation, ATP synthesis, solute transport). Cells must continuously replenish the $\Delta p$ as it is used for doing work. One should view the $\Delta p$ as a force pulling protons across the membrane into the cell toward their lower electrochemical potential. In order to replenish the $\Delta p$, an equal but opposite force must be exerted to push protons out of the cell toward the higher electrochemical potential (against the $\Delta p$). As will be discussed later, the force that generates the $\Delta p$ (translocates protons out of the cell) can be due to several exergonic reactions (reactions that give up energy), the most widely used being oxidation–reduction reactions in the membrane that occur during respiration, and ATP hydrolysis. Bacteria maintain an average $\Delta p$ of approximately −140 to −200 mV. The values for respiring bacteria tend to be a little higher than those for fermenting bacteria.[6] Equation 3.2 is a fundamental equation in cell biology.

It is derived in more detail later. (See eqs. 3.3–3.10.)

## Units: What is meant by volts, electron volts, and joules

It is important to distinguish between volts (V), electron volts (eV), and joules (J). Potential energy differences (e.g., $\Delta E$, $\Delta \Psi$, or $\Delta p$) are expressed as volts or millivolts. As long as charges are not moving, then these remain as potential energy and work is not done. When charges are moving, work is being done, either on the charges or by the charges, depending upon whether the charges are moving toward a higher potential or a lower potential. The quantity of work that is done is proportional to the product of the amount of charge that moves and the potential difference over which the charge moves. The units of work are either joules or electron volts. One joule is defined as the energy required to raise a charge of one coulomb (C) through a potential difference of one volt (i.e., J = C × V). (The older literature used calorie units instead of joules. One calorie = 4.184 J.) When calculating the change in joules when a *mole* of monovalent ions or electrons travels over a voltage gradient, then one multiplies volts by the faraday (F), since the faraday is the number of coulombs of charge *per mole* of electrons or monovalent ions. (See Note 7.) One electron volt is the increase in energy of a single electron or monovalent ion (rather than a coulomb) when raised through a potential difference of one volt. The charge on the electron or monovalent ion is approximately $1.6 \times 10^{-19}$ C. Therefore, one electron volt is equal to $1.6 \times 10^{-19}$ J. If the electrons are moving toward a lower energy level (i.e., toward a higher electrode potential), then work can be obtained from the system (e.g., the generation of a $\Delta p$). The energy available from the electron flow is $n \, \Delta E$ eV or $nF \, \Delta E$ J (if $n$ refers to moles of electrons). An equivalent amount of work must be done to move the electrons to a higher energy level. Similarly, if $y$ protons move over a potential of $\Delta p$ volts, then $y \, \Delta p$ eV of work are done. If $y$ moles of protons move, then $yF \, \Delta p$ J of work are done. Often work units are converted to volts or millivolts when one wishes to express the potential energy in a concentration gradient. For example, 17,400 joules are required to move one mole of solute against a concentration gradient of 1000. This can be expressed as the potential energy in the concentration gradient, 0.180 V (17,400/F). Another way of thinking about volts or millivolts is that it is a *force* that *pushes* a molecule down its electrical, electrochemical, or chemical gradient. The $\Delta p$ mV is the force that pushes protons; thus it is called the *proton motive force.*

## A more detailed explanation of and derivation of eq. 3.2

### 1. The electrical component of the $\Delta\mu_{H^+}$

A membrane potential, $\Delta\Psi$, exists across the cell membrane where $\Delta\Psi = \Psi_{in} - \Psi_{out}$. By convention, the $\Delta\Psi$ is negative when the inner membrane surface is negative. The units of $\Delta\Psi$ are volts. The work done on or by the electric field when charges traverse the membrane potential is equal to the total charges carried by the ions or electrons multipled by $\Delta\Psi$. If a single electron or monovalent ion such as the proton moves across the membrane, then the work done is $\Delta\Psi$ eV. (For a divalent ion the work done would be $2\Delta\Psi$ eV.) The amount of work that is done *per mole* of protons that traverses the $\Delta\Psi$ is:

$$\Delta G = F \Delta\Psi \quad J \qquad (3.3)$$

Bearing in mind that coulombs × volts = joules (i.e., V = J/F), eq. 3.3 is often expressed as electrical potential energy (or force) in volts by dividing both sides of the equation by the faraday:

$$\Delta G/F = \Delta\Psi \quad V \qquad (3.4)$$

### 2. The chemical component of the $\Delta\mu_{H^+}$

Of course, if a concentration gradient of protons exists, then we must add the chemical energy to the electrical energy in eq. 3.3 in order to obtain the expression for the electrochemical energy, $\Delta\mu_{H^+}$. The chemical energy of the proton (or any solute) as a function of its concentration is (see Note 8 for a more complete discussion):

$$G = G_0 + RT \ln[H^+] \quad J \qquad (3.5)$$

The free energy change accompanying the transfer of one mole of protons between a solution of protons outside $[H^+]_{out}$ and inside $[H^+]_{in}$ the cell is the difference between the free energies of the two solutions:

$$\Delta G = RT \ln[H^+]_{in} - RT \ln[H^+]_{out} \quad J \qquad (3.6)$$

or

$$\Delta G = RT \ln[H^+]_{in}/[H^+]_{out} \quad J$$

Equation 3.6 refers to the free energy change when one mole of protons moves from one concentration to another where the concentration gradient does not change (i.e., as applies to a steady state). It does *not* refer to the total energy released when the concentration of protons comes to equilibrium. Equation 3.6 can be used for the movement of *any solute* over a concentration gradient, not simply protons, and we will see this equation again when we discuss solute transport. (When describing the movement of solutes other than protons, the symbol $S$ may be substituted for $H^+$. Otherwise, the equation used is identical to eq. 3.6.)

Usually eq. 3.6 is expressed in electrical units of potential (volts). In order to do this, one substitutes 8.3144 J deg$^{-1}$ mole$^{-1}$ for $R$, 303 K (30°C) for $T$, converts ln to $\log_{10}$ by multiplying by 2.303, and divides by the faraday to convert joules to volts, thus deriving:

$$\Delta G/F = 0.06 \log_{10}([H^+]_{in}/[H^+]_{out}) \quad V$$

or

$$= 60 \log_{10}([H^+]_{in}/[H^+]_{out}) \quad mV \qquad (3.7)$$

Since $\log_{10}([H^+]_{in}/[H^+]_{out}) = pH_{out} - pH_{in}$, eq. 3.7 can be written as:

$$\begin{aligned} \Delta G/F &= 60(pH_{out} - pH_{in}) \\ &= -60(pH_{in} - pH_{out}) \\ &= -60 \Delta pH \quad mV \qquad (3.8) \end{aligned}$$

### 3. Proton electrochemical energy.

$\Delta\mu_{H^+}$ is the sum of the electrical (1) and chemical energies (2) of the proton. We are now ready to derive an expression for the proton motive force. The total energy change accompanying the movement of one mole of protons through the membrane is the sum of the energy due to the membrane potential (eq. 3.3) and the energy due to the concentration gradient (eq. 3.6). This sum is called the electrochemical energy, $\Delta\mu_{H^+}$.

$$\Delta\tilde{\mu}_{H^+} = F\,\Delta\Psi$$
$$+ RT\,\ln([H^+]_{in}/[H^+]_{out}) \quad J \quad (3.9)$$

One can also express the electrochemical energy as a potential in volts or millivolts (proton motive force, or $\Delta p$) by dividing by the faraday (or by adding eqs. 3.4 and 3.8):

$$\Delta p = \Delta\Psi - 60\,\Delta pH \quad mV \text{ at } 30°C \quad (3.10)$$

The same equation is used to express the electrochemical potential for any ion (e.g., $Na^+$)

$$\Delta\tilde{\mu}_{Na^+}/F = \Delta\Psi - 60\,\Delta pNa \quad mV, \quad (3.11)$$

where $p$Na is $-\log_{10}(Na^+)$. By convention, the values of the potentials are always negative when the cell membrane is energized for that particular ion.

### 3.2.2 Generating a $\Delta\Psi$ and a $\Delta pH$

We now consider some biophysical aspects of generating the proton motive force. First we will examine the formation of the $\Delta\Psi$, and then how the capacitance of the membrane affects the $\Delta\Psi$ that develops. Finally, we will consider the establishment of a $\Delta pH$.

### Electrogenic flow creates a $\Delta\Psi$

The movement of an uncompensated charge creates the membrane potential. When this happens, the charge movement is said to be *electrogenic*. The moving charge can be either a proton or an electron. For example, a membrane potential is generated when a proton is translocated through the membrane to the outer surface leaving behind a negative charge on the inner surface.

$$H^+_{in} \longrightarrow H^+_{out}$$

A membrane potential can also develop if a molecule is reduced on the inner membrane surface, picking up cytoplasmic protons in the process, and then diffuses across the membrane, where it is oxidized on the outer surface, releasing the protons to the exterior,

while the electrons return electrogenically across the membrane to the inner surface:

$$A + 2e^- + 2H^+_{in} \longrightarrow AH_2$$
$$AH_2 \longrightarrow A + 2e^- + 2H^+_{out}$$

This is also electrogenic flow, but in this case it is the electron that is the moving charge rather than the proton. The work done is the same because the electron carries the same charge as the proton (i.e., $1.6 \times 10^{-19}$ C). It makes no difference from the point of view of calculating the $\Delta p$ whether one thinks in terms of protons moving as positive charges across the membrane from the inside to the outside, or simply being carried as hydrogen atoms in a reduced organic compound (e.g., $AH_2$). There is a net translocation of protons in either case. In both instances, the total energy necessary to move $y$ protons against the $\Delta p$ is equal to $y\,\Delta p$ eV. We will see that bacteria use both electrogenic electron flow and electrogenic proton flow to create a membrane potential.

### The size of the membrane potential depends upon the capacitance of the membrane

The membrane potential that develops when even a small number of protons move across the membrane can be more than 100 mV. This can be understood by considering the membrane to be a capacitor. The membrane is a capacitor because the membrane lipids prevent the protons from rapidly leaking back into the cell, and so the membrane stores positively charged protons on one surface, just like a capacitor stores charges on one surface.

The relationship between the charge that accumulates on one face of a capacitor and the voltage across the capacitor is

$$\Delta V = \Delta Q/C \quad (3.12)$$

where $\Delta Q$ is the charge (in coulombs), $\Delta V$ is the voltage, and $C$ is a proportionality constant called the capacitance. The value of $C$ for biological membranes is low, about 1 microfarad ($\mu$F) cm$^{-2}$ of membrane. The equation can be rewritten with different symbols and used to predict the theoretical membrane potential:

$$\Delta\Psi(\text{volts}) = en/C \qquad (3.13)$$

where $e$ is the charge per proton ($1.6 \times 10^{-19}$ C), $n$ is the number of protons, and $C$ is the capacitance. Assuming a membrane area of about $3 \times 10^{-8}$ cm$^2$ (for a spherical cell the size of a typical bacterium), then $C =$ about $3 \times 10^{-14}$ F. Therefore, only 40,000 protons translocated to the cell surface is sufficient to generate a membrane potential of −200 millivolts.[9, 10] (By convention, the $\Delta\Psi$ is said to be negative when the inside potential is negative.) The membrane potential that actually develops varies in magnitude from approximately −60 to about −200 mV depending upon the bacterium and the growth conditions.[6] The membrane potential that develops when a relatively small number of protons are translocated limits electrogenic proton pumping and ultimately the size of the membrane potential itself. This is because the membrane potential, which is negative on the inside, pulls protons back into the cell.

## Generating a ΔpH

When a proton is translocated across the membrane, a $\Delta\Psi$ and a $\Delta$pH cannot be created simultaneously. This is summarized in Fig. 3.3. Let us first discuss creating a $\Delta$pH (i.e., accumulating protons in the external bulk phase). One should remember that the only way the bulk external medium can become acidified during proton translocation is if electrical neutrality is conserved (i.e., each proton in the bulk phase must have a negative counterion). This can happen if the proton is pumped out with an anion (i.e., H$^+$/R$^-$), or if a cation enters the cell from the external bulk phase in exchange for the proton (i.e., H$^+$ in exchange for R$^+$) (Fig. 3.3). However, these would be electroneutral events and a charge separation, hence a $\Delta\Psi$, would not develop. Therefore, *in the absence of compensating ion flow*, the protons that are pumped out of the cell remain on or very close to the membrane, and a $\Delta\Psi$ rather than a $\Delta$pH (measurable with a pH electrode) is created. (Theoretically, however, a $\Delta$pH and a $\Delta\Psi$ could develop if a cation in the external medium exchanged for the proton on the membrane.)

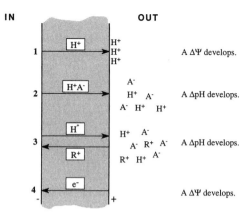

Fig. 3.3 A $\Delta$pH in the bulk phase and a $\Delta\Psi$ cannot develop simultaneously by the movement of a proton. (1) Electrogenic movement of a proton; (2) establishment of a $\Delta$ by the extrusion of both protons and counterions. 3. Establishment of a $\Delta$pH by the exchange of protons for cations in the medium. Theoretically, a $\Delta$pH could develop if the proton on the outer surface of the membrane exchanged with a cation from the bulk phase. But even under these circumstances the membrane potential (positive outside) would limit further efflux of protons. (4) A $\Delta\Psi$ can also develop via electrogenic flow of an electron.

Of course, some of the protons could be electroneutrally released into the bulk phase to create a $\Delta$pH, and some might remain at the outer membrane surface to create a $\Delta\Psi$, but even under these circumstances, a large $\Delta$pH cannot be generated in the face of a large $\Delta\Psi$. This is because the positive charge on the outside surface inhibits further efflux of protons. In fact, to demonstrate the formation of a $\Delta$pH experimentally as a result of proton pumping, a large $\Delta\Psi$ must not be allowed to develop. This is done experimentally by making the membrane permeable either to a cation so that the incoming cations can compensate electrically for the outgoing protons, or to an anion that moves in the same direction as the proton. For example, in many experiments the K$^+$ ionophore valinomycin is added to make the membrane permeable to K$^+$ (Section 3.4). When this is done, K$^+$ exchanges for H$^+$, and a $\Delta$pH can develop.

Although bacteria cannot make both a $\Delta\Psi$ and a $\Delta$pH with the same proton, they can create a $\Delta\Psi$ during proton translocation, and then convert it to a $\Delta pH$. Suppose a $\Delta\Psi$ is created because a few protons are translocated to the cell membrane outer surface. This cannot proceed for very long because a membrane potential develops quickly, which limits further efflux of protons. However, a cation such as $K^+$ might enter the cell electrogenically. This would result in a lowering of the membrane potential because of the positive charge moving in. Now more protons can be translocated out of the cell. The protons can leave the outer membrane surface and accumulate in the external phase because they are paired with the anion that was formerly paired with the $K^+$. Thus a membrane potential can form during proton translocation and be converted into a $\Delta pH$ by the influx of $K^+$. This is an important way for bacteria to maintain a $\Delta pH$ as described in Section 15.1.3.

## 3.3 The Contributions of the $\Delta\Psi$ and the $\Delta pH$ to the Overall $\Delta p$ in Neutrophiles, Acidophiles, and Alkaliphiles

Partly for the reasons stated in Section 3.2.2, the contributions of the $\Delta\Psi$ and the $\Delta pH$ to the $\Delta p$ are never equal. Additionally, the relative contributions of the $\Delta pH$ and the $\Delta\Psi$ to the $\Delta p$ vary, depending upon the pH of the environment where the bacteria naturally grow. Below are summarized the relative contributions of the $\Delta\Psi$ and the $\Delta pH$ to the $\Delta p$ in neutrophiles, acidophiles, and alkaliphiles.[11] Notice that in acidophiles and alkaliphiles the $\Delta\Psi$ (acidophiles) or the $\Delta pH$ (alkaliphiles) has the wrong sign and actually detracts from the $\Delta p$.

### Neutrophilic bacteria

For neutrophilic bacteria (i.e., those that grow with a pH optimum near neutrality), the $\Delta\Psi$ contributes approximately 70–80% to the $\Delta p$, with the $\Delta pH$ contributing only 20–30%. This is reasonable when one considers that the intracellular pH is near neutrality, and therefore the $\Delta pH$ cannot be very large.

### Acidophilic bacteria

For acidophilic bacteria (i.e., those that grow between pH 1 and 4, and not at neutral pH) the $\Delta\Psi$ is positive rather than negative (below an external pH of 3, which is where they are usually found growing in nature), and thus *lowers* the $\Delta p$. Under these conditions the force in the $\Delta p$ is due entirely to the $\Delta pH$. Let us examine an example of this situation. *Thiobacillus ferroxidans* is an aerobic acidophile that can grow at pH 2.0 (see Section 12.4.1). Because it has an intracellular pH of 6.5, the $\Delta pH$ is 4.5, which is very large (remember, $\Delta pH = pH_{in} - pH_{out}$). However, the aerobic acidophiles have a positive $\Delta\Psi$ when growing at low pH, and for *T. ferroxidans* the $\Delta\Psi$ is +10 mV.[12] (See Note 13.) The contribution of the $\Delta pH$ to the $\Delta p$ is $-60\ \Delta pH$ or $-270$ mV. Since $\Delta p = \Delta\Psi - 60\ \Delta pH$, the actual $\Delta p$ would be $-260$ mV. In this case, the $\Delta\Psi$ lowered the $\Delta p$ by 10 mV. As discussed in Section 15.1.3, the inverted membrane potential is necessary for maintaining the large $\Delta pH$ in the acidophiles.

### Alkaliphilic bacteria

An opposite situation holds for the aerobic alkaliphilic bacteria (i.e., those that grow above pH 9 often with optima between pH 10 and 12). For these organisms the $\Delta pH$ is one to two units *negative* (because the cytoplasmic pH is less than 9.6) and consequently *lowers* the $\Delta p$, by 60–120 mV.[14] Therefore, in the alkaliphiles, the potential of the $\Delta p$ may come entirely from the $\Delta\Psi$. In fact, as explained in Section 3.10, because of the large negative $\Delta pH$, the calculated $\Delta p$ in these organisms is so low that it raises conceptual problems regarding whether there is sufficient energy to synthesize ATP.

## 3.4 Ionophores

Before we continue with the discussion of the proton motive force, we must explain ionophores and their use because reference will be made to these important molecules later. Ionophores are important research tools for investigating membrane bioenergetics and their use has contributed to an understanding

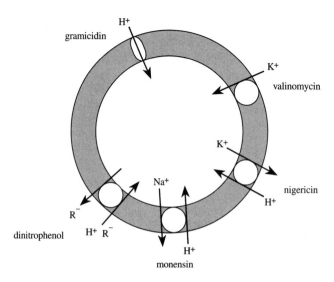

**Fig. 3.4** Examples of ionophores and the transport processes that they catalyze. All the reactions are reversible. *Valinomycin*: Transports K$^+$ and Rb$^+$. Valinomycin transports only one cation at a time. Since K$^+$ is positively charged, valinomycin carries out electrogenic transport, i.e., creates a membrane potential. If valinomycin is added to cells with high intracellular concentrations of K$^+$, then the K$^+$ will rush out of the cell ahead of counterions, and create a transient membrane potential, outside positive, predicted by the Nernst equation. In the presence of excess extracellular K$^+$ and valinomycin, the K$^+$ will rapidly diffuse into the cell, collapsing the existing potential. *Nigericin*: Carries out an electroneutral exchange of K$^+$ for H$^+$. When nigericin is added to cells, one can expect a collapse of the pH gradient as the internal K$^+$ exchanges for the external H$^+$, but the membrane potential should not decrease. The combination of nigericin and valinomycin will collapse both the $\Delta$pH and the membrane potential. *Monensin*: Carries out an electroneutral exchange of Na$^+$ or K$^+$ for H$^+$. There is a slight preference for Na$^+$. *Gramicidin*: Carries out electrogenic transport of H$^+$ > Rb$^+$, K$^+$, Na$^+$. Gramicidin differs from the other ionophores in that it forms polypeptide channels in the membrane. That is to say, it is not a diffusible carrier. Since the addition of gramicidin results in the equilibration of protons across the cell membrane, it will collapse the $\Delta\Psi$ and the $\Delta$pH i.e. the $\Delta p$. *Dinitrophenol*: This is an anion (R$^-$). It carries out electroneutral influx of H$^+$ and R$^-$ into the cell and returns to the outside without H$^+$, i.e., R$^-$. It will therefore collapse both the $\Delta$pH and the $\Delta\Psi$. This is the classic uncoupler of oxidative phosphorylation. *Carbonyl cyanide-p-trifluoromethylhydrazone (FCCP)*: (Not shown.) This is a lipophilic weak acid that exists as the nonprotonated anion (FCCP$^-$) and the protonated form (FCCPH), both of which can travel through the membrane. Protons are carried into the cell in the form of FCCPH. Inside the cell the FCCPH ionizes to FCCP$^-$, which exits in response to the membrane potential, outside positive. The result is a collapse of both the $\Delta$pH and the $\Delta\Psi$.

of the role of electrochemical ion gradients in membrane energetics. As mentioned earlier, membranes are poorly permeable to ions, and this is why the membrane can maintain ion gradients. Ionophores perturb these ion gradients. Most ionophores are organic compounds that form lipid-soluble complexes with cations (e.g., K$^+$, Na$^+$, H$^+$), and rapidly equilibrate these across the cell membrane (Fig. 3.4). The incorporation of an ionophore into the membrane is equivalent to short-circuiting an electrical device with a copper wire. Another way of saying this is that the ionophore causes the electrochemical potential difference of the ion to approach zero. Since some ionophores are specific for certain ions, it is sometimes possible to identify the ion current that is performing the work with the appropriate use of ionophores. For example, if ATP synthesis is prevented by a proton ionophore, then this implies that a current of protons carries the energy for ATP synthesis. One can also preferentially collapse the $\Delta\Psi$ or $\Delta$pH with judicious use of ionophores, and perhaps gain information regarding the driving force for the ion current. For example, nigericin, which catalyzes an electroneutral exchange of $K^+$ for $H^+$, will dissipate the pH gradient but not the $\Delta\Psi$.[15] Valinomycin plus $K^+$, on the other hand, will initially dissipate the $\Delta\Psi$ because it electrogenically carries $K^+$ into the cell, thus setting up a potential opposite to the membrane potential. It is also possible to *create* a $\Delta\Psi$ using valinomycin. The addition of valinomycin to starved cells or vesicles loaded with $K^+$ will induce a temporary $\Delta\Psi$ predicted by the Nernst equation, $\Delta\Psi = -60 \log([S]_{in}/[S]_{out})$ mV. (This is the concentration gradient expressed as mV at 30°C.) What happens is that the $K^+$ moves from the high internal concentration to the lower external concentration in the presence of valinomycin (Fig. 3.5). Because the $K^+$ moves faster than its counterion, a temporary diffusion potential, outside positive is created. The diffusion potential is temporary because of the movement of the counterions. The use of ionophores has helped researchers to determine which ions are carrying the primary current that is doing the work, and to investigate the relative importance of the membrane potential and ion concentration gradients in providing the energy for specific membrane functions.

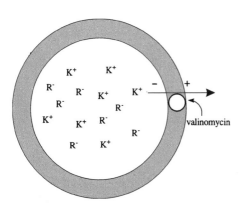

**Fig. 3.5** Valinomycin-stimulated $K^+$ efflux can impose a temporary membrane potential. When valinomycin is added to cells or membrane vesicles loaded with $K^+$, the valinomycin dissolves in the membrane and carries the $K^+$ out of the cell. The efflux of $K^+$ creates a diffusion potential, since the negative counterions lag behind the $K^+$. The membrane potential is predicted by the Nernst equation, $\Delta\Psi = -60 \log[K^+]_{in}/[K^+]_{out}$. The membrane potential is transient because it is neutralized by the movement of counterions.

### 3.4.1 The effect of uncouplers on respiration

Uncouplers are ionophores that have the following effects:

1. They collapse the $\Delta p$ and thereby inhibit ATP synthesis coupled to electron transport.

2. They stimulate respiration.

Why should uncouplers stimulate respiration? The flow of electrons through electron carriers in the membrane is *obligatorily* coupled to a flow of protons in a closed circuit (Section 4.5). This occurs at the coupling sites discussed in Section 3.7.1. Protons are translocated to the outer surface of the membrane and then re-enter via the ATP synthase. If re-entry is blocked by inhibitors of the ATP synthase (e.g., dicyclohexylcarbodiimide, DCCD) or slowed by depletion of ADP, then respiration is slowed. One possible explanation for the slowing of respiration is that, as the protons are translocated to the outside surface, the $\Delta p$ rises and approaches the $\Delta G$ of the oxidation–reduction reactions. This might slow respiration since the oxidation–reduction reactions are reversible. One can view this as the $\Delta p$ producing a "back-pressure." The re-entry of the protons via the ATP synthase can be viewed, in this context, as placing a limit on the rise of the $\Delta p$ and thereby promoting respiration. Uncouplers might also stimulate respiration because they collapse the $\Delta p$. In the presence of uncouplers such as dinitrophenol, $H^+$ rapidly enters the cell on the uncoupler rather than through the ATPase.

### 3.5 Measurement of the $\Delta p$

Measurements of the size of the $\Delta p$ are a necessary part of analyzing the role that the $\Delta p$ plays in the overall physiology of the cell. The two components of the $\Delta p$, the $\Delta\Psi$ and the $\Delta$pH, are measured separately.[16–18]

### 3.5.1 Measurement of $\Delta\Psi$

How is the membrane potential measured? Bacteria are too small for the insertion of electrodes; therefore, the membrane potential is measured indirectly. Suppose a membrane potential exists and an ion was allowed to come to its electrochemical equilibrium in response to the potential difference. Then at equilibrium the electrochemical energy of the ion is zero and, using the equation for electrochemical energy (for a monovalent ion),

$$\Delta \tilde{\mu}_S / F = 0 = \Delta \Psi$$
$$+ 60 \log_{10}([S]_{in}/[S]_{out}) \quad mV \quad (3.14)$$

Solving for $\Delta \Psi$,

$$\Delta \Psi = - 60 \log_{10}([S]_{in}/[S]_{out})$$
$$mV \text{ at } 30°C \quad (3.15)$$

Equation 3.15 is one form of the Nernst equation. It states that the measurement of the intracellular and extracellular concentrations of a permeant ion at equilibrium allows one to calculate the membrane potential. The bacterial cell membrane is relatively nonpermeable to ions; therefore, to measure the membrane potential one must use either an ion plus an appropriate ionophore (e.g., $K^+$ and valinomycin), or a lipophilic ion (i.e., one that can dissolve in the lipid membrane and pass freely into the cell). When the inside is negative with respect to the outside, a cation ($R^+$) is chosen because it accumulates inside the cells or vesicles (Fig. 3.6). When the inside is positive, an anion ($R^-$) is used. It is important to use a small concentration of the ion to prevent the collapse of the membrane potential by the influx of the ion.

Cationic or anionic fluorescent dyes have also been used to measure the $\Delta \Psi$. The dyes partition between the cells and the medium in response to the membrane potential, and the fluorescence is quenched. The fluorescent dye will monitor relative changes in membrane potential (Fig. 3.7). When using a fluorescent dye to measure the absolute membrane potential, it is necessary to produce a standard curve. This involves measuring the fluorescence quenching in a sample where the membrane potential is known (i.e., has been measured by independent means such as the distribution of a permeant ion, Fig 3.8).

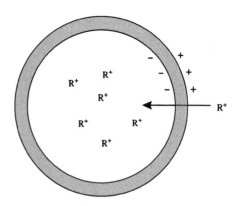

Fig. 3.6 Measurement of membrane potential by the accumulation of a lipophilic cation. The cation accumulates in response to the membrane potential until the internal concentraton reaches a point where efflux equals influx; i.e., equilibrium has been reached. At this time the electrochemical energy of the cation is zero and $\Delta \Psi$ = $-60 \log [(R^+)_{in}/(R^+)_{out}]$ mV at 30°C. Any permeant ion can be used as long as it diffuses passively, is used in small amounts, is not metabolized, and accumulates freely in the bulk phase on either side of the membrane. Permeant ions that have been used include lipophilic cations such as tetraphenylphosphonium ($TPP^+$).

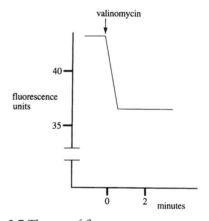

Fig. 3.7 The use of fluorescence to measure the membrane potential. An example of what one might expect if valinomycin were added to bacteria in buffer containing low concentrations of potassium ion in the presence of a fluorescent cationic lipophilic probe. The potassium inside would rush out, creating a diffusion potential predicted by the Nernst equation. The cationic probe would enter in response to the membrane potential, and the fluorescence would be quenched. It has been suggested that the quenching of fluorescence is due to the formation of dye aggregates with reduced fluorescence inside the cell.

## 3.5.2 Measuring the ΔpH

A common way to measure the ΔpH is to measure the distribution of a weak acid or weak base at equilibrium between the inside and outside of the cell. The assumption is that the uncharged molecule freely diffuses across the membrane but that the ionized molecule cannot. Inside the cell the acid becomes deprotonated or the base becomes protonated, the extent of which depends upon the intracellular pH. For example,

$$AH \longrightarrow A^- + H^+$$

or

$$B + H^+ \longrightarrow BH^+$$

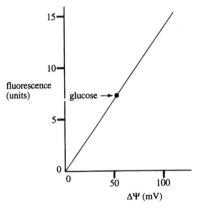

One uses a weak acid (e.g., acetic acid) when $pH_{in} > pH_{out}$ because it will ionize more extensively at the higher pH, and a weak base (e.g., methylamine) when $pH_{in} < pH_{out}$ because it will become more protonated at the lower pH. At equilibrium, for a weak acid

$$K_a = [H^+]_{in}[A^-]_{in}/[AH]_{in}$$
$$= [H^+]_{out}[A^-]_{out}/[AH]_{out}$$

where, $K_a$ is the dissociation constant of the acid, HA.

If $pH_{in}$ and $pH_{out}$ are at least 2 units higher than the $pK$, then most of the acid is ionized on both sides of the membrane and there is no need to take into account the unionized acid. Therefore, the ΔpH can be calculated using eq. 3.16.

$$pH_{in} - pH_{out} = \Delta pH$$
$$= \log_{10}[A^-]_{in}/[A^-]_{out} \quad (3.16)$$

Thus the $\log_{10}$ of the ratio of concentrations inside to outside of the weak acid is equal to the ΔpH. In practice one uses radioactive acids or bases as probes and measures the amount of radioactivity taken up by the cells. A more complex equation must be used when the concentration of the unionized acid (AH) cannot be ignored.[18] Cytoplasmic pH is sometimes also measured by $^{31}P$ nuclear magnetic resonance ($^{13}P$-NMR) of phosphate, whose spectrum is pH dependent. (See Note 19.)

**Fig. 3.8** Relationship between the membrane potential and fluorescence change of 1,1′-dihexyl-2,2′-oxacarbocyanine (CC₆). On the ordinate are plotted the changes in fluorescence (continuous line) corresponding to different membrane potentials imposed by the addition of valinomycin to *Streptococcus lactis* cells. The membrane potentials were calculated using the Nernst equation from potassium concentration ratios (IN/OUT) in parallel experiments where the intracellular $K^+$ concentrations were about 400 mM and the extracellular concentrations were varied. Also shown is the membrane potential caused by the addition of glucose (rather than valinomycin) to the cells. *Source*: Adapted from Maloney, P. C., E. R. Kashket, and T. H. Wilson. 1975. Methods for studying transport in bacteria. In: *Methods in Membrane Biology*, Vol. 5, Chap. 1, pp. 1–49. Korn, E. D. (Ed.). Plenum Press, New York.

## 3.6 Use of the Δp to Do Work

The Δp provides the energy for several membrane functions, including solute transport and ATP synthesis, discussed in this section. (See Chapter 16 for a more complete discussion of solute transport.)

### 3.6.1 Use of the Δp to drive solute uptake

As an example of how the Δp can be used for doing work, consider symport of an uncharged solute, S, with protons (Fig. 3.2, 4). The total driving force on S at 30°C is

$$y \,\Delta p + 60 \log_{10}([S]_{in}/[S]_{out}) \quad mV \quad (3.17)$$

where $y$ is the ratio of $H^+/S$. [$60 \log_{10}([S]_{in}/[S]_{out})$ is the force involved in moving S from one concentration to another. It is the same as eq. 3.7 but written for the solute, S, rather than for the proton.]

At equilibrium, the sum of the forces equals 0, and therefore,

$$y \,\Delta p = -60 \log_{10}([S]_{in}/[S]_{out}) \quad (3.18)$$

For $y = 1$ and $\Delta p = -0.180$ V

$$3 = \log_{10}([S]_{in}/[S]_{out})$$

and

$$[S]_{in}/[S]_{out} = 10^3$$

Therefore, a $\Delta p$ of $-180$ mV could maintain a $10^3$ concentration gradient of $[S_{in}]/[S_{out}]$ if the ratio of $H^+/S$ were one. If the ratio of $[H^+]/[S]$ were two, then a concentration gradient of $10^6$ could be maintained. It can be seen that very large concentation gradients can be maintained using the $\Delta p$.

### 3.6.2 Use of the $\Delta p$ to drive ATP synthesis

Built into the cell membranes of prokaryotes and mitochondrial and chloroplast membranes is an enzyme complex that couples the translocation of protons down a proton potential gradient to the phosphorylation of ADP to make ATP. It is called the proton-translocating ATP synthase, or simply ATP synthase. As discussed in the Section 3.7.2, the ATP synthase is reversible and can pump protons out of the cell, generating a $\Delta p$.

### Description of the ATP synthase

The ATP synthase consists of two regions (i.e., the $F_0$ and $F_1$) (Fig. 3.9).[20] The $F_0$ region spans the membrane and serves as a proton channel through which the protons are pumped. The $F_1$ region is located on the inner surface of the membrane and is the catalytic subunit responsible for the synthesis

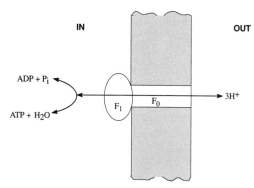

Fig. 3.9 ATP synthase. $F_0$ is a proton channel that spans the membrane. $F_1$ is the catalytic subunit on the inner membrane surface that catalyzes the reversible hydrolysis of ATP. Under physiological conditions the ATP synthase reaction is poised to proceed in either direction. When $\Delta p$ levels decrease relative to ATP, they can be restored by ATP hydrolysis. When the ATP levels decrease relative to $\Delta p$, then more ATP can be made.

and hydrolysis of ATP. The $F_1$ unit is also called the "coupling factor." The $F_1$ subunit from *E. coli* is made from five different polypeptides with the following stoichiometry: $\alpha_3$, $\beta_3$, $\gamma$, $\delta$, and $\varepsilon$. The $\beta$ polypeptides each contain a single catalytic site for ATP synthesis and hydrolysis, although the sites are not equivalent at any one time. The $F_0$ portion has three different polypeptides, which in *E. coli* are $a_1$, $b_2$, and $c_{10}$, all of which occur in multiples. All in all, *E. coli* uses 8 different types of polypetides to construct a 22-polypeptide machine that acts as a *reversible* pump driven by the proton potential or by ATP hydrolysis. As discussed more completely in Section 3.7.2, the amount of energy to synthesize one mole of ATP is given by $\Delta G_p$. Assume that $\Delta G_p$ is equal to approximately 50,000 J or (dividing by the faraday) 518 mV. Thus $y\Delta p$ must be $\geq$ $-518$ mV, where $y$ is the number of entering protons. If $y = 3$, then the $\Delta p$ must be at least $-518/3$ or $-173$ mV to synthesize 1 ATP. (By convention the $\Delta p$ is negative when energy is available to do work; see Note 21.)

### *Evidence that either a $\Delta \Psi$ or a $\Delta pH$ can drive the synthesis of ATP via proton influx through the ATP synthase*

It is possible to impose a $\Delta \Psi$ or a $\Delta pH$ on cells or membrane vesicles and demonstrate that the influx of protons through the ATP synthase results in the synthesis of ATP. A $\Delta\Psi$ can be imposed by using valinomycin and creating a potassium ion diffusion potential. (The use of valinomycin to impose a membrane potential is described in Section 3.4.) A $\Delta pH$ can be created by adding acid to the medium in which cells or membrane vesicles are suspended. These critical experiments were done with the lactic acid bacterium *Streptococcus lactis*.[22, 23] Washed cells of *Streptococcus lactis* containing a high intracellular concentration of $K^+$ (300–400 mM) were suspended in a medium containing a low concentration of $K^+$ (0.3–0.4 mM). The initial $\Delta p$ was close to zero. Valinomycin was then added. The valinomycin caused the rapid efflux of $K^+$, causing a temporary potassium diffusion potential, outside positive, predicted by the Nernst equation.

The cells responded to the valinomycin by making ATP (Fig. 3.10). This indicates that a $\Delta\Psi$ can drive ATP synthesis. A similar experiment was done, but this time in the presence of N,N'-dicyclohexylcarbodiimide (DCCD), which is an inhibitor of the ATP synthase. The DCCD inhibited ATP synthesis *and* proton influx (Fig. 3.11). *This indicates that a major route of proton influx is through the ATP synthase and that the ATP synthase is responsible for ATP synthesis.* Notice that proton entry continued even in the presence of DCCD. It was assumed that proton entry into DCCD-treated cells resulted from passive inflow or leakage by routes other than the ATP synthase.

To demonstrate that a $\Delta pH$ can drive ATP synthesis, washed cells of *S. lactis* were suspended in media to which a small volume of sulfuric acid was added to lower the

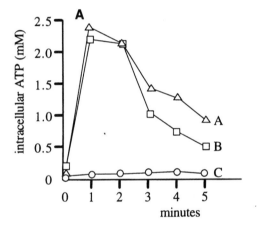
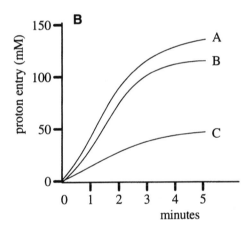

**Fig. 3.10** Proton entry and ATP synthesis in response to an imposed membrane potential. A dense suspension of *Streptococcus lactis* cells was incubated in buffer A, B, or C. Average internal potassium was about 360 mM. Buffer B, external $K^+$ was 0.4 mM. Buffers A and C, KCl added to a final concentration of 3 mM. External pH was set at pH 5 (sample A) or pH 6 (samples B and C), and proton entry was measured as the amount of acid added to maintain the external pH at the initial value, using a pH stat. Valinomycin was added at 0 time and intracellular ATP levels (A) and proton entry (B) were measured. Upon addition of valinomycin the cells immediately made ATP during proton influx (curves A and B). For sample A, the $\Delta p$ was 200 mV, of which 125 mV was due to the measured membrane potential and 75 mV due to the measured $\Delta pH$. For sample B, the $\Delta p$ was also about 200 mV despite the smaller contribution from the $\Delta pH$ (15 mV). This was because the external $K^+$ concentration in sample B was 0.4 mM instead of 3 mM, thus raising the membrane potential. Notice that the amounts of ATP made and proton influx in samples A and B were approximately the same, despite the differences in the membrane potential and the $\Delta pH$. This suggests that proton influx and ATP synthesis depend upon the $\Delta p$ rather than on the individual values of the $\Delta\Psi$ or the $\Delta pH$. Sample C had a $\Delta p$ of only 140 mV (membrane potential, 125 mV; $\Delta pH$, 15 mV) and made no ATP. This suggests a threshold value for $\Delta p$ for ATP synthesis. Presumably, below the threshold value of $\Delta p$, the ATP synthase pumps protons out of the cell at the expense of ATP. *Source*: From Maloney, P. C. Obligatory coupling between proton entry and the synthesis of adenosine 5'-triphosphate in *Streptococcus lactis*. 1977. *J. Bacteriol.* **132**:564–575.

external pH to about 3.5. The internal pH was initially about 7.6, giving a ΔpH of 4.1 and a driving force (−60 ΔpH) of about −246 mV.[24] The cells responded to the imposed ΔpH by making ATP (Fig. 3.12B) during proton influx (Fig. 3.12A). In these experiments it was necessary to add valinomycin, to the samples in order to stimulate proton influx and consequent ATP synthesis. This is because, in the absence of valinomycin, the entering protons established a membrane potential, inside positive, that impeded further net influx

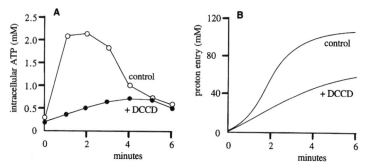

**Fig. 3.11** Effect of DCCD on proton entry and ATP synthesis. Similar conditions as for Fig. 3.10. Cells were suspended in buffer A, ($[K_{in}^+]/[K_{out}^+]$ about 120) and valinomycin added at 0 time. Some samples were incubated with DCCD, which is an inhibitor of the ATP synthase. DCCD blocked both ATP synthesis and proton entry, suggesting that a major route of influx of protons is through the ATP synthase, and that this flow results in ATP synthesis. *Source*: From Maloney, P. C. Obligatory coupling between proton entry and the synthesis of adenosine 5′-triphosphate in *Streptococcus lactis*. 1977. *J. Bacteriol.* 132:564–575.

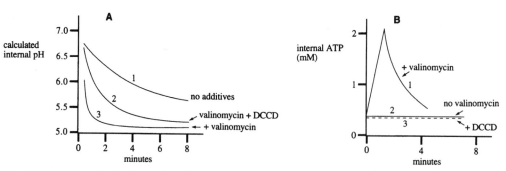

**Fig. 3.12** ATP synthesis driven by the ΔpH. *Streptococcus lactis* cells were placed in buffer with various additions. At zero time the external pH was lowered from pH 8 to 3.5 with sulfuric acid. (*A*) The internal pH dropped very slowly (curve 1) unless valinomycin was added (curve 3). This can be explained by assuming that in the absence of valinomycin the entering H⁺ imposed a membrane potential, inside positive, that limited the uptake of protons. Valinomycin allowed the exit of internal K⁺, thus diminishing the buildup of an internally positive potential. The addition of DCCD markedly inhibited proton uptake in the first minute, suggesting that most of the protons that entered rapidly were entering via the ATP synthase (curve 2). (*B*) The addition of the sulfuric acid at zero time resulted in an immediate synthesis of ATP (curve 1). No such increase in ATP was observed if valinomycin was omitted (curve 2), presumably because net proton influx was slow in the absence of valinomycin. The ATP synthase inhibitor DCCD also prevented ATP synthesis (curve 3). *Source*: Adapted from Maloney, P. C., and F. C. Hansen III. 1982. Stoichiometry of proton movements coupled to ATP synthesis driven by a pH gradient in *Streptococcus lactis*. *J. Memb. Biol.* 66:63–75.

of protons. Valinomycin allowed $K^+$ to exit the cell in exchange for the proton, thus electrically compensating for the influx of the protons. In this case the exiting $K^+$ did not produce a membrane potential because excess $K^+$ (0.1 M) was added to the medium to lower the $[K_{in}^+]/[K_{out}^+]$. Both ATP synthesis and the influx of protons were inhibited by the ATP synthase inhibitor DCCD, indicating that proton entry was primarily through the ATPase.

## Model of the mechanism of ATP synthase

The mechanism of ATP synthesis/hydrolysis by the ATP synthase is complex. One model is called the "binding change mechanism." It is based upon the finding that purified $F_1$ will synthesize tightly bound ATP in the absence of a $\Delta p$.[25] (The $K_d$ for the high-affinity site is $10^{-12}$ M.) The equilibrium constant between enzyme-bound ATP and bound hydrolysis products (ADP and $P_i$) is

approximately one, using soluble $F_1$.[26] The tight binding of ATP and the equilibrium constant of approximately one suggests that the energy requirement for net ATP synthesis is for the release of ATP from the enzyme rather than for its synthesis. The model proposes that, when protons move down the electrochemical gradient through the $F_0F_1$ complex, a conformational change in $F_1$ occurs that results in the release of newly synthesized ATP from the high-affinity site. This is illustrated in Fig. 3.13. The model is discussed in more detail, and with some revisions, in Ref. 27.

## 3.7 Exergonic Reactions that Generate a $\Delta p$

Section 3.6 describes the influx of protons down a proton potential gradient coupled to the performance of work (e.g., solute trans-

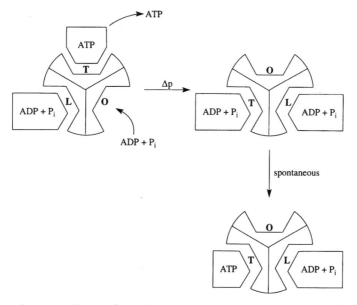

**Fig. 3.13** Binding-change mechanism for ATP synthase. Because there are three $\beta$ subunits encoded by a single gene, and because there is only one catalytic site per $\beta$ subunit, a maximum of three catalytic sites are possible. However, the three catalytic sites are not functionally equivalent and are thought to cycle through conformational changes driven by the electrochemical proton gradient. As a result of the conformational changes, newly synthesized ATP is released from the catalytic site. The three conformational states are O (open, very low affinity for substrates), L (loose binding,), and T (tight binding, the active catalytic site). ATP is made spontaneously at site T, which converts to site O in the presence of a $\Delta p$ and releases the ATP. The $\Delta p$ also drives the conversion of the pre-existing site O to site L, which binds ADP and $P_i$, as well as the conversion of the pre-existing site L to site T. In this model, the only two catalytic sites that bind substrates are L and T. *Source:* Adapted from Cross, L. R., D. Cunningham, and J. K. Tamura. 1984. *Curr. Top. Cell. Regul.* 24:335–344.

port or ATP synthesis). These activities dissipate the $\Delta p$. Let us now consider driving forces that generate a $\Delta p$ (i.e., move protons out of the cell toward a higher potential). The major driving reactions in most prokaryotes are the redox reactions in the cell membranes of respiring organisms (electron transport), ATP hydrolysis in fermenting organisms, and several other less frequently used driving reactions considered in Section 3.8. At this time, let us consider some general thermodyamic features common to all the driving reactions that generate the proton gradient. The translocation of protons out of cells is an energy-requiring process that can be written as

$$yH^+_{in} \longrightarrow yH^+_{out},$$

$$\Delta G = yF \, \Delta p \quad \text{J or } y \, \Delta p \text{ eV}$$

That is to say, the amount of energy required to translocate $y$ moles of protons out of the cell is $yF \, \Delta p$ J, or for $y$ protons, $y \, \Delta p$ electron volts. (This is the same energy, but of opposite sign, that is released when the protons enter the cell.) Therefore, the $\Delta G$ of the driving reaction must be equal to or greater than $yF \, \Delta p$ J or $y \, \Delta p$ eV. If the reaction is near equilibrium, which is the case for the major driving reactions, then the energy available from the driving reaction is approximately equal to the energy available from the proton gradient and $\Delta G_{driving\ reaction} = yF \, \Delta p$. This relationship allows one to calculate the $\Delta p$ if $y$ and $\Delta G_{driving\ reaction}$ are known. The most common classes of driving reactions will be discussed first. They are oxidation–reduction reactions that occur during respiration and photosynthetic electron transport, and ATP hydrolysis. In all cases we will write the $\Delta G$ for the driving reaction. Then we will equate the $\Delta G_{driving\ reaction}$ with $yF \, \Delta p$. Finally, we will solve for $\Delta p$.

### 3.7.1 Oxidation–reduction reactions as driving reactions

*Electrode potentials and energy changes during oxidation–reduction reactions*

The tendency of a molecule (A) to accept an electron from another molecule is given by its electrode potential $E$, also called the reduction potential. Under standard conditions (1 M for all solutes and gas at 1 atmosphere), the electrode potential has the symbol $E_0$ and is related to the actual electrode potential $E_h$, as follows: $E_h = E_0 + RT/nF \, \ln[\text{ox}]/[\text{red}]$. Usually standard potentials at pH 7 are quoted, and the symbol is $E'_0$. You can see that it is important to know the concentrations of the oxidized and the reduced forms in order to know the actual electrode potential. Molecules with more negative $E_h$ values are better donors (i.e., electrons spontaneously flow to molecules with more positive $E_h$ values). It is important to understand that when an electron moves over a $\Delta E_h$ (which is $E_{h,acceptor} - E_{h,donor}$) to a higher electrode potential, the electron is actually moving toward a lower potential energy, and as a consequence, energy is released. The energy released is proportional to the potential difference ($\Delta E_h$ volts) over which the electron travels. When $n$ electrons move over a potential difference of $\Delta E_h$ volts, the energy released is equal to $n \, \Delta E_h$ eV. (Recall that potential differences are given in volts but work units are in electron volts.) Since the total charge carried by one mole of electrons is $F$ coulombs, the *work done per n moles* of electrons is $\Delta G = -nF \, \Delta E_h$ J, where $\Delta G$ is the Gibbs free energy. The negative sign is placed in the equation to show that energy is released (work is done) when the $\Delta E_h$ is positive. As described next, the movement of electrons during an oxidation–reduction reaction toward a lower potential energy level (higher electrode potential) can be coupled to the extrusion of protons toward a higher potential energy level (higher $\Delta p$), that is, to the generation of a $\Delta p$.

*Oxidation–reduction reactions can generate a $\Delta p$*

An important method of generating a $\Delta p$ is to couple proton translocation to oxidation–reduction reactions that occur during electron transport. The details of electron transport are discussed in Chapter 4, but here we will simply compute the $\Delta p$ that can be created. Energy released from oxidation–

reductions can generate a $\Delta p$ because the oxidation–reduction reactions are coupled to proton translocation. The mechanism of coupling is described in Section 4.5, but here we describe only the energetic relationship between the $\Delta E_h$ and the $\Delta p$. Let us do a simple calculation to illustrate how to estimate the size of the $\Delta p$ that can be generated by an oxidation–reduction reaction that is coupled to proton translocation. Consider what happens in mitochondria and in many bacteria. A common oxidation–reduction reaction in the respiratory chain is the oxidation of reduced ubiquinone, $UQH_2$, by cytochrome $c_1$(cyt $c_1$) in an enzyme complex called the *bc$_1$ complex*. The oxidation–reduction is coupled to the translocation of protons across the membrane, hence the creation of a $\Delta p$. As stated earlier, the total energy change during an oxidation–reduction reaction is $\Delta G = -nF \Delta E_h$ J, where $n$ equals the number of moles of electrons. The total energy change during proton translocation is $yF \Delta p$ J, where $y$ equals the number of moles of translocated protons. [One can convert to electrical potential (volts) and describe these reactions in terms of forces by dividing by the faraday.] This is summarized below:

$$\Delta G(J)$$

$$UQH_2 + 2cyt\ c_1(ox) \longrightarrow UQ \qquad (1)$$
$$+\ 2cyt\ c_1(red) + 2H^+ \qquad -nF\Delta E_h$$
$$yH^+_{in} \longrightarrow yH^+_{out} \qquad yF\Delta p \qquad (2)$$

Reactions (1) and (2) are coupled (i.e., one cannot proceed without the other). What is the size of the $\Delta p$ that can be generated? These reactions are close to equilibrium and can proceed in either direction. Therefore, one can write that the total force available from the redox reaction is equal to the total force of the proton potential:

$$\Delta G/F = -n\ \Delta E_h = y\ \Delta p \qquad (3.19)$$

Equation 3.19 summarizes an important relationship between the $\Delta E_h$ of an oxidation–reduction reaction during respiration and the $\Delta p$ that can be generated at a coupling site. We will return to this point, and to this equation, in Chapter 4 when coupling sites are discussed in more detail.

Now we solve for $\Delta p$. Four protons ($y = 4$) are translocated for every two electrons ($n = 2$) that travel from reduced quinone to cytochrome $c_1$. The $\Delta E_h$ between quinone and cytochrome $c_1$ is approximately $+0.2$ V (200 mV). Substituting this value for $\Delta E_h$, and using $n = 2$ and $y = 4$, we obtain a $\Delta p$ of $-0.1$ V ($-100$ mV). That is to say, when two electrons travel down a $\Delta E_h$ of 200 mV and four protons are translocated, $-100$ mV are stored in the $\Delta p$. Another way of looking at this is that when two electrons travel down a redox gradient of 0.2 V, 0.4 eV are available for doing work (2 × 0.2 V). The 0.4 eV are used to raise each of four protons to a $\Delta p$ of 0.1 V. Oxidation–reduction reactions in the respiratory chain that are coupled to proton translocation are called coupling sites, of which the bc$_1$ complex is an example. Coupling sites are discussed in more detail in Section 4.5.

Coupling of redox reactions during electron transport to proton translocation is the main process by which a $\Delta p$ is created in respiring bacteria, in phototrophic prokaryotes, in chloroplasts, and in mitochondria.

### Reversed electron transport

The fact that some of the redox reactions in the respiratory pathway are in equilibrium with the $\Delta p$ has important physiological consequences. One of these is that it can be expected that the $\Delta p$ can drive electron transport in reverse. That is to say, protons driven into the cell by the $\Delta p$ at coupling sites can drive electrons to the more negative electrode potential (e.g., the reversal of reactions 2 and 1 for the oxidation of $UQH_2$ by cyt $c_1$ coupled to the extrusion of protons). In fact, one test for the functioning of reversed electron transport is its inhibition by ionophores that collapse the $\Delta p$ (Section 3.4). Reversed electron transport commonly occurs in bacteria that use inorganic compounds such as ammonia, nitrite, sulfur, and so on, as a source of electrons to reduce $NAD^+$ for biosynthesis of cell material (chemolithotrophs), because these electron donors are at a potential higher than $NAD^+$. (See Table 4.1 for a list of electrode potentials of biological molecules.) The chemolithotrophs are discussed in Chapter 12.

## Respiration coupled to sodium ion efflux

Although respiratory chains coupled to proton translocation appears to be the rule in most bacteria, a respiration-linked $Na^+$ pump (a $Na^+$-dependent NADH–quinone reductase) has been reported in several halophilic marine bacteria that require high concentrations of $Na^+$ (0.5 M) for optimal growth.[28, 29] The situation has been well studied with *Vibrio alginolyticus*, an alkalotolerant marine bacterium that uses a $\Delta\tilde{\mu}_{Na^+}$ for solute transport, flagella rotation, and ATP synthesis (at alkaline pH). *V. alginolyticus* creates the $\Delta\tilde{\mu}_{Na^+}$ in two ways. At pH 6.5, a respiration-driven $H^+$ pump generates a $\Delta\tilde{\mu}_{H^+}$, which drives a $Na^+$–$H^+$ antiporter that creates the $\Delta\tilde{\mu}_{Na^+}$. The antiporter creates the sodium ion gradient by coupling the influx of protons (down the proton electrochemical gradient) with the efflux of sodium ions. However, at pH 8.5, the $\Delta\tilde{\mu}_{Na^+}$ is created directly by a respiration-driven $Na^+$ pump, that is, the $Na^+$-dependent NADH–quinone reductase. In agreement with this conclusion, the generation of the membrane potential at alkaline pH (pH 8.5) is resistant to the proton ionophore, *m*-carbonylcyanide phenylhydrazone (CCCP), which short-circuits the proton current (Section 3.4), but is sensitive to CCCP at pH 6.5.[30] It has been suggested that switching to a $Na^+$-dependent respiration at alkaline pH may be energetically economical.[25] The reasoning is that when the external pH is more alkaline than the cytoplasmic pH, the only part of the $\Delta p$ that contributes energy to the antiporter is the $\Delta\Psi$, and therefore the antiporter must be electrogenic. That is to say, the $[H^+]/[Na^+]$ must be greater than one. If one assumes that the antiporter is electroneutral when the external pH is acidic, then the continued use of the antiporter at alkaline pH would necessitate increased pumping of protons out of the cell by the primary proton-linked respiration pumps. Rather than do this, the cells simply switch to a $Na^+$-dependent respiration pump to generate the $\Delta\tilde{\mu}_{Na^+}$. This argument assumes that the ratios

$[Na^+]/[e^-]$ and $[H^+]/[e^-]$ are identical so that the energy economies of the respiration-linked cation pumps are the same. (See Note 31 for additional information regarding proton and sodium ion pumping in *V. alginolyticus*.) Several other bacteria have recently been found to generate sodium ion potentials by a primary process. For example, primary $Na^+$ pumping is also catalyzed by sodium-ion translocating decarboxylases in certain anaerobic bacteria described in Section 3.8.1. It must be pointed out, however, that the $\Delta\tilde{\mu}_{Na^+}$, although relied upon for solute transport and motility in many other $Na^+$-dependent bacteria, is usually created by $Na^+/H^+$ antiport rather than by a primary $Na^+$ pump. For example, the marine sulfate reducer *Desulfovibrio salexigens*, which uses a $\Delta\tilde{\mu}_{Na^+}$ for sulfate accumulation, generates the $\Delta\tilde{\mu}_{Na^+}$ by electrogenic $Na^+/H^+$ antiport driven by the $\Delta\tilde{\mu}_{H^+}$, which is created by a respiration-linked proton pump.[32] Similarly, nonmarine aerobic alkaliphiles belonging to the genus *Bacillus*, which rely on the $\Delta\tilde{\mu}_{Na^+}$ for most solute transport and for flagella rotation, create the $\Delta\tilde{\mu}_{Na^+}$ using a $Na^+/H^+$ antiporter driven by a $\Delta\tilde{\mu}_{H^+}$ that is created by a respiration-linked proton pump (Section 3.10). Furthermore, in *D. salexigens* as well as the alkalophilic *Bacillus* species, the membrane-bound ATP synthase is $H^+$-linked rather than $Na^+$-linked.

## 3.7.2 ATP hydrolysis as a driving reaction for creating a $\Delta p$

Electron transport reactions are the major energy source for creating a $\Delta p$ in respiring organisms, but not in fermenting bacteria. (However, see Note 33.) A major energy source for the creation of the $\Delta p$ in fermenting bacteria is ATP hydrolysis catalyzed by the membrane-bound proton-translocating ATP synthase, which yields considerable energy (Fig. 3.2, 7). Consider the coupled reactions summarized below:

$$ATP + H_2O \longrightarrow ADP + P_i, \qquad \Delta G_p \quad (1)$$

$$yH^+_{in} \longrightarrow yH^+_{out}, \qquad yF\,\Delta p \quad (2)$$

When referring to the energy of ATP hydrolysis or synthesis using physiological concentrations of ADP, $P_i$, and ATP, the term $\Delta G_p$

(phosphorylation potential) is used instead of $\Delta G$.

$$\Delta G_p = \Delta G^{0'}$$
$$+ 2.303RT\log_{10}[ATP]/[ADP][P_i] \quad (3.20)$$

The ATP synthase reaction is close to equilibrium and can operate in either direction. Therefore, the total force available from the proton potential equals the force available from ATP hydrolysis:

$$\Delta G_p/F = y\,\Delta p \quad V \quad\quad (3.21)$$

The value of $\Delta G_p$ is about $-50,000$ J (518 mV), and the consensus value for $y$ is 3. Under these circumstances, the hydrolysis of one ATP would generate a maximum $\Delta p$ of $-173$ mV.

## 3.8 Other Mechanisms for Creating a $\Delta\Psi$ or a $\Delta p$

Redox reactions and ATP hydrolysis are the most common driving reactions for creating a proton potential. However, other mechanisms exist for generating proton potentials, and even sodium ion potentials. These driving reactions are not as widespread among the prokaryotes as the others. Nevertheless, they are very important for certain groups of prokaryotes, especially anaerobic bacteria and halophilic archaea. In some instances, they may be the only source of ATP. Some of these driving reactions are considered next.

### 3.8.1 Sodium transport decarboxylases can create a sodium potential

Although chemical reactions directly linked to $Na^+$ translocation (primary transport of $Na^+$) are not widespread among the bacteria, as most primary transport is of the proton, they can be very important for certain bacteria.[34-39] An example of primary $Na^+$ transport is the $Na^+$ pump coupled to respiration in *Vibrio alginolyticus* that was discussed in Section 3.7.1. There is also the decarboxylation of organic acids coupled to sodium ion efflux. This occurs in anaerobic bacteria that generate a sodium gradient by coupling

the decarboxylation of a carboxylic acid to the electrogenic efflux of sodium ions. The decarboxylases include methylmalonyl–CoA decarboxylase from *Veillonella alcalescens* and *Propionigenium modestum*, glutaconyl–SCoA decarboxylase from *Acidaminococcus fermentans*, and oxaloacetate decarboxylase from *Klebsiella pneumonia* and *Salmonella typhimurium*. A description of these bacteria and their metabolism will explain the importance of the sodium-translocating decarboxylases as a source of energy.

*Propionigenium modestum* is an anaerobe isolated from marine and freshwater mud, and from human saliva.[40] It grows only on carboxylic acids (i.e., succinate, fumarate, L-aspartate, L-malate, oxaloacetate, and pyruvate). The carboxylic acids are fermented to propionate and acetate, forming methylmalonyl–CoA as an intermediate. (See the description of the propionic acid fermentation in Section 14.7.1 for an explanation of these reactions.) The methylmalonyl–CoA is decarboxylated to propionyl–CoA and $CO_2$ coupled to the electrogenic translocation of 2 moles of sodium ions per mole of methylmalonyl–CoA decarboxylated.[41]

$$\text{methylmalonyl–CoA} + 2Na^+_{in}$$
$$\longrightarrow \text{propionyl–CoA} + CO_2 + 2Na^+_{out}$$

*P. modestum* differs from most known bacteria in that it has a $Na^+$-dependent ATP synthase rather than a $H^+$-dependent ATP synthase and thus relies on a $Na^+$ current to make ATP. The consequence of this is that, when it grows on succinate, the decarboxylation of methylmalonyl–CoA is its only source of ATP.

Another example is *V. alcalescens*. This is an anaerobic gram-negative bacterium that grows in the alimentary canal and mouth of humans and other animals. Like *P. modestum*, it is unable to ferment carbohydrates, but does ferment the carboxylic acids lactate, malate, and fumarate to propionate, acetate, $H_2$, and $CO_2$. During the fermentation, pyruvate and methylmalonyl–CoA are formed as intermediates, as in *P. modestum*.[42] Some of the pyruvate is oxidized to acetate, $CO_2$, and $H_2$ with the formation of one ATP via substrate-level phosphorylation (Section 7.3). Pyruvate can also be converted to methylmalonyl–CoA that is decarboxylated to propionyl–CoA coupled to the extrusion of sodium ions. The sodium ion gradient that is created might be used as a source of energy for solute uptake via a process called Na$^+$-coupled symport, described in Chapter 16.

*Acidaminococcus fermentans* is an anaerobe that ferments the amino acid glutamate to acetate and butyrate. During the fermentation glutaconyl–CoA is decarboxylated to crotonyl–CoA by a decarboxylase that is coupled to the translocation of sodium ions (eq. 3.21).[43]

$$\text{glutaconyl–CoA} + y\text{Na}^+_{\text{in}}$$

$$\longrightarrow \text{crotonyl–CoA} + CO_2 + y\text{Na}^+_{\text{out}}$$

### The oxaloacetate decarboxylase in Klebsiella pneumoniae

We will examine the Na$^+$-dependent decarboxylation of oxaloacetate by oxaloacetate decarboxylase from the facultative anaerobe *K. pneumoniae* because this has been well studied. A substantial sodium potential develops because the decarboxylation of oxaloacetate is coupled to the electrogenic efflux of sodium ion. (The standard free energy for the decarboxylation reaction is approximately $-29,000$ J.) The enzyme from *K. pneumoniae* translocates two Na$^+$ out of the cell per oxaloacetate decarboxylated according to the following reaction:

$$2\text{Na}^+_{\text{in}} + \text{oxaloacetate}$$

$$\longrightarrow 2\text{Na}^+_{\text{out}} + \text{pyruvate} + CO_2$$

#### 1. Evidence that oxaloacetate decarboxylase is a sodium pump

Inverted membrane vesicles (inside out) were prepared from *Klebsiella* and incubated with $^{22}$Na$^+$ and oxaloacetate. (See Note 44 for a description of how to prepare the vesicles.) As seen in Fig. 3.14, oxaloacetate-dependent sodium ion influx took place. The uptake of sodium ion was prevented by avidin, an inhibitor of the oxaloacetate decarboxylase.

#### 2. The structure of the oxaloacetate decarboxylase

Oxaloacetate decarboxylase consists of two parts, a peripheral catalytic portion attached

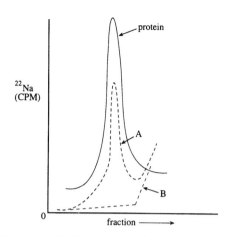

Fig. 3.14 Sodium ion influx into vesicles driven by oxaloacetate decarboxylation. Inside-out vesicles were prepared from *Klebsiella pneumoniae* and incubated with $^{22}$Na$^+$ in the presence (A) and absence (B) of oxaloacetate. After 1 minute of incubation the vesicles were isolated by Sephadex chromatography, which separates the vesicles from the $^{22}$Na$^+$ in the medium. The fractions from the Sephadex columns that contained vesicles were detected by absorbance at 280 nm (protein). Only when oxaloacetate was present did the vesicle fraction contain $^{22}$Na$^+$ (A). This demonstrates the dependency of sodium uptake upon oxaloacetate. In separate experiments it was demonstrated that (1) oxaloacetate decarboxylase was present in the vesicles, (2) the oxaloacetate was decarboxylated, and (3) inhibition of the oxaloacetate decarboxylase by avidin prevented uptake of $^{22}$Na$^+$. Adapted from Dimroth, P. 1980. A new sodium-transport system energized by the decarboxylation of oxaloacetate. *FEBS Lett.* 122:234–236.

to the inner surface of the membrane and an integral membrane portion that serves as a Na$^+$ channel.[37] The integral membrane protein also takes part in the decarboxylation step along with the peripheral membrane protein. How the decarboxylase translocates Na$^+$ to the cell surface is not understood.

#### 3. Use of the sodium current

What does *Klebsiella* do with the energy conserved in the sodium potential? It uses the energy actively to transport the growth substrate, citrate, into the cell as well as drive the reduction of NAD$^+$ by ubiquinol, that is, via a Na$^+$-dependent NADH:ubiquinone oxidoreductase. The latter is an example of reversed electron transport driven by the influx of Na$^+$. This all occurs during the fermentation of citrate and is depicted in Fig. 3.15.

### 3.8.2 Oxalate:formate exchange can create a $\Delta p$

*Oxalobacter formigenes*, an anaerobic bacterium that is part of the normal flora in mammalian intestines, uses dietary oxalic acid as its sole source of energy for growth. The organism has evolved a method for creating a proton potential at the expense of the free energy released from the decarboxylation of oxalic acid

to formic acid and carbon dioxide.[45–47] What is especially interesting is that the enzyme is not in the membrane and therefore cannot act as an ion pump, yet a $\Delta\Psi$ is created. The reaction catalyzed by oxalate decarboxylase is:

$$^-OOC–COO^- + H^+ \longrightarrow CO_2 + HCOO^-$$
$$\text{oxalate} \qquad\qquad\qquad \text{formate}$$

For every mole of oxalate that enters the cell one mole of formate leaves (Fig. 3.16). Since a dicarboxylic acid crosses the cell membrane in exchange for a monocarboxylic acid, there is net movement of negative charge toward the inside, thus creating a $\Delta\Psi$, inside

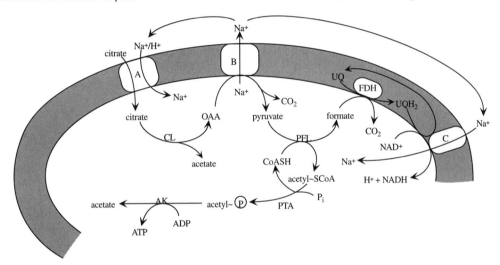

**Fig. 3.15** Citrate fermentation and proposed Na$^+$ ion currents in *Klebsiella pneumoniae*. (A) Na$^+$-citrate symporter. Citrate enters the cell via symport with Na$^+$. Uptake of citrate is not dependent upon the $\Delta\Psi$ and is therefore electroneutral and dependent upon the $\Delta p$Na$^+$. (It has been suggested that only two of the carboxyl groups of citrate are neutralized with sodium ions and the third is protonated.) The citrate is then cleaved by citrate lyase (CL) to oxaloacetate and acetate. (B) Na$^+$-pumping oxaloacetate decarboxylase. Na$^+$ is pumped out of the cell during the decarboxylation of oxaloacetate to pyruvate and CO$_2$, creating a sodium ion electrochemical gradient that is used for citrate uptake (Na$^+$-citrate symporter) and NADH production (Na$^+$-pumping NADH:ubiquinone oxidoreductase). The pyruvate is used to make ATP via substrate-level phosphorylation. First the pyruvate is cleaved by pyruvate–formate lyase (PFL) to acetyl–CoA and formate. Then the acetyl–CoA is converted to acetyl phosphate by phosphotransacetylase (PTA). Finally, the acetyl phosphate donates the phosphoryl group to ADP to form ATP and acetate in a reaction catalyzed by acetate kinase (AK). NADH is produced as follows: Some of the formate is oxidized to CO$_2$ by a membrane-bound formate dehydrogenase (FDH). The electron acceptor is ubiquinone. Ubiquinol is reoxidized by NAD$^+$ to form NADH + H$^+$ in a reaction catalyzed by a Na$^+$-pumping NADH:ubiquinone oxidoreductase. This is a case of reversed electron transport driven by the electrochemical Na$^+$ gradient. There is therefore a Na$^+$ circuit. Sodium ion is pumped out of the cell via the decarboxylase and enters via the citrate transporter and the NADH:ubiquinone oxidoreductase. (A) Na$^+$-citrate symporter. (B) Na$^+$-pumping oxaloacetate decarboxylase. (C) Na$^+$-pumping NADH–ubiquinone oxidoreductase. *Source:* Adapted from Dimroth, P. 1997. Primary sodium ion translocating enzymes. *Biochim. Biophys. Acta* 1318:11–51.

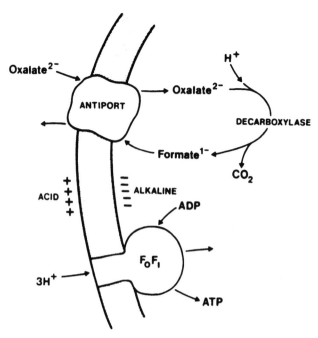

**Fig. 3.16** The electrogenic oxalate:formate exchange and the synthesis of ATP in *Oxalobacter formigenes* using a proton current. Oxalate$^{2-}$ enters via an oxalate$^{2-}$/formate$^{1-}$ antiporter. The oxalate$^{2-}$ is decarboxylated to formate$^{1-}$ and $CO_2$ while a cytoplasmic proton is consumed. (The oxalate is first converted to oxalyl–coenzyme A, which is decarboxylated to formyl–coenzyme A. The formyl–coenzyme A transfers the coenzyme A to incoming oxalate, thus forming formate and oxalyl–coenzyme A.) An electrogenic exchange of oxalate$^{2-}$ for formate$^{1-}$ creates a membrane potential, outside positive. It is suggested that the stoichiometry of the ATP synthase is 3H$^+$/ATP. Since the decarboxylation of one mole of oxalate results in the consumption of one mole of protons, the incoming current of protons during the synthesis of one mole of ATP is balanced by the decarboxylation of three moles of oxalate. In other words, a steady-state current of protons requires that $\frac{1}{3}$ of a mole of ATP is made per mole of oxalate decarboxylated. *Source*: From, Anantharam, V., M. J. Allison, and P. C. Maloney. 1989. Oxalate:formate exchange. *J. Biol. Chem.* **264**:7244–725. (Also see Refs. 46 and 47.)

negative. Also, a proton is consumed in the cytoplasm during the decarboxylation, which can contribute to a ΔpH, inside alkaline. Recently the antiporter has been purified and shown to be a 38 kD hydrophobic polypeptide that catalyzes the exchange of oxalate and formate in reconstituted proteoliposomes.[48] (For a discussion of proteoliposomes, see Note 49 and Section 16.1.)

### ATP synthesis

Assuming that the stoichiometry of the ATP synthase reaction is 3H$^+$/ATP and for oxalate decarboxylation it is 1H$^+$/oxalate, then a steady proton current requires the decarboxylation of 3 moles of oxalate per

mole of ATP synthesized. In other words, $\frac{1}{3}$ mole of ATP can be synthesized per mole of oxalate decarboxylated. This is apparently the only means of ATP synthesis in *Oxalobacter formigenes*.

### The decarboxylation of other acids may also create a Δp

In principle the influx and decarboxylation of any dicarboxylic acid coupled to the efflux of the monocarboxylic acid can create a proton potential (Fig. 3.17). For example, a decarboxylation and electrogenic antiport was reported for the decarboxylation of malate to lactate acid during the malolactate fermentation in *Lactobacillus lactis*.[50]

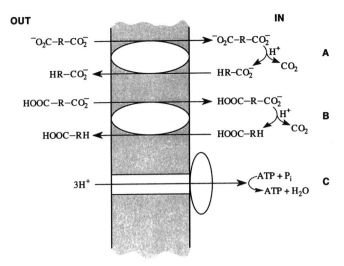

**Fig. 3.17** How the decarboxylation of a dicarboxylic acid can be coupled to the development of a proton potential. (A, B) A dicarboxylic acid enters as a divalent anion (A) or a monovalent anion (B). In the cytoplasm a decarboxylase cleaves off $CO_2$ consuming a proton in the process and producing a monocarboxylic acid with one fewer negative charge than the original dicarboxylic acid. Exchange of the dicarboxylic acid and the monocarboxylic acid via the antiporter is electrogenic and produces a membrane potential, inside negative. The consumption of the proton during the decarboxylation creates a $\Delta pH$. (C) Influx of protons via the ATP synthase completes the proton circuit and results in ATP synthesis. The energetics can be understood in terms of the decarboxylase removing the dicarboxylic acid from the inside, thus maintaining a concentration gradient that stimulates influx.

### 3.8.3 End-product efflux as the driving reaction

Theoretically, it is possible to couple the excretion of fermentation end products down a concentration gradient to the translocation of protons out of the cell, thereby creating a $\Delta p$. This is the reverse of solute uptake by proton-coupled transport systems and is called the "energy recycling model."[51, 52] In other words, the direction of solute transport depends upon which is greater, the energy from the proton potential, $\Delta p$, which drives solute uptake, creating an electrochemical solute gradient, $\Delta \tilde{\mu}/F$, or the electrochemical solute gradient, which drives solute efflux, creating a $\Delta p$. Solute efflux coupled to proton translocation can spare ATP, because in fermenting bacteria the hydrolysis of ATP (catalyzed by the ATP synthase) is used to pump protons to the outside to create the $\Delta p$. As the $\Delta p$ rises, ATP hydrolysis is diminished.

### Energetics

Consider solute/proton symport as reversible. Under these circumstances, the $\Delta p$ can drive the uptake of a solute against a concentration gradient, or the efflux of a solute down a concentration gradient can create a $\Delta p$. The driving force for solute transport in symport with protons is the sum of the proton potential and the electrochemical potential of the solute; that is,

$$\text{driving force (mV)} = y \, \Delta p + \Delta \tilde{\mu}_s/F \quad (3.22)$$

where $y$ is the number of protons in symport with the solute, S. The term $\Delta \tilde{\mu}_s/F$ represents the electrochemical potential of S; that is,

$$\Delta \tilde{\mu}_s/F = m \Delta \Psi + 60 \, \log_{10}([S]_{in}/[S]_{out}) \text{ mV} \quad (3.23)$$

where $m$ is the charge of the solute. Therefore, the overall driving force (substituting for $\Delta \tilde{\mu}_s/F$ and $\Delta p$) is driving force (mV)

$$= 60 \log_{10}([S]_{in}/[S]_{out})$$
$$+ (y + m) \Delta\Psi - y60 \Delta pH \quad (3.24)$$

Substituting $m = -1$ for lactate, eq. 3.24 becomes

driving force

$$= (y - 1) \Delta\Psi - y60 \Delta pH$$
$$+ 60 \log_{10}([S]_{in}/[S]_{out}) \quad (3.25)$$

During growth lactate transport is near equilbrium, and the net driving force is close to zero. Thus by setting eq. 3.25 equal to zero, one can solve for $y$:

$$y = \{(\Delta\Psi - 60 \log_{10}([S_{in}]/[S]_{out})\}/\Delta p \quad (3.26)$$

Note that eq. 3.25 states that when $y$, which is the number of moles of $H^+$ translocated per mole of lactate, is one, then the translocation is electroneutral and a $\Delta\Psi$ does not develop, only a small $\Delta pH$ develops due to the acidification of the external medium by the lactic acid. We will return to these points later when we discuss the physiological significance of energy conservation via end-product efflux. Let us now consider some data that support the hypothesis that lactate/$H^+$ symport and succinate/$Na^+$ symport can create a membrane potential in membrane vesicles derived from cells.

## Symport of protons and sodium ions with fermentation end products

Coupled translocation of protons and lactate can be demonstrated in membrane vesicles prepared from lactic acid bacteria belonging to the genus *Streptococcus*.[53-56] Similarly, coupled translocation of sodium ions and succinate is catalyzed by vesicles prepared from a rumen bacterium belonging to the genus *Selenemonas*.[57] In both cases a transporter exists in the membrane that simultaneously translocates the organic acid ($R^-$) with either protons or sodium ions out of the cell (symport) (Fig. 3.18). (See Note 58.)

## Lactate efflux

L-Lactate-loaded membrane vesicles were prepared from *Streptococcus cremoris* in the following way: A concentrated suspension of

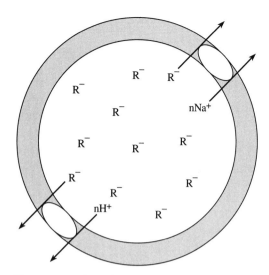

**Fig. 3.18** End-product efflux in symport with protons or sodium ions. The high intracellular concentrations of $R^-$ may drive the efflux of $Na^+$ or $H^+$ via symporters. If the ratio of protons or sodium ions to carboxyl, i.e., $n$/carboxyl, is > 1, then the symport is electrogenic and a membrane potential develops.

cells was treated with lysozyme in buffer to degrade the cell walls. The resulting cell suspension was gently lysed (broken) by adding potassium sulfate. The cell membranes spontaneously resealed into empty vesicles. The vesicles were purified and then incubated for 1 h with 50 mM L-lactate, which equilibrated across the membrane, thus loading the vesicles with L-lactate. Then the membrane vesicles loaded with L-lactate were incubated with the lipophilic cation, tetraphenylphosphonium ($Ph_4P^+$), in solutions containing: buffer, Fig. 3.19, curve 1; buffer +50 mM L-lactate, curve 2. Samples were filtered to separate the vesicles from the external medium, and the amount of $Ph_4P^+$ that accumulated was measured. The $Ph_4P^+$ accumulated inside the cells when the $[lactate]_{in}/[lactate]_{out}$ was high.

The accumulation of $Ph_4P^+$ implies that the efflux of lactate along its concentration gradient imposed a membrane potential on the vesicles.

The membrane potential was calculated using the Nernst equation and the $Ph_4P^+$ accumulation ratio, as explained in Section 3.5.1.

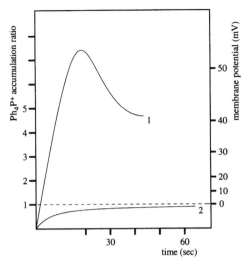

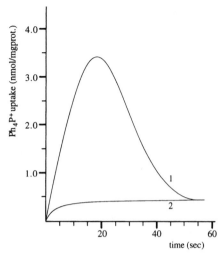

**Fig. 3.19** Lactate efflux can produce a membrane potential. Lactate-loaded membrane vesicles from *Streptococcus cremoris* were incubated with the lipophilic cation $Ph_4P^+$ without lactate (curve 1) and with 50 mM lactate in the external medium (curve 2). In the absence of external lactate the lipophilic probe accumulated inside the vesicles, suggesting that a membrane potential developed as lactate left the cells. The addition of lactate to the external medium prevented the formation of the membrane potential, because the lactate concentration (IN/OUT) was lowered. Additional experiments showed that the electrogenic ion was the proton. *Source*: Adapted from Otto, R., R. G. Lageveen, H. Veldkamp, and W. N. Konings. 1982. Lactate efflux-induced electrical potential in membrane vesicles of *Streptococcus cremoris*. J. Bacteriol. 149:733–738.

**Fig. 3.20** Succinate efflux can produce a membrane potential. Succinate-loaded membrane vesicles from *Selenomonas ruminantium* were incubated with the lipophilic cation $Ph_4P^+$ without succinate (curve 1) or with succinate (curve 2) in the external medium. The uptake of $Ph_4P^+$ indicates that a membrane potential developed when the $(succinate)_{in}/(succinate)_{out}$ was high. *Source*: Adapted from Michel, T. A., and J. M. Macy. 1990. Generation of a membrane potential by sodium-dependent succinate efflux in *Selenomonas ruminantium*. J. Bacteriol. 172:1430–1435.

## Succinate efflux

A similar experiment was done with vesicles from the rumen bacterium *Selenomonas ruminantium* loaded with succinate (Fig. 3.20).

The $Ph_4P^+$ was taken up by the vesicles when there was a concentration gradient of succinate, indicating that a membrane potential, inside negative, developed during succinate efflux (Fig. 3.20, curve 1).

When the external buffer contained a high concentration of succinate, the membrane potential did not develop, indicating that the energy to establish the potential was derived from the efflux of succinate along its concentration gradient, $[succ]_{in}/[succ]_{out}$ (Fig. 3.20, curve 2). The molar growth yields of

*S. ruminantium* are higher when succinate production is at a maximum, implying that more ATP is available for biosynthesis as a result of the succinate efflux.

## Physiological significace of end-product efflux as a source of cellular energy

In order to demonstrate the generation of a membrane potential due to end-product efflux, cells or membrane vesicles must be "loaded" with high concentrations of the end product by incubating de-energized cells or membrane vesicles for several hours before dilution into end-product-free media. These are not physiological conditions, and a question has been raised as to the conditions under which the electrogenic extrusion of lactate coupled to protons is physiologically relevant. Assuming a given value of *y*, one can use eq. 3.26 to calculate the necessary lactate concentration gradient given experimentally determined values of $\Delta\Psi$ and $\Delta pH$. When one substitutes $y = 2$ (for electrogenic lactate extrusion), a major problem appears. Substituting a measured external lactate concentration of 30 mM, a $\Delta\Psi$ of $-100$ mV, a $-60$ $\Delta pH$ of $-30$ mV (at pH 7.0), and $y = 2$, the calculated internal lactate concentration would have to be 14 M in order for lactate secretion to occur.[59] This is a nonphysiological concentration. (Internal lactate concentrations reach concentrations of about 0.2 M.) In fact, the values of $y$ were calculated by Brink et al. using experimentally measured values of lactate concentrations, membrane potential, and $\Delta pH$ (see Ref. 55). Only when the external lactate concentrations were very low was lactate/proton efflux electrogenic (i.e., $y > 1$). For example, when the cells were grown at pH 6.34 and the lactate accumulated in the medium over time, the value of $y$ decreased from about 1.44 to 0.9 when the external lactate concentrations increased from 8 to 38 mM. This means that only under certain growth conditions (low external lactate and high external pH so that the $\Delta pH$ is 0 or inverted) would one expect lactate efflux effectively to generate a $\Delta p$. This may occur in the natural habitat where growth of the producer may be stimulated by a population of bacteria that utilize lactate, thus keeping the external concentrations low.

### 3.8.4 Light absorbed by bacteriorhodopsin can drive the creation of a $\Delta p$

Certain archaea (i.e., the extremely halophilic archaea) have evolved a way to produce a $\Delta p$ using light energy directly (i.e., without the intervention of oxidation–reduction reactions, and without chlorophyll).[60–62] (See Note 63 for a more complete discussion of the extreme halophiles.) (It has been reported that under the proper conditions they can be grown photoheterotrophically, i.e., on organic carbon with light as a source of energy, but these conclusions have been questioned.[64, 65])

Halophilic archaea are heterotrophic organisms that carry out an ordinary aerobic respiration creating a $\Delta p$ driven by oxidation–reduction reactions during electron transport.

(See Note 66.) The $\Delta p$ is used to drive ATP synthesis via a membrane ATP synthase. (See Sections 3.7.1 and 3.6.2.) However, conditions for respiration are not always optimal, and, in the presence of light and low oxygen levels, the halophiles adapt by making photopigments (rhodopsins), one of which (bacteriorhodopsin) functions as a proton pump that is energized directly by light energy. (See Note 67.) Whereas photosynthetic electron flow is an example of an indirect transformation of light energy into an electrochemical potential (via redox reactions), bacteriorhodopsin illustrates the direct transformation of light energy into an electrochemical potential. We will first consider data that demonstrate the light-dependent pumping of protons. This will be followed by a description of the proton pump, the photocycle, and a model for the mechanism of pumping protons. It should be added that bacteriorhodopsin is the best-characterized ion pump, and, for this reason, it will be examined in detail.

### Evidence that halophiles can use light energy to drive a proton pump

As shown in Fig. 3.21, a suspension of *Halobacterium halobium* was subjected to light of different intensities and for different periods of time, and the pH of the external medium was monitored. As illustrated in Fig. 3.21A, the light produced an efflux of protons from the cell. Higher light intensities produced a greater rate of proton efflux. (Compare $I_c$ to $I_a$.) In Fig. 3.21B, the rate of proton efflux (ng/sec) as determined from the slopes in Fig. 3.21A is plotted as a function of the light intensity (neinstein/sec). (An einstein is equal to a "mole" of photons.) From data such as these, a quantum yield (i.e., protons ejected per photon absorbed) was calculated to be 0.52 proton/photon absorbed.[68] The reported values for the *maximum* quantum yield for proton efflux is a little higher (i.e., 0.6–0.7 proton/photon absorbed). The quantum yield for the photocycle (i.e., the fraction of bacteriorhodopsin molecules absorbing light that undergoes the photocycle described below) is $0.64\pm0.04$.[69, 70] These values suggest that one proton is pumped per photocycle.

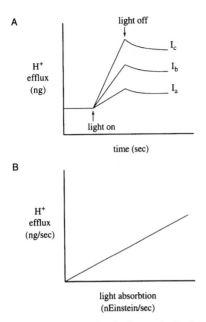

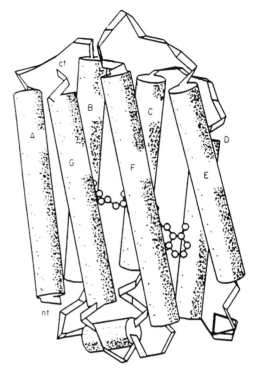

Fig. 3.21 Proton pumping by bacteriorhodopsin. Expected results if *Halobacterium* cells were illuminated by light. In order to measure proton outflow accurately, proton inflow through the ATP synthase must be blocked either with uncouplers or nigericin, which collapse the proton electrochemical potential, or an ATP synthase inhibitor such as DCCD. The extruded protons can be quantified with a pH meter. (A) Proton efflux measured as a function of time at increasing light intensities (neinsteins/sec), $I_a < I_b < I_c$. The slope of each line is the rate of proton efflux. (B) The rate of proton efflux is plotted as a function of the light intensity. From these data one can calculate the quantum yield, i.e., the number of protons extruded per photon absorbed. *Source*: Adapted from data by Bogomolni, R. A., R. A. Baker, R. H. Lozier, and W. Stoeckenius. 1980. Action spectrum and quantum efficiency for proton pumping in *Halobacterium halobium*. *Biochemistry* 19: 2152–2159. Abbreviation: ng, nanograms.

Fig. 3.22 Diagram of bacteriorhodopsin. The seven helices are shown as solid rods. The helices form a central channel where the retinal is attached to a lysine residue on helix G. *Source*: From Henderson, R., J. M. Baldwin, T. A. Ceska, F. Zemlin, E. Beckmann, and K. H. Downing. 1990. Model for the structure of bacteriorhodopsin based on high-resolution electron cryo-microscopy. *J. Mol. Biol.* 213:899–929.

the bacteriorhodopsin, is a pigment called retinal (a $C_{20}$-carotenoid), which is attached via a Schiff base to a lysine residue on the protein. (See Note 72 for a description of a Schiff base.) When the retinal absorbs light, the bacteriorhodopsin remarkably translocates protons out of the cell, and a $\Delta p$ is created.NP

## Bacteriorhodopsin is the proton pump

Built into the cell membrane of the halophilic archaebacteria is a pigment protein called bacteriorhodopsin, which is a pump responsible for the light-driven electrogenic efflux of protons. It consists of one large polypeptide (248 amino acids, 26,486 D) folded into seven a helices that form a transmembrane channel (Fig. 3.22). (See Ref. 71 for a review.) Located in the middle of the channel, and attached to

## The photocycle and a model for proton pumping

The photoevents occurring in halophiles can be followed spectroscopically, because when bacteriorhodopsin absorbs light, it loses its absorption peak at 568 nm (bleaches) while it is converted in the dark to a series of pigments that have absorption peaks at different wavelengths. This is called the photocycle. Also, site-specific mutagenesis of bacteriorhodopsin is being used to identify the amino acid side chains that transfer protons across the membrane.[73–76] The photocycle is shown in Fig. 3.23. The retinal is attached via a Schiff base to the epsilon amino group of lysine-216 ($K_{216}$). Before absorbing light, the retinal is protonated at the Schiff base and exists in the all-trans (13-trans) configuration ($bR_{568}$). The subscript refers to the absorption maximum. Upon absorbing a photon of light, the retinal isomerizes to the 13-cis form ($K_{625}$). All subsequent steps do not require light and represent the de-energization of the bacteriorhodopsin via a series of intermediates that have different absorbance maxima to the original unex-cited molecule. The transitions are very fast and occur in the nanosecond and millisec-ond range. These intermediates are, in or-der of their appearance, K, L, M, N, and

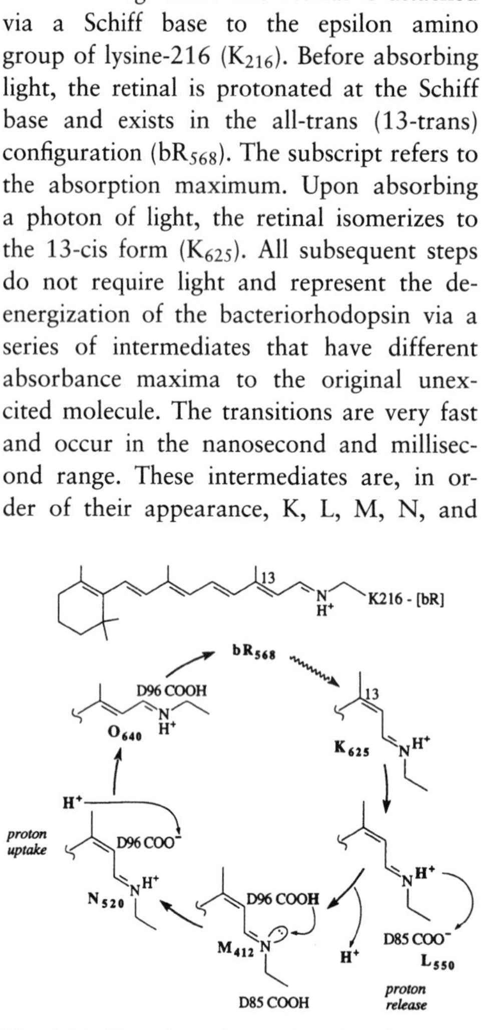

Fig. 3.23 The photochemical cycle of bacteri-orhodopsin. Upon absorption of a photon of light, $bR_{568}$ undergoes a *trans*-to-*cis* isomeriza-tion and is converted to a series of intermediates with different absorbance maxima. The Schiff base becomes deprotonated during the L to M transition and reprotonated during the M to N transition. A recent photocycle postulates two M states, i.e., L to $M_1$ to $M_2$ to N. Although not indicated, some of the steps are reversible. (See Lanyi, J. K. 1992. Proton transfer and en-ergy coupling in the bacteriorhodpsin photocy-cle. *J. Bioenerg. and Biomemb.* 24:169–179.) *Source*: From Krebs, M. P., and H. Gobind Kho-rana. 1993. Mechanism of light-dependent pro-ton translocation by bacteriorhodopsin. *J. Bac-teriol.* 175:1555–1560.

O. In converting from L to M the Schiff base loses its proton to aspartate-85 but re-gains a proton from aspartate-96 in going from M to N. Aspartate-96 acquires a proton from the cytoplasm in going from N to O. Thus a proton has moved from the cytoplas-mic side through aspartate-96 to the Schiff base to aspartate-85. From aspartate-85, the proton moves to the outside membrane sur-face. (See Ref. 77.) Precisely how the proton travels through the bacteriorhodopsin chan-nel from the cytoplasmic side to the Schiff base in the center of the channel, and from there to the external surface of the mem-brane, is not known. Probably the proton is passed from one amino acid side group that can be reversibly protonated to another. Two of these amino acids have been suggested from site-specific mutagenesis experiments to be aspartate-85 and aspartate-96. The proto-natable residues would extend along the pro-tein from the cytoplasmic surface to the out-side. In the case of bacteriorhodopsin, where the three-dimensional structure is known, the protonatable residues are oriented toward the center of the channel. Any bound wa-ter might also participate in proton transloca-tion. For example, there might be a chain of water and hydrogen-bonded protons con-necting protonatable groups. Since the Schiff base gives up its proton to the extracellular side of the channel (transition L to M) and becomes protonated with a proton from the cytoplasm (transition M to N), it has been assumed that there is a switch that reori-ents the Schiff base so that it alternatively faces the cytoplasmic and extracellular sides of the channel. Logically, the switch would be at M. The mechanism for the switch is unknown, although it has been suggested to be a conformational change in the bacteri-orhodopsin.[78] It is not obviously correlated with the retinal isomerizations because the retinal is in the cis configuration through-out most of the photocycle, including the protonation and deprotonation of the Schiff base. The transfer of protons along proto-natable groups has been suggested for several proton pumps besides bacteriorhodopsin, in-cluding cytochrome oxidase and the proton-translocating ATP synthase.[79]

## 3.9 Halorhodopsin, a Light-Driven Chloride Pump

The halophiles have a second light-driven electrogenic ion pump, but one that does not energize the membrane. The second pump, called halorhodopsin, is structurally similar to bacteriorhodopsin and is used to accumulate $Cl^-$ intracellularly in order to maintain osmotic stability.[60, 74, 80] Recall that the halophiles live in salt water, where the extracellular concentrations of NaCl can be 3–5 M. The osmotic balance is preserved by intracellular concentrations of KCl that match the extracellular $Cl^-$ concentrations. (As discussed in Chapter 15, $K^+$ is important for osmotic homeostasis in eubacteria as well.) Since the membrane is negatively charged on the inside, energy must be used to bring the $Cl^-$ into the cell, and halorhodopsin accomplishes this purpose. In the dark the halophiles use another energy source, probably ATP, to accumulate chloride ions actively.[81]

## 3.10 The $\Delta p$ and ATP Synthesis in Alkaliphiles

Alkaliphilic bacteria grow in habitats where the pH is very basic, usually around pH 10. These habitats include soda lakes, dilute alkaline springs, and desert soils, where the alkalinity is usually due to sodium carbonate. A definition of *obligate* alkaliphiles are organisms that cannot grow at pH values of 8.5 or less, and usually have an optimum around 9. (See Note 82.) These include *Bacillus pasteurii*, *B. firmus*, and *B. alcalophilus*. The $\Delta p$ has been measured in obligate aerobic alkaliphiles that grow optimally at pH 10–12, and appears to be too low to drive the synthesis of ATP. The problem is that because the external pH is so basic, the internal pH is generally at least two pH units more acid. This gives the $\Delta$pH a sign opposite to that found in the other bacteria, that is, negative. A $\Delta$pH of two is equivalent to 60 × 2 or 120 mV. Thus the $\Delta p$ can be lowered by about 120 mV in these organisms. Typical membrane potentials

for aerobic alkaliphiles are approximately −170 mV (positive out). Therefore, the $\Delta p$ values can be as low as −170 + 120 or −50 mV.[83] The ATPase in the alkaliphiles is a proton translocating enzyme as in most bacteria. Even if the ATPase translocated as many as four protons, this would generate only 0.2 eV, far short of the approximately 0.4 to 0.5 eV required to synthesize an ATP under physiological conditions. The energy to synthesize an ATP (i.e., $\Delta G_p$) is 40,000 to 50,000 J. Dividing this number by the faraday gives 0.4–0.5 eV. How can this dilemma be resolved? It is possible that the protons in the bulk extracellular phase may not be as important for the $\Delta p$ as protons on the membrane or a few angstrom units away from the membrane. One suggestion is that proton pumps may be in frequent contact with the ATPases due to random collisons within the membrane, and that as soon as a proton is pumped out of the cell it may re-enter via an adjoining ATPase without entering the pool of bulk protons.[84] This suggestion emphasizes the activity of protons at the face of the membrane rather than the $\Delta$pH, which is due to the concentration of protons in the bulk phase, and raises questions about the details of the proton circuit. An important tenet of the chemiosmotic theory may not apply in this situation, and that is that a delocalized $\Delta p$ is used. That is to say, the chemiosmotic theory postulates that proton currents couple any exergonic reaction with *any* endergonic reaction (i.e., proton circuits are delocalized over the entire membrane). In that way, proton extrusion during respiration can provide the energy for several different reactions, not simply the ATP synthase (Fig. 3.2). However, it may be that in some alkaliphiles there is direct transfer of protons extruded during respiration to the ATP synthase (i.e., *localized* proton circuits).

## 3.11 Summary

The energetics of bacterial cell membranes can be understood for most bacteria in terms of an electrochemical proton potential established by exergonic chemical reactions or light. The protons are raised from a low electrochemical potential on the inside of the cell to a high electrochemical potential on the outside of the cell. When the protons circulate back into the cell through appropriate carriers, work can be done (e.g., the synthesis of ATP via the membrane ATP synthase, solute transport, and flagellar rotation).

The proton potential is due to a combination of a membrane potential ($\Delta\Psi$), outside positive, and a $\Delta$pH, outside acid. Because of the low capacitance of the membrane, not very many protons need be extruded before a large membrane potential develops. The membrane potential seems to be the dominant component in the $\Delta p$ for most bacteria, except for acidophiles, which can have a reversed membrane potential. Other cations, especially sodium ions, can use the established membrane potential for doing work, principally solute accumulation. The sodium ions must be returned to the outside of the cell, and $Na^+/H^+$ antiporters serve this purpose in most bacteria. Thus, although the major ion circuit is a proton circuit, the sodium circuit is also important.

A $\Delta p$ is created when an exergonic chemical reaction is coupled to the electrogenic flow of charge across the cell membrane and the liberation of protons on the outer membrane surface. Energy input of at least $yF \, \Delta p$ joules is necessary in order to raise the electrochemical potential of $y$ moles of protons to $\Delta p$ V. The three most widespread reactions that provide the energy to create the $\Delta p$ are oxidation–reduction reactions during electron transport in membranes (respiration), oxidation–reduction reactions during electron transport stimulated by light absorption (photosynthesis), and ATP hydrolysis via the membrane ATP synthase. Respiration and ATP hydrolysis are reversible, and the $\Delta p$ can drive reversed electron transport as well as the synthesis of ATP. During reversed electron transport protons enter the cells rather than leave the cells. The ATP synthase is an enzyme complex that reversibly hydrolyzes ATP and pumps protons out of the cell. When protons enter via the ATP synthase, ATP is made.

Light energy can also be used directly to create a $\Delta p$ without the establishment of a redox potential. This occurs in the halophilic archaea. They create a $\Delta p$ using a light-driven proton pump called bacteriorhodopsin, which forms a proton channel through the membrane. Bacteriorhodopsin is being studied as a model system to investigate the mechanism of ion pumping across membranes.

Fermenting bacteria have evolved additional ways to generate a $\Delta p$. Lactic acid bacteria can create a $\Delta p$ via coupled efflux of protons and lactate (in addition to ATP hydrolysis) under certain growth conditions. Another anaerobic bacterium, *Oxalobacter*, creates a $\Delta p$ by the oxidation of oxalic acid to formic acid coupled with the electrogenic exchange of oxalate for formate and the consumption of protons during the decarboxylation. Other examples similar to these will no doubt be discovered in the future.

Sodium potentials are also important in the prokaryotes, especially for solute transport. Although most sodium potentials are created secondarily from the proton potential via antiporters, there are some prokaryotes that couple a chemical reaction to the creation of a sodium potential. For example, some marine bacteria, as exemplified by *Vibrio alginolyticus*, couple respiration to the electrogenic translocation of sodium ions out of the cell at alkaline pH. These bacteria can couple the sodium potential to the membrane ATP synthase and therefore rely on a sodium current rather than a proton current when growing in basic solutions. When growing in slightly acidic conditions, they use a proton potential.

Several different fermenting bacteria can create a sodium potential by coupling the decarboxylation of organic acids with the translocation of sodium ions out of the cell, or by coupling the efflux of end products of fermentation with the translocation of sodium ions to the outside. In some bacteria (e.g., *Klebsiella pneumoniae*), the sodium potential is used to drive the influx of the growth substrate into the cell but not for the generation of ATP. ATP is synthesized by a substrate-level phosphorylation during the conversion of pyruvate to formate and acetate. These bacteria may also generate a membrane potential by coupled efflux of the end products of citrate degradation (formate and acetate) with protons.

It should be emphasized that in fermenting bacteria ATP is hydrolyzed via the ATP synthase to create the $\Delta p$ that is necessary for membrane activities (e.g., solute transport and flagellar rotation). A decrease in the $\Delta p$ should result in even more ATP hydrolysis. Thus reactions that create a membrane potential (e.g., electrogenic efflux of sodium ions or protons in symport with end products of fermentation, or during decarboxylation reactions) are expected to conserve ATP.

Measurements of the $\Delta \Psi$ and the $\Delta pH$ are necessarily indirect because of the small size of the bacteria. The $\Delta \Psi$ is measured using cationic or anionic fluorescent dyes that equilibrate across the membrane in response to the potential. The distribution of the dyes is monitored by fluorescence quenching. A second way to measure the membrane potential is by the equilibration of a permeant ion, which achieves electrochemical equilibrium with the membrane potential. The membrane potential is computed using the Nernst equation and the intracellular and extracellular ion concentrations. The $\Delta pH$ is measured using a weak acid or weak base whose log ratio of concentrations inside the cell to outside the cell is a function of the $\Delta pH$.

## Study Questions

1. The $E_0'$ for ubiquinone (ox)/ubiquinone (red) is $+100$ mV and for $NAD^+/NADH$ it is $-320$ mV. What is the $\Delta E_0'$?

   *ans.* 420 mV

2. In the electron transport chain, oxidation–reduction reactions with a $\Delta E_h$ of about 200 mV appear to be coupled to the extrusion of protons. Assume that 100% of the oxidation–reduction energy is converted to the $\Delta p$; for a two-electron transfer and the extrusion of two protons, what is the expected $\Delta p$?

   *ans.* $-200$ mV

3. What is the maximum $\Delta p$ (i.e., 100% energy conversion) when the hydrolysis of one mole of ATP is coupled to the extrusion of four moles of protons? Three moles of protons? Assume the free energy of hydrolysis of ATP is $-50,000$ J.

   *ans.* $-130$ mV, $-173$ mV

4. Design an experimental approach that can show that the efflux of an organic acid along its concentration gradient is coupled to proton translocation and can generate a membrane potential. (You must not only be able to demonstrate the membrane potential but also show that the proton is the conducting charge.)

5. Explain how the decarboxylation of oxalic acid by *Oxalobacter* creates a $\Delta pH$ and a $\Delta \Psi$.

6. What is the reason for stating that light creates a $\Delta p$ indirectly in photosynthesis but directly in the extreme halophiles?

7. Lactate efflux was in symport with protons, whereas succinate efflux was in symport with sodium ions. Which ionophores might you use to distinguish which cations are involved?

8. A reasonable figure for the actual free energy of hydrolysis of ATP inside cells is $-50,000$ J mole$^{-1}$. It is believed that the hydrolysis of ATP is coupled to the extrusion of three protons in many systems. If a $\Delta p$ of $-150$ mV were generated, what would be the efficiency of utilization of ATP energy to create the $\Delta p$?

   *ans.* 87%

9. Assume a reduction potential of $+400$ mV for an oxidant and a potential of $-100$ mV for a reductant. How many joules of energy are released when two moles of electrons flow from the reductant to the oxidant?

   *ans.* 96,500 J

10. Assume that 45 kJ are required to synthesize one mole of ATP. What would be the required $\Delta p$, assuming that three moles of $H^+$ entered via the ATPase per mole of ATP made?

    *ans.* $-155$ mV

11. Assume that the $\Delta p$ is $-225$ mV. If the $\Delta pH$ at 30°C is 1.0, what is the membrane potential?

    *ans.* $-165$ mV

12. How much energy in joules is required to move a mole of uncharged solute into the cell against a concentration gradient of 1000 at 30°C? If transport were driven by the $\Delta p$, what would be the minimal value of the $\Delta p$ required if one $H^+$ were cotransported?

*ans.* 17,370 J, 180 mV

13. Assume membrane vesicles loaded with $K^+$ so that $[K_{in}^+]/[K_{out}^+]$ is 1,000. The temperature is 30°C. Upon adding valinomycin, what is the predicted initial membrane potential in mV?

*ans.* −180 mV

14. What is the rationale for adding valinomycin and $K^+$ to an experimental system where the number of protons being pumped out of the cell is measured?

15. Briefly, what does the chemiosmotic theory state?

16. What is meant by the $\Delta E_h$? What is the relationship between the $\Delta E_h$ and the $\Delta p$ that is generated at coupling sites?

17. Assume that a concentration gradient of an uncharged solute, S, exists across the cell membrane. What is the formula that expresses the driving force in the concentration gradient in mV at 30°C? What is the common expression used for the force due to the concentration gradient when S is a proton?

18. Assume that an ion traverses a charged membrane with a potential of $\Delta \Psi$ volts. Further assume that there is no concentration gradient. What is the expression that denotes the driving force? What will the sign be if the ion moves toward the side of opposite charge?

19. What is the expression, in mV at 30°C, for the $\Delta p$? What is the expression in joules?

## NOTES AND REFERENCES

1. Harold, F. M. 1986. *The Vital Force: A Study of Bioenergetics.* W. H. Freeman and Co., New York.

2. Nichols, D. G., and S. J. Ferguson. 1992. *Bioenergetics 2.* Academic Press, London.

3. Mitchell, P. 1961. Coupling of phosphorylation to electron and hydrogen transfer by a chemiosmotic type of mechanism. *Nature (London)* 191:144–148.

4. Mitchell, P. 1966. Chemiosmotic coupling in oxidative and photosynthetic phosphorylation. *Biol. Rev. Cambridge Philos. Soc.* 41:445–502.

5. Mitchell, P. 1979. Compartmentation and communication in living systems. Ligand conduction: A general catalytic principle in chemical, osmotic and chemiosmotic reaction systems. *Eur. J. Biochem.* 95:1–20.

6. Kashket, E. R. 1985. The proton motive force in bacteria: A critical assessment of methods. *Ann. Rev. Micro.* 39:219–242.

7. A single proton (or any monovalent ion, or electron) carries $1.6 \times 10^{-19}$ C of charge. If we multiply this by Avogadro's number, we arrive at the charge carried by a mole, which is approximately 96,500 C, or the faraday (F).

8. The actual equation is $G = G_0 + RT \ln a$, where $a$ is the activity. The activity is a product of the molal concentration ($c$) and the activity coefficient ($\gamma$) for the particular compound, $a = \gamma c$. In practice, concentrations are usually used instead of activities, and the concentrations are in molar units instead of molal. The symbol $T$ is the absolute temperature in degrees kelvin (273 + °C). The symbol $R$ is the ideal gas constant, and $G_0$ is the standard free energy (i.e., when the concentration of all reactants is 1 M). When one uses 8.3144 J deg$^{-1}$ mole$^{-1}$ as the units of $R$, then the free energy ($G$) is given in J mole$^{-1}$.

9. Cecchini, G., and A. L. Koch. 1975. Effect of uncouplers on "downhill" $\beta$-galactoside transport in energy-depleted cells of *Escherichia coli. J. Bacteriol.* 123:187–195.

10. Gould, J. M., and W. A. Cramer. 1977. Relationship between oxygen-induced proton efflux and membrane energization in cells of *Escherichia coli. J. Biol. Chem.* 252:5875–5882.

11. Padan, E., D. Zilberstein, and S. Schuldner. 1981. pH homeostasis in bacteria. *Biochim. et Biophys. Acta* 650:151–166.

12. Reviewed in Cobley, J. G., and J. C. Cox. 1983. Energy conservation in acidophilic bacteria. *Microbiol. Rev.* 47:579–595.

13. The pumping of protons out of the cell or the electrogenic influx of electrons will create a membrane potential, positive outside. However, in the aerobic acidophilic bacteria (i.e., those bacteria that live in environments of extremely low pH, pH 1 to 4) other events act to reverse the membrane potential. These bacteria have positive membrane potentials (i.e., inside positive with respect to outside, at low pH). It

is not clear why the aerobic acidophiles have a positive $\Delta\Psi$. One possibility is that they have an energy-dependent $K^+$ pump that brings $K^+$ into the cells at a rate sufficient to establish a net influx of positive charge, creating an inside positive membrane potential. This point is discussed further in Section 15.1.3.

14. Krulwich, T. A., and A. A. Guffanti. 1986. Regulation of internal pH in acidophilic and alkalophilic bacteria, pp. 352–365. In: *Methods in Enzymology*. S. Fleischer and B. Fleischer (Eds.). Vol. 125. Academic Press, New York.

15. Actually, what happens in the presence of nigericin is that an equalization of the $K^+$ and $H^+$ gradients occurs.

16. Padan, E., D. Zilberstein, and S. Schuldiner. 1981. pH homeostasis in bacteria. *Biochim. Biophys. Acta* **650**:151–166.

17. Rottenberg, H. 1979. The measurement of membrane potential and $\Delta$pH in cells, organelles, and vesicles. *Methods in Enzymology* LV:547–569.

18. Bakker, E. P. The role of alkali-cation transport in energy coupling of neutrophilic and acidophilic bacteria: An assessment of methods and concepts. 1990. *FEMS Microbiol. Rev.* **75**:319–334.

19. This is because the resonance frequency of inorganic phosphate or of the $\gamma$-$\rho$hosphate of ATP in a high magnetic field is a function of the degree to which the phosphate is protonated. (Ferguson, S. J., and M. C. Sorgato. 1982. Proton electrochemical gradients and energy-transduction processes. *Ann. Rev. Bioch.* **51**:185–217.)

20. Cross, R. L. 1992. The reaction mechanism of $F_0F_1$-ATP synthases, pp. 317–330. In: *Molecular Mechanisms in Bioenergetics*, L. Ernster (Ed.). Elsevier Science Publishers, Amsterdam.

21. Another way of stating this is in terms of the total force. The total force is equal to $y \Delta p + \Delta G_p/F$, where $\Delta G_p/F = 518$ mV. At equilibrium the total force is 0; therefore $y \Delta p = -518$ mV and when $y = 3$, $\Delta p = -173$ mV.

22. Maloney, P. C. 1977. Obligatory coupling between proton entry and the synthesis of adenosine 5′-triphosphate in *Streptococcus lactis*. *J. Bacteriol.* **132**:564–575.

23. Maloney, P. C., and F. C. Hansen III. 1982. Stoichiometry of proton movements coupled to ATP synthesis driven by a pH gradient in *Streptococcus lactis*. *J. Memb. Biol.* **66**:63–75.

24. The actual force was about 239 mV because the temperature was 20–21°C rather than 30°C.

25. Kandpal, R. P., K. E. Stempel, and P. B. Boyer. 1987. Characteristics of the forma-tion of enzyme-bound ATP from medium inorganic phosphate by mitochondrial $F_1$ adenosinetriphosphatase in the presence of dimethyl sulfoxide. *Biochemistry* **26**:1512–1517.

26. Grubmeyer, C., R. L. Cross, and H. S. Penefsky. 1982. Mechanism of ATP hydrolysis by beef heart mitochondrial ATPase: Rate constants for elementary steps in catalysis at a single site. *J. Biol. Chem.* **257**:12092–12100.

27. Weber, J., and A. E. Senior. 1997. Catalytic mechanism of $F_1$-ATPase. *Biochim. Biophys. Acta* **1319**:19–58.

28. Reviewed in Unemoto, T., H. Tokuda, and M. Hayashi. 1990. Primary sodium pumps and their signficance in bacterial energetics, pp. 33–54. In: T. A. Krulwich (Ed.). *The Bacteria*, Vol. XII. Academic Press, Inc., New York.

29. Reviewed in, Skulachev, V. P. 1992. Chemiosmotic systems and the basic principles of cell energetics, pp. 37–73. In: *Molecular Mechanisms in Bioenergetics*. Ernster, L. (Ed.). *New Comprehensive Biochemistry: Molecular Mechanisms in Bioenergetics*, Vol. 23. Elsevier, Amsterdam.

30. Tokuda, H., and T. Unemoto. 1982. Characterization of the respiration-dependent $Na^+$ pump in the marine bacterium *Vibrio alginolyticus*. *J. Biol. Chem.* **257**:10007–10014.

31. In his review of primary sodium ion translocating enzymes, Dimroth points out that *V. alginolyticus* has two different NADH:ubiquinone oxidoreductases, i.e., NQR1, which is $Na^+$ dependent and functions at pH 8.5 but not at pH 6.5, and NQR2, which is $Na^+$ independent and is not a coupling site. There is apparently no $H^+$-dependent NADH:ubiquinone oxidoreductase. However, these bacteria do have a cytochrome *bo* oxidase that oxidizes the quinol, is not $Na^+$ dependent, and is believed to be a proton pump as in other bacteria. The presence of both pumps can explain how *V. alginolyticus* operates a $Na^+$-dependent respiratory pump at pH 8.5 and a $H^+$-dependent respiratory pump at pH 6.5. The cytochrome *bo* proton pump must function at both acidic and basic pHs because mutants lacking the $Na^+$-dependent NADH:ubiquinone oxidoreductase extrude $Na^+$ at pH 8.5 using a $Na^+/H^+$ antiporter in combination with a primary proton pump, and the wild type is known to extrude $Na^+$ at pH 6.5 using the $Na^+/H^+$ antiporter in combination with the primary proton pump. Dimroth, P. 1997. Primary sodium ion translocating enzymes. *Biochim. Biophys. Acta* **1318**:11–51.

32. Kreke, B., and H. Cypionka. 1994. Role of sodium ions for sulfate transport and energy metabolism in *Desulfovibrio salexigens*. *Arch. Microbiol.* **161**:55–61.

33. Many nonfermenting anaerobic bacteria

carry out electron transport using either organic compounds such as fumarate, or inorganic compounds such as nitrate as electron acceptors. Thus electron flow in these bacteria can be coupled to proton efflux and the establishment of a $\Delta p$. Furthermore, even fermenting bacteria can carry out some fumarate respiration, generating a $\Delta p$. However, the major source of energy for the $\Delta p$ in most fermenting bacteria is ATP hydrolysis.

34. Dimroth, P. 1997. Primary sodium ion translocating enzymes. *Biochim. Biophys. Acta* **1318**:11–51.

35. Dimroth, P. 1990. Energy transductions by an electrochemical gradient of sodium ions, pp. 114–127. In: *The Molecular Basis of Bacterial Metabolism.* Springer-Verlag. G. Hauska and R. Thauer (Eds.). Springer-Verlag, Berlin.

36. Dimroth, P. 1980. A new sodium-transport system energized by the decarboxylation of oxaloacetate. *FEBS Lett.* **122**:234–236.

37. Dimroth, P., and A. Thomer. 1988. Dissociation of the sodium-ion-translocating oxaloacetate decarboxylase of *Klebsiella pneumoniae* and reconstitution of the active complex from the isolated subunits. *Eur. J. Biochem.* **175**:175–180.

38. Hilpert, W., and P. Dimroth. 1983. Purification and characterization of a new sodium transport decarboxylase. Methylmalonyl–CoA decarboxylase from *Veillonella alcalescens. Eur. J. Biochem.* **132**:579–587.

39. Buckel, W., and R. Semmler. 1983. Purification, characterization and reconstitution of glutaconyl–CoA decarboxylase. *Eur. J. Biochem.* **136**:427–434.

40. Schink, B., and N. Pfennig. 1982. *Propionigenium modestum* gen. nov. sp. nov. A new strictly anaerobic nonsporing bacterium growing on succinate. *Arch. Microbiol.* **133**:209–216.

41. Hilpert, W., B. Schink, and P. Dimroth. 1984. Life by a new decarboxylation-dependent energy conservation mechanism with Na$^+$ as coupling ion. *The EMBO Journal* **3**:1665–1670.

42. De Vries, W., R. Theresia, M. Rietveld-Struijk, and A. H. Stouthamer. 1977. ATP formation associated with fumarate and nitrate reduction in growing cultures of *Veillonella alcalescens. Antonie van Leeuwenhoek* **43**:153–167.

43. Buckel, W., and R. Semmler. 1982. A biotin-dependent sodium pump: Glutaconyl–CoA decarboxylase from *Acidaminococcus fermentans. FEBS Lett.* **148**:35–38.

44. Inside-out vesicles are prepared by sonicating whole cells, or shearing them with a French pressure cell. Right-side-out vesicles are prepared by first removing the cell wall with lysozyme in a hypertonic medium, and then osmotically lysing the protoplasts or spheroplasts in hypotonic medium.

45. Anantharam, V., M. J. Allison, and P. C. Maloney. 1989. Oxalate:formate exchange. *J. Biol. Chem.* **264**:7244–7250.

46. Baetz, A. L., and M. J. Allison. 1990. Purification and characterization of oxalyl–coenzyme A decarboxylase from *Oxalobacter formigenes. J. Bacteriol.* **171**:2605–2608.

47. Baetz, A. L., and M. J. Allison. 1990. Purification and characterization of formyl–coenzyme A transferase from *Oxalobacter formigenes. J. Bacteriol.* **171**:3537–3540.

48. Ruan, Z, V. Anantharam, I. T. Crawford, S. V. Ambudkar, S. Y. Rhee, M. J. Allison, and P. C. Maloney. 1992. Identification, purification, and reconstitution of OxlT, the oxalate:formate antiport protein of *Oxalobacter formigenes. J. Biol. Chem.* **267**:10537–10543.

49. To prepare proteoliposomes one disperses phospholipids (e.g., those isolated from *E. coli*) in water, where they spontaneously aggregate to form spherical vesicles called liposomes consisting of concentric layers of phospholipid. The liposomes are then subjected to high-frequency sound waves (sonic oscillation), which breaks them into smaller vesicles surrounded by a single phospholipid bilayer resembling the lipid bilayer found in natural membranes. Then purified protein (e.g., the OxlT antiporter) is mixed with the sonicated phospholipids in the presence of detergent, and the suspension is diluted into buffer. The protein becomes incorporated into the phospholipid bilayer and membrane vesicles called proteoliposomes are formed. When the proteoliposomes are incubated with solute, they catalyze uptake of the solute into the vesicles provided the appropriate carrier protein has been incorporated. In addition, one can "load" the proteoliposomes with solutes (e.g., oxalate) by including these in the dilution buffer.

50. Poolman, B., D. Molenaar, E. J. Smid, T. Ubbink, T. Abee, P. P. Renault, and W. N. Konings. 1991. Malolactic fermentation: Electrogenic malate uptake and malate/lactate antiport generate metabolic energy. *J. Bacteriol.* **173**:6030–6037.

51. Konings, W. N. 1985. Generation of metabolic energy by end-product efflux. *TIBS*, August, 317–319.

52. Konings, W. N., J. S. Lolkema, and B. Poolman. 1995. The generation of metabolic energy by solute transport. *Arch. Microbiol.* **164**:235–242.

53. Michels, J. P. J. Michel, J. Boonstra, and W. N. Konings. 1979. Generation of an electrochemical proton gradient in bacteria by the excretion of metabolic end-products. *FEMS Microbiol. Letts.* **5**:357–364.

54. Otto et al. 1982. Lactate efflux-induced electrical potential in membrane vesicles of *Streptococcus cremoris*. *J. Bacteriol.* **149**:733–738.

55. Brink, B. T., and W. N. Konings. 1982. Electrochemical proton gradient and lactate concentration gradient in *Streptococcus cremoris* cells grown in batch culture. *J. Bacteriol.* **152**:682–686.

56. Driessen, A. J. M., and W. N. Konings. 1990. Energetic problems of bacterial fermentations: Extrusion of metabolic end products, pp. 449–478. In: *Bacterial Energetics*, T. A. Krulwich (Ed.) Academic Press, N.Y.

57. Michel, T. A., and J. M. Macy. 1990. Generation of a membrane potential by sodium-dependent succinate efflux in *Selenomonas ruminantium*. *J. Bacteriol.* **172**:1430–1435.

58. The lactic and succinic acids are presumed to be in the ionized form because the intracellular pH is much larger than the $pK_a$ values.

59. Brink, B. T., R. Otto, U. Hansen, and W. N. Konings. 1985. Energy recycling by lactate efflux in growing and nongrowing cells of *Streptococcus cremoris*. *J. Bacteriol.* **162**:383–390.

60. Oesterhelt, D., and J. Tittor. 1989. Two pumps, one principle: Light-driven ion transport in halobacteria. *TIBS* **14**:57–61.

61. Bogomoleni, R. A., R. A. Baker, R. H. Lozier, and W. Stoeckenius. 1980. Action spectrum and quantum efficiency for proton pumping in *Halobacterium halobium*. *Biochemistry* **19**:2152–2159.

62. Henderson, R., J. M. Baldwin, T.A. Ceska. 1990. Model for the structure of bacteriorhodopsin based on high-resolution electron cryo-microscopy, *J. Mol. Biol.* **213**:899–929.

63. The extreme halophiles require unusually high external NaCl concentrations (at least 3–5 M, i.e., 17–28%) in order to grow. They inhabit hypersaline environments such as the solar salt evaporation ponds near San Francisco and salt lakes (e.g., the Great Salt Lake in Utah and the Dead Sea). There are now six recognized genera, two of them being the well-known *Halobacterium* and *Halococcus*. The best studied is *Hb. salinarium* (*halobium*). The other four genera are *Haloarcula*, *Haloferax*, *Natronobacterium*, and *Natronococcus*. The majority of the known halophilic archaea are aerobic chemo-organotrophs and can grow on simple carbohydrates as well as long-chained saturated hydrocarbons. They generally grow best at pH values between 8 and 9. However, *Natronobacterium* and *Natronococcus* are also alkalophilic and grow well at pH values up to 11. When oxygen is not present, the halobacteria will grow anaerobically using several electron acceptors in place of oxygen. These include fumarate, dimethylsulfoxide (DMSO), and trimethyamine N-oxide (TMAO). Members of the genera *Haloarcula* and *Haloferax* can grow on nitrate as the terminal electron acceptor. Some of the halobacteria reduce the nitrate to nitrite, and some reduce it completely to nitrogen gas. Some halobacteria can also grow fermentatively in the absence of oxygen. These include *Hb. salinarium* (*halobium*) which can ferment arginine to citrulline.

64. Oesterhelt, D., and G. Krippahl. 1983. Phototrophic growth of halobacteria and its use for isolation of photosynthetically-deficient mutants. *Ann. Microbiol. (Inst. Pasteur)* **134B**:137–150.

65. Gest, H. 1993. Photosynthetic and quasi-photosynthetic bacteria. *FEMS Microbiol. Lett.* **112**:1–6.

66. Some can carry out anaerobic respiration using nitrate as an electron acceptor.

67. Respiration can be severely limited under certain growth conditions, since the oxygen content of hypersaline waters, the normal habitat of these organisms, is usually 20% or less than is found in normal sea water, and in unstirred ponds oxygen becomes even more scarce. The halobacteria can derive energy from the fermentation of amino acids, but in the absence of a fermentable carbon source and respiration, light is the only source of energy.

68. Bogomoleni, R. A., R. A. Baker, R. H. Lozier, and W. Stoeckenius. 1980. Action spectrum and quantum efficiency for proton pumping in *Halobacterium halobium*. *Biochemistry* **19**:2152–2159.

69. Tittor, J., and D. Oesterhelt. 1990. The quantum yield of bacteriorhodopsin. *FEBS Lett.* **263**:269–273.

70. Govindjee, R., Balashov, S. P., and T. G. Ebrey. 1990. Quantum efficiency of the photochemical cycle of bacteriorhodopsin. *Biophys. J.* **58**:597–608.

71. Lanyi, J. K. 1997. Mechanism of ion transport across membranes. *J. Biol. Chem.* **272**:31209–31212.

72. A Schiff base is an imine and has the following structure: R–CH=N–R'. It is formed between a carbonyl group and a primary amine, i.e.,

$$R\text{–}CHO + H_2N\text{–}R' \longrightarrow R\text{–}CH{=}N\text{–}R' + H_2O$$

In bacteriorhodopsin, the amino group is donated by lysine in the protein and the carbonyl is donated by the retinal.

73. Lanyi, J. K. 1992. Proton transfer and energy coupling in the bacteriorhodopsin photocycle. *J. Bioenerg. and Biomemb.* **24**:169–179.

74. Oesterhelt, D., J. Tittor, and E. Bamberg. 1992. A unifying concept for ion translocation by retinal proteins. *J. Bioenerg. and Biomemb.* **24**:181–191.

75. Fodor, S. P. A., J. B. Ames, R. Gebhard, E. M. M. van den Berg, W. Stoeckenius, J. Lugtenburg, and R. A. Mathies. 1988. Chromophore structure in bacteriorhodopsin's N intermediate: Implications for the proton-pumping mechanism. *Biochemistry* **27**:7097–7101.

76. Krebs, M. P., and H. G. Khorana. 1993. Mechanism of light-dependent proton translocation by bacteriorhodopsin. *J. Bacteriol.* **175**:1555–1560.

77. Aspartate-85 remains protonated while a proton is released into the aqueous phase, and therefore it must be concluded that the immediate source of the released proton is a different amino acid residue.

78. Fodor, S. P. A., J. B. Ames, R. Gebhard, E. M. M. van den Berg, W. Stoeckenius, J. Lugtenburg, and R. A. Mathies. 1988. Chromophore structure in bacteriorhodopsin's N intermediate: Implications for the proton-pumping mechanism. *Biochemistry* **27**:7097–7101.

79. Senior, A. E. 1990. The proton-translocating ATPase of *Escherichia coli. Ann. Rev. Biophys. Chem.* **19**:7–41.

80. Lanyi, J. K. 1990. Halorhodopsin, a light-driven electrogenic chloride-transport system. *Physiol. Rev.* **70**:319–330.

81. Duschl, A., and G. Wagner. 1986. Primary and secondary chloride transport in *Halobacterium halobium. J. Bacteriol.* **168**:548–552.

82. There are many nonobligate alkalophilic bacteria whose optimal growth pH is 9 or greater, but which can grow at pH 7 or below. These are widely distributed among the bacterial genera, and include both bacteria and archaea. There are also *alkalophilic-tolerant* bacteria that can grow at pH values of 9 or more, but whose optimal growth pH is around neutrality.

83. Reviewed in Krulwich, T. A., and D. M. Ivey. 1990. Bioenergetics in extreme environments, pp. 417–447. In: *The Bacteria*, Vol. XII. T. A. Krulwich (Ed.). Academic Press, Inc., New York.

84. Krulwich, T. A., and A. A. Guffanti. 1989. Alkalophilic bacteria. *Ann. Rev. Microbiol.* **43**:435–463.

# 4

# Electron Transport

The main method by which energy is generated for growth-related physiological processes such as biosynthesis and solute transport in *respiring* prokaryotes is by coupling the flow of electrons in membranes to the creation of an electrochemical proton gradient (Fig. 4.1). (See Note 1.) As discussed in Section 3.7.1, electrons flow spontaneously down a potential energy gradient toward acceptors that have a more positive electrode potential. (The student should review Section 3.2.2 for a discussion of the generation of membrane potentials, and Section 3.7.1 for a discussion of oxidation–reduction reactions coupled to the generation of a $\Delta p$.) The electrons flow from primary electron donors to terminal electron acceptors through a series of *electron carrier proteins* and a class of lipids called *quinones*. One refers to electron flow via electron carriers in membranes as *respiration*. If the terminal electron acceptor is oxygen, then electron flow is called *aerobic respiration*. If it is not oxgyen, then it is called *anaerobic respiration*. Proton translocation takes place during respiration, and a $\Delta p$ is created at coupling sites described in Sections 3.7.1 and 4.5. Mitochondria have three coupling sites, whereas bacteria can have one to three coupling sites depending upon the bacterium and growth conditions. In other words, prokaryotic cell membranes (and mitochondria and chloroplasts) convert an electrode potential difference, $\Delta E_h$, into a proton

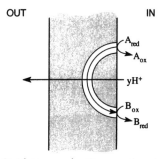

**Fig. 4.1** Oxidation–reduction reactions in the cell membrane result in a proton potential. Electrons flow from A to B through a series of electron carriers in the membrane from a low potential toward a higher potential. The intermediate redox reactions between A and B are not shown. Certain of the redox reactions in the series are coupled to the translocation of protons across the cell membrane. These are called coupling sites. In this way a redox potential ($\Delta E_h$) is converted into a proton potential ($\Delta p$). n $\Delta E_h$ = y $\Delta p$, where *n* is the number of electrons transferred and *y* is the number of protons extruded.

electrochemical potential difference ($\Delta p$). (As discussed in Section 3.7.1, $E_h$ is the actual electrode potential of a compound at the specified concentrations of its oxidized and reduced forms.) The proton potential is then used to drive solute transport, ATP synthesis, flagella rotation, and other membrane activities. In mitochondria, electron transport pathways are pretty much all the same. However, prokaryotes are diverse creatures, and their electron transport pathways differ depending upon the primary donor and terminal acceptor. This chapter describes electron transport pathways in mitochondria and bacteria and how they are coupled to the formation of a $\Delta p$. It appears that archaeal electron transport chains are similar to (eu)bacterial electron transport chains.[2]

## 4.1 Aerobic and Anaerobic Respiration

There is a steady current of electrons through electron carriers in prokaryotic cell membranes from low potential electron donors (the primary donors or *reductants*) to high potential electon acceptors (the terminal acceptor or *oxidant*). Electron acceptors can be oxygen or some other inorganic acceptor such as nitrate or sulfate. An example of an organic electron acceptor is fumarate. Thus there is oxygen respiration, nitrate respiration, sulfate respiration, fumarate respiration, and so on.

## 4.2 The Electron Carriers

The electrons flow through a series of electron carriers. These are:

1. Flavoproteins (hydrogen and electron carrier)

2. Quinones (hydrogen and electron carrier)

3. Iron–sulfur proteins (electron carrier)

4. Cytochromes (electron carrier)

The quinones are lipids, whereas the other electron carriers are proteins, which exist in multiprotein enzyme complexes called *oxidoreductases*. (See Note 3.) The electrons are not carried in the protein per se, but in a nonprotein molecule bound to the protein. The nonprotein portion that carries the electron is called a *prosthetic group*. (See Note 4 for useful definitions.) The prosthetic group in iron–sulfur proteins is a cluster of iron sulfide, which is abbreviated as FeS. The prosthetic group in flavoproteins (Fp) is a flavin, which can be either FAD or FMN. The prosthetic group in cytochromes is heme. The chemistry of the prosthetic groups is described in Section 4.2.1. Some of the prosthetic groups (flavins) carry hydrogen as well as electrons, and they are referred to as hydrogen carriers. The quinones are also hydrogen carriers. Some of the prosthetic groups (FeS and heme) carry only electrons, and they are referred to as electron carriers.

Each of the electron carriers has a different electrode potential, and the electrons are transferred sequentially to a carrier of a higher potential.

The standard potentials at pH 7 of the electron carriers and some electron donors and acceptors are listed in Table 4.1.

**Table 4.1** Standard electrode potentials (pH 7)

| Couple | $E'$ (mV) |
|---|---|
| $Fd_{ox}/Fd_{red}$ (spinach) | −432 |
| $CO_2$/formate | −432 |
| $H^+/H_2$ | −410 |
| $Fd_{ox}/Fd_{red}$ (*Clostridium*) | −410 |
| $NAD^+$/NADH | −320 |
| FeS(ox/red) in mitoch | −305 |
| Lipoic/dihydrolipoic | −290 |
| $S^o/H_2S$ | −270 |
| $FAD/FADH_2$ | −220 |
| Acetaldehyde/ethanol | −197 |
| $FMN/FMNH_2$ | −190 |
| Pyruvate/lactate | −185 |
| OAA/malate | −170 |
| Menaquinone(ox/red) | −74 |
| Cyt $b_{558}$(ox/red) | −75 to −43 |
| Fum/succ | +33 |
| Ubiquinone(ox/red) | +100 |
| Cyt $b_{556}$(ox/red) | +46 to +129 |
| Cyt $b_{562}$(ox/red) | +125 to +260 |
| Cyt d(ox/red) | +260 to +280 |
| Cyt c(ox/red) | +250 |
| FeS(ox/red) in mitochondria | +280 |
| Cyt a(ox/red) | +290 |
| Cyt $c_{555}$(ox/red) | +355 |
| Cyt $a_3$(ox/red) in mitochondria | +385 |
| $NO_3^-/NO_2^-$ | +421 |
| $Fe^{3+}/Fe^{2+}$ | +771 |
| $O_2$ (1 atm)/$H_2O$ | +815 |

*Key:* Fd, ferredoxin; OAA, oxaloacetate; Fum, fumarate; succ, succinate.
*Source:* Thauer, R. K., K. Jungermann, and K. Decker. 1977. Energy conservation in chemotrophic anaerobic bacteria. *Bacteriol Rev.* 41:100–180; Metzler, D. E. 1977. *Biochemistry: The Chemical Reactions of Living Cells*. Academic Press, New York.

### 4.2.1 Flavoproteins

Flavoproteins (Fp) are electron carriers that have as their prosthetic group an organic molecule called flavin. (Flavin is derived from the Latin word *flavius*, which means yellow, in reference to the color of flavins. They are synthesized by cells from the vitamin riboflavin, vitamin $B_2$.) There are two flavins, flavin mononucleotide (FMN) and flavin adenine dinucleotide (FAD) (Fig. 4.2). Phosphorylation of riboflavin at the ribityl $5'$-OH yields FMN, and adenylylation of FMN yields FAD. As Fig. 4.2 illustrates, when flavins are reduced, they carry 2H (two electrons and two hydrogens), one on each of two ring nitrogens. There are many different flavoproteins, and they catalyze diverse oxidation–reduction reactions in the cytoplasm, not merely those of the electron transport chain in the membranes. Although all the flavoproteins have FMN or FAD as their prosthetic group, they differ in which oxidations they catalyze, as well as in their redox potentials. These differences are due to differences in the protein component of the enzyme, not in the flavin itself.

### 4.2.2 Quinones

Quinones are lipid electron carriers. Due to their hydrophobic lipid nature, some are believed to be highly mobile in the lipid phase of the membrane, and carry hydrogen and electrons to and from the complexes of protein electron carriers that are not mobile. Their structure and oxidation–reduction reactions are shown in Fig. 4.3. All quinones have hydrophobic isoprenoid side chains that contribute to their lipid solubility. The number of isoprene units varies but is typically six to ten. Bacteria make two types of quinones that function during respiration, ubiquinone (UQ), a quinone also found in mitochondria, and menaquinone (MQ or sometimes MK). Menaquinones, which are derivatives of vitamin K, differ from ubiquinones in being naphthoquinones where the additional benzene ring replaces the two methoxy groups present in ubiquinones (Fig. 4.3). They also have a much lower electrode potential than ubiquinones and are used predominantly during anaerobic respiration, where the electron acceptor has a low potential (e.g., during fumarate respiration). A third type of quinone, plastoquinone, occurs in chloroplasts and cyanobacteria, and functions in photosynthetic electron transport. In plastoquinones, the two methoxy groups are replaced by methyl groups.

### 4.2.3 Iron–sulfur proteins

Iron–sulfur proteins contain nonheme iron and usually acid labile sulfur (Fig. 4.4). The term acid-labile sulfur means that, when the pH is lowered to approximately one, $H_2S$ is released from the protein. This is because there is sulfide attached to iron by bonds that are ruptured in acid. Generally, the proteins contain clusters in which iron and acid-labile sulfur are present in a ratio of 1:1. However,

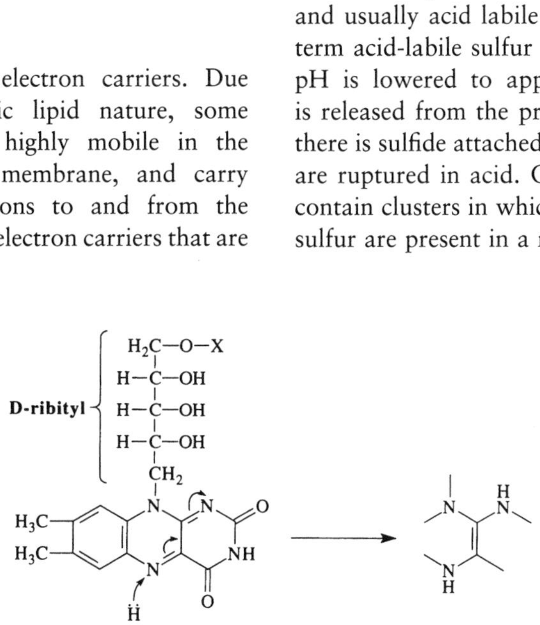

Fig. 4.2 Structures of riboflavin, FMN, and FAD. Riboflavin: X = H; FMN: X = $PO_3H_2$; FAD: X = ADP. For the sake of convenience, the reduction reaction is drawn as proceeding via a hydride ion even though this need not be the actual mechanism in all flavin reductions.

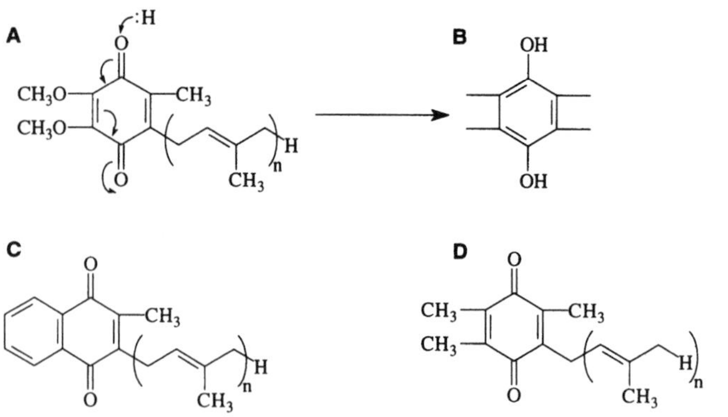

**Fig. 4.3** The structure of quinones: (A) oxidized ubiquinone; (B) reduced ubiquinone; (C) oxidized menaquinone; (D) oxidized plastoquinone. The value of $n$ can be 4–10 and is 8 for both quinones in *E. coli*. In *E. coli* ubiquinone plays a major role in aerobic and nitrate respiration, whereas menaquinone is dominant during fumarate respiration. One reason for this is that ubiquinone has a potential ($E_0'$) of +100, mV compared to + 30 mV for fumarate. It is therefore at too high a potential to deliver electrons to fumarate. Menaquinone has a low potential, −74 mV, and is thus able to deliver electrons to fumarate. Plastoquinone is used in chloroplast and cyanobacterial photosynthetic electron transport.

there may be more than one iron–sulfur cluster per protein. For example, in mitochondria the enzyme complex that oxidizes NADH has at least four FeS clusters (see Fig. 4.9). The FeS clusters have different $E_h$ values, and the electron travels from one FeS cluster to the next toward the higher $E_h$. It appears that the electron may not be localized on any particular iron atom and the entire FeS cluster should be thought of as carrying one electron, regardless of the number of Fe atoms. These proteins also contain cysteine sulfur, which is not acid labile, and which bonds the iron to the protein. There are several different types of iron–sulfur proteins, and these catalyze numerous oxidation–reduction reactions in the cytoplasm as well as in the membranes. (See Note 5 for more information on iron–sulfur proteins.) The iron–sulfur proteins have characteristic electron spin resonance spectra (EPR) because of an unpaired electron in either the oxidized or reduced form of the FeS cluster in different FeS proteins. (See Note 6 for a description of electron spin resonance.) The iron–sulfur proteins cover a very wide range of potentials from approximately −400 to +350 mV. They therefore can carry out oxidation-reduction reactions at both the low-potential end and the high-potential end of the electron transport chain, and indeed are found in several locations. An example of an FeS cluster is shown in Fig. 4.4. Note that each Fe is bound to two acid-labile S and two cysteine S. This would be called an $Fe_2S_2$ cluster.

### 4.2.4 Cytochromes

Cytochromes are electron carriers that have heme as the prosthetic group. Heme consists of four pyrrole rings attached to each other by methene bridges (Fig. 4.5). Because hemes have four pyrroles, they are called *tetrapyrroles*. Each of the pyrrole rings is substituted by a side chain. Substituted tetrapyrroles are called *porphyrins*. Therefore, hemes are also called porphyrins. (An unsubstituted tetrapyrrole is called a porphin.)

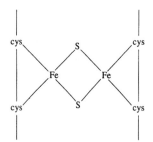

**Fig. 4.4** FeS cluster. This is an $Fe_2S_2$ cluster. More than one cluster may be present per protein. The sulfur atoms held only by the iron are acid labile. The iron is bonded to the protein via sulfur in cysteine residues.

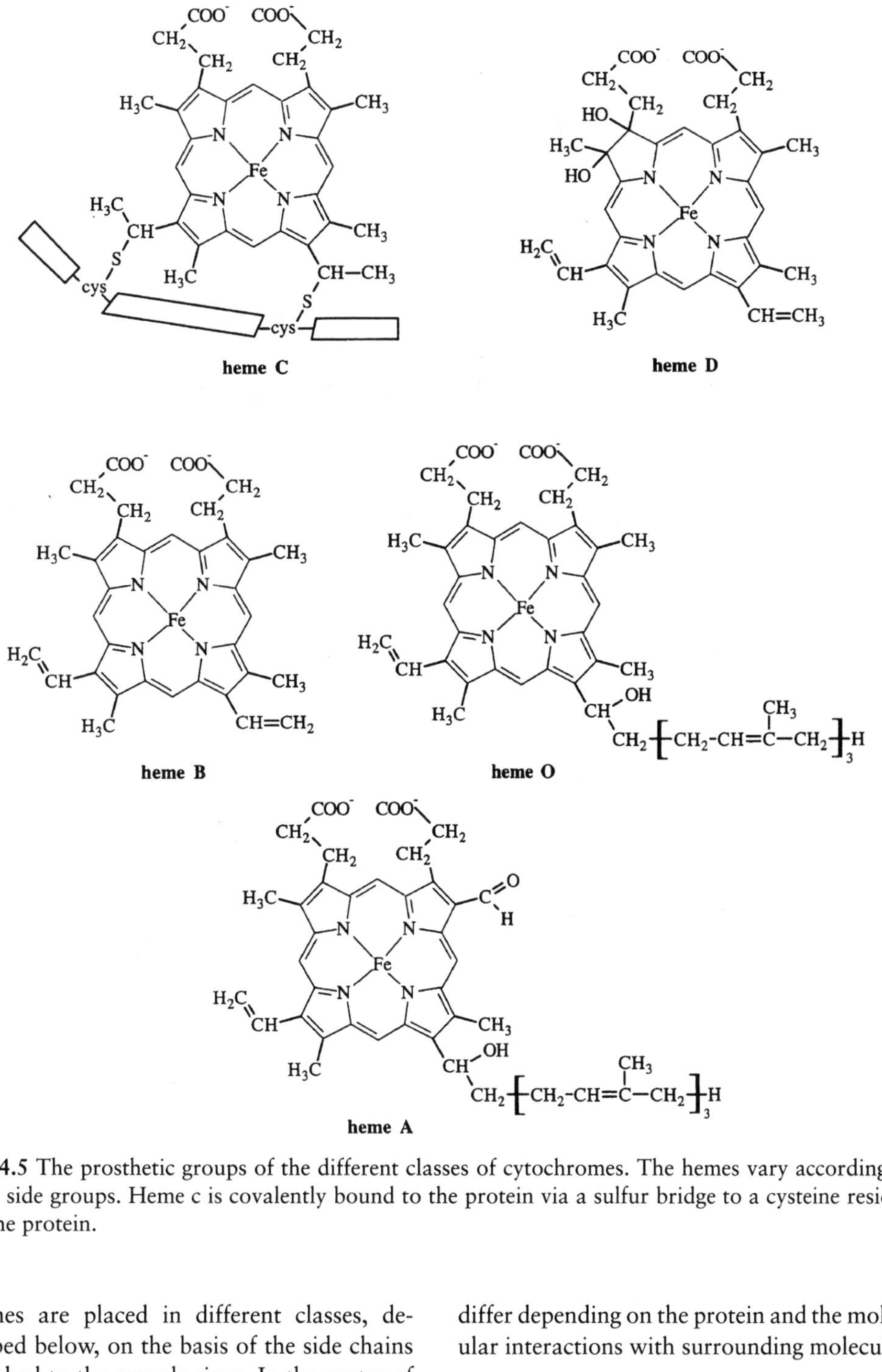

**Fig. 4.5** The prosthetic groups of the different classes of cytochromes. The hemes vary according to their side groups. Heme c is covalently bound to the protein via a sulfur bridge to a cysteine residue on the protein.

Hemes are placed in different classes, described below, on the basis of the side chains attached to the pyrrole rings. In the center of each heme there is an iron atom that is bound to the nitrogen of the pyrrole rings. The iron is the electron carrier and is oxidized to ferric or reduced to ferrous ion during electron transport. Cytochromes are therefore one-electron carriers. The $E_h$ of the different cytochromes differ depending on the protein and the molecular interactions with surrounding molecules.

## Classes of cytochromes

There are five classes of heme that distinguish the cytochromes (Fig. 4.5). These are hemes a, b, c, d, and o. Hemes d and o have been found only in the prokaryotic cytochrome oxidases.

Bacterial cytochromes include cytochromes bd (also called cytochrome d) and bo (also called cytochrome $bo_3$ or cytochrome o), which are quinol oxidases that reduce oxygen. (In naming cytochromes, sometimes the $O_2$-binding heme is given the subscript 3.[7]) As the names imply, they each have two types of heme, one being heme b and the other being heme d or o.[7] As mentioned previously, the hemes can be distinguished according to the side groups that they possess, and this is summarized in Fig. 4.5. For example, heme o differs from heme b in having an hydroxyethylfarnesyl group substituted for a vinyl group. However, the only difference between heme b and heme c is that the latter is covalently bound to protein by thioether linkages between the two vinyl groups and cysteine residues in the protein. Hemes can usually be distinguished spectrophotometrically. When cytochromes are in the reduced state, they have characteristic light absorption bands in the visible range due to absorption by the heme. These are the $\alpha$, the $\beta$, and the $\gamma$ bands. The $\alpha$ bands absorb light between 500 and 600 nm, the $\beta$ absorb at a lower wavelength, and the $\gamma$ bands are in the blue region of the spectrum. The spectrum for a cytochrome c is shown in Fig. 4.6. Cytochromes are distinguished, in part, by the position of the maximum in the $\alpha$ band. For example, cyt $b_{556}$ and cyt $b_{558}$ differ because the former has a peak at 556 nm and the latter a peak at 558 nm.

## Reduced minus oxidized spectra

It is very difficult to resolve the different peaks of individual cytochromes in whole cells because of light scattering and nonspecific absorption unless one employs difference spectroscopy. For difference spectroscopy, the cells are placed into two cuvettes in a split-beam spectrophotometer, and monochromatic light from a single monochromator scan is split to pass through both cuvettes. In one cuvette the cytochromes are oxidized by adding an oxidant, and in the second cuvette they are reduced by adding a reductant. The spectrophotometer subtracts the output of one cuvette from the other to give a reduced minus oxidized difference spectrum. In this way nonspecific absorption and light scattering are eliminated from the spectrum, and the cytochromes in the preparation are identified.

## Dual-beam spectroscopy

To follow the kinetics of oxidation or reduction of a particular cytochrome, a dual-beam spectrophotometer is used. In a dual-beam spectrophotometer there are two monochromators. Light from one monochromator is set at a wavelength at which absorbance will change during oxidation or reduction, and the second beam of light is at a nearby wavelength for which absorbance will not change. The light is sent alternatively from both monochromators through the sample cuvette, and the difference in absorbance between the two wavelengths is automatically plotted as a function of time.

## 4.2.5 Standard electrode potentials of the electron carriers

Table 4.1 shows standard electrode potentials at pH 7 ($E_0'$) of some electron donors, acceptors, and electron carriers. Notice that redox couples are generally written in the form oxidized/reduced. Many of the oxidation–reduction reactions in the electron transport chain can be reversed by the $\Delta p$, as discussed in Section 3.7.1. This means that the ratio ox/red for several of the electron carriers (flavoproteins, cytochromes, quinones, FeS proteins) must be close to one. Thus for these

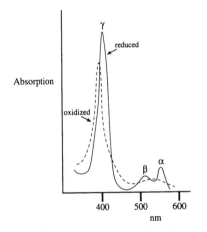

**Fig. 4.6** Absorption spectra of oxidized and reduced cytochrome c. The $\alpha$ band in the reduced form is used to identify cytochromes.

reactions the $E_h$ (actual potential at pH 7) of the redox couples are close to their midpoint potentials, $E'_m$, which is the potential at pH 7 when the couple is 50% reduced (i.e., [ox] = [red]).

## 4.3 Organization of the Electron Carriers in Mitochondria

The electron carriers are organized as an electron transport chain that transfers electrons from electron donors at a low electrode potential to electron acceptors at a higher electrode potential (Fig. 4.7). Electrons can enter at the level of flavoprotein, quinone, or cytochrome, depending upon the potential of the donor. The carriers are organized in the membrane as individual complexes. The complexes can be isolated from each other by appropriate separation techniques after mild detergent extraction, which removes the lipids but does not destroy the protein–protein interactions. The separated complexes can be analyzed for their components, and also can be incorporated into proteoliposomes in order to study the oxidation–reduction reactions that they each catalyze, in addition to proton translocation. (Proteoliposomes are artificial constructs of purified lipids and proteins. They are described in Section 16.1.) Four complexes can be recognized in mitochondria. They are complex I (NADH–ubiquinone oxidoreductase), complex II (succinate dehydrogenase), complex III (ubiquinol–cytochrome c oxidoreductase, also called the $bc_1$ complex), and complex IV (cytochrome c oxidase, which is cytochrome $aa_3$). Complexes I, III, and IV are coupling sites (Section 4.5). Each complex can have several proteins. The most intricate is complex I from mammalian mitochondria, which has about 40 polypeptide subunits, at least four iron–sulfur centers, one flavin mononucleotide (FMN), and one or two bound ubiquinones. Analogous complexes have been isolated from bacteria, but in some cases (e.g., NADH–ubiquinone oxidoreductase and the $bc_1$ complex), they have fewer protein components.[8–11] Note the pattern in the arrangement of the electron carriers; a dehydrogenase complex accepts electrons from a primary donor and transfers the electrons to a quinone. The quinone then transfers the electrons to an oxidase complex via intervening cytochromes. As described next, the same general pattern exists in bacteria.

## 4.4 Organization of the Electron Carriers in Bacteria

Bacterial electron transport chains vary among the different bacteria, and also according to the growth conditions. These variations will be discussed later (Section 4.7). First, the common features in bacterial electron transport schemes will be described and comparisons will be made with mitochondrial

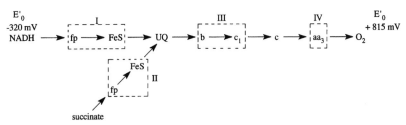

Fig. 4.7 Electron transport scheme in mitochondria. Electrons travel in the electron transport chain from a low to a high electrode potential. Complexes I–IV are bracketed by dotted lines. Complex I is NADH dehydrogenase, also called NADH–ubiquinone oxidoreductase. Complex II is succinate dehydrogenase, also called succinate–ubiquinone oxidoreductase. Complex III is the $bc_1$ complex, also called ubiquinol–cytochrome c oxidoreductase. Complex IV is the cytochrome $aa_3$ oxidase, also called cytochrome c oxidase. There are several FeS clusters in complexes I and II, and an FeS protein in complex III. Complex II also has a cytochrome b. Symbols: fp, flavoprotein; FeS, iron–sulfur protein; UQ, ubiquinone; b, cytochrome b; $c_1$, cytochrome $c_1$; c, cytochrome c; $aa_3$, cytochrome $aa_3$.

electron transport. As with the mitochondrial electron transport chain, the bacterial chains are organized into dehydrogenase and oxidase complexes connected by quinones (Fig. 4.8). The quinones accept electrons from dehydrogenases and transfer these to oxidase complexes that reduce the terminal electron acceptor. Bacteria are capable of using electron acceptors other than oxygen (e.g., nitrate and fumarate) during anaerobic respiration. The enzyme complexes that reduce electron acceptors other than oxygen are called *reductases*, rather than oxidases. Some of the dehydrogenase complexes are NADH and succinate dehydrogenase complexes analogous to complex I and II in mitochondria. In addition to these dehydrogenases, there are several others that reflect the diversity of substrates oxidized by the bacteria. For example, there are $H_2$ dehydrogenases (called hydrogenases), formate dehydrogenase, lactate dehydrogenase, methanol dehyhdrogenase, methylamine dehydrogenase, and so on. Depending upon the source of electrons and electron acceptors, bacteria can synthesize and substitute one dehydrogenase complex for another, or reductase complexes for oxidase complexes. For example, when growing anaerobically, *E. coli* makes the reductase complexes instead of the oxidase complexes, represses the synthesis of some dehydrogenases, and stimulates the synthesis of others. (This is discussed more fully in Chapter 18.) For this reason, the electron carrier complexes in bacteria are sometimes referred to as modules, since they can be synthesized and "plugged" into the respiratory chain when needed.

### 4.4.1 The different terminal oxidases

A word should be said about the many different terminal oxidases found in bacteria.[12] Whereas mitochondria all have the same cytochrome c oxidase (cytochrome $aa_3$), bacteria have a variety of terminal oxidases, often 2 to 3 different ones in the same bacterium (i.e., 2 to 3 branches to oxygen). Some of the bacterial terminal oxidases oxidize quinols (quinol oxidases), and some oxidize cytochrome c (cytochrome c oxidases). The terminal oxidases differ in their affinities for oxygen and whether or not they are proton pumps, as well as in the types of hemes and metals they contain. However, despite these differences, most

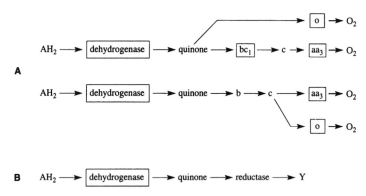

Fig. 4.8 Generalized electron transport pathways found in bacteria. The details will vary depending upon the bacterium and the growth conditions. (A) Aerobic respiration: A dehydrogenase complex removes electrons from an electron donor and transfers these to a quinone. The electrons are transferred to an oxidase complex via a branched pathway. Depending upon the bacterium, the pathway may branch at the quinone or at the cytochrome. Many bacteria have $bc_1$, cytochrome c, and cytochrome $aa_3$ in one of the branches and in this way resemble mitochondria. Other bacteria do not have a $bc_1$ complex, and may or may not have cytochrome $aa_3$. (B) Anaerobic respiration: Under anaerobic conditions the electrons are transferred to reductase complexes, which are synthesized anaerobically. Several reductases exist, each one specific for the electron acceptor. Y represents either an inorganic electron acceptor other than oxygen, e.g., nitrate, or an organic electron acceptor, e.g., fumarate. More than one reductase can simultaneously exist in a bacterium.

of them belong to a superfamily of oxidases called the heme–copper oxidase superfamily, to which the mitochondrial cytochrome c oxidase also belongs. An exception is the cytochrome bd oxidase, discussed later, which is not a member of the heme–copper oxidase superfamily. All members of the heme–copper oxidase superfamily share a protein subunit that is homologous to one of the subunits (subunit I) of the mitochondrial cytochrome c oxidase. This subunit has a bimetallic (binuclear) center that binds oxygen and reduces it to water, and pumps protons. (Several of the heme–copper oxidases are known to be proton pumps, whereas in the case of others it is not yet known whether they pump protons.) The bimetallic center contains a heme iron and copper. There is a second heme in all these oxidases, but it is not part of the bimetallic center and probably functions in transferring electrons to the bimetallic center. Heme–copper oxidases that are mentioned later in the context of bacterial respiratory systems include the $cbb_3$-type oxidases, cytochrome $aa_3$ oxidase, cytochrome $bo_3$ oxidase (cytochrome o oxidase), and cytochrome $bb_3$ oxidase. It should be emphasized that some of these are quinol oxidases, and some are cytochrome c oxidases.

## 4.4.2 Bacterial electron transport chains are branched

Two major difference between mitochondrial and bacterial electron transport chains are that (1) the routes to oxygen in the bacteria are branched, the branch point being at the quinone or cytochrome, and (2) many bacteria can alter their electron transport chains depending upon growth conditions (Fig. 4.8). Under aerobic conditions there are often two or three branches leading to different terminal oxidases. (See Section 4.4.1.) For example, a two-branched electron transport chain might contain a branch leading to cytochrome o oxidase (quinol oxidase) and a branch leading to cytochrome $aa_3$ oxidase (cytochrome c oxidase). Other bacteria (e.g., E. coli) have cytochrome o and d oxidase branches (both are quinol oxidases) but do not have the cytochrome $aa_3$ branch. The ability to synthesize

branched electron transport pathways to oxygen confers flexibility on the bacteria, because the branches may differ not only in the $\Delta p$ that can be generated (because they may differ in the number of coupling sites), but their terminal oxidases may also differ with respect to affinities for oxygen. For example, in E. coli cytochrome o has a low affinity for oxygen, whereas cytochrome d has a higher affinity for oxygen. Switching to an oxidase with a higher affinity for oxygen allows the cells to continue to respire even when oxygen tensions fall to very low values. This is important to ensure the reoxidation of the reduced quinones and NADH so that cellular oxidations such as the oxidation of glucose to pyruvate or the oxidation of pyruvate to $CO_2$ may continue. It has also been suggested that under microaerophilic conditions the use of an oxidase with a high affinity for oxygen would remove traces of oxygen that might damage oxygen-sensitive enzymes that are made under microaerophilic or anaerobic growth conditions.[13] As an example of a protective role that an oxidase might play, consider the situation with the strict aerobe Azotobacter vinelandii. This organism has a branched respiratory chain leading to cytochromes o and d as the terminal oxidases. It also fixes nitrogen aerobically. However, as with other nitrogenases, the Azotobacter nitrogenase is inactivated by oxygen. Azotobacter employs two mechanisms to protect its nitrogenase, respiratory and conformational. (See Section 12.3.2 for a discussion of this point.) It is thought that the rapid consumption by oxygen by a terminal oxidase maintains the intracellular oxygen levels sufficiently low so that the nitrogenase is not inactivated. In agreement with this suggestion, mutants of A. vinelandii that are deficient in the cytochrome d complex failed to fix nitrogen in air, although they did fix nitrogen when the oxygen tension was sufficiently reduced.[14] (Mutants in cytochrome o can fix nitrogen in air.)

The adaptability of the bacteria with respect to their electron transport chains can also be seen with many bacteria that can respire either aerobically or anaerobically. Under anaerobic conditions they do not make the oxidase complexes but instead synthesize reductases. For example, during anaerobic growth, E. coli synthesizes fumarate reductase, nitrate reductase, and tri-methylamine-N-oxide (TMAO) reductase. (The regulation of synthesis of these reductases is discussed in Chapter 17.) The different reductases enable the bacteria to utilize alternative electron acceptors under anaerobic conditions. (Some facultative anaerobes will ferment when a terminal electron acceptor is unavailable. Fermentation is discussed in Chapter 14.)

## 4.5 Coupling Sites

Sites in the electron transport pathway where redox reactions are coupled to proton extrusion creating a $\Delta p$ are called *coupling sites*.[15] Each coupling site is also a site for ATP synthesis, since the protons extruded re-enter via ATP synthase to make ATP. In mitochondria there are three coupling sites, and they are called sites 1, 2, and 3. These are shown in Fig. 4.7. Site 1 is the NADH dehydrogenase complex (complex I), site 2 is the $bc_1$ complex (complex III), and site 3 is the cytochrome $aa_3$ complex (complex IV). The succinate dehydrogenase complex (complex II) is not a coupling site. The ratio of protons translocated per two electrons varies, depending upon the complex. A consensus value of 10 protons are extruded per two electrons that travel from NADH to oxygen. The $bc_1$ complex translocates 4 protons per two electrons, and depending upon the reported value, complexes I and IV translocate from two to four protons per two electrons. (The consensus value for mitochondrial complex I is $4H^+/2e^-$.) During reversed electron flow protons enter the cell through coupling sites 1 and 2, driven by the $\Delta p$, and the electrons are driven toward the lower redox potential. This creates a positive $\Delta E$ at the expense of the $\Delta p$. (See eq. 3.19.) Coupling site 3 is not physiologically reversible. Thus water cannot serve as a source of electrons for $NAD^+$ reduction using reversed electron flow. However, during oxygenic photosynthesis, light energy can drive electrons from water to $NADP^+$. The mechanism of photoreduction of $NADP^+$ is different from reversed electron flow and is discussed in Chapter 5.

### 4.5.1 The identification of coupling sites

For an understanding of the physiology of energy metabolism during electron transport, it is necessary to study the mechanism of proton translocation, and for this the coupling sites must be identified and isolated. The coupling sites can be identified by the use of electron donors that feed electrons into the chain at different places and measuring the amount of ATP made per $2e^-$ transfer through the respiratory chain. The number of ATPs made per $2e^-$ transfer to oxygen is called the *P/O ratio*. It is equal to the number of ATP molecules formed per atom of oxygen taken up. When an electron acceptor other than oxygen is used, then $P/2e^-$ is substituted for P/O. The P/O ratio is approximately equal to the number of coupling sites. (However, see Section 4.5.2.) In mitochondria, the oxidation of NADH results in a P/O ratio of about three, indicating that three coupling sites exist between NADH and $O_2$. The use of succinate as an electron donor results in a P/O ratio of approximately two. Since electrons from succinate enter at the ubiquinone level, this indicates that coupling site 1 occurs between NADH and ubiquinone (i.e., the NADH–ubiquinone oxidoreductase reaction, Fig. 4.7). The other two coupling sites must occur between ubiquinone and oxygen. When electrons enter the respiratory chain after the $bc_1$ complex, the P/O ratio is reduced to one, indicating that the $bc_1$ complex is the second coupling site and that site 3 is cytochrome $aa_3$ oxidase. Site 3 can be demonstrated by bypassing the $bc_1$ complex. The $bc_1$ complex can be bypassed using an artificial electron donor to reduce cytochrome c (e.g., ascorbate and tetramethylphenylenediamine, TMPD), thus channeling electrons from ascorbate to TMPD to cytochrome c to cytochrome $aa_3$. Alternatively, one can simply use reduced cytochrome c as an electron donor directly to reduce cytochrome $aa_3$. Each of these sites is characterized by a drop in midpoint potential of about 200 mV, which is sufficient for generating the $\Delta p$ (Fig. 4.9). The size of the $\Delta p$ that can be generated with respect to the $\Delta E$ is discussed in Section 3.7.

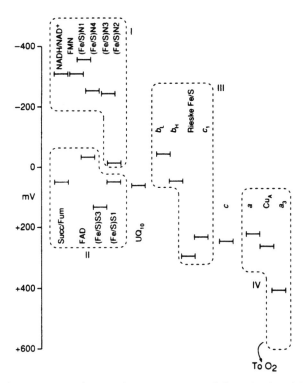

**Fig. 4.9** Average midpoint potentials, $E'_m$, for components of the mitochondrial respiratory chain. The complexes are in dotted boxes. The actual potentials ($E_h$) for most of the components are not very different from the midpoint potentials. An exception is cytochrome aa3, whose $E_h$ is much more positive than its midpoint potential. There are three sites where there are changes in potential of 200 mV or more. These drive proton translocation. One site is within complex I, a second within complex III, and the third between complex IV and oxygen. *Source:* Adapted from Nicholls, D. G., and S. J. Ferguson. 1992. *Bioenergetics 2.* Academic Press, London.

### 4.5.2 The actual number of ATPs that can be made per two electrons traveling through the coupling sites

According to the chemiosomotic theory, the number of ATPs made at a coupling site need not be a whole number. This is because the amount of ATP that can be made per coupling site is equal to the ratio of protons extruded at the coupling site to protons that re-enter via the ATP synthase (Fig. 4.10). For example, if two protons are translocated at a coupling site and three protons enter through the ATP synthase, then $\frac{2}{3}$ of an ATP can be made when two electrons pass through the coupling site. The ratio of protons translocated per two electrons traveling through all the coupling sites to $O_2$ is called the $H^+/O$ ratio. It can be measured by administering a pulse of a known amount of oxygen to an anaerobic suspension

of mitochondria or bacteria and measuring the initial efflux of protons with a pH electrode as the small amount of oxygen is used up. The experiment requires that valinomycin plus $K^+$ or a permeant anion such as thiocyanate, $SCN^-$, be in the medium to prevent a $\Delta\Psi$ from developing. (See Section 3.2.2 for a discussion of this point.) The reported values for $H^+/O$ for NADH oxidation vary. However, there is a consensus that the true value is probably around 10. The ratio of protons entering via the ATP synthase to ATP made is called the $H^+/ATP$. It can be measured using inverted submitochondrial particles prepared by sonic oscillation. These particles have the ATP synthase on the outside and will pump protons into the interior upon addition of ATP. Similar inverted vesicles can be made from bacteria by first enzymatically weakening or removing the cell wall and breaking the spheroplasts

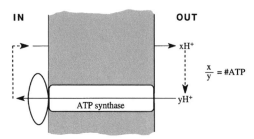

Fig. 4.10 The ratio of protons extruded to protons translocated through the ATPase determines the amount of ATP made.

or protoplasts by passage through a French press at high pressures. (See Note 16 for how to make right-side-out vesicles.) Values of H$^+$/ATP from two to four have been reported, and a consensus value of three can be used for calculations. For intact mitochondria, an additional H$^+$ is required to bring P$_i$ electroneutrally from the cytosol into the mitochondrial matrix in symport with H$^+$; so H$^+$/ATP would be four. A value of 10 for H$^+$/O predicts a maximum P/O ratio of 2.5 (10/4) for mitochondria. This means that the often-stated value of three for a P/O ratio for NADH oxidation by mitochondria may be too high. The number of protons ejected per 2 $e^-$ traveling between succinate and oxygen in mitochondria is six. Therefore, the maximal P/O ratio for this segment of the electron transport chain may be 6/4 or 1.5. The P/O ratios in bacteria can be higher since a proton is not required to bring P$_i$ into the cell. One significant aspect of branched aerobic respiratory chains in bacteria is that the number of coupling sites, and therefore the ratio of H$^+$/O, can differ in the branched chains. Thus the different branches are not equally efficient in generating a $\Delta p$ or making ATP. We will return to this point later.

Of course, an ATP can be made when three protons enter via the ATP synthase only if the $\Delta p$ is sufficiently large. As an excercise, we can ask how large the $\Delta p$ must be. Recall that $y \Delta p$ is the work that can be done in units of electron volts when $y$ protons traverse a proton potential of $\Delta p$ volts. If H$^+$/ATP is three, then the number of electron volts made available by proton influx through the ATP synthase is $3 \Delta p$. How many electron volts

are needed to synthesize an ATP? The free energy of formation of ATP at physiological concentrations of ATP, ADP, and P$_i$ is the phosphorylation potential, $\Delta G_p$, which is approximately 45,000–50,000 J. (See Note 17 for an explanation of the phosphorylation potential.) Dividing by the faraday (96,500 C) expresses the energy required to synthesize an ATP in electron volts. For 45,000 J this is 0.466 eV. Therefore, 3 $\Delta p$ must be greater than or equal to 0.466 eV. Thus the minimum $\Delta p$ is 0.466/3 or $-0.155$ V. Values of $\Delta p$ approximate to this are easily generated during electron transport. (However, see the discussion in Section 3.10 regarding the low $\Delta p$ in alkaliphiles.)

## 4.6 Q Loops, Q Cycles, and Proton Pumps

The previous discussion points out that proton translocation takes place at coupling sites when electrons travel "downhill" over a potential gradient of at least 200 mV. (See Fig. 4.9.) However, the mechanism by which the redox reaction is actually coupled to proton translocation was not explained. There are two ways in which this is thought to occur: (1) a Q loop or Q cycle, and (2) a proton pump. In the Q loop or Q cycle, reduced quinone carries hydrogen across the membrane and becomes oxidized, releasing protons on the external face of the membrane as the electrons return electrogenically via electron transport carriers to the inner membrane surface. On the cytoplasmic side, the same number of protons are taken up as were released on the outside as the terminal oxidant (e.g., oxygen) is reduced. The result is the net translocation of protons from the inside to the outside, although protons per se do not actually traverse the membrane. Proton translocation in the Q loop or Q cycle is referred to as *scalar* translocation. Proton pumps are electron carrier proteins that couple electron transfer to the electrogenic translocation of protons through the membrane. Such translocation of protons through the membrane is referred to as *vectorial*. Although these two mechanisms are fundamentally different, the result is the same, that is, the net translocation of protons across the membrane with the establishment of a $\Delta p$. The $\Delta p$ is the same regardless of whether the moving charges are electrons or protons, since they both carry the same charge. The relationship between the $\Delta p$ and the $\Delta E_h$ is given by eq. 3.19.

## 4.6.1 The Q loop

The essential feature of the Q loop model is that the electron carriers alternate between those that carry both hydrogen and electrons (flavoproteins and quinones) and those that carry only electrons (iron–sulfur proteins and

cytochromes). This is illustrated in Fig. 4.11. The electron carriers and their sequence are: flavoprotein (H carrier), FeS protein (e$^-$ carrier), quinone (H carrier), and cytochromes (e$^-$ carriers). The flavoprotein and FeS protein comprise the NADH dehydrogenase, which can be a coupling site (Section 4.6.3), although the mechanism of proton translocation by the NADH dehydrogenase is not understood. When electrons are transferred from the FeS protein to quinone (Q) on the inner side of the membrane, two protons are acquired from the cytoplasm. According to the model, the reduced quinone (QH$_2$), called quinol, then diffuses to the outer membrane surface and

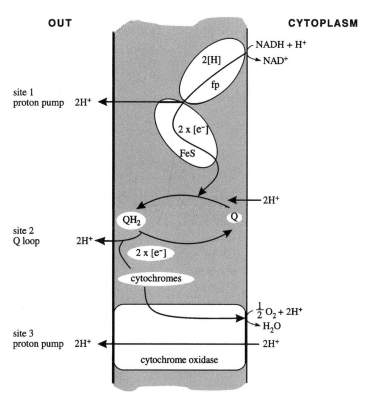

**Fig. 4.11** Proton translocation showing a Q loop and a proton pump. It is proposed that the electron carriers exist in an alternating sequence of hydrogen [H] and electron [e$^-$] carriers. This would be flavoprotein (fp), iron–sulfur protein (FeS), quinone (Q), and cytochromes. Oxidation of the flavoprotein deposits two protons on the outer membrane surface. The electrons return to the inner membrane surface, where a quinone is reduced, taking up two protons from the cytoplasm. The reduced quinone diffuses to the outer surface of the membrane, where it is oxidized, depositing two more protons on the surface. The electrons return to the cytoplasmic surface via cytochromes, where they reduce oxygen in a reaction that consumes protons. Some cytochrome oxidases function as proton pumps. During anaerobic respiration the cytochrome oxidase is replaced by a reductase, and the electrons reduce some other electron acceptor, e.g., nitrate or fumarate. It should be noted that electrons can also enter at the level of quinone, e.g., from succinate dehydrogenase.

becomes oxidized, releasing the two protons. The electrons then return via cytochromes to the inner membrane surface, where they reduce oxygen. This would create a $\Delta\Psi$ because the protons are left on the outer surface of the membrane as the electrons move electrogenically to the inner surface. Thus quinol oxidation is a second coupling site. As mentioned previously, the energy to create the membrane potential is derived from the $\Delta E_h$ between the oxidant and the reductant. One way to view this is that the energy from the $\Delta E_h$ "pushes" the electron to the negative membrane potential on the inside surface. Note that the role of the quinone is to ferry the hydrogens across the membrane, presumably by diffusing from a reduction site on the cytoplasmic side of the membrane to an oxidation site on the outer side. For the purpose of calculating the expected $\Delta p$ generated at a coupling site, it makes no difference whether one postulates the transmembrane movement of protons in one direction or electrons in the opposite direction, since they both carry the same charge (eq. 3.19).

## 4.6.2 The Q cycle

Although the linear Q loop as described above for the oxidation of quinol may accurately describe quinol oxidation in *E. coli* and some other bacteria, it is inconsistent with experimental observations of electron transport in mitochondria, chloroplasts, and many bacteria. For example, the Q loop predicts that the ratio of $H^+$ released per $QH_2$ oxidized is two, whereas the measured ratio in mitochondria and many bacteria is actually four. In order to account for the extra two protons, Peter Mitchell suggested a new pathway for the oxidation of quinol called the Q cycle.[18] The Q cycle operates in an enzyme complex called the $bc_1$ complex (complex III). The $bc_1$ complex from bacteria contains three polypeptides. These are cytochrome b with two b-type hemes, an iron–sulfur protein containing a single 2Fe–2S cluster (the Rieske protein), and cytochrome $c_1$ with one heme. The complex spans the membrane and has a site for binding reduced ubiquinone, $UQH_2$,

on the outer surface of the membrane called site P (for positive), and a second site on the inner surface for binding UQ, site N (for negative) (Fig. 4.12). At site P, $UQH_2$ is oxidized to the semiquinone anion, $UQ^-$ (Fig. 4.12A), and the two protons are released to the membrane surface. The electron travels to the FeS protein, and from there to cytochrome $c_1$ on its way to the terminal electron acceptor. The $UQ^-$ is then oxidized to UQ by the removal of the second electron, which is transferred to $b_{556}$, also called $b_L$ because of its relatively low $E'_m$. The electron is transferred across the membrane to $b_{560}$, also called $b_H$ because of its relatively high $E'_h$. Heme $b_H$ transfers the electron to UQ bound at site N, reducing it to the semiquinone anion, $UQ^-$. *Electron flow from the $Q_P$ site to the $Q_N$ site is transmembrane and creates a membrane potential.* A second $UQH_2$ is oxidized at the P site, and the $UQ_N^-$ is reduced to $UQH_2$, picking up two protons from the cytoplasm (Fig. 4.12B). The $UQH_2$ enters the quinone pool. Thus, for every two $UQH_2$ that are oxidized releasing four protons, one $UQH_2$ is regenerated. The net result is the oxidation of one $UQH_2$ to UQ, with the release of four protons. The situation can be summarized as the following:

$$\text{P site}: 2UQH_2 + 2\text{cyt } c_{ox} \longrightarrow 2UQ$$
$$+ 2\text{cyt } c_{red} + 4H^+_{out} + 2e^-$$

$$\text{N site}: UQ + 2e^- + 2H^+_{in} \longrightarrow UQH_2$$

$$UQH_2 + 2\text{cyt } c_{ox} + 2H^+_{in} \longrightarrow UQ$$
$$+ 2\text{cyt } c_{red} + 4H^+_{out}$$

In the pre-steady state, $UQH_2$ can be oxidized at the N site (by reversal) as well as the P site. The P site is inhibited by myxathiozol and by stigmatellin, whereas the N site is inhibited by antimycin. Therefore, as discussed later, a combination of inhibitors is required completely to inhibit the oxidation of $UQH_2$ in the pre-steady state.

### Bioenergetics of the Q cycle

Since the Q cycle translocates two protons per electron that flows to the terminal electron acceptor, it generates a larger proton current than the Q loop that translocates only one

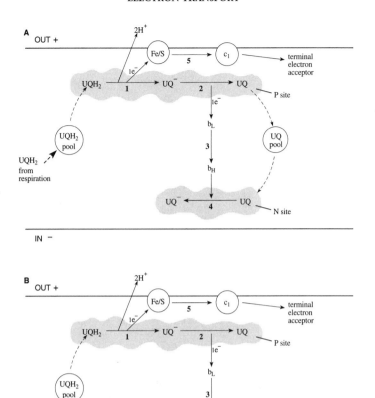

**Fig. 4.12** The $bc_1$ complex. The $bc_1$ complex isolated from bacteria contains three polypeptides, which are a cytochrome b containing two heme b groups, $b_H$ and $b_L$, an iron–sulfur protein, Fe/S (the Rieske protein), and cytochrome $c_1$. It is widely distributed among the bacteria, including the photosynthetic bacteria. (Mitochondrial $bc_1$ complexes are similar but contain an additional 6–8 polypeptides without prosthetic groups.) The iron–sulfur protein and cytochrome $c_1$ are thought to be located on the outside (positive or p) surface of the membrane, and the cytochrome b is believed to span the membrane acting as an electron conductor. On the outer surface there is a binding site in the $bc_1$ complex for ubiquinol, $UQH_2$ (the P site). On the inner surface (negative or n) there is a binding site in the $bc_1$ complex for ubiquinone, UQ (the N site). (A) Reduced ubiquinone binds to the P site, and one electron is removed, forming the semiquinone anion, $UQ^-$ (reaction 1). At this time two protons are released on the outer membrane surface. The electron that is removed is transferred to the iron–sulfur protein and from there to cytochrome $c_1$ (step 5). (Cytochrome $c_1$ transfers the electron to the terminal electron acceptor via a series of electron carriers in other reactions.) In reaction 2 the second electron is removed from the semiquinone anion, producing the fully oxidized quinone, UQ. The second electron is transferred transmembrane via cytochrome b to ubiquinone at site N (steps 3 and 4). Because the electron travels transmembrane, a membrane potential is created, outside positive. B. A second reduced ubiquinone is oxidized at the P site, releasing two more protons, and the sequence of electron transfers is repeated. The $UQ^-$ at site N becomes reduced to $UQH_2$, having acquired two protons from the cytoplasm. Note that for every two $UQH_2$ molecules that are oxidized, four protons are released, and one $UQH_2$ is regenerated. Therefore, four protons are released per one $UQH_2$ oxidized. Another way of saying this is that the ratio, $H^+/e^-$, is two.

proton per electron. This can result in more ATP synthesis. Consider the situation where an ATP synthase requires the influx of three protons to make one ATP. If the transfer of an electron through the electron transport pathway resulted in the translocation of one proton, then $\frac{1}{3}$ of an ATP could be made per electron. On the other hand, if electron transport resulted in the translocation of two protons per electron, then $\frac{2}{3}$ of an ATP could be made per electron. In other words, the size of the proton current generated by respiration determines the upper value of the amount of ATP that can be made.

## Distinguishing the Q loop from the Q cycle

When examining the physiology of electron transport in particular bacteria, it is important to learn whether a Q loop or a Q cycle is operating. There are several features of the Q cycle that help to distinguish it from the linear pathway of electron flow in the Q loop. These include:

1. The ratio of $H^+$ extruded in the Q cycle per $e^-$ is two rather than one.

2. Cytochrome b can be reduced by quinol at two sites, site P or site N (by reversal), whereas in a linear respiratory chain there would be only one site, because there is only one site for $UQH_2$ oxidation. These sites can be distinguished using inhibitors, as described next.

Examine Fig. 4.12A. Notice that there are two sites for the oxidation of $UQH_2$: site P (steps 1 and 2) and site N (steps 6 and 4). (Site N will oxidize $UQH_2$ in a pre-steady state by reversing electron flow.) One can follow the oxidation of $UQH_2$ by measuring the reduction of cytochrome b spectrophotometrically (see Section 4.2.4). These two sites for the reduction of cytochrome b can be demonstrated with the use of inhibitors. Myxothiazol and stigmatellin block the oxidation of ubiquinol at site P, whereas antimycin blocks the oxidation of ubiquinol at site N. These inhibitions are shown in Fig. 4.13. Therefore, antimycin alone does not inhibit the reduction of cytochrome $c_1$ or cytochrome b in the pre-steady state, since the electrons are coming from site

P. Also, myxothiazol and stigmatellin do not inhibit the reduction of cytochrome b, since electrons can come from site N. Thus the presence of both myxothiazol (or stigmatellin) and antimycin are necessary to block the reduction of cytochrome b.

The fact that one requires two different inhibitors of $UQH_2$ oxidation to block the reduction of cytochrome b in the pre-steady state indicates that there are two routes for cytochrome b reduction by $UQH_2$, and is evidence for a Q cycle.

### 4.6.3 Pumps

Proton pumps also exist. These catalyze the electrogenic translocation of protons across the membrane rather than electrons (Fig. 4.11). For example, proton extrusion accompanies the cytochrome $aa_3$ oxidase reaction when cytochrome c is oxidized by oxygen in mitochondria and some bacteria. This can be observed by feeding electrons into the respiratory chain at the level of cytochrome c, thus bypassing the quinone. The experimental procedure is to incubate the cells in lightly buffered anaerobic media with a reductant for cytochrome c and a permeant anion (e.g., $SCN^-$ or valinomycin plus $K^+$). Changes in pH are measured with a pH electrode. Upon the addition of a pulse of oxygen, given as an air-saturated salt solution, a sharp, transient acidification of the medium occurs, and the $H^+/O$ can be calculated. If the electron donor itself releases protons, then these scalar protons are subtracted from the total protons translocated to determine the number of vectorially translocated protons (due to proton pumping by the cytochrome c oxidase). For example, when ascorbate plus TMPD are used to reduce cytochrome c, the ascorbate is oxidized to dehydroascorbate with the release of one scalar proton per two electrons. Any additional protons released are due to proton pumping by the cytochrome c oxidase. When similar experiments were done with *Paracoccus denitrificans*, it was found that the *P. denitrificans* cytochrome $aa_3$ oxidase translocates protons with a stoichiometry of $1H^+/1e^-$.[19, 20] Cytochrome oxidase pumping activity can also be demonstrated

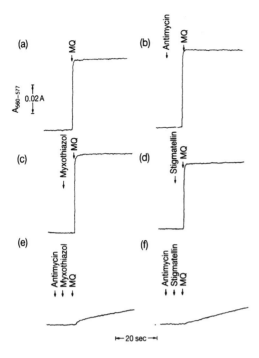

$A_{560-577}$  0.02 A

(a) MQ

(b) ↓ Antimycin MQ

(c) ↓ Myxothiazol MQ

(d) ↓ Stigmatellin MQ

(e) Antimycin Myxothiazol MQ ↓ ↓ ↓

(f) Antimycin Stigmatellin MQ ↓ ↓ ↓

⊢— 20 sec —⊣

Fig. 4.13 Inhibitor studies can provide evidence for a Q cycle. One distinguishing feature of the Q cycle is that there are two sites for cytochrome b reduction, i.e., site P and N. At site P quinol is oxidized by the iron–sulfur protein generating the semiquinone anion ($Q^-$). This oxidation is inhibited by stigmatellin or myxothiazol. The semiquinone anion, $Q^-$, reduces cytochrome b. Thus stigmatellin or myxothiazol prevents the oxidation of quinol and the reduction of cytochrome b at site P. The reduced cytochrome b is reoxidized by Q or $Q^-$ in an antimycin sensitive step at site N. If site P is blocked, e.g., by myxothiazol or stigmatellin, or by removal of the iron–sulfur protein, then cytochrome b can be reduced by quinol by a reversal of the antimycin-sensitive step at site N. There are therefore two routes for the reduction of cytochrome b, as opposed to a single route in a linear pathway between quinol and cytochrome b. This experiment was done with a purified bc₁ complex isolated from the bacterium *Paracoccus denitrificans*, which was incorporated into liposomes. The reductant was menaquinol (MQ). Cytochrome b reduction is reflected by a rise in the trace, which is a measure of the absorbance of reduced cytochrome b. The addition of antimycin, myxothiazol, or stigmatellin alone did not prevent the reduction of cytochrome b by MQ. However, the addition of both antimycin and stigmatellin, or antimycin and

myxothiazol, blocked cytochrome b reduction. Trace a, control without inhibitors. Traces b, c, d, antimycin, myxothaizol, and stigmatellin, respectively. Traces e and f, antimycin with myxothiazol and antimycin with stigmatellin. *Source:* From Yang, X, and B. L. Trumpower. 1988. Protonmotive Q cycle pathway of electron transfer and energy transduction in the three-subunit ubiquinol–cytochrome c oxidoreductase complex of *Paracoccus denitrificans*. *J. Biol. Chem.* **263**:11962–11970.

in proteoliposomes made with purified cytochrome aa₃. Because the proton pumps move a positive charge across the membrane, leaving behind a negative charge, a membrane potential, outside positive, develops. The membrane potential should be the same as when an electron moves inward, since the proton and the electron carry the same charge (i.e., $1.6 \times 10^{-19}$ C). The mechanism of pumping is not known but probably requires conformational changes in cytochrome oxidase resulting from its redox activity. Conformational changes probably also occur in bacteriorhodopsin during proton pumping (Section 3.8.4).

The mitochondrial NADH dehydrogenase complex (NADH:ubiquinol oxidoreductase) translocates four protons per NADH oxidized.[21] However, the mechanism has not yet been elucidated. Two types of NADH:ubiquinol oxidoreductases exist in bacteria.[8–11] One of these, called NDH-1, is similar to the mitochondrial complex I in that it is a multisubunit enzyme complex (approximately 14 polypeptide subunits) consisting of FMN and FeS clusters, and translocates protons across the membrane during NADH oxidation ($4H^+/2e^-$). The mechanism of proton translocation is not known, but it is referred to as a proton pump. [The marine bacterium, *Vibrio alginolyticus*, has a sodium-translocating NDH (Na-NADH). See Section 3.7.1.] A second NADH:ubiquinol oxidoreductase, called NDH-2, may also be present. NDH-2 differs from NDH-1 in consisting of a single polypeptide and FAD, and not being an energy-coupling site. In *E. coli*, NDH-1 and

NDH-2 are simultaneously present.[22] How *E. coli* regulates the partitioning of electrons between the two NAD dehydrogenases is not known, but clearly has important energetic consequences.

## 4.7 Patterns of Electron Flow in Individual Bacterial Species

Although the major principles of electron transport as outlined above apply to bacteria in general, several different patterns of electron flow exist in particular bacteria, often within the same bacterium grown under different conditions.[23] The patterns of electron flow reflect the different sources of electrons and electron acceptors that are used by the bacteria. For example, bacteria may synthesize two or three different oxidases in the presence of air and several reductases anaerobically. In addition, certain dehydrogenases are made only anaerobically because they are part of an anaerobic respiratory chain. Furthermore, whereas electron donors such as NADH and $FADH_2$ generated in the cytoplasm are oxidized inside the cell, there are many instances of oxidations in the periplasm in gram-negative bacteria. Examples of substances that are oxidized in the periplasm include hydrogen gas, methane, methanol, methylamine, formate, perhaps ferrous ion, reduced inorganic sulfur, and elemental sulfur. In most of these instances periplasmic cytochromes c accept the electrons from the electron donor and transfer them to electron carriers in the membrane. The discussion that follows is not complete, but is meant to convey the diversity of electron transport systems found in bacteria. The electron transport chains for the respiratory metabolism of ammonia, nitrite, inorganic sulfur, and iron are described in Chapter 12.

### 4.7.1 Escherichia coli

*E. coli* is a gram-negative heterotrophic facultative anaerobe. It can be grown aerobically using oxygen as an electron acceptor, anaerobically (e.g., using nitrate or fumarate as the electron acceptor) or anaerobically via fermentation (Chapter 14). The bacteria adapt to their surroundings, and the electron transport system that the cells assemble reflects the electron acceptor that is available. (See Ref. 24 for a review.) All the electron transport chains present in *E. coli* branch at the level of quinone, which connects the different dehydrogenases with the various terminal reductases and oxidases that are present under different growth conditions. *E. coli* makes three quinones, and their relative amounts depend upon the nature of the electron acceptor. These are ubiquinone (UQ), menaquinone (MQ), and demethylmenaquinone (DMQ). When growing aerobically UQ accounts for 60% of the total quinone, DMQ is 37% of the total, and MQ is only about 3%. However, when growing anaerobically, very little UQ is made. When growing anaerobically on nitrate, the major quinones are DMQ (70%) and MQ (30%). When growing anaerobically on fumarate or DMSO, the major quinone is MQ (74%), with DMQ (16%) and UQ (10%) contributing the rest.[24] There is no $bc_1$ complex and no cytochrome c, which may serve as additional branch points in other bacterial respiratory chains (e.g., see Section 4.7.2 on *Paracoccus denitrificans*).

### Aerobic respiratory chains

When grown aerobically, *E. coli* makes two different quinol oxidase complexes, cytochrome bo complex (has heme b and heme o and is also called $bo_3$ or cytochrome o) and cytochrome bd complex (has heme b and heme d and is also called cytochrome d), resulting in a branched respiratory chain to oxygen (Fig. 4.14).[25-29] In these pathways electrons flow from ubiquinol to the terminal oxidase complex, which is why the oxidases are called (ubi)quinol oxidases. The cytochrome bo complex from *E. coli* is a proton pump, with a stoichiometry of $1H^+/e^-$. Cytochrome bo complex is the predominant oxidase when the oxygen levels are high. When the oxygen tensions are lowered, *E. coli* makes more cytochrome bd relative to cytochrome bo. The cytochrome bd complex has a higher affinity for oxygen (lower $K_m$) than does the cytochrome bo complex, suggesting a rationale for why it is the dominant oxidase

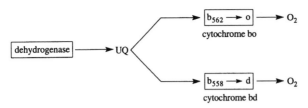

**Fig. 4.14** Aerobic respiratory chain in *E. coli*. The chain branches at the level of ubiquinone (UQ) to two alternate quinol oxidases, cytochrome bo and bd. Cytochrome bd complex has a higher affinity for oxygen and is synthesized under low oxygen tensions, where it becomes the major route to oxygen. Coupling sites, i.e., sites where protons are translocated to the outer surface, are indicated by the $H^+/e^-$ ratio. Proton translocation occus during NADH oxidation, catalyzed by NADH:ubiquinone oxidoreductase, also called NDH-1, and quinol oxidation, catalyzed by cytochrome bo complex or cytochrome bd complex. Additionally, cytochrome bo is a proton pump. *E. coli* has a second NADH dehydrogenase, called NADH-2, which does not translocate protons. Therefore, the number of protons translocated per NADH oxidized can vary from two (NADH-2 and bd complex) to 8 (NADH-1 and bo complex).

under low oxygen tensions. Cytochrome bd is not a proton pump. In the same sense that the $bc_1$ complex is a coupling site because it catalyzes the oxidation of quinol resulting in the scalar extrusion of protons, both the bo and bd complexes are coupling sites. The difference between the two, as mentioned, is that the bo complex is also a proton pump and catalyzes the vectorial extrusion of protons. The oxidation of quinol by the bo complex results in 2 protons translocated per electron (one scalar and one vectorial), whereas the oxidation of quinol by the bd complex results in only 1 proton translocated per electron (scalar). *E. coli* has two NADH dehydrogenases, NDH-1 and NDH-2, encoded by the *nuo* operon and *ndh*, respectively. Only NDH-1 is a coupling site (proton pump). It is a complex enzyme consisting of 14 subunits and translocates 2 protons per electron. As discussed in Section 4.6.3, NDH-1 is similar to complex I of mitochondria. It has been shown that NDH-1 is used during fumarate respiration, whereas NDH-2 is used primarily during aerobic and nitrate respiration.[30] The regulation of transcription of the NADH dehydrogenase genes in *E. coli* is complex.[30, 31]

*Physiological significance of alternate electron routes that differ in the number of coupling sites*

Since the NDH-1 dehydrogenase may translocate as many as 4 protons per $2\ e^-$, whereas the NDH-2 dehydrogenase translocates 0, the number of protons translocated per NADH oxidized can theoretically vary from 2 (NDH-2 and bd complex) to 8 (NDH-1 and bo complex). Assuming a $H^+/ATP$ of 3 (i.e., the ATP synthase translocates inwardly 3 protons per ATP made), the ATP yields per NADH oxidized can vary fourfold from $\frac{2}{3}$, or 0.67, to $\frac{8}{3}$, or 2.7. It also means that *E. coli* has great latitude in adjusting the $\Delta p$ generated during respiration. Since a large $\Delta p$ can drive reversed electron transport and thus slow down oxidation of NADH and quinol, it may be an advantage to be able to direct electrons along alternate routes that bypass coupling sites and translocate fewer protons. This could ensure adequate rates of reoxidation of NADH and quinol.

*Anaerobic respiratory chains*

In the absence of oxygen, *E. coli* can use either nitrate or fumarate as an electron acceptor. The nitrate ($NO_3^-$) is reduced to nitrite ($NO_2^-$), which is further reduced to ammonia ($NH_4^+$), and the fumarate is reduced to succinate, all of which are excreted into the medium.[32] (See Note 33 for a further description of fumarate and nitrate reduction.) The anaerobic respiratory chains consists of a dehydrogenase, a reductase, and a diffusible quinone to mediate the transfer of electrons between the dehydrogenase and the reductase (Fig. 4.15). The number of coupling sites

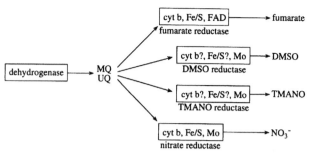

**Fig. 4.15** Anaerobic respiratory chains in *E. coli*. When oxygen is absent, *E. coli* synthesizes any one of several membrane-bound reductase complexes depending upon the presence of the electron acceptors. Nitrate induces the synthesis of nitrate reductase and represses the synthesis of the other reductases. Menaquinone ($E'_0 = -74$ mV) (or demethylmenaquinone, $E'_0 = -40$ mV) must be used to reduce some of the reductases, e.g., fumarate reductase, because it has a sufficiently low midpoint potential. Ubiquinol ($E'_0 = +100$ mV) or menaquinone can reduce nitrate reductase because nitrate has a $E'_0$ of 421 mV. Each reductase may be a complex of several proteins and prosthetic groups through which the electrons travel to the terminal electron acceptor. The transfer of electrons from the dehydrogenases to the reductases results in the establishment of a proton potential. If the dehydrogenase has site 1 activity, there can theoretically be two coupling sites, one at the dehydrogenase step and one linked to quinol oxidation at the reductase step. Abbreviations: cyt b, cytochrome b; Fe/S, nonheme iron–sulfur protein; FAD, flavoprotein with flavin adenine dinucleotide as the prosthetic group; Mo, molybdenum; TMANO, trimethylamine N-oxide; DMSO, dimethylsulfoxide; MQ, menaquinone; UQ, ubiquinone.

depends upon whether the electron acceptor is nitrate or fumarate.[24] When nitrate is the electron acceptor, then there can be two coupling sites, one at the dehydrogenase step (site 1) and one at the quinol oxidation step (the nitrate reductase, scalar). The latter means that the oxidation of one quinol and release of two protons takes place on the periplasmic side of the membrane, and the uptake of two protons during the reduction of one nitrate to nitrite takes place on the cytoplasmic side. Thus the oxidation of quinol by nitrate catalyzed by nitrate reductase yields a $H^+/e^-$ ratio of one.[24] See the discussion of the reaction for nitrate reductase in Section 4.7.2.

However, the $H^+/e^-$ ratio for quinol oxidation by fumarate in *E. coli* or *W. succinogenes* is zero, indicating that the release of protons when $MQH_2$ is oxidized and the uptake of protons when fumarate is reduced both take place on the cytoplasmic side of the membrane. (It does not appear that either fumarate reductase or any of the nitrate reductases are proton pumps.) Thus, in order for *E. coli* to make ATP during the ox-

idation of NADH by fumarate, it must use NDH-1 rather than NDH-2, because only the former is a coupling site. In agreement with this conclusion, it has been found that the genes for NDH-1 (*nuo* genes) are essential for anaerobic respiration with fumarate as the electron acceptor. (Reviewed in Ref. 34.) Anaerobic respiration occurs only in the absence of oxygen because oxygen represses the synthesis of the nitrate and fumarate reductases. When presented with nitrate anaerobically, only the nitrate reductase is made, because nitrate represses the synthesis of the fumarate reductase. (The regulation by oxygen nitrate and nitrite of the electron transport chain in *E. coli* is discussed in Sections 18.2 and 18.3.) Thus there exists a hierarchy of electron acceptors for *E. coli* (i.e., oxygen > nitrate > fumarate). This may reflect the energy yields of the electron transport pathways, since oxygen has the highest electrode potential and fumarate the lowest (Table 4.1). When *E. coli* grows in the absence of oxygen, the citric acid cycle does not operate as an oxidative pathway (Section 18.2).

## 4.7.2 Paracoccus denitrificans

*P. denitrificans* is a nonfermenting gramnegative facultative anaerobe that can obtain energy from either aerobic respiration or nitrate respiration.[18, 35, 36] It is found primarily in soil and sewage sludge. The bacteria can grow heterotrophically on a wide variety of carbon sources, or autotrophically on $H_2$ and $CO_2$ under anaerobic conditions using nitrate as the electron acceptor. It can be isolated from soil by anaerobic enrichment with media containing $H_2$ as the source of energy and electrons, $Na_2CO_3$ as the source of carbon, and nitrate as the electron acceptor. Electron transport in *P. denitrificans* receives a great deal of research attention because certain features closely resemble electron transport in mitochondria.

### Aerobic pathway

*P. denitrificans* differs from *E. coli* in that it has a $bc_1$ complex and a cytochrome $aa_3$ oxidase (cytochrome c oxidase), and in this way resembles mitochondria. In addition to the cytochrome $aa_3$, there are two other terminal oxidases in the aerobic pathway.[20, 36] These are a different cytochrome c oxidase (cytochrome $cbb_3$) and a ubiquinol oxidase (cytochrome $bb_3$). (Cytochromes $cbb_3$ are heme–copper oxidases found in several bacteria, including *Thiobacillus*, *Rhodobacter*, *Paracoccus*, and *Bradyrhizobium*.[37] Previously, cytochrome $bb_3$ was called cytochrome o or $b_0$, but no heme O was detected when the hemes were extracted from membranes and analyzed by reversed-phase HPLC.[38, 39, 20]) The cytochrome $aa_3$ and cytochrome $bb_3$ are proton pumps. (Whether or not cytochrome $cbb_3$ pumps protons apparently depends upon the assay conditions.[36]) The aerobic pathways as well as the sites of proton translocation are shown in Fig. 4.16. Electrons traveling from NADH to oxygen can pass through as many as three coupling sites (NDH-1, $bc_1$ complex, cyt $aa_3$ or perhaps cyt $cbb_3$) or as few as two couping sites (NDH-1 and cyt $bb_3$). Recall that the $bc_1$ complex and cyt $bb_3$ are coupling sites because they oxidize quinol, whereas the NDH-1 and cytochrome c oxidases are proton pumps. As described later, *P. denitrificans* also oxidizes methanol, and in this case the electrons enter at a cytochrome c, thus bypassing the $bc_1$ site.

### Anaerobic pathway

*P. denitrificans* can also grow anaerobically using nitrate as an electron acceptor, reducing it to nitrogen gas in a process called *denitrification* (Fig. 4.16).[32, 40] During anaerobic growth on nitrate *P. denitrificans* has a complete citric acid cycle, which donates electrons to the electron transport chain, but the electron transport chain is very different than during aerobic growth. The cells contain cytochrome $bb_3$, nitrate reductase, nitrite reductase, nitric oxide reductase, and nitrous oxide reductase. (For a more complete description of these enzymes, see Note 41.) The levels of cytochrome $aa_3$ are very low. The cytochrome c shown in Fig. 4.16 is periplasmic, although there is also cytochrome c in the membrane associated with some of the electron carriers. The nitrate ($NO_3^-$) is reduced to nitrite ($NO_2^-$) in a two-electron transfer via a membrane-bound nitrate reductase. The $NO_2^-$ is reduced to nitric oxide (NO) in a one-electron transfer via a periplasmic nitrite reductase. The NO is reduced to one-half a mole of nitrous acid ($\frac{1}{2}$ $N_2O$) in a one-electron step via a membrane-bound nitric oxide reductase. And the $\frac{1}{2}$ $N_2O$ is reduced to one-half a mole of dinitrogen ($\frac{1}{2}$ $N_2$) in a one-electron step by a periplasmic nitrous oxide reductase. Thus a total of five electrons flow in and out of the cell membrane through membraneous electron and periplasmic electron carriers from ubiquinol to the various reductases in order to reduce one mole of $NO_3^-$ to $\frac{1}{2}$ $N_2$. As shown in Fig. 4.16, the electron transport pathway includes several branches to the individual reductases. The first branch site is at $UQH_2$, where electrons can flow either to nitrate reductase or to the $bc_1$ complex. Then there are three branches to the three other reductases after the $bc_1$ complex at the level of cytochrome c. In agreement with the model, electron flow to nitrate reductase is not sensitive to inhibitors of the $bc_1$ complex, whereas electron flow to the other reducatses is sensitive to the inhibitors. The nitrate reductase spans the membrane, and it has been proposed that it creates a $\Delta p$ via a Q loop,

**aerobic**

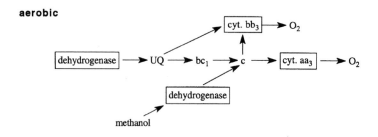

**anaerobic**

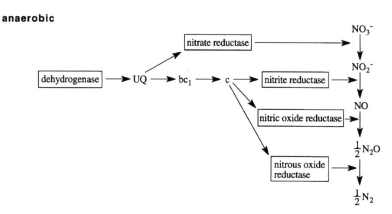

**Fig. 4.16** A model for electron transport pathways in *Paracoccus denitrificans*. *Aerobic*. The pathway has two branch points. One branch is at the level of ubiquinone leading to one of two ubiquinol oxidases, i.e., the cyt $bc_1$ complex or cyt $bb_3$. Both of these quinol oxidases are coupling sites. The $bc_1$ complex extrudes 2 protons per electron via the Q cycle. Cyt $bb_3$ is a proton pump and extrudes one proton per electron vectorially, whereas the second proton is extruded via a Q loop. A second branch point occurs at the level of the $bc_1$ complex. Electrons can flow either to cyt $aa_3$, which is a proton pump, or to cyt $cbb_3$, which has been reported to pump protons under certain experimental conditions. NDH 1 is also a coupling site. *Anaerobic*. When the bacteria are grown anaerobically using nitrate as the electron acceptor, the cytochrome $aa_3$ levels are very low, and the electrons travel from ubiquinone to nitrate reductase and also through the $bc_1$ complex to nitrite reductase, nitric oxide reductase, and nitrous oxide reductase.

similar to that described in Section 4.6.1. It is proposed that the nitrate reductase accepts electrons from $UQH_2$ on the periplasmic side of the membrane and reduces nitrate on the cytoplasmic side. This would create a $\Delta\Psi$ as two electrons from $UQH_2$ flow electrogenically across the membrane, leaving two $H^+$ on the outside. In the cytoplasm protons are taken up during nitrate reduction according to the following reaction:

$$2H^+ + 2e^- + NO_3^- \longrightarrow NO_2^- + H_2O$$

The result is the net translocation of protons to the outside, although only electrons moved across the membrane, not protons. This is the same as the Q loop described in Section 4.6.1 except that the terminal electron acceptor is nitrate instead of oxygen. The electrons to the other reductases flow from $UQH_2$ through the $bc_1$ complex, and a $\Delta p$ is generated via the Q cycle catalyzed by the $bc_1$ complex, which, as described in Section 4.6.1, is a modification of the Q loop that results in the translocation of two protons per electron rather than one.

In *P. denitrificans, Pseudomonas aeruginosa*, as well as many other facultative anaerobes, both the synthesis and activity of the denitrifying enzymes are prevented by oxygen. However, in certain other facultative anaerobes, including *Comamonas* sp., certain species of *Pseudomonas*, *Thiosphaera pantotropha*, and *Alcaligenes faecalis*, denitrifying enzymes are made and are active in the presence of oxygen.[42] In these systems, both oxygen and nitrate are used simultaneously as electron acceptors, although aeration can significantly decrease the rate of nitrate reduction. The advantage of corespiration using both oxygen and nitrate is not obvious.

## Periplasmic oxidation of methanol

Many gram-negative bacteria oxidize substances in the periplasm and transfer the electrons to membrane-bound electron carriers, often via periplasmic cytochromes c. (See Section 1.2.4 for a description of the periplasm.) An example is *P. denitrificans*, which can grow aerobically on methanol ($CH_3OH$) by oxidizing it to formaldehyde (HCHO) and $2H^+$ with a periplasmic dehydrogenase, methanol dehydrogenase (Fig. 4.17).

$$CH_3OH \longrightarrow HCHO + 2H^+$$

(Growth on methanol is autotrophic, since the formaldehyde is eventually oxidized to $CO_2$, which is assimilated via the Calvin cycle.) The electrons are transferred from the dehydrogenase to c-type cytochromes in the periplasm. The cytochromes c transfer the electrons to cytochrome $aa_3$ oxidase in the membrane, which reduces oxygen to water on the cytoplasmic surface. A $\Delta p$ is established due to the inward flow of electrons and the outward pumping of protons by the cytochrome $aa_3$ oxidase, as well as the release of protons in the periplasm during methanol oxidation and consumption in the cytoplasm during oxygen reduction. Since the oxidation of methanol bypasses the $bc_1$ coupling site, the ATP yields are lower. In addition to methanol, the bacteria can grow on methylamine ($CH_3NH_3^+$), which is oxidized by methylamine dehydrogenase to formaldehyde, $NH_4^+$, and $2H^+$.

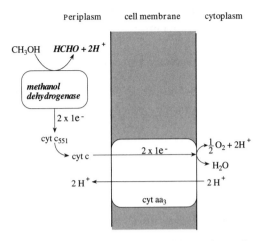

**Fig. 4.17** Oxidation of methanol by *P. denitrificans*. Methanol is oxidized to formaldehyde by a periplasmic methanol dehydrogenase. The electrons are transferred to periplasmic cytochromes c and to a membrane-bound cytochrome $aa_3$, which is also a proton pump. A $\Delta p$ is created due to the electrogenic influx of electrons and the electrogenic efflux of protons, accompanied by the release of protons in the periplasm (methanol oxidation) and uptake in the cytoplasm (oxygen reduction).

$$CH_3NH_3^+ + H_2O \longrightarrow HCHO + NH_4^+ + 2H^+$$

Methylamine dehydrogenase is also located in the periplasm and donates electrons via cytochromes c to cytochrome $aa_3$, bypassing the $bc_1$ complex.

## 4.7.3 Electron transport chain in Rhodobacter sphaeroides

*Rhodobacter sphaeroides* is a purple photosynthetic bacterium that can be grown photoheterotrophically under anaerobic conditions or aerobically in the dark. When growing aerobically in the dark, the bacteria obtain energy from aerobic respiration. The respiratory chain resembles the mitochondrial and *P. denitrificans* respiratory chains, in that it consists of NADH:ubiquinone oxidoreductase (coupling site 1), a $bc_1$ complex (coupling site 2), and a cytochrome $aa_3$ complex (coupling site 3). As in *E. coli*, there are two NADH:ubiquinone oxidoreductases: NDH-1 and NDH-2.

### 4.7.4 Fumarate respiration in Wolinella succinogenes

Fumarate respiration occurs in a wide range of bacteria growing anaerobically.[43] This is probably because fumarate itself is formed from carbohydates and protein during growth. The following is a description of the electron transport pathway in *W. succinogenes*, a gram-negative anaerobe isolated from the rumen. *W. succinogens* can grow at the expense of $H_2$ or formate, both produced in the rumen by other bacteria. The electron transport pathway is shown in Fig. 4.18A. The active sites for both the hydrogenase and formate dehydrogenase are periplasmic, whereas the active site for the fumarate reductase is cytoplasmic. An examination of the topology of the components of the respiratory chain reveals how a $\Delta p$ is generated.

### Topology of the components of the electron transport pathway

The electron transport chain consists of a periplasmic enzyme that oxidizes the electron donor, a membrane-bound menaquinone (MQ) that serves as an intermediate electron carrier, and membrane-bound fumarate reductase, which accepts the electrons from the menaquinone and reduces fumarate on the cytoplasmic side of the membrane (Fig. 4.18B). Both the hydrogenase and formate dehydrogenase are made of three polypeptide subunits, two facing the periplasm and one an integral membrane protein (cytochrome b). Note that cytochrome b serves not only as a conduit for electrons, but also binds the dehydrogenases into the membrane. The two periplasmic subunits of the hydrogenase are a Ni-containing protein subunit and an

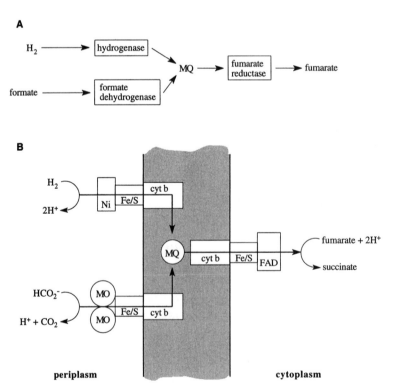

**Fig. 4.18** A model for the electron transport system of *W. succinogenes*. (A) Electrons flow from $H_2$ and formate through menaquinone (MQ) to fumarate reductase. (B) Illustration showing that the catalytic portions of hydrogenase and formate dehydrogenase are periplasmic, whereas fumarate reductase reduces fumarate on the cytoplasmic side. Electrons flow electrogenically to fumarate. A $\Delta p$ is created because of electrogenic influx of electrons together with the release of protons in the periplasm, and their consumption in the cytoplasm. *Source:* Modified from Kroger, A., V. Geisler, E. Lemma, F. Theis, and R. Lenger. 1992. Bacterial fumarate respiration. *Microbiol.* **158**:311–314.

iron–sulfur protein. The two periplasmic subunits of the formate dehydrogenase are a Mo-containing protein subunit and an iron–sulfur protein. The fumarate reductase is a complex containing three subunits. One subunit of the fumarate reductase is a flavoprotein with FAD as the prosthetic group (subunit A). A second subunit has several FeS centers (subunit B). And the third subunit has two hemes of the b type (subunit C), which binds the fumarate reductase to the membrane. Fumarate reductase is similar in structure to succinate dehydrogenase isolated from several different sources, which catalyzes the oxidation of succinate to fumarate in the citric acid cycle.

### Electron flow and the establishment of a $\Delta p$

Electron flow is from the dehydrogenase or hydrogenase to cytochrome b to menaquinone to fumarate reductase. Two electrons are electrogenically transferred across the membrane to fumarate per $H_2$ or formate oxidized, leaving two $H^+$ on the outside, thus establishing a $\Delta p$. A value of 1.1 was obtained for the ratio of $H^+/e^-$ during fumarate reduction when studying whole cells, suggesting perhaps that a mechanism of proton translocation through the membrane may also exist.[43] If one assumes that 1.1 is a correct number, and also assumes a stoichiometry for the ATP synthase of $3H^+/ATP$, then the theoretical maximum number of ATPs that can be formed from the transfer of two $e^-$ to fumarate is 2.2/3, or 0.73. The actual number measured by experimentation was 0.56. Note that the quinone functions as an electron carrier between the cytochromes b, but does not take part in hydrogen translocation across the membrane as in a Q loop or Q cycle.

## 4.8 Summary

All electron transport schemes can be viewed as consisting of membrane-bound dehydrogenase complexes, such as NADH dehydrogenase or succinate dehydrogenase, that remove electrons from their substrates, and transfer the electrons to quinones that in turn transfer the electrons to oxidase or reductase complexes. The latter complexes reduce the terminal electron acceptors. In contrast to mitochondria, which all have the same electron transport scheme, bacteria differ in the details of their electron transport pathways, although the broad outlines of all the electron transport schemes are similar. In bacteria, the dehydrogenase, oxidase, and reductase complexes are sometimes referred to as modules because specific ones are synthesized under certain growth conditions and "plugged" into the respiratory pathway. For example, in facultative anaerobes such as E. coli, the oxidase modules are synthesized in an aerobic atmosphere and the reductase modules under anaerobic conditions.

Other dehydrogenases besides NADH dehydrogenase and succinate dehydrogenase exist. These oxidize various electron donors (e.g., methanol, hydrogen, formate, $H_2$, glycerol, and so on), and are located in the periplasm or the cytoplasm. The coenzyme or prosthetic group for these soluble dehydrogenases vary (e.g., they may be $NAD^+$ or flavin). The electrons from the various dehydrogenases are transferred to one of the electron carriers (e.g., quinone or cytochrome) and from there to a terminal reductase or oxidase.

An important difference between electron transport chains in bacteria and those in mitochondria is that the former are branched. Branching can occur at the level of quinone or cytochrome. The branches lead to different oxidases or reductases, depending upon whether the bacterium is growing aerobically or anaerobically. Many bacteria, including E. coli, transfer electrons from reduced quinone to cytochrome o, which is the major one used when oxygen levels are high, and to cytochrome d, which is used when oxygen becomes limiting. Other bacteria may have in addition, or instead of, an electron transport pathway in which electrons travel from reduced quinone through a $bc_1$ complex, to cytochrome $aa_3$, which reduces oxygen. This is the same as the electron transport pathway found in mitochondria, although the carriers may not be identical. The alternative branches may differ in the number of coupling sites, and this could have important regulatory significance regarding the rates of oxidation of reduced electron carriers, as well as ATP yields.

Another difference from mitochondria is that bacteria can have either aerobic or anaerobic electron transport chains, or as is the case with facultative anaerobes such as *E. coli*, either can be present, depending upon the availability of oxgyen or alternative electron acceptors. A hierarchy of electron acceptors is used. For *E. coli*, oxygen is the preferred acceptor, followed by nitrate, and finally fumarate.

With respect to cytochrome c oxidase, there are two classes. Cytochrome $aa_3$ is the major class and has been reported in many bacteria, including *Paracoccus denitrificans*, *Nitrosomonas europaea*, *Pseudomonas AM1*, *Bacillus subtilis*, and *Rhodobacter sphaeroides*. A different cytochrome c oxidase has been reported for *Azotobacter vinelandii*, *Rhodobacter capsulata*, *Rhodobacter sphaeroides*, *Rhodobacter palustris*, and *Pseudomonas aeruginosa*, as well as *P. denitrificans*.[20, 44] Apparently, both classes of cytochrome c oxidase coexist in the same organism and serve as alternate routes to oxygen.

The main energetic purpose of the respiratory electron transport pathways is to convert a redox potential ($\Delta E_h$) into a proton potential ($\Delta p$). This is done at coupling sites. A membrane potential is created by electrogenic influx of electrons, leaving the positively charged proton on the outside, or electrogenic efflux of protons during proton pumping, leaving a negative charge on the inside. Influx of electrons occurs when oxidations take place on the periplasmic membrane surface or in the periplasm, and electrons move vectorially across the membrane to the cytoplasmic surface, where reductions take place. There are two situations where this occurs. One is when the substrate (e.g., $H_2$ or methanol, and so on) is oxidized by dehydrogenases in the periplasm, and the electrons move across the membrane to reduce the electron acceptor. The second situation is when quinones are reduced on the cytoplasmic side of the membrane, diffuse across the membrane, and are oxidized on the outside membrane surface. Quinone oxidation can be via a Q loop or a Q cycle.

A Q loop is an electron transport pathway in which a reduced quinone carries hydrogen to the outside surface of the cell membrane and releases $2H^+$ as a result of oxidation. The $2e^-$ return to the inner surface, via cytochromes, where they reduce an electron acceptor. The inward transfer of the electron is electrogenic (i.e., a membrane potential is created.) In the Q loop, which is a linear pathway of electron flow, $1H^+$ is released per $e^-$. A more complicated pathway is called the Q cycle and results in $2H^+$ released per $e^-$. In the Q cycle, $QH_2$ gives up two $H^+$ and two $e^-$, but one of the electrons is recycled back to oxidized quinone. Thus the ratio of $H^+/e^-$ is two, rather than one. The Q cycle is more efficient since it results in more protons available for influx through the ATP synthase per electron, and thus can increase the yield of ATP.

A second method of coupling oxidation-reduction reactions to the establishment of a proton potential relies on some of the electron carriers acting as proton pumps. A membrane potential is created when protons are pumped electrogenically out of the cell. A well-established example is cytochrome $aa_3$ oxidase. However, there is evidence that other oxidases are also proton pumps, including cytochrome bo in *E. coli* and cytochromes $bb_3$ and $cbb_3$ in *P. denitrificans*. Mitochondrial and certain bacterial NADH dehydrogenases are also coupling sites, and these may function as proton pumps.

## Study Questions

1. What is it about the solubility properties and electrode potentials of quinones that make them suitable for their role in electron transport?

2. Design an experiment that can quantify the $H^+/O$ for the cytochrome $aa_3$ reaction in proteoliposomes.

3. What are two features that distinguish the Q cycle from linear flow of electrons? How can you verify that the Q cycle is operating?

4. What is the relationship between $H^+/O$, $H^+/ATP$, and the number of ATPs that can be synthesized?

5. What is the relationship between the $\Delta p$ and number of ATPs that can be synthesized?

6. Draw a schematic outline of three ways that a cell might create a membrane potential during respiration. You can include both periplasmic and cytoplasmic oxidations.

7. Assume a $\Delta p$ of $-200$ mV and a $H^+/ATP$ of 3, calculate what the $\Delta G_p$ must be in joules/mole if $\Delta G_p$ is in equilibrium with $\Delta p$. Assume that the temperature is 30°C, and that $\Delta G'_o$ is 37 kcal/mole. What will be the ratio of $[ATP]/[ADP][P_i]$?

8. What are the similarities and differences between bacterial respiratory pathways and the mitochondrial respiratory pathway?

9. Calculate the maximum number of ATPs that can be made per NADH oxidized by *E. coli* when using the NDH-1 and bo combination, and the NDH-2 and bd combination.

10. What drives reversed electron transport? What experiment can be done to support this conclusion?

11. What is it about the sequential arrangement of electron carriers that make proton translocation possible in a Q loop or Q cycle?

12. Consider the following data concerning proton translocation by *P. denitrificans*:

Observed $H^+/2e^-$ ratios for whole cells of *Paracoccus denitrificans*[a]

| Substrate | $H^+/O$ | $H^+/2Fe(CN)_6^{3-}$ |
|---|---|---|
| Endogenous | $7.12 \pm 0.24$ | $5.26 \pm 0.34$ |
| ascorbate + TMPD | $2.81 \pm 0.14$ | $1.04 \pm 0.05$ |

[a]Adapted from Van Verseveld, H. W., et al. 1981. *Biochim. Biophys. Acta* 635:525–534.

Refer to Fig. 4.16. Ferricyanide is an oxidant that allows the terminal oxidase to be bypassed. Ascorbate feeds electrons via TMPD into cytochrome c and releases one scalar proton for every two electrons removed from it. Calculate the $H^+/2e^-$ for the cytochrome c oxidase pump.

## NOTES AND REFERENCES

1. As discussed in Section 3.7.1, some marine bacteria couple electron transport to the creation of a sodium ion gradient.

2. Lubben, M. 1995. Cytochromes of archaeal electron transfer chains. *Biochim. Biophys. Acta* 1229:1–22.

3. For example, the enzyme complex that oxidizes NADH and reduces quinone is an NADH–quinone oxidoreductase. It consists of a flavoprotein and iron–sulfur proteins. (Reduced quinone is called quinol.) An enzyme complex that oxidizes quinol and reduces cytochrome c is called a quinol–cytochrome c oxidoreductase. It consists of cytochromes and an iron–sulfur protein.

4. Here are some definitions that are useful to know. *Cofactors* are nonprotein molecules, either metals or organic, bound with varying degrees of affinity to enzymes and required for enzyme activity. *Coenzymes* are organic cofactors that shuttle back and forth between enzymes carrying electrons, hydrogen, or organic moieties (e.g., acyl groups). *Prosthetic groups* are cofactors (organic or inorganic) that are tightly bound to the protein and do not dissociate from the protein. Thus one difference between a coenzyme and a prosthetic group is that the former shuttles between enzymes and the latter remains tightly bound to the enzyme. Coenzyme A (carrier of acyl groups) and NAD$^+$ (electron and hydrogen carrier) are examples of coenzymes, whereas FAD (electron and hydrogen carrier) and heme (electron carrier) are examples of prosthetic groups.

5. The first iron–sulfur protein identified was a soluble protein isolated from bacteria and called ferredoxin. A protein similar to ferredoxin with a low redox potential can be isolated from chloroplasts and mediates electron transfer to NADP$^+$ during noncyclic electron flow. Iron–sulfur proteins that have no other prosthetic groups are divided into four classes based upon the number of iron atoms per molecule and whether or not the sulfur is acid labile. Examples from the four different classes are: rubredoxin (no acid-labile sulfur), isolated from bacteria, high-potential iron protein (HiPIP), isolated from photosynthetic bacteria, chloroplast ferredoxin, and various bacterial ferredoxins.

6. Electron spin resonance detects unpaired electrons such as the ones that exist in the FeS cluster. Monochromatic microwave radiation is absorbed by the unpaired electron when a magnetic field is applied. The size of the magnetic field required for radiation absorption depends upon the molecular environment of the unpaired electron. A spectrum is obtained by varying the

magnetic field and keeping the frequency of the microwave radiation constant. The spectra differ for different FeS clusters. A spectroscopic constant, called the g value, is obtained that is characteristic of the FeS cluster in the protein.

7. Puustinen, A., and M. Wikstrom. 1991. The heme groups of cytochrome o from Escherichia coli. Proc. Natl. Acad. Sci. USA 88:6122–6126.

8. Yagi, T., X. Xu, and A. Matsuno-Yagi. 1992. The energy-transducing NADH–quinone oxidoreductase (NDH-1) of Paracoccus denitrificans. Biochim. Biophys. Acta 1101:181–183.

9. Finel, M. 1993. The proton-translocating NADH:ubiquinone oxidoreductase: A discussion of selected topics. J. Bioenerg. Biomemb. 25:357–366.

10. Ohnishi, T. 1993. NADH–quinone oxidoreductase, the most complex complex. J. Bioenerg. Biomemb. 25:325–329.

11. Yagi, T., Yano, T., and A. Matsuno-Yagi. 1993. Characteristics of the energy-transducing NADH–quinone oxidoreductase of Paracoccus denitrificans as revealed by biochemical, biophysical, and molecular biological approaches. J. Bioenerg. Biomemb. 25:339–345.

12. García-Horsman, J. A., B. Barquera, J. Rumbley, J. Ma, and R. B. Gennis. 1994. The superfamily of heme–copper respiratory oxidases. J. Bacteriol. 176:5587–5600.

13. Hill, S., S. Viollet, A. T. Smith, and C. Anthony. 1990. Roles for enteric d-type cytochrome oxidase in $N_2$ fixation and microaerobiosis. J. Bacteriol. 172:2071–2078.

14. Kelly, M. J. S., R. K. Poole, M. G. Yates, and C. Kennedy. 1990. Cloning and mutagenesis of genes encoding the cytochrome bd terminal oxidase complex in Azotobacter vinelandii: Mutants deficient in the cytochrome d complex are uanble to fix nitrogen in air. J. Bacteriol. 172:6010–6019.

15. Reviewed in J. Bioenerg. Biomemb. 25(4), 1993.

16. Right-side-out vesicles are made by osmotic lysis of the spheroplasts or protoplasts.

17. The phosphorylation potential, $\Delta G_p$, is dependent upon the standard free energy of formation of ATP and the actual ratios of ATP, ADP, and $P_i$, according to the following equation, $\Delta G_p = \Delta G_0' + RT \ln[\text{ATP}]/[\text{ADP}][P_i]$ J/mole. The free energy of formation is proportional to the equilibrium constant (i.e., $\Delta G_0' = -RT \ln K_{eq}'$).

18. Reviewed in Trumpower, B. L. 1990. Cytochrome $bc_1$ complexes of microorganisms. Microbiol. Rev. 54:101–129.

19. Verseveld, H. W. Van, K. Krab, and A. H. Stouthamer. 1981. Proton pump coupled to cytochrome c oxidase in Paracoccus denitrificans. Biochem. Biophys. Acta 635:525–534.

20. de Gier, J.-W. L., M. Lubben, W. M. Reijnders, C. A. Tipker, D-J. van Slotboom, R. J. M. Spanning, A. H. Stouthamer, and J. van der Oost. 1994. The terminal oxidases of Paracoccus denitrificans. Molec. Microbiol. 13:183–196.

21. Hinkle, P. C., M. a. Kumar, A. Resetar, and D. L. Harrs. 1991. Mechanistic stoichiometry of mitochondrial oxidative phosphorylation. Biochemistry 30:3576–3582.

22. Matsushita, K., T. Ohnishi, and H. R. Kaback. 1987. NADH–ubiquinone oxidoreductases of the Escherichia coli aerobic respiratory chain. Biochemistry 26:7732–7737.

23. Sled, V. D., T. Freidrich, H. Leif, H. Weiss, S. W. Meinhardt, Y. Fukumori, M. W. Calhoun, R. B. Gennis, and T. Ohnishi. 1993. Bacterial NADH–quinone oxidoreductases: Iron–sulfur clusters and related problems. J. Bioenerg. Biomemb. 25:347–356.

24. Unden, G., and J. Bongaerts. 1997. Alternative respiratory pathways of Escherichia coli: Energetics and transcriptional regulation in response to electron acceptors. Biochim. Biophys. Acta 1320:217–234.

25. van der Oost, J., A. P. N. de Boer, J.-W. L. de Gier, W. G. Zumft, A. H. Stouthamer, and R. J. M. van Spanning. 1994. The heme–copper oxidase family consists of three distinct types of terminal oxidases and is related to nitric oxide reductase. FEMS Microbiol. Lett. 121:1–10.

26. Minghetti, K. C., V. C. Goswitz, N. E. Gabriel, J. Hill, C. A. Barassi, C. D. Georgiou, S. I. Chan, and R. B. Gennis. 1992. A modified, large-scale purification of the cytochrome o complex of Escherichia coli yields a two heme/one copper terminal oxidase with high specific activity. Biochemistry 31:6917–6924.

27. Gennis, R. B., and V. Stewart. 1996. Respiration, pp. 217–261. In: Escherichia coli and Salmonella, Cellular and Molecular Biology, Vol. 1. F. C. Neidhardt (Ed. in chief). ASM Press. Washington, D.C.

28. Anraku, Y., and R. B. Dennis. 1987. The aerobic respiratory chain of Escherichia coli. Trends Biochem. Sci. 12:262–266.

29. Calhoun, M. W., K. L. Oden, R. B. Gennis, M. J. Teixeira de Mattos, and O. M. Neijssel. 1993. Energetic efficiency of Escherichia coli: Effects of mutations in components of the aerobic respiratory chain. J. Bacteriol. 175:3020–3025.

30. Tran, Q. H., J. Bongaerts, D. Vlad, and G. Unden. 1997. Requirement for the proton-pumping NADH dehydrogenase I of *Escherichia coli* in respiration of NADH to fumarate and its bioenergetic implications. *Eur. J. Biochem.* **244**:155–160.

31. Green, J., M. F. Anjum, and J. R. Guest. 1997. Regulation of the *ndh* gene of *Escherichia coli* by integration host factor a novel regulator. *Arr. Microbiology* **143**:2865–2875.

32. Reviewed in: Berks, B. C., S. J. Ferguson, J. W. B. Moir, and D. J. Richardson. 1995. Enzymes and associated electron transport systems that catalyze the respiratory reduction of nitrogen oxides and oxyanions. *Biochem. Biophys. Acta* **1232**:97–173.

33. Fumarate is reduced to succinate using the membrane-bound enzyme fumarate reductase. Nitrate is reduced to nitrite using the membrane-bound nitrate reductase, which resembles the dissimilatory nitrate reductase in *P. denitrificans*. Both fumarate and nitrate reductases are coupled to the generation of a $\Delta p$, probably via a quinone loop. Nitrite is reduced to ammonia using an NADH-linked cytoplasmic enzyme (NADH–nitrite oxidoreductase). Energy is not conserved in the reaction. If $NO_3^-$ is the only source of nitrogen, then some of the ammonia is assimilated and the excess is excreted. Nitrite is toxic to bacteria, and probably the reduction to ammonia functions to reduce toxic levels.

34. Unden, G., and J. Schirawski. 1997. The oxygen-responsive transcriptional regulator FNR of *Escherichia coli*: The search for signals and reactions. *Molec. Microbiol.* **25**:205–210.

35. Verseveld, H. W., and A. H. Stouthamer. 1992. The genus *Paracoccus*, pp. 2321–2334. In: *The Prokaryotes*, Vol. III, 2nd ed. A. Balows, H. G. Truper, M. Dworkin, W. Harder, and K.-H. Schleifer (Eds.). Springer-Verlag, Berlin.

36. Van Spanning, R. J. M., A. P. N. de Boer, W. N. M. Reijnders, J.-W. L. de Gier, C. O. Delorme, A. H. Stouthamer, H. V. Westerfoff, N. Harms, and J. van der Oost. 1995. Regulation of oxidative phosphorylation: The flexible respiratory network of *Paracoccus denitrificans*. *J. Bioenerg. Biomembr.* **27**:499–512.

37. Visser, J. M., A. H. de Jong, S. de Vries, L. A. Robertson, and J. G. Kuenen. 1997. *cbb₃*-type cytochrome oxidase in the obligately chemolithoautotrophic *Thiobacillus* sp. W5. *FEMS Microbiol. Lett.* **147**:127–132.

38. Cox, J. C., W. J. Ingledew, B. A. Haddock, and H. G. Lawford. 1978. The variable cytochrome content of *Paracoccus denitrificans* grown aerobically under different conditions. *FEBS Lett.* **93**:261–265.

39. Puustinen, A., M. Finel, M. Virkki, and M. Wikstrom. 1989. Cytochrome *o* (*bo*) is a proton pump in *Paracoccus denitrificans* and *Escherichia coli*. *FEBS Lett.* **249**:163–167.

40. Ferguson, S. J. 1987. Denitrification: A question of the control and organization of electron and ion transport. *TIBS* **12**:354–357.

41. Nitrate reductase is a membrane-bound molybdenum protein with three subunits: $\alpha$, $\beta$, and $\gamma$. The $\gamma$ subunit is a cytochrome b. A proposed route for electrons is from ubiquinol to the FeS centers of the $\beta$ subunit, and from there to the molybdenum center at the active site in the $\alpha$ subunit. The $\alpha$ subunit then reduces the nitrate. The molybdenum is part of a cofactor (prosthetic group) called the molybdenum cofactor (Moco), which consists of molybdenum bound to a pterin. The Moco prosthetic group is in all molybdenum-containing proteins, except nitrogenase. (The molybdenum cofactor from nitrogenase contains nonheme iron and is called FeMo cofactor or FeMoco; Section 12.3.2.) Nitrite reductase is a periplasmic enzyme that reduces nitrite to nitric oxide. Nitrite reductase from *P. denitrificans* has two identical subunits. These have hemes c and d₁. Nitric oxide reductase is a membrane-bound enzyme that reduces nitric oxide to nitrous oxide. It has heme b and heme c. Nitrous oxide reductase is a periplasmic enzyme that reduces nitrous oxide to nitrogen gas. It has two identical subunits containing Cu.

42. Patureau, D., N. Bernet, and R. Moletta. 1996. Study of the denitrifying enzymatic system of *Comamonas* sp. strain SGLY2 under various aeration conditions with a particular view on nitrate and nitrite reductases. *Curr. Microbiol.* **32**:25–32.

43. Reviewed in Kroger, A., V. Geisler, E. Lemma, F. Theis, and R. Lenger. 1992. Bacterial fumarate respiration. *Microbiology* **158**:311–314.

44. Anraku, Y. 1988. Bacterial electron transport chains, pp. 101–132. In: *Ann. Rev. Biochem.*, Vol. 57. C. C. Richardson, P. D. Boyer, I. B. Dawid, and A. Meister (Eds.). Annual Reviews, Inc. Palo Alto.

# 5

# Photosynthesis

Photosynthesis is the conversion of light energy into chemical energy used for growth. Organisms that obtain most or all of their energy in this way are called phototrophic or photosynthetic. Depending upon the photosynthetic system, light energy can do one or both of the following: (1) In all photosynthetic systems light energy can drive the phosphorylation of ADP to make ATP (called photophosphorylation). (2) In some photosynthetic systems light can also drive the transfer of electrons from $H_2O$ ($\Delta E_{m,7} = +820$ mV) to NADP+ ($\Delta E_{m,7} = -320$ mV). The latter is referred to as the photoreduction of NADP+. During photoreduction of NADP+ the water becomes oxidized to oxygen gas, and the NADP+ becomes reduced to NADPH. The student will recognize that this is the reverse of the spontaneous direction in which electrons flow during aerobic respiration (Chapter 4). During both ATP and NADPH synthesis electromagnetic energy is absorbed by photopigments in the photosynthetic membranes and converted into chemical energy. This chapter explains how that is done. At the heart of the process is the light-driven oxidation of chlorophyll or bacteriochlorophyll. This initiates electron transport, which results in the generation of a $\Delta p$ and subsequent synthesis of ATP, and in the case of chlorophyll, NADP+ reduction. It is of interest that (bacterio)chlorophyll-based photosynthesis is widely distributed, being found in bacteria, green plants, and algae, but is not present in the known archaea. (Interestingly, bacteriochlorophyll has also been found in nonphototrophic bacteria, i.e., rhizobia.[1,2]) This chapter begins with backround information on the different groups of phototrophic prokaryotes, and then describes photosynthetic electron transport in all the photosynthetic systems.

## 5.1 The Phototrophic Prokaryotes

The phototrophic prokaryotes are a diverse assemblage of organisms that share the common feature of being able to use light as a source of energy for growth. Their classification is based upon physiological differences, including whether they produce oxygen, and what they use as a source of electrons for biosynthesis. For example, there are both oxygenic phototrophs (produce oxygen) and anoxygenic phototrophs. Oxygenic phototrophs include the well-known *cyanobacteria* (formerly called blue-green algae) and members of the genera *Prochloron*, *Prochlorothrix*, and *Prochlorococcus*.[3] The latter three genera consist of organisms phylogenetically related to the cyanobacteria but having chlorophyll b as a light-harvesting pigment instead of phycobilins. The anoxygenic phototrophs are the *purple photosynthetic bacteria*, the *green photosynthetic bacteria*, and the *heliobacteria*.

(For a more complete discussion of the taxonomy of the photosynthetic bacteria, see Note 4.) These photosynthesize only under anaerobic conditions. The purple and green photosynthetic bacteria are further subdivided according to whether they use sulfur as a source of electrons. Thus there are purple sulfur and nonsulfur photosynthetic bacteria and green sulfur and nonsulfur photosynthetic bacteria. The various phototrophic prokaryotes and some of their properties are summarized in Table 5.1.

### 5.1.1 Oxygenic phototrophs

The oxygenic prokaryotic phototrophs use $H_2O$ as the electron donor for the photosynthetic reduction of $NADP^+$. Most of the species belong to the cyanobacteria which are widely distributed in nature, occurring in fresh and marine waters and terrestial habitats. They have only one type of chlorophyll (chlorophyll a) and light-harvesting pigments called phycobilins. Another kind of oxygenic microbial phototroph is the prochlorophytes, which are a diverse group of photosynthetic prokaryotes that are evolutionarily related to the cyanobacteria but differ from the latter in having chlorophyll b rather than phycobilins as the light-harvesting pigment.[5] There are three genera among the prochlorophytes: *Prochloron*, *Prochlorothrix*, and *Prochlorococcus*. Members of the genus *Prochloron* are obligate symbionts of certain ascidians (sea squirts). *Prochlorothrix* is a flilamentous, free-living microorganism that lives in freshwater lakes. *Prochlorococcus* is a free-living marine microorganism.

### 5.1.2 Anoxygenic phototrophs

The other phototrophic prokaryotes do not produce oxygen (i.e., $H_2O$ is not a source of electrons). Instead, they use organic compounds, inorganic sulfur compounds, or hydrogen gas as a source of electrons. *These organisms will grow phototrophically only anaerobically or when oxygen tensions are low.* There are four major groups of anoxygenic phototrophs (see Table 5.1):

1. Purple photosynthetic bacteria, which includes both the purple sulfur and the purple nonsulfur bacteria;

2. Green sulfur photosynthetic bacteria;

3. Green nonsulfur photosynthetic bacteria, also called green gliding bacteria;

4. Heliobacteria.

### Purple sulfur phototrophs

The purple sulfur phototrophic bacteria grow photoautotrophically in anaerobic environments using hydrogen sulfide as the electron donor and $CO_2$ as the carbon source. Their natural habitats are freshwater lakes and ponds, or marine waters where the sulfide content is high due to sulfate-reducing bacteria. They oxidize sulfide to elemental sulfur, which accumulates as granules intracellularly in all known genera except in the genus *Ectothiorhodospira*, which deposits sulfur extracellularly. However, some can grow photoheterotrophically, and several have been grown chemoautotrophically under low partial pressures of oxygen with reduced inorganic sulfur as an electron donor and energy source. There are many genera, including the well-studied *Chromatium*.

### Purple nonsulfur phototrophs

The purple nonsulfur phototrophic bacteria are extremely versatile with regard to sources of energy and carbon. Originally it was thought that these bacteria were not able to utilize sulfide as a source of electrons; hence the name "nonsulfur." However, it turns out that the concentrations of sulfide used by the purple sulfur bacteria are toxic to the nonsulfur purples, and that, provided the concentrations are sufficiently low, some purple nonsulfur bacteria (i.e., *Rhodopseudomonas* and *Rhodobacter*) can use sulfide and/or thiosulfate as a source of electrons during photoautotrophic growth. The purple nonsulfur phototrophs can grow photoautotrophically or photoheterotrophically in anaerobic environments. When growing photoautotrophically, they use $H_2$ as the electron donor and $CO_2$ as the source of carbon. Photoheterotrophic growth uses simple organic acids such as

**Table 5.1** Phototrophic prokaryotes

| Type | Pigments[a] | Electron donor | Carbon source | Aerobic dark growth |
|---|---|---|---|---|
| **Oxygenic** | | | | |
| Cyanobacteria | Chl a, phycobilins | $H_2O$ | $CO_2$ | No |
| *Prochloron* | Chl a, Chl b | $H_2O$ | $CO_2$ | No |
| *Chlorothrix* | Chl a, Chl b | $H_2O$ | $CO_2$ | No |
| *Prochlorothrix* | Chl a, Chl b | $H_2O$ | $CO_2$ | No |
| **Anoxygenic** | | | | |
| Purple sulfur | Bchl a or Bchl b | $H_b S^b$, $S°$, $S_2O_3^=$, $H_2$, organic | $CO_2$, organic | Yes[d] |
| Purple nonsulfur | Bchl a or Bchl b | $H_2$, organic, $H_2S$ (some) | $CO_2$, organic | Yes |
| Green sulfur | Bchl a and c, d, or e | $H_2S^c$, $S°$, $S_2O_3^=$, $H_2$ | $CO_2$, organic | No |
| Green gliding | Bchl a and c or d | $H_2S$, $H_2$, organic | $CO_2$, organic | Yes |
| Heliobacteria | Bchl g | organic | organic | No |

[a] Carotenoids are usually present.

[b] Members of the genus *Chromatium* accumulate S° intracellularly. Members of the genus *Ectothiorhodospira* accumulate extracellular sulfur.

[c] Accumulates S° extracellularly.

[d] In the natural habitat growth occurs primarily anaerobically in the light. However, several purple sulfur species (Chromatiaceae) can be grown continuously in the laboratory aerobically in the dark at low oxygen concentrations. *Source:* Overmann, J., and N. Pfennig. 1992. Continuous chemotrophic growth and respiration of Chromoatiaceae species at low oxygen concentrations. *Arch. Microbiol.* **158**:59–67.

malate or succinate as the electron donor and source of carbon, and light as the source of energy. If these organisms are placed in the dark in the presence of oxygen, they will carry out an ordinary aerobic respiration and grow chemoheterotrophically (e.g., on succinate or malate). A few can even grow fermentatively very slowly in the dark. Because of their physiological versatility, the purple nonsulfur photosynthetic bacteria have received much attention in research. They are found in lakes and ponds with low sufide content. Representative genera are *Rhodobacter* and *Rhodospirillum*.

## Green sulfur phototrophs

For a review of photosynthesis in the green sulfur and green nonsulfur photosynthetic bacteria, see Ref. 6. The green sulfur phototrophic bacteria are strict anaerobic photoautotrophs that use $H_2S$, $S°$, $S_2O_3^{2-}$ (thiosulfate), or $H_2$ as the electron donor and $CO_2$ as the source of carbon. The green sulfur bacteria coexist with the purple sulfur phototrophs in sulfide-rich anaerobic aquatic habitats, although they are found in a separate layer.

Their light-harvesting pigments are located in special inclusion bodies called *chlorosomes,* which will be described later, and their reaction centers (that part of the photosynthetic membrane where the bacteriochlorophyll is oxidized) differ from those of the purple photosynthetic bacteria. There are several genera, including the well-studied *Chlorobium.*

## Green nonsulfur phototrophs (green gliding bacteria)

The best-studied green nonsufur photosynthetic bacterium is *Chloroflexus,* which is a thermophilic filamentous, gliding, green phototroph. *Chloroflexus* can be isolated from alkaline hot springs, whose pH can be as high as 10. Most isolates have a temperature optimum for growth from 52 to 60°C. It can be grown photoheterotrophically, photoautotrophically with either $H_2$ or $H_2S$ as the electron donor, or chemoheterotrophically in the dark in the presence of air. However, it really grows best as a photoheterotroph. *Chloroflexus* has chlorosomes, but its reaction center is a quinone type, resembling that of the purple photosynthetic bacteria. It differs from the reaction center in the green sulfur bacteria, which is an iron–sulfur type resembling the reaction center of heliobacteria and reaction center I of cyanobacteria and chloroplasts.

## Heliobacteria

The heliobacteria are recently discovered strictly anaerobic photoheterotrophs that differ from the other phototrophs in having bacteriochlorophyll g as their reaction center chlorophyll, containing little carotenoids, and having no internal photosynthetic membranes or chlorosomes. The two known genera are *Heliobacterium* and *Heliobacillus*. The tetrapyrrole portion of bacteriochlorophyll g is similar to chlorophyll a, and in fact isomerizes into chlorophyll a in the presence of air. Both 16S rRNA analyses and the lack of lipopolysaccharide in the cell wall place these organisms with the gram-positive bacteria, and it has been suggested that they are phylogenetically related to the clostridia. In agreement with this, most species of heliobacteria make endospores that resemble those produced by *Bacillus* or *Clostridium*.

## 5.2 The Purple Photosynthetic Bacteria

In all photosynthetic systems, light is absorbed by light-harvesting pigments (also called accessory pigments), and the energy is transferred to reaction centers where bacteriochlorophyll is oxidized, creating a redox potential, $\Delta E$. The light-harvesting pigments, the photosynthetic pigments in the reaction center, and energy transfer to the reaction center are considered in Sections 5.6 and 5.7. The structure of the photosynthetic membrane systems in the prokaryotes as revealed by electron microscopy is described in Section 5.8. The generation of the $\Delta E$ and photosynthetic electron transport are described below.[7, 8] The reaction center of green nonsulfur filamentous photosynthetic bacteria is similar. See Note 9 for a comparison between the two reaction centers.

### 5.2.1 Photosynthetic electron transport

This section discusses the oxidation of bacteriochlorophyll, the generation of the $\Delta E$, and electron transport. Photosynthetic electron flow in the purple photosynthetic bacteria is illustrated in Fig. 5.1A and summarized below. A detailed discussion of the reaction center is given in Section 5.2.2.

**Step 1.** A dimer of two bacteriochlorophyll molecules ($P_{870}$) in the reaction center absorbs light energy, and one of its electrons becomes excited to a higher energy level ($P_{870}^*$). $P_{870}^*$ has a very low reduction potential ($E_m'$) in contrast to $P_{870}$.

**Step 2.** The electron is transferred to an acceptor molecule within the reaction center, thus oxidizing the $P_{870}$ and reducing the acceptor molecule. The acceptor molecule is bacteriopheophytin (Bpheo), which is bacteriochlorophyll without magnesium. The light creates a $\Delta E_m'$ of about 1 V between $P_{870}^+/P_{870}$, which is around $+0.45$ V, and Bpheo$^+$/Bpheo, which is around $-0.6$ V.

**Step 3.** The electron is transferred from the bacteriopheophytin to a quinone called ubiquinone A ($UQ_A$). The $E_m'$ for $UQ_A/UQ_A^-$ is about $-0.2$ V.

**Step 4.** $UQ_A$ transfers the electron to a ubiquinone in site B, called $UQ_B$. $UQ_B$ becomes reduced to $UQ_B^-$. Steps 1–4 are repeated so that $UQ_B$ accepts a second electron and becomes reduced to $UQ_B^{2-}$.

**Step 5.** $UQ_B^{2-}$ picks up two protons from the cytoplasm and is released from the reaction center as $UQH_2$. The $UQH_2$ joins the quinone pool in the membrane.

**Steps 6 and 7.** $UQH_2$ transfers the electrons to a $bc_1$ complex, and the $bc_1$ complex reduces cytochrome $c_2$. The $bc_1$ complex translocates protons to the cell surface, thus generating a $\Delta p$. (See Section 4.6.2 for a discussion of the $bc_1$ complex and the Q cycle.)

**Step 8.** The cytochrome $c_2$ returns the electron to the oxidized bacteriochlorophyll

**A**

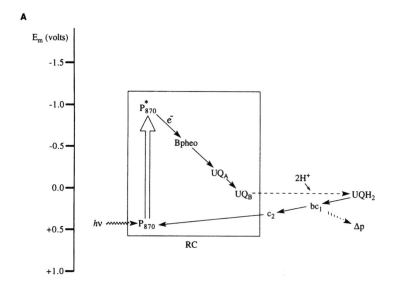

**B**

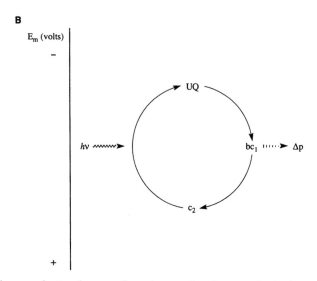

**Fig. 5.1** Photosynthetic electron flow in purple photosynthetic bacteria. (A) Light energizes bacteriochlorophyll, here shown as $P_{870}$ (bacteriochlorophyll a) in the reaction center (RC). (The number in the subscript refers to the major long-wavelength absorption peak.) Some purple photosynthetic bacteria have bacteriochlorophyll b instead, which absorbs at 1020–1035 nm. The energized bacteriochlorophyll reduces bacteriopheophytin (Bpheo), which is bacteriochlorophyll without $Mg^{++}$. The electron then travels through two ubiquinones, $UQ_A$ and $UQ_B$, in the reaction center. (Some species use menaquinone.) After a second light reaction $UQ_B$ becomes reduced to $UQH_2$ and leaves the reaction center. The electron returns to the reaction center via a proton-translocating $bc_1$ complex and cytochrome $c_2$. ATP is synthesized via a membrane ATP synthase driven by the $\Delta p$ (not shown). Midpoint potentials at pH 7 (approximate): $Bchl_{870}$ ($P_{870}$), +450 mV; Bpheo, −600 mV; $UQ_A$, −200 mV; $UQ_B$, +80 mV; c cytochromes, +380 mV. (B) A simplified version of A that emphasizes the cyclic flow of electrons. *Source:* Data from Mathis, P. 1990. *Biochem. et Biophys. Acta* 1018:163–167.)

molecule in the reaction center. Thus light energy causes a cyclic electric current in the membrane.

Steps 1–8 describe cyclic electron flow. We can follow the path of the electron starting with cytochrome $c_2$ (Fig. 5.1B). Notice that an electron moves "uphill" (to a lower electrode potential) from cytochrome $c_2$ to ubiquinone when "boosted" or "pushed" in the reaction center by the energy from a quantum of light. As pointed out above, in order for the reaction center to reduce UQ to $UQH_2$, two electrons and therefore two light reactions are required. This can be written as follows:

$$2\text{cyt } c_2(\text{red}) + 2H^+ + 2UQ \xrightarrow{\text{2 quanta of light}}$$

$$2 \text{ cyt } c_2(\text{ox}) + UQH_2 \qquad (5.1)$$

Because two electrons are required before reduced quinone leaves the reaction center, $UQ_B$ is referred to as a *two-electron gate*.

As illustrated in Fig. 5.1, the reduced quinone transfers the electrons to cytochrome $c_2$ via a $bc_1$ complex outside the reaction center. The $bc_1$ complex translocates protons across the membrane via a Q cycle:

$$UQH_2 + 2H^+_{in} + 2\text{cyt } c_2(\text{ox}) \xrightarrow{bc_1}$$

$$UQ + 4H^+_{out} + 2\text{cyt } c_2(\text{red}) \qquad (5.2)$$

The $\Delta p$ that is generated is used to drive the synthesis of ATP via a membrane-bound ATP synthase:

$$ADP + P_1 + 3H^+_{out} \xrightarrow{\text{ATP synthase}}$$

$$ATP + H_2O + 3H^+_{in} \qquad (5.3)$$

Thus the net result of photosynthesis by purple photosynthetic bacteria is the synthesis of ATP. This is called *cyclic photophosphorylation*.

The actual amount of ATP made depends upon the number of protons translocated by the $bc_1$ complex and the number of protons that enter via the ATP synthase. For example, if four protons per two electron transfers are translocated out of the cell by the $bc_1$ complex and three protons re-enter via the ATP synthase, then the maximum number of ATP molecules made per two electrons is 4/3.

Figure 5.2 illustrates the topographical relationships between the reaction center, the $bc_1$ complex, and the ATP synthase in photosynthetic membranes and can be compared to Fig. 5.1, which illustrates the sequential steps in electron transfer. In the reaction center, electrons travel across the membrane from a periplasmic cytochrome $c_2$ to the quinone at site A. This creates a membrane potential, which is negative inside. After a second electron is transferred, the quinone at site B accepts two protons from the cytoplasm and leaves the reaction center as $UQH_2$. The $UQH_2$ diffuses through the lipid matrix to the $bc_1$ complex where it is oxidized. The $bc_1$ complex translocates four

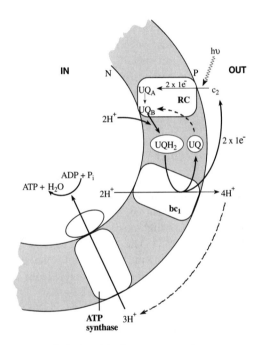

Fig. 5.2 Relationship between reaction center, $bc_1$ complex, and ATP synthase in photosynthetic membranes. The reaction center (RC) takes two protons from the cytoplasm and two electrons from a periplasmic cytochrome c to reduce ubiquinone (UQ) to $UQH_2$. The $UQH_2$ leaves the reaction center and diffuses to the $bc_1$ complex. The $bc_1$ complex oxidizes $UQH_2$ and translocates four protons to the periplasmic surface in a Q cycle, which is described in Section 4.6.2. The electrons travel back to cytochrome $c_2$ (cyclic flow). The translocated protons re-enter via the ATP synthase, which makes ATP.

protons to the outside per $UQH_2$ oxidized via a Q cycle, creating a $\Delta p$. The electrons are transferred from the $bc_1$ complex to a periplasmic cytochrome $c_2$, which returns them to the reaction center. The proton circuit is completed when the protons are returned to the cytoplasmic side via the ATP synthase accompanied by the synthesis of ATP. *Thus light is used to drive an electron circuit, which drives a proton circuit, which drives ATP synthesis.* This successful method for harnessing light energy to do chemical work is used in all phototrophic organisms from bacteria to plants.

### 5.2.2 A more detailed examination of the reaction center and what happens there

The description given above for electron flow in the reaction center is incomplete in that it does not include all the components and electron carriers, or the structure of the reaction center. A more complete description is given next.

### Structure and composition of the reaction center

The reaction centers from *Rhodopseudomonas viridis* and *Rhodobacter sphaeroides* have been crystallized, and the structures determined from high-resolution X-ray diffraction studies.[10, 11] The reaction centers are similar. The following proteins and pigments are found in the reaction center from *R. sphaeroides*:

1. *Reaction center protein.* The reaction center protein has 11 membrane-spanning alpha helices, and a globular portion on the cytoplasmic side of the membrane. It consists of three polypeptides: H, L, and M. The reaction center protein serves as a scaffolding to which the bacteriochlorophyll and bacteriopheophytin are attached, and to which the quinones and nonheme iron are bound.

2. *Four bacteriochlorophyll molecules (Bchl).* There are four bacteriochlorophyll molecules in the reaction center. Two of the bacteriochlorophyll molecules exist as

a dimer $(Bchl)_2$, and two are monomers ($Bchl_A$ and $Bchl_B$).

3. *Two molecules of bacteriopheophytin ($Bpheo_A$ and $Bpheo_B$), which is Bchl without $Mg^{2+}$.*

4. *Two molecules of ubiquinone ($UQ_A$ and $UQ_B$).*

5. *One molecule of nonheme ferrous iron (Fe).*

6. *One carotenoid molecule.* The function of the carotenoid is to protect the reaction center pigments from photodestruction.[12] (For a further explanation, see Note 13.)

### Electron transfer in the reaction center

The sequence of redox reactions in the reaction center summarized in Fig. 5.1 (boxed area) is shown in more detail in Fig. 5.3, along with the proposed time scale for the electron transfer events. As shown in Fig. 5.3, the arrangement of the pigment molecules and the quinones shows twofold symmetry with a right and left half very similar. On the periplasmic side of the membrane there sits a pair of bacteriochlorophyll molecules $(Bchl)_2$ (Fig. 5.3). These bacteriochlorophyll molecules are $P_{870}$ in Fig. 5.1. When the energy from a quantum of light is absorbed by $(Bchl)_2$, an electron becomes excited [i.e., $(Bchl)_2$ becomes $(Bchl)_2^*$] (Fig. 5.3, step 1). The redox potential of $(Bchl)_2^*$ is sufficiently low so that it reduces bacteriochlorophyll A ($Bchl_A$), forming $(Bchl)_2^+$ and $Bchl_A^-$ (Fig. 5.3, step 2). The electron then moves to bacteriopheophytin$_A$ ($Bpheo_A$), forming $Bpheo_A^-$ (Fig. 5.3, step 3). (As reflected by the question mark in step 2 of Fig. 5.3, the exact route of the electron is not known. It has not been unequivocally demonstrated that the electron moves through the $Bchl_A$ monomer. Some researchers suggest that the electron travels from $Bchl_2^*$ to $Bchl_A$ and very quickly moves on to $Bpheo_A$, whereas other investigators postulate that the electron moves directly from $(Bchl^*)_2$ to $Bpheo_A$.) Then the electron moves to ubiquinone bound to site A on the cytoplasmic side of the reaction center ($UQ_A$), forming the semiquinone, $UQ_A^-$, Fig. 5.3, step 4. At this time an electron is returned to $(Bchl)_2^+$ from reduced cytochrome

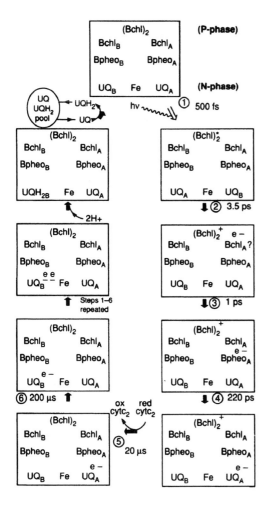

Fig. 5.3 The reaction center in purple photosynthetic bacteria. The reaction center, which spans the membrane, is represented by the boxed area. The absorption of light creates a transient membrane potential, positive (P-phase) on the periplasmic side and negative (N-phase) on the cytoplasmic side. Step 1: Absorption of a photon of light energizes a bacteriochlorophyl dimer $(Bchl)_2$. Step 2: The energized $(Bchl)_2$ reduces $Bchl_A$. The question mark refers to the fact that this has not been unequivocally demonstrated. Step 3: Bacteriopheophytin A $(Bpheo_A)$ is reduced. Step 4: The electron is transferred to a quinone $(UQ_A)$. Step 5: The oxidized $(Bchl)_2$ is reduced via cytochrome $c_2$. Step 6: The electron moves from $UQ_A$ to $UQ_B$. Steps 1–6 are repeated, forming $UQ_B^-$. Two protons are acquired from the cytoplasm to produce $UQH_2$, which enters the reduced quinone pool in the membrane and returns the electrons to oxidized cytochrome $c_2$ via the $bc_1$ complex. It may be that one proton is acquired by $UQ_B^-$ and the second proton is acquired when the second electron arrives. *Source:* From Nicholls, D. G., and S. J. Ferguson. 1992. Bioenergetics 2. Academic Press, London.

$c_2$, which is periplasmic (step 5). Then the electron is transferred from $UQ_A^-$ to $UQ_B$, to form $UQ_B^-$ (Fig. 5.3, step 6). Electron flow is therefore across the membrane from periplasmic $c_2$ to $UQ_B$ located on the cytoplasmic side. This generates a membrane potential, outside positive. A second light reaction occurs and steps 1 through 6 in Fig. 5.3 are repeated so that $UQ_B$ has two electrons ($UQ_B^{2-}$). Two protons are picked up from the cytoplasm to form $UQH_2$, which leaves the reaction center to join the quinone pool in the membrane. (Although Fig. 5.3 indicates that two protons enter after both electrons arrive at $UQ_B$, it has been suggested that one proton enters after $UQ_B$ receives the first electron.) Eventually, the protons carried by $UQH_2$ are released on the periplasmic side during oxidation of $UQH_2$ by the $bc_1$ complex. Thus the reaction center and the $bc_1$ complex cooperate to translocate protons to the outside. As previously explained, because $UQ_B$ cannot leave the reaction center until it has accepted two electrons, it is called a *two-electron gate*. The length of time that it takes the electron to travel from $(Bchl)_2$ to $UQ_A$ is a little more than 200 psec. (One psec is one trillionth of a second, or $10^{-12}$ sec.) The rates of subsequent steps are slower (but still very fast) and measured in microseconds. (One microsecond is a millionth of a second.) Determining the pattern and timing of electron transfer in the reaction center requires the use of picosecond laser pulses and very rapid recording of the absorption spectra of the electron carriers. Interestingly, only one side of the reaction center (the A side) appears to be involved in electron transport. For example, there is photoreduction of only one of the bacteriopheophytin molecules. The reason why only one branch appears to function in electron transport is not known.

## The contribution to the $\Delta p$ by the reaction center

When reaction centers absorb light energy, a $\Delta\Psi$ and a proton gradient are created. The reason for the $\Delta\Psi$ is that the electrons move electrogenically from cytochrome $c_2$ to $UQ_B$ across the membrane from outside to inside (Fig. 5.2). (This will produce a $\Delta\Psi$, outside positive, because the inward movement of a negative charge is equivalent to the outward movement of a positive charge.) The creation of the membrane potential is detected by a shift in the absorption spectra of membrane carotenoids. (For a further explanation, see Note 14.) The value of this technique is that very rapid changes in membrane potential can be monitored. The $\Delta\Psi$ produced by the reaction center is delocalized and increases the $\Delta\Psi$ made in the $bc_1$ complex (Section 5.2.). Additionally, two protons are taken from the cytoplasm to reduce $UQ_B$ to $UQH_2$. Eventually, the two protons will be released on the outside during the oxidation of $UQH_2$ by the $bc_1$ complex. Thus the reaction center contributes to both the $\Delta\Psi$ and the $\Delta pH$ components of the $\Delta p$, as well as creating a $\Delta E$ between $UQ/UQH_2$ and $c_{2,ox}/c_{2,red}$.

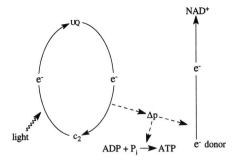

**Fig. 5.4** Relationship between cyclic electron flow and reverse electron transport in the purple photosynthetic bacteria. The electron is driven by light from a cytochrome $c_2$ ($c_2$) to a ubiquinone (UQ), and then returns via electron carriers to the cytochrome $c_2$ with production of a $\Delta p$. The $\Delta p$ is used to drive ATP synthesis as well as reverse electron transport.

## Summary of photosynthesis by purple photosynthetic bacteria

$$ADP + P_i \xrightarrow{\text{light and bchl}} ATP + H_2O$$

### 5.2.3 Source of electrons for growth

In order to grow, all organisms must reduce $NAD(P)^+$ to $NAD(P)H$. That is because NADH and NADPH are the electron donors for almost all the biosynthetic reactions in the cell, including the reduction of carbon dioxide to carbohydrate. When the electron donor is of a higher potential than the $NAD^+/NADH$ couple, then energy is required to reduce $NAD^+$. For example, this is the case for the purple photosynthetic bacteria that use certain inorganic sulfur compounds or succinate as a source of electrons. The purple photosynthetic bacteria drive electron transport in reverse, using the $\Delta p$ created by light energy (Fig. 5.4). During reversed electron flow, ubiquinol reduces $NAD^+$ via the NADH:ubiquinone oxidoreductase. (See Section 3.7.1 for a discussion of reversed electron transport.) Figure 5.4 illustrates the situation where electrons from the electron donor do not pass through the reaction center but simply enter the electron transport chain and travel directly to $NAD^+$ via ubiquinone. That is to say, there is no noncyclic electron transport. Many investigators hold this view. However,

another scenario has been suggested, which is useful to consider because it illustrates the role that an increase in the $\Delta p$ can play in slowing the rate of electron transfer. It should be recalled that ubiquinone can also accept electrons from bacteriopheophytin during cyclic electron transport and reduce the $bc_1$ complex, which generates a $\Delta p$ (Fig. 5.1). It has been suggested that as the $\Delta p$ grows larger, it might exert "backpressure" on the oxidation of ubiquinol by the $bc_1$ complex, thus slowing down the oxidation of ubiquinol via this route and making it available for $NAD^+$ reduction via reversed electron transport.[15] To the extent that this might occur, the electron donor (succinate or inorganic sulfur compounds) would replenish electrons to the bacteriochlorophyll via either ubiquinone or cytochrome c.

## 5.3 The Green Sulfur Bacteria

### 5.3.1 Electron transport

The green sulfur photosynthetic bacteria have reaction centers that are distinguished from the reaction centers of the purple photosynthetic bacteria by:

1. Reducing $NAD(P)^+$ instead of quinone;

2. Possessing iron–sulfur centers as reaction center intermediate electron carriers.

The reaction centers of green sulfur bacteria are similar to the reaction centers I of cyanobacteria and chloroplasts. A similar reaction center is found in heliobacteria.

However, the principles underlying the transformation of electrochemical energy into a $\Delta E$ and a membrane potential are the same as for the purple photosynthetic bacteria.[16, 17] Light energizes a bacteriochlorophyll a molecule ($P_{840}$), which reduces a primary acceptor ($A_0$), establishing a redox potential difference greater than 1 V (Fig. 5.5). The primary electron acceptor ($A_0$) in the green sulfur bacteria has recently been reported to be an isomer of chlorophyll a called bacteriochlorophyll 663.[18] It might be added that the primary acceptor of heliobacteria is also a chlorophyll a derivative, hydroxychlorophyll a, reflecting the similarities known to exist between the reaction centers of the green

sulfur bacteria, heliobacteria, and photosystem I of chloroplasts and cyanobacteria. (See Note 19.) The electron flows from $A_0$ to a quinonelike acceptor called $A_1$. The electron is then transferred from $A_1$ through three iron–sulfur centers to ferredoxin. There can be both cyclic and noncyclic electron flow. In cyclic flow the electron returns to the reaction center via menaquinone (MQ) and a $bc_1$ complex, creating a $\Delta p$. In noncyclic electron flow inorganic sulfur donates electrons, which travel through the reaction center to $NAD^+$. According to the scheme in Fig. 5.5, electrons from inorganic sulfur enter at the level of cytochrome c, and the $bc_1$ complex is bypassed. This is a widely held view based upon available data using isolated oxidoreductases. From an energetic point of view, it is wasteful because a coupling site is bypassed, even though the $E'_m$ values of some of the sulfur couples are sufficiently low to reduce menaquinone. For example, the $E'_m$ for sulfur/sulfide ($n = 2$) is $-0.27$, for sulfite/sulfide ($n = 6$) it is $-0.11$ V, and

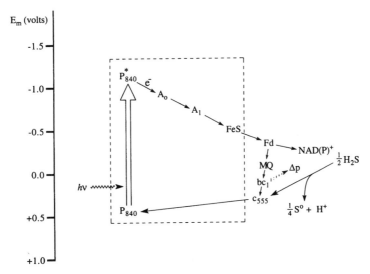

**Fig. 5.5** A model for electron flow in the green sulfur bacteria. Both cyclic and noncyclic flow of electrons are possible. Reaction center bacteriochlorophyll a (P-840) becomes energized and reduces $A_0$, which is bacteriochlorophyll 663. The electron then travels to $A_1$, a quinonelike molecule, and then through two or three iron–sulfur centers (FeS). The first reduced product outside the reaction center is the iron–sulfur protein ferredoxin (Fd). The ferredoxin reduces $NAD(P)^+$ in the noncyclic pathway. Cyclic flow occurs when the electron reduces menaquinone (MQ) instead of $NAD(P)^+$, and returns to the reaction center via a $bc_1$ complex and cytochrome $c_{555}$. A $\Delta p$ is created in the $bc_1$ complex. In the noncyclic pathway the electron donor is a reduced inorganic sulfur compound, here shown as hydrogen sulfide. Elemental sulfur or thiosulfate can also be used. The inorganic sulfur is oxidized by cytochrome $c_{555}$, which feeds electrons into the reaction center.

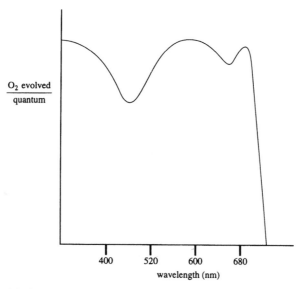

**Fig. 5.6** Quantum yield of photosynthesis. When the rate of $O_2$ evolution per quantum absorbed is plotted against wavelength, the rate drops off sharply above 680 nm. *Source:* After Stryer, L. 1988. *Biochemistry,* W. H. Freeman and Co., New York.

for sulfate/sulfite ($n = 2$) it is $-0.54$ V. They are all at a potential low enough to reduce menaquinone, which has an $E'_m$ of $-0.074$ V. However, the point of entry of the electron is still an unresolved issue.[20]

*Summary of photosynthesis by green sulfur bacteria (Chlorobiaceae)*

$$H_2S + NAD^+ + ADP + P_i \xrightarrow{\text{light and bchl}}$$

$$S^0 + NADH + H^+ + ATP$$

(These organisms can also use elemental sulfur, oxidizing it to sulfate.)

## 5.4 Cyanobacteria and Chloroplasts

Photosynthesis in cyanobacteria and chloroplasts differs from photosynthesis discussed thus far in three important respects:

1. $H_2O$ is the electron donor, and oxygen is evolved.

2. There are two light reactions in series, hence two different reaction centers.

3. Electron flow is primarily noncyclic, producing both ATP and NADPH. (Noncyclic

flow also occurs in the green sulfur bacteria, as described in Section 5.3.1)

### 5.4.1 Two light reactions

As we shall see, chloroplasts and cyanobacteria have essentially combined the light reactions of purple photosynthetic bacteria and green sulfur photosynthetic bacteria in series, so that two light reactions energize a single electron that energizes ATP synthesis and reduces NADP$^+$. The initial evidence for two light reactions came from early studies of photosynthesis performed with algae by Emerson and his colleagues.[21, 22] They observed that the efficiency of photosynthesis, measured as the moles of oxygen evolved per einstein absorbed (i.e., the *quantum yield*) is high over all the wavelengths absorbed by chlorophyll and the light-harvesting pigments, but drops off sharply at 685 nm despite the fact that chlorophyll continues to absorb light between 680 and 700 nm (Fig. 5.6). This became known as the "red drop" effect because 700 nm light is red. One can restore the efficiency of photosynthesis of 700 nm light by supplementing it with light at shorter wavelengths (e.g., 600 nm light). The explanation for this is that there are two reaction centers, one that is energized by light at a wavelength of around 700 nm, called reaction center I, and one that is energized by lower wavelengths of light, called reaction center II. These are also called photosystems I and II or PS I and PS II. Photosystems I and II operate in series, and therefore both must be energized to maintain the electron flow from water to NADP$^+$. The lower wavelengths of light can energize both reaction centers, but 700 nm light can energize only reaction center I. It is for this reason that 700 nm light is effective only when given in combination with supplemental doses of shorter wavelengths.

### 5.4.2 Electron transport

A schematic drawing of the overall pattern of electron flow in cyanobacteria and chloroplasts is shown in Fig. 5.7. These systems have two reaction centers, PS I and PS II, connected by a short electron transport chain that includes the analog to the bc$_1$ complex (i.e., the b$_6$f complex). (For a description of the b$_6$f complex, see Note 23.) Let us begin with PS II, which functions like the reaction center in the

purple photosynthetic bacteria. Chlorophyll a, with a major absorption peak at 680 nm (i.e., P$_{680}$—probably a dimer), becomes energized and reduces an acceptor molecule, which is pheophytin (pheo). Having lost an electron, the P$_{680}^+$/P$_{680}$ couple is of a sufficiently high redox potential (estimated at about +1.1 V) to replace the lost electron with one from water, thus oxidizing $\frac{1}{2}$ H$_2$O to $\frac{1}{4}$ O$_2$. The energized electron travels to pheophytin, which has an $E'_m$, of about $-0.6$ V. The pheophytin reduces plastoquinone (PQ), which is structurally similar to ubiquinone (Fig. 4.3). A two-electron gate, similar to the two-electron gate in the reaction center of the purple photosynthetic bacteria, operates during quinone reduction. Electrons leave the reaction center in PQH$_2$ and are transferred through a b$_6$f complex (structurally and functionally similar to the bc$_1$ complex, except that cytochrome f, which has a c-type heme, replaces cytochrome c$_1$ and there are some differences in the cytochrome b) to a copper-containing protein called plastocyanin (Pc). A $\Delta p$ is created by the b$_6$f complex, but there is some controversy whether or not

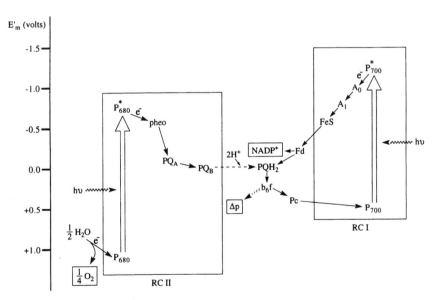

**Fig. 5.7** Photosynthesis in cyanobacteria and chloroplasts. Light stimulates electron flow from water through reaction center II (RC II) to reaction center I (RC I) to NADP$^+$. Cyclic flow is possible using RC I from ferredoxin through the b$_6$f complex. Abbreviations: P$_{680}$, chlorophyll a with a major absorption peak at 680 nm; Pheo, chlorophyll pheophytin; PQ, plastoquinone; b$_6$f, cytochrome b$_6$f complex; Pc, plastocyanin; P$_{700}$, chlorophyll a with a major absorption peak at 700 nm; A$_o$, chlorophyll a; A$_1$, phylloquinone; FeS, one of several iron–sulfur centers; Fd, ferredoxin.

the $b_6f$ complex catalyzes a Q cycle.[24] The plastocyanin reduces $P_{700}^+$ in reaction center I, previously oxidized by a light reaction described next.

Photosystem I is similar to the reaction center in the green photosynthetic bacteria. A photon of light energizes $P_{700}$, which is chlorophyll a with a major absorption peak at 700 nm. The $P_{700}^*$ reduces a chlorophyll a molecule, called $A_0$. For this initial redox reaction the energized electron travels from an $E_m'$ of about +0.5 V ($P_{700}^+/P_{700}$) to an $E_m'$ of about −1.0 V ($A_0/A_0^-$). Note that this is a far more negative potential than that generated in reaction center II. From $A_0$ the electrons travel to $A_1$, which is a phylloquinone. Phylloquinones have a structure similar to menaquinone, but with only one double bond in the isoprenoid chain. (See Fig. 4.3 for the structure of menaquinone.) The electron is transferred from the quinone through several iron–sulfur centers (FeS), which reduce the iron–sulfur protein, ferredoxin (Fd), that is outside of the reaction center. Ferredoxin in turn reduces $NADP^+$. Thus the two light reactions in series energize electron flow from $H_2O$ to $NADP^+$, which is over a net potential difference of about 1.1 V. This is called noncyclic electron flow because the electron never returns to the reaction center. However, cyclic electron flow is possible. There is a branch point at the ferredoxin step, and it is possible for the electron to cycle back to reaction center I via the $b_6f$ complex, augmenting the $\Delta p$, rather than reducing $NADP^+$. This may be a way to increase the amounts of ATP made relative to NADPH. The Calvin cycle, which is the pathway for reducing $CO_2$ to carbohydrate in oxygenic phototrophs, requires three ATPs per two NADPHs in order to reduce one $CO_2$ to the level of carbohydrate.

*Summary of photosynthesis by green plants, algae, cyanobacteria*

$$H_2O + NADP^+ + ADP + P_i \xrightarrow{\text{light and chl}}$$

$$\tfrac{1}{2} O_2 + NADPH + H^+ + ATP$$

## 5.5 Efficiency of Photosynthesis

The efficiencies of photosynthesis based upon input light energy and products of photosynthesis are calculated below as an exercise. The calculated efficiencies are only approximations based upon assumptions regarding ATP yields, actual redox potentials, standard free energies, and so on.

### 5.5.1 ATP synthesis

Basically, photosynthesis is work done by energized electrons. The work is the phosphorylation of ADP and the reduction of $NAD(P)^+$. Each electron is energized by a photon (quantum) of light, and each mole of electrons by an einstein ($6.023 \times 10^{23}$ quanta) of light. It is instructive to ask how much of this energy is conserved in ATP and NADPH. Let us consider the synthesis of ATP. Assume that each energized electron results in the translocation of two protons (e.g., during the Q cycle in the $bc_1$ complex) but that three protons must re-enter through the ATP synthase to make one ATP. As mentioned before, a value of $3H^+$/ATP is reasonable in light of experimental data. Thus each energized electron (or each photon) results in the synthesis of $\tfrac{2}{3}$ of an ATP. How much energy is required to synthesize $\tfrac{2}{3}$ of an ATP? The $\Delta G_p$ for the phosphorylation of one mole of ADP to make ATP is about 45 kJ. Therefore, $\tfrac{2}{3}$ of a mole of ATP should require approximately $+45(\tfrac{2}{3})$ kJ = 30 kJ. An einstein of 870 nm light, which corresponds to the absorption maximum of bacteriochlorophyll a (found in purple phototrophs), has 138 kJ of energy. Therefore, the efficiency is (30/138)(100) or about 22%. One can also calculate the efficiency using electron volts instead of joules. The energy in a photon of light at 870 nm is 1.43 eV. The synthesis of $\tfrac{2}{3}$ of a mole of ATP requires 30,000 J. Dividing this number by the faraday gives the energy in electron volts (i.e., 0.31 eV).

### 5.5.2 ATP and NADPH synthesis

What about photosystems that reduce $NADP^+$ as well as make ATP (i.e., photosystems I and II)? The $E_m'$ for $O_2/H_2O$ is +0.82

V, and for NADP$^+$/NADPH it is $-0.32$ V. Therefore, the energized electron must have $0.82 - (-0.32)$ or $1.14$ eV to move from water to NADP$^+$. (The answer would not be very different if $E_h$ values were used instead of $E'_m$.) If $\frac{2}{3}$ of an ATP are made, then a total of $0.31 + 1.14 = 1.45$ eV would be required to make $\frac{2}{3}$ of an ATP and $\frac{1}{2}$ of an NADPH. A photon of light at wavelength 680 nm (the major long-wave absorption peak of chlorophyll a in reaction center II) has 1.82 eV. Two photons, or the equivalent of about 3.6 eV, are used. Therefore, approximately 40% of the light energy is conserved as ATP and NADPH.

### 5.5.3 Carbohydrate synthesis and oxygen production

One can also estimate the approximate efficiency by considering the number of light quanta required to produce oxygen:

$$6CO_2 + 12H_2O \longrightarrow C_6H_{12}O_6 + 6H_2O$$
$$+ 6O_2, \qquad \Delta G'_0 = 2870 \text{ kJ}$$

Therefore, per mole of $O_2$ produced, the standard free energy requirement at pH 7 is 2870/6, or 478 kJ. The number of einsteins of light required to produce one mole of $O_2$ is eight (Fig. 5.7). An einstein of 680 nm light carries 176 kJ of energy. Therefore, 176 $\times$ 8 or 1408 kJ of light energy are used to produce one mole of $O_2$. The efficiency is thus (478/1408)(100), or 34%.

## 5.6 Photosynthetic Pigments

### 5.6.1 Light-harvesting pigments

The photosynthetic pigments are divided into two categories. They are *reaction center pigments* (primarily chlorophylls) and *light-harvesting pigments* (carotenoids, phycobilins, chlorophylls). The light-harvesting pigments are sometimes called accessory pigments or antennae pigments. By far, most of the photosynthetic pigments are light-harvesting pigments. The light-harvesting pigments are critical to photosynthesis because they absorb light of different wavelengths and funnel the energy to the reaction center. Whole cell absorption spectra of a purple photosynthetic bacterium and a cyanobacterium showing the absorption wavelengths of the pigments are shown in Fig. 5.8, and the various pigments and their absorption peaks are listed in Table 5.2. The nature of the light-harvesting pigments will vary with the type of organism, but some important generalizations about them can be made:

1. They absorb light at wavelengths different from reaction center chlorophyll and therefore extend the range of wavelengths over which photosynthesis is possible because they transfer energy to the reaction center (Table 5.2).

2. In the purple photosynthetic bacteria they are embedded in the membrane as pigment–protein complexes that are in close physical association with the reaction center.

3. In the green sulfur bacteria and Chloroflexus, they exist in chlorosomes, which are separate "organelles," also called inclusion bodies, attached to the inner surface of the cell membrane.

4. In the cyanobacteria (and eukaryotic red algae), the light-harvesting pigments (phycobilins) that transfer energy to PS II are localized in granules called phycobilisomes that are attached to the photosynthetic membranes (also called thylakoids). It should be pointed out that as in other photosynthetic organisms, cyanobacteria also use carotenoids as light-harvesting pigments. Carotenoids have been found in isolated complexes of photosystems I and II from cyanobacteria, and fluorescence data suggest that carotenoids absorb light and transfer energy to reaction center chlorophyll in PS I.[25] (See Section 5.7.2.) Carotenoids in the reaction centers also serve to protect the chlorophyll from photooxidation. (See Note 13.)

### Light-harvesting pigments of the purple photosynthetic bacteria

The light-harvesting complexes of the purple photosynthetic bacteria are *bacteriochlorophyll–protein–carotenoid complexes* localized

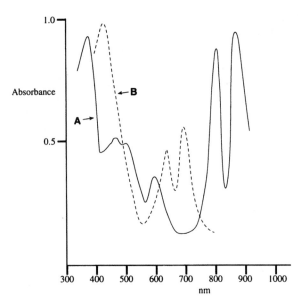

**Fig. 5.8** Absorption spectra of a purple photosynthetic bacterium, *Rhodopseudomonas palustris*, (A) and a cyanobacterium (B). (A) Bacteriochlorophyll a peaks at 360, 600, 805, and 870 nm. Carotenoid peaks at 450–525 nm. (B) Chlorophyll a peaks at 440 and 680 nm. The phycocyanin peak is around 620 nm. *Source:* Adapted from Brock, T. D., and M. T. Madigan. 1988. *Biology of Microorganisms.* Reprinted by permission of Prentice Hall, Englewood Cliffs, New Jersey.

**Table 5.2** In vivo long wavelength absorption maxima of photosynthetic pigments of phototrophic prokaryotes

| Pigment | Wavelength (nm) |
|---|---|
| Chlorophyll | |
| a | 680 |
| b | 675 |
| Bacteriochlorophyll | |
| a | 800–810, 850–910 |
| b | 835–850, 1020–1035 |
| c | 745–760 |
| d | 725–745 |
| e | 715–725 |
| g | 670, 788 |
| Carotenoid | |
| Chlorobactene | 458 |
| Isorenieratene | 517 |
| Lycopene, rhodopin | 463, 490, 524 |
| Okenone | 521 |
| Rhodopinal | 497, 529 |
| Spheroidene | 450, 482, 514 |
| Spirilloxanthin | 486, 515, 552 |
| $\beta$ carotene | 433, 483 |
| $\gamma$ carotene | 433, 483 |
| Phycobilins | |
| Phycocyanin | 620–650 |
| Phycoerythrin | 560–566 |

*Source:* Stolz, J. F. 1991. *Structure of Phototrophic Prokaryotes.* CRC Press, Boston.

in the cell membrane in close association with the reaction centers. The pigment–protein complexes absorb light in the range of 800–1000 nm (i.e., in the near infrared range). Upon treatment of the photosynthetic membranes with a mild detergent (e.g., lauryl dimethyl-amine-*N*-oxide, LDAO), the light-harvesting complexes can be separated from the reaction centers and analyzed. In *Rhodobacter sphaeroides* there are two light-harvesting complexes (LH 1 and LH 2), each containing two polypeptides ($\alpha$ and $\beta$) to which the pigments are attached (Fig. 5.9). Analysis of the hydrophobic domains of the polypeptides suggests that they span the membrane. It should be emphasized that, in order for effective energy transfer to take place between pigment molecules, they must be positioned correctly with respect to one another (Section 5.7). Organization on protein molecules probably accomplishes this. When light is absorbed by the light-harvesting pigments, the energy is quickly transferred to the reaction center (Fig. 5.9).

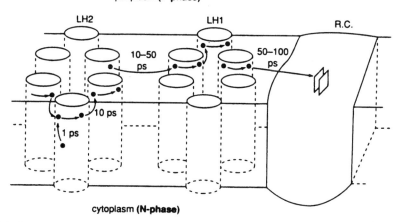

**Fig. 5.9** Light-harvesting complexes, LH 1 and LH 2, in the purple nonsulfur photosynthetic bacterium, *R. sphaeroides*. The arrow shows transfer of energy to reaction center (RC). Each cylinder represents a polypeptide unit. The small filled circles symbolize bound pigments. Excitation energy begins in the LH 2 units and then travels to the LH 1 units. The LH 1 units transfer the excitation energy to the bacteriochlorophyll dimer in the reaction center. Abbreviations: ps, picoseconds. *Source:* From Nicholls, D. G., and S. J. Ferguson. 1992. *Bioenergetics 2.* Academic Press, London.

## Light-harvesting pigments in the green photosynthetic bacteria

The major light-harvesting pigment of the green photosynthetic bacteria is either bacteriochlorophyll c, d, or e, depending upon the species (see Fig. 5.14). (The green sulfur bacteria may contain Bchl c and/or d or e, but the green nonsulfur photosynthetic bacteria have only bacteriochlorophyll c.) In addition, there is bacteriochlorophyll a. As with the purple photosynthetic bacteria, the bacteriochlorophylls are associated in some way with carotenoids and protein. In contrast to the purple bacteria, most of the light-harvesting pigments are not in the cell membrane but rather in *chlorosomes*. The chlorosomes are interesting inclusion bodies, which are about $150 \times 70 \times 30$ nm, attached to the inner membrane surface by a baseplate, which is attached to the reaction centers (Fig. 5.10). The chlorosomes contain all the bacteriochlorophylls c, d, or e, much of the carotenoid, and some of the bacteriochlorophyll a. Within the chlorosomes are rod elements of about 10 nm in diameter in *Chlorobium*, which run lengthwise through the chlorosome. The rods are composed of a polypeptide to which the bacteriochlorophyll c, d, or e is thought

to be attached. However, some investigators have suggested that the bacteriochlorophyll exists as self-aggregated clusters in the chlorosomes rather than attached to protein. Bacteriochlorophyll a exists as a protein–pigment complex in the baseplate. In addition, there is bacteriochlorophyll a in the membrane that is bound to the reaction centers. Most of the light is absorbed primarily by the bacteriochlorophyll c, d, or e in the chlorosome, and the energy is transferred to bacteriochlorophyll a in the baseplate. The bacteriochlorophyll in the baseplate then transfers the energy to membrane bacteriochlorophyll a, which transfers the energy to reaction center bacteriochlorophyll a. The green sulfur photosynthetic bacteria have an iron–sulfur-based reaction center resembling reaction center I of cyanobacteria and chloroplasts, and the green nonsulfur (filamentous) photosynthetic bacteria have a quinone-based reaction center resembling that of the purple photosynthetic bacteria.

## Light-harvesting pigments in the cyanobacteria

The photosynthetic membranes of cyanobacteria are actually intracellular membrane-bound sacs called thylakoids that have both reaction centers as well as light-harvesting complexes called *phycobilisomes*.[262, 27] The phycobilisomes are numerous granules covering the thylakoids and attached to PS II (Fig. 5.11). They can be removed from the photosynthetic membranes with nonionic detergent (Triton X-100) for purification and analysis. The phycobilisomes consist of proteins called *phycobiliproteins* that absorb light in the range of 450–660 nm due to open-chain tetrapyrroles (phycobilins) covalently bound to the protein via thioether linkages to a cysteine residue in the phycobiliprotein (Fig. 5.12). Examination of the absorption spectra of chlorophylls and carotenoids reveals that the absorbance is relatively poor in the 500–600 nm range, which is where the phycobiliproteins absorb. There are three classes of phycobiliproteins: *phycoerythrin, phycocyanin,* and *allophycocyanin.* Allophycocyanin is usually at a much lower concentration than phycoerythrin and phycocyanin, and as described below is considered to be an energy funnel to the reaction center. Although most species of cyanobacteria contain both phycoerythrin (red) and phycocanin (blue), the phycocyanins generally predominate. The phycobiliproteins are arranged in layers around each other with phycoerythrin on the outside, phycocyanin in the middle, and allophycocyanin in the inside, closest to the reaction center. A simplified drawing is shown in Fig. 5.11A. Figure 5.11B represents a detailed model at the molecular level of the phycobilisomes from the cyanobacterium *Synechococcus*, which does not contain phycoerythrin. These are fan-shaped (hemidiscoidal) phycobilisomes that consist of a dicylindrical core (A and B) containing allophycocyanin and six cylindrical phycocyanin rods radiating from the core. (In other cyanobacteria the core may consist of three cylinders.) If phycoerythrin cylinders were present, they would be an extension of the phycocyanin rods. The shorter wavelengths of light are absorbed by the phycoerythrin, the longer wavelengths by phycocyanin, and the still longer wavelengths (650 nm) by the allophycocyanin. These pigments are thus well positioned to transfer energy via inductive resonance to the reaction center. (See Section 5.7 for a discussion of energy transfer.)

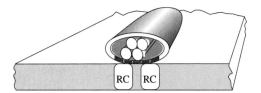

Fig. 5.10 Structure of the chlorosome. The chlorosome is surrounded by a galactolipid layer, which contains some protein, and is attached to the reaction centers in the cell membrane via a baseplate (shaded area). There is bacteriochlorophyll a in the baseplate and in the membrane. One model for the inside of the chlorosomes postulates rod-shaped proteins to which is attached the major light-harvesing pigment, bacteriochlorophyll c, d, or e. An alternative model is that the bacteriochlorophyll exists as an aggregate not attached to protein. It is suggested that light is absorbed by bacteriochlorophyll in the interior of the chlorosome and the energy is transmitted to bacteriochlorophyll a in the baseplate, then to bacteriochlorophyll a attached to the reaction centers, and from there to reaction center bacteriochlorophyll a.

A second kind of light-harvesting complex is located in the thylakoids and transmits energy to the closely associated PS I. It consists of chlorophyll a, carotenoids, and protein.

## Light-harvesting pigments in algae

All the algae contain chlorophyll a and the carotenoid β-carotene. In addition, there are more than 60 other carotenoids found among algae, some of which are shown in Fig. 5.13.

1. The *green algae* (Chlorophyta) contain chlorophyll b and xanthophylls (modified carotenes) called lutein.

2. The *brown algae* (Chromophyta) have chlorophyll $c_1$ and $c_2$ and xanthophylls called fucoxanthin and peridinin.

3. The *red algae* (Rhodophyta) have phycobilins (phycoerythrin) (Fig. 5.12), chlorophylls, carotenes, and xanthophylls.

The pigment–protein complexes can be divided into two classes. One class is called the

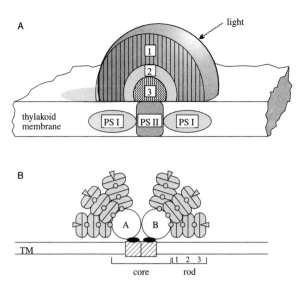

**Fig. 5.11** Structure of phycobilisome. (A) Hemispherical phycobilisome common in red algae. They are approximately 48 × 32 nm high × 32 nm. The major pigment is phycoerythrin. Light energy absorbed by the pigments in the outer region is transferred through an inner core to the reaction center: (1) phycoerythrin; (2) phycocyanin; (3) allophycocyanin. (B) Phycobilisomes from cyanobacteria are generally fan shaped with dimensions about 60 nm wide, 30 nm high, and 10 nm thick. The major pigment is phycocyanin. Detailed model of phycobilisome from the cyanobacterium *Synechococcus*. The outermost region of the phycobilisome consists of six cylindrical rods of phycocyanin. Each rod is made from three hexamers of phycocyanin, 1, 2, and 3. (*Synechococcus* does not have phycoerythrin.) The innermost region of the phycobilisome (the core) consists of two closely associated cylinders (A and B) of allophycocyanin. Each cylinder of allophycocyanin consists of two hexamers (not shown). The cross-hatched rectangle within the thylakoid membrane (TM) represents the reaction center and associated chlorophylls. The solid filled ovals between the core and the thylakoid membrane represent high-molecular-weight polypeptides thought to anchor the phycobilisome to the thylakoid membrane. There is a linker polypeptide associated with each phycocyanin hexamer that is drawn as an open circle between the hexamers. The open triangle at the terminus of each rod represents a polypeptide that may terminate the rod substructure. *Source:* Adapted from Gantt, E. 1986. *Phycobilisomes*, pp. 260–268, in *Photosynthesis III: Photosynthetic Membranes and Light Harvesting Systems*. Staehelin, L. A., and C. J. Arntzen (Eds.). Springer-Verlag, Berlin; and Grossman, A. R., M. R. Schaefer, G. G. Chiang, and J. L. Collier. 1993. The phycobilisome, a light-harvesting complex responsive to environmental conditions. *Microbiol. Rev.* 57:725–749.

inner antenna $\beta$-carotene–Chl a–protein complex, which exists as part of photosystems I and II. The second class is called the outer antenna pigment–protein class that forms light-harvesting complex I (LHC I) associated with PS I and light-harvesting complex II (LHC II) associated with PS II. The inner antenna complex that is part of the reaction centers is the same in all algae and higher plants, but the composition and structure of LHC I and LHC II varies. The main function of LHC I and II is to harvest light and transmit the energy to the pigment complex in the reaction center.

## 5.6.2 Structures of the chlorophylls, bacteriochorophylls, and carotenoids

### Chlorophylls and bacteriochlorophylls

The chlorophylls are substituted tetrapyrroles related to heme (Figs. 4.5 and 5.14). The differences are that chlorophylls have $Mg^{2+}$ in the center of the tetrapyrrole rather than $Fe^{2+(3+)}$, and a fifth ring (ring V) is present. The different substitutions on the pyrrole rings and the number of double bonds in ring II distinguish the chlorophylls from each other. Recently a Zn-bacteriochlorophyl a has been

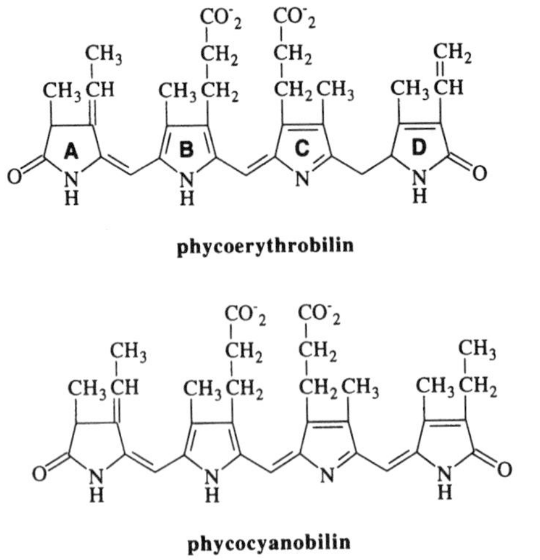

**phycoerythrobilin**

**phycocyanobilin**

**Fig. 5.12** Structures of phycoerythrobilin and phycocyanobilin. The protein is covalently bonded via a thio-ether linkage between a cysteine residue in the phycobiliprotein and C2 of $CH_3$–$CH=$ in ring A.

discovered.[28] It is present in an acidophilic photosynthetic bacterium called *Acidiphilium rubrum*.

## Carotenoids

Carotenoids are long *isoprenoids*, usually $C_{40}$ tetraterpenoids made of eight isoprene units (Fig. 5.13). (For other isoprenoids, see Note 29.) Carotenoids have a system of conjugated double bonds, which accounts for their absorption spectrum in the visible range at wavelengths poorly absorbed by chlorophylls, and makes them valuable as light-harvesting pigments. Light energy absorbed by carotenoids is transferred to the reaction center chlorophyll (Section 5.7). Carotenoid pigments with six-membered rings at both ends of the molecule are called carotenes (e.g., $\beta$-carotene) (Fig. 5.14). Modification of

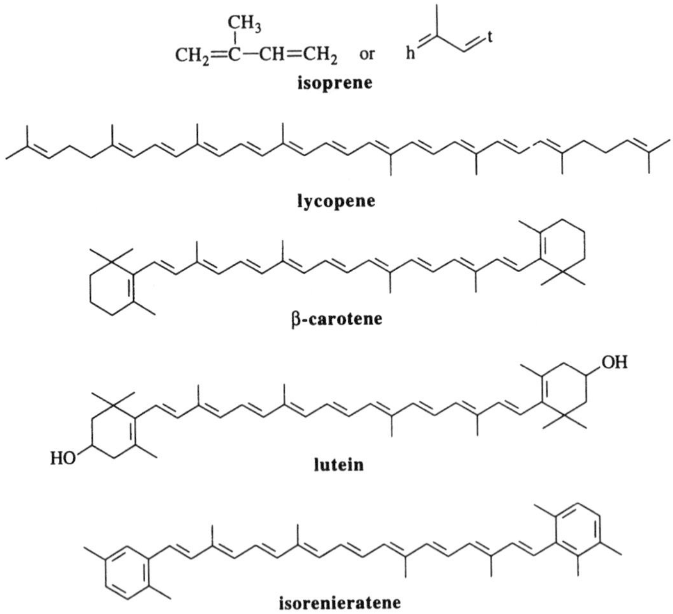

**isoprene**

**lycopene**

**β-carotene**

**lutein**

**isorenieratene**

**Fig. 5.13** Structure of carotenoids. An isoprene molecule is shown at the top. One end can be called the head (h) and the other the tail (t). All carotenoids are made of isoprene subunits with a "tail-to-tail" connection in the middle. Carotenoids in which the ends have circularized into rings are called carotenes. Modified carotenes, e.g., by hydroxylation, are called xanthophylls. Lycopene is the carotenoid responsible for the red color in tomatoes. $\beta$-Carotene is part of the inner antenna complex in all plants and algae. Lutein is a xanthophyll found in algae. Isorenieratene is a carotenoid found in green sulfur bacteria.

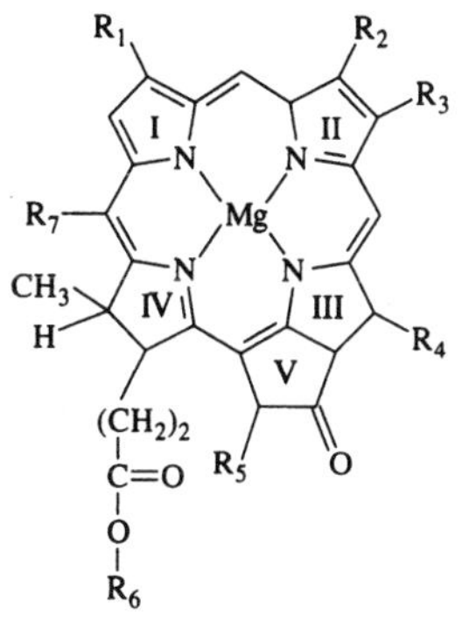

Fig. 5.14 Structure of bacteriochlorophyll. There is no double bond in ring II of bacteriochlorophyll a, b, and g between C3 and C4. (Nitrogen is position 1.)

| R | bchl a | bchl b | bchl c | bchl d | bchl e | bchl g | chl a |
|---|--------|--------|--------|--------|--------|--------|-------|
| 1 | $COCH_3$ | $COCH_3$ | $CHOHCH_3$ | $CHOHCH_3$ | $CHOHCH_3$ | $CH=CH_2$ | $CH=CH_2$ |
| 2 | $CH_3$ | $CH_3$ | $CH_3$ | $CH_3$ | $CHO$ | $CH_3$ | $CH_3$ |
| 3 | $CH_2CH_3$ | $=CHCH_3$ | $CH_2CH_3$ | $CH_2CH_3$ | $CH_2CH_3$ | $=CHCH_3$ | $CH_2CH_3$ |
| 4 | $CH_3$ | $CH_3$ | $CH_2CH_3$ | $CH_2CH_3$ | $CH_2CH_3$ | $CH_3$ | $CH_3$ |
| 5 | $COOCH_3$ | $COOCH_3$ | H | H | H | $COOCH_3$ | $COOCH_3$ |
| 6 | phytyl | phytyl | farnesyl | farnesyl | farnesyl | farnesyl | phytyl |
| 7 | H | H | $CH_3$ | H | $CH_3$ | H | H |

carotenes (e.g., by hydroxylation) produces xanthophylls, which comprise a major portion of the light-harvesting pigments in algae. Carotenoids perform a function in addition to serving as a light-harvesting pigment. As described in Section 5.2.2 and Note 13, they protect against photooxidation.

## 5.7 The Transfer of Energy from the Light-Harvesting Pigments to the Reaction Center

### 5.7.1 Mechanism of energy transfer

When light is absorbed in one region of the pigment system, excitation energy is transferred to other regions and eventually to the reaction center. The mechanism that probably accounts for the transfer of excitation energy from one pigment complex to another throughout the pigment system is called *inductive resonance transfer.*

### Inductive resonance transfer

In an unexcited molecule, all the electrons occupy molecular orbitals having the lowest available energy (the ground state). When a photon of light is absorbed, an electron becomes energized and occupies a higher-energy orbital that had been empty. This usually occurs without a change in the spin quantum number (Fig. 5.15). The electron is now in the excited singlet state. The excitation can be transferred from one pigment complex to another via *inductive resonance.* During resonance transfer, electrons do not travel between complexes, but energy is transferred. Essentially what happens is that, when the electron in the excited molecule drops back to its ground state, the energy is transferred to a molecule close by, resulting in an electron in that molecule being raised to the excited singlet state. This can occur if there is an overlap between the fluorescence spectrum of the donor molecule and the absorption spectrum of the acceptor, and if the molecules are in the proper orientation to each other. That is why the steric relationships of the light-harvesting pigments to each other and to the reaction center is of critical importance. It must be emphasized that transfer by this

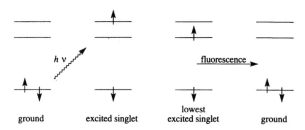

**Fig. 5.15** Light raises an electron to a higher energy level in an unoccupied orbital. The electron is then said to be in the excited singlet state. As the diagram shows, the electron may drop down to a lower excited singlet state. The electron may return to the ground state, releasing energy as light (fluorescence) or heat. Under appropriate conditions, the energy made available when the electron returns to the ground state is not released but instead is used to raise an electron in a nearby pigment to an excited singlet state, thereby transferring the energy by a process called inductive resonance. In the reaction center, electrons raised to a higher-energy orbital are transferred to a nearby acceptor molecule, creating a redox gradient.

mechanism does not occur as a result of emission and reabsorption of light.

### Delocalized excitons

A second method of excitation transfer occurs over very short distances (i.e., less than 2 nm). When the excited electron is raised to the singlet state, it is said to leave a positive "hole" in the ground state (Fig. 5.15). The combination of the excited electron and the positive hole is called an *exciton*. If two or more similar molecules are very close to each other (< 2 nm), the exciton migrates over the molecular orbitals belonging to both molecules (i.e., the exciton becomes *delocalized*). Thus the excitation energy is actually *shared* by the group of interacting molecules. However, delocalized excitons in the pigment complexes extend only over very short distances and cannot connect the outer regions of the light-harvesting complex with the reaction center. On the other hand, the sharing of excitation energy by delocalized excitons is more rapid than inductive resonance transfer and is therefore expected to be an important factor for molecules at small intermolecular distances.

### 5.7.2 Evidence that energy absorbed by the light-harvesting pigments is transferred to the reaction center

When isolated chloroplasts are irradiated, fluorescence from reaction center chlorophyll (chlorophyll a) can be measured, but not from the light-harvesting pigments (chlorophyll b, carotenoids). This means that the energy absorbed by the light-harvesting pigments is efficiently transferred to reaction center chlorophyll, whereupon an electron in the reaction center chlorophyll becomes excited to a higher energy level. In the absence of fluorescence, the energized electron can reduce the primary electron acceptor. It has been suggested that the initial redox reaction is initiated in reaction center chlorophyll by an electron in the lowest excited singlet state.[30]

### 5.8 The Structure of Photosynthetic Membranes in Bacteria

The anoxygenic phototrophic bacteria have a variety of types of photosynthetic membrane structures, depending upon the organism (Fig. 5.16). For example, the purple photosynthetic bacteria have numerous intracellular photosynthetic membranes where the light-harvesting pigments and the reaction centers are closely associated. The green bacteria separate most of the light-harvesting pigments into chlorosomes.

### 5.9 Summary

When the energy from a photon of light reaches reaction center chlorophyll, an electron in the chlorophyll becomes energized and

GREEN BACTERIA (CHLOROSOMES)

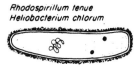

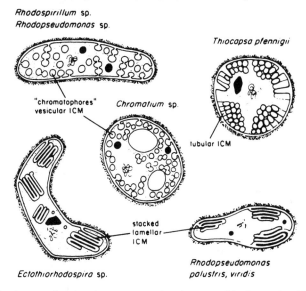

SIMPLE PHOTOSYNTHETIC BACTERIA
( NO CHLOROSOMES OR INTERNAL MEMBRANES )

PURPLE BACTERIA (INTRACYTOPLASMIC MEMBRANES)

**Fig. 5.16** Photosynthetic membranes in anoxygenic phototrophic bacteria. The green bacteria (*Chlorobium, Chloroflexus,* and *Prosthecochloris*) contain chlorosomes attached to the cytoplasmic surface of the cell membrane. Most of the purple photosynthetic bacteria have extensive invaginations of the cell membrane that can take the form of stacked lamellae (*Ectothiorhodospira* sp., *Rhodopseudomonas palustris, viridis*), vesicles (*Rhodospirillum* sp., *Rhodopseudomonas* sp.), or tubules (*Thiocapsa pfennigii*). A few photosynthetic bacteria do not have internal membranes or chlorosomes (*Rhodospirillum tenue, Heliobacterium chlorum*). *Source:* From Sprague, S. G., and A. R. Varga. 1986. Membrane architecture of anoxygenic photosynthetic bacteria, pp. 603–619. In L. A. Staehelin and C. J. Arntzen (Eds.). *Photosynthesis III: Photosynthetic Membranes and Light Harvesting Systems,* Springer-Verlag, Berlin.

is transferred to a primary acceptor molecule within the reaction center. This leaves the chlorophyll in an oxidized form. The fate of the electron lost by the primary donor differentiates two different types of reaction centers. One type of reaction center reduces quinone. Quinone-reducing reaction centers are the reaction centers in purple photosynthetic bacteria, and reaction center II in chloroplasts and cyanobacteria. In the photosynthetic bacteria, the quinone is ubiquinone. In chloroplasts and cyanobacteria, it is plastoquinone. A second type of reaction center reduces an FeS protein. The latter reaction centers are reaction center I in chloroplasts and cyanobacteria, and the reaction center in green sulfur bacteria and heliobacteria.

In the purple photosynthetic bacteria the reduced quinone transfers the electrons to a $bc_1$ complex. The $bc_1$ complex reduces cytochrome $c_2$, which returns the electron to the reaction center. This is called cyclic electron flow. A $\Delta p$ is generated by the $bc_1$ complex.

In cyanobacteria and chloroplasts the reduced quinone leaves reaction center II and reduces the $b_6f$ complex, which is similar to the $bc_1$ complex. The $b_6f$ complex reduces plastocyanin, which transfers the electron to reaction center I. A $\Delta p$ is established by the $b_6f$ complex. Having lost an electron, reaction center II accepts an electron from water, thus producing oxygen as a byproduct of the light reactions in the cyanobacteria and in chloroplasts. Reaction center I is also energized by light and transfers the electron to ferredoxin. The ferredoxin reduces $NADP^+$. The combination of reaction centers I and II is called noncyclic electron flow. In noncyclic flow, both ATP and NADPH are synthesized. Cyclic flow is also possible as the electron returns to reaction center I from reduced ferredoxin.

The green sulfur photosynthetic bacteria have a reaction center similar to reaction center I that reduces ferredoxin and $NAD^+$. The electron donors are inorganic sulfur compounds (e.g., sulfide and elemental sulfur). The electron can return to the reaction center via a quinone and a $bc_1$ complex in cyclic flow rather than reduce $NAD^+$.

Most of the light energy used in photosynthesis is not absorbed by reaction centers directly but is instead absorbed by light-harvesting pigments organized within the membrane and closely positioned to the reaction centers, in special inclusions called chlorosomes attached to the membrane, or in granules called phycobilisomes attached to the membrane. There is considerable variability in the composition and structure of the light-harvesting complexes. However, they all consist of pigment–protein complexes that transfer energy to the reaction center. The mechanism of energy transfer between the pigment molecules demands that they be positioned very close to each other in a precise orientation. The proteins in the light-harvesting complexes and in the reaction centers fulfill this function.

## Study Questions

1. If one measures oxygen production per quantum of absorbed light, the efficiency falls when using monochromatic light of 700 nm. The efficiency can be restored if the 700 nm light is supplemented with low-intensity light of shorter wavelength (e.g., 600 nm). Explain these results in terms of the light reactions of oxygenic photosynthesis.

2. Describe the similarities and differences between reaction centers I and II.

3. When bacteriochlorophyll absorbs a photon of 870 nm light (about 1.4 eV of energy), an electron travels out of the reaction center, through a $bc_1$ complex, and back to the reaction center (cyclic flow) with the extrusion of two protons to the outer membrane surface at the site of the $bc_1$ complex. Protons return to the cytoplasm through the ATP synthase with a $H^+$/ATP of three. Assuming that the $\Delta G_p$ is 45 kJ, what is the efficiency of photophosphorylation?

4. For all anoxygenic photosynthetic bacteria except the green sulfur bacteria, electrons from the reductant do not pass through the reaction center. This means that light does not stimulate electron flow from the reductant to $NAD^+$. But many of the reductants have reduction potentials more positive than that of the $NAD^+$/NADH couple. What energizes electron flow to $NAD^+$? Devise an experiment using ionophores that would support your conclusion.

5. Energy from light absorbed by the accessory pigments is transferred to the reaction center. What is the evidence for that?

6. Explain how the absorption of light by photosynthetic membranes creates a $\Delta E$.

Explain how the $\Delta E$ is converted into a $\Delta p$. Explain how the $\Delta p$ is used to drive ATP synthesis.

7. Why are two light reactions required for oxygenic photosynthesis but only one light reaction for anoxygenic photosynthesis?

## NOTES AND REFERENCES

1. Fleischman, D. E., W. R. Evans, and I. M. Miller. 1995. In: *Anoxygenic photosynthetic bacteria*. Blankenship, R. E., M. T. Madigan, and C. E. Bauer (Eds.). pp. 123–136. Kluwer Academic, Dordrecht.

2. Kramer, D. M., A. Kanazawa, and D. Fleischman. 1997. Oxygen dependence of photosynthetic electron transport in a bacteriochlorophyll-containing rhizobium. *FEBS Letts.* 417:275–278.

3. Reviewed in: *The Molecular Biology of Cyanobacteria*. 1995. Bryant, D. A. (Ed.). Kluwer Academic Publishers, Boston, MA.

4. These groupings do not reflect the complex taxonomy of the photosynthetic bacteria, as revealed by ribosomal RNA sequencing. There are five distinct evolutionary lines of photosynthetic prokaryotes (see Fig. 1.1 and Table 1.1). The purple photosynthetic bacteria are a heterogeneous assemblage that are part of the evolutionary line of prokaryotes known as the *purple bacteria*. The purple bacteria are subdivided into four subdivisions (i.e., the alpha, beta, gamma, and delta groups). Photosynthetic bacteria, along with nonphotosynthetic bacteria, are in the alpha, beta, and gamma subdivisions, whereas the delta subdivision contains only nonphotosynthetic bacteria. The purple nonsulfur photosynthetic bacteria are in the alpha and beta subdivision, whereas the purple sulfur bacteria are in the gamma subdivision. In the alpha group are *Rhodospirillum, Rhodopseudomonas, Rhodobacter, Rhodomicrobium*, and *Rhodopila*, all of which are nonsulfur purple bacteria. Another nonsulfur purple bacterium, *Rhodocyclus*, is in the beta group. The gamma group contains purple sulfur bacteria, including *Chromatium* and *Thiospirillum*. The *green sulfur bacteria*, including *Chlorobium* and *Chloroherpeton*, are only distantly related to the purple bacteria and are in a distinctly separate evolutionary line. The *green nonsulfur bacteria* (the *Chloroflexus* group) are in a separate evolutionary line phylogenetically distinct from the purple photosynthetic bacteria and the green sulfur bacteria. The *cyanobacteria* occupy a fourth evolutionary line. The photosynthetic *heliobacteria*, which consist of *Heliobacterium, Heliospirillum*, and *Heliobacillus*, are grouped in the evolutionary line that consists of the gram-positive bacteria.

5. Bullerjahn, G. S., and A. F. Post. 1993. The prochlorophytes: Are they more than just chlorophyll a/b-containing cyanobacteria? *Crit. Rev. Microbiol.* 19:43–59.

6. Olson, J. M., 1998. Chlorophyll organization and function in green photosynthetic bacteria. *Photochemistry and Photobiology* 67:61–75.

7. Dutton, P. L. 1986. Energy transduction in anoxygenic photosynthesis, pp. 197–237. In: *Photosynthesis III: Photosynthetic Membranes and Light Harvesting Systems*. Staehelin, L. A., and C. J. Arntzen (Eds.). Springer-Verlag, Berlin.

8. Mathis, P. 1990. Compared structure of plant and bacterial photosynthetic reaction centers. Evolutionary implications. *Biochim. Biophys. Acta* 1018:163–167.

9. Green nonsulfur photosynthetic bacteria have a reaction center resembling that of the purple photosynthetic bacteria. The primary donor of the electron is a dimer of Bchl a called $P_{865}$. The initial electron acceptor is bacteriopheophytin (Bpheo). The two quinones ($Q_A$ and $Q_B$) are menaquinones rather than ubiquinones.

10. Deisenhofer, J., H. Michel, and R. Huber. 1985. The structural basis of photosynthetic light reactions in bacteria. *TIBS* 10:243–248.

11. Komiya, H., T. O. Yeates, D. C. Rees, J. P. Allen, and G. Feher. 1988. Structure of the reaction center from *Rhodobacter sphaeroides* R-26 and 2.4.1: Symmetry relations and sequence comparisons between different species. *Proc. Natl. Acad. Sci. USA* 85:9012–9016.

12. Cogdell, R. J. 1978. Carotenoids in photosynthesis. *Phil. Trans. R. Soc. Lond. B* 284:569–579.

13. During photosynthesis under high light intensities an electron in bacteriochlorophyll can go from the excited singlet state to the lower-energy triplet state (Section 5.7). In the triplet state, the spin of the excited electron has changed, so that there are now two unpaired electrons as opposed to paired electrons in the singlet state. (Phosphorescence occurs from the triplet state.) When triplet-state bacteriochlorophyll reacts with oxgyen, the oxygen becomes energized to its first excited state, singlet oxygen, $^1O_2$. Singlet oxygen is very reactive and combines with cell molecules such as unsaturated fatty acids, amino acids, and purines, thus causing oxidative damage to various cell components, including lipids, enzymes, and nucleic acids. The carotenoid in the reaction center quenches the triplet state in bacteriochlorophyll. Thus carotenoids prevent photooxidation under high light intensities. When the carotenoid absorbs the energy from the triplet bacteriochlorophyll, the carotenoid itself becomes energized to the triplet state

(triplet–triplet energy transfer). However, the energy in triplet-state carotenoid is dissipated harmlessly as heat to the medium. Because the triplet-state energy of carotenoids is below the singlet-state energy of oxygen, carotenoids will also quench singlet oxygen.

14. Carotenoids are isoprenoid membrane pigments. They are useful in detecting changes in membrane potential. The energy levels of electrons in the conjugated double bond system of the carotenoids is altered by the electric field of the membrane potential. When the membrane potential changes, the carotenoid undergoes a rapid shift in its absorption spectrum peaks. This can be followed using split-beam spectroscopy similar to that discussed for examining spectral changes of cytochromes during oxidation–reduction reactions (Section 4.2.4).

15. Dutton, P. L. 1986. Energy transduction in anoxygenic photosynthesis, pp. 197–237. In: *Photosynthesis III: Photosynthetic Membranes and Light Harvesting Systems*. Staehelin, L. A. and C. J. Arntzen (Eds.). Springer-Verlag, Berlin.

16. Nitschke, W., U. Feiler, and A. W. Rutherford. 1990. Photosynthetic reaction center of green sulfur bacteria studied by EPR. *Biochemistry* 29:3834–3842.

17. Miller, M., X. Liu, S. W. Snyder, M. C. Thurnauer, and J. Biggins. 1992. Photosynthetic electron-transfer reactions in the green sulfur bacterium *Chlorobium vibrioforme*: Evidence for the functional involvement of iron–sulfur redox centers on the acceptor side of the reaction center. *Biochemistry* 31:4354–436.

18. Meent, van de E. J., M. Kobayashi, C. Erkelens, P. A. van Veelen, S. C. M. Otte, K. Inoue, T. Watanabe, and J. Amesz. 1992. The nature of the primary electron acceptor in green sulfur bacteria. *Biochim. Biophys. Acta* 1102:371–378.

19. The reaction centers of the green sulfur bacteria, heliobacteria, and photosystem I of chloroplasts differ from reaction center II and the reaction center in purple photosynthetic bacteria in producing low-potential reduced FeS proteins, rather than reducing quinone.

20. Reviewed in Dutton, 1986. Energy transduction in anoxygenic photosynthesis, pp. 197–237, In: *Photosynthesis III: Photosynthetic Membranes and Light Harvesting Systems*. Staehelin, L. A. and C. J. Arntzen (Eds.). Springer-Verlag, Berlin.

21. Emerson, R., and C. M. Lewis. 1943. The dependence of the quantum yield of Chlorella photosynthesis on wave length of light. *Am. J. Botany* 30:165–178.

22. Emerson, R., R. Chalmers, and C. Cederstrand. 1957. Some factors influencing the longwave limit of photosynthesis. *Proc. Natl. Acad. Sci. USA* 43:133–143.

23. The $b_6f$ complex is similar to the $bc_1$ complex. It contains four proteins, i.e., cyt $b_6$, subunit IV, the Rieske Fe-S protein, and cyt f. There are four prosthetic groups, i.e., two b-hemes, one c-type heme (cytochrome f), and a 2Fe–2S center. The $b_6f$ complex functions in respiration as well as in photosynthetic electron transport in the cyanobacteria. See the review by Kallas, T. 1995. The cytochrome $b_6f$ complex, pp. 259–317. In: *The Molecular Biology of Cyanobacteria*. Bryant, D. A. (Ed.). Kluwer Academic Publishers. Boston, MA.

24. Knaff, D. B. 1993. The cytochrome $bc_1$ complexes of photosynthetic purple bacteria. *Photosynthesis Res.* 35:117–133.

25. Hirschberg, J., and D. Chamovitz. 1995. Carotenoids in cyanobacteria, pp. 559–579. In: *The Molecular Biology of Cyanobacteria*. Bryant, D. A. (Ed.). Kluwer Academic Publishers, Boston, MA.

26. Grossman, A. R., M. R. Schaefer, G. G. Chiang, and J. L. Collier. 1993. The phycobilisome, a light-harvesting complex responsive to environmental conditions. *Microbiol. Rev.* 57:725–749.

27. Gantt, E. 1986. Phycobilisomes, pp. 260–268. In: *Photosynthesis III: Photosynthetic Membranes and Light Harvesting Systems*. Staehelin, L. A., and C. J. Arntzen (Eds.). Springer-Verlag, Berlin.

28. Wakao, N., Yokoi, N., Isoyama, N., Hiraishi, A., Shimada K., Kobayashi, M., Kise, H., Iwaki, M., Itoh, S., Takaichi, S., and Y. Sakurai. 1996. Discovery of natural photosynthesis using Zn-containing bacteriochlorophyll in an aerobic bacterium *Acidiphilium rubrum*. *Plant Cell Physiol.* 37:889–893.

29. Isoprenoids are an important class of molecule. Other isoprenoids are the side chains of quinones, the phytol chain in chlorophyll, and the retinal chromophore in bacteriorhodopsin, halorhodopsin, and rhodopsin.

30. Diner, B. A. 1986. Photosystems I and II: Structure, proteins, and cofactors, pp. 422–436. In: *Photosynthesis III: Photosynthetic Membranes and LIght Harvesting Systems*. Staehelin, L. A., and C. J. Arntzen (Eds.). Springer-Verlag, N.Y.

# 6

# The Regulation of Metabolic Pathways

Bacteria catalyze a very large number of chemical reactions (approximately 1000–2000) that are organized into interconnecting metabolic pathways. It is of the utmost importance that there is coordination between the pathways to avoid inefficiency, if not chaos. Pathways are regulated by adjusting the rate of one or more regulatory enzymes that govern the overall rate of the pathway. The rates that regulatory enzymes operate are modified in two ways. One method is by the noncovalent binding to the enzyme of certain biochemical intermediates of the pathways (called *allosteric effectors*) that either stimulate or inhibit the regulatory enzyme, and in this way signal to the regulatory enzyme whether the pathway is producing optimal amounts of intermediates (Sections 6.1.1 and 6.1.2). Regulatory enzymes are also called *allosteric* enzymes because in addition to having binding sites for their substrates, they also have separate binding sites for the effector molecules. (*Allo* is a Greek term meaning "other." It refers to the fact that allosteric enzymes have a second, "other," site besides the substrate site.) A second method of altering enzyme activity is by the *covalent modification* of the enzyme (Section 6.4). Covalent modifications include the attachment and removal of chemical groups such as phosphate and nucleotides.

## 6.1 Patterns of Regulation of Metabolic Pathways

In addition to learning the mechanisms of enzyme regulation (allosteric or covalent modification), the student should be aware of the *patterns* of regulation. What is meant by "patterns of regulation" will be made clear below. Patterns of regulation will vary, *even with the same pathway in different organisms*. Nevertheless, some common patterns of metabolic regulation have evolved. These are described next.

### 6.1.1 Feedback inhibition by an end product of the pathway

For biosynthetic pathways the end product is usually a negative allosteric effector for a branch point enzyme. For example, in Fig. 6.1, F and J are negative effectors for the regulatory enzymes 1 and 2, respectively. Such control is called *end-product inhibition* or *feedback inhibition* by an end product. Its role is to maintain a steady state where the end product is utilized as rapidly as it is synthesized. When the rates of utilization of the end product increase, then the concentration of the end product decreases. The decrease in concentration relieves the inhibition, and the regulatory enzyme speeds up, resulting in more end-product synthesis to meet the

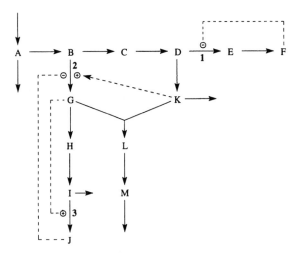

**Fig. 6.1** Branched metabolic pathways. The pathways provide precursors to other pathways at the branch points. Both positive and negative regulation occurs. In the figure, F and J are negative effectors that prevent their own overproduction by inhibiting the regulatory enzymes, 1 and 2. K is a positive effector that stimulates the production of G (in a different pathway), which is needed to react with K to form L. G is a precursor that activates a later reaction (enzyme 3) in its own pathway.

demands dictated by more rapid utilization (e.g., during rapid growth). There are three recognized patterns of feedback inhibition in biosynthetic pathways. They are *simple*, *cumulative*, and *concerted* (Fig. 6.2). Simple feedback inhibition describes the situation where the regulatory enzyme is inhibited by a single end product (Fig. 6.2, A). It is encountered in linear biosynthetic pathways. Cumulative and concerted inhibition refers to situations where more than one end product inhibits the enzyme (Fig. 6.2, B). They are seen in branched pathways. Concerted inhibition refers to a situation where both end products must bind to the regulatory enzyme simultaneously to achieve any inhibition

(Fig. 6.2, B). In cumulative inhibition the enzyme is not completely inhibited by any single end product. For example, one end product might inhibit the enzyme by 25%, and a second might inhibit the enzyme by 45%. Both end products together might inhibit the enzyme by 60% (not 70%) (Fig. 6.2, B). Thus the inhibition is cumulative, but not necessarily additive. The necessity for cumulative or concerted inhibition is that branched pathways may share a common regulatory enzyme prior to the point that leads to the separate branches of the pathway. Under these circumstances it is important that a single end product does not shut down the activity of the common enzyme, and therefore

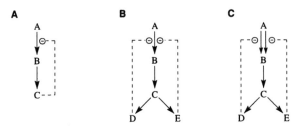

**Fig. 6.2** Patterns of feedback inhibition. (A) Simple feedback inhibition in an unbranched pathway. (B) Concerted or cumulative inhibition. In concerted inhibition, there is no inhibition unless both D and E bind. In cumulative inhibition D and E each exert partial inhibition, but the combination is less than additive. (C) Simple inhibition in a branched pathway using isoenzymes.

the synthesis of all the end products. However, sometimes branched biosynthetic pathways are regulated by simple feedback inhibition. This may occur when a reaction shared by the various branches uses enzymes that have the same catalytic sites but different effector sites (isoenzymes) (Fig. 6.2, C). (Thus isoenzymes catalyze the same reaction but are responsive to different end products.) It can be seen that different regulatory patterns (cumulative, concerted, isoenzymes) can produce similar results (i.e., the control of the rate of the pathway by the levels of one or more of the end products).

It should not be concluded that feedback inhibition applies only to biosynthetic pathways. Catabolic pathways, including those for the breakdown of sugars and carboxylic acids, may also be subject to feedback inhibition by an end product.

## 6.1.2 Positive regulation

A metabolic pathway can also be positively regulated, sometimes by an intermediate in a second pathway. A situation where this occurs is shown in Fig. 6.1. One pathway produces an intermediate, G, which combines with K in a second pathway to produce L. If insufficient G is formed, then K accumulates and stimulates the enzyme that produces G. An example of this is the activation of PEP carboxylase by acetyl-CoA described in Section 8.9.1. Another pattern of positive regulation is sometimes called "precursor activation." The latter refers to a situation where a precursor intermediate stimulates a regulatory enzyme "downstream" in the same pathway. This ensures that the rate of the downstream reactions matches that of the upstream reactions. In Fig. 6.1, G is a precursor that activates enzyme 3, which converts I to J. An example of precursor activation is the activation of pyruvate kinase by fructose-1,6-bisphosphate described in Section 8.1.2. We will encounter other examples of both positive and negative regulation by multiple effectors when we discuss the regulation of the enzymes of central metabolism in Chapter 8.

## 6.1.3 Regulatory enzymes catalyze irreversible reactions at branch points

There are certain generalizations that one can make about regulatory enzymes and the reactions that they catalyze. The reactions catalyzed by regulatory enzymes are usually at a metabolic branch point. For example, in Fig. 6.1, the intermediates B, D, and I are at branch points. Also, regulatory enzymes often catalyze reactions that are physiologically irreversible (i.e., they are far from equilibrium). Thus they are poised to accelerate in one direction when stimulated.

## 6.2 Kinetics of Regulatory and Nonregulatory Enzymes

To appreciate how effector molecules alter the activities of regulatory enzymes, one must understand enzyme kinetics and the enzyme kinetic constants, $K_m$ and $V_{max}$. This is best done by beginning with the kinetics of nonregulatory enzymes.

### 6.2.1 Nonregulatory enzymes

A plot of the rate of formation of product (or disappearance of substrate, S) as a function of substrate concentration for most enzyme-catalyzed reactions generates a hyperbolic curve similar to the one shown in Fig. 6.3. As the substrate concentration [S] is increased, the rate of the reaction approaches a maximum, $V_{max}$, because the enzyme becomes saturated with substrate. For each enzyme there is a substate concentration that gives $\frac{1}{2}V_{max}$. This is called the *Michaelis–Menten constant*, or $K_m$. The equation that describes the kinetics of enzyme activity is called the Michaelis–Menten equation:

$$v = (V_{max}S)/(K_m + S) \qquad (6.1)$$

When the substrate concentration, S, is very small compared to $K_m$, then the initial velocity, $v$, is proportional to S (actually to $V_{max}S/K_m$). However, when S becomes much larger than $K_m$, then S cancels out and $v$ approaches $V_{max}$. There are two units for $V_{max}$ (i.e., specific activity and turnover number). The

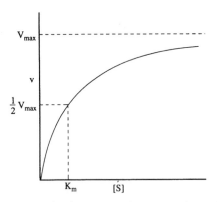

Fig. 6.3 Michaelis–Menten kinetics. When substrate concentrations [S] are plotted against initial velocity ($v$), a hyperbolic curve is obtained that approaches a maximum rate ($V_{max}$). The substrate concentration that yields $1/2 V_{max}$ is a constant for each enzyme and is called the Michaelis–Menten constant, or $K_m$.

units of specific activity are micromoles of substrate converted per minute per milligram of protein, and the units for turnover number are micromoles of substrate converted per minute per micromole of enzyme.

## Derivation of the Michaelis–Menten equation

In 1913 L. Michaelis and M. L. Menten formulated a theory of enzyme action that explained the kinetics shown in Fig. 6.3. The following derivation of the Michaelis–Menten equation was developed later by Briggs and Haldane. Consider the situation where a substrate (S) is converted to a product (P) in an enzyme-catalyzed reaction. During the reaction, S combines with the enzyme (E) at the substrate site on the enzyme to form an enzyme–substrate complex [(ES)], which then breaks down to enzyme and product. For the Briggs and Haldane derivation it is necessary to assume a steady state (i.e., the enzyme is present in catalytic amounts, S ≫ total enzyme). Under these conditions, the rate of formation and breakdown of ES are equal, which results in a steady-state level of ES. The situation can be summarized as follows:

$$S + E \text{ (free)} \underset{k_2}{\overset{k_1}{\rightleftharpoons}} (ES) \underset{k_4}{\overset{k_3}{\rightleftharpoons}} E + P$$

Notice that there are four rate constants, $k_1$, $k_2$, $k_3$, and $k_4$. If one were to measure the initial velocity ($v$), then the rate of the reaction (e.g., the rate of formation of P or disappearance of S) would be proportional to $k_3(ES)$. In this treatment we are ignoring the formation of (ES) from E and P and $k_4$. This would be small, in any event, if the initial rates were measured before P accumulated. If all the enzyme were bound to S, then all of E would be in the form of (ES), and the rate would be the maximum rate. These relationships are shown in eqs. 6.2 and 6.3.

$$v = \frac{dP}{dt} = k_3(ES) \tag{6.2}$$

and, substituting total enzyme, $E_t$, for E, when all of the enzyme is saturated with substrate:

$$V_{max} = k_3(E_t S) \tag{6.3}$$

(For enzymes that may have complex reaction pathways that depend upon more rate constants than simply $k_3$, the symbol $k_{cat}$ is used for the maximal catalytic rate, $V_{max} = k_{cat}(E_t S)$.[1]

The instantaneous rate of formation of (ES) is:

$$\frac{d(ES)}{dt} = k_1(E_t - ES)(S) \tag{6.4}$$

The concentration of free enzyme (E) is $E_t - (ES)$. S represents the substrate concentration.

Equation 6.4 describes the instantaneous rate of a bimolecular reaction between the substrate whose concentration is (S), and the free enzyme whose concentration is $E_t - ES$. Now we must consider the rate of breakdown of ES. This is given by:

$$\frac{-d(ES)}{dt} = k_2(ES) + k_3(ES) \tag{6.5}$$

Let us assume the reaction reaches a steady state at the time of measurement. This is ensured by using substrate concentrations that are in large excess over the total amount of enzyme. In the steady state, the rate of formation of ES is equal to its rate of breakdown; therefore,

$$k_1(E_t - ES)(S) = k_2(ES) + k3(ES) \tag{6.6}$$

Rearranging the above equation gives:

$$(S)(E_t - ES)/(ES) = [k_2 + k_3]/k_1 = K_m \quad (6.7)$$

The constant, $K_m$, is called the Michaelis–Menten constant. Note that if $k_3$ is small compared to $k_2$ (i.e., if the rate of product formation is small with respect to the rate that the substrate dissociates from the enzyme), then the $K_m$ is approximately equal to $k_2/k_1$ (i.e., the dissociation constant for the enzyme, $K_D$). This is true for some enzyme-substrate combinations. However, the student should recognize that it is not always true, and it is a mistake to assume (as is often done) that $1/K_m$ is a direct measure of the affinity of the substrate for the enzyme. It is best to refer to the $K_m$ as being equal to the substrate concentration that gives $\frac{1}{2}$ maximal velocity, as described in Fig. 6.3.

Now we can solve for $(ES)$:

$$(ES) = (E_t)(S)/(K_m + S) \quad (6.8)$$

where $(E_t)$ is total enzyme. Since the initial rate $(v)$ is proportional to $k_3(ES)$, we can write:

$$v = k_3(E_t)(S)/(K_m + S) \quad (6.9)$$

$V_{max} = k_3(ES)$ when $(ES) = E_t$; therefore, $V_{max} = k_3(E_t)$. Equation 6.9 then becomes:

$$v = (V_{max}S)/(K_m + S) \quad (6.10)$$

Equation 6.10 is the Michaelis–Menten equation that describes the kinetics in Fig. 6.3. Typical $K_m$ values range from $10^{-4}$ M (100 $\mu$M) and lower.

### The $K_m$ is the substrate concentration that gives $\frac{1}{2}V_{max}$

If one substitutes $\frac{1}{2}V_{max}$ for $v$ in the Michaelis–Menten equation and solves for $S$, then $S$ is equal to $K_m$. That is to say, the $K_m$ is equal to the substrate concentration that gives $\frac{1}{2}V_{max}$. This is shown as:

$$V_{max}/2 = V_{max}(S)/(K_m + S) \quad (6.11)$$

Divide both sides by $V_{max}$:

$$\tfrac{1}{2} = S/(K_m + S) \quad (6.12)$$

Rearrange:

$$K_m + S = 2S \quad (6.13)$$

$$K_m = S \quad (6.14)$$

Because the $V_m$ and therefore the $K_m$ can only be approximated from the plot shown in Fig. 6.3, eq. 6.10 is frequently written as the reciprocal:

$$1/v = 1/V_{max} + (K_m/V_{max})(1/S) \quad (6.15)$$

$1/v$ is plotted against $1/[S]$ to give a *double reciprocal* or *Lineweaver–Burk* plot (Fig. 6.4). One can find $K_m/V_{max}$ from the slope, or $-1/K_m$ from the intercept on the x axis, and $1/V_{max}$ from the intercept on the y axis.

### 6.2.2 Regulatory enzymes

Regulatory enzymes may not follow simple Michaelis–Menten kinetics. Instead, they typically show sigmoidal kinetics (Fig. 6.5). An explanation for substrate-dependent sigmoidal kinetics is that the binding of one substrate molecule increases the affinity of the enzyme for a second substrate molecule or increases the rate of formation of product from sites already occupied. This is called *positive cooperativity*. Positive cooperativity makes sense for a regulatory enzyme because it makes enzyme catalysis very sensitive to small changes in substrate-level concentrations when the substrate concentrations are very small compared to the $K_m$. (Notice in Fig. 6.5 how the enzyme rate rapidly changes with small changes

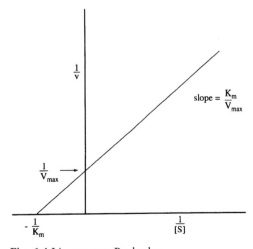

**Fig. 6.4** Lineweaver–Burk plot.

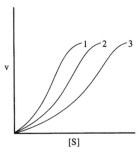

**Fig. 6.5** Sigmoidal kinetics of regulatory enzymes. Initial velocities ($v$) are plotted against substrate concentrations [S] in the presence of a positive effector (curve 1); no effector (curve 2); or a negative effector (curve 3). The positive effector decreases the $K_m$ as well as decreasing the sigmoidicity of the curve. The negative effector increases the $K_m$ and the sigmoidicity of the curve. In other cases the effectors may change the $V_{max}$.

of substrate at critical concentrations.) Allosteric effectors (positive and negative) can also bind cooperatively to enzymes, making the enzyme more sensitive to small changes in effector concentration. Even though sigmoidal kinetics are not explained by the Michaelis–Menten equation, one can still measure the $K_m$ and $V_{max}$ because these are defined operationally. When one measures these "constants" for regulatory enzymes, one finds that they are subject to change, depending upon the presence or absence of effector molecules. Some of the effectors raise the $K_m$, and some lower the $K_m$. An effector that raises the $K_m$ will decrease the velocity of the reaction, and one that lowers the $K_m$ will increase the velocity of a reaction, provided that the enzyme is not saturated with substrate. This is shown in Fig. 6.5. Notice that the negative effector (curve 3) increases the sigmoidicity of the curve and also increases the $K_m$. The positive effector (curve 1) decreases the sigmoidicity of the curve and lowers the $K_m$. Effectors can also change the $V_{max}$. However, in Fig. 6.5 the $V_{max}$ is not changed.

## 6.3 Conformational Changes in Regulatory Enzymes

When the effector binds to the allosteric site (the effector site), the protein undergoes a

conformational change, and this changes its kinetic constants. Many of the regulatory enzymes are *oligomeric* or *multimeric* (i.e., they have multiple subunits). However, for illustrative purposes we will first consider a monomeric polypeptide (Fig. 6.6). Assume the polypeptide has three binding sites: one for the substrate (substrate site), a second for a positive effector, and a third for a negative effector. Further assume that the enzyme exists in two states, A and B. When the enzyme is in conformation A, it binds the substrate and also binds the positive effector, which locks the enzyme in state A. You might say that state A has a low $K_m$. State B has a high $K_m$ and also binds the positive effector poorly. But state B does bind the negative effector, which locks it into

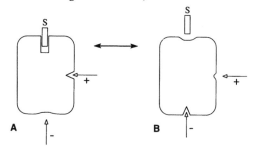

**Fig. 6.6** Conformational changes in a regulatory enzyme. Shown is a monomeric enzyme with three binding sites: one for the substrate (S), one for the positive effector (+), and one for the negative effector (−). (A) When the enzyme binds to the positive effector, a conformational change occurs that lowers the $K_m$. (B) When the enzyme binds to the negative effector, there is an increase in the $K_m$ and loss of affinity for the positive effector. In the presence of a sufficient concentration of positive effector, essentially all the enzyme will be in conformation A and the enzyme will be turned "on." On the other hand, if sufficient negative effector is present, then the enzyme will be mostly in conformation B and turned "off." The fraction of enzyme in the "on" or "off" conformation will depend upon the relative concentrations of substrate, positive effector, and negative effector. Multimeric enzymes are regulated in a similar fashion except that the effector binding sites can be on a subunit (regulatory subunit) separate from the substrate binding site (the catalytic subunit). Conformational changes in one subunit induce conformational changes in the attached subunit.

state B. Thus the positive effector lowers the $K_m$, and the negative effector raises the $K_m$. In the cell, there is an interplay of positive and negative effectors that adjust the ratio of active and less active enzyme.

### Regulatory subunits

Many regulatory enzymes are multimeric. Some consist of a regulatory subunit and one or more catalytic subunits. The catalytic subunits bear the active sites and bind the substrate. The regulatory subunits bind the effectors. When the regulatory subunits bind the effector, the polypeptide undergoes a conformational change induced by the binding. The regulatory subunit then either inhibits or activates the catalytic subunit, resulting in changes in the kinetic constants.

## 6.4 Regulation by Covalent Modification

Although most regulatory proteins appear to be regulated by conformational changes induced by the binding of allosteric effectors as described above, there are many important instances in both prokaryotes and eukaryotes of regulation by covalent modification of the protein. The enzyme may also be regulated by allosteric interactions. Covalent modification occurs by the reversible attachment of chemical groups such as acetyl groups, phosphate groups, methyl groups, adenyl groups, and uridyl groups. The covalent attachment of a chemical group can activate or inhibit

the protein. Table 6.1 summarizes some examples of covalent modification of proteins, several of which are discussed in later chapters.

## 6.5 Summary

Metabolism is regulated by key regulatory enzymes that help to keep metabolism in a steady state where concentrations of intermediates do not change. Generally, the regulatory enzymes catalyze physiologically irreversible reactions at metabolic branch points. The enzymes can be regulated by negative effectors, positive effectors, both negative and positive effectors, and by covalent modification. In biosynthetic pathways, the negative effectors are the end products. This is called feedback or end-product inhibition. Positive regulation by a precursor intermediate can speed up a subsequent reaction in the same metabolic pathway to avoid buildup of the precursor. Also, an intermediate in one pathway may regulate a reaction in a second pathway positively, if the second pathway provides a precursor metabolite for the first pathway. Enzymes that are regulated by effector molecules that bind to effector sites are called allosteric enzymes, and the effector molecules are called allosteric effectors.

Nonregulatory enzymes display simple saturation kinetics called Michaelis–Menten kinetics. Two kinetic constants are the $V_{max}$ and the $K_m$. The $V_{max}$ is the velocity of the enzyme when it is saturated with substrate. The $K_m$ is the substrate concentration yield-

**Table 6.1** Covalent modifications of bacterial enzymes and other proteins

| Enzyme | Organism | Modification |
|---|---|---|
| Glutamine synthetase | E. coli and others | Adenylylation |
| Isocitrate lyase | E. coli and others | Phosphorylation |
| Isocitrate dehydrogenase | E. coli and others | Phosphorylation |
| Chemotaxis proteins | E. coli and others | Methylation |
| $P_{II}$ | E. coli and others | Uridylylation |
| Ribosomal protein L7 | E. coli and others | Acetylation |
| Citrate lyase | Rhodopseudomonas gelatinosa | Acetylation |
| Histidine protein kinase | Many bacteria | Phosphorylation |
| Phosphorylated response regulators | Many bacteria | Phosphorylation |

Source: Neidhardt, F. C., J. L. Ingraham, and M. Schaechter. 1990. *Physiology of the Bacterial Cell*. Sinauer Associates, Inc., Sunderland, MA.

ing $\frac{1}{2}V_{max}$. Regulatory enzymes usually show sigmoidal kinetics. Nevertheless a $V_{max}$ and $K_m$ can be measured. A positive effector can raise the $V_{max}$ and lower the $K_m$, but not necessarily both. A negative effector does just the opposite. Regulatory enzymes are frequently multimeric, and consist of catalytic subunits and regulatory subunits. The allosteric effector binds to the regulatory subunit and effects a conformational change in the catalytic subunit that alters its kinetic constants.

## Study Questions

1. Plot the following data and derive a $K_m$ and $V_{max}$:

| Time (min) | Substrate (M × 10−6) | | | |
|---|---|---|---|---|
| | 16 | 24 | 48 | 144 |
| | Product (arbitrary units) | | | |
| 0.4 | 0.55 | 0.70 | 0.70 | 1.00 |
| 0.8 | 1.10 | 1.30 | 1.35 | 1.70 |
| 1.2 | 1.65 | 1.85 | 2.00 | 3.00 |
| 1.6 | 2.05 | 2.35 | 2.55 | 3.15 |
| 2.0 | 2.50 | 2.85 | 3.10 | 3.80 |

2. What is a rationale for having regulatory enzymes catalyze physiologically irreversible reactions at branch points?

3. What are three patterns of feedback inhibition for the regulation of a branched biosynthetic pathway?

4. Under what circumstances might you expect to see positive regulation of enzyme activity?

5. What is meant by positive cooperativity? What is the physiological advantage to it?

## NOTES AND REFERENCES

1. The turnover number is equal to $k_{cat}$ ($k_3$), and is related to the concentration of active sites ($E_t$) [i.e., $V_{max} = k_{cat}(E_t)$]. The turnover number is measured with a pure enzyme whose molecular weight is known.

# 7

# Bioenergetics in the Cytosol

Two different kinds of energy drive all cellular reactions. One of these is electrochemical energy, which refers to ion gradients (in bacteria primarily the proton gradient) across the cell membrane (Chapter 3). Electrochemical energy energizes solute transport, flagella rotation, ATP synthesis, and other membrane activities. The second type of energy is chemical energy in the form of high-energy molecules (e.g., ATP) in the soluble part of the cell. High-energy molecules are important because they drive biosynthesis in the cytoplasm, including the synthesis of nucleic acids, proteins, lipids, and polysaccharides. Additionally, the uptake into the cell of certain solutes is driven by high-energy molecules rather than by electrochemical energy. This chapter introduces the various high-energy molecules used by the cell and explains why "high energy" is applied to them, how they are made, and how they are used. The major high-energy molecules include ATP as well as other nucleotide derivatives, phosphoenolpyruvate, acyl phosphates, and acyl–CoA derivatives. The chapter begins with a discussion of the chemistry of the major high-energy molecules (Section 7.1), followed by an explanation of how high-energy molecules are used to drive biosynthetic reactions (Section 7.2), and how the high-energy molecules are synthesized (Section 7.3). The information in this chapter will be referred to in later chapters when the metabolic roles of

the high-energy molecules are considered in more detail.

## 7.1 High-Energy Molecules and Group Transfer Potential

High-energy molecules such as ATP have bonds that have a high free energy of hydrolysis that are sometimes depicted by a "squiggle" ($\sim$). These bonds are sometimes called "high-energy" bonds, but, as will be discussed below, this is a misnomer because the bonds have normal bond energy. The important point is that the chemical group attached to the "squiggle" is readily transferred to acceptor molecules. Therefore, the high-energy molecules are said to have a *high group transfer potential*. When the chemical groups are transferred, new linkages between molecules (e.g., ester linkages, amide linkages, glycosidic linkages, and ether linkages) are made. This results in the synthesis of the different small molecules in the cell (e.g., complex lipids, nucleotides, and so on), as well as polymers such as nucleic acids, proteins, polysaccharides, and fatty acids. We will first consider group transfer reactions in general, and then explain why certain molecules have a high group transfer potential.[1] Finally, some examples of how group transfer reactions are used in biosynthesis will be discussed.

## 7.1.1 Group transfer potential

A common chemical group that is transferred between molecules is the phosphoryl group, and phosphoryl group transfer will be used as an example of a group transfer reaction. Let us consider a generic phosphoryl group transfer reaction (shown in Fig. 7.1). We will examine both the chemical mechanism of transfer and the thermodynamics. Notice that the phosphorus in all phosphate groups carries a positive charge (i.e., the P=O bond is drawn as the semipolar $P^+-O^-$ bond). This is because phosphorus forms double bonds poorly, and the electrons in the bond are shifted toward the electron-attracting oxygen. During the phosphoryl group transfer reaction, the phosphorus atom is attacked by a nucleophile (an attacking atom with a pair of electrons seeking a positive center) shown in Fig. 7.1. The chemical group Y is displaced with its bonding electrons, and the phosphoryl group is transferred to the hydroxyl-forming ROP. This is a general scheme for group transfer reactions, not simply phosphoryl group transfers. That is to say, a nucleophile bonds to an electropositive center and displaces a leaving group with its bonding electrons. The reactions are called *nucleophilic displacements* or $S_N2$ reactions (substitution nucleophilic bimolecular). As we shall see later, various molecules such as ATP, acyl phosphates, and phosphoenolpyruvate have a high phosphoryl group transfer potential and undergo similar nucleophilic displacement reactions. But how can one compare the phosphoryl group transfer potential of all these molecules since the acceptors (the attacking nucleophile) differ? A scale is used, where the standard nucleophile is the hydroxyl group of water and the phosphoryl donors are all compared with respect to the tendency to donate the phosphoryl group to water. The group transfer potential is thus defined as the negative of the standard free energy of hydrolysis at pH 7. It is a quantitative assessment of the tendency of a molecule to donate the chemical group to a nucleophile. For example, suppose the standard free energy of hydrolysis of the phosphate ester bond in YOP is $-29,000$ J/mol. Then its phosphoryl group transfer potential is the negative of this number, or $+29,000$ J/mol. Bonds that have

a standard free energy of hydrolysis at pH 7 equal to or greater than $-29,000$ J are usually called "high-energy" bonds, although as discussed later, they have normal bond energy. The group transfer potential is not really a potential in an electrical sense, but a free energy change per mole of substrate hydrolyzed. However, the word *potential* is widely used in this context, and the convention will be followed here. The molecules with high phosphoryl group transfer potentials that we will revisit in the ensuing chapters are listed in Table 7.1. Also listed is glucose-6-phosphate, which has a low phosphoryl group transfer potential.

Group transfer potentials are a convenient way of estimating the direction in which a reaction will proceed. For example, the phosphoryl group transfer potential of ATP at pH 7 is 35 kJ/mol, and for glucose-6-phosphate it is only 14 kJ/mol. This means

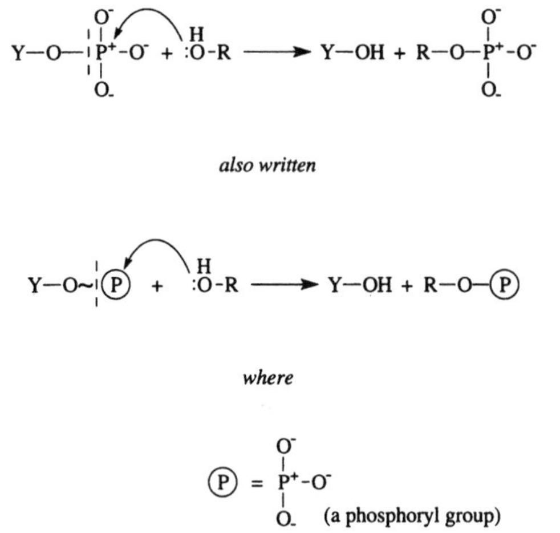

Fig. 7.1 Phosphoryl group transfer reaction. The phosphate group is shown as ionized. Because phosphorus is a poor double bond former, the phosphorus–oxygen bond exists as a semipolar bond. The positively charged phosphorus is attacked by the electronegative oxygen in the hydroxyl. The leaving group is YOH. If ROH is water, then the reaction is a hydrolysis and the product is inorganic phosphate and YOH. One can compare the tendency of different molecules to donate phosphoryl groups by comparing the free energy released when the acceptor is water, i.e., the free energy of hydrolysis. The group transfer potential is the negative of the free energy of hydrolysis.

**Table 7.1** Group transfer potentials

| Compound | $G'_{ohyd}$ (kJ/mol) | Phosphoryl group transfer potential (kJ/mol) |
|---|---|---|
| PEP + $H_2O \longrightarrow$ pyruvate + $P_i$ | −62 | +62 |
| 1,3-BPGA + $H_2O \longrightarrow$ 3-PGA + $P_i$ | −49 | +49 |
| Acetyl-P + $H_2O \longrightarrow$ acetate + $P_i$ | −47.7 | +47.7 |
| ATP + $H_2O \longrightarrow$ ADP + $P_i$ | −35 | +35 |
| Glucose-6-P + $H_2O \longrightarrow P_i$ | −14 | +14 |

that ATP is a more energetic donor of the phosphoryl group than is glucose-6-phosphate. It also means that ATP will transfer the phosphoryl group to glucose to form glucose-6-phosphate with the release of 35 − 14 or 21 kJ/mol under standard conditions, pH 7.

$$\text{ATP} + \text{glucose} \longrightarrow \text{glucose-6-phosphate}$$
$$+ \text{ADP}, \qquad \Delta G'_0 = -21\text{kJ/mol}$$

This can also be seen by summing the hydrolysis reactions because their sum equals the transfer of the phosphoryl group from ATP to glucose. This can be done for thermodynamic calculations, even though glucose-6-phosphate is not synthesized by hydrolysis reactions as written below, because the overall energy change is independent of the path of the reaction.

$$\text{ATP} + \text{H}_2\text{O} \longrightarrow \text{ADP} + \text{P}_i,$$
$$\Delta G'_0 = -35\text{kJ/mol}$$

$$\text{glucose} + \text{P}_i \longrightarrow \text{glucose-6-phosphate} + \text{H}_2\text{O},$$
$$\Delta G'_0 = +14\text{kJ/mol}$$

| | |
|---|---|
| ATP + glucose $\longrightarrow$ glucose-6-phosphate | |
| + ADP, | $\Delta G'_0 = -21\text{kJ/mol}$ |

The release of 21 kJ per mole means that the equilibrium lies far in the direction of glucose-6-phosphate. Because $\Delta G'_0 = -RT \ln K'_{eq} = -5.80 \log_{10} K'_{eq}$ kJ (at 30°C), the equilibrium constant is $4.2 \times 10^3$ in favor of glucose-6-phosphate and ADP. However, whether a particular reaction will proceed in the direction as written depends upon the actual free energy change and not the equilibrium constant. The actual free

energy change at pH 7 is $\Delta G'$, which is a function of the physiological concentrations of products and reactants and in this case is equal to:

$$\Delta G' = \Delta G'_0$$
$$+ RT\ln[\text{glucose-6-phosphate}][\text{ADP}]/[\text{ATP}][\text{glucose}]$$

Reactions proceed only in the direction of a negative $\Delta G'$. In the cell, the above reaction proceeds only in the direction of glucose-6-phosphate and ADP. This is because the ratio [glucose-6-phosphate][ADP]/[ATP][glucose] would have to be greater than $4.2 \times 10^3$ in order to change the sign of the $\Delta G'$ so that the reaction would proceed in the direction of ATP and glucose. This does not occur, and therefore under physiological conditions the direction of phosphoryl flow is always from ATP to glucose. We will now consider why ATP and the other molecules in Table 7.1 have such high group transfer potentials (high free energies of hydrolysis).

### 7.1.2 Adenosine triphosphate (ATP)

As seen in Table 7.1, the standard free energy of hydrolysis of the phosphate ester bond in ATP at pH 7 is −35 kJ. To understand why so much energy is released during the hydrolysis reaction, we must consider the structure of the ATP molecule (Fig. 7.2). Notice that at pH 7 the phosphate groups are ionized. (Actually, at pH 7 most of the ATP is a mixture of $\text{ATP}^{3-}$ and $\text{ATP}^{4-}$.) This produces electrostatic repulsion between the negatively charged phosphates, which accounts for much of the free energy of hydrolysis. Reactions during which phosphate is removed from

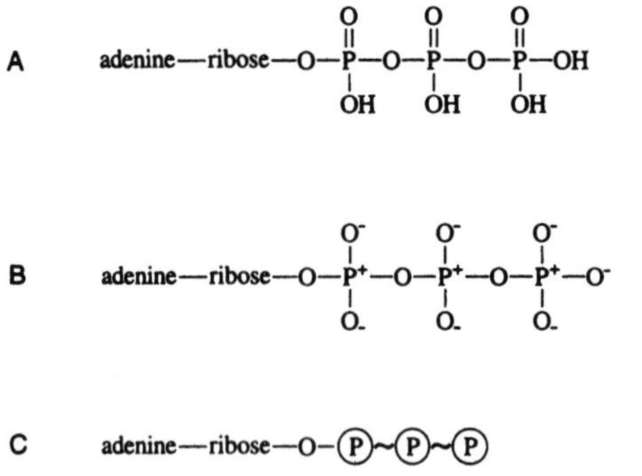

A     adenine—ribose—O—P(=O)—O—P(=O)—O—P(=O)—OH, with OH, OH, OH below

B     adenine—ribose—O—P$^+$(O$^-$)—O—P$^+$(O$^-$)—O—P$^+$(O$^-$)—O$^-$

C     adenine—ribose—O—(P)~(P)~(P)

**Fig. 7.2** Structure of ATP. ATP has three phosphate groups. (A) Unionized. (B) Although all the phosphate groups are shown ionized, giving them a net negative charge, the actual number at pH 7 is 3 or 4 for most of the ATP. Note that the phosphate–oxygen double bond is drawn as a semipolar bond, which takes into account the electronegativity of the oxygen and the low propensity of phosphorus to form double bonds. The structure predicts strong electrostatic repulsion between the phosphate groups. The electrostatic repulsion favors the transfer of phosphate to a nucleophile, e.g., water. (C) ATP drawn with squiggles to show the bonds with high free energy of hydrolysis. The phosphates are drawn as phosphoryl groups, as illustrated in Fig. 7.1.

ATP will be favored because the electrostatic repulsion is decreased as a result of the hydrolysis.

### Transfer of a phosphoryl group to other acceptors besides water

Any group that is electronegative (e.g., the hydroxyl groups in sugars) can attack the electropositive phosphorus shown in Fig. 7.2, resulting in phosphoryl group transfer, provided the appropriate enzyme is present to catalyze the reaction. In this way, ATP can phosphorylate many different compounds. Enzymes that catalyze phosphoryl group transfer reactions are called *kinases*. In summary, then, the high free energy of hydrolysis of ATP is a good predictor for the tendency of ATP to donate a phosphoryl group to nucleophiles such as the hydroxyl groups in sugars. Much of the biochemistry of ATP is directly related to this tendency. We will return to this point later, but first we must further discuss the "squiggle" and the "high-energy bond."

### The "squiggle"

As mentioned, one denotes a high negative free energy of hydrolysis, and thus a greater group transfer potential, with the symbol of a "squiggle" ($\sim$). ATP is usually drawn with two squiggles because there are two phosphate ester bonds with a high free energy of hydrolysis (Fig. 7.3). Note: The squiggle does not refer to the energy in the phosphate bond but rather to the free energy of hydrolysis. To make the distinction clear, consider the definition of bond energy. Bond energy is the energy required to break a bond. It is not the energy released when a bond is broken. In fact, the P–O bond energy is about +413 kJ (100 kcal). Compare this to the −35 kJ of hydrolysis energy.

### 7.1.3 Phosphoenolpyruvic acid

Another high-energy phosphoryl donor is phosphoenolpyruvate (PEP). In fact, PEP is a more energetic phosphoryl donor than ATP and will donate the phosphoryl group to ADP to make ATP with release of 62 − 35 or 27 kJ/ mole (under standard conditions, pH 7)

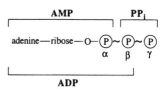

**Fig. 7.3** ATP drawn with squiggles. A high free energy of hydrolysis, i.e., high group transfer potential, is denoted by a squiggle. In ATP, two of the phosphate ester bonds have a high free energy of hydrolysis due to electrostatic repulsion. The phosphates are labelled $\alpha$, $\beta$, and $\gamma$, starting with the one nearest to the ribose. If the a phosphate is attacked, then AMP is transferred and pyrophosphate (PP$_i$) is displaced. If the g phosphate is attacked, then the phosphoryl group is transferred and ADP is displaced. Less frequently the $\beta$ phosphate is attacked and the pyrophosphoryl group is transferred, displacing the AMP. The reaction that take place depends upon the specificity of the enzyme.

(Table 7.1). As discussed in Section 7.3.4, this is an important source of ATP. PEP also donates the phosphoryl group to sugars during sugar transport in the phosphotransferase (PTS) system (Chapter 16).

## Why does PEP have a high phosphoryl group transfer potential?

The reason why PEP has a high phosphoryl group transfer potential is different from that for ATP. Consider the reaction written below and illustrated in Fig. 7.4:

$$PEP + H_2O \longrightarrow pyruvic\ acid + P_i,$$

$$\Delta G'_0 \approx -62\,kJ/mol$$

The hydrolysis removes the phosphate and allows the enol form of pyruvic acid to tautomerize into the keto form. Energy is released because the keto form is more stable than the enol form. One can account for the energy released during hydrolysis by the difference in bond energies between the keto and enol forms of pyruvic acid. A summation of bond energies reveals that the keto form has 76 kJ/mol (18 kcal/mol) more bond energy than the enol form. (Recall that bond energy is the energy required to break a bond, and

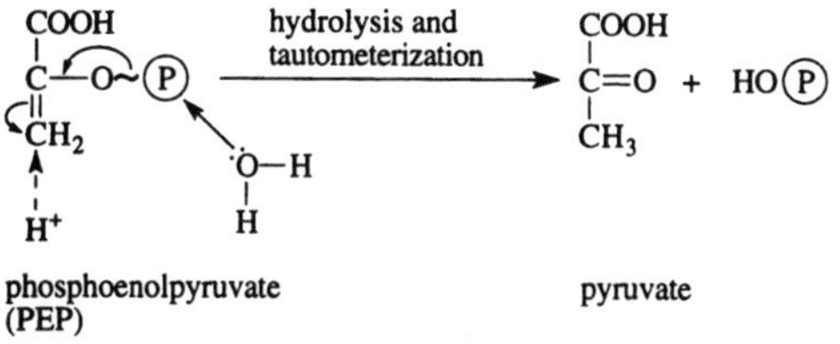

therefore it is equal to the energy released when the bond is formed. Hence the formation of a molecule with higher bond energy will result in the *release* of energy.) Thus the hydrolysis of the phosphate ester bond results in the release of energy because it promotes the enol–keto tautomerization.

## 7.1.4 Acyl derivatives of phosphate and coenzyme A

Acyl derivatives of phosphate and coenzyme A also have a high free energy of hydrolysis. An acyl group is a derivative of a carboxylic acid and has the structure shown in Fig. 7.5, where R may be an alkyl or aryl group. The acyl derivatives of coenzyme A (CoA) and of phosphate are also shown in Fig. 7.5.

## Why do acyl derivatives have high group transfer potentials?

The acyl derivatives of phosphate esters and thioesters have high group transfer potential because they do not resonate well. Consider the resonance forms of a normal ester and a phosphate ester illustrated in Fig. 7.6. In a normal ester, when two electrons shift from the oxygen to form a double bond during resonance, oxygen acquires a positive charge. In acyl–CoA and acyl–phosphate derivatives the oxygen attached to R′ is replaced by a phosphorus or sulfur atom (Fig. 7.6). Resonance is hindered in phosphate esters because the positive charge on the phosphorus

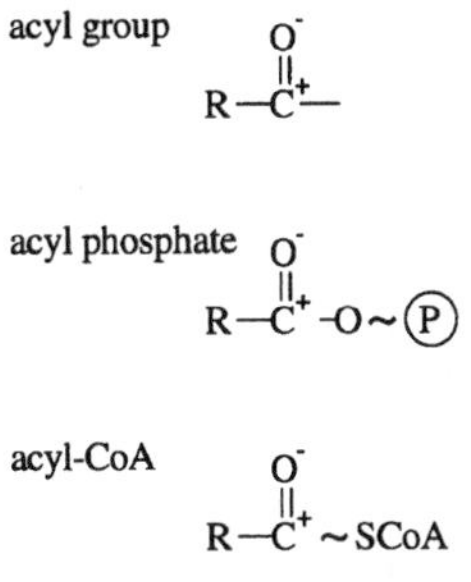

**Fig. 7.4** Conversion of PEP to pyruvate. This is an enol–keto tautomerization. The removal of the phosphoryl group allows the electrons to shift into the keto form. Energy is released because the keto form has less free energy and is more stable than the enol form. A comparison of the difference in the bond energies of pyruvic acid and enolpyruvic acid shows that pyruvic acid has 76 kJ more bond energy than enolpyruvic acid.

| enolpyruvic acid | | Pyruvic | |
|---|---|---|---|
| C–O | 293 kJ | C–C | 247 kJ |
| O–H | 460 kJ | C–H | 364 kJ |
| C=C | 418 kJ | C=O | 636 kJ |
| | 1171 kJ | | 1247 kJ |

**Fig. 7.5** Structures of acyl group, acyl–CoA and acyl–phosphates. Notice that the carbonyl group is polarized and the carbon is subject to nucleophilic attack during acyl group transfer reactions. The phosphate ester and thioester bonds in the acyl phosphates and acyl–CoAs have a high free energy of hydrolysis.

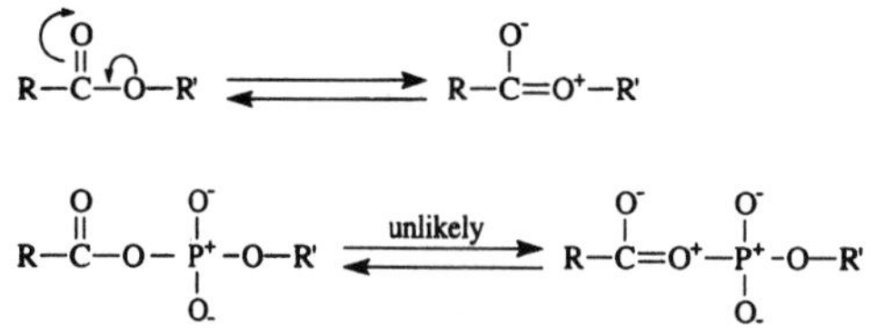

**Fig. 7.6** Resonance of an oxygen ester. Note that electrons shift from the oxygen in the C–OR' to form a double bond. This is unlikely in phosphate esters because the phosphorus atom bears a positive charge and prevents electrons from shifting in from the oxygen atom to form the double bond. Resonance is made less likely in thioesters because the sulfur atom does not form double bonds.

atom prevents electrons from shifting in from the oxygen to form a double bond, which would leave two adjacent positive centers. Thioesters also do not resonate as well as normal esters. The reason for this is that sulfur forms double bonds poorly. The poor resonance of the phosphate and thioesters is in sharp contrast to the high resonance of the free carboxylate group that forms when the phosphoryl group or CoA is transferred to an acceptor molecule. Thus the hydrolysis of the acyl derivatives of phosphate and coenzyme A is energetically favored because they lead to products stabilized by resonance with respect to the reactants.

### The importance of acyl derivatives in group transfer reactions

Acyl derivatives are very versatile. Depending on the specificity of the enzyme catalyzing the reaction, they donate *acyl groups, CoA groups,* or *phosphoryl groups.* For example, fatty acids and proteins are synthesized via acyl group transfer reactions, CoA transfer takes place during fermentative reactions in bacteria, and acyl phosphates donate their phosphoryl groups to ADP to form ATP. Enzymes that catalyze the transfer of acyl groups are called *transacylases.* Those that transfer CoA or phosphoryl groups are called *CoA transferases* and *kinases,* respectively.

## 7.2 The Central Role of Group Transfer Reactions in Biosynthesis

Group transfer reactions are central to all of metabolism because biological molecules such as proteins, lipids, carbohydrates, and nucleic acids, are synthesized as a result of group transfer reactions using high-energy donors such as ATP or other nucleotide derivatives, acyl–CoAs, and acyl–phosphates.

### 7.2.1 How ATP can be used to form amide linkages, thioester bonds, and ester bonds

#### ATP as a donor of AMP, phosphoryl groups, or pyrophosphoryl groups

ATP can donate other parts of the molecule besides the phosphoryl group (e.g., AMP, $PP_i$). This is very important for the synthesis of many polymers (e.g., proteins, polysaccharides, and nucleic acids) as well as other biochemical reactions. That is because when these groups are transferred to an acceptor molecule, the acceptor molecule itself becomes a high-energy donor for subsequent biosynthetic reactions. In other words, the energy from ATP can be transferred to other molecules that can then drive biosynthetic reactions (i.e., the formation of new covalent bonds).

Let us look again at the structure of ATP and examine some of the group transfer reactions that it can undergo (Fig. 7.3). The phosphates are labelled $\alpha$, $\beta$, and $\gamma$, when counting from the ribose moiety, with $\alpha$ being the phosphate closest to the ribose. If the $\alpha$ phosphate is attacked, then AMP is the group transferred and $PP_i$ is the leaving group. For example, this occurs during protein synthesis, discussed in Section 7.2.2. Some enzymes catalyze the transfer of the pyrophosphoryl group when the $\beta$ phosphate is attacked. For example, enzymes that synthesize phosphoenolpyruvate from pyruvate make an enzyme–pyrophosphate derivative by transferring the pyrophosphoryl group from ATP to the enzyme (Section 8.13.2). Another example is the synthesis of phosphoribosylpyrophosphate (PRPP), during which the pyrophosphoryl group is transferred from ATP to the C1 of ribose-5-phosphate (Section 9.2.2). In a subsequent reaction, the ribose-5-phosphate is transferred from PRPP to an appropriate acceptor molecule in the synthesis of purines and pyrimidines. If the $\gamma$ phosphate is attacked, then the phosphoryl group is transferred, forming phosphorylated derivatives and ADP as the leaving group. This is a very common reaction in metabolism (e.g., the synthesis of glucose-6-phosphate). As stated, all the above reactions can occur. The reaction that takes place depends upon which enzyme is the catalyst, because it is the enzyme that determines the specificity of the attack. As examples, we will consider the cases where ATP is used to form amide linkages, thioesters, and esters. In all these reactions, a carboxyl group accepts either AMP or phosphate from ATP in a group transfer reaction to form a *high-energy intermediate*. The high-energy intermediate is an acyl derivative with a high group transfer potential. The acyl–AMP or acyl–phosphate can donate the acyl group in a subsequent reaction forming an amide, ester, or thioester bond, depending upon whether the attacking nucleophile is N:, O:, or S:, respectively. Consider the reactions in Fig. 7.7. Notice the sequence of reactions. First there is a displacement on either the $\alpha$ or $\gamma$ phosphate of ATP to form the acyl derivative. Then there is a displacement on the carbonyl carbon to transfer the acyl moiety to the nucleophile to form the ester or amide bond. Notice that a pyrophosphatase is associated with reactions in which pyrophosphate is displaced. Because of the high free energy of hydrolysis of the pyrophosphate bond, the group transfer reaction is driven to completion.

### 7.2.2 How ATP is used to form peptide bonds during protein synthesis

As an example of the principles described in Sections 7.2 and 7.2.1, we will consider the formation of peptide bonds during protein synthesis. The acyl donor is made by derivatizing the $\alpha$-carboxyl group on the amino acid with AMP using ATP as the AMP donor (Fig. 7.8). The acyl–AMP is the high-energy in-

termediate. In the next reaction the acyl group is transferred to transfer RNA (tRNA) so that the carboxyl group becomes derivatized with tRNA. The acyl–tRNA also has a high group transfer potential. Thus energy has flowed from ATP to aminoacyl–AMP to aminoacyl–tRNA. The aminoacyl–tRNA is attacked by the nucleophilic nitrogen of an amino group from another amino acid, and the acyl portion is transferred to the amino group forming the peptide bond. This last reaction takes place on the ribosome. The reaction goes to completion because of the large difference in group transfer potential between the aminoacyl–tRNA and the peptide that is formed. These reactions exemplify the principle that the energy to make covalent bonds (in this case a peptide bond) derives from a series of group transfer reactions starting with ATP, in which ATP provides the energy to make high-energy intermediates that serve as group donors. In this way, all the large complex molecules (i.e., proteins, nucleic acids, polysaccharides, lipids, and so on) are synthesized.

## 7.3 ATP Synthesis by Substrate-Level Phosphorylation

We have seen how ATP can drive the synthesis of biological molecules via a coupled series of group transfer reactions. But how is ATP itself made? The answer depends upon whether the ATP is synthesized in the membranes or in the cytosol. In the membranes the phosphorylation of ADP is coupled to oxidation–reduction reactions via the generation of an electrochemical gradient of protons ($\Delta p$), which is then used to drive the phosphorylation of ADP via the membrane ATP synthase. That process is called oxidative phosphorylation or electron transport phosphorylation and is discussed in Sections 3.6.2 and 3.7.1. In electron transport phosphorylation electrons travelling over a potential difference of $\Delta E_h$ volts provide the energy to establish the $\Delta p$ ($n\Delta E_h = y\Delta p$, where $n$ is the number of electrons transferred and $y$ is the number of protons extruded). ATP in the soluble part of the cell

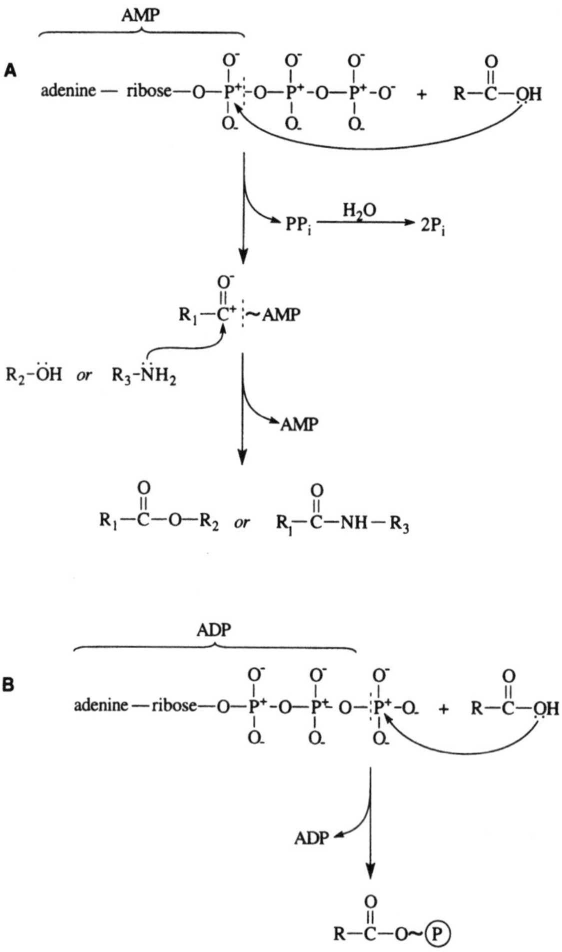

Fig. 7.7 ATP provides the energy to make ester and amide linkages. (A) A carboxyl attacks the $\alpha$ phosphate of ATP, displacing pyrophosphate, and forming the AMP derivative. The hydrolysis of pyrophosphate catalyzed by pyrophosphatase drives the reaction to completion. The AMP is then displaced by an attack on the carbonyl carbon by a hydroxyl or amino group, resulting in the transfer of the acyl group to form the ester or the substituted amide. (B) Similar to (A), except that the attack is on the $\gamma$ phosphate of ATP, displacing ADP, and forming the acyl phosphate.

is made by phosphorylating ADP by a process called *substrate-level phosphorylation*. We can define a substrate-level phosphorylation as the phosphorylation of ADP in the soluble part of the cell using a high-energy phosphoryl donor. Substrate-level phosphorylations are catalyzed by enzymes called *kinases*:

$$Y \sim P + ADP \xrightarrow{\text{kinase}} Y + ATP$$

During a substrate-level phosphorylation an oxygen in the $\beta$ phosphate of ADP acts as a nucleophile and bonds to the phosphate phosphorus in the high-energy donor. The

phosphoryl group is transferred to ADP, making ATP. (This is a phosphoryl group transfer reaction similar to that shown in Fig. 7.1.) Consider the phosphoryl group transfer from an acyl phosphate to ADP (Fig. 7.9). Phosphoryl donors for ATP synthesis during substrate-level phosphorylations include 1,3-bisphosphoglycerate (BPGA), phosphoenolpyruvate (PEP), acetyl-phosphate, and succinyl-CoA plus inorganic phosphate. These high-energy phosphoryl donors are listed in Table 7.1. The four major substrate-level phosphorylations are listed in Table 7.2. The metabolic pathways in which they occur are:

**Table 7.2** Four substrate level phosphorylations

| | |
|---|---|
| 1,3-BPGA + ADP | $\longrightarrow$ 3-PGA + ATP |
| PEP + ADP | $\longrightarrow$ Pyruvic acid + ATP |
| Acetyl-P + ADP | $\longrightarrow$ Acetic acid + ATP |
| Succinyl-CoA + $P_i$ | $\longrightarrow$ Succinic acid |
| + ADP | + ATP + CoASH |

1. The substrate-level phosphorylations using BPGA and PEP take place during glycolysis.

2. The succinyl–CoA reaction is part of the citric acid cycle.

3. Acetyl-phosphate is formed from acetyl–CoA, itself formed from pyruvate. This is an important source of ATP in anaerobically growing bacteria.

The synthesis of ATP via substrate-level phosphorylation first requires the synthesis of one of the high-energy phosphoryl donors listed in Table 7.2. All the reactions that synthesize a high-energy molecule are oxidations, with one exception. The single exception is the synthesis of phosphoenopyruvate, which results from a dehydration. During the oxidation-reduction reaction, $-nF\,\Delta E$ J is used to create a molecule with a high phosphoryl group transfer potential. The synthesis of the phosphoryl donors and the substrate-level phosphorylations of ADP are described next.

### 7.3.1 1,3-Bisphosphoglycerate

During the degradation of sugars in the metabolic pathway called glycolysis, the 6-carbon sugar glucose is cleaved into two 3-carbon fragments called phosphoglyceraldehyde (PGALD) (Chapter 8). The phosphoglyceraldehyde is then oxidized to 1,3-bisphosphoglycerate using inorganic phosphate as the source of phosphate and $NAD^+$ as the electron acceptor (Fig. 7.10)

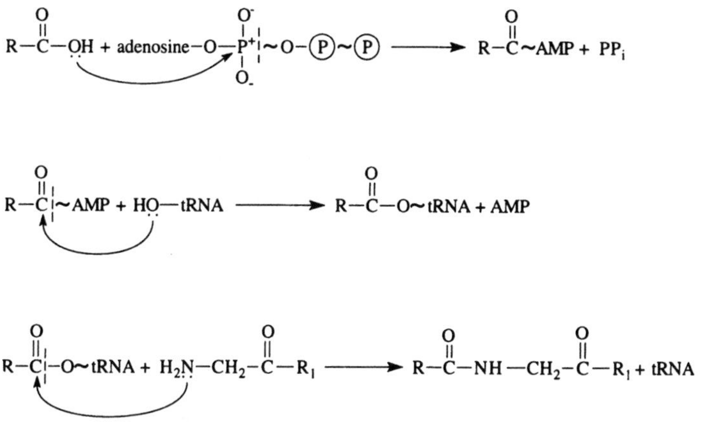

**Fig. 7.8** Formation of a peptide bond (an amide linkage between two amino acids). Peptide bonds are formed as a result of a series of group transfer reactions. The first reaction is the transfer of AMP from ATP to the carboxyl group of the amino acid. In this reaction the $\alpha$ phosphorus of ATP is attacked by the OH in the carboxyl group. Recall that the P=O bonds in ATP are semipolar and the phosphorus is an electropositive center. Most of the group transfer potential of ATP is trapped in the product, and the reaction is freely reversible. However, the reaction is driven to completion by the hydrolysis of the pyrophosphate (not shown). The second reaction is the displacement of the AMP by tRNA. This is not done for energetic reasons but rather because the tRNA is an adaptor molecule that aids in placing the amino acid in the correct position with respect to the mRNA on the ribosome. The synthesis of the aminoacyl–tRNA is reversible, indicating that the group transfer potential of the aminoacyl–tRNA is similar to that of the aminoacyl–AMP. The third reaction, which takes place on the ribosome, is the displacement of the tRNA by the amino group of a second amino acid, resulting in the synthesis of a peptide bond. This reaction proceeds with the release of a relatively large amount of free energy and is irreversible. (On the ribosome, it is the amino group of the incoming aminoacyl–tRNA at the "a" site that attacks the carbonyl of the resident aminoacyl–tRNA at the "p" site. The polypeptide is thus transferred from the "p" site to the "a" site as the tRNA at the "p" site is released. In the figure $R_1$ is the tRNA of the incoming amino acid at the "a" site, and the depicted incoming amino acid is glycine.)

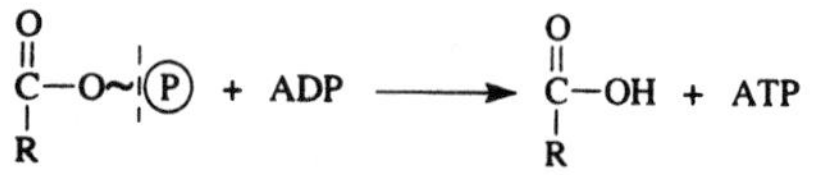

**Fig. 7.9** A substrate-level phosphorylation. This is an example of an acyl phosphate donating a phosphoryl group to ADP. The carboxylic acid is displaced. An example is the phosphorylation of ADP by 1,3-bisphosphoglycerate.

The reaction is catalyzed by *phosphoglyceraldehyde dehydrogenase*. The energy that would normally be released as heat from the oxidation $(-2F\Delta\ E'_0\ =\ -44,000$ J) is used to drive the synthesis of 1,3-bisphosphoglycerate, which has a phosphoryl group transfer potential of 49,000 J $(-\Delta G'_{o,hyd})$. (These are standard free energy changes. The actual free energy changes depend upon the ratios of concentrations of products to reactants.) The 1,3-bisphosphoglycerate then donates a phosphoryl group to ADP, in a substrate-level phosphorylation, to form ATP and 3-phosphoglycerate in a reaction catalyzed by *phosphoglycerate kinase* (Fig. 7.11). Thus the oxidation of phosphoglyceraldehyde by $NAD^+$ drives ATP synthesis.

### 7.3.2 Acetyl-phosphate

Acetyl-phosphate, another high-energy phosphoryl donor, can be made from pyruvate via acetyl–CoA. The sequence is

Pyruvate $\longrightarrow$ acetyl–CoA $\longrightarrow$ acetyl-phosphate

The acetyl-phosphate donates the phosphoryl group to ADP (to form ATP), and the acetate that is produced from the acetyl-phosphate

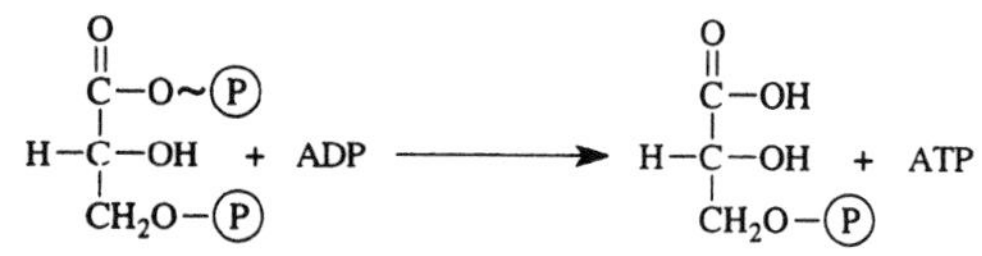

**1,3-BPGA**          **3-PGA**

**Fig. 7.11** The phosphoglycerate kinase reaction. ADP carries out a nucleophilic attack on the phosphoryl group of 1,3-bisphosphoglycerate (BPGA), displacing the free carboxylic acid (3-phosphoglycerate, 3-PGA).

is excreted into the medium. These reactions are extremely important for fermenting bacteria and accounts for the acetic acid that is produced during certain fermentations. The oxidation of pyruvate to acetyl–CoA will be discussed first (the *pyruvate dehydrogenase*, *pyruvate–ferredoxin reductase*, and *pyruvate–formate lyase* reactions). This will be followed by a description of the conversion of acetyl–CoA to acetyl-phosphate (the *phosphotransacetylase* reaction). Then the synthesis of ATP and acetate from acetyl-phosphate and ADP will be described (the *acetate kinase* reaction).

### Formation of acetyl–CoA from pyruvate

Acetyl–CoA is usually made by the oxidative decarboxylation of pyruvate, which is a key intermediate in the breakdown of sugars. There are three well-characterized enzyme systems in the Bacteria that decarboxylate pyruvate to acetyl–CoA. One is found in aerobic bacteria (and mitochondria) and is called *pyruvate dehydrogenase*. It is usually not present in anaerobically growing bacteria. The pyruvate dehydrogenase reaction is an oxidative decarboxylation of

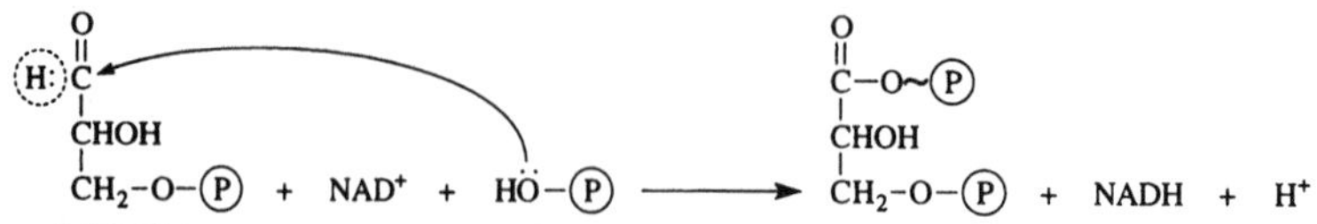

**3-PGALD**          **P_i**          **BPGA**

**Fig. 7.10** Oxidation of phosphoglyceraldehyde (PGALD). The incorporation of inorganic phosphate ($P_i$) into a high-energy phosphoryl donor occurs during the oxidation of phosphoglyceraldehyde (PGALD). The product is the acyl phosphate, 1,3-bisphosphoglycerate (BPGA). Energy that would normally be released as heat is trapped in the 1,3- bisphosphoglycerate because inorganic phosphate rather than water is the nucleophile. As a consequence, an acyl phosphate rather than a free carboxylic acid is formed.

pyruvate to acetyl–CoA, where the electron acceptor is NAD$^+$. Acetyl–CoA formed aerobically using pyruvate dehydrogenase is not a source of acetyl-phosphate but usually enters the citric acid cycle, where it is oxidized to $CO_2$ (Chapter 8). The other two enzyme systems that oxidize pyruvate to acetyl–CoA are found only in bacteria growing anaerobically. These are *pyruvate–ferredoxin oxidoreductase* and *pyruvate–formate lyase* (Fig. 7.12). (Pyruvate-ferredoxin oxidoreductase has been found in

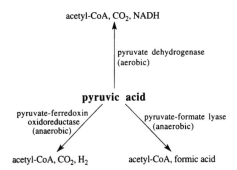

acetyl-CoA, $CO_2$, NADH

pyruvate dehydrogenase
(aerobic)

**pyruvic acid**

pyruvate-ferredoxin
oxidoreductase
(anaerobic)

pyruvate-formate lyase
(anaerobic)

acetyl-CoA, $CO_2$, $H_2$          acetyl-CoA, formic acid

**Fig. 7.12** Three enzyme systems that decarboxylate pyruvic acid to acetyl–CoA. Aerobically growing bacteria and mitochondria use the pyruvate dehydrogenase complex. The acetyl-CoA that is produced is oxidized to carbon dioxide in the citric acid cycle. Some anaerobic bacteria may also have a pyruvate dehydrogenase. Anaerobically growing bacteria generally use pyruvate–ferredoxin oxidoreductase or pyruvate–formate lyase instead of pyruvate dehydrogenase. The acetyl–CoA that is produced anaerobically (during fermentations) is converted to acetyl phosphate via the phosphotransacetylase reaction. The acetyl phosphate serves as a phosphoryl donor for ATP synthesis (the acetate kinase reaction), and the product acetate is excreted.

several archaea, but pyruvate–formate lyase and pyruvate dehydrogenase have not been found in any of the archaea thus far examined.[2]) The pyruvate–ferredoxin oxidoreductase reaction is an oxidative decarboxylation of pyruvate to acetyl–CoA, where the electron acceptor is ferredoxin. The pyruvate–formate lyase is an oxidative decarboxylation of pyruvate to acetyl–CoA, where the electrons remain with the carboxyl group (rather than being transferred to an acceptor such as NAD$^+$ or ferredoxin), which is released as formate.

An important difference between pyruvate dehydrogenase and the other two enzymes is that the pyruvate dehydrogenase reaction produces NADH. This can be disadvantageous to fermenting bacteria because there is often no externally provided electron acceptor to reoxidize the NADH. This may be why pyruvate dehydrogenase is usually not found in fermenting bacteria.

### 1. The pyruvate dehydrogenase reaction
Bacteria that are respiring aerobically use pyruvate dehydrogenase[3] to decarboxylate pyruvic acid to acetyl–CoA. This is an enzyme reaction that is also found in mitochondria. The reaction is shown in Fig. 7.13. A more detailed description of the pyruvate dehydrogenase reaction can be found in Section 8.7.

### 2. Pyruvate–ferredoxin oxidoreductase
Most anaerobically growing bacteria do not use pyruvate dehydrogenase to oxidize pyruvate to acetyl–CoA and $CO_2$. Instead, they use *pyruvate–ferredoxin oxidoreductase* or *pyruvate–formate lyase*. Pyruvate–ferredoxin oxidoreductase is found in the clostridia, sulfate-reducing bacteria, and some other anaerobes. The enzyme catalyzes a reaction

$$\begin{array}{l} COOH \\ | \\ C=O \\ | \\ CH_3 \end{array} + NAD^+ + CoASH \longrightarrow \begin{array}{l} O \\ || \\ C{\sim}SCoA \\ | \\ CH_3 \end{array} + H^+ + NADH + CO_2$$

**Fig. 7.13** The pyruvate dehydrogenase reaction. In this reaction, the two electrons that bond the carboxyl group to the rest of the molecule are transferred by the enzyme to NAD$^+$. At the same time, coenzyme A attaches to the carbonyl group to form the acylated coenzyme A derivative. If the oxidation were to take place using the :OH from water instead of the :SH from CoASH to supply the fourth bond to the carbonyl carbon, then the product would be acetic acid, and a great deal of energy would be lost as heat. But the thioester of the carboxyl group cannot resonate as well as the free carboxyl group, and for this reason, the energy that would have normally been released during the oxidation is "trapped" in the acetyl–SCoA, a molecule with a high group transfer potential.

similar to pyruvate dehydrogenase except that the electron acceptor is not $NAD^+$. Instead, it is an iron–sulfur protein called ferredoxin (Fig. 7.14). An important feature of the pyruvate–ferredoxin oxidoreductase in fermenting bacteria is that the enzyme is coupled to a second enzyme called *hydrogenase*. The hydrogenase catalyzes the transfer of electrons from reduced ferredoxin to $H^+$ to form hydrogen gas, accounting for much of the hydrogen gas produced during fermentations. The importance of the hydrogenase reaction is that it reoxidizes the reduced ferredoxin, thus allowing the continued oxidation of pyruvate. The ferredoxin-linked decarboxylation of pyruvate to acetyl–CoA is reversible and is used for autotrophic $CO_2$ fixation in certain anaerobic bacteria (Sections 13.1.4 and 13.1.6)

### 3. Pyruvate–formate lyase

Pyruvate–formate lyase is an enzyme found in some fermenting bacteria (e.g., the enteric bacteria and certain lactic acid bacteria). In the reaction the electrons stay with the carboxyl that is removed, and therefore formate is formed instead of carbon dioxide. One of the advantages to using pyruvate–formate lyase is that reduced ferredoxin or NADH is not produced, and the electrons are disposed of as part of the formate. The reaction is illustrated in Fig. 7.15.

### Formation of acetyl-phosphate from acetyl–CoA

The acetyl–CoA that is made using either pyruvate–ferredoxin oxidoreductase or pyruvate–formate lyase is converted to acetyl-phosphate by the displacement of the CoASH by inorganic phosphate in a reaction catalyzed

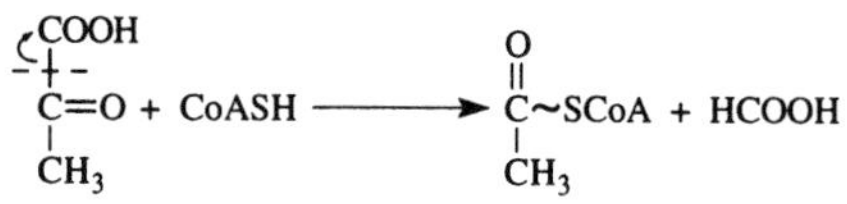

$$\begin{array}{c} \text{COOH} \\ | \\ \text{C=O} + \text{CoASH} \\ | \\ \text{CH}_3 \end{array} \longrightarrow \begin{array}{c} \text{O} \\ \| \\ \text{C}\sim\text{SCoA} + \text{HCOOH} \\ | \\ \text{CH}_3 \end{array}$$

**Fig. 7.15** The pyruvate–formate lyase reaction. Part of the molecule becomes oxidized, and part becomes reduced. The part that becomes reduced is the carboxyl group that leaves as formic acid.

by *phosphotransacetylase*. It is sometimes referred to as the PTA enzyme.

$$\text{acetyl–CoA} + P_i \xleftrightarrow{\text{phosphotransacetylase}}$$

$$\text{acetyl-P} + \text{CoASH}$$

(Acetyl-P can also be made directly from pyruvate and inorganic phosphate using *pyruvate oxidase*. This is a flavoprotein enzyme found in certain *Lactobacillus* species. See Section 14.9.)

### Formation of ATP from acetyl–phosphate

The acetyl-phosphate then donates the phosphoryl group to ADP in a substrate-level phosphorylation catalyzed by *acetate kinase*, sometimes referred to as the ACK enzyme. The formation of acetate from acetyl–CoA using ACK and PTA is referred to as the ACK–PTA pathway. Fermenting bacteria oxidize pyruvate to acetate using both the phosphotransacetylase and the acetate kinase, and derive an ATP from the process while excreting acetate into the medium. We will return to this subject in Chapter 14, where fermentations are discussed in more detail.

$$\text{acetyl-P} + \text{ADP} \xleftrightarrow{\text{acetate kinase}} \text{acetate}$$
$$+ \text{ATP}$$

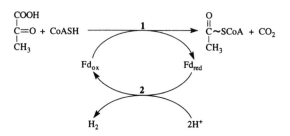

**Fig. 7.14** The pyruvate–ferredoxin oxidoreductase reaction and the hydrogenase. The electrons travel from the ferredoxin (Fd) to protons via the enzyme hydrogenase. Enzymes: 1, pyruvate–ferredoxin oxidoreductase; 2, hydrogenase.

## Making acetyl–CoA during growth on acetate

Note that acetate kinase and phosphotransacetylase catalyze reversible reactions and can be used during growth on acetate to make acetyl–CoA, which is then incorporated into cell material or oxidized to $CO_2$. Another means to make acetyl–CoA when growing on acetate is by use of the enzyme acetyl–CoA synthetase (the ACS enzyme).

$$acetate + ATP + CoA$$

$$\xleftrightarrow{\text{AMP-dependent acetyl–CoA synthetase}}$$

$$acetyl–CoA + AMP + PP_i$$

## An ADP-dependent acetyl–CoA synthetase

Although essentially all acetyl–CoA synthetases that have been studied generate AMP and $PP_i$ (the AMP-dependent acetyl–CoA synthetases) and have been isolated from bacteria, methanogenic archaea, and eukaryotes, there also exist the so-called ADP-dependent acetyl–CoA synthetases that generate ADP and $P_i$. The latter enzyme has thus far not been found to be widely distributed and has been reported in *Entamoeba histolytica* and *Giardia lamblia*, two eukaryotic microorganisms, and in certain archaea, including the hypothermophilic archaeon *Pyrococcus furiosus*, which ferments carbohydrates and peptides and grows at temperatures as high as 105°C.[4] The latter three organisms excrete acetate into the medium and appear to use the ADP-dependent acetyl–CoA synthetase as a means of generating ATP. Note that the AMP-dependent acetyl–CoA synthetase is generally used for acetate utilization, whereas the ADP-dependent acetyl–CoA synthetase may be involved primarily with acetate production (and ATP synthesis), and can

be viewed as an alternative to the ACK–PTA pathway.

$$acetate + ATP + CoA$$

$$\xleftarrow{\text{ADP-dependent acetyl–CoA synthetase}}$$

$$acetyl–CoA + ADP + P_i$$

### 7.3.3 Succinyl–CoA

Succinyl–CoA is made by the oxidative decarboxylation of $\alpha$-ketoglutarate, a reaction that occurs in the citric acid cycle (Fig. 7.16). It is strictly analogous in its mechanism and cofactor requirements to the oxidative decarboxylation of pyruvate by pyruvate dehydrogenase. The enzyme that carries out the oxidation of $\alpha$-ketoglutarate is called *$\alpha$-ketoglutarate dehydrogenase*. Succinyl–CoA then drives the synthesis of ATP from inorganic phosphate and ADP in a reaction catalyzed by the citric acid cycle enzyme *succinate thiokinase*. The reaction is:

$$succinyl–CoA + ADP + P_i \longrightarrow succinate + CoASH + ATP$$

Succinyl-phosphate is not a free intermediate. Perhaps the CoASH is transferred from succinyl–CoA to the enzyme, where it is displaced by phosphate. The phosphorylated enzyme would then be the phosphoryl donor for ADP in ATP synthesis. It should be pointed out that, whereas bacteria and plants produce ATP from succinyl–CoA, animals produce GTP instead. Succinyl–CoA is important not only for ATP synthesis but also as a precursor for heme synthesis.

### 7.3.4 Phosphoenolpyruvate

When cells are growing on sugars using the glycolytic pathway, they make phospho-

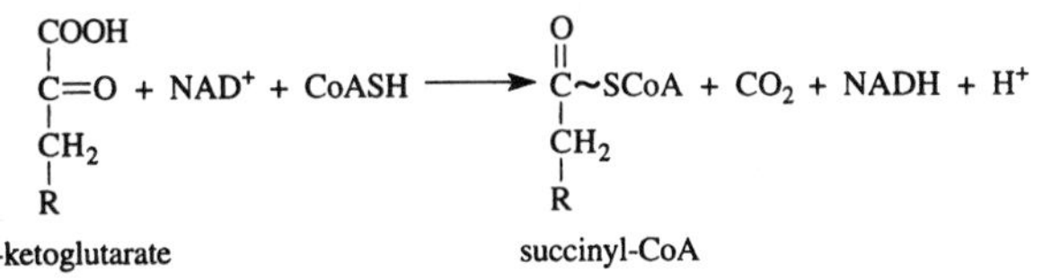

**Fig. 7.16** The $\alpha$-ketoglutarate dehydrogenase reaction. Notice that the substrate molecule resembles pyruvic acid. The difference is that in pyruvic acid the R group is H, whereas in $\alpha$-ketoglutaric acid, it is $CH_2$–COOH.

enolpyruvate from 2-phosphoglycerate (2-PGA), an intermediate in the breakdown of the sugars. The reaction is a dehydration and is catalyzed by the enzyme *enolase* (Fig. 7.17). Phosphoenolpyruvate then phosphorylates ADP in a reaction catalyzed by *pyruvate kinase*:

$$PEP + ADP \longrightarrow pyruvate + ATP$$

The enolase and pyruvate kinase reactions are important in energy metabolism because they serve to regenerate the ATP that is used to phosphorylate the sugars during the initial stages of sugar catabolism.[5] However, they cannot account for the synthesis of net ATP from ADP and inorganic phosphate, since the phosphate in the PEP originated from ATP (or PEP), rather than from inorganic phosphate. Phosphoenolpyruvate is also necessary for the synthesis of muramic acid and certain amino acids (Chapters 9 and 11).

## 7.4 Summary

Biochemical reactions in the cytosol are driven by high-energy molecules. There are several high-energy molecules (e.g., ATP, BPGA, PEP, acetyl-P, acetyl–CoA, and succinyl–CoA). They are called high-energy molecules because they have a bond with a high free energy of hydrolysis. The reasons for this depend upon the structure of the whole molecule, not on any particular bond. High free energies of hydrolysis can be due to elec-

trostatic repulsion between adjacent phosphate groups and/or diminished resonance. The term *high-energy bond* is a misnomer because the free energy of hydrolysis is not bond energy. The term *group transfer potential* refers to the negative of the free energy of hydrolysis and is a useful concept when comparing the tendency of chemical groups to be transferred to attacking nucleophiles. Thus ATP has a high phosphoryl group transfer potential (i.e., around 35 kJ standard conditions, pH 7), whereas glucose-6-phosphate has a low phosphoryl group transfer potential (i.e., around 14 kJ). Hence ATP will transfer the phosphoryl group to glucose with the release of 35−14 = 21 kJ of energy.

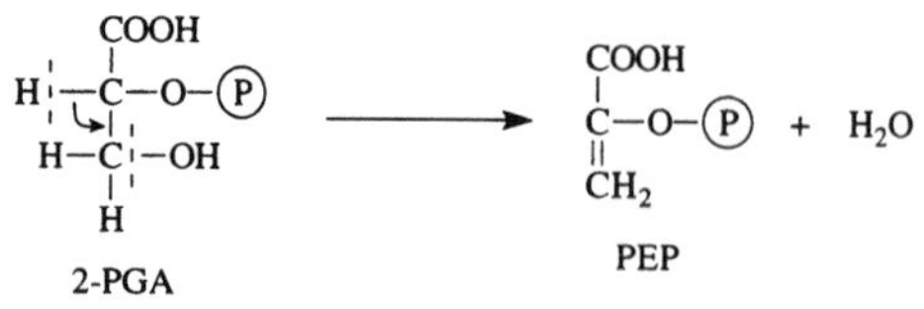

Group transfer reactions can occur with conservation of energy to form high-energy intermediates, which themselves can be group donors in coupled reactions. This explains how ATP can provide the energy to drive a series of coupled chemical reactions that result in the synthesis of nucleic acids, proteins, polysaccharides, lipids, and so on.

The formation of a high-energy molecule usually involves an oxidation of an aldehyde (3-phosphoglyceraldehyde) or the oxidative decarboxylation of a $\beta$-keto carboxylic acid (pyruvate or succinate). In substrate-level phosphorylation the energy of the the redox reaction is trapped in an acyl-phosphate or an acyl–CoA derivative. This is in contrast to respiratory phosphorylation, where the energy from the redox reaction is trapped in a $\Delta p$ that drives ATP synthesis.

Phosphoenolpyruvate is not formed as a result of an oxidation–reduction reaction but from a dehydration. However, the phosphate in phosphoenolpyruvate was already present in 2-phosphoglycerate, having been previously donated by ATP or phosphenolpyruvate during the sugar phosphorylations. Therefore, synthesis of ATP from phosphenolpyruvate does not represent the formation of ATP from ADP and inorganic phosphate, but rather the regeneration of ATP that was used previously to phosphorylate the sugars.

Fig. 7.17 The enolase reaction. 2-Phosphoglycerate is dehydrated to phosphoenolpyruvate. Isotope exchange studies suggest that the first step is the removal of a proton from C2 to form a carbanion intermediate, which loses the hydroxyl and becomes phosphoenolpyruvate. Abbreviations: 2-PGA, 2-phosphoglycerate; PEP, phosphoenolpyruvate.

## Study Questions

1. What is the definition of group transfer potential? Why do ATP, acyl phosphates, and PEP have a high phosphoryl group transfer potential?

2. Write a series of hypothetical reactions in which PEP drives the synthesis of A–B from A, B, and ADP.

3. What are two features that distinguish substrate-level phosphorylations from electron transport phosphorylation?

4. What features do the synthesis of BPGA, acetyl–CoA, and succinyl–SCoA have in common with each other but not with PEP?

5. What do the syntheses of acetyl–CoA and succinyl–CoA have in common?

6. Write a series of reactions that result in the synthesis of a substituted amide or an ester from a carboxylic acid. Use ATP as the source of energy. How is protein synthesis a modification of this reaction?

## NOTES AND REFERENCES

1. For a discussion of high-energy molecules, see Ingraham, L. L. 1962. *Biochemical Mechanisms*. John Wiley and Sons, Inc., New York.

2. Selig, M., and P. Schonheit. 1994. Oxidation of organic compounds to $CO_2$ with sulfur or thiosulfate as electron acceptor in the anaerobic hyperthermophilic archaea *Thermoproteus tenax* and *Pyrobaculum islandicum* proceeds via the citric acid cycle. *Arch. Microbiol.* **162**:286–294.

3. Dehydrogenases are enzymes that catalyze oxidation–reduction reactions in which hydrogens as well as electrons are transferred. They are named after one of the substrates (e.g., pyruvate dehydrogenase).

4. Mail, X., and M. W. W. Adams. 1996. Purification and characterization of two reversible and ADP-dependent acetyl–coenzyme A synthetases from the hyperthermophilic archaeon *Pyrococcus furiosus. J. Bacteriol.* **178**:5897–5903.

5. With rare exceptions, sugars must be phosphorylated in order to be metabolized.

# 8

# Central Metabolic Pathways

The central metabolic pathways are those pathways that provide the precursor metabolites to all the other pathways. They are the pathways for the metabolism of carbohydrates and carboxylic acids, such as $C_4$ dicarboxylic acids and acetic acid. The major carbohydrate pathways are the *Embden–Meyerhof–Parnas pathway* (also called the EMP pathway or glycolysis), the *pentose phosphate pathway* (PPP), and the *Entner–Doudoroff pathway* (ED). The Entner–Doudoroff pathway has been found to be restricted almost entirely to the prokaryotes, being present in gram-negative and gram-positive bacteria, as well as archaea. (It has been reported in one amoeba, i.e., *Entamoeba histolytica* and two fungi, i.e., *Aspergillus niger* and *Penicillium notatum*.[1] ) The three pathways differ in many ways, but two generalizations can be made:

1. All three pathways convert glucose to phosphoglyceraldehyde, albeit by different routes.

2. The phosphoglyceraldehyde is oxidized to pyruvate via reactions that are the same in all three pathways.

From an energetic point of view, the reactions that convert phosphoglyceraldehyde to pyruvate are extremely important because they generate ATP from inorganic phosphate and ADP. This is because there is an oxidation in which inorganic phosphate is incorporated into an acyl-phosphate (i.e., the oxidation of phosphoglyceraldehyde to 1,3-bisphosphoglycerate). The 1,3-bisphosphoglycerate then donates the phosphoryl group to ADP in a substrate-level phosphorylation. (See Section 7.3.1.)

The fate of the pyruvate that is formed during the catabolism of carbohydrates depends on whether the cells are respiring. If the organisms are respiring, then the pyruvate that is formed by the carbohydrate catabolic pathways is oxidized to acetyl–CoA, which is subsequently oxidized to carbon dioxide in the *citric acid cycle*. The latter generally operates only during aerobic respiration. (However, see Section 8.8.4.) If fermentation rather than respiration is taking place, then the pyruvate is converted to fermentation end products such as alcohols, organic acids, and solvents, rather than oxidized in the citric acid cycle. Fermentations are discussed in Chapter 14.

An overview of the carbohydrate catabolic pathways and their relationship to one another and to the citric acid cycle is shown in Fig. 8.1. Several points can be made about this figure. Notice that there are three substrate-level phosphorylations, two during carbohydrate catabolism and one in the citric acid cycle. Furthermore, there are six oxidation reactions, one in glycolysis, one in the pyruvate dehydrogenase reaction, and four in the citric acid cycle. These

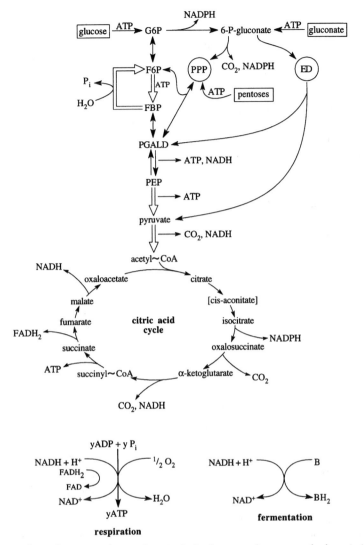

**Fig. 8.1** Relationships between the major carbohydrate pathways and the citric acid cycle. The pathway from glucose-6-phosphate to pyruvate is the Embden–Meyerhof–Parnas pathway (glycolysis). The pentose phosphate pathway (PPP) and the Entner–Doudoroff pathway (ED) branch from 6-phosphogluconate. Both of these pathways intersect with the glycolytic pathway at phosphoglyceraldehyde. All the carbohydrate pathways produce pyruvate, which is oxidized to acetyl–CoA. In aerobically growing organisms, the acetyl–CoA is oxidized to $CO_2$ in the citric acid cycle. The electrons from NAD(P)H and $FADH_2$ are transferred to the electron transport chain in respiring organisms with the formation of ATP. In fermenting cells, the NADH is reoxidized by an organic acceptor (B) that is generated during catabolism. The citric acid cycle does not operate as an oxidative pathway during fermentative growth. Abbreviations: G6P, glucose-6-phosphate; F6P, fructose-6-phosphate; FBP, fructose-1,6-bisphosphate; PGALD, 3-phosphoglyceraldehyde; PEP, phosphoenolpyruvate; PPP, pentose phosphate pathway; ED, Entner–Doudoroff pathway.

oxidations produce NADH (primarily) and $FADH_2$. The NADH and $FADH_2$ must be reoxidized in order to regenerate the $NAD^+$ and FAD that are required for the oxidations. The route of reoxidation and the energy yield depend upon whether the organism is respiring or fermenting. During respiration, the NADH and $FADH_2$ are reoxidized via electron transport with the formation of a $\Delta p$. (The $\Delta p$ is used for ATP synthesis via respiratory phosphorylation as explained in Chapter 4.) In fermenting cells, most of the NADH is reoxidized in the cytosol by an organic acceptor, but ATP is not made. (However, see Note 2.) The different pathways for the reoxidation of NADH in fermenting bacteria are discussed in Chapter 14. The student will notice that the citric acid cycle generates a great deal of NADH and $FADH_2$. The reoxidation of the NADH and $FADH_2$ requires adequate amounts of electron acceptor, such as is provided to respiring organisms. In fact, the oxidative citric acic cycle as illustrated in Fig. 8.1 is coupled to respiration, and during fermentative growth it becomes modified into a reductive pathway (Section 8.10).

## 8.1 Glycolysis

It is best to think of glycolysis as occurring in two stages:

**Stage 1.** This stage catalyzes the splitting of the glucose molecule ($C_6$) into two phosphoglyceraldehyde ($C_3$) molecules. It consists of four consecutive reactions. Two ATPs are used per glucose metabolized; and these donate the phosphoryl groups that become the phosphates in phosphoglyceraldehyde. (The phosphoglyceraldehyde eventually becomes phosphoenolpyruvate in stage two, and the phosphate that originated from ATP is returned to ATP in the pyruvate kinase step, thus regenerating the ATP.)

**Stage 2.** This stage catalyzes the oxidation of phosphoglyceraldehyde to pyruvate. It consists of five consecutive reactions. Stage 2 generates four ATPs per glucose metabolized, hence the net yield of ATP is two. Stage 2 reactions are not unique to glycolysis and also occur when pyruvate is formed from phosphoglyceraldehyde in the pentose phosphate pathway and the Entner–Doudoroff pathway, accounting for ATP synthesis in these pathways.

Stage 1: Glucose $+ 2ATP \longrightarrow 2PGALD + 2ADP$

Stage 2: $2PGALD + 2P_i + 4ADP + 2NAD^+$
$\longrightarrow 2$ pyruvate $+ 4ATP + 2NADH + 2H^+$

Sum: Glucose $+ 2ADP + 2P_i + 2NAD^+$
$\longrightarrow 2$ pyruvate $+ 2ATP + 2NADH + 2H^+$

The reactions are summarized in Fig. 8.2. The pathway begins with the phosphorylation of glucose to form glucose-6-phosphate (reaction 1). The phosphoryl donor is ATP in a reaction catalyzed by hexokinase. The ATP is regenerated from phosphoenolpyruvate in stage 2. Some bacteria phosphorylate glucose during transport into the cell via the phosphotransferase (PTS) system, in which case the phosphoryl donor is phosphoenolpyruvate. (See Section 16.3.4 for a discussion of the phosphotransferase system.) The glucose-6-phosphate (G6P) isomerizes to fructose-6-phosphate (F6P) in a reaction catalyzed by the enzyme isomerase (reaction 2). The isomerization is an electron shift where two electrons from the C2 carbon reduce the C1 aldehyde of the glucose-6-phosphate molecule to an alcohol (Section 8.1.3). The fructose-6-phosphate is phosphorylated at the expense of ATP to fructose-1,6-bisphosphate (FBP) by the enzyme fructose-6-phosphate kinase (reaction 3). The ATP used to phosphorylate fructose-6-phosphate is also regenerated from phosphoenolpyruvate in Stage 2. The fructose-1,6-bisphosphate is split into phosphoglyceraldehyde (PGALD) and dihydroxyacetone phosphate (DHAP) by fructose-1,6-bisphosphate aldolase (reaction 4). The splitting of fructose-1,6-bisphosphate is facilitated by the electron attracting keto group at C2, thus rationalizing the isomerization of glucose-6-phosphate to fructose-6-phosphate (Section 8.1.3). The dihydroxyacetone phosphate is isomerized to phosphoglyceraldehyde (reaction 5), in a reaction similar to the earlier isomerase reaction (reaction 2). Thus stage 1 produces two moles of phosphoglyceraldehyde per mole of glucose.

In Stage 2, both moles of phosphoglyceraldehyde are oxidized to 1,3-bisphosphoglycerate (also called disphosphoglycerate, DPGA) (reaction 6). The bisphosphoglycerate serves as the phosphoryl donor for a substrate-level phosphorylation catalyzed by the enzyme phosphoglycerate kinase (reaction 7). At this point, two ATPs are made, one from each of the two bisphosphoglycerates. The product of the phosphoglycerate

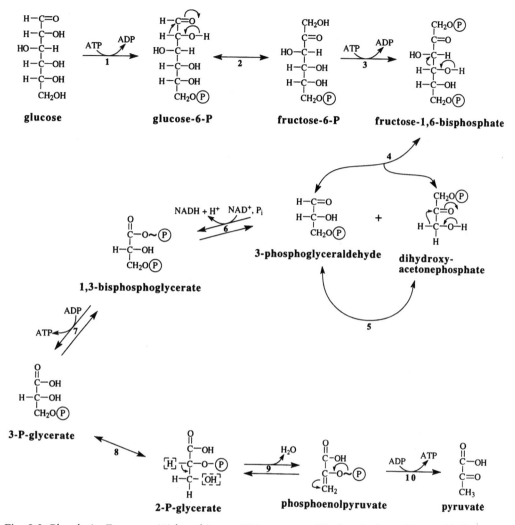

**Fig. 8.2** Glycolysis. Enzymes: (1) hexokinase; (2) isomerase; (3) phosphofructokinase; (4) fructose-1,6-bisphosphate aldolase; (5) triosephosphate isomerase; (6) triosephosphate dehydrogenase, (7) phosphoglycerate kinase; (8) mutase; (9) enolase; (10) pyruvate kinase.

kinase reaction is 3-phosphoglycerate (3-PGA). The two moles of 3-phosphoglycerate are converted to two moles of 2-phosphoglycerate (2-PGA) (reaction 8), which are dehydrated to two moles of phosphoenolpyruvate (PEP) (reaction 9). The phosphoenolpyruvate serves as the phosphoryl donor in a second site substrate-level phosphorylation to form two more moles of ATP and two moles of pyruvate (reaction 10). Notice that the phosphate in the 2-phosphoglycerate originated from ATP during the phosphorylations in stage 1 (reactions 1 and 3). Thus the net synthesis of ATP from ADP and inorganic phosphate in

glycolysis is coupled to the oxidation of phosphoglyceraldehyde to 3-phosphoglycerate (reactions 6 and 7). For a more detailed discussion of substrate-level phosphorylations, see Section 7.3.

### 8.1.1 Glycolysis as an anabolic pathway

The glycolytic pathway serves not only to oxidize carbohydrate to pyruvate and to phosphorylate ADP, but it also provides precursor metabolites for many other pathways. Figure 8.3 summarizes the glycolytic reactions and points out only a few of the branch points to other pathways. For example, glucose-6-phosphate is a precursor to polysaccharides, pentose phosphates, and aromatic amino acids; fructose-6-phosphate is a precursor to amino sugars (e.g., muramic acid and glucosamine found in the cell wall), dihydroxyacetone phosphate is a precursor to phospholipids, 3-phosphoglycerate is a precursor to the amino acids glycine, serine, and cysteine; and phosphoenolpyruvate is a precursor to aromatic amino acids and to the lactyl portion of muramic acid. When the organisms are not growing on carbohydrate, they must synthesize these glycolytic intermediates from other carbon sources. Figure 8.3 shows that some of the carbon from amino acids, carboxylic acids (organic acids), and lipids is converted to phosphoenolpyruvate from which the glycolytic intermediates can be synthesized. Also, pyruvate can serve as a carbon source and therefore must be converted to glycolytic intermediates. However, it can be seen that the glycolytic pathway can be reversed from phosphoenolpyruvate only to fructose-1,6-bisphosphate (FBP), and not at all from pyruvate. This is because the pyruvate kinase and phosphofructokinase reactions are physiologically irreversible due to the high free energy in the phosphoryl donors with respect to the phosphorylated products. Therefore, to reverse glycolysis the kinase reactions are bypassed. The conversion of fructose-1,6-bisphosphate to fructose-6-phosphate requires fructose-1,6-bisphosphate phosphatase (Fig. 8.4). There are also alternative ways to convert pyruvate directly to phosphoenolpyruvate without using pyruvate kinase. These are discussed in Section 8.13.

## 8.1.2 Regulation of glycolysis

Figure 8.3 also illustrates the regulation of glycolysis in E. coli. Two key enzymes in regulating the *directionality* of carbon flow are phosphofructokinase (reaction 1) and fructose-1,6-bisphosphate phosphatase (reaction 2), which catalyze physiologically irreversible steps. The kinase catalyzes the phosphorylation of fructose-6-phosphate

to fructose-1,6-bisphosphate, whereas the phosphatase catalyzes the dephosphorylation of fructose-1,6-bisphosphate to fructose-6-phosphate. Models for the regulation of glycolysis are based primarily on in vitro studies of the allosteric properties of the enzymes. Important effector molecules are AMP and ADP. When both of these are high, ATP is low, since they are both derived from ATP. That is to say,

$$ATP \longrightarrow ADP + P_i$$

$$ATP \longrightarrow AMP + PP_i$$

Thus high ADP and AMP concentrations are a signal that the ATP levels are low. (Allosteric activation and inhibition are discussed in Chapter 6.) Since glycolysis produces ATP, it makes sense to stimulate glycolysis when the ATP levels are low. E. coli accomplishes this by allosterically activating the phosphofructokinase with ADP, which, as mentioned, is at a higher concentration when the ATP levels are low. At the same time that glycolysis is stimulated by ADP, the reversal of glycolysis is slowed by AMP, which is also at a higher concentration when the ATP levels are low. The reason for this is that AMP inhibits the fructose-1,6-bisphosphate phosphatase reaction. The student may notice that the sum of the phosphofructokinase and fructose-1,6-bisphosphatase reaction is the hydrolysis of ATP (i.e., ATPase activity). The stimulation of the phosphofructokinase by ADP and the inhibition of the phosphatase by AMP prevents the unnecessary hydrolysis of ATP when the ATP levels are low. Glycolysis is not only regulated by AMP and ADP in E. coli, but also by phosphoenolpyruvate and fructose-6-phosphate. As indicated in Fig. 8.3, the phosphofructokinase is feedback inhibited by phosphoenolpyruvate. This can be considered an example of end-product inhibition. The pyruvate kinase, another physiologically irreversible reaction, is positively regulated by fructose-1,6-bisphosphate, which is an example of a precursor metabolite activating a later step in the pathway. (Feedback inhibition and precursor activation are discussed in Sections 6.1.1 and 6.1.2.)

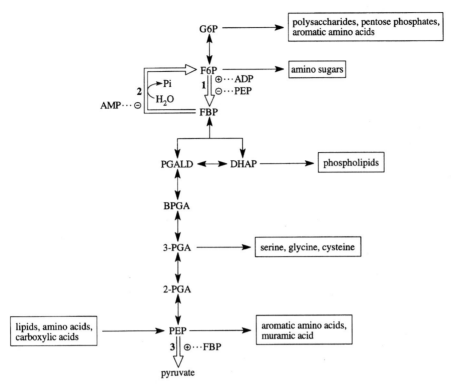

**Fig. 8.3** Glycolysis as an anabolic pathway and its regulation in *E. coli*. The rationale for the pattern of regulation is that when the ADP and AMP levels are high, the ATP levels are low and therefore glycolysis is stimulated. Steady-state levels of intermediates are maintained by positive and negative feedback inhibition.

### 8.1.3 The chemical bases for the isomerization and aldol cleavage reactions in glycolysis

It is important to understand the chemistry of metabolic reactions as well as to learn the pathways and their physiological role. To this end, the isomerization and aldol cleavage reactions will be explained because they are common reactions that we will see later in other pathways. Consider the isomerization of glucose-6-phosphate to fructose-6-phosphate. The rationale for this isomerization is that

it creates an electron attracting keto group at C2 of the sugar, and the electron attracting keto group is necessary to break the bond between C3 and C4 in the aldolase reaction. These reactions are shown in Fig. 8.5. The isomerization can be viewed as the oxidation of C2 by C1, because two electrons shift from C2 to C1. This happens in two steps. A hydrogen dissociates from C2 and two electrons shift in to form the cis-enediol. Then the hydrogen in the C2 hydroxyl dissociates and two electrons shift in, forcing the two electrons in the double bond to go

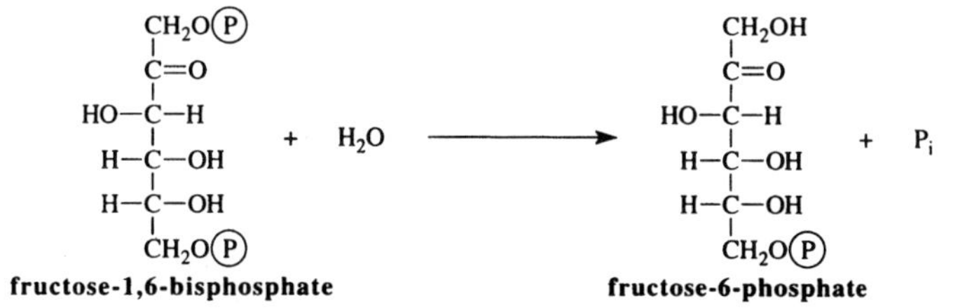

CH₂O(P)
|
C=O
|
HO—C—H          +   H₂O
|
H—C—OH
|
H—C—OH
|
CH₂O(P)

CH₂OH
|
C=O
|
HO—C—H          +   Pᵢ
|
H—C—OH
|
H—C—OH
|
CH₂O(P)

**fructose-1,6-bisphosphate**          **fructose-6-phosphate**

**Fig. 8.4** The fructose-1,6-bisphosphatase reaction

to the C1. The result is fructose-6-phosphate. The fructose-6-phosphate becomes phosphorylated to fructose-1,6-bisphosphate. The fructose-1,6-bisphosphate is split by the aldolase when the keto group on C2 pulls electrons away from the C–C bond between C3 and C4, as two electrons shift in from the hydroxyl on C4 (Fig. 8.5). The products of the split are phosphoglyceraldehyde and dihydroxyacetone phosphate. A second isomerization converts the dihydroxyacetone phosphate to phosphoglyceraldehyde via the same mechanism as the isomerization between glucose-6-phosphate and fructose-6-phosphate. In this way, two phosphoglyceraldehydes can be formed from glucose-6-phosphate.

### 8.1.4 Why are the glycolytic intermediates phosphorylated?

In glycolysis, the phosphorylation of ADP by inorganic phosphate is due to two reactions that take place in Stage 2 (i.e., the oxidation of the C1 aldehyde of 3-phosphoglyceraldehyde

to the acyl phosphate and the subsequent transfer of the phosphoryl group to ADP). The first reaction is catalyzed by triose-phosphate dehydrogenase, and the second reaction is catalyzed by phosphoglycerate kinase. Given that these are the steps in which net ATP is made from ADP and inorganic phosphate, one can ask why the other intermediates are phosphorylated. Phosphorylation of the intermediates requires the use of two ATPs in Stage 1, which are simply regenerated in the pyruvate kinase step. There is probably more than one reason why all of the intermediates are phosphorylated. One reason may be because the kinase reactions in Stage 1 are irreversible and therefore drive the reactions rapidly in the direction of pyruvate. Another reason has to do with the physiological role of the pathway. Glycolysis is not simply a pathway for the oxidation of glucose and the provision of ATP. Very importantly, the glycolytic pathway also provides phosphorylated precursors to many other pathways. In subsequent chapters, we will study these interconnections with other pathways.

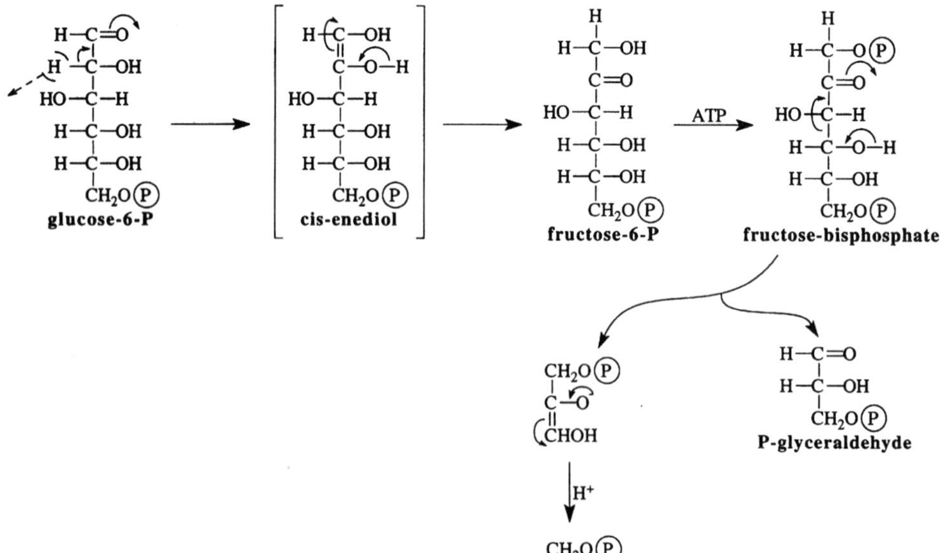

Fig. 8.5 Making two phosphoglyceraldehydes from glucose-6-phosphate. Glucose-6-phosphate itself cannot be split because there is no electron attracting group to withdraw the electrons from the C–C bond between carbons #3 and #4. An electron-withdrawing keto group is created on C2 when glucose-6-phosphate is isomerized to fructose-6-phosphate.

## 8.2 The Fate of NADH

If the NADH were not reoxidized to $NAD^+$, then all pathways (including glycolysis) that require $NAD^+$ would stop. Clearly, glycolysis must be coupled to pathways that re-oxidize NADH back to $NAD^+$. Bacteria have three ways to reoxidize NADH: respiration, fermentation, and the hydrogenase reaction. Respiration (aerobic or anaerobic):

$$NADH + H^+ + B + yADP + yP_i$$
$$\longrightarrow NAD^+ + BH_2 + yATP$$

where $y$ is approximately equal to the number of coupling sites and B is the terminal electron acceptor. (See Section 4.5.2 for a discussion of the number of ATP molecules formed per coupling site according to the chemiosmotic theory.)

Fermentation (anaerobic):

$$NADH + H^+ + B \text{ (organic)} \longrightarrow$$
$$NAD^+ + BH_2$$

Hydrogenase (anaerobic):

$$NADH + H^+ \longrightarrow H_2 + NAD^+$$

Aerobic and anaerobic respiration are discussed in Chapter 4. In the absence of respiration, NADH can be reoxidized in the cytosol via fermentation discussed in Chapter 14. A third way to reoxidize NADH is via the enzyme hydrogenase in the cytosol. Hydrogenases that use NADH as the electron donor are found in fermenting bacteria. However, the oxidation of NADH with the production of hydrogen gas generally proceeds only when the hydrogen gas concentration is kept low (e.g., during growth with hydrogen gas utilizers). This is because the equilibrium favors the reduction of $NAD^+$. Interspecies hydrogen transfer is discussed in Section 14.4.1.

## 8.3 Why Write $NAD^+$ instead of NAD, and NADH instead of $NADH_2$?

Oxidized nicotinamide adenine dinucleotide is written as $NAD^+$, and the reduced form is written NADH, not $NADH_2$. To understand why, we must examine the structures (Fig. 8.6). Notice that the molecule can accept two electrons but only one hydrogen. That is why it is written as $NADH + H^+$. The oxidized molecule is written $NAD^+$ because the nitrogen carries a formal positive charge.

## 8.4 A Modified EMP Pathway in the Hyperthermophilic Archaeon Pyrococcus furiosus

The study of archaeal metabolism has recently been receiving wider attention, compared to the long history of research with bacteria, and there have been many rewarding findings that suggest the presence of several metabolic features distinct from those of the bacteria.[3] Examples include the novel ether-linked lipids described in Chapter 1 and their biosynthesis, discussed in Chapter 9. Additional features of archaeal metabolism that appear to be unique to these microorganisms include the synthesis of methane, and the presence of novel coenzymes for acetate and methane metabolism discussed in Chapter 13. A modified Embden–Meyerhof–Parnas pathway has been proposed for Pyrococcus furiosus.[4] This organism has a growth temperature optimum of 100°C and ferments carbohydrates and peptides to acetate, $H_2$, and $CO_2$. Monosaccharides such as

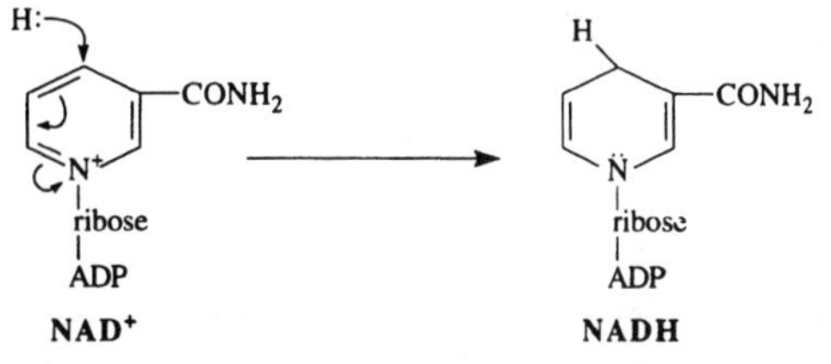

Fig. 8.6 The structures of $NAD^+$, NADH, and nicotinamide. $NAD^+$ is a derivative of nicotinamide, to which ADP–ribose is attached to the nitrogen of nicotinamide. In the oxidized form the nitrogen has four bonds, and hence carries a positive charge. $NAD^+$ accepts two electrons, but only one hydrogen, (hydride ion) to become NADH. The second hydrogen removed from the electron donor (the reductant) is released into the medium as a proton.

glucose and fructose do not support growth, but other carbohydrates such as maltose are transported into the cell and converted to glucose. If S° is present in the medium, it is used as an electron sink and reduced to $H_2S$. A pathway proposed for the catabolism of glucose to acetate is shown in Fig. 8.7.[5] The postulated pathway resembles the classic EMP pathway but differs in several respects. Very interestingly, the phosphoryl donor in the hexokinase and fructokinase reactions appears not to be ATP but rather ADP. Another difference is that the enzyme that oxidizes glyceraldehyde-3-phosphate to 3-phosphoglycerate is suggested to be a ferredoxin-linked enzyme rather than an $NAD^+$-linked enzyme, and it appears that 1,3-bisphosphoglycerate is not an intermediate. Pyruvate is oxidized to acetyl–CoA and $CO_2$ by pyruvate–ferredoxin oxidoreductase. (See Section 7.3.2 for a description of this reaction.) Finally, an ADP-dependent acetyl–CoA synthetase catalyzes a reaction in which ADP is phosphorylated and acetate is formed. See Section 7.3.2 for a description of the ADP-dependent acetyl–CoA synthetase and its distribution. All the enzymes proposed for the modified EMP pathway have been detected in extracts of *P. furiosus*, and in vivo NMR studies of the products formed from [$^{13}C$]glucose are in agreement with the pathway drawn in Fig. 8.7.[5]

## 8.5 The Pentose Phosphate Pathway

Another important pathway for carbohydrate metabolism is the pentose phosphate pathway. The pentose phosphate pathway is important first because it produces the pentose phosphates, which are the precursors to the ribose and deoxyribose in the nucleic acids, and second because it provides erythrose phosphate, which is the precursor to the aromatic amino acids, phenylalanine, tyrosine, and tryptophan. Also, the NADPH produced in the pentose phosphate pathway is a major source of electrons for biosynthesis in most of the pathways in which reductions occur. (See Note 6 for an alternative means of generating

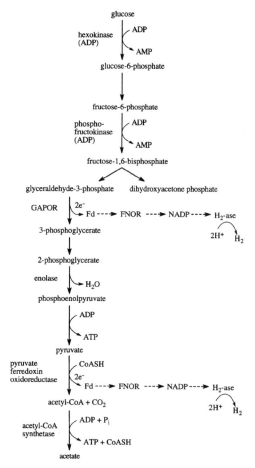

Fig. 8.7 Proposed pathway for maltose fermentation to acetate, $CO_2$, and $H_2$ in *Pyrococcus furiosus*. GAPOR, glyceraldehyde-3-phosphate ferredoxin oxidoreductase; Fd, ferredoxin; FNOR, ferredoxin NADP oxidoreductase; $H_2$ase, hydrogenase. *Source:* Mukund, S., and M. W. W. Adams. 1995. Glyceraldehyde-3-phosphate ferredoxin oxidoreductase, a novel tungsten-containing enzyme with a potential glycolytic role in the hyperthermophilic archaeon *Pyrococcus furiosus*. J. Biol. Chem. **270**:8389–8392.

NADPH.) The pathway is important to learn for yet one more reason. Several of the reactions of the pentose phosphate pathway are the same as the reactions in the Calvin cycle, which is used by many autotrophic organisms to incorporate $CO_2$ into organic carbon (Chapter 13).

The overall reaction of the pentose phosphate pathway is

$$G6P + 6\,NADP^+ \longrightarrow 3CO_2 + PGALD$$
$$+ 6\,NADPH + 6\,H^+ \qquad (8.1)$$

## 8.5.1 The reactions of the pentose phosphate pathway

The pentose phosphate pathway is complex and can be best learned by dividing the reactions into three stages. Stage 1 consists of oxidation–decarboxylation reactions. The $CO_2$ and NADPH are produced in Stage 1. Stage 2 consists of isomerization reactions that make the precursors for Stage 3. Stage 3 reactions are sugar rearrangements. The phosphoglyceraldehyde is produced in Stage 3.

### Stage 1: Oxidation–decarboxylation reactions

The oxidation–decarboxylation reactions are shown in Fig. 8.8. These reactions oxidize the C1 in glucose-6-phosphate to a carboxyl group and remove it as carbon dioxide. Glucose actually exists as a ring structure, which forms because the aldehyde group at C1 reacts with the C5 hydroxyl group, forming a hemiacetal. The C1 is therefore not a typical aldehyde in that it does not react in the Schiff test and does not form a bisulfite addition product. Nevertheless, it is easily oxidized. The glucose-6-phosphate is oxidized by $NADP^+$ to 6-P-gluconolactone by glucose-6-phosphate dehydrogenase (reaction 1). The lactone is then hydrolyzed to 6-P-gluconate by gluconolactonase (reaction 2). In the oxidation of glucose-6-phosphate to 6-phosphogluconate, water contributes the second oxygen in the carboxyl group, and an acyl-phosphate intermediate is not formed. This means that the energy from the oxidation is lost as heat, and the reaction is physiologically irreversible, as is often the case for the first reaction in a metabolic pathway. (Recall that during the oxidation of phosphoglyceraldehyde, inorganic phosphate is added and 1,3-bisphosphoglycerate is formed. The energy of oxidation is trapped in the acyl-phosphate rather than being lost as heat. A subsequent substrate-level phosphorylation

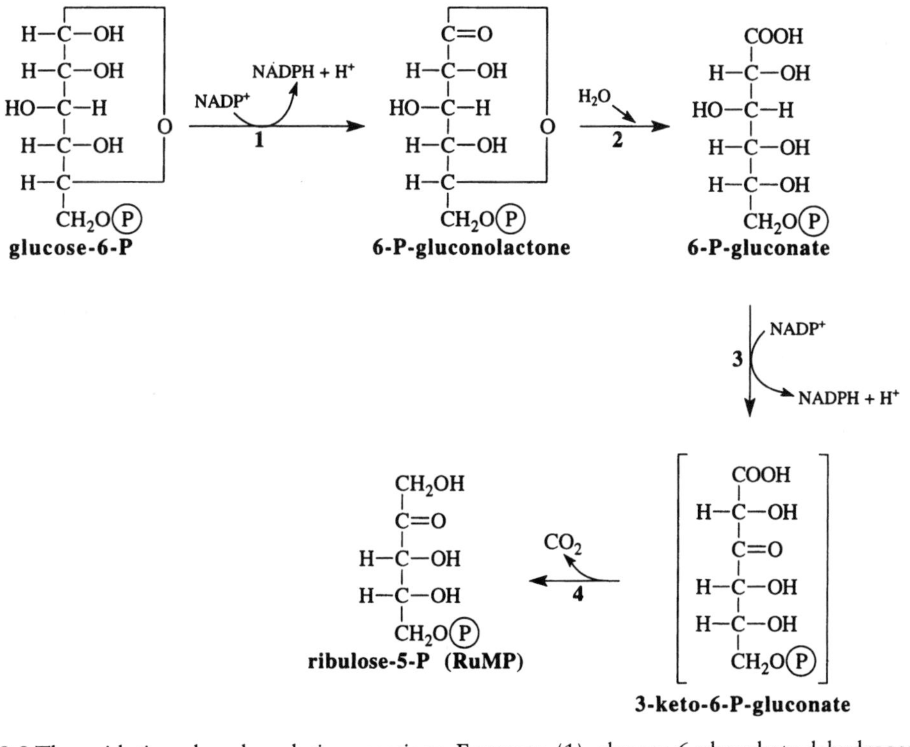

Fig. 8.8 The oxidation–decarboxylation reactions. Enzymes: (1), glucose-6-phosphate dehydrogenase; (2) gluconolactonase; (3) and (4) 6-phosphogluconate dehydrogenase.

recovers the energy of oxidation in the form of ATP.) The product of the oxidation, 6-P-gluconate, is then oxidized on the C3 to generate a keto group $\beta$ to the carboxyl (reaction 3). A $\beta$-decarboxylation then occurs, generating ribulose-5-phosphate (reaction 4). (The mechanism of $\beta$-decarboxylations is described in Section 8.11.2.) Therefore, the products of Stage 1 are carbon dioxide, 2 NADPH, and the five-carbon sugar phosphate ribulose-5-phosphate. The rest of the pathway continues with ribulose-5-phosphate.

## Stage 2: The isomerization reactions

During the second stage, some of the ribulose-5-phosphate is isomerized to ribose-5-phosphate and to xylulose-5-phosphate. Isomers are molecules having the same chemical formula but different structural formulae. That is to say, their parts have been switched around. For example, the chemical formula for ribulose-5-phosphate is $C_5H_{11}O_8P$. Ribose-5-phosphate and xylulose-5-phosphate have the same chemical formula. However, their structures are different (Fig. 8.9).

One of the isomerases in an epimerase. The epimerase catalyzes a movement of the hydroxyl group from one side of the C3 in ribulose-5-phosphate to the other. The product is xylulose-5-phosphate (the epimer[7] of ribulose-5-phosphate). The other isomerase converts ribulose-5-phosphate to ribose-5-phosphate.

## Stage 3: The sugar rearrangement reactions

Stage 3 of the pentose phosphate pathway involves sugar rearrangement reactions. There are two basic types of reactions. One kind transfers a *two*-carbon fragment from a ketose to an aldose. The enzyme that catalyzes the transfer of the two-carbon fragment is called a *transketolase* (TK). A second kind of reaction transfers a *three*-carbon fragment from a ketose to an aldose. The enzyme that catalyzes the transfer of a three-carbon fragment is called a *transaldolase* (TA). The rule is that the donor is always a ketose (with the OH group of the third carbon "on the left," as in xylulose-5-phosphate) and the acceptor is always an aldose. This rule is important to learn because we shall see other transketolase and transaldolase reactions later. Knowing the requirements will make it easier to remember the reactions. The transketolase and transaldolase reactions are summarized in Fig. 8.10.

## Summarizing the pentose phosphate pathway

Figure 8.11 summarizes the pentose phosphate pathway. Reactions 1–3 comprise the oxidative decarboxylation reactions of Stage 1. Three moles of glucose-6-phosphate must be oxidized in order to produce three moles of $CO_2$ and one mole of phosphoglyceraldehyde. Therefore, Stage 1 produces three moles of ribulose-5-phosphate. Reactions 4 and 5 are the isomerization reactions of Stage 2, in which the three moles of ribulose-5-phosphate are converted to one mole of ribose-5-phosphate and two moles of xylulose-5-phosphate. Reactions 6, 7, and 8 comprise Stage 3. Reaction 6 is a transketolase reaction in which a xylulose-5-phosphate ($C_5$) transfers a two-carbon moiety to ribose-5-phosphate ($C_5$) with the formation of sedoheptulose-7-phosphate ($C_7$) and phosphoglyceraldehyde ($C_3$). The two-carbon moiety is highlighted as a boxed area. Reaction 7 is a transaldolase reaction in which the sedoheptulose-7-phosphate transfers a three-

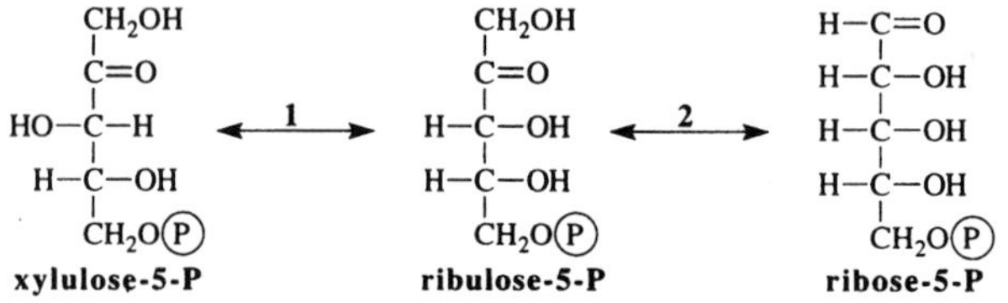

Fig. 8.9 The isomerization reactions. Enzymes: 1, RuMP epimerase; 2, ribose-5-phosphate isomerase.

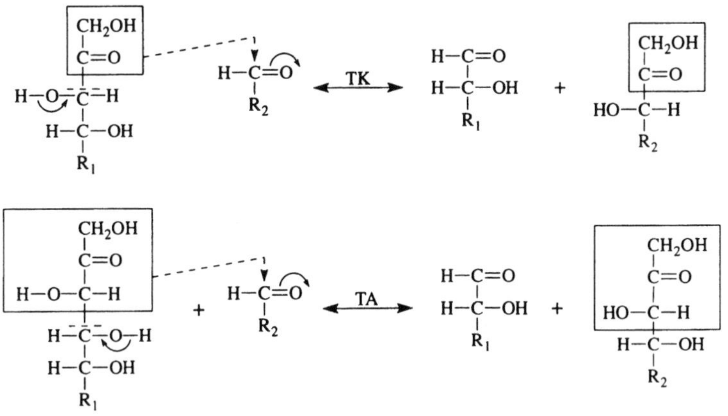

**Fig. 8.10** The transketolase and transaldolase reactions. Enzymes: TK, transketolase; TA, transaldolase. The donor is always a ketose with the keto group on C2, and the hydroxyl on C3 on the "left." In the transketolase reaction a $C_2$ unit is transferred with its bonding electrons to the carbonyl group on an aldehyde acceptor. The transaldolase transfers a three-carbon fragment. In the transketolase reaction, the newly formed alcohol group is on the "left," which means that the products of both the transketolase and transaldolase reactions can act as donors in a subsequent transfer.

carbon moiety to the phosphoglyceraldehyde to form erythrose-4-phosphate ($C_4$) and fructose-6-phosphate ($C_6$). The three-carbon moiety is highlighted as a dashed box. Reaction 8 is a transketolase reaction in which xylulose-5-phosphate transfers a two-carbon moiety to the erythrose-4-phosphate, forming phosphoglyceraldehyde and fructose-6-phosphate. You will notice that the sequence of reactions is:

Transketolase $\longrightarrow$ transaldolase

$\longrightarrow$ transketolase

The result is that three moles of glucose-6-phosphate are converted to two moles of fructose-6-phosphate and one mole of phosphoglyceraldehyde. The two moles of fructose-6-phosphate become glucose-6-phosphate by isomerization, and the net result is the conversion of one mole of glucose-6-phosphate to one mole of phosphoglyceraldehyde, three moles of carbon dioxide, and six moles of NADPH. This is shown in the carbon balance below.

The carbon balance for the pentose phosphate pathway is:

| | |
|---|---|
| Oxidative decarboxylation | 3 glucose-6-P $\longrightarrow$ 3 ribulose-5-P + $3CO_2$ |
| | $3C_6$ $\qquad\qquad$ $3C_5$ $\qquad$ $3C_1$ |
| Isomerizations | 3 ribulose-5-P $\longrightarrow$ 2 xylulose-5-P + ribose-5-P |
| | $3C_5$ $\qquad\qquad$ $2C_5$ $\qquad$ $C_5$ |
| Transketolase | xylulose-5-P + ribose-5-P $\longrightarrow$ sedoheptulose-7-P + phosphoglyceraldehyde |
| | $C_5$ $\qquad$ $C_5$ $\qquad\qquad$ $C_7$ $\qquad\qquad$ $C_3$ |
| Transaldolase | sedoheptulose-7-P + phosphoglyceraldehyde $\longrightarrow$ fructose-6-P + erythrose-4-P |
| | $C_7$ $\qquad\qquad$ $C_3$ $\qquad\qquad$ $C_6$ $\qquad$ $C_4$ |
| Transketolase | xylulose-5-P + erythrose-4-P $\longrightarrow$ fructose-6-P + phosphoglyceraldehyde |
| | $C_5$ $\qquad$ $C_4$ $\qquad\qquad$ $C_6$ $\qquad\qquad$ $C_3$ |
| Sum: | glucose-6-P $\longrightarrow$ phosphoglyceraldehyde + $3CO_2$ |
| | $C_6$ $\qquad\qquad$ $C_3$ $\qquad\qquad$ $3C_1$ |

Later we will study the Calvin cycle, which is a pathway by which many organisms can grow on $CO_2$ as the sole source of carbon. In the Calvin cycle, $CO_2$ is first reduced to phosphoglyceraldehyde. The phosphoglyceraldehyde is then converted via sedoheptulose-7-phosphate to pentose phosphates using essentially a reversal of the pentose phosphate pathway. However, the Calvin cycle has no transaldolase and synthesizes sedoheptulose-7-phosphate by an alternate route that runs irreversibly in the direction of pentose phosphates (Section 13.1.1).

The pentose phosphate pathway serves important biosynthetic functions. Notice that Stage 1 (the oxidative decarboxylation reactions) and Stage 2 (the isomerization reactions) generate the pentose phosphates required for nucleic acid synthesis. Stage 1 also produces NADPH, which is used in several biosynthetic pathways. Stage 3 generates the erythrose-4-phosphate necessary for aromatic amino acid biosynthesis.

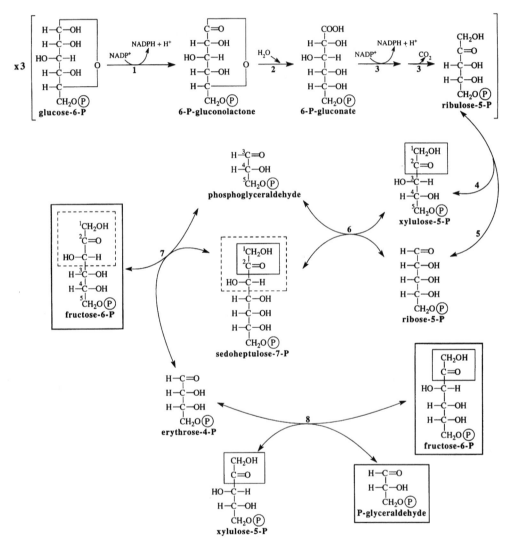

Fig. 8.11 The pentose phosphate pathway. Enzymes: 1, glucose-6-phosphate dehydrogenase; 2, lactonase; 3, 6-phosphogluconate dehydrogenase; 4, ribulose-5-phosphate epimerase; 5, ribose-5-phosphate isomerase; 6, 8, transketolase; 7, transaldolase. The two-carbon moiety transferred by the transketolase is shown in the boxed area. The three-carbon fragment transferred by the transaldolase is shown in the dashed box. The product of the glucose-6-phosphate dehydrogenase reaction is the lactone, which is unstable and hydrolyzes spontaneously to the free acid. However, there exists a specific lactonase that catalyzes the reaction.

## Some bacteria rely completely on the pentose phosphate pathway for sugar catabolism

*Thiobacillus novellus* and *Brucella abortus* lack both Stage 1 of the Embden–Meyerhof–Parnas pathway and the enzymes of the Entner–Doudoroff pathway. These organisms use only an oxidative pentose phosphate pathway to grow on glucose. They oxidize the glucose to phosphoglyceraldehyde via the pentose phosphate pathway. The phosphoglyceraldehyde is then oxidized to pyruvate via reactions that are the same as Stage 2 of the EMP pathway, and then the pyruvate is oxidized to $CO_2$ via the citric acid cycle.

## Relationship of the pentose phosphate pathway to glycolysis

The pentose phosphate pathway and glycolysis interconnect at phosphoglyceraldehyde and fructose-6-phosphate (Fig. 8.1). Thus organisms growing on pentoses can make hexose phosphates. Furthermore, because Stages 2 and 3 of the pentose phosphate pathway are reversible, it is possible to synthesize pentose phosphates from phosphoglyceraldehyde and avoid the oxidative decarboxylation reactions of Stage 1. This would uncouple pentose phosphate synthesis from NADPH production, and confer a possibly advantageous metabolic flexibility on the cells.

## 8.6 The Entner–Doudoroff Pathway

Many prokaryotes have another pathway for the degradation of carbohydrates called the Entner–Doudoroff or ED pathway. The other pathways that we have been studying are common to all cells, whether they be prokaryotes or eukaryotes, but the Entner–Doudoroff pathway has been found almost entirely among the prokaryotes, including bacteria and archaea (reviewed in Ref. 1). The pathway is widespread, particularly among the aerobic gram-negative bacteria. It is usually not found among anaerobic bacteria, perhaps because of the low ATP yields discussed below. Most bacteria degrade sugars via the Embden–Meyerhof–Parnas pathway, but when grown on certain compounds

(e.g., gluconic acid), they use the Entner–Doudoroff pathway. However, some strictly aerobic bacteria cannot carry out Stage 1 of the Embden–Meyerhof–Parnas pathway and rely entirely on the Entner–Doudoroff pathway for sugar degradation (Table 8.1). The overall reaction for the Entner–Doudoroff pathway is

$$glucose + NADP^+ + NAD^+ + ADP + P_i$$
$$\longrightarrow 2 \text{ pyruvic acid} + NADPH + 2H^+$$
$$+ NADH + ATP$$

It can be seen that the pathway catalyzes the same overall reaction as the Embden–Meyerhoff–Parnas pathway, (i.e., the oxidation of one mole of glucose to two moles of pyruvic acid), except that only one ATP is made, and one NADPH and one NADH are made instead of two NADHs. The reason why only one ATP is made is that only one phosphoglyceraldehyde is made from glucose (Fig. 8.12).

### 8.6.1 The Entner–Doudoroff reactions

The first oxidation is the oxidation of the C1 in glucose-6-phosphate to the carboxyl in 6-P-gluconate (Fig. 8.12, reaction 2). These are the same enzymatic reactions that oxidize glucose-6-phosphate in the pentose phosphate pathway, and proceed through the gluconolactone.

**Table 8.1** Distribution of the Embden-Meyerhoff-Parnas (EMP) and Entner-doudoroff (ED) pathways in certain bacteria.

| Bacterium | EMP | ED |
|---|---|---|
| *Arthrobacter species* | + | − |
| *Azotobacter chroococcum* | + | − |
| *Alcaligenes eutrophus* | − | + |
| *Bacillus* species | + | − |
| *Escherichia coli* and other enteric bacteria[a] | + | − |
| *Pseudomonas* species | − | + |
| *Rhizobium* species | − | + |
| *Thiobacillus* species | − | + |
| *Xanthomonas* species | − | + |

[a] Organisms such as *E. coli* synthesize the enzymes of the ED pathway when growing on gluconate.
*Source:* Gottschalk, G. 1986. *Bacterial Metabolism.* Springer-Verlag, New York, Berlin.

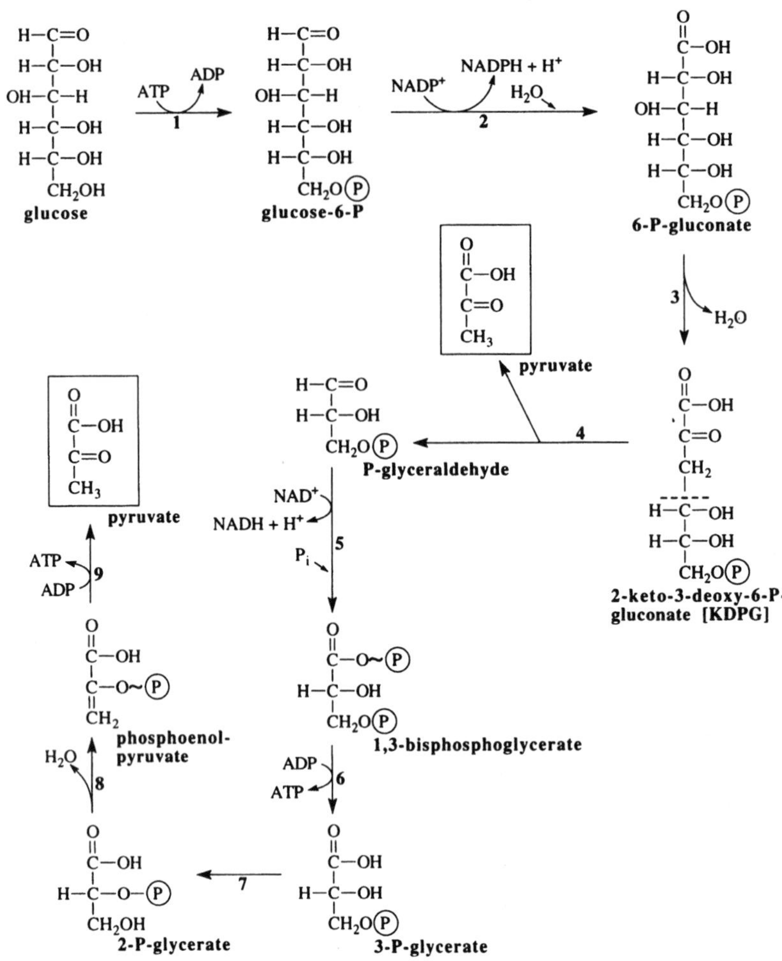

**Fig. 8.12** The Entner–Doudoroff pathway. Because there is only one PGALD formed, there is only one ATP made. The enzymes unique to this pathway are the 6-phosphogluconate dehydratase (reaction 3) and the KDPG aldolase (reaction 4). The other enzymes are present in the pentose phosphate pathway and the glycolytic pathway. Enzymes: 1, hexokinase; 2, glucose-6-phosphate dehydrogenase; 3, 6-phosphogluconate dehydratase; 4, KDPG aldolase; 5, triose phosphate dehydrogenase; 6, PGA kinase; 7, mutase; 8, enolase; 9, pyruvate kinase.

The pathway diverges from the pentose phosphate pathway at this point because some of the 6-P-gluconate is dehydrated to 2-keto-3-deoxy-6-P-gluconate (KDPG), rather than being oxidized to ribulose-5-phosphate (reaction 3). The KDPG is split by KDPG aldolase to pyruvate and phosphoglyceraldehyde (reaction 4). The phosphoglyceraldehyde is oxidized to pyruvate in a sequence of reactions identical to those in Stage 2 of the EMP pathway (reactions 5–9).

Reaction 3, which is the dehydration of 6-phosphogluconate to KDPG, takes place via an enol intermediate that tautomerizes to

KDPG. This is illustrated in Fig. 8.13. In this way, it is similar to the dehydration of 2-phosphoglycerate to phosphoenolpyruvate by enolase described in Section 7.3.4. However, in phosphoenolpyruvate, the enol derivative is stabilized by the phosphate group, and the tautomerization does not take place until the phosphoryl group is transferred.

## 8.6.2 Physiological role for the Entner–Doudoroff pathway

Since the Entner–Doudoroff pathway produces only one ATP, one can ask why the

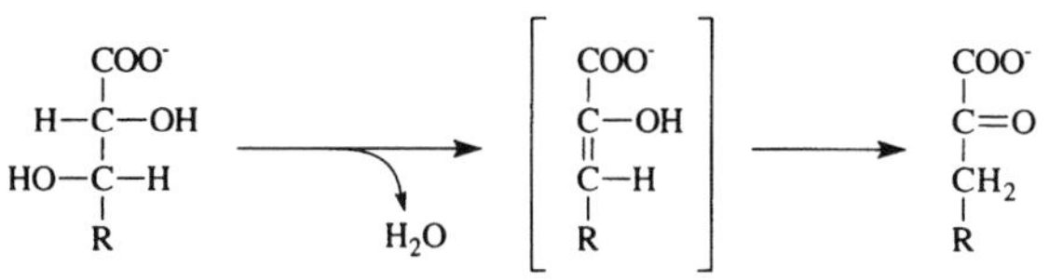

**Fig. 8.13** Dehydration of a carboxylic acid with hydroxyl groups in the $\alpha$ and $\beta$ position. The dehydration of a carboxylic acid with hydroxyl groups in both the $\alpha$ and $\beta$ position leads to the formation of an enol, which tautomerizes to the keto compound. That is because the hydroxyl on the C3 leaves with its bonding electrons and the electrons bonded to the hydrogen on the C2 shift in to form the double bond. This happens when 6-phosphogluconate is dehydrated to 2-keto-3-deoxy-6-phosphogluconate in the Entner–Doudoroff pathway, and when 2-phosphoglycerate is dehydrated to phosphoenolpyruvate during glycolysis. The phosphoenolpyruvate tautomerizes to pyruvate when the phosphate is removed during the kinase reaction.

pathway is so common in the bacteria. Whereas hexoses are readily degraded by the Embden–Meyerhoff–Parnas pathway, aldonic acids (aldoses oxidized at the aldehydic carbon) such as gluconate are not, but can be degraded via the Entner–Doudoroff pathway. (Aldonic acids occur in nature and can be an important nutrient.) An example of this occurs when *E. coli* is transferred from a medium containing glucose as the carbon source to one in which gluconate is the source of carbon. Growth on gluconate results in the induction of three enzymes: a gluconokinase that makes 6-P-gluconate from the gluconate (at the expense of ATP), 6-P-gluconate dehydratase, and KDPG aldolase. Thus *E. coli* uses the Entner–Doudoroff pathway to grow on gluconate and the Embden–Meyerhoff–Parnas pathway to grow on glucose.

### Some bacteria do not have a complete Embden–Meyerhoff–Parnas pathway and rely on the Entner–Doudoroff pathway for hexose degradation

Several prokaryotes (e.g., the pseudomonads in general) do not make phosphofructokinase. Hence they cannot carry out glucose oxidation using the Embden–Meyerhoff–Parnas pathway (EMP). Instead, they use the Entner–Doudoroff pathway (ED). See Table 8.1 for the distribution of the EMP and ED pathways. Some of the pseudomonads even oxidize glucose to gluconate before degrading it via the ED pathway, instead of making glucose-6-phosphate. The oxidation of glucose to

gluconate may confer a competitive advantage, since it removes glucose, which is more readily utilizable by other microorganisms.

### 8.6.3 A partly nonphosphorylated Entner–Doudoroff pathway

A modified ED pathway has been found in the archaeon *Halobacterium saccharovorum*, and in several bacteria, including members of the genera *Clostridium*, *Alcaligenes*, *Achromobacter*, and in *Rhodopseudomonas sphaeroides*. The pathway is characterized as having nonphosphorylated intermediates prior to 2-keto-3-deoxygluconate. The first reaction is the oxidation of glucose to gluconate via an $NAD^+$-dependent dehydrogenase. The gluconate is then dehydrated to 2-keto-3-deoxygluconate by a gluconate dehydratase. The 2-keto-3-deoxygluconate is phosphorylated by a special kinase to form KDPG, which is metabolized to pyruvate using the ordinary ED reactions. Those bacteria that have a modified ED pathway can use it for the catabolism of gluconate, or glucose if the glucose dehydrogenase is present.

### 8.7 The Oxidation of Pyruvate to acetyl–CoA: The Pyruvate Dehydrogenase Reaction

Pyruvate is the common product of sugar catabolism in all the major carbohydrate catabolic pathways (Fig. 8.1). We now examine the metabolic fate of pyruvate. One of the fates of pyruvate is to be oxidized to acetyl–CoA. As described in Section 7.3.2, this is catalyzed in anaerobically growing bacteria by either pyruvate–ferredoxin oxidoreductase or pyruvate–formate lyase. The acetyl–CoA thus formed is converted to acetyl–phosphate via the enzyme phosphotransacetylase, and the acetyl–phosphate donates the phosphoryl group to ADP in a substrate-level phosphorylation catalyzed by acetate kinase. The acetate thus formed is excreted. (However, certain anaerobic archaea oxidize pyruvate to acetyl–CoA using pyruvate–ferredoxin oxidoreductase and oxidize the acetyl–CoA to $CO_2$ via the citric acid cycle.[10]) The oxidation of pyruvate to acetyl–CoA during aerobic growth is carried out by the enzyme complex *pyruvate dehydrogenase*, which is widespread in both prokaryotes and eukaryotes. (The acetyl–CoA formed during aerobic growth is oxidized to $CO_2$ in the citric acid cycle.) The overall reaction for pyruvate dehydrogenase is:

$$CH_3COCOOH + NAD^+ + CoASH$$
$$\longrightarrow CH_3COSCoA + CO_2 + NADH + H^+$$

The pyruvate dehydrogenase complex is a very large enzyme complex (in *E. coli* about 1.7 times the size of the ribosome) located in the mitochondria of eukaryotic cells and in the cytosol of prokaryotes. The pyruvate dehydrogenase from *E. coli* consists of 24 molecules of enzyme E1 (pyruvate dehydrogenase), 24 molecules of enzyme E2 (dihydrolipoate transacetylase), and 12 molecules of enzyme E3 (dihydrolipoate dehydrogenase). Several very important cofactors are involved. The cofactors are thiamine pyrophosphate (TPP) derived from the vitamin thiamine, flavin adenine dinucleotide (FAD) derived from the vitamin riboflavin, lipoic acid ($RS_2$), nicotinamide adenine dinucleotide ($NAD^+$) derived from the vitamin nicotinamide, and coenzyme-A, derived from the vitamin pantothenic acid. (See Note 8 for a more complete discussion of vitamins.) The large size of the complex is presumably designed to process the heavy stream of pyruvate that is generated during the catabolism of sugars and other compounds. As described below, the pyruvate dehydrogenase complex catalyzes a short metabolic pathway rather than simply a single reaction.

The individual reactions carried out by the pyruvate dehydrogenase complex are as follows (Fig. 8.14):

**Step 1.** Pyruvate is decarboxylated to form "active acetaldehyde" bound to TPP (Fig. 8.14). The reaction is catalyzed by pyruvate dehydrogenase ($E_1$). (The mechanism of this reaction is described in Section 14.10.)

**Step 2.** The "active acetaldehyde" is oxidized to the level of carboxyl by the disulfide in lipoic acid. The disulfide of the lipoic acid is reduced to a sulfhydryl. During the reaction, TPP is displaced and the acetyl group is transferred to the lipoic acid. The reaction is also catalyzed by pyruvate dehydrogenase.

**Step 3.** A transacetylation occurs where lipoic acid is displaced by CoASH, forming acetyl–CoA and reduced lipoic acid. The reaction is catalyzed by dihydrolipoate transacetylase, $E_2$.

**Step 4.** The lipoic acid is oxidized by dihydrolipoate dehydrogenase, $E_3$-FAD.

**Step 5.** The $E_3$-$FADH_2$ transfers the electrons to $NAD^+$.

All the intermediates remain bound to the complex and are passed from one active site to another. Presumably, this has the advantage inherent in all multienzyme complexes (i.e., there is no dilution of intermediates in the cytosol, and side reactions are minimized). The student should refer to Section 1.2.6 for a discussion of multienzyme complexes in the cytoplasm.

## 8.7.1 Physiological control

The pyruvate dehydrogenase reaction, which is physiologically irreversible, is under metabolic control by several allosteric effectors (Fig. 8.15). The *E. coli* pyruvate dehydrogenase is feedback inhibited by the products it forms, acetyl–CoA and NADH. This can be rationalized as ensuring that the enzyme produces only as much acetyl–CoA and NADH as can be used immediately. It is also stimulated by phosphoenolpyruvate (the precursor to pyruvate), presumably signaling the dehydrogenase that more pyruvate is on the way. It is also stimulated by AMP, which

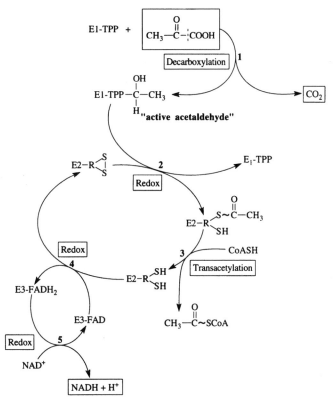

**Fig. 8.14** The pyruvate dehydrogenase complex (PDH). *Step 1.* Pyruvate is decarboxylated to "active acetaldehyde." The decarboxylation requires thiamine pyrophosphate (TPP). *Step 2.* The "active acetaldehyde" is oxidized to an acylthioester with a high acyl group transfer potential. The oxidant is reduced lipoic acid (R–S$_2$). *Step 3.* The lipoylacylthioester transfers the acetyl group to coenzyme A (CoASH) to form acetyl–CoA. *Step 4.* The reduced lipoic acid is reoxidized by FAD, which in turn is reoxidized by NAD$^+$ (*Step 5*). The products of the reaction are acetyl–CoA, CO$_2$, and NADH. Enzymes: E$_1$, pyruvate dehydrogenase; E$_2$, dihydrolipoate transacetyase; E$_3$, dihydrolipoate dehydrogenase.

signals low ATP. The stimulation by AMP probably reflects the fact that the oxidation of the product acetyl–CoA in the citric acid cycle is a major source of ATP (via respiratory phosphorylation).

## 8.8 The citric acid cycle

The acetyl–CoA that is formed by pyruvate dehydrogenase is oxidized to CO$_2$ in the citric acid cycle (Fig. 8.1). The overall reaction is:

$$\text{acetyl–CoA} + 2H_2O + ADP + P_i + FAD$$
$$+ NADP^+ + 2NAD^+$$
$$\rightarrow 2CO_2 + ATP + FADH_2 + NADPH$$
$$+ 2NADH + 3H^+ + CoASH$$

Notice that there are four oxidations per acetyl–CoA producing two NADH, one NADPH, and one FADH$_2$, and one substrate-level phosphorylation producing ATP. The cycle usually operates in conjunction with respiration that reoxidizes the NAD(P)H and FADH$_2$. Other names for this pathway are the tricarboxylic acid (TCA) cycle and the Krebs cycle. The latter name honors Sir Hans Krebs, who did much of the pioneering work and proposed the cycle in 1937.

### 8.8.1 The individual reactions of the citric acid cycle

The pathway is outlined in Fig. 8.16. Reaction 1 is the addition of the acetyl group from

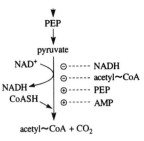

Fig. 8.15 Regulation of pyruvate dehydrogenase in *E. coli*. The activity of the enzyme in vitro is modified by several effector molecules. NADH and acetyl–CoA are negative effectors, and PEP and AMP are positive effectors.

acetyl–CoA to oxaloacetate to form citrate. In this reaction the methyl group of acetyl–CoA acts as a nucleophile and bonds to the carbon in the keto group of oxaloacetate (OAA). The reaction is driven to completion by the hydrolysis of the thioester bond of acetyl–CoA, which has a high free energy of hydrolysis. Reaction 1 is catalyzed by citrate synthase. This enzyme operates irreversibly in the direction of citrate. The oxaloacetate acts catalytically in the cycle, and, if it is not regenerated or replenished, the pathway stops. Examine the structure of citrate shown in Fig. 8.16. In the subsequent reactions of the citric acid cycle, the C3 and C5 carbons will be removed as $CO_2$ and the C1–C5 carbons will regenerate the oxaloacetate. In reaction 2, catalyzed by aconitase, the citrate is dehydrated to *cis*-aconitate, which remains bound to the enzyme. Reaction 3 (also catalyzed by aconitase) is the rehydration of *cis*-aconitate to form isocitrate, an isomer of citrate. In reaction 4 (isocitrate dehydrogenase) the isocitrate is oxidized to oxalosuccinate. This oxidation creates a keto group $\beta$ to the carboxyl group. The creation of the keto group is necessary for the decarboxylation that takes place in the next reaction ($\beta$-keto decarboxylations are explained in Section 8.11.2).

Reaction 5 (isocitrate dehydrogenase) is the decarboxylation of oxalosuccinate to $\alpha$-ketoglutarate. Reaction 6 ($\alpha$-ketoglutarate dehydrogenase) is the oxidative decarboxylation of $\alpha$-ketoglutarate to succinyl–CoA. This is an $\alpha$-decarboxylation, in contrast to a $\beta$-decarboxylation. It is a complex reaction and requires the same cofactors as does the

decarboxylation of pyruvate to acetyl–CoA. Reaction 7 (succinate thiokinase) is a substrate-level phosphorylation resulting in the formation of ATP from ADP and inorganic phosphate. This reaction was described in Section 7.3.3. Reaction 8 (succinate dehydrogenase) is the oxidation of succinate to fumarate, catalyzed by a flavin enzyme. Succinate dehydrogenase is the only citric acid cycle enzyme that is membrane bound. In bacteria it is part of the cell membrane and transfers electrons directly to quinone in the respiratory chain. (See the description of the electron transport chain in Section 4.3.) Reaction 9 (fumarase) is the hydration of fumarate to malate. Finally, the oxaloacetate is regenerated by oxidizing the malate to oxaloacetate in reaction 10 (malate dehydrogenase). The citric acid cycle proceeds in the direction of acetyl–CoA oxidation because of two irreversible steps: the citrate synthase and the $\alpha$-ketoglutarate dehydrogenase reactions.

## Summing up the citric acid cycle

When one examines the reactions in the citric acid cycle, it can be seen that there is no net synthesis. In other words, all the carbon that enters the cycle exits as $CO_2$. This is made clear by writing a carbon balance. In the carbon balance written below, $C_2$ represents the two-carbon molecule acetyl–CoA, $C_6$ represents citrate or isocitrate, $C_5$ represents $\alpha$-ketoglutarate, and $C_4$ represents either succinate fumarate, malate, or oxaloacetate. Of course, $C_1$ represents carbon dioxide.

Carbon balance for citric acid cycle:

$$
\begin{aligned}
C_2 + C_4 &\longrightarrow C_6 \\
C_6 &\longrightarrow C_5 + C_1 \\
C_5 &\longrightarrow C_4 + C_1 \\
\hline
\text{sum}: \quad C_2 &\longrightarrow 2\,C_1
\end{aligned}
$$

## 8.8.2 Regulation of the citric acid cycle

The citric acid cycle is feedback inhibited by several intermediates that can be viewed as end products of the pathway. In gram-negative bacteria, the citrate synthase is allosterically inhibited by NADH, and in

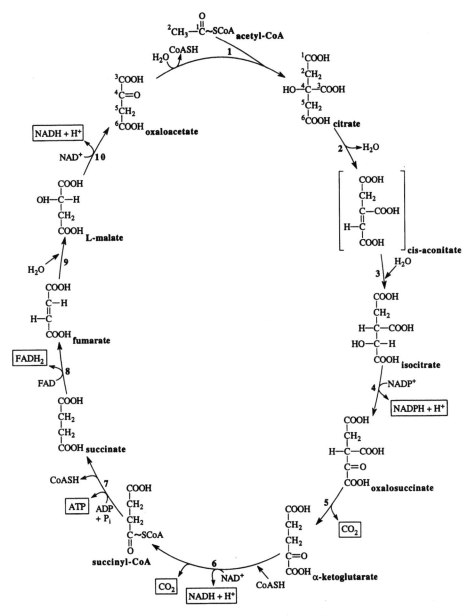

**Fig. 8.16** The citric acid cycle. Enzymes: (1) citrate synthase; (2, 3) aconitase; (4, 5) isocitrate dehydrogenase; (6), α-ketoglutarate dehydrogenase; (7), succinate thiokinase; (8), succinate dehydrogenase; (9), fumarase; (10), malate dehydrogenase. *cis*-Aconitate is drawn in parentheses because it is an enzyme-bound intermediate.

facultative anaerobes such as *E. coli*, also by α-ketoglutarate. The inhibition of the citrate synthase by NADH may be a way to prevent oversynthesis of NADH. The inhibition by α-ketoglutarate can also be viewed as an example of end-product inhibition, in this case to prevent overproduction of the amino acid glutamate, which is derived from α-ketoglutarate.[9] The citrate synthase from gram-positive bacteria and eukaryotes is not sensitive to NADH and α-ketoglutarate but is inhibited by ATP, another end product of the citric acid pathway. Recall the discussion in Chapter 6 emphasizing that the pattern of regulation of a particular pathway need not be the same in different bacteria.

## 8.8.3 The citric acid cycle as an anabolic pathway

The citric acid cycle reactions provide precursors to 10 of the 20 amino acids found in proteins. It is therefore a multifunctional pathway and is not used simply for the oxidation of acetyl–CoA. Succinyl–CoA is necessary for the synthesis of the amino acids L-lysine and L-methionine. Succinyl–CoA is also a precursor to tetrapyrroles, which are the prosthetic group in several proteins, including cytochromes and chlorophylls. Oxaloacetate is a precursor to the amino acid aspartate, which itself is the precursor to five other amino acids. In some bacteria, fumarate is also a precursor to aspartate. $\alpha$-Ketoglutarate is the precursor to the amino acid glutamate, which itself is the precursor to three other amino acids. The biosynthesis of amino acids is described in Chapter 9. However, since the citric acid cycle requires a constant level of oxaloacetate in order to function, net synthesis of these molecules requires replenishment of the oxaloacetate. This is discussed in Section 8.9.

## 8.8.4 Distribution of the citric acid cycle

The citric acid cycle is present in most heterotrophic bacteria growing aerobically. However, not all aerobic bacteria have a complete citric acid cycle. For example, organisms that grow on $C_1$ compounds (methane, methanol, and so on described in Chapter 13) lack $\alpha$-ketoglutarate dehydrogenase and carry out a reductive pathway as described in Section 8.9. An oxidative citric acid cycle is

not necessary for these organisms since acetyl–CoA is not an intermediate in the oxidation of the $C_1$ compounds. Although the oxidative citric acid cycle is a pathway usually associated with aerobic bacteria, it is also present in certain anaerobes that oxidize organic compounds completely to $CO_2$. These include the group II sulfate reducers (discussed in Section 12.2.2) and certain archaea. The latter are anaerobic hyperthermophilic archaea that use sulfur or thiosulfate as the terminal electron acceptor (Ref. 10).

## 8.9 Carboxylations that Replenish Oxaloacetate: The Pyruvate and Phosphoenolpyruvate Carboxylases

Because the citric acid cycle intermediates are constantly being removed to provide precursors for biosynthesis, they must be replaced (Section 8.8.3). Failure to do this would decrease the level of oxaloacetate that is necessary for the citrate synthase reaction, and thus for the continuation of the cycle. If the organism is growing on amino acids or organic acids (e.g., malate), then replenishment of oxaloacetate is not a problem, since these molecules are easily converted to oxaloacetate (Fig. 8.27). If the carbon source is a sugar (e.g., glucose), then the carboxylation of pyruvate or phosphoenolpyruvate replenishes the oxaloacetate (Fig. 8.17). Two enzymes that carry out the carboxylation of phosphoenolpyruvate and pyruvate are *PEP carboxylase* and *pyruvate carboxylase*, which are widespread among the bacteria. A bacterium will have one

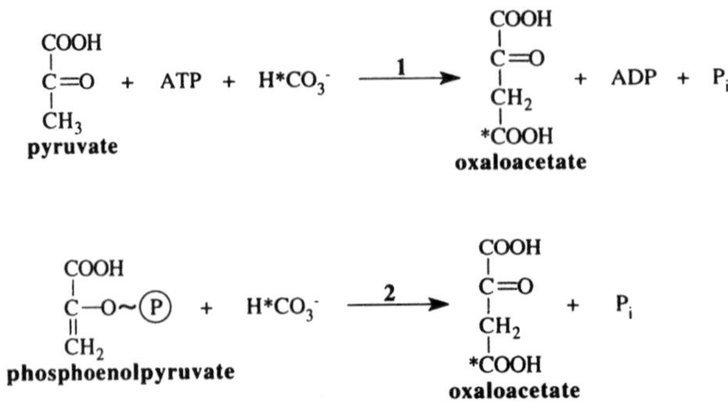

**Fig. 8.17** Carboxylation reactions that replenish the supply of oxaloacetate. Enzymes: 1, pyruvate carboxylase; 2, PEP carboxylase. Bacteria may have one or the other.

or the other. PEP carboxylase is not found in animal tissues or in fungi.

## 8.9.1 Regulation of PEP carboxylase

In *E. coli* PEP carboxylase is an allosteric enzyme that is positively regulated by acetyl–CoA and negatively regulated by aspartate (Fig. 8.18). Presumably, if the oxaloacetate levels drop, then acetyl–CoA will accumulate and result in the activation of the PEP carboxylase. This should produce more oxaloacetate. Aspartate can slow down its own synthesis by negatively regulating PEP carboxylase.

## 8.10 Modification of the Citric Acid Cycle into a Reductive (Incomplete) Cycle during Fermentative Growth

In the presence of air, the citric acid cycle operates as an oxidative pathway coupled to aerobic respiration in respiratory organisms. Since fermenting organisms are not carrying out aerobic respiration, it seems best not to have an oxidative pathway that produces so much NADH and FADH$_2$. On the other hand, the reactions that make oxaloacetate, succinyl–CoA, and $\alpha$-ketoglutarate are necessary because these molecules are required for the

biosynthesis of amino acids and tetrapyrroles. (See Section 8.8.3 and Fig. 8.27.) The solution to the problem is to convert the citric acid cycle from an oxidative into a reductive pathway. The reductive pathway is also referred to as an incomplete citric acid cycle. Fermenting bacteria have little or no activity for the enzyme $\alpha$-ketoglutarate dehydrogenase. Thus the pathway is blocked between $\alpha$-ketoglutarate and succinyl–CoA, and cannot operate in the oxidative direction (Fig. 8.19). Succinyl–CoA is made by reversing the reactions between oxaloacetate and succinyl–CoA, using the enzyme fumarate reductase instead of succinate dehydrogenase. The latter enzyme is replaced by fumarate reductase under anaerobic conditions. These reactions consume 4H. If one includes the reactions from citrate to $\alpha$-ketoglutarate that produce 2H, the net result is that the reductive pathway consumes 2H. The reductive citric acid pathway is found not only in fermenting bacteria, but also in some other bacteria (including the enteric bacteria) that are carrying out anaerobic respiration using nitrate as the electron acceptor.[11] The reason for this is that oxygen induces the synthesis of $\alpha$-ketoglutarate dehydrogenase in certain facultative anaerobes, and under anaerobic conditions, the enzyme levels are very low. (However, some nitrate respirers do have an oxidative citric acid cycle (e.g., *Pseudomonas stutzeri* grown under denitrifying conditions.[12])

One of the consequences of an incomplete citric acid cycle is that acetate is excreted as a by-product of sugar metabolism during anaerobic growth. That is because some of the acetyl–CoA is converted to acetate concomitant with the formation of an ATP. These reactions, which are an important source of ATP, are discussed in Chapter 14. It should be pointed out that some strict anaerobes (e.g., the green photosynthetic sulfur bacteria) have a reductive citric acid pathway that is "complete" in that it reduces oxaloacetate to citrate. The pathway, called the reductive carboxylic acid pathway, is a $CO_2$ fixation pathway used for autotrophic growth and differs in some key enzymological reactions from the pathways discussed here. The reductive carboxylic acid pathway is described in Section 13.1.9.

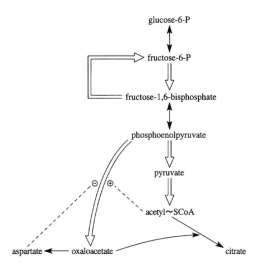

**Fig. 8.18** The regulation of PEP carboxylase in *E. coli*. The carboxylase is positively regulated by acetyl–CoA and negatively regulated by aspartate.

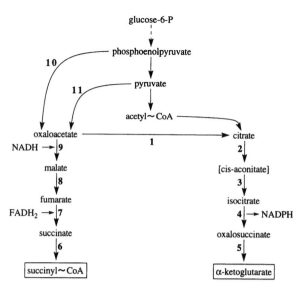

**Fig. 8.19** The reductive citric acid pathway in fermenting bacteria. There are two "arms." One route oxidizes citrate to α-ketoglutarate. A second route reduces oxaloacetate to succinyl-SCoA. The enzyme a-ketoglutarate dehydrogenase is missing. The enzyme fumarate reductase replaces succinate dehydrogenase. Enzymes: 1, citrate synthase; 2, 3, aconitase; 4, 5, isocitrate dehydrogenase; 6, succinate thiokinase; 7, fumarate reductase; 8, fumarase; 9, malate dehydrogenase; 10, PEP carboxylase; 11, pyruvate carboxylase.

## 8.11 Chemistry of Some of the Reactions in the Citric acid Cycle

This section presents a rational basis for understanding the chemistry of some of the key reactions in the citric acid cycle. Similar reactions are seen in other pathways.

### 8.11.1 Acetyl–CoA condensation reactions

Acetyl–CoA is a precursor for the biosynthesis of many different molecules besides citrate. These include lipids (Chapter 9) and various fermentation end products (Chapter 14). The reason why acetyl–CoA is so versatile is that it undergoes condensations at both the methyl and carboxyl end of the molecule. Condensations at both ends of acetyl–CoA can be understood in terms of the chemistry of the polarized carbonyl group. The electrons in the carbonyl group are not shared equally by the carbon and oxygen; rather, the oxygen is much more electronegative than the carbon and pulls the electrons in the double bond closer to itself. That is to say, the C=O group is polarized, making the carbon slightly positive (Fig. 8.20).

Because the oxygen in the carbonyl group is electron attracting, there is a tendency to pull electrons away from the C–H bond in the carbon adjacent to the carbonyl. This results in the formation of an enolate ion, which acts as a nucleophile (Fig. 8.21). The enolate anion seeks electrophilic centers, (e.g., the carbon atoms in carbonyl groups). So acetyl–CoA can be a nucleophile at its methyl end and attack other carbonyl groups, even other acetyl–CoA molecules. In the formation of citric acid, the methyl group of acetyl–CoA attacks a carbonyl group in oxaloacetate to form citric acid. At the same time, the thioester linkage to coenzyme A is hydrolyzed, driving the reaction to completion. Later, we will examine other pathways in which the methyl carbon of acetyl–CoA attacks a carbonyl.

Another result of the polarization of the carbonyl group is that the carbon in the carbonyl is electron deficient and subject to

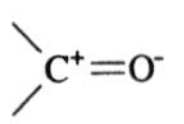

$$\overset{\diagdown}{\underset{\diagup}{C^+}}=O^-$$

**Fig. 8.20** The carbonyl group is polarized. Electrons are attracted by the oxygen, leaving a partial positive charge on the carbon.

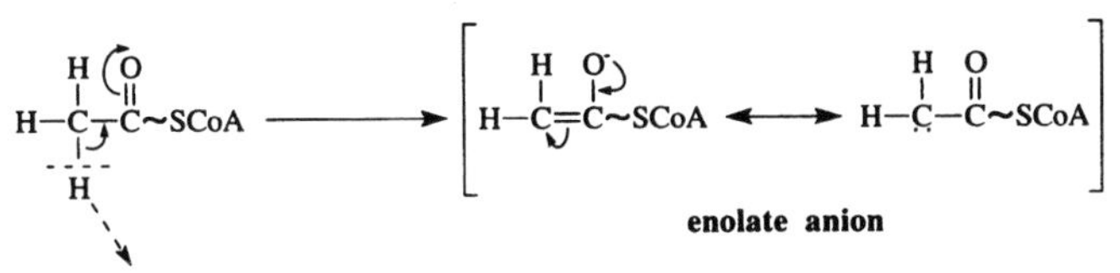

Fig. 8.21 The methylene carbon of acetyl–CoA can act as a nucleophile. Because of the electron-attracting ability of the carbonyl group, a hydrogen dissociates from the methylene carbon, forming an enolate anion, which resonates. Electrons can shift to the methylene carbon, which then seeks a positive center, e.g., a carbonyl group. Because of this, acetyl–CoA will form covalent bonds to the carbon of carbonyl groups in condensation reactions.

attack by nucleophiles that seek a positive center (i.e., it is electrophilic). Thus acetyl–CoA undergoes condensations at the carboxyl end in reactions in which the CoASH is displaced during a nucleophilic displacement and the acetyl portion is transferred to the nucleophile (Fig. 8.22). For example, this occurs during fatty acid synthesis and during butanol fermentations (Chapters 9 and 14), as well as in several other pathways. Because of the reactivity of acetyl–CoA at both the methyl and carboxyl end, the molecule is widely used in building larger molecules.

### 8.11.2 Decarboxylation reactions

Oxalosuccinate is a $\beta$-ketocarboxylic acid (i.e., the carboxyl group is $\beta$ to a keto group). The decarboxylation of $\beta$-ketocarboxylic acids occurs throughout metabolism. Another $\beta$-decarboxylation occurs in the pentose phosphate pathway. When 6-phosphogluconate is oxidized by NADP$^+$, a 3-keto intermediate is formed (Fig. 8.8). The decarboxylation of $\beta$-ketocarboxylic acids is relatively straightforward. A single enzyme is required, and the cofactor requirements are met by Mn$^{2+}$ or Mg$^{2+}$. A $\beta$-decarboxylation is shown in Fig. 8.23. The $\beta$-keto group attracts electrons, facilitating the breakage of the bond holding the carboxyl group to the molecule. Note the similarity to the aldolase reaction, where the keto group facilitates the breakage of a C–C bond that is beta to the keto group (Fig. 8.5).

### Decarboxylation of $\alpha$-ketoglutarate

$\alpha$-Ketoglutarate is an $\alpha$-keto carboxylic acid, and its decarboxylation is more complex than that of a $\beta$-keto carboxylic acid. The mechanism is the same as for the decarboxylation of pyruvate, another $\alpha$-keto

Fig. 8.22 Acetyl group transfer. Because the carbonyl group in acetyl–CoA is polarized, it is subject to nucleophilic attack. The result is that the acetyl group is transferred to the nucleophile and CoASH is displaced.

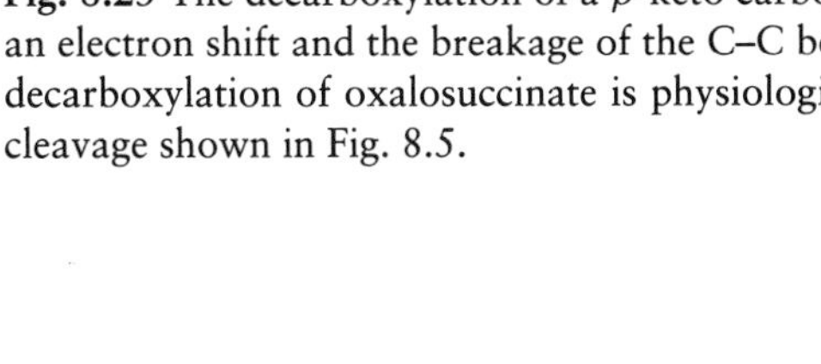

Fig. 8.23 The decarboxylation of a $\beta$-keto carboxylic acid. The keto group attracts electrons causing an electron shift and the breakage of the C–C bond holding the carboxyl group to the molecule. The decarboxylation of oxalosuccinate is physiologically reversible. Notice the resemblance to the aldol cleavage shown in Fig. 8.5.

carboxylic acid, and is described in more detail in Section 14.10. Pyruvate and $\alpha$-ketoglutarate dehydrogenases are similar in that they can be separated into three components (i.e., the TPP-containing decarboxylase, the FAD-containing dihydrolipoyl dehydrogenase, and the diydrolipoyl transacetylase). Other $\alpha$-ketocarboxylic acid dehydrogenases exist (e.g., for the catabolism of $\alpha$-ketoacids derived from the degradation of the branched-chain amino acids, leucine, isoleucine, and valine).

## 8.12 The Glyoxylate Cycle

We now come to a second pathway central to the metabolism of acetyl–CoA: the glyoxylate cycle, also called the glyoxylate bypass (Fig. 8.24). The glyoxylate cycle is required by aerobic bacteria to grow on fatty acids and acetate. (Plants and protozoa also have the glyoxylate cycle. However, it is absent in animals.) The glyoxylate cycle resembles the citric acid cycle except that it bypasses the two decarboxylations in the citric acid cycle. For this reason the acetyl–CoA is not oxidized to $CO_2$. Examine the summary of the glyoxylate cycle shown in Fig. 8.24. The glyoxylate cycle shares with the citric acid cycle the reactions that synthesize isocitrate from acetyl–CoA. The two pathways diverge at isocitrate. In the glyoxylate cycle the isocitrate is cleaved to succinate and glyoxylate by the enzyme *isocitrate lyase* (reaction 1). The glyoxylate condenses with acetyl–CoA to form malate, in a reaction catalyzed by *malate synthase* (reaction 2). The malate synthase reaction is of the same type as the citrate synthase (i.e., the methylene carbon of acetyl–CoA attacks the carbonyl group in glyoxylate). The reaction is driven to completion by the hydrolysis of the coenzyme A thioester bond just as during citrate synthesis. The malate replenishes the oxaloacetate, leaving one succinate and NADH as the products. The cells incorporate the succinate into cell material by first oxidizing it to oxaloacetate from which phosphoenolpyruvate is synthesized (Section 8.12).

### Carbon balance for the glyoxylate cycle

The carbon balance is written below. The net result of the glyoxylate cycle is the condensation of two molecules of acetyl–CoA to form succinate. Carbon balance for the glyoxylate pathway:

$$
\begin{aligned}
C_2 + C_4 &\longrightarrow C_6 \\
C_6 &\longrightarrow C_4 + C_2 \\
\underline{C_2 + C_2} &\longrightarrow \underline{C_4} \\
C_2 + C_2 &\longrightarrow C_4
\end{aligned}
$$

### 8.12.1 Regulation of the glyoxylate cycle

As Fig. 8.24 illustrates, isocitrate is at a branch point for both the citric acid cycle and the glyoxylate cycle. That is to say, it is a substrate for both the isocitrate lyase and the isocitrate dehydrogenase. What regulates the fate of the isocitrate? In *E. coli* the isocitrate dehydrogenase activity is partially inactivated by phosphorylation when cells are grown on acetate.[13] (The regulation of enzyme activity by covalent modification is discussed in Chapter 6, and other examples are listed in Table 6.1.) Acetate also induces the enzymes of the glyoxylate cycle. Therefore, in the presence of acetate, isocitrate lyase activity increases while the isocitrate dehydrogenase is partially inactivated. However, the $K_m$ for the isocitrate lyase is high relative to the concentrations of isocitrate. This means that the isocitrate lyase requires a high intracellular concentration of isocitrate. It is presumed that the partial inactivation of isocitrate dehydrogenase increases the concentration of isocitrate, resulting in an increase in flux through the glyoxylate cycle.

## 8.13 Formation of Phosphoenolpyruvate

Organic acids such as lactate, pyruvate, acetate, succinate, malate, amino acids, and so on can be used for growth because pathways exist to convert them to phosphoenolpyruvate, which is a precursor to the glycolytic

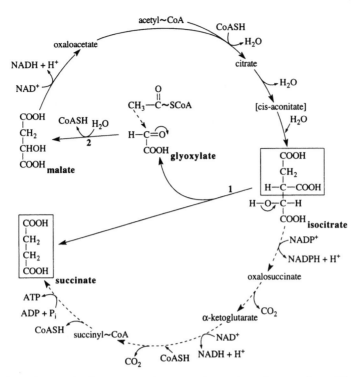

**Fig. 8.24** The glyoxylate cycle. Enzymes: 1, isocitrate lyase; 2, malate synthase. The dotted arrows represent the reactions of the citric acid cycle that are bypassed.

intermediates. The phosphoenolpyruvate is generally made in two ways. It can be made via the decarboxylation of oxaloacetate or via the phosphorylation of pyruvate. These two reactions are described next.

### 8.13.1 Formation of phosphoenopyruvate from oxaloacetate

This reaction is an ATP-dependent decarboxylation catalyzed by PEP carboxykinase, as shown in Fig. 8.25. The enzyme is widespread and accounts for phosphoenolpyruvate synthesis in both eukaryotes and prokaryotes. This enzyme, along with malic enzyme, is important for growth on succinate and malate (Section 8.14). (Although PEP carboxykinase is an important enzyme for the synthesis of phosphoenolpyruvate from oxaloacetate, it catalyzes the synthesis of oxaloacetate from phosphoenolpyruvate in some anaerobic bacteria. For example, anaerobic bacteria that ferment glucose to succinate may use this enzyme to carboxylate phosphoenolpyruvate to

oxaloacetate and then in other reactions reduce the oxaloacetate to succinate.[14])

### 8.13.2 Formation of phosphoenolpyruvate from pyruvate

Prokaryotes are able to synthesize phosphoenolpyruvate by phosphorylating pyruvate, instead of converting the pyruvate to oxaloacetate and then decarboxylating the oxaloacetate. This is necessary for growth on pyruvate for those bacteria that cannot synthesize oxaloacetate from pyruvate. Prokaryotes that

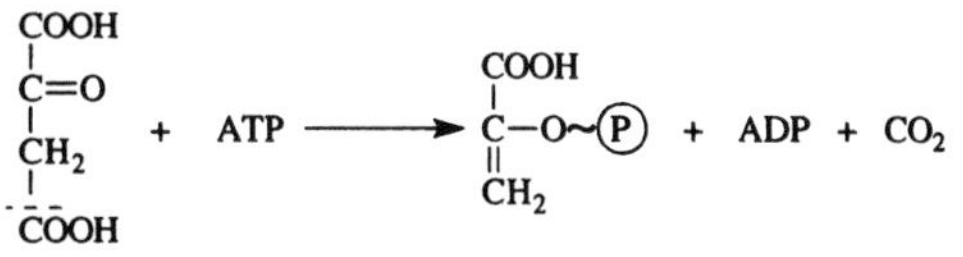

$$\begin{array}{c} COOH \\ | \\ C=O \\ | \\ CH_2 \\ | \\ COOH \end{array} + ATP \longrightarrow \begin{array}{c} COOH \\ | \\ C-O\sim\textcircled{P} \\ || \\ CH_2 \end{array} + ADP + CO_2$$

**Fig. 8.25** The PEP carboxykinase reaction. The PEP carboxykinase generally operates in the direction of PEP synthesis, although during fermentation in anaerobes it can work in the direction of oxaloacetate, which is subsequently reduced to the fermentation end product, succinate.

fall into the latter category do not have a glyoxylate pathway and also lack pyruvate carboxylase. For example, *E. coli* does not have a glyoxylate cycle unless growing on acetate, and it has PEP carboxylase instead of pyruvate carboxylase. Furthermore, there are strict anaerobes (e.g., methanogens and green sulfur photosynthetic bacteria) that grow autotrophically converting $CO_2$ to acetyl–CoA, which is carboxylated to pyruvate (Sections 13.1.2 and 13.1.3). They do not have a glyoxylate cycle and must phosphorylate the pyruvate to form phosphoenolpyruvate. The phosphorylation of pyruvate to phosphoenolpyruvate is described next.

## PEP synthetase and pyruvate–phosphate dikinase reactions

Prokaryotes that convert pyruvate directly to phosphoenolpyruvate use one of two enzymes. They are *PEP synthetase* and *pyruvate–phosphate dikinase*. These enzymes are found in prokaryotes and plants, but not in animals. The reactions are illustrated in Fig. 8.26.

$$\text{pyruvate} + H_2O + ATP \xrightarrow{\text{PEP synthetase}}$$
$$PEP + AMP + P_i$$

$$\text{pyruvate} + ATP + P_i \xrightarrow{\text{pyruvate–phosphate dikinase}}$$
$$PEP + AMP + PP_i$$

$$PP_i + H_2O \xrightarrow{\text{pyrophosphatase reaction}} 2P_i$$

In the PEP synthetase reaction, a pyrophosphoryl group is transferred from ATP to the enzyme. (See Fig. 7.3 for a description of

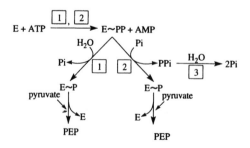

**Fig. 8.26** The PEP synthetase and pyruvate–phosphate dikinase reactions. Enzymes: 1, PEP synthetase; 2, pyruvate–phosphate dikinase; 3, pyrophosphatase.

pyrophosphoryl group transfer reactions.) One phosphate is removed by hydrolysis, leaving a phosphorylated enzyme. The phosphorylated enzyme then donates the phosphoryl group to pyruvate to form PEP. The reaction catalyzed by the pyruvate–phosphate dikinase is similar except that, instead of the phosphate being hydrolytically removed from the pyrophosphorylated enzyme, it is transferred to inorganic phosphate to form pyrophosphate. The pyrophosphate is hydrolyzed by a pyrophosphatase, pulling the reaction to completion. The net result from both reactions is the same (i.e., the sum of the pyruvate–phosphate dikinase and pyrophosphatase reactions is the same as the PEP synthetase reaction). In either case, the synthesis of phosphoenolpyruvate from pyruvate requires the hydrolysis of two phosphodiester bonds with a high free energy of hydrolysis.

## 8.14 Formation of Pyruvate from Malate

Bacteria and mitochondria can convert malate directly to pyruvate using the *malic enzyme*:

$$\text{L-malate} + NAD^+ \underset{\text{Malic enzyme}}{\rightleftharpoons}$$
$$\text{pyruvate} + NADH + CO_2$$

Bacteria and mitochondria actually possess two malic enzymes, one specific for $NAD^+$ and the other for $NADP^+$. The latter provides NADPH for reductions that occur during biosynthesis, such as the biosynthesis of fatty acids (Chapter 9). Malic enzyme, along with PEP carboxykinase (Section 8.13.1), is an important enzyme for growth on citric acid cycle intermediates such as succinate or malate. *E. coli* mutants that lack both PEP carboxykinase and malic enzyme will not grow on succinate or malate, although strains lacking only one of these enzymes will grow.

## 8.15 Summary of the Relationships between the Pathways

Figure 8.27 illustrates the relationship between the different pathways discussed in this chapter. Reactions 1 and 2 are key enzymes in

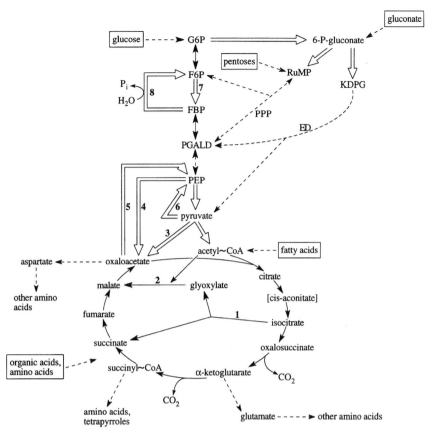

**Fig. 8.27** Relationship between the glyoxylate cycle, the citric acid cycle, and the major carbohydrate pathways. Enzymes: 1, isocitrate lyase; 2, malate synthase; 3, pyruvate carboxylase; 4, PEP carboxylase; 5, PEP carboxykinase; 6, PEP synthetase or pyruvate phosphodikinase; 7, phosphofructokinase; 8, fructose-1,6-bisphosphatase. Not shown is malic enzyme (Section 8.14).

the glyoxylate cycle. Reaction 1 is isocitrate lyase. Reaction 2 is malate synthase. Fatty acids and acetate are converted to acetyl–CoA that enters the citric acid cycle and the glyoxylate cycle. Dicarboxylic acids and amino acids are eventually degraded to citric acid cycle intermediates. Sugars can be catabolized via the Embden–Meyerhof–Parnas pathway, the pentose phosphate pathway, or the Entner–Doudoroff pathway. All the sugar pathways intersect at phosphoglyceraldehyde.

## 8.16 Summary

The nearly ubiquitous pathway for glucose degradation is the Embden–Meyerhof–Parnas pathway, also called the glycolytic pathway. The pathway oxidizes one mole of glucose to two moles of pyruvate, and pro-

duces two moles of NADH and two moles of ATP. There is only one oxidation, and that is the oxidation of phosphoglyceraldehyde to phosphoglycerate. An intermediate, 1,3-bisphosphoglycerate, is formed, which donates a phosphoryl group to ADP in a substrate-level phosphorylation. Since two phosphoglyceraldehydes are formed from one glucose, two ATPs are made. The pathway can be considered as two stages. Stage 1 generates two phosphoglyceraldehydes. Stage 2 is the oxidation of phosphoglyceraldehyde to pyruvate. Glycolysis not only provides ATP, NADH, and pyruvate, but its intermediates are used for biosynthesis in other pathways. This will become more evident in subsequent chapters when we examine other metabolic pathways. The pathway cannot be reversed from pyruvate to glucose-6-phosphate because of two irreversible reactions, the pyruvate kinase reaction (phospho-enolpyruvate to pyruvate) and the phospho-fructokinase reaction (fructose-6-phosphate to fructose-1,6-bisphosphate). Both these reactions proceed so far in the direction of products that they are not physiologically reversible. However, the pathway can be reversed from phosphoenolpyruvate provided fructose-1,6-bisphosphate phosphatase is present to convert fructose-1,6-bisphosphate to fructose-6-phosphate. A modification of the EMP pathway exists in the archaea, where ADP serves as the phosphoryl donor in the kinase reactions and ferredoxin is the electron acceptor in the oxidation reactions.

A second sugar-catabolizing pathway is the Entner–Doudoroff (ED) pathway. The ED pathway is widespread among prokaryotes, where it can be used for growth on gluconate. An intermediate of this pathway is phospho-glyceraldehyde, which is oxidized to pyruvate using the same reactions that occur in Stage 2 of glycolysis. Because only one phosphoglyceraldehyde is produced from glucose-6-phosphate in the ED pathway, only one ATP is made. Several bacteria lack key enzymes in the first stage of the EMP pathway (i.e., phosphofructokinase and fructose-1,6-bisphosphate aldolase) and rely on the ED pathway for growth on glucose. Interestingly, modifications of the ED pathway exist in a few bacteria and archaea in which some of the intermediates are not phosphorylated.

Among the many pathways that intersect with the second stage of glycolysis is the pentose phosphate pathway. The pentose phosphate pathway oxidizes glucose-6-phosphate to pentose phosphates. The pentose phosphates can be used for the synthesis of nucleic acids, or be converted via the transaldolase and transketolase reactions to phosphoglyceraldehyde. The pathway also produces NADPH for biosynthetic reductions, and erythrose-4-phosphate for aromatic amino acid biosynthesis. The pentose phosphate pathway is also used when pentoses are the source of carbon and energy.

In aerobic prokaryotes pyruvate is oxidatively decarboxylated via pyruvate dehydrogenase to acetyl-CoA, $CO_2$, and NADH. The acetyl–CoA enters the citric acid cycle, where it is oxidized completely to $CO_2$. The electrons generated during the oxidations are transferred to the electron transport chain, where a $\Delta p$ is created. The citric acid cycle is also an anabolic pathway. Several biosynthetic pathways draw citric acid cycle intermediates out of the pools. For example, oxaloacetate and $\alpha$-ketoglutarate are used for amino acid biosynthesis, and succinyl–CoA is used for tetrapyrrole, lysine, and methionine synthesis. Oxaloacetate is also used to synthesize phosphoenolpyruvate for glucogenesis. Since oxaloacetate is used catalytically in the cycle, it must be replenished during growth. When organisms are growing on carbohydrates, the oxaloacetate is replenished by carboxylating pyruvate or phosphoenolpyruvate. Growth on proteins or amino acids poses no problem since they are degraded to citric acid cycle intermediates.

In order to use pyruvate as a source of cell carbon, it must be converted to phosphoenolpyruvate. The reason for this is that phosphoenopyruvate is a precursor to several metabolites, including the intermediates of glycolysis. One route is the ATP-dependent carboxylation of pyruvate to oxaloacetate (pyruvate carboxylase) combined with the ATP-dependent decarboxylation to phosphoenolpyruvate using the enzyme PEP carboxykinase. Two ATPs are therefore used to convert one pyruvate to phosphoenolpyruvate. A second route is the direct phosphorylation of pyruvate to phosphoenolpyruvate using either PEP synthetase or pyruvate–phosphate dikinase combined with a pyrophosphatase. The products are phosphoenolpyruvate, AMP, and $P_i$. Thus, regardless of the pathway used, two ATPs are required to synthesize phosphoenolpyruvate from pyruvate.

Bacteria growing aerobically on fatty acids or acetate use the glyoxylate cycle for net incorporation of $C_2$ units into cell material. This pathway is a modification of the citric acid cycle, in which the two decarboxylation reactions are bypassed. The decarboxylation reactions are bypassed using two enzymes unique to the glyoxylate cycle (i.e., isocitrate lyase and malate synthetase). The net result is that acetyl–CoA is converted to succinate rather than to carbon dioxide.

The citric acid cycle is usually present only during aerobic growth. Under anaerobic conditions, prokaryotes have a modified pathway called the reductive citric acid pathway. However, some sulfate reducers, which are anaerobic bacteria that use sulfate as an electron acceptor, have an oxidative citric acid cycle. In the reductive citric acid pathway there are two major changes. The $\alpha$-ketoglutarate dehydrogenase activity is low or missing, and fumarate reductase replaces succinate dehydrogenase. The result is that oxaloacetate is reduced to succinyl–CoA, and citrate is oxidized to $\alpha$-ketoglutarate. Thus the intermediates necessary for biosynthesis of amino acids and tetrapyrroles are made, but the number of reductions exceed the number of oxidations.

## Study Questions

1. Suppose you isolated a mutant that did not grow on glucose as the sole source of carbon. However, the mutant did grow on the glucose when it was supplemented with succinate, fumarate, or malate. An examination of broken cell extracts revealed that all the citric acid cycle enzymes were present. What might be the metabolic defect in the mutant?

2. There are two physiologically irreversible reactions in the EMP pathway starting with glucose-6-phosphate and ending with pyruvate. Which ones are they? How are they regulated in *E. coli*?

3. Do you think that the glyoxylate cycle and the citric acid cycle can operate at the same time in an organism growing on acetic acid? How is this regulated in *E. coli*?

4. Fermenting bacteria have little or no $\alpha$-ketoglutarate dehydrogenase activity, and therefore have a block in the citric acid cycle. Under these conditions, the pathway operates reductively instead of oxidatively. Draw a reductive citric acid pathway showing how these organisms make oxaloacetate, succinyl–CoA, and $\alpha$-ketoglutarate. Why is it important to be able to synthesize these compounds under all growth conditions?

5. Which nonglycolytic enzyme is necessary to reverse glycolysis from PEP to G6P?

6. Some bacteria (e.g., *Brucella abortus*) lack key enzymes in the EMP and the ED pathway, but they can still grow on glucose. *Brucella abortus* is an animal pathogen that causes spontaneous abortion in cattle and can also infect humans, causing fever, headache, and joint pains. It has a citric acid cycle. Write a series of reactions showing how the pentose phosphate pathway can result in glucose oxidation completely to $CO_2$ without the involvement of Stage 1 of the EMP pathway or the ED pathway.

7. The ED pathway is only 50% as efficient as the EMP pathway in making ATP from glucose. Why is that?

8. *E. coli* mutants lacking G6P dehydrogenase can be grown on glucose as the sole source of carbon. Since this is the enzyme that catalyzes the entrance of G6P into the pentose phosphate pathway, how can these mutants make NADPH or pentose phosphates?

9. Show how the pentose phosphate pathway might oxidize glucose completely to $CO_2$ without using the citric acid cycle.

10. The combination of phosphofructokinase (fructose-6-phosphate kinase) and fructose-1,6-bisphosphatase has ATPase activity. Write these reactions and sum them to satisfy yourself that this is so. How does *E. coli* ensure that these reactions do not use up all the ATP in the cell?

11. When *E. coli* is grown on pyruvate as its sole source of carbon, it does not synthesize the glyoxylate cycle enzymes. How is the pyruvate converted to PEP? [Hint: For this question and the ones that follow, start by examining Fig. 8.27.]

12. Suppose you are growing a bacterial culture on ribose as its only source of carbon. Write a series of reactions that will convert ribose to glucose-6-phosphate. Why is it important to be able to synthesize glucose-6-phosphate?

13. Suppose you are growing a culture on gluconate. How would the cells convert the gluconate to phosphoglyceraldehyde if they lacked 6-phosphogluconate dehydrogenase? How might they make ribose-5-phosphate and erythrose-4-phosphate? Why is it necessary to have a supply of the latter two compounds?

14. Suppose you are growing a bacterial culture on succinate as the source of carbon. Write the reactions by which succinate is converted to PEP. Why is it important to be able to synthesize PEP?

## NOTES AND REFERENCES

1. Reviewed in Conway, T. 1992. The Entner–Doudoroff pathway: History, physiology, and molecular biology. *FEMS Microbiol. Rev.* 103:1–28.

2. During some fermentations a portion of the NADH is reoxidized by fumarate via membrane-bound electron carriers and a $\Delta p$ is created. This is called fumarate respiration. An example of fumarate respiration occurs during mixed acid fermentation discussed in Section 14.10 and propionate fermentation discussed in Section 14.7.

3. Danson, M. J. 1988. Archaebacteria: The comparative enzymology of their central metabolic pathways, pp. 166–231. In: *Advances in Microbial Physiology*, Vol. 29. Rose, A. H., and D. W. Tempest (Eds.). Academic Press, New York.

4. Kengen, S. W. M., F. A. M. deBok, N-D. van Loo, C. Dijkema, A. J. M. Stams, and W. M. de Vos. 1994. Evidence for the operation of a novel Embden–Meyerhof pathway that involves ADP-dependent kinases during sugar fermentation by *Pyrococcus furiosus. J. Biol. Chem.* 269:17537–17541.

5. Mukund, S., and M. W. W. Adams. 1995. Glyceraldehyde-3-phosphate ferredoxin oxidoreductase, a novel tungsten-containing enzyme with a potential glycolytic role in the hyperthermophiulic archaeon *Pyrococcus furiosus. J. Biol. Chem.* 270:8389–8392.

6. However, some bacteria (e.g., *E. coli*) also contain an enzyme called transhydrogenase, which catalyzes the reduction of $NADP^+$ by NADH. Therefore, for those bacteria the pentose phosphate pathway is not essential for NADPH synthesis.

7. Sugars that differ in the configuration at a single asymmetric carbon are called *epimers*. For

example, xylulose-5-phosphate and ribulose-5-phosphate are epimers of each other at C3.

8. Vitamins are growth factors that cannot be made by animals. During the early years of vitamin research, vitamins were divided into two groups: The fat-soluble vitamins were placed into one group and the water-soluble vitamins (B complex and C) into another. Vitamins are generally assayed according to specific diseases that they cure in animals fed on a vitamin-deficient diet. We now know that the water-soluble vitamin most responsible for stimulating growth in rats is riboflavin ($B_2$). Vitamin $B_6$ (pyridoxine) prevents rat facial dermatitis. Pantothenic acid cures chick dermatitis. Nicotinic acid (a precursor to nicotinamide) cures human pellagra. (The symptoms of pellagra are weakness, dermatitis, diarrhea, mental disorder, and death.) Vitamin C (ascorbic acid) prevents scurvy. Folic acid and vitamin $B_{12}$ (cobalamin) prevent anemia. Vitamin D or calciferol (fat-soluble) prevents rickets. Vitamin E or tocopherol (fat-soluble) is required for rats for full-term pregnancy and prevents sterility in male rats. Vitamin K (fat-soluble) is necessary for normal blood clotting. A deficiency in vitamin A (fat-soluble) leads to dry skin, conjunctivitis of the eyes, night blindness, and retardation of growth. Male rats fed a diet deficient in vitamin A do not form sperm and become blind. Vitamin A (retinol) is an important component of the light-sensitive pigment in the rod cells of the eye. In addition, animals cannot make polyunsaturated fatty acids ("essential fatty acids") because they lack the enzyme to desaturate monounsaturated fatty acids. A diet deficient in polyunsaturated fatty acids leads to poor growth, skin lesions, impaired fertility, and kidney damage. Lipoic acid is synthesized by animals and is therefore not a vitamin. However, certain bacteria and other microorganisms require lipoic acid for growth. It was soon recognized that lipoic acid was required for the oxidation of pyruvic acid. The chemical identification of lipoic acid followed the purification of 30 mg from 10 tons (!) of water-soluble residue of liver. It was done by Lester Reed at the University of Texas in 1949, with collaboration from scientists at Ely Lilly and Co. Lipoic acid is a growth factor for some bacteria. It is an eight-carbon saturated fatty acid in which carbons 6 and 8 are joined by a disulfide bond to form a ring when the molecule is oxidized. The reduced molecule has two sulfhydryl groups. The coenzyme functions in the oxidative decarboxylation of $\alpha$-keto acids.

9. Danson, M. J., S. Harford, and P. D. J. Weitzman. 1979. Studies on a mutant form of *Escherichia coli* citrate synthase desensitized to allosteric effectors. *Eur. J. Biochem.* 101:515–521.

10. Selig, M., and P. Schonhet. 1994. Oxidation of organic compounds to $CO_2$ with sulfur or thiosulfate as electron acceptor in the anaerobic hyperthermophilic archaea *Thermoproteus tenax* and *Pyrobaculum islandicum* proceeds via the citric acid cycle. *Arch. Microbiol.* **162**:286–294.

11. Stewart, V. 1988. Nitrate respiration in relation to facultative metabolism in Enterobacteria. *Microbiol. Rev.* **52**:190–232.

12. Spangler, W. J., and C. M. Gilmour. 1966. Biochemistry of nitrate respiration in *Pseudomonas stutzeri*. Aerobic and nitrate respiration routes of carbohydrate catabolism. *J. Bacteriol.* **91**:245–250.

13. LaPorte, D. C. 1993. The isocitrate dehydrogenase phosphorylation cycle: Regulation and enzymology. *J. Cell. Biochem.* **51**:14–18.

14. Podkovyrov, S. M., and J. G. Zeikus. 1993. Purification and characterization of phosphoenolpyruvate carboxykinase, a catabolic $CO_2$-fixing enzyme, from *Anaerobiospirillum succiniciproducens*. *J. Gen. Microbiol.* **139**:223–228.

# 9

# Metabolism of Lipids, Nucleotides, Amino Acids, and Hydrocarbons

The central pathways (i.e., glycolysis, the pentose phosphate pathway, the Entner–Doudoroff pathway, and the citric acid cycle) provide precursors to all the other metabolic pathways. This chapter focuses on some of those metabolic pathways that feed off of the central pathways and are necessary for lipid, protein, and nucleic acid metabolism.

## 9.1 Lipids

Lipids are a structurally heterogeneous group of substances that share the common property of being highly soluble in nonpolar solvents (e.g., methanol, chloroform, and so on) and relatively insoluble in water. Essentially all the lipids in prokaryotes are in the membranes. The major membrane lipids in bacteria and eukaryotes are phospholipids consisting of fatty acids esterified to glycerol phosphate derivatives and are called phosphoglycerides. Archaea can also have phosphoglycerides, but their structure and mode of synthesis are different. The structures of archaeal lipids are summarized in Section 1.2.5 and their synthesis in Section 9.1.3. The metabolism of fatty acids will be discussed first.

### 9.1.1 Fatty acids

#### Types found in bacteria

Fatty acids are chains of methylene carbons with a carboxyl group at one end. They can be branched (methyl), saturated, unsaturated, or hydroxylated. Some examples are shown in Table 9.1.

Fatty acids differ in the number of carbon atoms, the number of double bonds they contain, where the double bonds are placed in the molecule, and whether the molecule is branched, as is tuberculostearic, or whether it is cyclopropane (e.g., lactobacillic). However, some generalizations can be made about the fatty acids. They are usually 16 or 18 carbons long and are either saturated or have one double bond. Generally, gram-positive bacteria are richer in branched-chain fatty acids than are gram-negative bacteria.

#### The role of fatty acids

Fatty acids do not occur free in bacteria but are covalently attached to other molecules. Most of the fatty acids are esterified to glycerolphosphate derivatives to make *phosphoglycerides*, which are an important structural component of membranes. Fatty acids may also be esterified to carbohydrate. For example,

**Table 9.1** Some fatty acids

| Number of carbon atoms | Fatty Acid | |
|---|---|---|
| 16 | $CH_3(CH_2)_{14}COOH$ | Palmitic |
| 18 | $CH_3(CH_2)_{16}COOH$ | Stearic |
| 18 | $CH_3(CH_2)_7CH{=}CH(CH_2)_7COOH$ | Oleic |
| 18 | $CH_3(CH_2)_4CH{=}CHCH_2CH{=}CH(CH_2)_7COOH$ | Linoleic |
| 9 | $CH_3(CH_2)_5CH-CH(CH_2)_9COOH$ <br> $\backslash \ /$ <br> $CH_2$ | Lactobacillic |
| 19 | $CH_3(CH_2)_7CH(CH_2)_8COOH$ <br> \| <br> $CH_3$ | Tuberculostearic |

the *lipid A* portion of lipopolysaccharide consists of fatty acids esterified to glucosamine (Section 11.2.1). Fatty acids can also be esterified to protein. For example, gram-negative bacteria have a *lipoprotein* that is covalently attached to the peptidoglycan and protrudes into the outer membrane. The lipoprotein apparently is important for the stability of the outer membrane (Section 1.2.3). Because fatty acids are nonpolar, they anchor the molecules to which they are attached to the membrane. Furthermore, whether a fatty acid is unsaturated, saturated, or branched has important significance for the physical properties of the membrane. Unsaturated fatty acids and branched-chain fatty acids make the membrane more fluid, which is a necessary condition for membrane function.

### β-Oxidation of fatty acids

Many bacteria can grow on long-chain fatty acids (e.g., pseudomonads, various bacilli, and *E. coli*). The fatty acids are oxidized to acetyl–CoA via a pathway called *β-oxidation* (Fig. 9.1). For this to take place the fatty acid is first converted to the acyl–CoA derivative in a reaction catalyzed by *acyl–CoA synthetase*. This takes place in two steps. First, pyrophosphate is displaced from ATP to make the AMP derivative of the carboxyl group (acyl adenylate) which remains tightly bound to the enzyme (Fig. 9.1, reaction 1). Then the AMP is displaced by CoASH to make the fatty acyl–CoA. The activation of a carboxyl group by formation of an acyl adenylate is common in metabolism and is discussed in Section 7.2.1. (See Fig. 7.7.) For example, amino acids are activated for protein

synthesis in this way. As with other reactions in which pyrophosphate is displaced from ATP, the reaction is driven to completion by hydrolysis of the pyrophosphate. A keto

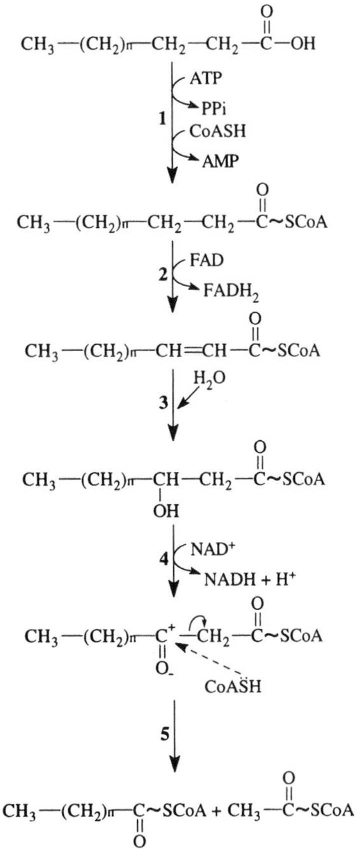

**Fig. 9.1** β-oxidation of fatty acids. Enzyme: 1, acyl–CoA synthetase; 2, fatty acyl–CoA dehydrogenase; 3, 3-hydroxyacyl–CoA hydrolyase; 4, L-3-hydroxyacyl–CoA dehydrogenase; 5, β-ketothiolase.

group is then generated $\beta$ to the carboxyl. The sequence of reactions leading to the keto group are: (1) an oxidation to form a double bond (reaction 2) catalyzed by acyl–CoA dehydrogenase; (2) hydration of the double bond to form the hydroxyl (reaction 3) catalyzed by 3-hydroxyacyl–CoA hydrolyase; and (3) oxidation of the hydroxyl to form the keto goup (reaction 4), catalyzed by L-3-hydroxyacyl–CoA dehydrogenase. The chemistry is the same as the conversion of succinate to oxaloacetate in the citric acid cycle (Fig. 8.16). The carbonyl of the $\beta$-keto acyl–CoA is then attacked by CoASH, displacing an acetyl–CoA (reaction 5) catalyzed by $\beta$-ketothiolase. Usually the bacterium is growing aerobically and the acetyl–CoA is oxidized to carbon dioxide in the citric acid cycle. The acetyl–CoA can also be a source of cell carbon when it is assimilated via the glyoxylate cycle (Section 8.12). The displacement of the acetyl–CoA results in the generation of a fatty acid acyl–CoA that is recycled through the $\beta$-oxidation pathway.

If the fatty acid has an even number of carbon atoms, then the entire chain is degraded to acetyl–CoA. However, if the fatty acid is an odd-chain fatty acid, then the last fragment is propionyl–CoA rather than acetyl–CoA. There is a variety of possible pathways for the oxidation of propionyl–CoA to acetyl–CoA. Two of these are shown in Figs. 9.2 and 9.3. E. coli forms pyruvate from propionate via methylcitrate (Fig. 9.3).[1,2] Review the glyoxylate cycle in Section 8.12 and Fig. 8.24. Notice the similarity to the methylcitrate pathway.

## The synthesis of fatty acids

An examination of metabolic pathways reveals that catabolic pathways are usually different from synthetic pathways. In other words, biosynthesis is not simply due to reversing the reactions of catabolism. This is true because there is usually at least one irreversible step in the catabolic pathway. For example, glycolysis has two irreversible reactions, the phosphofructokinase and pyruvate kinase reactions, and these must be bypassed in order to reverse glycolysis. Often, an entirely different set of reactions is used for biosynthesis. This is seen in fatty acid metabolism, where the biosynthetic pathway consists of reactions completely different from the $\beta$-oxidation pathway.[3] *The biosynthetic pathway differs from the $\beta$-oxidation pathway in the following ways:*

1. The reductant is NADPH, rather than NADH.

2. Biosynthesis requires $CO_2$ and proceeds via a carboxylated derivative of acetyl–CoA, called malonyl–CoA.

3. The acyl carrier is not CoA, as it is in the degradative pathway, but a protein called the acyl carrier protein (ACP). The acyl carrier protein is a small protein (MW = 10,000 in E. coli) having a residue similar to CoASH attached to one end. (See Note 4 for comparison of the chemical structures of CoA and ACP.)

In eukaryotic cells, fatty acid biosynthesis takes place in the cytosol, and the degradation takes place in the matrix of the mitochondria. Since prokaryotes do not have similar organelles, both synthesis and degradation take place in the same compartment (i.e., the cytosol). The fatty acid biosynthetic enzymes exist as multienzyme complexes in eukaryotes, including yeast. However, in E. coli they are present in the cytosol as separate enzymes.

The sequence of reactions begins with the carboxylation of acetyl–CoA to make malonyl–CoA (Fig. 9.4A, reaction 1). The carboxylation requires ATP, a vitamin, biotin, and is catalyzed by acetyl–CoA carboxylase. (See Note 5 for more information on

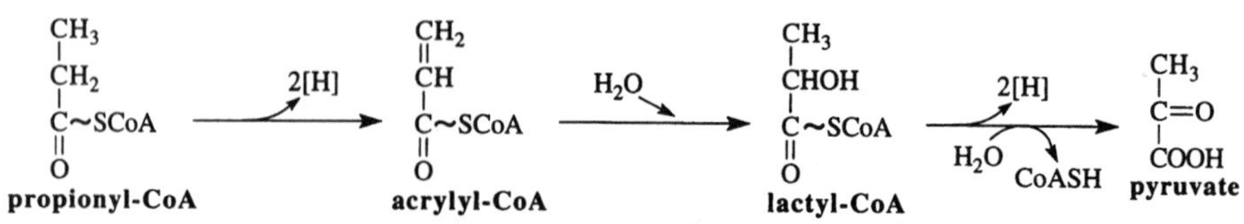

Fig. 9.2 Oxidation of propionyl–CoA to pyruvate via acrylyl–CoA.

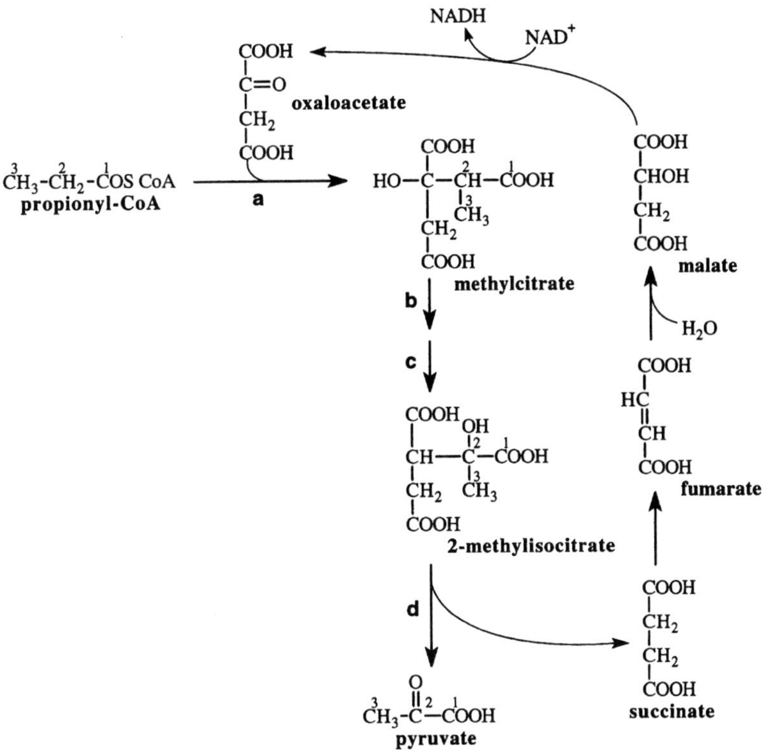

**Fig. 9.3** Oxidation of propionyl–CoA to pyruvate via the methylcitrate pathway. This pathway seems to operate in *E. coli.* Enzymes: (a) 2-methylcitrate synthase. (b and c) 2-methylaconitase. (d) 2-methylisocitrate lyase. *Source:* Adapted from Textor, S., V. F. Wendisch, A. A. De Graff, U. Müller, M. I. Linder, D. Linder, and W. Buckel. 1997. Propionate oxidation in *Escherichia coli:* Evidence for operation of a methylcitrate cycle in bacteria. *Arch. Microbiol.* **168:**428–436.

acetyl–CoA carboxylase.) Then the CoA is displaced by the acyl carrier protein, ACP, to form malonyl-ACP in a reaction catalyzed by malonyl–CoA:ACP transacetylase (reaction 2). A similar reaction, catalyzed by acetyl–CoA:ACP transacetylase, displaces CoA from another molecule of acetyl–CoA to form the acetyl-ACP derivative (reaction 3). (See Note 6.) The methylene carbon in malonyl-ACP acts as a nucleophile and displaces the ACP from acetyl-ACP to form the β-ketoacyl-ACP derivative in a reaction catalyzed by β-ketoacyl-ACP synthase (reaction 4). (See Section 8.11.1 for a discussion of these types of condensation reactions.) At the same time, the newly added carboxyl is removed, driving the reaction to completion. (In *E. coli*, either acetyl–CoA or acetyl–ACP can be used for the condensation with malonyl-ACP because there is more than one condensing enzyme. See Note 7 for more information on fatty

acid synthesis in *E. coli.*) The β-keto acyl-ACP is then reduced to the hydroxy derivative by an NADPH-dependent β-ketoacyl-ACP reductase (reaction 5), and dehydrated to the a-β derivative by the β-hydroxylacyl-ACP dehydrase to form the unsaturated acyl-ACP derivative (reaction 6). (The double bond is between C2, the carbon α to the carboxyl carbon, and C3, the carbon β to the carboxyl carbon.) The unsaturated acyl-ACP derivative is then reduced by enoyl-ACP reductase to the saturated acyl-ACP (reaction 7). The acyl-ACP chain is elongated by a series of identical reactions initiated by the attack of malonyl-ACP on the carboxyl end of the growing acyl-ACP chain, displacing the ACP. Fatty acid synthesis must be regulated because long-chain acyl-ACPs do not accumulate. The mechanism of regulation has not been elucidated, but it is reasonable to suggest that it involves feedback inhibition of a regulatory

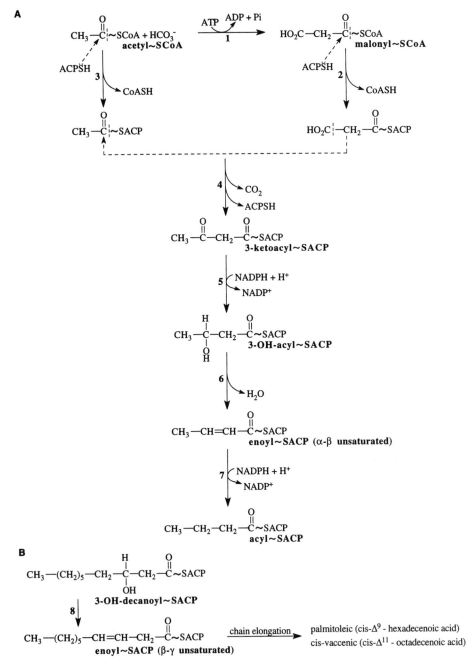

**Fig. 9.4** Biosynthesis of fatty acids. Enyzmes: 1, acetyl–CoA carboxylase; 2, malonyl transacetylase; 3, acetyl transacetylase; 4, 3-ketoacyl–ACP synthase; 5, 3-ketoacyl–ACP reductase; 6, $\beta$-hydroxyacyl–ACP dehydrase; 7, enoyl–ACP reductase; 8, 3-hydroxydecenoyl–ACP dehydrase.

fatty acid biosynthetic enzyme by long-chain acyl ACPs. (See Section 6.1.1 for a discussion of feedback inhibition.)

When the acyl-ACP chain is completed, the acyl portion is immediately transferred to membrane phospholipids by the glycerol phosphate acyltransferase reactions described in Section 9.1.2. Of course, not all the fatty acids are incorporated into phospholipids. For example, some of the acyl-ACPs are used for lipid A biosynthesis (Section 11.2.2). These include $\beta$-hydroxymyristoyl-ACP, lauroyl-ACP, and myristoyl-ACP.

## Synthesis of unsaturated fatty acids

If the fatty acid is unsaturated, it is almost always monounsaturated (one double bond). Polyunsaturated fatty acids are more typical of eukaryotic organisms. (As explained in Note 8, polyunsaturated acids are required in the diets of mammals.) Unsaturated fatty acids are formed in two different ways depending upon the organism, the *anaerobic* pathway and the *aerobic* pathway. The anaerobic pathway is restricted to prokaryotes. It is widespread in bacteria, being found in *Clostridium*, *Lactobacillus*, *Escherichia*, *Pseudomonas*, the cyanobacteria, and the photosynthetic bacteria. The aerobic pathway is found in *Bacillus*, *Mycobacterium*, *Corynebacterium*, *Micrococcus*, and eukaryotes.

In the anaerobic pathway, a special dehydrase desaturates the $C_{10}$ hydroxyacyl-ACP intermediate to the *trans-α,β*-decenoyl-ACP, which is isomerized while bound to the enzyme to *cis-β,γ*-decenoyl-ACP (Fig. 9.4B). In the $\beta$-$\gamma$ derivative the double bond is between C3 and C4, whereas in the $\alpha$-$\beta$ derivative the double bond is between C2 and C3. Very importantly, the *cis-β,γ* does not serve as a substrate for the enoyl reductase. The consequence of this fact is that the *cis-β,γ* derivative is not reduced. Instead, it is elongated, leading to a fatty acid with a double bond. Two common monounsaturated fatty acids in bacteria synthesized in this way are palmitoleic, which has 16 carbons (*cis-$\Delta^9$*-hexadecenoic acid), and *cis*-vaccenic acid, which has 18 carbons (*cis-$\Delta^{11}$*-octadecenoic acid). (See Note 9 for a description of *cis* and *trans* fatty acids.) In palmitoleic acid the double bond is between $C_9$ and $C_{10}$ in the final product. This is because six carbons are added to the C1 of the $C_{10}$ $\beta$-$\gamma$ derivative. Since vaccenic acid is two carbons longer, the double bond is between $C_{11}$ and $C_{12}$.

The aerobic pathway makes use of special desaturases (oxidases) that introduce a double bond into the completed fatty acyl–CoA or fatty acyl-ACP derivative. The enzyme system requires molecular oxygen and NADPH.

In a multienzyme reaction sequence two electrons are removed from the fatty acyl derivative, forming the double bond, and two electrons are removed from NADH. The four electrons are transferred to $O_2$, forming two $H_2O$ molecules. Bacteria that use the aerobic pathway make oleic from its $C_{18}$ saturated progenitor, stearic acid, rather than *cis*-vaccenic. Oleic acid has a double bond between C9 and C10 and is called *cis-$\Delta^9$*-octadecenoic acid.

## 9.1.2 Phospholipid synthesis in bacteria

Phospholipids are lipids with covalently attached phosphate groups. They are an important constituent of cell membranes. Because the phosphate groups are ionized, phospholipids always have negatively charged groups. There may also be positively charged groups (e.g., the protonated amino group in phosphatidyl enthanolamine). There are several different types of phospholipids. However, the ones that concern us here are the major phospholipids in bacteria. These are phospholipids that contain fatty acids and phosphate esterified to glycerol, and usually some other molecule (e.g., an amino acid, an amine, or a sugar covalently bound to the phosphate). They are called *phosphoglycerides*. The structure of typical phosphoglycerides is shown in Fig. 9.5. The fatty acids in positions C1 and C2 within the same molecule need not be the same.

## Phospholipids and the structure of the cell membrane

A major structural feature of phospholipids is that they are amphibolic. That is to say, one part of the molecule is hydrophobic (apolar) and another part is hydrophilic (polar). The hydrophobic area (tail) is the part where the fatty acids are located, and the hydrophilic area (head) is the end containing the phosphate and its attached group. The amphibolic nature of the phospholipids explains their orientation in membranes. Phospholipids are oriented in membranes as a bilayer with the charged head groups pointing out into the aqueous phase and the hydrophobic fatty

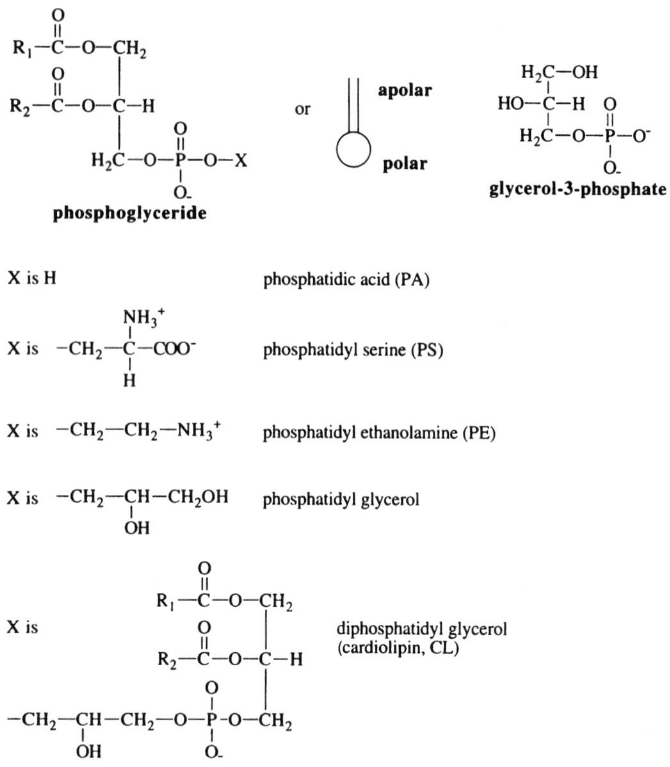

**Fig. 9.5** Some common phosphoglycerides in bacteria. R represents a fatty acyl moiety esterified to the glycerol phosphate. Note that phospholipids have an apolar and a polar end. The configuration of glycerol-3-phosphate and the glycerol-3-phosphate moiety of phosphoglycerides is called sn-glycerol-3-phosphate and belongs to the L-stereochemical series. Horizontal bonds are projected to the front and vertical bonds behind the plane of the page. Thus the H and OH are in front of the plane of the page, and C1 and C3 are behind.

acid tails interacting with each other in a hydrophobic interior. The cell membrane is completed by proteins (Fig. 1.15). Some properties of phospholipid membranes that should be remembered are: (1) They are permeable only to water, gases, and small hydrophobic molecules; (2) they have a low ionic conductance and are capable of doing work when ions (especially protons and sodium ions) are transported through them via special carriers along an electrochemical gradient; (3) the lipid portion of membranes must remain fluid in order for the membrane to function; and (4) the fluidity of the membrane is due to the presence of unsaturated fatty acids or branched-chain (methyl) fatty acids in the phospholipids.

## The synthesis of the phosphoglycerides

Phosphoglyceride synthesis starts with dihydroxyacetone phosphate (DHAP), an intermediate in glycolysis (Fig. 9.6). *Step 1* is the reduction of dihydroxyacetone phosphate to glycerol phosphate by the enzyme *glycerol phosphate dehydrogenase*. *Step 2* is the transfer to glycerol phosphate of fatty acids from fatty acyl-S-ACP (newly synthesized, as described in Section 9.1.1). The reactions are catalyzed by membrane-bound enzymes called *G3P acyl transferases*. (Some bacteria are also capable of using acyl–SCoA derivatives as the acyl donor.) The first phospholipid made is phosphatidic acid, which is glycerol phosphate esterified to two fatty acids. The other phospholipids are made from phosphatidic acid. *Step 3* is a reaction in which the phosphate on the phosphatidic acid reacts with cytidine triphosphate (CTP) and displaces PP$_i$ to form CDP-diacylglycerol. The reaction is catalyzed by *CDP–diglyceride synthase*. The formation of CDP-diacylglycerol is driven to completion

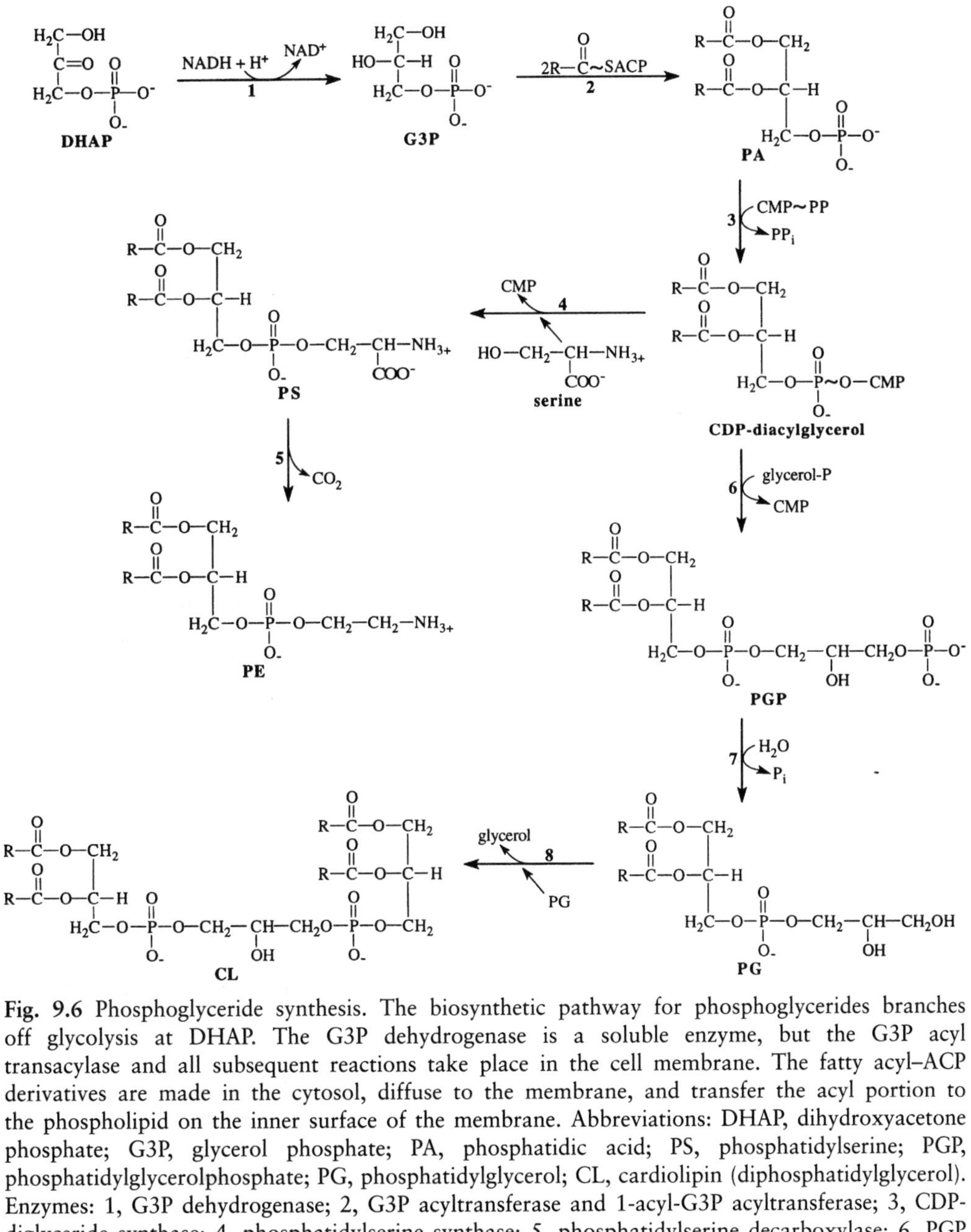

**Fig. 9.6** Phosphoglyceride synthesis. The biosynthetic pathway for phosphoglycerides branches off glycolysis at DHAP. The G3P dehydrogenase is a soluble enzyme, but the G3P acyl transacylase and all subsequent reactions take place in the cell membrane. The fatty acyl–ACP derivatives are made in the cytosol, diffuse to the membrane, and transfer the acyl portion to the phospholipid on the inner surface of the membrane. Abbreviations: DHAP, dihydroxyacetone phosphate; G3P, glycerol phosphate; PA, phosphatidic acid; PS, phosphatidylserine; PGP, phosphatidylglycerolphosphate; PG, phosphatidylglycerol; CL, cardiolipin (diphosphatidylglycerol). Enzymes: 1, G3P dehydrogenase; 2, G3P acyltransferase and 1-acyl-G3P acyltransferase; 3, CDP-diglyceride synthase; 4, phosphatidylserine synthase; 5, phosphatidylserine decarboxylase; 6, PGP synthase; 7, PGP phosphatase; 8, CL synthase.

by the hydrolysis of the $PP_i$ catalyzed by a pyrophosphatase. *Step 4* is the displacement of CMP by serine catalyzed by *phosphatidylserine synthase* (PS synthase). *Step 5* is the decarboxylation of phosphatidylserine to yield phosphatidylethanolamine, which is the major phospholipid in several bacteria.

Bacteria make other phospholipids from CDP-diglyceride. The displacement of CDP by $\alpha$-glycerolphosphate yields phosphatidylglycerol phosphate (*step 6*). Phosphatidylglycerol phosphate is dephosphorylated to yield phosphatidylglycerol (*step 7*). The latter displaces glycerol from a second molecule of phosphatidylglycerol to form diphosphatidyl-glycerol (cardiolipin) (*step 8*). Note that modification of the head groups in all these cases occurs via a nucleophilic attack by a hydroxyl on the electropositive phosphorous in the phosphate group.

### Stringent response

The student may recall the discussion of the stringent response in Section 2.2.2, where it was pointed out that the stringent response, which is a response to starvation for a required amino acid or a carbon and energy source, is correlated with an accumulation of intracellular (p)ppGpp and is accompanied by an inhibition of phospholipid synthesis as well as stable RNA synthesis. During the stringent response imposed by amino acid starvation in *E. coli* long-chain acyl-ACPs accumulate, suggesting that the acyl-transferase is inhibited, which can account for the inhibition of phospholipid biosynthesis.[10] How (p)ppGpp might be involved in the inhibition of the acyltransferase is not known.

### 9.1.3 Synthesis of archaeal lipids

The student should review the structure of archaeal lipids and membranes described in Section 1.2.5. The lipids of archaea differ in the following two ways from those of bacteria:

1. Instead of fatty acids, archaeal lipids have long-chain alcohols called isopranyl alcohols.

2. The linkage to glycerol is via an ether bond rather than an ester bond.

The metabolic pathway for the synthesis of the glycerol ethers in the archaea is not well understood at all. Figure 9.7 illustrates a recent proposal for *Halobacterium halobium*. The glycerol backbone is synthesized from either glycerol-3-phosphate or dihydroxyacetone phosphate. The alcohol is believed to be derived from geranylgeranyl-pyrophosphate, which is synthesized from acetyl–CoA via mevalonic acid in a well-known pathway that is widespread among the bacteria and

higher organisms. (The alcohol group is esterified to the pyrophosphate.) The C1 hydroxyl on glycerol-3-phosphate or the dihydroxy-acetone phosphate nucleophilically attacks the geranylgeranyl-pyrophosphate, displacing pyrophosphate and forming the ether linkage. If dihydroxyacetone phosphate is used, then the monoalkenyl-DHAP is formed. If glycerol-3-phosphate is used, then the monoalkenylglycerol-1-phosphate is formed. The monoalkenylglycerol-1-phosphate, which is formed either directly or in two steps, reacts with another geranylgeranyl-pyrophosphate to form the dialkenyl-glycerol-1-phosphate. It is proposed that in *Methanobacterium* and *Sulfolobus* glycerol-1-phosphate, rather than dihydroxacetone phosphate or glycerol-3-phosphate, attacks the geranylgeranyl-pyrophosphate. The rest of the pathway is unknown but involves a reduction of the double bonds in the hydrocarbons, the attachment of the polar head groups, and the formation of the tetraether lipids. It is not known whether the condensation between the alkyl groups to form the tetraethers occurs before or after substitution with polar head groups.

## 9.2 Nucleotides

### 9.2.1 Nomenclature and structures

A nucleotide is a molecule containing a pyrimidine or purine, a sugar (ribose or deoxyribose), and phosphate. If there were no phosphate, it would be called a nucleoside. Thus nucleotides are nucleoside phosphates. If one phosphate is present, then the molecule is called a nucleoside monophosphate. If two phosphates are present, then it is called a nucleoside diphosphate, and so on. The three major pyrimidines are cytosine, thymine, and uracil. The corresponding nucleosides are called cytidine, thymidine, and uridine, respectively. The cytidine nucleotides are called cytidine monophosphate (CMP), cytidine diphosphate (CDP), and cytidine triphosphate (CTP). A similar naming system is used for the uridine nucleotides (UMP, UDP, UTP) and the

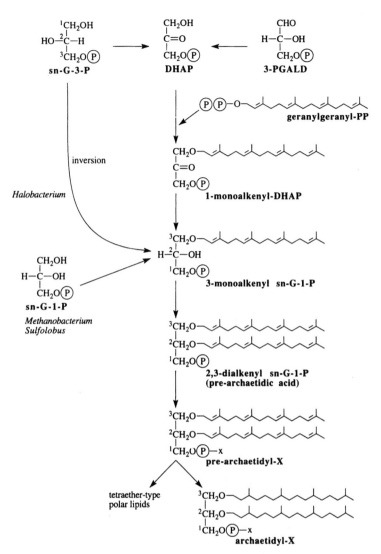

**Fig. 9.7** Postulated pathway for the synthesis of glycerol ether lipids in archaea. Two pathways are depicted, one for *Halobacterium* and one for *Methanobacterium* and *Sulfolobus*. It is proposed that a nucleophilic displacement takes place on geranylgeranyl-PP, displacing the pyrophosphate and forming the ether linkage. The molecule that condenses with the geranylgeranyl-PP might be glycerol-3-phosphate, glycerol-1-phosphate, or dihydroxyacetone phosphate. The rest of the pathway is unknown. *Source:* From Koga et al. 1993. Ether polar lipids of methanogenic bacteria: structures, comparative aspects, and biosynthesis. *Microbiol. Rev.* 57:164–182.

thymidine nucleotides (TMP, TDP, TTP). If the sugar is deoxyribose rather than ribose, then the nucleotide is written dCTP and so on. The two major purines are adenine and guanine. The nucleosides are adenosine and guanosine. The nucleotides are adenosine mono-, di-, and triphosphate and guanosine mono-, di-, and triphosphate. There are even separate names for the nucleoside monophosphates. For example, adenosine monophosphate is also called adenylic acid. Guanosine monophosphate is guanylic acid. We also have cytidylic acid, uridylic acid, and thymidylic acid. The structures of the purines, pyrimidines, and sugars are shown in Fig. 9.8. Notice that deoxyribose differs from ribose in having a hydrogen substituted for the hydroxyl group on C2 of the sugar.

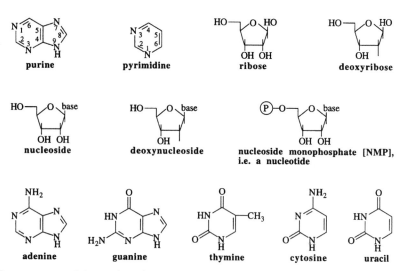

**Fig. 9.8** The structures of the nucleotides and their subunits. The bases are attached to the C1 of the sugars and the phosphates to the C5. Not shown are the nucleoside di- and triphosphates.

## 9.2.2 The synthesis of the pyrimidine nucleotides

The pyrimidine part of the nucleotide is made from aspartic acid, ammonia, and carbon dioxide (Fig. 9.9). *Step 1* is the biotin-dependent synthesis of carbamoyl phosphate from $HCO_3^-$, glutamine, and ATP (Fig. 9.10). The enzyme that catalyzes the reaction is called *carbamoyl phosphate synthetase*. (See Note 11.) It requires two ATPs. One ATP is used to form a carboxylated biotin intermediate, and the second ATP is used to phosphorylate the carboxyl group. Notice that carbamoyl phosphate is an acyl phosphate and therefore has high group transfer potential. In *step 2*, the carbamoyl group is transferred to aspartate as the phosphate is displaced by the nitrogen atom of aspartate. The product is N-carbamoylaspartate. The enzyme that catalyzes step 2 is *aspartate transcarbamylase* (ATCase), which is feedback inhibited by the end product, CTP. *Step 3* is the cyclization of N-carbamoylaspartate with loss of water to form the first pyrimidine, dihydroorotate. *Step 4* is the oxidation of dihydroorotate to orotate. In *step 5* the orotate displaces $PP_i$ from phosphoribosyl pyrophosphate (PPRP) to form the first nucleotide, orotidine-5′-phosphate, also called orotidylate. The reaction is driven to completion by the hydrolysis of $PP_i$ catalyzed by a pyrophosphatase. Note that the PRPP

itself is synthesized from ribose-5-phosphate and ATP via a pyrophosphoryl group transfer reaction from ATP to the C1 carbon of ribose-5-phosphate. Orotodine-5′-phosphate is decarboxlated in *step 6* to uridine monophosphate (UMP). Phosphorylation of UMP using ATP as the donor yields UDP and UTP (*steps 7 and 8*). CTP is synthesized from UTP by an ATP-dependent amination using $NH_4^+$ (*step 9*). (See Note 12.)

The CTP can be dephosphorylated to CDP, which is the precursor to the deoxyribonucleotides dCTP and dTTP. The sequence of reactions is:

$$CDP \longrightarrow dCDP \longrightarrow dCTP \longrightarrow dUTP \longrightarrow dUMP$$
$$\longrightarrow dTMP \longrightarrow dTDP \longrightarrow dTTP$$

Also, UDP can be reduced to dUDP, which can be converted to dUTP. Thus there are two routes to dUTP, one from CDP and the other (a minor route) from UDP.

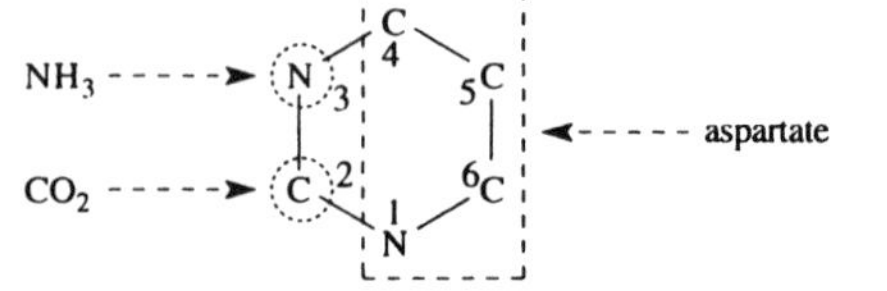

**Fig. 9.9** Origins of atoms in the pyrimidine ring. C2 is derived from $CO_2$, and N3 comes from ammonia, via carbamoyl phosphate. The rest of the atoms are derived from aspartate.

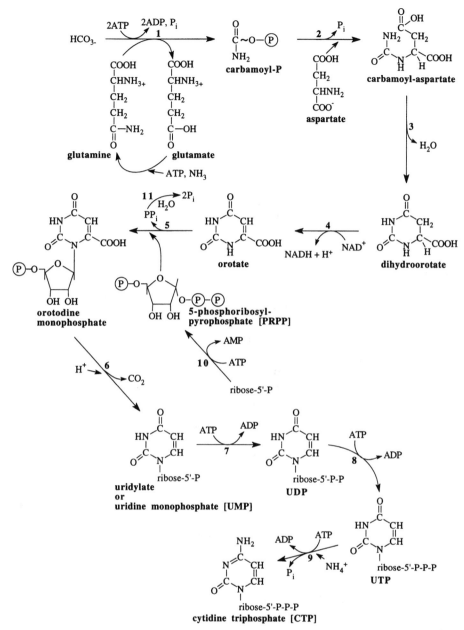

**Fig. 9.10** The biosynthesis of pyrimidine nucleotides. Enzymes: 1, carbamoyl phosphate synthetase; 2, aspartate transcarbamoylase; 3, dihydroorotase; 4, dihydroorotate dehydrogenase; 5, orotate phosphoribosyl transferase; 6, orotidine-5-phosphate decarboxylase; 7, nucleoside monophosphate kinase; 8, nucleoside diphosphate kinase; 9, CTP synthetase; 10, PRPP synthetase; 11, pyrophosphatase.

## 9.2.3 The synthesis of purine nucleotides

A purine is drawn in Fig. 9.11 showing the origin of the atoms. The synthesis of the purine ring requires, as precursors, glutamine and as-partic acid (to donate amino groups), glycine, carbon dioxide, and a $C_1$ unit at the oxidation state of formic acid. The latter three donate all the carbon atoms in the purine. The $C_1$ unit at the oxidation level of formic acid can come from formic acid itself, or from serine.

223

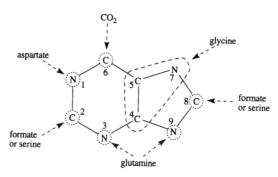

**Fig. 9.11** Metabolic origins of atoms in purines.

## Enzymatic reactions

The biosynthetic pathway for purines is shown in Fig. 9.12. The purine molecule is synthesized in stages while attached to ribose phosphate, which in turns comes from PRPP. *Step 1* is the attachment of an amino group to PRPP, donated by glutamine. The product is 5-phosphoribosylamine. This is a reaction in which the amide nitrogen of glutamine displaces the pyrophosphate in a nucleophilic displacement. The pyrophosphate that is released has a high free energy of hydrolysis, and the amination is driven to completion by the hydrolysis of the pyrophosphate by pyrophosphatase. In *step 2*, glycine is added to the amino group of phosphoribosylamine to form 5-phosphoribosyl-glycineamide in an ATP-dependent step. In this reaction the carboxyl group of glycine is added to the amino group of phosphoribosylamine to make the amide bond. The ATP first phosphorylates the carboxyl group, forming an acyl phosphate. Then the nitrogen in the amino group displaces the phosphate, forming the amide bond. The student should review these kinds of reactions in Section 7.2.1. *Step 3* is the formylation of the amino group on the glycine moiety. This requires a molecule to donate the formyl group (i.e., a molecule with high formyl group transfer potential). The donor of the formyl group is formyl-tetrahydrofolic acid (formyl-THF). The product is 5-phosphoribosyl-*N*-formylglycineamide. In *step 4* the amide group is changed into an amidine, the nitrogen being donated by glutamine in an ATP-dependent reaction. The product is 5-phosphoribosyl-*N*-formylglycineamidine. In *step 5* the 5-phosphoribosyl-*N*-formylglycineamidine cy-

clizes to 5-phosphoribosyl-5-aminoimidazole in a reaction that requires ATP. Probably the carbonyl is phosphorylated and the phosphate is displaced by the nitrogen to form the C–N bond. The cyclized product tautomerizes to form the amino group on aminoimidazole ribonucleotide. *Reaction 6* is the carboxylation of 5-phosphoribosyl-5-aminoimidazole to form 5-phosphoribosyl-5-aminoimidazole-4-carboxylic acid. In *step 7* aspartic acid combines with 5-phosphoribosyl-5-aminoimidazole-4-carboxylic acid to form 5-phosphoribosyl-4-(*N*-succinocarboxamide)-5-aminoimidazole in an ATP-dependent step. In *step 8* fumarate is removed, leaving the nitrogen, to form 5-phosphoribosyl-4-carboxamide-5-aminoimidazole, which is formylated in *step 9* to form 5-phosphoribosyl-4-carboxamide-5-formaminoimidazole. The latter is cyclized in *step 10* to inosinic acid (IMP). The purine itself is called hypoxanthine. Inosinic acid is the precursor to all the purine nucleotides.

## Biosynthesis of AMP and GMP from IMP

IMP is converted to AMP by substitution of the carbonyl oxygen at C6 with an amino group (Fig. 9.13). The amino donor is aspartate, and the reaction requires GTP. AMP can then be phosphorylated to ADP using ATP as the phosphoryl donor. The ADP can be reduced to dADP or phosphorylated to ATP via substrate-level phosphorylation or respiratory phosphorylation. GMP is synthesized by an oxidation of IMP at C2 followed by an amination at that carbon. The oxidation produces a carbonyl group, which accepts an amino group from glutamine.

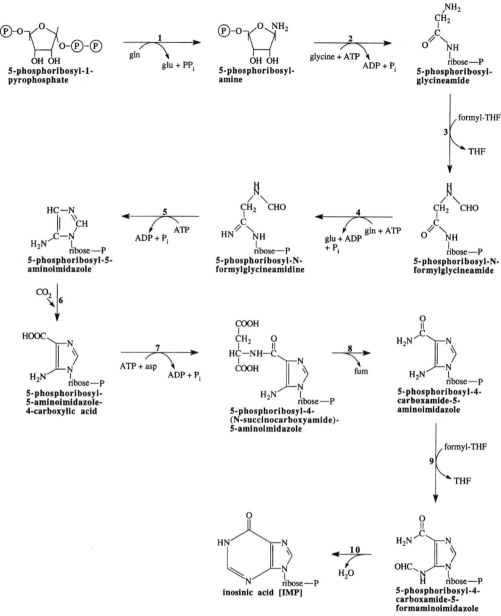

**Fig. 9.12** Biosynthesis of purine nucleotides. Enzymes: 1, PRPP amidotransferase; 2, phosphoribosylglycineamide synthetase; 3, phosphoribosylglycineamide formyltransferase; 4, phosphoribosylformylglycineamidine synthetase; 5, phosphoribosyl-aminoimidazole synthetase; 6, phosphoribosylaminoimidazole carboxylase; 7, phosphoribosylaminoimidazole succinocarboxamide synthetase; 8, adenylosuccinate lyase; 9, phosphoribosylaminoimidazolecarboxamide formyltransferase; 10, IMP cyclohydrolase. Abbreviations: gln, glutamine; glu, glutamate; asp, aspartate; fum, fumarate; THF, tetrahydrofolate.

## 9.2.4 The role of tetrahydrofolic acid

Derivatives of tetrahydrofolic acid (THF; synthesized from the B vitamin folic acid) are coenzymes that carry single carbon groups and are very important in metabolism. For example, the $C_1$ units used in purine biosynthesis are carried by derivatives of tetrahydrofolic acid. For an explanation of how THF is used, examine Fig. 9.14. The $C_1$ units can

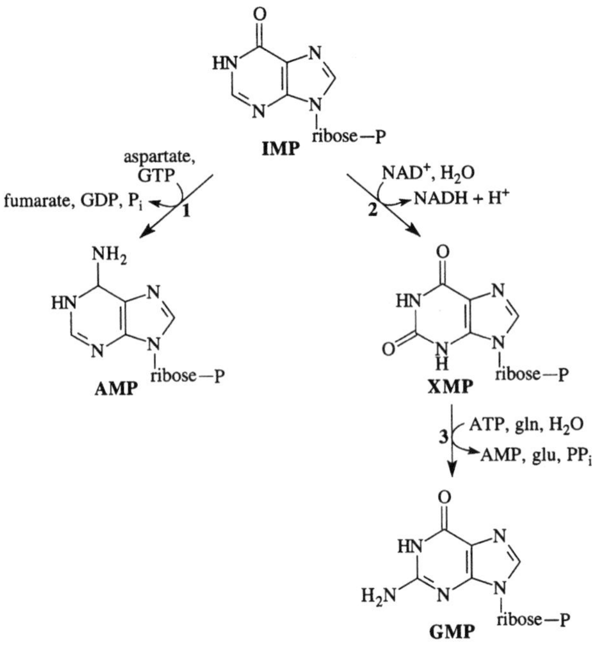

**Fig. 9.13** Synthesis of AMP and GMP from IMP. Enzymes: 1, adenylosuccinate synthetase and adenylosuccinate lyase; 2, IMP dehydrogenase; 3, GMP synthetase. Abbreviations: IMP, inosinic acid; AMP, adenylic acid; XMP, xanthylic acid; GMP, guanylic acid.

be donated to THF from serine, which donates its $\beta$-carbon at the oxidation level of formaldehyde to THF acid to form methylene-THF and glycine. (This is also the pathway for glycine biosynthesis, as described in Section 9.3.1.) Methylene-THF is then oxidized to *formyl-THF*, which donates the formyl group in purine biosynthesis. Bacteria can also use formate for purine biosynthesis. The formate is attached to THF acid to make formyl-THF in an ATP-dependent reaction. THF is also an important methyl carrier. Instead of methylene-THF being oxidized to formyl-THF, it is reduced to *methyl-THF*. Methyl-THF donates the methyl group in various biosynthetic reactions, including the biosynthesis of methionine. *Sulfanilamide* and its derivatives, called sulfonamides (the "sulfa drugs"), inhibit the formation of folic acid and therefore inhibit purine biosynthesis in bacteria. Sulfonamides resemble *p*-aminobenzoic acid, an intermediate in folic acid biosynthesis, and the sulfonamides inhibit the enzyme that utilizes *p*-aminobenzoic acid for folic acid synthesis. The sulfonamides are toxic to bacteria and not to animals because animals obtain their folic acid from their diet, whereas most bacteria synthesize folic acid.

### 9.2.5 Synthesis of deoxyribonucleotides

The deoxyribonucleotides are synthesized from the ribonucleoside diphosphates by a reductive dehydration catalyzed by *ribonucleoside diphosphate reductase* (Fig. 9.15). The electron donor is a sulfhydryl protein called thioredoxin that obtains its electrons from NADPH.

### 9.3 Amino Acids

There are 20 different amino acids in proteins. Inspection of Table 9.2 reveals that six of the 20 amino acids are synthesized from oxaloacetate and four from $\alpha$-ketoglutarate, two intermediates of the citric acid cycle. In addition, succinyl–CoA donates a succinyl group in the formation of intermediates in the biosynthesis of lysine, methionine, and diaminopimelic acid. (The succinate is removed at a later step and does not appear in the product.) This

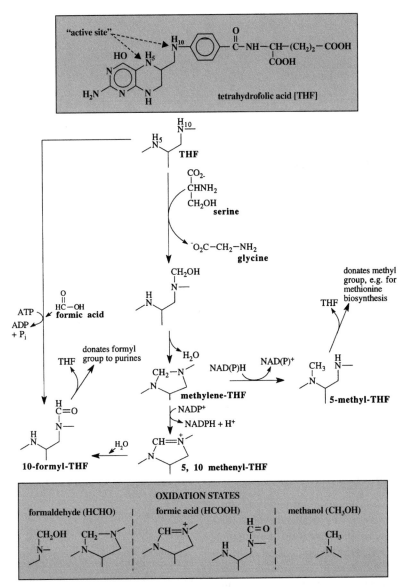

**Fig. 9.14** Tetrahydrofolic acid (THF) as a $C_1$ carrier. Tetrahydrofolic acid is the reduced form of the vitamin folic acid. Tetrahydrofolic acid and its derivatives are important coenzymes that function in many pathways to carry one-carbon units. The one-carbon units can be at the oxidation level of formic acid (HCOOH), formaldehyde (HCHO), or methyl ($CH_3-$). It is essential for purine biosynthesis because it transfers groups at the oxidation level of formic acid in two separate steps. The most common precursor of the $C_1$ unit is the amino acid serine, which donates the $C_1$ unit as formaldehyde to THF. The product can then be oxidized to the level of formate to form 10-formyl-THF. Some bacteria can use formic acid itself as the source of the $C_1$ unit. Methylene–THF can also be reduced to methyl–THF, which serves as a methyl donor.

again emphasizes that not only is the citric acid cycle important for energy generation, but also for biosynthesis. The table also points out that three glycolytic intermediates (i.e., pyruvate, 3-PGA, and PEP) are precursors to seven more amino acids. The pentose phosphate pathway provides erythrose-4-P, which is a precursor to aromatic amino acids. (It is actually used with PEP in the synthesis of the aromatic amino acids.) Finally, 5-phosphoribosyl-1-

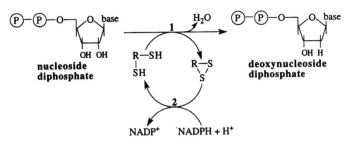

**Fig. 9.15** Formation of deoxynucleotides by ribonucleoside diphosphate reductase. The deoxynucleotides are made from the nucleoside diphosphates. The OH on C2 of the ribose leaves with its bonding electrons to form water. The OH is replaced by a hydride ion ($H:^-$) donated by reduced thioredoxin, a dithioprotein. Thioredoxin is rereduced by thioredoxin reductase, a flavoprotein. Thioredoxin reductase, in turn, accepts electrons from NADPH. Enzymes: 1, ribonucleotide diphosphate reductase; 2, thioredoxin reductase. Abbreviations: $R–(SH)_2$, thioredoxin.

pyrophosphate (PRPP), which donates the sugar for nucleotide synthesis, is also the precursor to the amino acid histidine. Therefore, the three central pathways (i.e., glycolysis, the pentose phosphate pathway, and the citric acid cycle) provide the precursors to all the amino acids. In this discussion, we will examine the biosynthesis and catabolism of a select group of amino acids in order to illuminate some principles of amino acid metabolism.

### 9.3.1 Synthesis

#### Glutamate and glutamine synthesis

The ability to synthesize glutamate and glutamine is of extreme importance because this is the only route for incorporation of inorganic nitrogen into cell material. All inorganic nitrogen must first be converted to ammonia, which is then incorporated as an amino group into glutamate and glutamine. (The synthesis of

**Table 9.2** Precursors for amino-acid biosynthesis

| Precursor | Amino Acid |
|---|---|
| Pyruvic acid | L-alanine, L-valine, L-leucine |
| Oxaloacetic acid | L-aspartate, L-asparagine, L-methionine, L-lysine, L-threonine, L-isoleucine |
| α-Ketoglutaric acid | L-glutamate, L-glutamine, L-arginine, L-proline |
| 3-PGA | L-serine, glycine, L-cysteine |
| PEP and erythrose-4-P | L-phenylalanine, L-tyrosine, L-tryptophan |
| PRPP and ATP | L-histidine |

ammonia from nitrate and from nitrogen gas is described in Chapter 12.) The amino group is then donated from these amino acids to all the other nitrogen-containing compounds in the cell. Glutamate is the amino donor for most of the amino acids, and glutamine is the amino donor for the synthesis of purines, pyrimidines, amino sugars, histidine, tryptophan, asparagine, NAD$^+$, and p-aminobenzoate.

Glutamate is synthesized by two alternate routes. One requires the enzyme *glutamate dehydrogenase*, and the second requires two enzymes: *glutamine synthetase* (GS) and *glutamate synthase*, which is also called *glutamine oxoglutarate aminotransferase* or the *GOGAT* enzyme.

Glutamate dehydrogenase catalyzes the reductive amination of α-ketoglutarate (Fig. 9.16). However, it should be pointed out that because of its high $K_m$ for ammonia, the glutamate dehydrogenase reaction can be used for glutamate synthesis only when ammonia concentrations are high (> 1 mM), otherwise ammonia is incorporated into glutamate via glutamine. This is the situation in many natural environments. Under conditions of low ammonia concentration, bacteria use two enzymes in combination for ammonia incorporation. One is L-glutamine synthetase (Fig. 9.17, reaction 1). This enzyme incorporates ammonia into glutamate to form glutamine. The glutamine synthetase reaction uses ATP, which drives the reaction to completion. The other enzyme is glutamate synthase (reaction 2). This enzyme transfers the newly incorporated ammonia from glutamine to α-

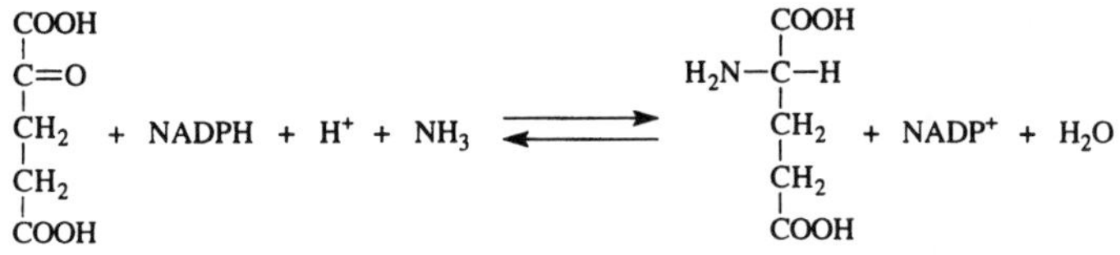

COOH
C=O
CH2
CH2
COOH
+ NADPH + H+ + NH3
COOH
H2N—C—H
CH2
CH2
COOH
+ NADP+ + H2O

**Fig. 9.16** The glutamate dehydrogenase reaction.

ketoglutarate to form glutamate. Notice in Fig. 9.17 that glutamine is used catalytically. This is also seen in the reactions written below. The glutamine synthetase and glutamate synthase reactions:

$$\text{glutamate} + \text{ATP} + NH_3 \longrightarrow \text{glutamine} + \text{ADP} + P_i$$

$$\text{glutamine} + \alpha\ \text{ketoglutarate} + \text{NADPH} + H^+ \longrightarrow 2\ \text{glutamate} + \text{NADP}^+$$

---

$$\alpha\ \text{ketoglutarate} + \text{ATP} + NH_3 + \text{NADPH} + H^+ \longrightarrow \text{glutamate} + \text{ADP} + P_i + \text{NADP}^+$$

The use of ATP drives the reaction to completion.

As mentioned, glutamine is an amino donor for many compounds, including pyrimidines and purines. Therefore, glutamine synthetase is an indispensable enzyme unless the medium is supplemented with glutamine. Glutamate dehydrogenase, however, is not necessary for growth, since the organism can synthesize glutamate from glutamine and $\alpha$-ketoglutarate using glutamate synthase. The regulation of glutamine synthetase activity is complex and discussed in Section 18.8.3.

COOH
H2N—C—H
CH2
CH2
C—OH
O
L-glutamate
ATP ADP + Pi
NH3
1
COOH
H2N—C—H
CH2
CH2
C—NH2
O
L-glutamine
2
NADP+
NADPH + H+
COOH
H2N—C—H
CH2
CH2
COOH
L-glutamate
COOH
C=O
CH2
CH2
COOH
α-ketoglutarate

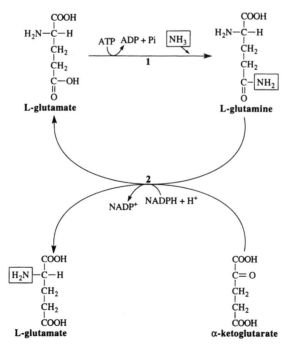

**Fig. 9.17** The GS and GOGAT reactions. Enzymes: 1, L-glutamine synthetase; 2, glutamine: $\alpha$-oxoglutarate aminotransferase, also called the GOGAT enzyme, or glutamate synthase.

## Transamination reactions from glutamate are required for the synthesis of the other amino acids

The synthesis of the other amino acids requires that glutamate donates an amino group to an $\alpha$-keto carboxylic acid. The enzyme that catalyzes the amino group transfer is called a *transaminase*. A generalized transamination reaction is shown in Fig. 9.18. Note that after the glutamate has donated its amino group it becomes $\alpha$-ketoglutarate again, which then becomes aminated once more to form another molecule of glutamate. Thus

glutamate can be thought of as a conduit through which ammonia passes into other amino acids. Transaminases use a coenzyme called pyridoxal phosphate (derived from vitamin $B_6$) to carry the amino group from glutamate to the $\alpha$-keto carboxylic acids.

### Synthesis of aspartate and alanine

Aspartate is synthesized via a transamination from glutamate to oxaloacetate, and alanine is made via a transamination from glutamate to pyruvate (Fig. 9.18).

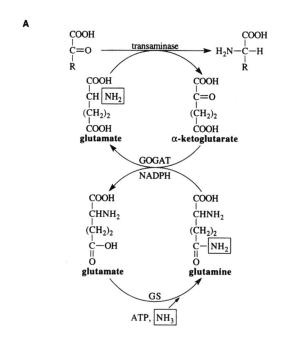

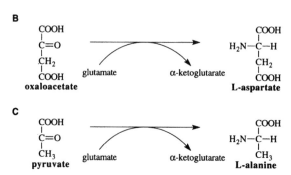

**Fig. 9.18** The transamination reaction. (A) Glutamate donates an amino group to an $\alpha$-keto carboxylic acid to form an $\alpha$-amino acid. The glutamate becomes $\alpha$-ketoglutarate, which is aminated either via the glutamate dehydrogenase or the GS and GOGAT reactions. (B) Formation of L-aspartate from oxaloacetate. (C) Formation of L-alanine from pyruvate.

## Synthesis of serine, glycine, and cysteine

Serine is synthesized from 3-phosphoglycerate (Fig. 9.19). The phosphoglycerate is first oxidized to the $\alpha$-keto acid, phosphohydroxypyruvate. A transamination, using glutamate as the donor, converts the phosphohydroxypyruvate to phosphoserine. Then the phosphate is hydrolytically removed to form L-serine. The hydrolysis of the phosphate ester bond drives the reaction to completion. Serine is a precursor to both glycine and cysteine. Glycine is formed by transfer of the $\beta$-carbon of serine to tetrahydrofolic acid (Fig. 9.14). The route to cysteine is initiated when serine accepts an acetyl group from acetyl–CoA to become O-acetylated. Sulfide then displaces the acetyl group from O-acetylserine forming cysteine (see Fig. 12.2).

## 9.3.4 Catabolism of amino acids

A scheme showing the overall pattern of carbon flow during amino acid catabolism is shown in Fig. 9.20. The first reaction is always the removal of the amino group to generate the $\alpha$-keto acid, which eventually enters the citric acid cycle. All 20 amino acids are degraded to seven intermediates that enter the citric acid cycle. These seven intermediates are: pyruvate, acetyl–CoA, acetoacetyl–CoA, $\alpha$-ketoglutarate, succinyl–CoA, fumarate, and oxaloacetate. Acetoacetyl–CoA itself is a precursor to acetyl–CoA.

## Removal of the amino group

There are several ways that amino groups can be removed from amino acids during their catabolism (Fig. 9.21). Usually amino acids are oxidatively deaminated to their corresponding keto acid. The oxidation may be catalyzed by a nonspecific *flavoprotein oxidase* that oxidizes any one of a number of amino acids and feeds the electrons directly into the electron transport chain (Fig. 9.21A). These oxidases can be D-amino oxidases as well as L-amino oxidases. The D-amino acid oxidases are useful because several biological molecules (e.g., peptidoglycan and certain

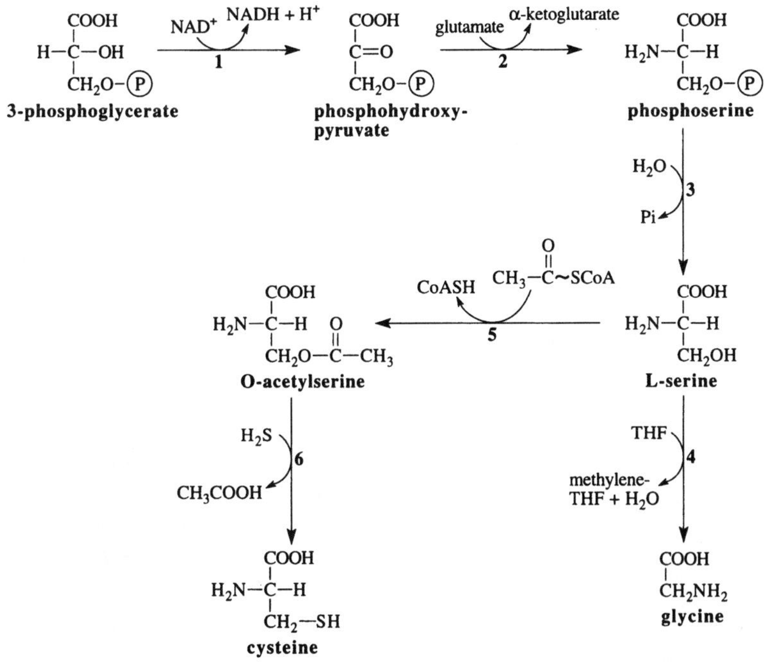

**Fig. 9.19** The synthesis of serine, glycine, and cysteine from 3-PGA. Enzymes: 1, phosphoglycerate dehydrogenase; 2, phosphoserine aminotransferase; 3, phosphoserine phosphatase; 4, serine hydroxymethyltransferase; 5, serine transacetylase; 6, O-acetylserine sulfhydrylase.

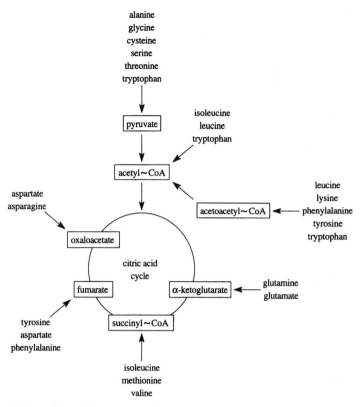

**Fig. 9.20** Fates of the carbon skeletons of amino acids. Amino acid carbon can be used to synthesize all the cell components or be oxidized to $CO_2$. In the absence of the glyoxylate cycle, carbon entering at acetyl–CoA cannot be used for net glucogenesis except in some strict anaerobic bacteria that can carboxylate acetyl–CoA to pyruvate (Chapter 13).

antibiotics) have D-amino acids. There are also *NAD(P)$^+$-linked dehydrogenases*, which are more specific than the flavoprotein oxidases (Fig. 9.21B). The dehydrogenases catalyze the reversible reductive amination of the keto acid. A transaminase coupled to a dehydrogenase can lead to the deamination of amino acids (Fig. 9.21C). In the example shown in Fig. 9.21C, a nonspecific transaminase transfers an amino group to pyruvate-forming L-alanine. The amino group is released as ammonia, and pyruvate is regenerated by L-alanine dehydrogenase. $\alpha$-Ketoglutarate and its dehydrogenase serve the same function in amino acid catabolism. Certain amino acids are deaminated by specific *deaminases* (Fig. 9.21D). In these cases, a redox reaction is not involved. These amino acids include serine, threonine, aspartate, and histidine. Some deaminations (i.e., the deamination of serine and threonine) proceed by the elimination of water.

## 9.4 Aliphatic Hydrocarbons

Many bacteria can grow on long-chain hydrocarbons, for example, alkanes ($C_{10}$–$C_{18}$). A few bacteria (mycobacteria, flavobacteria, *Nocardia*) grow on short-chain hydrocarbons ($C_2$–$C_8$). The hydrocarbons exist as droplets of oil outside of the cell, and a major problem in their utilization is to transfer these water-insoluble molecules from the oil layer across the cell wall to the cell membrane, where they can be metabolized. Some bacteria have cell walls containing glycolipids in which the hydrocarbons can dissolve and be transported to the membrane. *Acinetobacter* strains secrete particles resembling the outer membrane in which the hydrocarbons dissolve. The particles with the dissolved hydrocarbon fuse with the outer membrane, and the hydrocarbons are then transferred to the cell membrane. Once in the cell membrane, the degradation

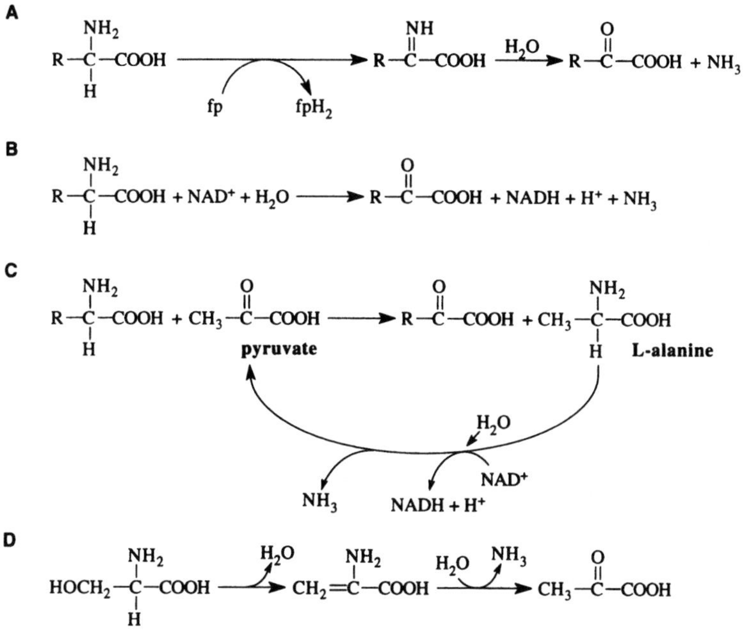

Fig. 9.21 The removal of amino groups from amino acids during catabolism. (A) Amino acid oxidases are flavoproteins that are specific for L or D amino acids, but generally not for the particular amino acid. Electrons are transferred from the flavoprotein to the electron transport chain. (B) An amino acid dehydrogenase. These are NAD(P) linked and more specific for the amino acid than are the flavoprotein oxidases. (C) Transamination catalyzed by a nonspecific transaminase that uses pyruvate as the amino acid acceptor. The amino group is removed from the alanine by alanine dehydrogenase. $\alpha$-Ketoglutarate can also accept amino groups and be deaminated by $\alpha$-ketoglutarate dehydrogenase. (D) The deamination of serine by serine dehydratase.

of the alkanes requires their hydroxylation using molecular oxygen and an enzyme called *monooxygenase*.

### 9.4.1 Degradative pathways

#### Hydroxylation at one end

After the hydrocarbon dissolves in the membrane, it is hydroxylated at one end using molecular oxygen in a reaction catalyzed by a monooxygenase. Membrane-bound enzymes then oxidize the long-chain alcohol to the carboxylic acid (Fig. 9.22A). The carboxylic acid is derivatized with CoASH in an ATP-dependent reaction and oxidized via $\beta$-oxidation in the cytoplasm. There must exist an alternative mechanism that does not require oxygen because aliphatic hydrocarbons can be oxidized to $CO_2$ by certain sulfate-reducing bacteria in the absence of oxygen.[13] However, the mechanism is not known. In the sequence shown in Fig. 9.21A, a short electron

transport chain carries electrons from NADH via flavoprotein and an FeS protein to a b-type cytochrome called P-450.[14] The P-450 is in a complex with the hydrocarbon substrate and is called a cosubstrate. The reduced P-450 binds $O_2$ and activates it, presumably converting it to bound $O_2^-$ (superoxide) or $O_2^{2-}$ (peroxide), which attacks the C1 carbon on the hydrocarbon. One of the oxygen atoms replaces a hydrogen on the hydrocarbon and becomes a hydroxyl group, while the other oxygen atom is reduced to $H_2O$. Some bacteria use an iron protein called rubredoxin instead of P-450 to activate the oxygen. Electrons flow from NADH through a rubredoxin:NADH oxidoreductase to rubredoxin.

#### Hydroxylation on the penultimate carbon

An alternative degradative pathway is found in *Nocardia* species (Fig. 9.22B). The

233

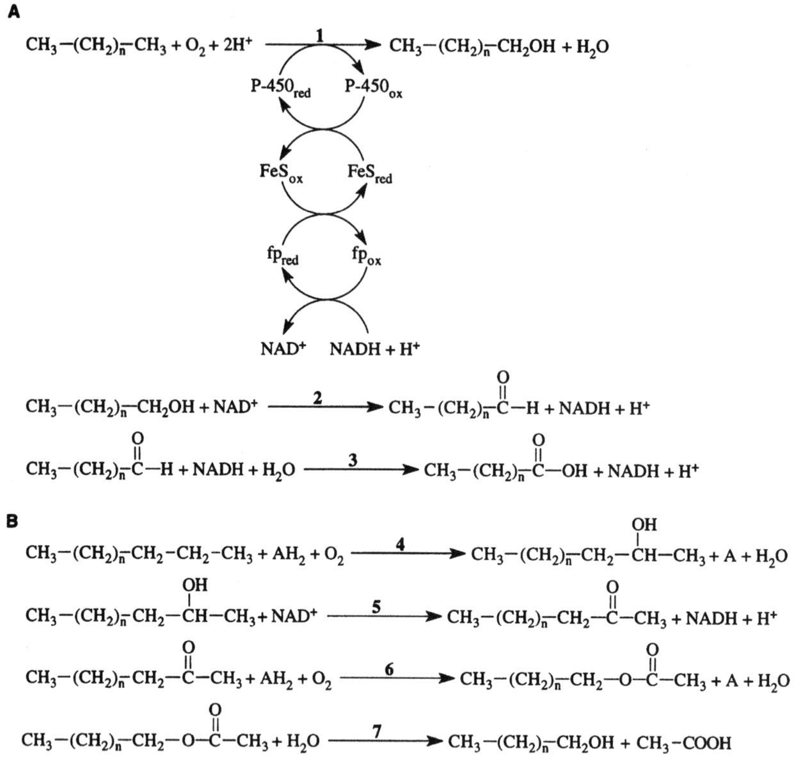

**A**

$$CH_3-(CH_2)_{\overline{n}}-CH_3 + O_2 + 2H^+ \xrightarrow{\quad 1 \quad} CH_3-(CH_2)_{\overline{n}}-CH_2OH + H_2O$$

P-450$_{red}$    P-450$_{ox}$

FeS$_{ox}$    FeS$_{red}$

fp$_{red}$    fp$_{ox}$

NAD$^+$    NADH + H$^+$

$$CH_3-(CH_2)_{\overline{n}}-CH_2OH + NAD^+ \xrightarrow{\quad 2 \quad} CH_3-(CH_2)_{\overline{n}}-\overset{O}{\overset{\|}{C}}-H + NADH + H^+$$

$$CH_3-(CH_2)_{\overline{n}}-\overset{O}{\overset{\|}{C}}-H + NADH + H_2O \xrightarrow{\quad 3 \quad} CH_3-(CH_2)_{\overline{n}}-\overset{O}{\overset{\|}{C}}-OH + NADH + H^+$$

**B**

$$CH_3-(CH_2)_{\overline{n}}-CH_2-CH_2-CH_3 + AH_2 + O_2 \xrightarrow{\quad 4 \quad} CH_3-(CH_2)_{\overline{n}}-CH_2-\overset{OH}{\overset{|}{CH}}-CH_3 + A + H_2O$$

$$CH_3-(CH_2)_{\overline{n}}-CH_2-\overset{OH}{\overset{|}{CH}}-CH_3 + NAD^+ \xrightarrow{\quad 5 \quad} CH_3-(CH_2)_{\overline{n}}-CH_2-\overset{O}{\overset{\|}{C}}-CH_3 + NADH + H^+$$

$$CH_3-(CH_2)_{\overline{n}}-CH_2-\overset{O}{\overset{\|}{C}}-CH_3 + AH_2 + O_2 \xrightarrow{\quad 6 \quad} CH_3-(CH_2)_{\overline{n}}-CH_2-O-\overset{O}{\overset{\|}{C}}-CH_3 + A + H_2O$$

$$CH_3-(CH_2)_{\overline{n}}-CH_2-O-\overset{O}{\overset{\|}{C}}-CH_3 + H_2O \xrightarrow{\quad 7 \quad} CH_3-(CH_2)_{\overline{n}}-CH_2OH + CH_3-COOH$$

**Fig. 9.22** Oxidation of an aliphatic hydrocarbon to a carboxylic acid. (A) Pathway in yeast and *Corynebacterium* species. The hydrocarbon is hydroxylated by oxygen, which is activated by cytochrome P-450. The second oxygen atom is reduced to water. Other bacteria, e.g., *Pseudomonas* species, use rubredoxin instead of P-450. In the latter case the electron transport is from NADH to rubredoxin:NADH oxidoreductase, to rubredoxin. The primary alcohol is oxidized to an aldehyde, which is then oxidized to a carboxylic acid. The carboxylic acid is oxidized via the β-oxidation pathway. (B) Pathway in *Nocardia* species. The C2 carbon is hydroxylated, forming a secondary alcohol, which is then oxidized to a ketone. A second monooxygenase reaction converts the ketone into an acetylester. The acetylester is hydrolyzed to acetate and a long-chain alcohol, which is oxidized as in (A). Enzymes: 1, 4, 6, monooxygenase; 2, 5 alcohol dehydrogenase; 3, 6, aldehyde dehydrogenase; 7, acetylesterase.

hydroxylation takes place on the second carbon, forming a secondary alcohol, which is then oxidized to form a ketone. Then, in an unusual reaction, a second monooxygenase reaction creates an acetylester, which is subsequently hydrolyzed to acetate and the long-chain alcohol. The alcohol is oxidized to the carboxylic acid, which is degraded via the β-oxidation pathway.

## 9.5 Summary

Many aerobic bacteria can grow on long-chain fatty acids. The major pathway for degradation is the β-oxidative pathway. First, the fatty acid is converted to the acyl–CoA derivative in an ATP-dependent reaction catalyzed by acyl–CoA synthetase. The first reaction is a displacement of pyrophosphate from ATP to form the acyl–AMP, which remains bound to the enzyme. Then the AMP is displaced by CoASH to form the acyl–CoA. The reaction is driven to completion by hydrolysis of the pyrophosphate catalyzed by pyrophosphatase. Thus the activation of fatty acids requires the equivalent of two ATP molecules. A keto group is then introduced β to the carboxyl via an oxidation, hydration, and oxidation to form the β-ketoacyl–CoA derivative. Finally, the β-ketoacyl–CoA derivative is cleaved with CoASH to yield acetyl–CoA. The process is repeated as two-carbon fragments are sequentially removed as acetyl–CoA from the fatty acid. If the fatty acid is an odd-chain fatty acid, then the last fragment is propionyl–CoA, which can be oxidized to pyruvate. Because fatty acids are degraded to acetyl–CoA, growth on fatty acids uses the same metabolic pathways as growth on acetyl–CoA.

Fatty acid synthesis occurs via an entirely different set of reactions. Acetyl–CoA is carboxylated to malonyl–CoA. A transacetylase then converts malonyl–CoA to malonyl–ACP. Acetyl–CoA is also transacylated to acetyl–ACP. Malonyl–ACP then condenses with acetyl–ACP (or acetyl–CoA, depending upon which synthase is used) to form a $\beta$-ketoacyl–ACP. The condensation with malonyl–ACP is accompanied by a decarboxylation reaction that drives the condensation. The $\beta$-ketoacyl–ACP is reduced via NADPH, dehydrated, and reduced again to form the saturated acyl–ACP. The acyl–ACP is elongated by condensation with malonyl–ACP.

There are two systems found among the bacteria for synthesizing unsaturated fatty acids: an anaerobic and an aerobic pathway. In the anaerobic pathway the $C_{10}$ hydroxyacyl–ACP is desaturated and elongated rather than being desaturated and reduced. In the aerobic pathway, which is found in eukaryotes and certain aerobic bacteria, unsaturated fatty acids are synthesized in a pathway requiring oxygen and NADPH. The double bond is introduced into the completed fatty acyl–ACP or CoA derivative by special desaturases (oxidases) that require $O_2$ and NAD(P)H as substrates. During the introduction of the double bond, four electrons are transferred to $O_2$ to form two $H_2O$. Two electrons are contributed by NAD(P)H, and two electrons are derived from the fatty acid when the double bond is formed. Bacteria that use this pathway make oleic acid rather than cis-vaccenic acid, which is made in the anaerobic pathway. They are both $C_{18}$ carboxylic acids, but in oleic acid the double bond is between C9 and C10, whereas in cis-vaccenic acid it is between C11 and C12.

The phosphoglycerides are synthesized from the fatty acyl–ACP derivatives and glycerol phosphate, the latter being the reduced product of dihydroxyacetone phosphate. The phosphoglycerides differ with respect to the group that is substituted on the phosphate. The major phospholipid in cell membranes from several bacteria is phosphatidylethanolamine. Substitution on the phosphate begins with the displacement of pyrophosphate from CTP by the phosphate on phosphatidic acid, the product being CDP-diglyceride. The pyrophosphate is hydrolyzed by a pyrophosphatase, which makes the pathway irreversible. CMP is then displaced by serine to form phosphatidylserine. A decarboxylation leads to phosphatidylethanolamine. Phosphatidylglycerol-phosphate is formed by displacing CMP from CDP-diacylglyceride with glycerol phosphate. A subsequent dephosphorylation produces phosphatidylglycerol. Diphosphatidylglycerol is made from two phosphatidylglycerol molecules. The transfer of the fatty acyl groups to the glycerol and all subsequent steps in phospholipid synthesis take place in the cell membrane. Archaeal phospholipids are synthesized from either dihydroxyacetone phosphate or glycerol phosphate. The precursor to the alcohol portion is geranylgeranyl-pyrophosphate, which forms an ether linkage to the glycerol backbone.

The ribose and deoxyribose moieties of the nucleotides are derived from 5-phosphoribosyl-1-pyrophosphate (PPRP), which is synthesized from ribose-5-phosphate and ATP. In this reaction, the OH on the C1 of ribose-5-phosphate displaces AMP from ATP to form the pyrophosphate derivative. It is therefore a pyrophosphoryl group transfer rather than the usual phosphoryl or AMP group transfer. The pyrophosphate itself is displaced from PPRP by orotic acid during pyrimidine synthesis or by an amino group from glutamine during purine biosynthesis. A pyrophosphatase hydrolyzes the pyrophosphate to inorganic phosphate, thus driving nucleotide synthesis. Whenever pyrophosphate is hydrolyzed to inorganic phosphate, the equivalent of two ATPs are necessary to restore both of the phosphates to ATP. Pyrimidine and purine biosynthesis differ in that the pyrimidines are made separately and then attached to the ribose phosphate, whereas the purine ring is built piece by piece while attached to the ribose phosphate. The deoxyribonucleotides are formed by reduction of the ribonucleotide diphosphates.

All the phosphates in the nucleotides are donated by ATP. The α phosphate is derived from ribose-5-phosphate, which in turn is synthesized from glucose-6-phosphate. In some bacteria (e.g., *E. coli*) the phosphate in glucose-6-phosphate may come from PEP during transport into the cell via the phosphotransferase system, whereas in other bacteria it is transferred to glucose from ATP via hexokinase or glucokinase. The β and γ phosphates are derived from kinase reactions in which ATP is the donor. The inorganic phosphate is incorporated into ATP via substrate-level phosphorylation or electron transport phosphorylation. Since the various nucleotide triphosphates provide the energy for the synthesis of nucleic acids, protein, lipids, and polysaccharides, as well as other reactions, it is clear that ATP fuels all the biochemical reactions in the cytosol.

Cells must synthesize both glutamate and glutamine in order to incorporate ammonia into cell material. The glutamate donates amino groups to the amino acids, and the glutamine donates amino groups to purines, pyrimidines, amino sugars, and some amino acids. Glutamine is synthesized by the ATP-dependent amination of glutamate to form glutamine. This is catalyzed by glutamine synthetase. The glutamine then donates the amino group to α-ketoglutarate to form glutamate. The latter reaction is catalyzed by glutamate synthase, also known as the GOGAT enzyme. Glutamate can also be synthesized by the reductive amination of α-ketoglutarate to glutamate by glutamate dehydrogenase, a reaction that requires high concentrations of ammonia.

The degradation of amino acids occurs by pathways different from the biosynthetic ones. The first step is the removal of the amino group either by an amino acid oxidase, an amino acid dehydrogenase, or a deaminase. The carbon skeleton eventually enters the citric acid cycle.

Hydrocarbon catabolism begins with a hydroxylation to form the alcohol, which is oxidized to the carboxylic acid. The carboxylic acid is degraded via the β-oxidative pathway. Some bacteria use an alternative pathway to initiate hydrocarbon degradation, in which the hydrocarbon is oxidized to a ketone, which is eventually hydrolyzed to acetate and a long-chain alcohol.

## Study Questions

1. Fatty acid synthesis is not simply the reverse of oxidation. What features distinguish the two pathways from each other?

2. What is it about the structure of phospholipids that causes them to form bilayers spontaneously?

3. Glycerol can be incorporated into phospholipids. Write a pathway showing the synthesis of phosphatidic acid from glycerol.

4. The C3 of glycerol becomes the C2 and −8 of purines. Write a sequence of reactions showing how this can occur.

5. Show how three carbons of succinic acid can become C4, C5, and C6 of pyrimidines. What happens to the fourth carbon from succinate?

6. What drives the condensation reaction in fatty acid synthesis?

7. Write a reaction sequence by which bacteria incorporate the nitrogen from ammonia into glutamate and glutamine when ammonia concentrations are low. How might this occur when ammonia concentrations are high? What is the fate of the nitrogen incorporated into glutamate, and that incorporated into glutamine?

8. ATP drives the synthesis of phospholipids. Write the reactions showing the incorporation of glycerol into phosphatidyl serine. Focus on those steps that require a high-energy donor. How is ATP involved? (You must account for the synthesis of CTP.)

9. Bacteria that utilize aliphatic hydrocarbons as a carbon and energy source are usually aerobes. What is the explanation for the requirement for oxygen?

## NOTES AND REFERENCES

1. Textor, S., V. F. Wendisch, A. A. De Graff, U. Müller, M. I. Linder, D. Linder, and W. Buckel.

1997. Propionate oxidation in *Escherichia coli*: Evidence for operation of a methylcitrate cycle in bacteria. *Arch. Microbiol.* **168**:428–436.

2. Gerike, U., D. W. Hough, N. J. Russell, M. L. Dyall-Smith, and M. J. Danson. 1998. Citrate synthase and 2-methylcitrate synthase: Structural, functional and evolutionary relationships. *Microbiology* **144**:929–935.

3. Magnunson, K., S. Jackowski, C. O. Rock, and J. E. Cronan, Jr. 1993. Regulation of fatty acid biosynthesis in *Escherichia coli*. *Microbiol. Rev.* **57**:522–542.

4. CoASH = P-AMP-pantothenic acid-$\beta$-mercaptoethylamine-SH. Acyl carrier protein = protein-pantothenic acid-$\beta$-mercaptoethylamine-SH. Coenzyme A has a phosphorylated derivative of AMP (AMP-3'-phosphate) attached via a pyrophosphate linkage to the vitamin pantothenic acid (a $B_2$ vitamin), which is covalently bound to $\beta$-mercaptoethylamine via an amide linkage. The $\beta$-mercaptoethylamine provides the SH group at the end of the molecule. In the acyl carrier protein, the AMP is missing and the pantothenic acid is bound directly to the protein. Therefore, the functional end of the acyl carrier protein is identical to CoASH, but the end that binds to the enzymes is different.

5. Acetyl–CoA carboxylase consists of four subunit proteins. These are biotin carboxylase, biotin carboxyl carrier protein, and two proteins that carry out the transcarboxylation of the carboxy group from biotin to acetyl–CoA.

6. Whether there is a separate acetyl–CoA:ACP transacetylase is not certain. Reaction 3 can be catalyzed by $\beta$-ketoacyl–ACP synthase III (acetoacetyl-ACP synthase), which has transacetylase activity.

7. There are actually three synthases and three possible routes to acetoacetyl–ACP in *E. coli*. In one pathway, $\beta$-ketoacyl–ACP synthase III catalyzes the condensation of acetyl–CoA with malonyl–ACP. A defect in synthase III leads to overproduction of 18-carbon fatty acids, whereas overproduction of synthase III leads to a decrease in the average chain lengths of the fatty acids synthesized. The decrease in the fatty acid chain lengths in strains overproducing synthase III has been rationalized by assuming that synthase III is primarily active in the initial condensation of acetyl–CoA and malonyl–ACP, and that the increased levels of synthase III stimulate the initial condensation reaction and divert malonyl–ACP from the terminal elongation reactions. In a second initiation pathway acetyl–ACP is a substrate for $\beta$-ketoacyl–ACP synthase I or II. In a third pathway, which is not thought to be physiologically significant under most growth conditions, malonyl–ACP is decarboxylated by synthase I, and the resultant acetyl–ACP condenses with malonyl–ACP. The different synthases appear to be involved in determining the types of fatty acids that are made. In particular, synthases I and II, which catalyze condensations in both saturated and unsaturated fatty acid synthesis, appear to have specific roles in the synthesis of unsaturated fatty acids. For example, mutants of *E. coli* that lack synthase I do not make any unsaturated fatty acids, suggesting that only synthase I is capable of catalyzing the elongation of *cis*-3-decenoyl–ACP. Mutations in synthase II lead to an inability to synthesize *cis*-vaccenate, in agreement with the finding that synthase II can elongate palmitoleoyl–ACP but synthase I cannot.

8. Mammals require linoleate (18:2 *cis*-$\Delta^9$, $\Delta^{12}$) and linoleate (18:3 *cis*-$\Delta^9$, $\Delta^{12}$, $\Delta^{15}$). Hence these are essential fatty acids and must be supplied in the diet of mammals. The reason why mammals cannot synthesize these fatty acids is that mammals do not have the enzymes to introduce double bonds in fatty acids longer than C9. Arachidonate, a C20 fatty acid with four double bonds can be synthesized by mammals from linolenate. Arachidonate, in turn, is a precursor to various signaling molecules such as prostaglandins.

9. Naturally occurring fatty acids are mostly *cis* with respect to the configuration of the double bond (i.e.,

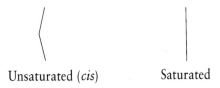

This *cis* configuration produces a bend of about 30° in the chain, whereas the *trans* configuration is a straight chain, as is the saturated.

Unsaturated (*cis*)    Saturated

10. Heath, R. J., S. Jackowski, and C. O. Rock. 1994. Guanosine tetraphosphate inhibition of fatty acid and phospholipid synthesis in *Escherichia coli* is relieved by overexpression of glycerol-3-phosphate acyltransferase (*plsB*). *J. Biol. Chem.* **266**:26584–26590.

11. In addition to the carbamoyl synthetase described here, verterbrates have a carbamoyl synthetase that combines $NH_4^+$, $CO_2$, 2 ATP, and $H_2O$ to form carbamoyl phosphate, which is used to convert $NH_4^+$ to urea in the urea cycle.

12. In mammals the amino group is donated from the amide group of glutamine. In *E. coli* the enzyme can use either $NH_4^+$ or glutamine. In both cases the requirement of ATP can be

explained by the formation of a phosphate ester with the carbonyl oxygen. The amino group then displaces the phosphate.

13. Aeckersberg, F., F. Bak, and F. Widdel. 1991. Anaerobic oxidation of saturated hydrocarbons to $CO_2$ by a new type of sulfate-reducing bacterium. *Arch. Microbiol.* **156**:5–14.

14. Munro, A. W., and J. G. Lindsay. 1996. Bacterial cytochromes P-450. *Molec. Microbiol.* **20**:1115–1125.

# 10

# Macromolecular Synthesis

From a biochemical point of view, nucleic acids and proteins are polymerized by donating the subunit (i.e., a nucleotide or amino acid) from a donor with a high group transfer potential to the growing chain via a nucleophilic displacement reaction. (See Section 7.2.) This condensation reaction is generally referred to as chain elongation. For nucleic acids the donors of the nucleotide monophosphates, (d)NMP, are the respective nucleotide triphosphates, (d)NTP. During the condensation reaction in nucleic acid biosynthesis the $\alpha$ phosphate in (d)NTP is attacked by the 3′ hydroxyl group on ribose or deoxyribose (at the 3′ end of the growing polynucleotide), and the released pyrophosphate ($PP_i$) is subsequently hydrolyzed by a pyrophosphatase. (See Figs. 7.3 and 7.7.) This pulls the reaction to completion. These condensation reactions are discussed in Sections 7.2.1 and 7.2.2. For protein synthesis the donors of the amino acids are the aminoacylated t-RNAs (Fig. 7.8). However, because of specific requirements that have to do with the structure of DNA as well as the need to use nucleic acid as a template during transcription (RNA synthesis) and translation (protein synthesis), the biosynthesis of DNA, RNA, and protein involves much more than simply the condensation reactions. This chapter discusses important features of the biosynthesis of DNA, RNA, and protein, including metabolic regulation of the pathways.

## 10.1 DNA Replication and Partitioning

### 10.1.1 Semiconservative replication

DNA consists of two strands wound around each other in a right-handed double helix (Fig. 10.1). The strands are held together by base pairing between A–T and G–C residues (Fig. 10.2). DNA is replicated via semiconservative replication. This means that each

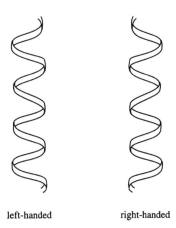

left-handed          right-handed

**Fig. 10.1** The direction of helical turns. In order to determine whether the molecule is in a right- or left-handed helix, sight down the molecule from one end. The right-handed helix turns clockwise, and the left-handed helix turns counterclockwise.

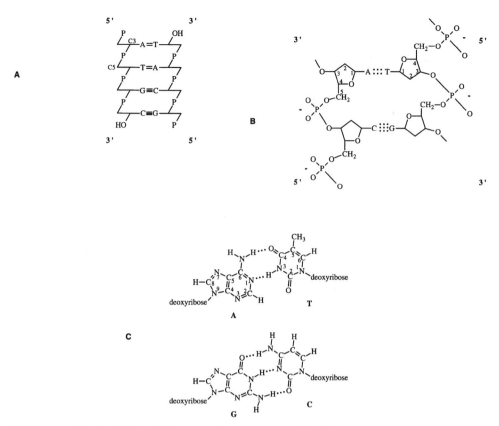

**Fig. 10.2** Base pairing in DNA. (A) Two complementary strands of the double helix are of opposite polarity and held together by hydrogen bonds between A–T and G–C pairs. The deoxyribose moieties are attached via phosphodiester linkages between the C3 hydroxyl of one sugar and the C5 hydroxyl of another. (B) A more detailed examination showing the structure of the deoxyribose connected by phosphodiester linkages. The phosphate–oxygen double bond is drawn as a semipolar bond because of the high electronegativity of oxygen and the low propensity of phosphorus to form double bonds. (C) The structures of the bases showing the hydrogen bonds. Note that two hydrogen bonds hold the A–T base pairs together, whereas three hydrogen bonds hold the G–C base pairs together. A, adenine; T, thymine; G, guanine; C, cytosine; P, phosphate.

strand acts as a template for the synthesis of a daughter strand. Semiconservative replication was demonstrated by the Meselson and Stahl experiment described in Fig. 10.3.

### 10.1.1 The topological problem

DNA replication is a complex topological problem because the DNA in a typical bacterium exists as a covalently closed circle of a right-handed double helix that may be 500–600 times longer than the cell and is tightly folded into supercoiled loops. (However, not all bacteria have circular chromosomes.[1, 2]) *Supercoiling* refers to the twisting of the DNA double helix around its central axis (Fig. 10.4). To visualize supercoiling, think of a telephone cord wound into secondary coils. (See Note 3 for a further explanation of supercoiling.) Pulling apart of the duplex strands at the replication fork in the closed circle makes the unreplicated portion ahead of the replication fork twist so that the helix becomes overwound and more tightly coiled into a positive supercoil (Fig. 10.5). Unless something were done about this, the unreplicated portion of the DNA helix would become bunched up in positive supercoils, and further unwinding would stop. As described next, DNA gyrase solves the problem.

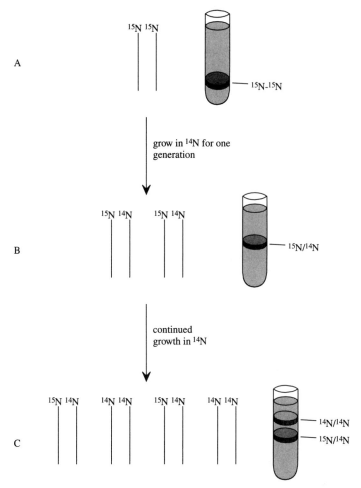

**Fig. 10.3** The Meselson–Stahl experiment showing that DNA is replicated semiconservatively. (A) Cells are grown in $^{15}N$ for many generations so that all the nitrogen in the DNA is heavy ($^{15}N$). The DNA is isolated and centrifuged to equilibrium in a cesium chloride (CsCl) density gradient that separates the molecules according to their density. If both strands have $^{15}N$, then the duplex DNA sediments to a position near the bottom of the tube. (B) The $^{15}N$ cells are then grown in $^{14}N$ (light) media, and after one generation the DNA is isolated and centrifuged in the CsCl gradient. Semiconservation replication predicts that "hybrid" DNA would be formed, one strand being labeled with $^{15}N$ and the other strand with $^{14}N$. The $^{15}N/^{14}N$ DNA occupies a position in the gradient higher than the $^{15}N/^{15}N$ DNA. C. Further growth in $^{14}N$ results in the formation of "light" DNA, i.e., $^{14}N/^{14}N$, which occupies the highest position in the gradient.

## Positive versus negative supercoiling

Positively supercoiled DNA has more than 10.5 base pairs per helical turn (i.e., it is overwound), and negatively supercoiled DNA has fewer than 10.5 base pairs per helical turn (i.e., it is underwound). The DNA duplex itself is a right-handed helix, and therefore positive supercoils are twisted in the same direction as the helix (overwound). If the twist of the coil is counterclockwise (left-handed), then it is negatively supercoiled (opposite to the helix) and underwound.

## DNA gyrase solves the problem of overwinding

The problem of overwinding the double helix during DNA replication is solved using an enzyme called topoisomerase II (DNA gyrase). DNA gyrase continuously removes the positive supercoils (overwound DNA)

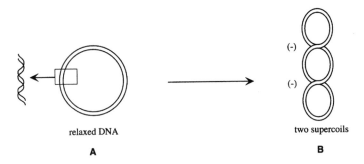

relaxed DNA

**A**

two supercoils

**B**

**Fig. 10.4** Supercoiled DNA. When there are 10.5 base pairs per helical turn, the DNA is relaxed (A). Native DNA is generally underwound; i.e., there are fewer than 10.5 base pairs per helical turn. This introduces a strain in the molecule, and as a result it twists upon itself to form supercoils in order to reduce the strain (B). Supercoiling resulting from underwound DNA is referred to as negative supercoiling. Supercoils will also form if DNA is overwound, i.e., more than 10.5 base pairs per turn. Supercoiling due to overwound DNA is called positive supercoiling. In negative supercoiling the DNA is twisted in a direction opposite to that of the right-handed double helix, and in positive supercoiling the DNA is twisted in the same direction as the right-handed double helix.

that form in the unreplicated DNA ahead of the replication fork and converts them into negative supercoils (underwound DNA). This is probably advantageous, since the duplex must unwind in order that DNA replication and RNA transcription take place. As shown in Fig. 10.6, DNA gyrase works by making a double-stranded break in the DNA, passing an unbroken portion through the gap, and resealing the break. Note 4 provides more information about topoisomerases and how they work.

*More problems to be solved*

Not only must the DNA be unwound and copied rapidly (in *E. coli* about 1000 nucleotides per second must be polymerized so that the chromosome is replicated in about 40 min), but the strands must be separated from each other without getting entangled, and finally, they must be partitioned into the daughter cells, a process that is not completely understood. At least 30 different proteins are required to replicate the DNA in *E. coli*. The following discussion describes the replication process and partitioning of bacterial DNA.[5–8]

### 10.1.2 Creating the replication fork

DNA replication takes place at a replication fork, which must be created each time

replication of the chromosome is initiated. (The student should review Section 2.2.3 for a discussion of the relationship between the timing of initiation of DNA replication and the growth rate.) This section explains how that is done. The first thing that must happen in order to make the replication fork is that the strands must unwind so that each strand can act as a template. The unwinding does not begin just any place but rather at a particular site in the DNA duplex termed the *origin*. (In *E. coli* this is called the *oriC* locus.) When the duplex is unwound, a Y is created where the arms of the Y are single stranded because the duplex has become unwound there, but the region downstream of the juncture where the two arms come together is still double stranded (Fig. 10.7). The juncture is called the *replication fork*. DNA is usually replicated bidirectionally; that is, there are two replication forks proceeding in opposite directions (Fig. 10.8). (Note 9 explains how this can be determined.) This has the advantage of halving the time it takes to replicate the DNA molecule and generally takes place with phages, plasmids, bacteria, and eukaryotic cells. The detailed steps for how the replication forks are created are described next.

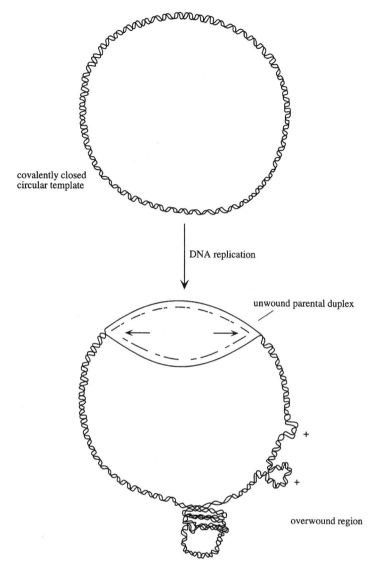

covalently closed
circular template

DNA replication

unwound parental duplex

+

+

overwound region

**Fig. 10.5** Supercoiling ahead of the replication fork. As the template strands in the closed circle are pulled apart, the duplex ahead of the replication fork overwinds as positive supercoils are formed. The twists in the overwound regions are removed by DNA gyrase, which produces negative supercoils and underwinds the duplex.

## Unwinding the duplex: Creation of the prepriming complex

The prepriming complex is formed first, and it is created in two stages.[8] In Stage 1 the *open complex* is formed, and in Stage 2 the open complex develops into the *prepriming complex*. DNA synthesis, which will be described later, actually begins with the prepriming complex.

### 1. Formation of the open complex

Creation of the open complex is initiated with ATP and two DNA binding proteins, DnaA and a histonelike protein called HU, at the origin of replication (*oriC*) (Fig. 10.9). (Two other proteins that will not be discussed, i.e., IHF and FIS, also bind to *oriC* and may help DnaA to unwind the duplex. See the discussion of these proteins and the nucleoid in Section 1.2.6.) Within the origin (*oriC*) there are five

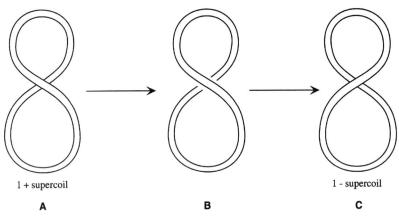

1 + supercoil

1 - supercoil

A                    B                    C

**Fig. 10.6** Action of DNA gyrase. DNA gyrase converts positive supercoils into negative supercoils. (A) A positively supercoiled node, i.e., the duplex is twisted around its central axis. For example, this happens during DNA replication as the duplex is being unwound by helicase at the replication fork and the duplex ahead of the replication fork spontaneously becomes overwound by being twisted in a clockwise direction. (B) Both strands are cut by DNA gyrase, and then the uncut portion is passed by the gyrase through the gap and the gap is sealed. (C) The duplex is now twisted in the opposite direction; i.e., it is negative supercoiled and underwound. DNA gyrase is an ATP-dependent enzyme. DNA gyrase is sometimes referred to as providing a "swivel" that allows the replication fork to continue.

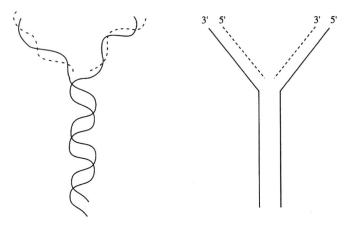

**Fig. 10.7** A replication fork. The two template strands have become unwound and are being copied. The unwinding is caused by DnaB protein, which is the replication fork helicase (not shown) and requires ATP. Note that there are single-stranded regions near the fork. A DNA-binding protein, called SSB, binds to the single-stranded regions, preventing them from coming together (not shown).

sites, called Dna boxes, that are recognized by DnaA. (See Note 10.) Approximately 30 DnaA molecules bind to the sites as the DNA wraps around a core of DnaA molecules. Then, in an ATP-dependent reaction that is aided by HU, the adjacent A + T-rich region at the 5' end of the origin sequence unwinds to form the 45-bp *open complex*. However, something must be done to prevent the single strands from coming together again to reform the duplex. This is the task of single-stranded binding proteins (SSB), which coat the strands.

*2. Formation of the prepriming complex*
The open complex unwinds into the *prepriming complex*. The unwinding is performed by a protein called helicase (DnaB), which must

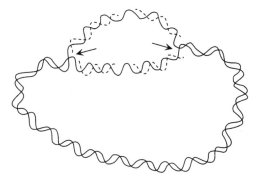

**Fig. 10.8** Bidirectional replication of DNA. In most organisms and viruses the DNA is replicated bidirectionally. This means that there are two replication forks proceeding in opposite directions (arrows). This reduces the time required to replicate the DNA. The template strands are the solid lines, and the daughter strands are the broken lines.

first bind to the DNA. However, DnaB does not bind to the DNA on its own but must be transferred to the open complex from a DnaB:DnaC:ATP complex. After binding to the DNA, DnaB further unwinds the strands bidirectionally to form the prepriming complex, with two replication forks ready for the initiation of DNA replication.

## Timing of initiation

Precisely what determines the timing of initiation of bacterial DNA replication is still a matter of speculation. As discussed in Section 2.2.3, the cell mass per *chromosomal* replication origin (as opposed to plasmid replication origins) at the time of initiation is constant, and all *oriC* origins (even plasmid origins) begin replication at the same time. This mass is called the initiation mass or the initiation volume, and some have suggested that it corresponds to the activity of DnaA. Another question is, what prevents the newly replicated origins from reinitiating in the same cell cycle? For a discussion of this last point, as well as what regulates the activity of DnaA, the sequestration of *oriC* in the membrane fraction, the role of methylation, and other aspects of the control of initiation timing, see Note 11, and the article by Messer and Weigel as well as the review by Funnel.[7, 12]

## More about the replication fork helicase (DnaB)

The replication fork helicase (DnaB) breaks the hydrogen bonds between the base pairs and moves along the DNA as it continues to unwind the duplex. Proteins such as the helicases that move along the DNA are said to be *processive*, and the farther they move while functioning before detaching, the higher the processivity. DnaB is highly processive. (Other processive enzymes include RNA polymerase and DNA polymerase III.) DnaB binds to the lagging strand template and moves in the 5′ to 3′ direction, displacing the leading strand template. The energy to unwind the duplex and to move is derived from the hydrolysis of ATP. (For a further description of DNA helicases and the unwinding process, refer to Refs. 13 and 14, and Note 15.)

## The necessity for DNA gyrase (topoisomerase II)

Because the ends of the closed circular DNA duplex are not free to rotate, the two strands of the duplex cannot unwind by freely rotating around each other. Thus, when the helicase pulls the strands apart at the replication fork, the duplex downstream of the replication fork winds tighter and tighter into *positive* supercoils, one for each turn of the unwound helix (Fig. 10.5). Unless something were done about this, the DNA would become wound so tightly in positive supercoils that further unwinding of the double helix would quickly stop. To see the magnitude of the problem, consider that *E. coli* replicates its DNA at a rate of approximately 1000 nucleotides per second. Given that there are about 10.5 base pairs per helical turn, this means about 95 turns are unwound per second or 95 supercoils would be introduced each second! The role of the gyrase is to change the positive supercoils into *negative* supercoils (underwound DNA) at the expense of ATP to relieve the torsional stress in the DNA that results from the unwinding activity of DnaB. Thus unwinding of the duplex requires DNA gyrase in addition to helicase, SSB, and ATP.

Now we can consider how the single-stranded DNA templates at the replication forks are copied into daughter strands of

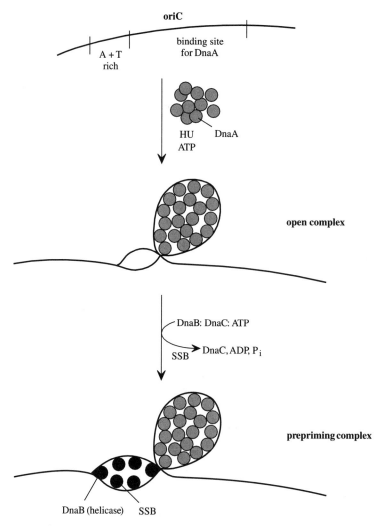

**Fig. 10.9** Creation of the prepriming complex. A multimer of about 20 DnaA proteins recognize specific sequences and binds to the duplex DNA at the origin adjacent to an A + T–rich region. The A + T-region adjacent to the DnaA proteins then unwinds to form the *open complex*. This is aided by the HU protein and requires ATP. DnaB protein (helicase) then binds in a DnaC-dependent reaction and further unwinds the duplex. Single-stranded DNA-binding proteins (SSB) bind to the single strands behind the helicase, preventing the single strands from coming together again.

DNA. As described next, this requires RNA polymerase (also called DnaG or primase) and DNA polymerase.

### 10.1.3 Replicating the DNA

#### DNA polymerases

The enzymes that synthesize DNA are called DNA polymerases. Two important characteristics of DNA polymerases are (1) they can only *extend* nucleic acid chains (i.e., they

cannot initiate new ones), and (2) they add mononucleotides to the 3′ OH of deoxyribose and therefore elongate nucleic acid only at the 3′ end. *E. coli* has three DNA polymerases. These are: DNA polymerase I encoded by *polA*, DNA polymerase II encoded by *polB*, and DNA polymerase III encoded by *polC* (*dna*E). Here is what they do:

1. DNA polymerase III replicates DNA at the fork.

2. DNA polymerase II is not important for

DNA replication, but does function in the repair of damaged DNA.

3. DNA polymerase I does have a role to play during DNA replication, which will be discussed later. (See the section on attaching the Okazaki fragments to each other.) DNA polymerase I is also important for DNA repair (e.g, see Sections 19.2.3 and 19.2.5).

## The problem in synthesizing strands of opposite polarity at the replication fork

At the replication fork the DNA templates are antiparallel, that is, one strand is 5' to 3', and the other strand is 3' to 5'. In other words, the 5' phosphate end of one strand is paired with the 3' hydroxyl end of the other strand (Fig. 10.7). Since the DNA copy strand is antiparallel to the template strand, this means that the new strands of DNA must be of opposite polarity too (i.e., one is 3'–5' whereas the other is 5'–3'). That is to say, at the replication fork one of the copy strands has its 3' end pointed in the direction of movement of the fork, whereas the other copy strand has its 3' end pointed away from the fork. This means that one strand must be synthesized in the direction that the replication fork is moving, whereas the other strand must be synthesized in the opposite direction.

## Okazaki fragments

DNA polymerase not only elongates both strands at their 3' ends but remains with the replication fork. How can the polymerase manage to remain at the replication fork and yet synthesize a strand of DNA whose 3' end keeps moving farther and farther away from the replication fork? As will be explained next, the answer lies in synthesizing the strand whose 3' end faces away from the fork in short fragments about 1000 nucleotides long called Okazaki fragments, after the investigator who discovered them, and then returning to the fork to initiate replication of another short fragment (Fig. 10.10). (See Note 16 for a description of the Okazaki experiments.) In this way the polymerase remains at least within 1000 nucleotides of the replication fork. The strand copied in short fragments is

called the *lagging strand* template. The other strand, which is copied in the direction of the replication fork, is synthesized in one piece and is called the *leading strand* template.

## Synthesis of leading strand

Recall that DNA polymerase cannot initiate new polynucleotide strands but can only elongate the 3' end of an existing strand. This means that the initiation of DNA replication must begin with the synthesis of a a relatively short RNA primer (5 to 60 nucleotides) by an RNA polymerase. It is not known whether the RNA polymerase that synthesizes the primer for the leading strand is the same RNA polymerase that synthesizes cellular RNA, or whether it is the RNA polymerase, called *primase* (DnaG), that synthesizes the RNA primer for the lagging strand. (See Note 17.) DNA polymerase III then extends the RNA primer at its 3' end and continues the elongation to complete the DNA as the replication fork moves forward (Fig. 10.10). The DNA polymerase that synthesizes the leading strand is said to be highly *processive* (i.e., it moves with the replication fork as it adds nucleotides and dissociates rarely, if at all).

## Synthesis of the lagging strand

The opposite template strand is called the *lagging* template strand. It cannot be copied processively in the same way as the leading template strand is copied because it is of opposite polarity and the 3' end of the RNA primer (i.e., the growing end), is facing away from the replication fork. What happens is that the polymerase periodically disengages from the template strand to return to a newly synthesized primer at the replication fork, thus synthesizing short polynucleotide fragments, the Okazaki fragments (Fig. 10.10).

## The primosome undwinds the DNA at the replication fork and synthesizes the RNA primer for the lagging strand

Replication begins at the replication fork with the synthesis of an RNA primer by the primase, DnaG (Fig. 10.11). The primase is associated with the helicase (DnaB) to form a mobile complex called the *primosome* that

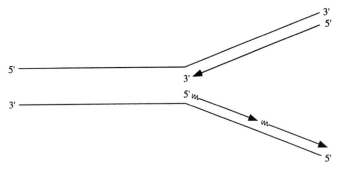

**Fig. 10.10** Synthesis of leading and lagging strands. Both template strands must be copied in the 3′ to 5′ direction so that the copy strands, which grow at the 3′ ends (arrows) are of opposite polarity to their respective templates. Since the template strands are of opposite polarity, the copy strands must be synthesized in opposite directions; i.e., one strand in the direction of replication fork movement, and the other strand in the opposite direction. However, it is known that the DNA polymerase, which is at the 3′ end of the growing strand, remains at the replication fork. The strand whose 3′ end faces the replication fork is synthesized by a polymerase that stays attached to the template and moves with the replication fork. This template is called the leading-strand template, and the strand being synthesized is called the leading strand. The strand whose 3′ end faces away from the replication fork is synthesized in short fragments called Okazaki fragments, which are about 1000 nucleotides long. Upon completion of an Okazaki fragment the DNA polymerase leaves the template and returns to the replication fork to synthesize another fragment. The template that is copied in short fragments is called the lagging-strand template, and the strand being synthesized is called the lagging strand. Each Okazaki fragment begins with the synthesis of an RNA primer that is subsequently elongated by the DNA polymerase. The RNA primers are drawn as wavy lines. Eventually, the RNA primers are removed, and the fragments are elongated by DNA polymerase and sealed. See text for details.

stays with the replication fork as the DNA is replicated. The primase must synthesize the primer in the direction opposite to the movement of the replication fork. (The primosome consists of several other proteins in addition to the primase and helicase, some of which function in primosome assembly.) The DnaB helicase unwinds the duplex. As this happens, the primosome moves forward with the replication fork. Then the DNA polymerase III elongates the RNA primer at its 3′ end to synthesize the Okazaki fragment (about 1000 nucleotides). Once every second or two, the DNA polymerase *disengages* from the template and returns to the replication fork to synthesize a new Okazaki fragment.

### A model to explain how the same DNA polymerase can synthesize both the leading and lagging strands

It is believed that the same polymerase synthesizes both the leading and lagging strand, but this cannot take place at the same time if the polymerase moves away from the replication fork for 1000 nucleotides while it synthesizes the lagging strand. One suggestion as to how the polymerase might stay with the replication fork and still synthesize the lagging strand postulates that the lagging strand template loops around the polymerase at the replication fork so that its 5′ end faces the replication fork rather than away from it (Fig. 10.11). If this were to occur, then the Okazaki fragments can be elongated at their 3′ ends in the direction of the replication fork and a dimeric DNA polymerase III could elongate both the leading and lagging strands at the same time. According to the model, after polymerizing approximately 1000 nucleotides, the polymerase would disengage from the lagging strand template and a new loop would form.

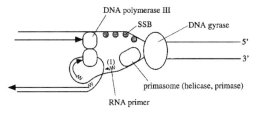

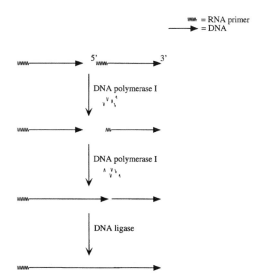

**Fig. 10.11** Replication fork. The model explains how DNA polymerase might synthesize the lagging strand and stay with the replication fork. It is suggested that the lagging-strand template loops around a dimeric polymerase so that the polymerase moves along it in the direction of the replication fork extending an RNA primer (wavy line) on the lagging strand and elongating the leading strand. Another RNA primer, labeled (1), is made by primase ahead of the polymerase at the replication fork. The polymerase keeps moving on the template until it encounters the 5′ end of a previously made Okazaki fragment (wavy line). Upon encountering the previously made Okazaki fragment, the polymerase disengages from the template and reinitiates at the new RNA primer with a new loop.

**Fig. 10.12** Attaching DNA fragments to each other in the lagging strand. When the growing Okazaki fragment reaches the 5′ end of the previously synthesized Okazaki fragment, then DNA polymerase III dissociates from the template DNA. DNA polymerase I then removes the RNA via 5′-3′ exonuclease activity from the 5′ end of the previously synthesized Okazaki fragment and replaces the RNA with DNA. DNA ligase then seals the break in the DNA. RNA is the wavy line; DNA is the straight line.

## Attaching the Okazaki fragments to each other

When DNA polymerase III reaches the RNA at the 5′ end of the previously synthesized segment of DNA in the lagging strand, it stops and leaves the 3′ end of the newly synthesized DNA. The result is a nick in the DNA strand between the DNA and RNA. The short RNA primer fragments (10–12 nucleotides long) in the Okazaki fragments are removed by the 5′ to 3′ activity of DNA polymerase I, and then the gap is filled with deoxyribonucleotides polymerized by DNA polymerase I. The formation of the phosphodiester bond between the 3′-hydroxyl at the end of one strand of DNA and the 5′-phosphate of the other strand after the gap is filled is catalyzed by *DNA ligase*. The overall scheme is shown in Fig. 10.12. The way that the ligase makes the covalent bond between the 3′ phosphate of the newly synthesized DNA and the 3′ hydroxyl of the previously synthesized DNA is to transfer AMP from either ATP or NAD$^+$ (depending upon the organism) to the phosphate (Fig. 10.13). This produces an AMP derivative with a high group transfer potential. The ligase then catalyzes the displacement of the AMP by the 3′ hydroxyl, and the phosphodiester bond is formed.

### 10.1.4 Termination

#### Ter sites and Tus protein

In *E. coli* the two replication forks meet in a region of the chromosome called the termination region. When the replication forks meet, replication stops and the daughter chromosomes are separated. The termination region contains specific sequences of nucleotides called Ter sequences (Fig. 10.14). The Ter sites are quite unusual because they allow the replication fork to pass through in only one direction! This can account for termination at the Ter sites. For example, assume a situation where

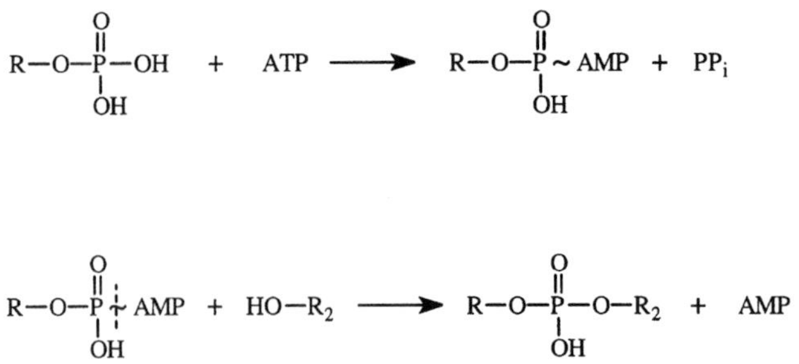

$$R-O-\overset{\overset{\displaystyle O}{\|}}{\underset{\underset{\displaystyle OH}{|}}{P}}-OH \quad + \quad ATP \quad \longrightarrow \quad R-O-\overset{\overset{\displaystyle O}{\|}}{\underset{\underset{\displaystyle OH}{|}}{P}}\sim AMP \quad + \quad PP_i$$

$$R-O-\overset{\overset{\displaystyle O}{\|}}{\underset{\underset{\displaystyle OH}{|}}{P}}\vdots AMP \quad + \quad HO-R_2 \quad \longrightarrow \quad R-O-\overset{\overset{\displaystyle O}{\|}}{\underset{\underset{\displaystyle OH}{|}}{P}}-O-R_2 \quad + \quad AMP$$

**Fig. 10.13** DNA ligase reaction. The enzyme catalyzes the adenylylation of the 5′-phosphate. Depending upon the ligase, the AMP can be derived either from ATP, in which case pyrophosphate is displaced, or NAD$^+$, in which case nicotinamide monophosphate (NMN) is displaced. *E. coli* DNA ligase uses NAD$^+$ as the AMP donor. Then the 3′-hydroxyl of the ribose attacks the AMP derivative and displaces the AMP. The result is a phosphodiester bond.

there are two Ter sites and these are located next to each other, with perhaps a short segment of DNA between them. Suppose Ter site 1 allows the replication fork moving clockwise to pass through but not the replication fork moving counterclockwise, and that Ter site 2 allows counterclockwise movement but not clockwise movement. Thus, if the clockwise moving replication fork gets to site 2, it will stall, and if the counterclockwise fork gets to site 1, it will stall. Consequently, the replication forks will meet at site 1 or site 2 or somewhere between them. As shown in Fig. 10.14, the situation in *E. coli* is a little bit more complicated because it has six Ter sites. The six Ter sites are divided into two groups of three. One group of three (i.e. Ter A, D, and E) prevents counterclockwise rotation of the fork, and the second group of three (i.e., Ter C, B, and F) prevents clockwise rotation of the fork. As explained in the legend to Fig. 10. 14, the presence of three Ter sites per fork direction provides a backup in case the fork gets by one of the Ter sites. A protein, which in *E. coli* is called the Tus protein (terminator utilization substance) binds to the Ter site and imposes the one-way travel. Tus does this by inhibiting the replicative helicase (DnaB).[18–20] Ter sites and the Ter-binding protein have been well studied in *B. subtilis*. See Note 21 for similarities and differences with respect to *E. coli*.

### 10.1.5 Chromosome partitioning

#### Overview

Chromosome partitioning refers to the separation and segregation of daughter chromosomes (nucleoids) to opposite cell poles in preparation for cell division. The experimental data are derived primarily from studies with *E. coli*, *B. subtilis*, and *Caulobacter crescentus*. (See Ref. 22 for a review.)

Chromosomal *separation* refers to the *detachment* of the newly replicated chromosomes from one another. If the linkage between the daughter chromosomes is noncovalent, then topoisomerases such as topoisomerase IV separate the daughter chromosomes. (See the section on topoisomerase IV later.) If an unequal number of recombinations have occurred between the daughter chromosomes, then the daughter chromosomes are covalently linked, and a site specific recombination must occur. (See the section on site-specific recombination at dif later.)

Chromosomal *segregation* refers to the *movement* of the detached chromosomes to opposite poles of the cell prior to cell division. Much has been learned about the separation, but relatively little is known about segregation. Segregation seems to involve several different proteins. Two of these proteins will be described later. They are the Par proteins and the Muk proteins. In some bacteria the Par proteins attach to

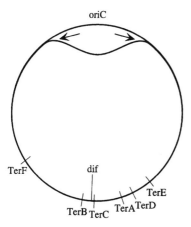

**Fig. 10.14** Termination of DNA replication in *E. coli*. A terminator region exists 180° opposite the origin. In the terminator region there are nucleotide sequences that allow only one-way travel of the replication fork. Thus the terminator sequences are said to be polar. *E. coli* has six terminator (Ter) sites. These are, in the sequence in which they exist in the DNA, TerE, TerD, TerA, TerC, TerB, and TerF. TerE, TerD, and TerA occupy one side of the terminator region and TerC, TerB, and TerF occupy the opposite side. TerC, TerB, and TerF inhibit clockwise-traveling forks, and TerA, TerD, and TerE inhibit counterclockwise-traveling forks. Therefore, there are three chances to trap each replication fork in the termination region. For example, if the counterclockwise-moving replication fork gets by TerA, it will be trapped at TerD or TerE. Consequently, the replication forks will meet at one of the termination sites or between them. A protein, called the Tus protein in *E. coli*, binds to Ter and prevents the helicase from proceeding in one direction, thus accounting for the polarity of inhibition of movement of the replication fork. Notice the site marked *dif*. It is here where recombination takes place to separate chromosomes that have undergone an unequal number of recombinations during replication.

the DNA and appear to align it. (See the section on how the Par proteins might position the chromosomal origins during segregation later.) The Muk proteins seem to be involved in moving the DNA to opposite poles of the cell in *E. coli*. (See the section on what moves the daughter chromosomes to the poles later.)

## Phenotypes of partition mutants

The genes involved in chromosome partitioning, including those encoding Par and Muk, have been discovered by analyzing mutants. Mutants defective in chromosome partitioning can be defective in any one of several genes, some of which are involved in detaching the chromosomes from each other, some of which are involved in the segregation of the detached chromosomes, and even genes involved in the replication of DNA, because only fully replicated DNA molecules can detach from one another. Depending upon whether the defect is in detachment or segregation of nucleoids, the phenotype differs. If the defect is in detaching the daughter chromosomes from each other, then the phenotype includes elongated cells with an unusually large nucleoid mass positioned near the cell center. Anucleate cells may also form. (See the discussion of topoisomerase IV next.) On the other hand, if the daughter chromosomes can detach but cannot segregate to opposite poles, then the phenotype is an increased production of *anucleate cells* of the size of a newborn cell, reflecting the fact that, because the daughter chromosomes do not segregate to opposite cell poles, they can both be in the same daughter cell upon cell division. Under these conditions, the nucleoids are single nucleoids and not larger than normal. (See Note 23 for a further discussion of partition mutants and proteins involved in partitioning.)

## Topoisomerase IV

When the two replication forks meet, which is about 180° from the start site, the two completed DNA molecules are linked as two circles in a chain and must be separated from each other (Fig. 10.15). In *E. coli* and *Salmonella typhymurium*, this requires the action of a DNA gyraselike enzyme called topoisomerase IV.[24] Topoisomerase IV mutants are temperature sensitive for growth; that is, they grow at 30°C but die at 42°C. At the restrictive temperature cell division is inhibited, and the mutants form elongated cells, often with unusually large nucleoids (revealed by DNA staining) in midcell or as several nucleoid masses unequally distributed in the elongated cells.[25] The large nucleoids are

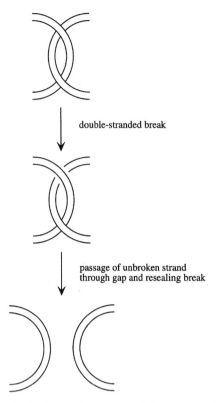

double-stranded break

passage of unbroken strand
through gap and resealing break

**Fig. 10.15** Separating catenated daughter circles after DNA replication using type IV topoisomerase. The reaction is catalyzed by Type IV topoisomerase (similar to DNA gyrase). The enzyme catalyzes a double-stranded cut in one of the duplexes. Then the unbroken duplex passes through the gap. The gap is sealed and the DNA molecules have become separated. Mutants defective in topoisomerase IV are temperature sensitive for growth and show abnormal nucleoid segration. The phenotype is called Par⁻. See Kato, J., Y. Nishimura, and R. Imamura. 1990. New topoisomerase essential for chromosome segregation in *E. coli*. *Cell* **63**:393–404.

due to the fact that the two daughter nucleoids are not able to separate. The filamentous cells may divide in regions where there is no DNA to produce anucleate cells. (Inhibition of cell division resulting in elongated or filamentous cells is typical of mutants that have a block in DNA replication or separation of daughter chromosomes. This reflects the coupling between cell division and DNA metabolism, as described in Note 26.) (Topoisomerase IV mutants have been isolated in *C. crescentus*, but the phenotype differs from that of *E. coli* and *S. typhimurium* mutants.[27])

Topoisomerase IV makes a double-stranded break in one of the DNA molecules, and the unbroken molecule is passed through the break (Fig. 10.15). This is followed by sealing the break. Unlike DNA gyrase, topoisomerase IV does not introduce negative supercoils.

### Site-specific recombination at dif

If an unequal number of recombinations have occurred between daughter *circular* chromosomes, then they are covalently linked to each other, and topoisomerases cannot separate them (Fig. 10.16). Under these circumstances, the DNA molecules must be separated by *site-specific recombination* using enzymes called *recombinases*. *E. coli* has a locus called *dif* (deletion- induced filamentation), which is located in the replication terminus (Ter) region.[28] It is here where the recombinational event that separates the two daughter chromosomes that have undergone an unequal number of recombinations *during replication* takes place (Fig. 10.16), The *site-specific* recombination is catalyzed by two recombinase proteins (i.e., XerC and XerD), both of which are required.[29] Mutants that are *dif* are viable, and most of the cells are normal. This is to be expected, since *dif* is required only when there has been an unequal number of recombinations. However, approximately 10% of the cells are filamentous with unusually large nucleoids, indicative of a failure of the newly replicated chromosomes to separate. As expected, *xerC* and *xerD* mutants show the same phenotype as *dif* mutants.

### Movement of chromosomal replication origins to cell poles in B. subtilis

There is microscopic evidence derived from experiments with *B. subtilis* that the daughter chromosomes become aligned prior to segregation such that replication origins move to the cell poles, whereas the termini remain near the cell center. This evidence was obtained using a fluoresecent protein to specifically label the chromosomal origins and termini. A cassette consisting of multiple copies of the *E. coli* lactose operon was inserted either near the origin of replication or at the terminus of the chromosome.[30] The operon could then be

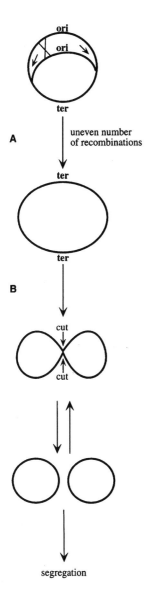

A

uneven number
of recombinations

B

cut

cut

segregation

Fig. 10.16 Site-specific recombination at termination to separate chromosomes that have undergone unequal numbers of recombinations. Simplified drawing of recombination at *dif* to resolve a DNA dimer into monomers. (A) An unequal number of recombinations have taken place covalently, linking the two circles to form a dimer. (B) Recombination takes place catalyzed by recombinases acting at the *dif* sites in the termination (Ter) regions. The chromosomes are then segregated to opposite cell poles. A *dif⁻* mutant produces filaments and anucleate cells as well as normal cells. As the figure indicates, it is believed that the recombination event at *dif* can proceed in both directions. However, monomer formation is favored. Perhaps the segregation of

visualized in dividing or sporulating cells using the lactose operon repressor (Lac I) fused to the green fluorescent protein (GFP) from the jellyfish *Aequorea victoria*. (See the discussion of GFP in Section 2.2.5, *ZipA*.) A *gfp–lacI* fusion was constructed that had a sporulation promoter that was active at the beginning of sporulation. The fused gene was introduced into *B. subtilis,* and the fusion protein was synthesized at the onset of sporulation. As GFP–LacI accumulated during the initial stages of sporulation, it became bound to the operators of the *lac* operons. It was possible to visualize the cluster of GFP–LacI proteins bound to the operons in whole cells using fluorescent microscopy. The fusion protein could also be localized in vegetatively dividing cells by placing the promoter of the fused gene under the control of a xylose-inducible sigma factor, and then inducing *gfp–lacI* with xylose. Using GFP–LacI to label the origin and terminus of the chromosome, it was found that in dividing or sporulating cells the chromosomal origins were located at the cell poles, whereas the termini were at mid-cell. The results of these experiments are summarized next in the model developed for *B. subtilis* and shown in Fig. 10.17.

### 1. Model for sporulation

The student should review the discussion of the partitioning of daughter chromosomes during *B. subtilis* sporulation, Section 18.15.2 and see the review cited in Ref. 31. It is proposed that at stage 0, prior to the initiation of sporulation, the two chromosomes are a single axial filament with their origins attached to opposite poles of the cell (Fig. 10.17, Ai). Septation occurs, trapping approximately $\frac{1}{3}$ of the forespore chromosome in the forespore (Fig. 10.17, Aii). DNA translocation across the forespore septum takes place, resulting in the forespore chromosome being moved

the monomers ensures that the newly replicated *dif* sites cannot recombine. In order for recombination at *dif* to resolve the dimers, *dif* must be at its original site in the terminus region. If a copy of *dif* is reinserted in a different chromosomal site in a *dif⁻* strain, the cells still show the *dif⁻* phenotype.

entirely into the forespore (Fig. 10.17, Aiii). Engulfment occurs, and the origins are detached from the poles (Fig. 10.17, Aiv).

## 2. Model for cell division during vegetative growth of B. subtilis

Prior to replication a single chromosome with one origin exists (Fig. 10.17, Bi). After replication the origins are moved to opposite poles (Fig. 10.17, Bii). The chromosomes condense (Fig. 10.17, Biii), and a septum forms (Fig. 10.17, Biv). It is not known whether the origin in a newborn cell is attached near a pole before replication begins and the new origin is moved to the opposite pole, or the origins move to the poles and

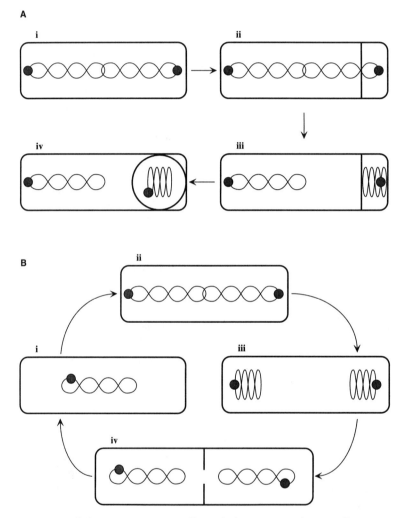

**Fig. 10.17** Partitioning of chromosomes in *Bacillus subtilis*. (A) Sporulating cell. (i) Stage 0. The two chromosomes are a single axial filament attached by their origins to opposite poles of the cell. (ii) A septum forms trapping the region proximal to the origin in the forespore. (iii) DNA translocation moves the rest of the chromosome into the forespore. (iv) After engulfment the origins detach from the poles. (B) Vegetative cell. (i) Prior to DNA replication there is one origin. It may or may not be attached to a pole. (ii) The DNA replicates, and both origins are attached to the poles. (iii) The chromosomes condense, moving them away from each other toward the poles. (iv) A septum forms in the middle, dividing the cell into two daughter cells. *Source:* Adapted from Webb, C. D., A. Teleman, S. Gordon, A. Straight, A. Belmont, D. C-H. Lin, A. D. Grossman, A. Wright, and R. Losick. 1997. Bipolar localization of the replication origin regions of chromosomes in vegetative and sporulating cells of *B. subtilis*. *Cell* 88:667–674.

attach after replication begins. Neither is it known what moves the origins to the poles. However, as discussed next, the Par proteins may be involved in positioning the daughter chromosomes.

## The Par proteins might position the chromosomal origins during segregation

The apparatus involved in positioning the daughter chromosomes and localizing (and perhaps attaching) the replication origins to opposite cell poles in some bacteria (as shown in Fig. 10.17) might involve the ParA and ParB proteins or homologues of these proteins. (For a review, see Ref. 32.) The *parAB* genes, originally discovered in plasmids and required for plasmid partitioning, but now known to be homologous to genes in *B. subtilis, C. crescentus,* and *Pseudomonas putida,* encode proteins required for chromosomal partitioning. (The ParA homologue in *B. subtilis* is Soj, and the ParB homologue is SpoOJ.) (See Note 23 for a summary of the Par proteins.) ParB is a DNA-binding protein that binds to specific sequences near the origin of replication (called *parS* sequences). It has been possible to follow the localization of the Par proteins during cell division in *C. cresentus.* The results of such experiments, described next, indicate that the Par proteins may be involved in localizing the chromosomal origins of replication to the cell poles. However, homologues of the *par* genes have not been discovered in the *E. coli* chromosome, although they are present in *E. coli* plasmids.

### 1. Involvement of Par proteins during chromosome segregation in Caulobacter

Normal cell division in *Caulobacter* produces two cell types, a motile swarmer cell with pili and a polar flagellum, and a stalked cell. (See Section 18.7.1 for a description of the life cycle of *Caulobacter.*) The stalked cell undergoes repeated divisions to give rise to swarmer cells, which do not synthesize DNA and do not divide. After a while the swarmer cell sheds the flagellum, grows a stalk, and becomes a dividing cell. During cell divison in *C. crescentus* the ParA and ParB proteins, both of which are required for viability in

*C. crescentus,* can be seen to localize to both cell poles when chromosomal replication is complete.[33] (See Note 34 for a discussion of the role of ParA and ParB in plasmid partitioning.) This can be demonstrated using immunofluorescence microscopy with anti-ParA or anti-ParB antibody. (See Note 35.) Thus it is reasonable to suppose that ParB attaches to the DNA near the origin of replication, and with the aid of ParA and some receptor in the membrane, attaches the origins to opposite poles. The question remains, however, as to the motor that moves the origins, and perhaps the Par proteins, to the poles, and what condenses the chromosomes once they arrive at the poles. The condensation is presumably important for leaving a gap in the center of the cell for the formation of the septum.

## What moves the daughter chromosomes to the poles?

In all bacteria, the newly replicated daughter chromosomes (nucleoids) are segregated to the cell poles prior to cell division. As discussed, there is microscopic evidence from studies with *B. subtilis* that the origins of replication of the daughter chromosomes segregate to the cell poles, and that in *C. cresecentus* the Par proteins may bind to the origins and be part of the machinery that localizes the origins to the cell poles. But what moves the nucleoids and the Par proteins? Nucleoid segregation is not understood. (See Ref. 36 for a review.) It seems that some sort of machinery must be present that attaches to the daughter chromosomes and moves them to opposite cell poles. However, what that machinery is, is a mystery. There is no obvious apparatus equivalent to a mitotic spindle that can be seen. An early suggestion was that the DNA was attached to the membrane and that membrane/wall growth between the nucleoid attachment sites in the membrane could separate the nucleoids to opposite cell poles.[37, 38] However, it has been shown that membrane and cell wall growth in *E. coli* and *B. subtilis* is diffuse, and this is not consistent with the model that proposes localized membrane/wall growth to segregate nucleoids.

### 1. Muk proteins

There is no obvious cytoskeleton or visible machinery that might move the daughter chromosomes to the poles during cell division (or sporulation). One protein from *E. coli* is frequently mentioned as possibly being part of a motor that moves the nucleoids, although the evidence for this is indirect. MukB is a DNA-binding protein in *E. coli* required for chromosome partitioning (but not plasmid partitioning) at temperatures higher than 22°C. It has been postulated to function as a motor that moves the nucleoids, at least in *E. coli*.[39] At the higher temperatures (e.g., 42°C) anucleate cells and multinucleate filaments with abnormal patterns of nucleoid segregation accumulate in *mukB*[−] cells. The nature of the "track" on which MukB might move as it pulls or pushes the daughter chromosomes to opposite poles is not known. The reason for suggesting that MukB is a motor is that its structure resembles the structure of dynamin, a protein motor in eukaryotic cells that moves along microtubules[40] (see Fig. 10.18). However, there is no direct evidence that MukB is actually a motor. Two other Muk proteins, MukE and MukF, are also required for chromosomal partitioning in *E. coli*.[41] The *muk* genes have been detected thus far only in *E. coli*, and in the case of *mukB*, in *Hemophilus influenzae*. (Reviewed in Ref. 42.) (Read Section 2.2.5 and Note 42 in Chapter 2 for a discussion of the Muk proteins and the roles that they might play in chromosome partitioning.)

### MinD seems to have a role in chromosome partitioning in E. coli

The *min* gene locus appears also to be involved in chromosome partitioning in *E. coli*.[43] (See Section 2.2.5 for a discussion of the role of the Min proteins in determining where the septum is placed in dividing cells.) The *minD* mutants produce not only anucleate minicells, but also anucleate rods of normal size, as well as nucleated rods. When filamentation is induced by inhibiting an enzyme (PBP3) required for peptidoglycan synthesis specifically in the septum, the nucleoids of *min* mutants can be seen to be irregularly spaced along the filaments, with no DNA at the poles, whereas filaments produced under the same circumstances by wild-type cells have regularly spaced masses of DNA, including DNA at the cell poles. Upon removal of the inhibitor of peptidoglycan synthesis, the *min* mutants divide to form anucleate rods in addition to nucleated cells, whereas the wild type forms essentially only nucleated cells. These experiments clearly show that the *min* locus affects DNA partitioning as well as regulating the site of septation. There is insufficient information available to know how the *min* locus affects both the site of septum formation and chromosome partitioning.

### The roles of Muk, Par, and MinD have not been established in all the bacteria where cell division is being studied

At this time there is no evidence that all bacteria whose cell division is being studied have proteins homologous to MukB that bind to the DNA and perhaps move the replication origins to opposite cell poles, or that they all have proteins homologous to the Par proteins that might attach the origins of replication to opposite poles and possibly align the chromosomes. For example, the *mukB* gene seems to be absent from most bacterial chromosomes that have been sequenced. Also, *E. coli* does not appear to have chromosomal *par* genes. Perhaps other proteins have evolved to take their place. Furthermore, the role that the *min* locus might play in chromosome partitioning is not understood, and in fact several bacteria lack one or more of the *min* genes. (See Section 2.2.5 and the description of a model for early events in cell division for a discussion of the presence of *min* and other cell division genes in various bacteria.) A missing feature that has not yet been elucidated is the "anchor" that tethers the chromosomes or some proteins attached to the chromosomes to the membrane/cell wall complex. Because the peptidoglycan is a strong, rigid structure in the cell wall, it is tempting to speculate that the chromosomes are directly or indirectly anchored to it. One must conclude, therefore, that a "universal" bacterial chromosome partitioning apparatus

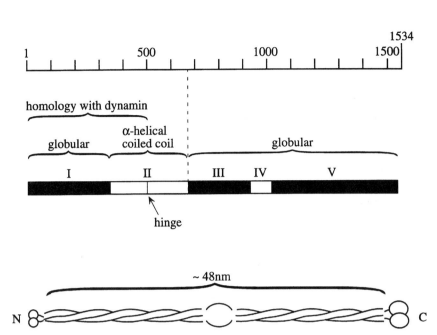

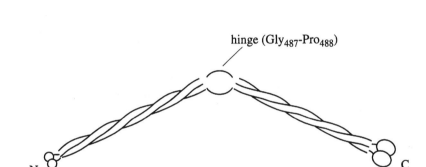

**Fig. 10.18** A model of MukB. Based upon amino acid sequence studies and electron microscopic data, the model postulates that MukB is a homodimer with a rod-and-hinge structure. At the C-terminal end there are a pair of large globular domains. The N-terminal end has a pair of smaller globular domains. Electron micrographs suggest that MukB can bend at a hinge site in the middle of the rod section. MukB binds to DNA, ATP, and GTP. The amino acid sequence of MukB indicates homology with a microtubule-associated protein called dynamin from rats. It has been speculated that MukB attaches to the DNA and to some cellular structure (cytoskeletal filaments?) and moves the DNA to opposite poles. *Source:* From Niki, H., R. Imamura, M. Kitaoka, K. Yamanaka, T. Ogura, and S. Hiraga. 1992. *E. coli* MukB protein involved in chromosome partition forms a homodimer with a rod-and-hinge structure having DNA binding and ATP/GTP binding activities. *The EMBO Journal* 11:5101–5109.

has not yet been described, and much remains to be learned about chromosome partitioning in bacteria.

### 10.1.6 Inhibitors of DNA replication

A variety of antibiotics and drugs are known that inhibit DNA replication itself or inhibit the synthesis of substrates for DNA replication. Two antibiotics produced by *Streptomyces* include novobiocin, which inhibits DNA gyrase, and mitomycin C, which crosslinks the guanine bases in DNA, sometimes in the same strand and sometimes in opposite strands. Crosslinking guanine in opposite strands prevents strand separation. Another antibiotic is nalidixic acid, a quinolone compound that inhibits DNA gyrase. There are also several chemically produced drugs that are widely used to inhibit DNA synthesis. These include acridine dyes. The acridine dyes, such as ethidium, proflavin, and chloroquine, insert between the bases in the same strand of DNA and distort the double helix. At low concentrations the dyes inhibit plasmid DNA replication but not chromosomal replication, although they can cause frameshift mutations. (See Note 44.) If the concentration is sufficiently high, then chromosomal replication is inhibited. Chemicals that inhibit the synthesis of precursors to DNA include the monophosphate of 5-fluorodeoxyuridine, aminopterin, and methotrexate. 5-Fluorodeoxyuridine-5′-phosphate (FdUMP, made by cells from fluorouracil, FU) inhibits thymidylate synthase, which is the enzyme that converts dUMP to dTMP (Section 9.2.2) (see Fig. 10.19). In this reaction a methylene group ($-CH_2$) and hydride (H:) are transferred from methylenetetrahydrofolate (methylene–THF) to d-UMP to form a methyl ($-CH_3$) and thus converts dUMP to dTMP. As a consequence of losing the hydride ion, the tetrahydrofolate is oxidized to dihydrofolate. The tetrahydrofolate (THF) is regenerated by reducing the dihydrofolate with NADPH in a reaction catalyzed by dihydrofolate reductase. The dihydrofolate analogues aminopterin and methotrexate inhibit dihydrofolate reductase. (For other reactions of tetrahydrofolate, see Fig. 9.14.)

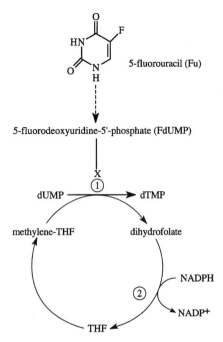

Fig. 10.19 Inhibition of dTMP synthesis by anticancer drugs. 5-Fluorouracil is converted to 5-fluorodeoxyuridine-5′-phosphate (fluorodeoxyuridylate or FdUMP), which inhibits thymidylate synthase, the enzyme that converts dUMP to dTMP. A methylene group ($-CH_2$) and hydride (H:) are transferred from methylenetetrahydrofolate (methylene–THF) to d-UMP to form a methyl ($-CH_3$), which converts dUMP to dTMP. As a consequence of losing the hydride, the THF is oxidized to dihydrofolate. The dihydrofolate is reduced back to THF by NADPH in a reaction catalyzed by dihydrofolate reductase. The dihydrofolate reductase is inhibited by dihydrofolate analogs such as methotrexate and aminopterin. 1, thymidylate synthase; 2, dihydrofolate reductase.

### 10.1.7 Repairing errors during replication

All cells have repair mechanisms that fix DNA that has been damaged or has suffered errors during replication. Repair systems in bacteria that operate during DNA replication and correct the error of inserting the wrong nucleotide are described next. These errors cause relatively minor topological changes in the DNA. Damaged DNA caused by ultraviolet light or mutagens, and which cause severe distortion of the double helix, are repaired by mechanisms described in Section 19.2.

### Editing repair

Occasionally the wrong base is inserted in the copy strand, for example, a T rather than a C opposite a G. This can occur if a base tautomerizes to a form that allows hydrogen bonding to the wrong partner. See Fig. 10.20 for an explanation. If the wrong nucleotide is added to the growing chain, for example, a T opposite a G, the T must be removed, or else when the strand containing the G–T pair is replicated, one of its progeny duplexes

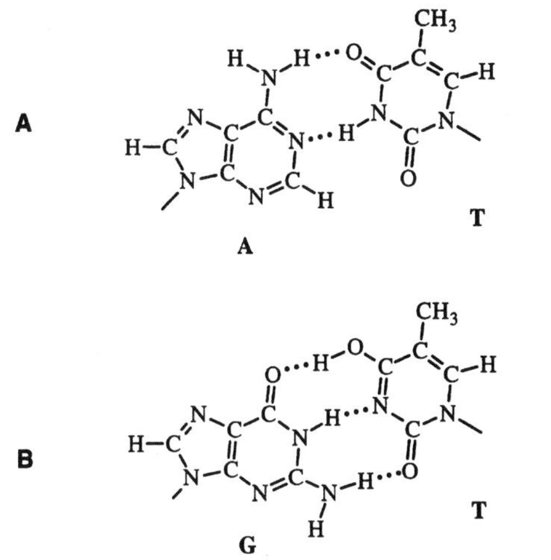

**Fig. 10.20** Tautomerization can lead to incorrect base pairing. Two isomers in equilibrium that differ in the arrangements of their atoms are called tautomers. Commonly, tautomers differ in the placement of hydrogen. For example, one isomer might exist in the enol form (–OH attached to a carbon–carbon double bond), whereas the other isomer might exist in the keto form (contains a C═O group). Keto–enol tautomerizations greatly favor the keto form. When thymine tautomerizes from the keto to the enol form, a proton dissociates from the nitrogen in the ring and moves to the oxygen in the keto group to form the hydroxyl. When this happens two electrons shift in from the nitrogen to form the C═N. (A) Adenine (A) correctly base pairs with the keto form of thymine (T). (B) Thymine has tautomerized to the enol form and base pairs with guanine (G) to form a mismatch.

will have an A–T pair in place of the G–C pair (i.e., a mutation will result). When the wrong base is inserted in the growing chain so that a mismatched pair results (e.g., G–T), DNA replication stops (i.e., the DNA polymerase does not continue to the next position). Presumably this occurs because of the resultant topological change in the DNA (distortion in the double-stranded helix). The incorrect nucleotide is then removed via a 3′-exonuclease. All DNA polymerases have 3′-exonuclease activity and are thought to function in *proofreading* and removing mismatched nucleotides when they are added. Mutations in the *dnaQ* gene (*mutD*) that codes for a 3′-exonuclease subunit in the DNA polymerase III holoenzyme result in greatly increased rates of spontaneous mutations. See

Note 45 for a description of the subunits in DNA polymerase III.

## Mismatch repair

The proofreading system is not perfect, and a certain number of wrong bases do get inserted into the newly replicated DNA. These can be removed by the system of mismatch repair, which improves the fidelity of DNA replication $10^2$- to $10^3$-fold (Fig. 10.21). It is called the methyl- directed mismatch repair (MMR) system. (Reviewed in Ref. 46.) The nucleotide that is removed is the incorrect nucleotide in the copy strand rather than the template strand, so that the template strand is not changed and a mutation does not occur. The reason why the repair system removes the nucleotide from the copy strand and not the template strand is because the template strand is marked by methylation. *E. coli* has an enzyme called deoxyadenosine methylase (*Dam methylase*) that methylates all adenines at the $N^6$ position within 5′-GATC-3′ sequences. (The sequence is a palindrome and therefore is present in both strands but in opposite orientation; i.e., 3′-CTAG-5′.) However, the enzyme does not begin to methylate the DNA until a short period after replication of the region of DNA containing the GATC sequence. Thus, for a few seconds or minutes after replication, the template strand is fully methylated, but the copy strand is undermethylated. During this brief period the copy strand can be repaired. What happens is that the copy strand with the incorrect base is cut at the 5′ side of the G in the unmethylated GATC, and the newly synthesized DNA is removed by an exonuclease to a point just beyond the mismatch (Fig. 10.21). The gap is then filled with DNA polymerase III and sealed with DNA ligase.

Several proteins are involved in mismatch repair. In *E. coli*, these include the products of *mutH*, *mutL,* and *mutS*. MutH, L, and S are thought to form a complex with the DNA. In this complex, MutH is the endonuclease. According to the model, MutS binds to the mismatch. Then MutL binds to MutS. MutH

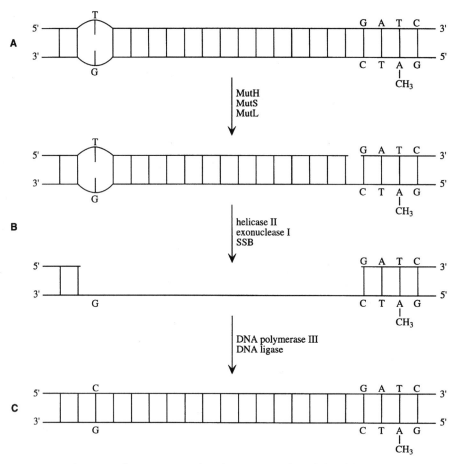

Fig. 10.21 Mismatch repair. This can occur when the wrong nucleotide has been inserted. For example, a T instead of a C might be inserted opposite a G. The template strand is methylated at an adenine in a CTAG sequence. The newly synthesized strand is not yet methylated, and this aids the repair enzymes in distinguishing between the newly synthesized strand and the template strand. (There is a slight delay in the methylation of newly synthesized DNA. However, the A in GATC eventually becomes methylated.) (A) An endonucleolytic cut is made either on the 5′ side or the 3′ side of the mismatch in the GATC sequence in the nonmethylated strand. (B) An exonuclease then removes the newly synthesized DNA past the point of the mismatch. This requires helicase II (MutU). (Helicase II is not identical to DnaB, which unwinds the strands during DNA replication.) If the mismatch is on the 5′ side of the cut, then exonuclease I degrades the DNA 3′ to 5′ through the mismatch. If the mismatch is on the 3′ side of the cut, then exonuclease VII or RecJ protein degrades the DNA 5′ to 3′ through the mismatch. (C) DNA polymerase III then fills in the missing DNA, and the gap is sealed with DNA ligase. Single-stranded DNA-binding protein is also required. The proteins involved in recognizing the mismatch and making the cuts are MutH, MutS, and MutL. The MutS protein recognizes the mismatch. The MutH protein is the endonuclease that makes the cut. It recognizes the GATC sequence and cleaves the unmethylated DNA on the 5′ side of the G in the GATC. MutS and MutH form a complex in which they are linked by MutL.

binds to a nearby GATC sequence, and its endonuclease activity is stimulated by the MutS/MutL complex. (Another *mut* gene, i.e., *mutD*, encodes the DNA polymerase III 3′-exonuclease that is important for editing as described earlier. See the section on editing repair.) DNA helicase II (the product of the *mutU* gene) is also required, and it is thought that its role is to unwind the cut strand so that it can be degraded by exonuclease. There is more than one exonuclease that can be used. If the endonucleolytic cut by MutH is made 3' to the mismatch, then exonuclease I (ExoI), which is a 3'-to-5' exonuclease, is used. If the cut is made 5' to the mismatch, then either exonuclease VII (ExoVII) or RecJ, which are 5'-to-3' exonucleases, are used. Mismatch repair can also be used to repair DNA damaged by the incorporation of base analogs or by certain types of alkylating agents as long as the distortion of the double helix is not severe. Otherwise, the repair mechanisms described in Section 19.2.1 are used. Null mutations in any of the *mut* genes involved in mismatch repair results in a 100- to 1000-fold increase in mutation frequency in *E. coli*. Homologues of the *mut* genes can be found in eukaryotes such as yeast, mice, and humans, where loss of function also results in increased mutation rates, and in the case of humans, is correlated with cancer in tumor cell lines from certain tissues (e.g., colorectal carcinomas). (For reviews, see Refs. 47–49.)

## 10.2 RNA Synthesis

RNA synthesis is called *transcription* and involves the use of one of the strands of DNA as a template to make an RNA copy. The template strand is sometimes called the coding strand or the minus (−) strand. The nontemplate strand is sometimes referred to as the plus (+) strand. The RNA is synthesized in opposite polarity to the template DNA strand; that is, the 3' end of the RNA (the growing end) faces in the same direction as the 5' end of the DNA.

In order for transcription to occur, the DNA must unwind at the site of transcription so that the enzyme that synthesizes RNA, DNA-dependent RNA polymerase, has access to the single-stranded DNA. (See Note 50 for more information on the different types of RNA polymerases.) There are three stages in transcription: initiation, chain elongation, and termination.

### 10.2.1 Initiation, chain elongation, and termination

#### Intiation begins at the promoter region

Initiation takes place at the promoter site, which is a region of DNA at the beginning of the gene where the RNA polymerase binds (Fig. 10.22).[51] Within the promoter region of most genes there exist short similar sequences of nucleotides centered approximately −10 and −35 base pairs upstream from the site where transcription is actually initiated, that is, where the first nucleotide is incorporated (the +1 position, or the *start site*). The polymerase is a very large enzyme and covers the entire promoter, not simply the −10 and −35 regions. However, the −10 and −35 regions are critical, as mutations that affect promoter function usually occur in one of these sequences. Furthermore, as discussed later, the frequency of initiation of transcription is specific for promoters and reflects variations in the sequences in the −35 and −10 regions. In *E. coli* the most common sequence in most of the genes at −10, sometimes called the Pribnow box, is 5'-TATAAT-3', and the sequence at

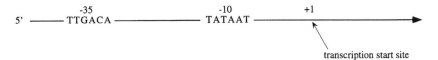

Fig. 10.22 Consensus promoter sequences for −10 and −35 regions in noncoding strand. Centered about 10 base pairs upstream of the mRNA start site (at position +1) is a six-base region that is usually 5'-TATAAT-3' in the noncoding strand or some minor variation thereof. It is called the −10 sequence or Pribnow box. A second consensus sequence is centered approximately 35 base pairs upstream of the transcription start site. It is generally referred to as the −35 sequence and is 5'-TTGACA-3'. The RNA polymerase binds to the entire promoter, not simply to the −35 and −10 regions. Unwinding of the DNA to initiate transcription occurs at the −10 region and extends for about 20 base pairs past the mRNA transcription start site. It is believed that the RNA polymerase itself unwinds the DNA.

−35 is 5 '-TTGACA-3'. These are called the consensus sequences, and although they are not identical in all promoters, there is sufficient similarity so that the investigator (and the major RNA polymerase called the sigma 70 polymerase) can recognize these sequences. Promoter sequences other than the consensus −35 and −10 sequences exist, but these are not recognized by the major RNA polymerase. Instead, they are recognized by RNA polymerases complexed with other sigma factors. (See the discussion of the sigma subunit later.) The binding of the RNA polymerase to these sequences and to the rest of the promoter region leads to the localized unwinding of the DNA strands, which allows the RNA polymerase to move down the template strand, making an RNA copy starting at the start site. In contrast to DNA replication, a helicase is not required.

## Steps in initiation and chain elongation

### 1. Formation of the closed complex
RNA polymerase holoenzyme (core plus sigma factor) binds to the promoter region. At this stage the DNA exists as a double helix and the DNA–polymerase complex is referred to as a "closed complex" (Fig. 10. 23B).

### 2. Formation of open complex
The DNA duplex unwinds at the promoter region, including the start site, to form the *open complex* (Fig. 10.23C). There is no requirement for a helicase or a nucleotide triphosphate such as ATP.

### 3. Binding of initiating ribonucleotides
Transcription is usually initiated at the start site (+1) with ATP or GTP, whose ribose provides the free 3'-hydroxyl that attacks the α phosphate in the incoming nucleotide, displacing the pyrophosphate. The newly synthesized RNA therefore has a triphosphate at its 5' end (Fig. 10.23D).

### 4. Elongation
RNA synthesis is said to be in its elongation stage after a few nucleotides (about 12) are added to the growing chain of RNA, the sigma subunit dissociates from the polymerase, and the polymerase moves forward from the promoter region as it elongates the RNA (Fig. 10.23E). There is an opening in the DNA

duplex called a *transcription bubble* about 18 bases long in which the elongating RNA forms an RNA/DNA hybrid with the template DNA strand. The RNA/DNA hybrid helps to keep the RNA polymerase attached to the DNA. The average rate of elongation of messenger RNA for *E. coli* is between 40 and 50 nucleotides per second, as compared to approximately 1000 nucleotides per second for elongation of DNA. This matches the rate of protein synthesis. That is because the rate at which amino acids are added to a growing polypeptide chain is approximately 16 per second. In order for this to occur, the messenger RNA must move through the ribosome at a rate of 48 nucleotides (i.e., 16 codons) per second, or approximately the same rate as its synthesis. (See the discussion of coupled transcription/translation in Section 10.4.5.) Ribosomal RNA in fast-growing cells is synthesized at a rate of about 90 nucleotides per second.

## Termination

At termination the RNA and RNA polymerase are released from the DNA, and the DNA duplex re-forms at that site.[52, 53] Transcription is terminated by a DNA sequence called a terminator. There are two patterns of termination. One requires an additional protein factor and is called factor-dependent termination. The best-studied termination factor is a protein called Rho. A second type of termination does not require a protein factor and is called factor independent, which will be described first.

### 1. Factor independent
Factor-independent termination occurs when the RNA polymerase transcribes a self-complementary sequence of bases that promotes the localized hybridization of the RNA to itself (formation of a hairpin loop) rather than to the DNA template (Fig. 10.24). The loop is followed by a short RNA–DNA duplex consisting of a string of A–U base pairs that is unstable. (The interactions between A and U are weaker than the other base pair interactions.) The consequence is that the RNA spontaneously dissociates from the template along the A–U region and from the polymerase. The RNA polymerase also dissociates from the DNA and the DNA duplex reforms.

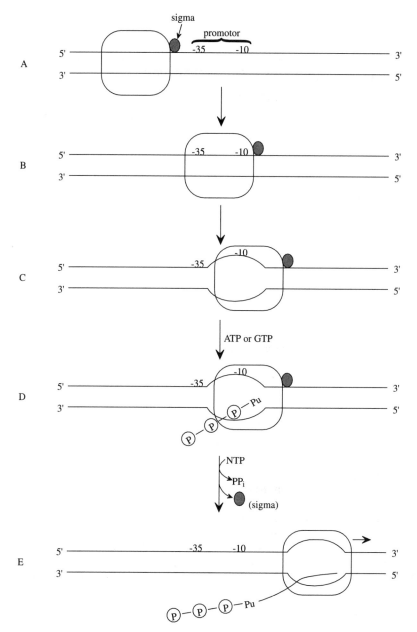

**Fig. 10.23** Initiation of transcription. (A) The RNA polymerase, with its attached sigma subunit, binds to the DNA upstream of the −35 region. (B) The RNA polymerase moves to cover the promoter. This is called the *closed complex*. (C) The polymerase moves past the 10 region as it unwinds the duplex beyond the transcription start site to form the *open complex*. (D) RNA synthesis is initiated at the start site with either ATP or GTP. (E) As the polymerase migrates past the promoter, the sigma subunit is released. There is an opening in the DNA duplex of about 18 bases called a transcription bubble in which the elongating RNA forms an RNA/DNA hybrid.

## 2. Rho dependent

Other terminator regions require protein factors for termination and do not rely on RNA–RNA hairpin loops to disrupt the RNA/DNA hybrid in the transcription bubble.

The nucleotide sequences in the terminator regions are not closely related to one another and are not always easy to recognize. There are three termination factors in *E. coli*. They are Rho, Tau, and NusA. The most well studied

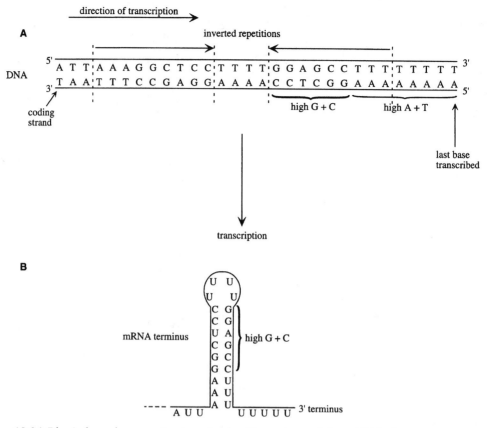

**Fig. 10.24** Rho-independent termination. At the 3′ terminus of the mRNA there is an inverted repeat separated by a nonrepeating sequence. For example, the sequence in the *E. coli tryp* mRNA (encoding enzymes to synthesize tryptophan) is: AAAGGCUCC–UUUU–GGAGCCUUU, which causes the mRNA to form a stem and loop structure (hairpin structure) near the 3′ end of the mRNA. Near the loop or within the stem there is a sequence rich in G + C base pairs. There are also 6 to 8 Us at the 3′ end of the mRNA (corresponding to a string of As in the template DNA). These Us are necessary for termination, as shown by mutants in which some of the A–T base pairs in this region of the DNA have been removed. The model is that when the RNA hairpin forms, the RNA hybridizes to itself rather than to the DNA and thus the RNA–DNA duplex is disrupted in the transcription complex. In addition, the U–A duplex after the hairpin is unstable. This leads to the dissociation of the RNA from the DNA.

is Rho, which is an RNA-dependent ATPase and an ATP-dependent RNA/DNA helicase. As we shall see, these two activities of Rho are necessary for termination. A model for how Rho works is that it binds to particular sequences, called *rut* (rho utilization sites) in the RNA directly behind the RNA polymerase. If a translating ribosome is present at that sequence, then Rho cannot bind. If, however, translation is terminated and Rho binds, then it moves along the mRNA behind the RNA polymerase in the 5′-to-3′ direction. The

energy to move along the mRNA is provided by the hydrolysis of ATP. When the RNA polymerase reaches a termination site, it stops moving. Rho reaches the polymerase, and the Rho helicase activity unwinds the RNA–DNA duplex in the transcription bubble. This results in the disengagement of the RNA polymerase from the DNA and the re-formation of the DNA/DNA duplex. Thus, when the ribosome stops translating at the translational stop site in the mRNA, Rho can bind at the next accessible *rut* site and transcription will stop at the next Rho-dependent termination site. Rho-dependent termination can also explain polarity, as discussed in Section 10.3.8.

## 10.2.2 Frequency of initiation

As stated, promoter regions are not identical, and they do differ in the extent to which they influence the frequency of initiation of transcription ("promoter strength"). Strong promoters cause initiation every few seconds and weak promoters every few minutes. This is related to the consensus sequences at the $-10$ and $-35$ regions because strong promoters have sequences that are more similar to the consensus sequences, whereas weak promoters have base substitutions in these regions. Various transcription factors, both positive and negative, also influence the frequency of transcription. The transcription factors bind to regions near the promoter and either promote RNA polymerase activity (e.g., by stimulating its binding or the formation of the open complex) or inhibit the activity of the polymerase. See Section 10.2.5 for a discussion of the regulation of RNA transcription.

## 10.2.3 Role of topoisomerases

Just as during DNA replication, the unwinding of the duplex produces positive supercoils downstream of the unwinding region. If nothing were done with these supercoils, DNA would stop unwinding, and transcription would stop. Topoisomerases change the positive supercoils to negative supercoils within which the DNA strands are more readily unwound.

## 10.2.4 The sigma subunit

The RNA polymerase in the Bacteria usually has 5 subunits $(\alpha_2\beta\beta'\sigma)$, one of them being a protein called sigma factor, $\sigma$, which is responsible for recognition of the promoter. After initiation, sigma falls off the polymerase; that is, it is not needed for chain elongation. The different sigma factors recognize different $-10$ and $-35$ consensus sequences. A bacterium can have more than one sigma factor, and the core polymerase can exchange one for another, thus contributing towards the specificity of gene transcription. For example, the main sigma factor in E. coli is $\sigma^{70}$ (because it has a molecular weight of 70 kD), which recognizes the consensus sequence in most of the promoters. But E. coli has at least six other sigma factors, and these bind to consensus sequences, which differ from the sequences recognized by $\sigma^{70}$. For example, when E. coli enters stationary phase as a result of starvation, it synthesizes $\sigma^s$ (also called $\sigma^{38}$), which recognizes promoters for genes that are expressed during stationary phase (Section 2.2.2). And, when E. coli is grown under limiting ammonia, it uses $\sigma^{54}$ to transcribe genes in the Ntr regulon (Section 18.8.1). When subjected to heat stress, E. coli increases the amounts of another sigma factor, $\sigma^{32}$, which stimulates the transcription of genes encoding heat-shock proteins (Section 19.1). The expression of genes whose activities increase during sporulation of Bacillus is due to increased synthesis of sporulation-specific sigma factors. Sigma factors determine whether or not a gene will be transcribed, and in cases where the activity or synthesis of a particular sigma factor increases, then this can result in increased transcription of target genes.

## 10.2.5 Regulation of transcription

Bacteria generally regulate protein synthesis at the transcriptional level. (See Ref. 54 for a review.) Since mRNA in bacteria is generally unstable (i.e., it is rapidly enzymatically degraded), the inhibition of transcription of a particular gene generally means that the synthesis of the protein encoded by that gene ceases quickly. Conversely, stimulation of transcription of a particular gene usually results in a rapid synthesis of the protein encoded by that gene. In this way bacteria modulate the mix of enzymes and other proteins in response to environmental challenges, such as the presence or absence of specific carbon and energy sources, or inorganic compounds such as inorganic phosphate or ammonium ion, and environmental stress such as starvation, changes in pH, changes in osmolarity, and changes in temperature. The responses to environmental challenges and the signal transduction pathways are discussed in Chapters 18 and 19.

## Operons

Frequently bacterial genes are arranged in *operons*. An operon consists of two or more genes that are cotranscribed into a polycistronic mRNA from a single promoter at the beginning of the first gene. All the genes in the operon are coordinately regulated from a single promoter.

## Transcription factors

Transcription is commonly regulated at the initiation step by proteins that affect in some way the binding of RNA polymerase to the promoter, the "melting" of the DNA to form the transcription bubble, or the movement of the polymerase along the DNA. The transcription factors bind to specific sequences in or near the promoter region or in regions upstream from the promoter, called *enhancer regions*. The transcription factors generally have a helix–turn–helix motif described next. *E. coli* has at least 100 such transcription factors. The regulation of transcription by transcription factors is especially important in situations where the levels of the sigma factor for the particular gene are not changed. Under these circumstances it is only the transcription factors that determine the rates of transcription of specific genes.

## Helix–turn–helix motif

Many prokaryotic DNA binding proteins, including activators and repressors of transcription, have a *helix–turn–helix motif* in part of the protein (Fig. 10.25A.)[55] In one part of the protein approximately 7 to 9 amino acids form an $\alpha$-helix, which is called helix 1. Helix 1 is connected to a second $\alpha$-helix of about the same size by about 4 amino acids. The helices are approximately at right angles to each other. Helix 1 lies across the DNA, and helix 2 (the recognition helix) lies in the major groove of the DNA (Fig. 10.25B). Because helix 2 lies in the major groove, the amino acid side chains that protrude from the surface of the protein can contact and form noncovalent bonds (e.g., hydrogen bonds) with base pairs in the DNA, and therefore can bind to specific nucleotide sequences (Fig. 10.25C). Hydrogen bonds can form between the carbonyl and amino groups in the protruding side chains of certain amino acids in the DNA-binding region of the protein, and amino groups, ring nitrogen, and carbonyl groups in specific bases. The amino acids that are involved in the bonding to the bases are generally asparagine, glutamine, glutamate, lysine, and arginine. For example, glutamine can bind to A–T base pairs because the carbonyl group and amino group in glutamine can hydrogen bond with the amino group on $^6C$ and the $^7N$ in adenine, respectively (Fig. 10.25C). This does not interfere with the hydrogen bonding of the bases to each other.

## How transcription factors work

Activators bind to specific sites in the DNA upstream of the RNA polymerase and make contact with the RNA polymerase while enhancing its activity. Repressors bind downstream of the RNA polymerase (in the operator region) and block progression of the RNA polymerase. Figure 10.26 illustrates general models of how transcription activators and repressors are believed to work. Some specific examples of positive and negative transcription factors are described next.

## Positive regulators

Positive transcription regulators bind to the promoter region or to sequences upstream from the promoter region (called *enhancer sites*). They may make contact with RNA polymerase and promote its binding to the promoter region, and some may facilitate the "melting" of DNA to form the transcription bubble. Consider the cyclic AMP receptor protein (CRP). When it binds to cyclic AMP (cAMP), it can then bind to specific sites in the DNA and stimulate transcription. For example, consider the stimulation by cAMP–CRP of the *lac* operon, which encodes proteins required for lactose uptake and metabolism. cAMP–CRP binds to specific nucleotide sequences upstream of the *lac* promoter as well as to an incoming RNA polymerase. The binding of cAMP–CRP to the

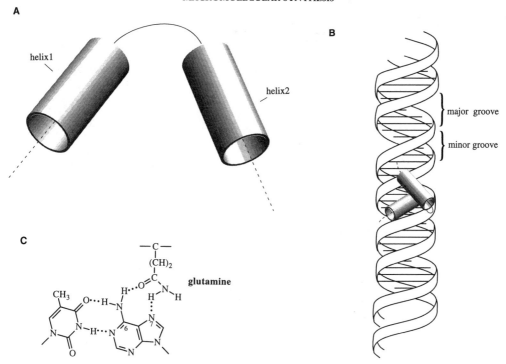

Fig. 10.25 Helix–turn–helix motif and DNA binding. (A) A DNA-binding protein typically has in one portion two α helices connected by a short segment of about four amino acids. (B) Helix 2 lies in the major groove of the DNA duplex, and helix 1 lies at approximately right angles to helix 2. (C) Because helix 2 lies in the major groove, it makes contact with the bases. Certain amino acids in helix 2 with amino or carbonyl groups in their side chains can hydrogen bond to specific bases. In the example shown, glutamine ($-CH_2-CH_2-CONH_2$) is shown hydrogen bonded to adenine in an A:T base pair. Other amino acids, e.g., asparagine ($-CH_2-CONH_2$), glutamate ($-CH_2-CH_2-COOH$), lysine ($-CH_2-CH_2-CH_2-CH_2-NH_2$), and arginine [$-CH_2-CH_2-CH_2-NH-C(NH_2)=N^+H_2$] can also hydrogen bond to specific base pairs, and in this way the DNA-binding protein can bind to specific nucleotide sequences. DNA-binding proteins are frequently dimers and bind to inverted repeats in the DNA.

RNA polymerase enhances its binding to the promoter and thus stimulates transcription. Many unrelated genes are positively regulated by cAMP–CRP and are said to be part of the CRP regulon. (They are also called cAMP-dependent genes.) cAMP-dependent genes may encode proteins required for the catabolism of several different carbon and energy sources such as lactose, galactose, arabinose, and maltose. In those bacteria where this is the case mutants unable to synthesize cAMP or CRP are unable to grow on these carbon sources. Because cAMP–CRP stimulates the transcription of many unrelated genes, it is called a *global regulator*. The levels of cAMP are lowered by glucose uptake in many bacteria. This lowers the transcription of cAMP–dependent genes, explaining glucose repression of genes required for the catabolism of carbon sources other than glucose. Glucose repression mediated by cAMP is discussed in Section 16.3.4. (Refer to Sections 18.13.1 and 18.13.2 for glucose repression that does *not* involve cAMP and for a general discussion of catabolite repression.)

Another example of a global transcription activator is NR$_I$-P, which is part of a two-component regulatory system that activates the *ntr regulon,* as discussed in Section 18.8.1. NR$_I$-P binds to enhancer sites 100 to 130 base pairs upstream from the promoter region of the target genes and interacts with the RNA polymerase to form the open complex, thus stimulating transcription. For this to happen, the DNA must bend in order to bring NR$_I$-P to the polymerase (Fig. 10.26).

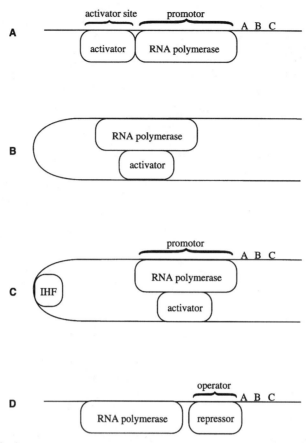

**Fig. 10.26** Models for how transcription activators and repressors work. Although not shown, several transcription factors are known to function as dimers. (A) Activator proteins can bind at specific sites, called activator sites, upstream of the promoter and make contact with the RNA polymerase facilitating its binding to the promoter. Some activators must first bind to a small effector molecule (inducer) before they can bind to the DNA. (B) If the activator binds a distance away upstream of the promoter, then the DNA must bend in order to bring the activator to the polymerase. (C) Sometimes a second protein, e.g., integration host factor (IHF), is required to bend the DNA in order to bring the activator site to the RNA polymerase. In the example shown, IHF binds to a site between the promoter region and the activator site and brings the activator to the RNA polymerase. (D) Repressor proteins bind to the operator region, which may or may not be in the promoter region. The repressors block progression of the RNA polymerase. Some repressor proteins (called aporepressors) must first bind to an effector molecule (corepressor) before they can bind to the DNA.

As shown in Fig. 10.26C, IHF promotes DNA bending. (See Section 18.8.2.) $NR_I$-P helps the RNA polymerase (sigma 54 polymerase) to separate the strands of DNA around the promoter so that the polymerase can gain access to the transcription start site. Several other positive transcriptional activators are described in Chapter 18.

There are other examples of the role of IHF in transcriptional regulation. For example, as discussed in Section 18.3, in order for the positive transcription regulator NarL-P to

stimulate transcription of the nitrate reductase gene, IHF, must also bind to the DNA. The model is that IHF bends the DNA to bring the bound NarL-P to the promoter, where it stimulates transcription (Fig. 10.26). IHF can also inhibit transcription if it binds to the promoter region. (See Section 18.10.1.) See the discussion of the nucleoid in Section 1.2.6 for a description of IHF and other DNA-binding proteins that bind to bent DNA.

Many operons are specifically activated by a protein (activator) only when the protein binds an inducer. Examples include the activation of the L-*ara* operon, which encodes proteins required for the conversion of L-arabinose into xylulose-5′-phosphate, an intermediate in the pentose phosphate pathway. The activator protein is AraC, which must first bind to L-arabinose before it can bind to the DNA and activate transcription of the operon. This is believed to be due to a conformational change in the activator increasing its affinity for the DNA sequence upon binding of the positive effector. As described later, some repressor molecules must also bind to small effector molecules before they bind to DNA.

## Negative regulators

Negative regulators are called *repressors*. They bind to nucleotide sequences in the DNA called *operator* regions and inhibit transcription (Fig. 10.26). For some genes the operator is within the promoter region, but for others it is outside of the promoter and may even be upstream of the start site. An important negative repressor is LexA, which is inactivated by proteolytic cleavage during the SOS response as discussed in Chapter 19. The SOS response is activated when DNA is damaged (e.g., by UV radiation). Single-stranded DNA that results from the damage activates RecA, which then activates the cleavage of LexA. As a consequence of the cleavage of LexA, many genes important for DNA repair are induced.

Another well-known repressor is the *lac* repressor, which binds to the operator region that overlaps the start site of the first gene in the *lac* operon. Transcription is inhibited when the repressor is bound because the repressor blocks RNA polymerase from proceeding into the first gene of the operon. The repressor is inactivated when it binds an isomeric product of lactose metabolism (allolactose), accounting for the induction of the lactose operon by lactose. (The allolactose binds to the repressor, causing a conformational change in the repressor protein, which lowers its affinity for the DNA. As a consequence, the repressor comes off the DNA, and the *lac* operon is induced.) Often investigators use isopropyl-β-D-thiogalactoside (IPTG) as

an inducer of the *lac* operon because IPTG is not metabolized.

Sometimes a repressor must bind to a small molecule before it can bind to the operator. This is the case for biosynthetic operons, that is, operons that encode enzymes for the biosynthesis of amino acids and other small molecules. As an example, consider the Trp repressor in *E. coli* that represses the *trp* operon. (The *trp* operon encodes the enzymes required to synthesize tryptophan from chorismic acid.) The repressor can bind to the operator only if it first binds to tryptophan, which is a *corepressor*. (In the absence of the corepressor, the inactive repressor is called an *aporepressor*.) In this way, tryptophan regulates the synthesis of its biosynthetic enzymes by feedback repression of transcription.

## Some transcription regulators can be activators or repressors

Some transcription regulators activate certain genes and repress others. Several examples are discussed in Chapter 18. For example, see the discussion of Cra described in Section 18.13.1.

## Attenuation of the trp operon

Whereas positive and negative regulators of transcription act at the promoter region to influence the initiation of transcription, attenuation refers to regulation of transcription *after* initiation. Attenuation refers to the termination of transcription before the first gene in the operon is transcribed. Operons can be regulated by attenuation as well as by positive and negative regulation of initiation of transcription. Several operons that encode enzymes required for the biosynthesis of amino acids are known to be regulated by attenuation, including the *trp* operon in *E. coli*. (See the previous discussion of the regulation of the *trp* operon by feedback repression by tryptophan.) Downstream of the promoter region in the *trp* operon there exists a region called the leader region (*trpL*), which lies between the promoter and the first gene of the *trp* operon (*trpE*). When the tryptophan concentrations rise, this results in an increased amount of aminoacylated tRNA[Trp], which terminates transcription in the leader region (trpL). As a consequence, the operon is not transcribed. How this occurs is described next.

The model for attenuation of the trp operon is diagrammed in Fig. 10.27. There are four regions in the trpL RNA: regions 1, 2, 3, and 4. They can form hairpin loops with one another: 1:2, 2:3, and 3:4. There are two adjacent tryptophan codons in trpL, and these are in region 1. When the ribosome reaches these codons, it will stall if there is insufficient aminoacylated tRNA^Trp. If the ribosome stalls at the trp codons in region 1, then a hairpin loop forms between the newly synthesized regions 2 and 3. (Region 4 has not yet been synthesized.) If, on the other hand, there is

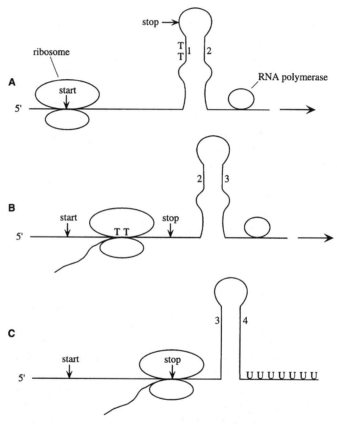

Fig. 10.27 Model for attenuation of the trp operon. There are four regions in the leader RNA, which precedes the transcript for the first gene in the trp operon. Region 1 has two trp codons (T). (A) The RNA polymerase transcribes trpL just past region 2 and pauses briefly. The reason why it pauses is that regions 1 and 2 form a hairpin loop. The translational start and stop sites are marked, as are the tryptophan codons (T). While the polymerase is paused, a ribosome begins translating the mRNA at the 5' end. As the ribosome moves, region 1 of the mRNA is drawn into the ribosome. As translation proceeds, the polymerase begins to move again and transcribes region 3. (B) Region 1 has been drawn into the ribosome and is being translated. This frees region 2 to pair with the newly transcribed region 3. If the ribosome stalls at region 1 because of an insufficiency of tryptophan, then region 2 remains paired with region 3. This prevents a termination loop between regions 3 and 4 from forming and transcription of the trp operon continues. (C) If the ribosome does not stall at region 1, then it will move to the end of trpL, where it stops at the UGA codon in region 2, which is now in the ribosome. While the ribosome occupies the UGA codon in region 2, it prevents region 2 from pairing with region 3. Meanwhile, the polymerase continues to move and transcribes region 4. A termination loop forms between regions 3 and 4. Thus, in the presence of excess tryptophan, transcription of the trp operon is terminated. See Fig. 10.24.

sufficient tryptophan so that the ribosome does not stall in region 1, then it moves to the stop codon in region 2. This frees region 3 to pair with region 4 of the RNA when it is synthesized. If regions 3 and 4 form a hairpin loop, then transcription stops because this hairpin is a factor-independent termination signal. (See Section 10.2.1 for a discussion of factor-independent termination.)

## 10.2.6 Processing of ribosomal and transfer RNAs

### Ribosomal RNA

In order to synthesize rRNA, bacteria first make a 30S preribosomal transcript from a single gene. The 30S transcript contains the 16S, 23S, and 5S rRNAs, as well as one or more tRNAs, and is processed according to the diagram in Fig. 10.28. There are actually seven sets of rRNA genes in *E. coli* that yield the 30S preribosomal transcript. The genes all code for

the same rRNAs but differ with respect to the number and type of tRNAs that are encoded. Eukaryotes also process rRNA transcripts, as described in Note 56.

### Transfer RNA

Transfer RNAs are extensively processed after transcription in both bacteria and eukaryotes (Fig. 10.29).[57] The initial transcript has segments at both the 3′ and at the 5′ end that are enzymatically removed. The removal of the 5′ end is catalyzed by RNase P, an endonuclease, which is present in both prokaryotes and eukaryotes. RNAse P is a ribonucleoprotein where the RNA has catalytic activity. It is therefore an example of a *ribozyme*. (The peptidyl transferase is another example of a ribozyme, Section 10.3.5.) The 3′ end is processed by an endonuclease that removes a terminal piece, followed by RNase D, an exonuclease, that trims the remaining nucleotides to generate the mature 3′ end. In addition to

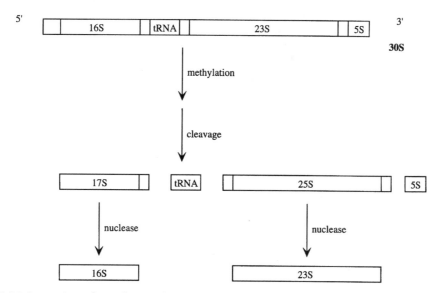

**Fig. 10.28** Processing of preribosomal RNA transcripts. *E. coli* has 7 genes that code for a 30S transcript containing the 16S, 23S, 5S rRNAs and 1 or more tRNAs. For example, one of the *E. coli* 30S transcripts is 5′-16S–tRNA[Ile]–tRNA[Ala]–23S–5S–tRNA[Asp]–tRNA[Trp]–3′. After methylation of specific bases, the 30S transcript is cleaved into 17S, 25S, and 5S rRNA transcripts and the tRNAs. Specific nucleases then generate the 16S rRNA from the 17S transcript and the 23S rRNA from the 25S transcript. Recall that each ribosome has 1 23S, 1 16S, and 1 5S rRNA. An advantage to cleaving the three rRNA transcripts from a single transcript rather than having the three genes transcribed separately is that it ensures that the rRNA molecules are produced in a 1:1:1 ratio. It should be pointed out that not all tRNA genes are present within the rRNA transcript. Several of the tRNA genes are encoded outside the rRNA genes.

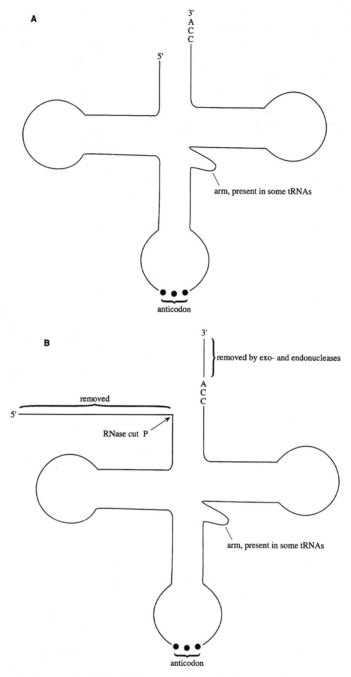

**Fig. 10.29** Processing of transfer RNA. (A) Processed transfer RNA molecules have between 73 and 93 nucleotides. They consist of a 3′ end, which ends in CCA. The amino acid is esterified to a hydroxyl in the ribose portion of the terminal AMP. There are four arms, three of which end in loops. Often a short fifth arm exists between two of the arms. The anticodon is present in the loop of the anticodon arm. (B) Transfer RNA is made as a larger precursor and processed at both the 3′ and 5′ ends. An exonuclease called RNase D removes the nucleotides at the 3′ end up to the CCA trinucleotide. Some tRNAs do not have CCA in the original transcript. Thus, after trimming the 3′ end, CCA-3′-OH is added using tRNA nucleotidyltransferase. The 5′ end is processed by a ribozyme called RNase P that trims the 5′ end to produce the 5′ terminus in mature tRNAs. After the 3′ and 5′ ends are trimmed, certain of the bases are enzymatically modified. (See text.)

exo- and endonucleolytic cleavage, additional processing may occur at the 3′ end. All eukaryotic tRNAs and some bacterial tRNAs lack the CCA-3′ terminus, and this must be added to the molecule after transcription by the enzyme tRNA nucleotidyltransferase. In addition, all tRNAs have bases at specific sites that are modified after transcription (Fig. 10.30).

## 10.2.7 Some antibiotic or other chemical inhibitors of transcription

Transcription in both prokaryotes and eukaryotes is inhibited by the antibiotic *actinomycin D*, which inserts into the DNA between base pairs and deforms the DNA. The result is that the RNA polymerase cannot continue to move down the template, and elongation is inhibited. Another inhibitor of elongation is *acridine*, which behaves in a similar way. A commonly used antibiotic to inhibit transcription in prokaryotes is *rifamycin* or the chemically modified derivative *rifampicin*, which is more effective. Rifamycin (and rifampicin) binds to a subunit in bacterial RNA polymerases (the ß subunit) and inhibits the initiation of transcription. It is not an inhibitor of RNA synthesis in eukaryotic nuclei. A related antibiotic that also binds to the β subunit of bacterial RNA polymerase is *streptolydigin*. *Amanitin* inhibits eukaryotic RNA polymerase II (synthesizes mRNA) but not bacterial RNA polymerase.

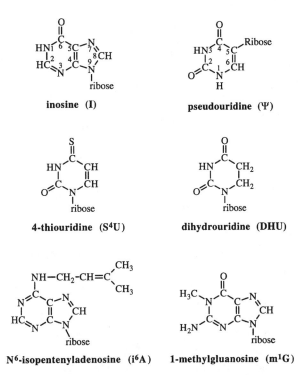

inosine (I)

pseudouridine (Ψ)

4-thiouridine (S⁴U)

dihydrouridine (DHU)

N⁶-isopentenyladenosine (i⁶A)    1-methylgluanosine (m¹G)

Fig. 10.30 Some modified nucleosides in tRNAs. The modifications are produced post-transcriptionally. Inosine differs from adenosine in having hypoxanthine as its base rather than adenine. The difference is that adenine has an amino group attached to C6, whereas hypoxanthine has an oxygen. Pseudouridine differs from uridine in having the ribose attached to C5 rather than N1. Thiouridine differs from uridine in having a sulfur attached to C4 rather than an oxygen. Dihydroxy-yuridine differs from uridine in having the ring reduced between C5 and C6. Isopentenyladenosine differs from adenosine in having an isopentenyl group attached to the amino group at the C6 position. Methylguanosine differs from guanosine in having a methyl group attached to N1.

## 10.3 Protein Synthesis

### 10.3.1 Overview

Protein synthesis (called translation because the sequence of bases in the mRNA is translated into a sequence of amino acids in the protein) occurs in three stages on the ribosome: (1) initiation, (2) elongation, and (3) termination.[58] A summary of each of the stages is given first, followed by a more detailed description (Sections 10.3.3–10.3.5).

### Initiation

During initiation an initiator transfer RNA (tRNA$_f$) carrying formylmethionine binds to the start site at the 5' end of the mRNA and at a tRNA binding site called the P site (peptidyl site) on the ribosome. Amino acids do not recognize or bind to mRNA, and it is the tRNA that is responsible for placing the correct amino acid into the polypeptide sequence. For this reason, rRNAs are called adaptor molecules. They each have an anticodon that base pairs with the codon in the mRNA, thus ensuring that the amino acid sequence in the completed protein reflects the nucleotide sequence in the mRNA (Fig. 10.29).

### Elongation

Elongation begins with the binding of an aminoacylated tRNA to a second tRNA binding site on the ribosome called the A site (aminoacyl site). Then the $\alpha$ amino group of the amino acid bound to the tRNA at the A site displaces the tRNA from the formylmethionyl–tRNA so that a dipeptidyl–tRNA is formed at the A site. (See Fig. 7.8.) This is followed by the transfer of the dipeptidyl–tRNA from the A site to the P site (peptidyl site) and the release of the initiator tRNA from the ribosome. In addition, the ribosome moves down the mRNA in the 3' direction so that the next codon is available at the A site for the next incoming aminoacylated tRNA. That part of chain elongation that includes the movement of the peptidyl–tRNA from the A site to the P site and the movement of the ribosome one codon down the mRNA is called *translocation*.

### Termination

The process of chain elongation is repeated until a stop signal on the mRNA arrives at the A site. At this time the elongation is terminated and the polypeptide is released from the ribosome. Termination requires specific protein factors called termination factors.

### Removal of the formyl group and of methionine

Most completed proteins do not have formyl methionine, or even methionine, at the N-terminus. Two enzymes remove these moieties. *Peptide deformylase* removes the formyl group, and *methionine aminopeptidase* removes methionine.

### Elongation rate

For *E. coli* growing at 37°C, the polypeptide elongation rate is about 16 amino acids per second. The average protein (MW = 40,000) has around 364 amino acids. This means that it takes only about 20 seconds to make an average-size protein.

### Comparison to cytosolic eukaryotic protein synthesis

Overall, protein synthesis by cytosolic eukaryotic ribosomes is quite similar to protein synthesis by bacterial ribosomes but differs in certain details, as described in Note 59.

### 10.3.2 Ribosomes

Ribosomes are ribonucleoprotein particles that catalyze protein synthesis (Section 1.2.6). Rapidly growing bacteria contain about 70,000 ribosomes per cell, all of which are engaged in the synthesis of protein. Each ribosome has a diameter of approximately 20 nm (200 Å ) and a sedimentation coefficient of 70S. The bacterial ribosome has two subunits named after their sedimentation coefficients. These are the 50S subunit and the 30S subunit. (See Section 1.1.1 and Table 1.2 for a comparison with archaeal and eukaryotic ribosomes.) Together they form the 70S ribosome, which makes the proteins. The 30S subunit has 21 different proteins and one 16S rRNA molecule. Its proteins are named S1, S2, etc., S standing for "small subunit." The 16S rRNA helps to align the mRNA so that it is positioned correctly on the ribosome for the initiation of protein synthesis. The 50S subunit has 31 different proteins, one 23S rRNA molecule, and one 5S rRNA molecule. Its proteins are named L1, L2, etc., L standing for "large subunit." The 23S rRNA catalyzes the formation of the peptide bonds. The ribosome should not be viewed as a static structure but rather as a machine that synthesizes peptide bonds and undergoes conformational changes that result in the threading of the mRNA through the ribosome (translocation) as the protein is being made.

### 10.3.3 Charging of the tRNA (making the aminoacyl–tRNA)

#### Overview

Prior to the initiation of protein synthesis on the ribosome each amino acid must be covalently bound to a tRNA. There are 20 different amino acids that must be activated in this way and the process is catalyzed by 20 different enzymes called aminoacyl–tRNA synthetases.[60] The number of tRNAs can be more than 20 since the code is degenerate; that is, there exist more than one codon for certain amino acids. For example, E. coli can have from 1 to 6 codons for each of its amino acids resulting in 61 codons. (However, there are fewer than 61 different tRNAs in E. coli. The reason for this is explained in Note 61.) The aminoacyl–tRNA synthetases recognize both the amino acid and certain structural features of the tRNA. Some synthetases recognize the anticodon, whereas others recognize nucleotide sequences in other parts of the tRNA. The process of attaching the amino acid to the tRNA is called charging the tRNA; it takes place in the cytosol, and the product is called aminoacyl–tRNA. Of particular importance is the fact that the amino acid must be attached to the right tRNA because it is the tRNA that recognizes the codon, and if the wrong amino acid is attached, then a protein with the wrong sequence of amino acids will be made.

### Transfer RNA (tRNA)

Transfer RNAs are small single-stranded RNAs that are folded with three hairpin loops resembling a cloverleaf (Fig. 10.29). The amino acid is esterified via its carboxyl group to the 3'-OH in the ribose part of the adenylyic acid residue at the 3' end, which is 5'-CCA-OH-3'. One of the loops holds the anticodon triplet that hybridizes to the codon in the RNA template and thus brings the correct amino acid to the growing polypeptide chain. All tRNAs have bases that are modified after transcription during the processing of the tRNA, and these bases appear in or near the hairpin loops. The structures of some of these modified bases are shown in Fig. 10.30.

### Attaching the amino acid to the tRNA (charging the tRNA)

The charging of the tRNA takes place in two steps, both of which are catalyzed by the same enzyme (Fig. 10.31). The enzyme is called aminoacyl-tRNA synthetase and each enzyme recognizes a specific amino acid and its corresponding tRNA. (There are some examples of the same aminoacyl synthetase charging two different tRNAs, followed by the appropriate modification of the amino acid in one of the tRNAs.[62] There are also examples of two different aminoacyl–tRNA synthetases activating the same amino acid and charging the same tRNA.[63] All these cases are summarized in Note 64.) Some aminoacyl–tRNA synthetases recognize simply the anticodon region of the tRNA that determines the specificity of the synthetase for the tRNA, whereas for other synthetases the specificity has little to do with the anticodon but is determined by other parts of the tRNA molecule. The synthesis of the aminoacyl-tRNA takes place via a route similar to the synthesis of acyl–CoA derivatives, that is, via an acyl–AMP intermediate. The first reaction is an attack of an oxygen atom of the carboxyl group of the amino acid on the $\alpha$-phosphate of ATP, displacing pyrophosphate from the ATP and forming an aminoacyl–AMP derivative. Then an oxygen atom of a hydroxyl group on the ribose of the tRNA attacks the aminoacyl–AMP, displacing the AMP and forming the

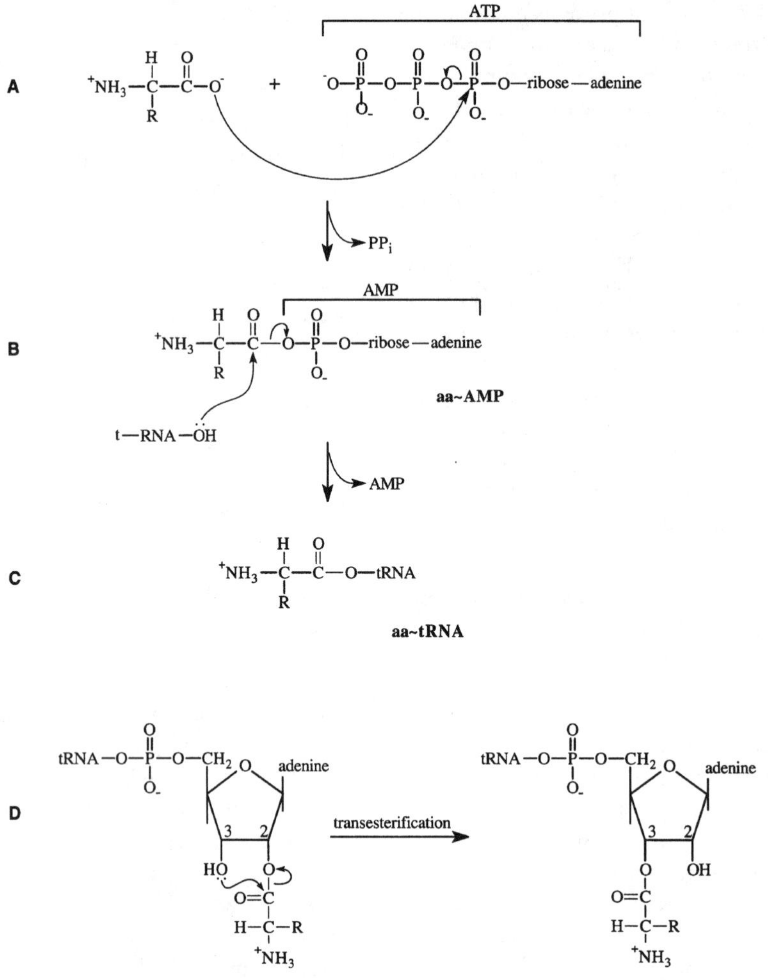

Fig. 10.31 The aminoacyl synthetase reaction. (A) The carboxyl group of the amino acid attacks the α-phosphate in ATP and displaces pyrophosphate to form the AMP derivative of the amino acid. (B) A hydroxyl on the ribose in the terminal adenylate residue attacks the carbonyl group of the amino acid and displaces the AMP. (C) The amino acylated tRNA. (D) Transesterification reaction. There are two classes of synthetases. Class I enzymes catalyze an attack by the 2′-OH of the terminal adenylate residue, and the amino acid is then transferred by a transesterification reaction to the 3′-OH. Class II enzymes catalyze an attack by the 3′-OH so that the ester linkage between the 3′-OH and the amino acid forms directly.

aminoacyl–tRNA derivative. The bond in the final product is an ester linkage between the carboxyl group of the amino acid and the 3′-hydroxyl group on the ribose portion of the tRNA.[65] (See Note 66 for a further description of this reaction.) The reaction is driven to completion by a pyrophosphatase, which hydrolyzes the pyrophosphate. (See Fig. 7.8.) Therefore, the equivalent of 2 ATPs are required to make one aminoacyl–tRNA.

### Making fMet–tRNA^fMet

The first amino acid in the growing polypeptide chain in bacteria is formylmethionine (fMet), and protein synthesis begins with the synthesis in the cytosol of fMet–tRNA^fMet. (The formyl group is removed after the protein is synthesized. Sometimes the methionine is also removed.) The fMet–tRNA^fMet is made in the cytosol when methionine is attached to a special initiator tRNA (tRNA^fMet). The reaction is catalyzed by the same aminoacyl–tRNA synthetase that attaches methionine to the tRNA (tRNA$^{Met}$) responsible for inserting methionine in internal positions in the protein. Then a *transformylase* (tRNA methionyl transformylase) binds to the methionine on the Met–tRNA$^{fMet}$ and catalyzes the transfer of a formyl group from N$^{10}$-formyltetrahydrofolate (10-formyl-THF) to the Met–tRNA$^{fMet}$ to form fMet–tRNA$^{fMet}$. (See Fig. 9.14 for a description of THF as a C$_1$ carrier.) The transformylase recognizes not only the methionine but the tRNA$^{fMet}$ as well

and is able to distinguish between tRNA$^{fMet}$ and tRNA$^{Met}$. The stage is now set for the initiation of protein synthesis on the ribosome.

### 10.3.4 Initiation

#### Making the preinitiation complex

A preinitiation complex is formed with the 30S ribosomal subunit, messenger RNA (mRNA), formylmethionyl–tRNA$^{fMet}$ (fMet–tRNA$^{fMet}$), and initiation factors IF-1, IF-2, and IF-3 (Fig. 10.32). First IF-3 binds to the 30S ribosomal

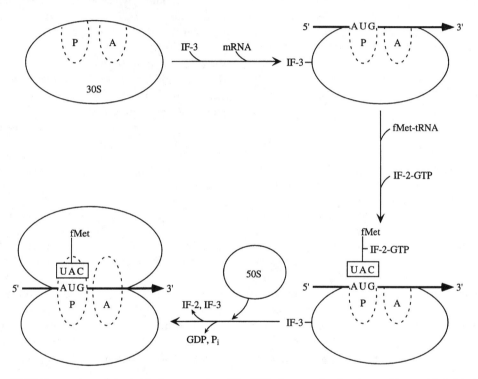

Fig. 10.32 Formation of the initiation complex. The 30S ribosomal subunit combines with initiating factor IF-3 and mRNA to form the 30S initiation complex. There are two sites in the 30S subunit, i.e., the A site and the P site, which are completed upon binding of the 50S subunit. The mRNA is positioned such that its start codon (AUG for formylmethionine) is at the P site. The mRNA binds correctly at the P site because near the 5′ end of the mRNA there is a consensus sequence called the Shine–Dalgarno sequence, and this sequence base pairs to a complementary sequence near the 3′ end of the 16S RNA in the 30S subunit. Then the initiating aminoacylated tRNA, i.e., fmet–tRNA$^{fmet}$, binds to initiating factor IF-2, which itself is complexed to GTP (IF2–GTP), and the ternary complex binds to the 30S subunit such that the anticodon in the fmet–tRNA hydrogen bonds to the AUG at the P site. The 30S initiation complex then binds to the 50S ribosomal subunit to form the 70S initiation complex. When this happens, the GTP is hydrolyzed to GDP and P$_i$, and the GDP, P$_i$, and initation factors leave the ribosome. The role of IF-3 is to prevent premature binding of the 50S subunit to the 30S subunit. IF-2 is required for correct binding of the fMet–tRNA$^{fMet}$ to the 30S subunit. A third initiation factor, IF-1, is also present. It stimulates the activities of IF-3 and IF-2. As mentioned, the A and P sites involve both the 30S and the 50S ribosomal subunits.

subunit, preventing premature association between the 30S and 50S subunits. Then IF-2–GTP helps the fMet–tRNA<sup>fMet</sup> to bind to the start codon via the anticodon on the tRNA$_f$ at what will become the P site when the 50S subunit binds. IF-1 is somehow involved in promoting optimal activity of the other initiation factors.

### Making the 70S initiation complex

After the mRNA, fMet–tRNA<sup>fMet</sup>, and 30S subunit form a complex, the 50S subunit joins to form the *70S initiation complex* as IF-1, IF-2, IF-3, GDP, and P$_i$ are released. It is very important that the mRNA binds to the 30S subunit in the correct position so that the start codon is at the P site. As described next, for many mRNAs, this is due to the Shine–Dalgarno sequence, which positions the mRNA on the ribosome.

### Shine–Dalgarno sequence

The Shine–Dalgarno sequence is a sequence of nucleotides in the initiation region of mRNA that specifies a ribosome binding site. It is a short string of bases (about 5 to 10) approximately 6 to 10 nucleotides downstream of the start codon in many mRNAs (Fig. 10.33). The Shine–Dalgarno sequence in the mRNA base pairs to a complementary sequence at the 3′ region of the 16S rRNA. This places the start codon in the mRNA at the P site in the ribosome, which is where fMet–tRNA<sup>fMet</sup> will bind. Not all mRNAs have a Shine–Dalgarno sequence,

and in fact the start codon may be at the extreme 5′ end of the mRNA. Presumably, under the latter circumstances the sequence in the mRNA that binds to the 16S rRNA to position the start codon in the mRNA at the P site is downstream of the start codon.

### 10.3.5 Chain elongation and termination

#### Start codon

Translation usually begins at an AUG codon (the start codon), which codes for methionine in the *initiator region of the mRNA*. [In a relatively small number of cases the fMet is encoded by the valine codon (GUG) or by the leucine codons (CUG, UUG) if they are at the start site. These codons encode valine or leucine, respectively, if they are downstream from the start site.] There may be internal (downstream) start codons, but these do not serve as translation initiation sites.

#### Binding of aminoacylated–tRNA to the A site

Chain elongation begins when an incoming aminoacylated tRNA binds to the A site on the ribosome as the anticodon on the tRNA base pairs with the codon in the mRNA. The correct positioning of the incoming aminoacyl–tRNA is facilitated by elongation factor Tu bound to GTP (i.e., EF-Tu-GTP), which forms a complex with the incoming aminoacyl–tRNA and guides the aminoacyl–tRNA to the A site (Fig. 10.34A).

**Fig. 10.33** Binding of initiator region of mRNA to 16S rRNA. Upstream from the initiation codon in the mRNA there is a short sequence of nucleotides called the Shine–Dalgarno sequence. The Shine–Dalgarno sequence hybridizes to a complementary sequence of nucleotides at the 3′ end of the 16S rRNA, and this places the initiation codon at the P site, where the fMet–tRNA<sup>fMet</sup> binds. Shine–Dalgarno sequences vary and can be complementary to different regions of the 16S rRNA. Some bacterial genes do not have a Shine–Dalgarno sequence; i.e., the initiation codon is at the very end of the mRNA, as revealed by sequencing the N-terminal region of the protein and comparing it to the nucleotide sequence at the 5′ end of the mRNA. Presumably, mRNAs that do not have a Shine–Dalgarno sequence upstream of the initiation codon have a 16S rRNA binding site downstream of the initiation codon.

## Recycling of EF-Tu-GTP

After the aminoacyl–tRNA is in the A site, GTP is hydrolyzed and EF-Tu-GDP is released from the ribosome. The EF-Tu-GDP then reacts with EF-Ts, which stimulates the exchange of GDP for GTP so that the EF-Tu-GTP is regenerated.

## Formation of the peptide bond

A peptide bond forms when the free $\alpha$-amino group on the amino acid bound to the tRNA at the A site displaces the $tRNA_f$ at the P site (Fig. 10.34B). The reaction is catalyzed by peptidyl transferase, which is part of the 50S ribosomal subunit. Very interestingly, the peptidyl transferase activity appears to reside in the 23S rRNA rather than in a protein.[67] (For other examples of RNA molecules acting as enzymes, see Note 68.) The result is a peptide attached to the tRNA at the A site. The formation of the peptide bond as the ester linkage is broken between the amino acid and the tRNA does not require any additional energy. Recall that ATP was used to form the ester linkage in the first place. Thus the energy to make the peptide bond is derived from ATP. In fact, the formation of the peptide bond uses the equivalent of two ATPs, since the formation of the aminoacyl–tRNA is driven to completion by the hydrolysis of pyrophosphate.

## Translocation

The formation of the peptide bond is followed by translocation (Fig. 10.34C). During translocation the uncharged tRNA leaves the P site, the peptidyl–tRNA moves from the A site to the P site, and the mRNA moves one codon with respect to the ribosome so that a new codon is positioned at the A site. The elongation factor EF-G (translocase) is required for translocation. During translocation GTP is hydrolyzed to GDP and $P_i$. The empty A site is now ready to receive a new aminoacyl–tRNA. Notice that the ribosome moves down the message from the $5'$ end to the $3'$ end and that the protein grows at the carboxy end.

## Termination

Termination occurs when a stop codon reaches the A site (Fig. 10.34D). The stop codons are UAA, UGA, and UAG. There are no tRNAs with anticodons for these codons, and because of this no aminoacyl–tRNA enters the A site when a stop codon is present. Release factors (proteins) bind to the stop codons and cause the release of the completed polypeptide from the ribosome. There are three release factors in E. coli: RF1, RF2, and RF3. RF1 binds to UAA and UAG, RF2 binds to UAA and UGA, and RF3 facilitates the activity of RF1 and RF2. It has been suggested that release factors cause the peptidyl transferase to catalyze an attack by water on the ester linkage between the peptide and the tRNA at the P site, releasing the protein. After the release of the protein the tRNA and mRNA are released, and the ribosome dissociates into its subunits. IF3 is important at this stage because it prevents the 30S and 50S from reassociating before the next initiation complex is formed.

## The requirement for GTP

GTP provides the energy for translocation. It is also necessary for the binding of certain proteins to either tRNA or to the ribosome. How does GTP act? GTP does not act in an analogous manner to ATP. In contrast to ATP, which drives the formation of covalent bonds, GTP is thought to activate proteins by causing a conformational change in the protein. One result of this conformation change is that the protein might then be in an active shape, enabling it to bind to certain targets. Thus the hydrolysis of GTP is associated with the release of GDP and $P_i$ from protein and its return to the inactive state, rather than with the activation of protein.

## 10.3.6 Polysomes, coupled transcription, and translation

Messenger RNAs are translated by several ribosomes in series (typically around 50), ensuring that several copies of the protein can be made from a single mRNA. The complex of several ribosomes attached to a mRNA is called a *polysome*. This increases the rate of protein synthesis, since the amount of protein made per unit time from a particular mRNA is proportional to the

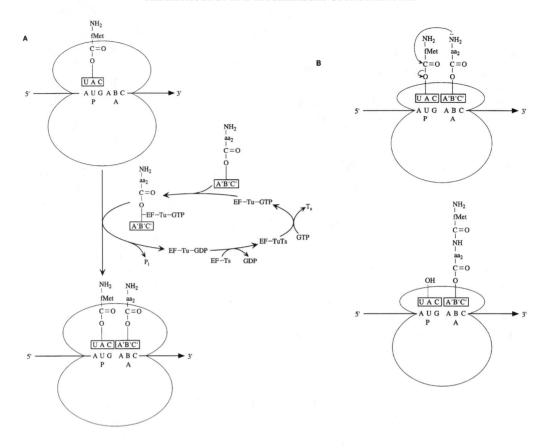

**Fig. 10.34** Chain elongation and termination. (A) Binding of aminoacylated tRNA to the A site. Elongation factor Tu bound to GTP forms a ternary complex with the aminoacylated tRNA, and this complex binds to the ribosome such that the anticodon of the tRNA base pairs with the codon at the A site. The GTP is hydrolyzed to GDP and $P_i$, and the Tu–GDP and $P_i$ are released from the ribosome. The GDP is then displaced from the Tu–GDP complex by elongation factor Ts, and the Ts is subsequently displaced from the complex by GTP to regenerate Tu–GTP, which can then bind to the next aminoacylated tRNA. (B) Formation of the peptide bond. The enzyme, peptidyl transferase, which is actually the 23S RNA in the 50S ribosomal subunit, catalyzes the displacement of the tRNA at the P site by the free amino group of the incoming amino acid. The result is that the growing peptide is transferred from the P site to the A site. (RNA molecules such as the 23S RNA that act as enzymes are called ribozymes.) (C) Translocation. The ribosome moves one codon toward the 3′ end of the mRNA as the tRNA in the P site is released. This positions the growing peptide at the P site and leaves an empty A site for the incoming aminoacylated tRNA. Translocation requires elongation factor G (EF-G) as well as energy from the hydrolysis of GTP to GDP and $P_i$. (D) Termination. When a stop codon reaches the A site, a protein release factor (RF) binds to the termination codon and stimulates peptidyl transferase to hydrolyze the linkage between the peptide and the tRNA at the P site. The ribosome dissociates into its subunits, and the newly synthesized protein is released. IF-3 binds to the 30S subunit to prevent its premature reassociation with the 50S subunit.

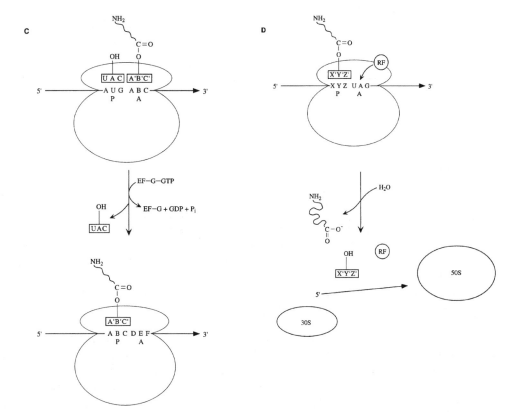

Fig. 10.34 (*continued*)

number of ribosomes translating the mRNA. Furthermore, translation and transcription are coupled in bacteria. What happens is that ribosomes attach to the 5′ end of the mRNA even before the message is completely transcribed, and then move along the mRNA toward the 3′ end as transcription continues (Fig. 10.35). Because translation is coupled to transcription, proteins can begin to be made even before the mRNA is completed, thus shortening the delay between the onset of transcription and the appearance of the protein. Coupled transcription and translation cannot take place in eukaryotes since transcription and translation occur in separate cell compartments (i.e., the nucleus and cytoplasm), respectively.

## 10.3.7 Polycistronic messages

Bacteria frequently make polycistronic messages. (Eukaryotes do not.) This means that several genes are transcribed in series into a single mRNA molecule. Initiation of transcription of the cotranscribed genes is from a single promoter upstream of the gene cluster, and the cluster of genes is called an *operon*. Usually genes that encode proteins that function in a common pathway are in the same operon. Operons in bacteria are the rule, and they may have as few as 2 genes and as many as 10 or more. In order for individual proteins to be made from a polycistronic message, there must be translational start sites at the beginning of each gene transcript and translational stop sites at the end (Fig. 10.36). Futhermore, there must be initiating regions (i.e., ribosome binding sites) at the beginning of each gene transcript. Thus the translation initiation complex forms at the beginning of each gene transcript in the polycistronic message, and the ribosome, mRNA, tRNA, and polypeptide dissociate at the termination site of each gene transcript in the polycistronic message.

## 10.3.8 Polarity

Polarity refers to the phenomenon in polycistronic mRNAs where a block in the translation of an upstream gene can prevent

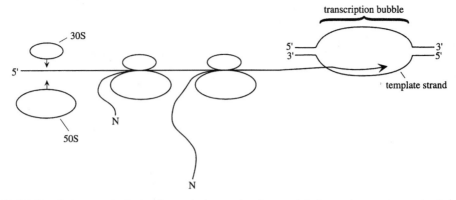

Fig. 10.35 Coupled transcription and translation and polysomes. In bacteria protein synthesis begins before the mRNA is completed. The result is a string of ribosomes on the same mRNA (called a polysome). There may be up to 100 ribosomes on the same mRNA.

*transcription* of a downstream gene. This can happen if, for example, a mutation results in a nonsense codon (i.e., a codon that does not bind to a tRNA) in the upstream gene and there is a Rho-dependent transcription termination site in the upstream gene after the translational stop site. Under these circumstances the RNA polymerase will disengage from the DNA and transcription will stop before the polymerase reaches the downstream gene. For a discussion of Rho-dependent termination of transcription, see Section 10.2.1.

### 10.3.9 Coupled translation

In some instances translation can be coupled. This means that in a polycistronic mRNA translation of the immediate upstream gene is required for translation of the downstream gene. For example, the translation of the genes in the operons coding for ribosomal proteins in *E. coli* is coupled. This affords the opportunity to regulate the translation of the entire operon by regulating the translation of one of the genes. (See Note 22 in Chapter 2 for a further discussion of the regulation of

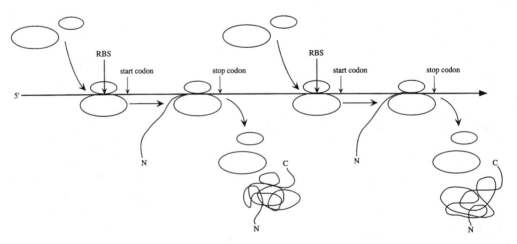

Fig. 10.36 Polycistronic mRNA. The polycistronic mRNA has more than one gene. At the beginning of each gene there is a ribosome binding site (RBS) where initiation begins. The ribosome then begins translating at the initiation codon and stops at the stop codon at the end of the gene. Upon termination, the ribosome and polypeptide are released and initiation begins anew at the beginning of the next gene. There exists an intercistronic spacer region that may comprise as few as 1 or as many as 30 to 40 bases between genes. The amino and carboxy terminals of the polypeptide are labeled N and C, respectively.

translation of the ribosomal protein operons.) A model for translational coupling postulates that the translational start region, including the start codon, for the downstream gene is inside a hairpin loop in the mRNA so that it cannot bind to a ribosome. When the ribosome translating the upstream gene comes to the stop codon of the upstream gene, then the ribosome may disrupt the secondary structure of the mRNA so that the translational start region for the downstream gene is accessible to a second ribosome (Fig. 10.37).

## 10.3.10 Folding of newly synthesized proteins: The role of chaperone proteins

### The role of chaperone proteins in folding of newly synthesized proteins

The information to fold into the active tertiary structure is contained in the sequence of amino acids, that is, the primary structure of the protein. (See Note 69 for definitions of primary, secondary, tertiary, and quaternary structures of proteins.) However, for many proteins proper folding of newly synthesized proteins needs help from other proteins called

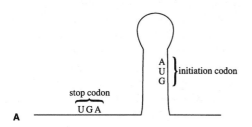

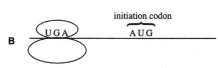

**Fig. 10.37** Model for translational coupling. (A) It is postulated that in a polycistronic mRNA the start codon for the downstream gene is in a hairpin loop so that it is not accessible to the ribosome. (B) When the ribosome translating the upstream gene arrives at the stop codon, it disrupts the secondary structure of the mRNA so that a second ribosome has access to the initiation site in the downstream gene.

*chaperone proteins* that transiently associate with the newly synthesized proteins and assist in the folding. One role of chaperone proteins is to prevent misfolding and aggregation. It has been estimated that at least 50% of newly synthesized proteins are aided in their folding by chaperone proteins. Some proteins, however, fold independently of chaperone proteins. These are probably rapidly folding small proteins. Chaperone proteins also bring SecB-independent preproteins to the membrane for secretion (Section 19.1.3).

### Measuring the flux of proteins through the chaperonins

One way to quantitate the flux of newly synthesized proteins through the chaperone proteins is to incubate the bacteria for a few seconds with a radioactive amino acid and then add a large excess of the nonradioactive amino acid. The addition of the nonradioactive amino acid lowers the specific activity of the radioactive amino acid sufficiently so that further incorporation of radioactivity stops. This is called a "pulse-chase" experiment. A specific chaperone protein can be immunoprecipitated from cell extracts at various times during the "chase," and the different radioactive proteins transiently associated with the chaperone protein, as well as the transit time, can be visualized by autoradiography after gel electrophoresis. The flux of proteins through the chaperone protein can be estimated by measuring the radioactivity in individual protein bands separated by gel electrophoresis, as well as the total amount of radioactivity that is immunoprecipitated during the "chase."

### GroEL/GroES

Experiments such as the ones just described were performed using *E. coli* and anti-GroEL antibody to monitor the flux of proteins through the GroEL chaperonin, which cooperates with GroES to produce proper folding.[70] (Chaperone proteins such as GroEL exist as multisubunit complexes. These complexes are called *chaperonins*. See Note 71 for a more complete description of GroEL and GroES.) The major conclusions from the pulse-chase experiment were that: (1) A variety of newly synthesized proteins (as judged by the number of bands that migrated during gel electrophoresis) transiently associate with GroEL, and this can account for 10% to 15% of the protein during normal growth temperatures. (2) Upon exposure to heat stress, where the relative amounts of GroEL and other chaperonins increase in the cell, at least 30% of the newly synthesized proteins associate with GroEL. (See the discussion of chaperonins and heat-shock proteins in Section 19.1.3.) It appears that GroEL binds to the protein *after* synthesis is complete. The reason for concluding this is that when the size distribution of polypeptides bound to GroEL was examined, it was observed that they were of distinct sizes, rather than a continuum, as might be expected if nascent polypeptide chains bound to the chaperonin. However, although GroEL becomes associated with polypeptides after their release from ribosomes, it appears that two other chaperone proteins, DnaK and DnaJ, associate with nascent polypeptides while on the ribosome. This is described next.

### A model: First DnaK and DnaJ, then GroEL.

There is evidence that the chaperone proteins DnaK and DnaJ can interact with a least some nascent polypeptides on ribosomes and that GroEL becomes associated with the polypepeptides after their release from ribosomes.[72] (See Note 73 for a description of the experiments that support this conclusion, and Section 19.1.3 for the role of these proteins, as well as the GroE system, in the heat-shock response.) Perhaps what happens is that DnaK and DnaJ prevent the polypeptides from premature folding and aggregation while they are being synthesized, but the actual folding of the completed polypeptide takes place when it is transferred to GroEL. (See Ref. 74 for a review.) Recently another ribosome-associated protein called *trigger factor* has been implicated in the folding of certain nascent proteins and may also take part in transferring nascent proteins to GroEL.[75] It is clear that much still needs to be learned about the roles that various chaperones play in the folding of nascent polypeptides.

### 10.3.11 Inhibitors

Some antibiotic inhibitors of protein synthesis and their sites of action are listed in Table 10.1. Note that chloramphenicol, erythromycin, and streptomycin inhibit bacterial but not archaeal or eukaryotic (cytosolic) protein synthesis, and that diphtheria toxin does not inhibit bacterial but does inhibit archaeal and eukaryotic protein synthesis. This reflects differences in the ribosomal and accessory proteins required for translation in these systems.

### 10.4 Summary

Several problems must be attended to in order to replicate DNA. These include unwinding the double helix without causing it to overwind in a positive supercoil in the unreplicated portion, creating a primer that can be extended by DNA polymerase III, and keeping the polymerase that synthesizes the lagging strand at the moving replication fork. These and other factors necessitate using about 20 different proteins to replicate the DNA. The replication fork is created at a specific sequence in the DNA (called the origin) when DNaA, HU, and ATP begin to unwind the helix. Further unwinding of the DNA during replication requires replication fork helicase (DnaB). Single-stranded binding protein (SSB) binds to the single strands to prevent them from coming together again so that they can serve as templates for new DNA. Meanwhile, DNA gyrase binds to the DNA downstream of the replication fork and continuously converts the positive supercoils into negative supercoils so that the DNA does not become overwound.

Replication of the DNA is initiated when the enzyme primase synthesizes a short RNA oligonucleotide at the replication fork on both the single strands. One copy strand, called the leading strand, is synthesized continuously by elongating the 3' end of the RNA primer, which faces the replication fork. The other copy strand, called the lagging strand, is synthesized in short fragments of about 1000 nucleotides called Okazaki fragments in a direction opposite to the movement of the

**Table 10.1** Antibiotics that inhibit protein synthesis

| Antibiotic | Site of action | Bacteria | Archaea | Eukaryotes |
|---|---|---|---|---|
| Erythromycin | Inhibits translocation | + | − | − |
| Streptomycin[a] | Inhibits initiation and causes mistakes in reading mRNA at low concentrations | + | − | − |
| Tetracycline[b] | Inhibits binding of aminoacyl–tRNA (blocks A site) | + | + | + |
| Chloramphenicol | Inhibits peptidyl transferase | + | − | |
| Puromycin[c] | Resembles charged tRNA and causes causes premature chain termination | + | − | − |
| Cycloheximide | Inhibits peptidyl transferase | + | + | + |
| Diphtheria toxin | Inhibits elongation factor 2 (eEF2), which is analogous to bacterial EF-G, by ADP-ribosylation | − | − | + |
| | | − | + | + |

[a] Streptomycin binds to the S12 protein in the 30S ribosomal subunit and inhibits the binding of fMet–tRNA$^{fMet}$ to the P site. Neomycin and kanamycin act in a manner similar to streptomycin.

[b] Tetracyclines inhibit bacterial protein synthesis by binding to the 30S ribosomal subunit and preventing attachment of aminoacylated tRNA. They also inhibit eukaryotic ribosomes but not at the concentrations used to treat bacterial infections. This is apparently because eukaryotic membranes are not very permeable to tetracycline.

[c] Puromycin resembles the aminoacyl portion of aminoacylated tRNA. It has a free amino group that attacks the ester linkage in the growing peptide in the P site, with the result that a peptidyl puromycin forms at the A site. However, the peptidyl puromycin cannot be transferred to the P site (i.e., translocation does not occur), and therefore chain elongation is terminated as the peptidyl puromycin dissociates from the ribosome.

replication fork. Each fragment begins with a short oligonucleotide RNA primer. The 3′ end of the lagging strand faces away from the replication fork. The same DNA polymerase would be able to synthesize both copy strands if the polymerase were dimeric and the template for the lagging strand looped around the polymerase so that its 5′ end faced the replication fork. The RNA primers as the 5′ end of the copy strands are removed by DNA polymerase I, which also fills the gap. The break is sealed by DNA ligase. Replication is usually bidirectional.

Termination occurs at *ter* sites and requires Tus protein. The enzyme topoisomerase IV separates the newly synthesized duplexes. In the event that an unequal number of recombinations have taken place between sister chromsomes during replication, then the daughter chromosomes are covalently linked, and a site-specific recombination must take place at the *dif* site, which is in the Ter region.

The partitioning of the nucleoids to opposite poles of the cells is not an understood process. However, proteins are being discovered that play a role in certain bacteria. These are the Par proteins and MukB. The Par proteins are involved in aligning DNA in some bacteria and perhaps attaching the origins of replication to the cell poles. The MukB protein may be involved in moving the nucleoids in *E. coli* to the cell poles.

Errors made during replication are repaired in two ways. One way is called editing repair and involves the immediate removal of the incorrect nucleotide by the 3′-exonuclease activity of DNA polymerase III. A second method of repair can occur if the polymerase fails to remove the incorrect nucleotide and proceeds beyond the error site. If this happens, then mismatch repair can take place. During mismatch repair an exonuclease removes the segment of copy DNA that includes the mismatched nucleotide pair and the gap is filled with DNA polymerase III and sealed with DNA ligase. The copy strand is recognized because for a short period after synthesis it is undermethylated.

RNA synthesis also requires localized unwinding of the DNA duplex, but the unwinding is done by RNA polymerase rather than helicase. Initiation takes place at promoter sites, which are nucleotide sequences where RNA polymerase binds upstream of the transcription start site. The promoter is recognized by a subunit in the polymeras called sigma factor. The sigma 70 polymerase recognizes concensus sequences at the $-10$ and $-35$ regions, where the transcription start site is $+1$. Other sigma factors exist, but these do not recognize the same sequence recognized by sigma 70. DNA gyrase is required for transcription to prevent the DNA from being overwound downstream from the transcribed region. Special termination codons exist that signal the end of the gene and the release of the RNA polymerase. Transcription is regulated by a variety of positive and negative transcription factors that bind to the DNA in the promoter region or upstream of the promoter (enhancer regions). The transcription factors can influence the binding of the RNA polymerase to the promoter or the rate at which the open complex forms. In many operons, RNA synthesis is also regulated by attenuation.

Protein synthesis begins with the aminoacylation of tRNA$^{fMet}$ in the cytosol. The enzymes that aminoacylate the tRNAs are called aminoacyl–tRNA synthetases. The tRNA is an adaptor molecule, and its anticodon base pairs with the codon in the mRNA, ensuring that the amino acid is placed in the correct sequence. The aminoacyl–tRNA synthetases, which attach the amino acids to the tRNA, distinguish between the different tRNAs and attach the amino acid only to the tRNA with the correct anticodon for that amino acid. Then, in the presence of initiation factors IF-1, IF-2, and IF-3, as well as mRNA and the 30S and 50S ribosomal subunits, the 70S initiation complex is formed. In this complex the fMet–tRNA$^{fMet}$ is positioned at the P site in the ribosome. This positioning is directed by a sequence of nucleotides called the Shine–Dalgarno sequence 6 to 10 nucleotides downstream of the transcription start codon. The Shine–Dalgarno sequence hybridizes with the 3′ end of the 16S rRNA in the 30S ribosomal subunit, and because of this the start codon is positioned at the P site on the ribosome. The next aminoacylated tRNA enters and is positioned at the A site. This requires EF-Tu and GTP. Once the P site and A site are occupied by aminoacylated tRNAs, the alpha amino group of the amino acid in the A site displaces the tRNA at the P site, and an aminoacylated peptide is made at the A site. After formation of the aminoacylated peptide, translocation takes place. During translocation the ribosome moves one codon with respect to the mRNA, and the aminoacylated peptide is transferred from the A site to the P site, thus enabling the A site to receive the next aminoacylated tRNA. Translocation requires elongation factor EF-G and GTP. The process is repeated until the ribosome reaches the end of the gene. At the end of the gene there is a termination codon for which there is no tRNA. When the A site of the ribosome reaches the termination codon, protein termination factors cause peptidyl transferase to hydrolyze the linkage between the protein and the tRNA in the P site, thus releasing the protein. The ribosome also dissociates into its subunits, which can then initiate a new round of translation at the beginning of a different gene.

Once the protein is synthesized, it folds into its tertiary structure. This is frequently aided by a class of proteins called chaperone proteins.

## Study Questions

1. Describe the roles that the following proteins play in DNA replication: helicase, DNA gyrase, DNA ligase, DNA polymerase III, DNA polymerase I, primase, DnaA, SSB.

2. Compare the biochemical reactions catalyzed by DNA polymerase and DNA ligase. How do they differ?

3. Describe the Meselson–Stahl experiment and explain how it proves that DNA is replicated semiconservatively.

4. DNA exists in a negative supercoil. How might this be advantageous to DNA replication and RNA synthesis? During DNA replication the unreplicated portion of the DNA winds tighter. Why is this the case?

5. What are Okazaki fragments, and why are they necessary?

6. Why is the 3-exonuclease activity of DNA polymerase III important?

7. What is the role of the sigma subunit in RNA synthesis? Under what circumstances might a bacterium use different sigma factors?

8. During RNA synthesis, what unwinds the DNA? How does this compare to the unwinding during DNA replication?

9. How might transcription factors increase or decrease the rates of transcription?

10. How is RNA synthesis terminated?

11. The addition of tryptophan to the growth medium of an *E. coli trpR* mutant lowers the expression of the *trp* operon. What is the explanation for this?

12. What is meant by a termination codon? How is it involved in the termination of protein synthesis?

13. What is meant by ribozyme? Give an example of a ribozyme in protein synthesis.

14. What ensures that the correct amino acid is added to the correct tRNA? If an incorrect amino acid is added, can this mistake be corrected?

15. Describe the roles of IF-1, IF-2, IF-3, and EF-G in protein synthesis.

16. What is the direction in which mRNA is translated (i.e., 5′ to 3′ or 3′ to 5′)? How is that relevant to the coupling between transcription and translation?

17. Explain the role of chaperone proteins in protein folding. What is the evidence that the binding to chaperone proteins is transient?

## NOTES AND REFERENCES

1. Most bacteria have circular chromosomes. Exceptions include *Borrelia burgdorferi*, which causes Lyme disease, *Rhodococcus fasciens*, and *Streptomyces lividans*. See Ref. 2.

2. Hinnebursch, J., and K. Tilly. 1993. Linear plasmids and chromosomes in bacteria. *Mol. Microbiol.* 10:917–922.

3. Supercoiling of DNA is an important aspect of its structure, in part because it makes the DNA more compact. Consider DNA as consisting of two strands coiled around a central axis. Supercoiling refers to the twisting or coiling of the central axis. This is analogous to a coiled telephone cord that twists on itself to form supercoils. Suppose the DNA exists as a closed double helical circle. If one strand is cut and unwound while the other strand is prevented from turning, and then the cut is sealed, there will be fewer helical turns in the circular DNA. The DNA will then be under structural strain and as a consequence will supercoil to relieve the strain. If the DNA is underwound (i.e., fewer helical turns), then it twists on itself into *negative* (left-handed) supercoils to relieve the stress. If the DNA is overwound (i.e., more helical turns are introduced before sealing the cut), it twists upon itself to produce *positive* (right-handed) supercoils to relieve the strain. Most cellular DNA is underwound (i.e., fewer helical turns) and negatively supercoiled. It is believed that the strands of the underwound DNA are more readily separated for replication and transcription. The production of negative supercoils is accomplished by a topoisomerase II enzyme called DNA gyrase and does not occur by unwinding the cut strands before rejoining them. (See text for how DNA gyrase introduces negative supercoils.)

4. Topoisomerases change the linking number of a covalently closed duplex of DNA. The linking number is the number of times that the chains in the duplex of a covalently closed circle cross one another; that is, it is the number of helical turns. DNA is said to be overwound when it has more than 10.5 base pairs per helical turn and underwound when it has fewer than 10.5 base pairs per helical turn. The original model proposed that the topoisomerase cut one strand of the duplex, and while holding both ends of the nicked strands, rotated a free end around the intact strand, thus changing the number of helical turns. This was followed by sealing the break. A more recent model suggests that rotation is not necessary and the number of helical turns can be changed simply by passing the unbroken strand through the break in the complementary strand. An example is Type I topoisomerase. Type I topoisomerases catalyze a break in one of the strands, which allows the passage through the break of the unbroken partner strand. The break is then sealed. This results in changing the linking number (number of helical turns) by multiples of one. Type II topoisomerases do it somewhat differently and change the linking number by multiples of two. Type II topoisomerases catalyze breakage of *both* strands of the duplex and allow the passage of *two* unbroken complementary strands from a nearby region of the molecule through the breaks before sealing. (The breaks are staggered breaks, i.e., not directly opposite each other.) Because type II topoisomerases change the number of times two strands are crossed per catalytic event, they alter the linking number

in multiples of two. An example of a type II topoisomerase is DNA gyrase from *E. coli*. It produces negative supercoils, unwinding the double helix in the following way: (1) It binds to the DNA, wrapping approximately 120 base pairs of duplex around itself. (2) It cleaves both strands with a four-base-pair stagger. When it does this, the enzyme covalently bonds to the 5′ ends of both strands. (3) It moves the uncut complementary strands through the break. The region that is moved is either within the region that is wrapped around the enzyme or close by. (4) It reseals the breaks. Passing the unbroken portion through the broken portion to produce negative supercoils unwinds the DNA. ATP is required for DNA gyrase activity.

To see how negative supercoiling can unwind a helix, take two pieces of rubber tubing and wrap them around each other in a right-hand helix. This would be clockwise sighting down the helix away from you. Clamp both ends. Now twist the linear helix so that the central axis twists counterclockwise. This produces negative supercoils. The helix will unwind. Twisting the helix so that the central axis turns clockwise (positive supercoils) makes the helix wind tighter.

Topoisomerase I and DNA gyrase adjust the supercoiling of DNA. Gyrase adds negative supercoils, and topoisomerase I removes negative supercoils. For a review of topoisomerases, see Luttinger, A. 1995. The twisted "life" of DNA in the cell: Bacterial topoisomerases. *Molec. Microbiol.* 15:601–608.

5. Marians, K. J. 1992. Prokaryotic DNA replication. *Ann. Rev. Biochem.* 61:673–719.

6. Baker, T. A., and S. H. Wickner. 1992. Genetics and enzymology of DNA replication in *Escherichia coli. Ann. Rev. Gen.* 26:447–477.

7. Messer, W., and C. Weigel. 1996. Initiation of chromosome replication, pp. 1579–1601. In: *Escherichia coli and Salmonella: Cellular and Molecular Biology.* Vol. 1. F. C. Neidhardt (Ed.). ASM Press, Washington, D.C.

8. Marians, K. J. 1996. Replication fork propagation, pp. 749–763. In: *Escherichia coli and Salmonella: Cellular and Molecular Biology.* Vol. 1. F. C. Neidhardt (Ed.). ASM Press, Washington, D.C.

9. Bidirectional replication can be demonstrated in two ways. One way is to insert into the chromosome prophage λ at the att site and prophage Mu within various mapped genes around the chromosome. The Mu can be located because the location of the gene is known, and when it inserts into a gene, it causes a mutation in that gene. If such mutants are also auxotrophic for certain amino acids, the experimenter can stop the initiation of new rounds of replication by removing the required amino acid. By adding the amino acid back again, the experimenter can initiate new rounds of replication at the replication origin. If new rounds of replication are begun with bromouracil in the medium, then bromouracil is incorporated into the DNA in place of thymine. The presence of the bromouracil makes the newly synthesized DNA denser than the parental strand. The newly synthesized strands can then be separated at various times after the initiation of replication by density gradient centrifugation and hybridized with both λ and Mu, and the ratio of Mu DNA to λ DNA can be measured. Knowing the map position of the various Mu inserts, it is possible to demonstrate that replication proceeds bidirectionally from the origin of replication. Bidirectional replication forks have also been observed using radioautography and electron microscopy after a pulse of tritiated thymidine. Both replication forks can be seen to be labeled when examining replication in several bacteria. If DNA replication were unidirectional, then only one replication fork would be labeled.

10. DnaA binds to a series of four 9-base-pair sites arranged as two inverted repeats called R1 to R4 as well as a fifth site called M in the duplex DNA The five sites are sometimes called *DnaA boxes.*

11. The concentration of DnaA does not vary very much during the cell cycle. However, it is thought that the *activity* of DnaA does vary and is responsible for the initiation of DNA replication. One line of evidence for the role of DnaA in initiation is derived from studying a temperature-sensitive *dnaA* mutant harboring a plasmid containing wild-type *dnaA* under the control of the inducible *lac* promoter. When the cells are grown at 42°C, replication is dependent upon the expression of the plasmid *dnaA* gene. The timing of initiation varies with the levels of expression of the gene. When DnaA levels are increased, initiation occurs earlier in the cell cycle. This has led to the suggestion that although the total amounts of DnaA do not vary very much during the cell cycle in wild-type cells, the activity of the protein may vary. It is not known what regulates the activity of DnaA, but it has been suggested that binding to phospholipids in the cell membrane may play a role. A fraction of the cellular DnaA can be isolated with the membrane fractions. DnaA also forms complexes with phospholipids, and the phospholipid can inhibit DnaA activity. Thus one model proposes that the phospholipids in the membrane keep sequestered DnaA in an inactive form until it is bound to *oriC*.

Another factor that may play a role in regulating the initiation of DNA synthesis is methylation of the DNA. The origin region is rich in GATC/CTAG sequences. These sequences are methylated at the adenine residues by Dam methylase shortly after a strand is copied. Therefore, for a short period of time after initiation (about 10 min) the origin is hemimethylated; i.e., the template strand is methylated, but the copy strand is not. DNA fragments containing *oriC* cosediment with the outer membrane fraction in cell lysates, and this has led to the conclusion that for at least part of the replication cycle, *oriC* is bound to the membrane. The model proposes that the hemimethylated origin is bound to the membrane right after replication with the help of a protein called SeqA. This sequesters the origin in the membrane and makes it unavailable for reinitiation in the same cell cycle. The release of *oriC* from the membrane occurs when it becomes fully methylated by Dam methylase. However, this takes place well before the next round of initiation, and therefore other controls besides sequestration in the membrane prevent early reinitiation. [What these other controls are in *E. coli* are a matter of speculation. However, there is evidence for a repressor (CtrA) of the initiation of replication in *Caulobacter crescentus*. See Section 18.16.2.] The evidence in favor of the model includes: (1) Hemimethylated *oriC* regions bind more readily to membranes than do fully methylated *oriC* regions. (2) Higher levels of Dam methylase cause premature initiation of replication. This was done by introducing a plasmid containing *dam*[+] into an *E. coli* dam[−] mutant. (Methylation is not required for viability.) The plasmid *dam* gene was expressed from a heat-inducible promoter ($\lambda$p$_L$ promoter and the temperature-sensitive *cI857* repressor gene). As the temperature is increased, more Dam is made.

The binding of *oriC* to the membrane has been demonstrated in several ways. These include isolation of the membrane fractions and showing the presence of *oriC* sequences using restriction mapping and Southern hybridization. The *oriC* region is generally isolated with the outer membrane fraction.

For a review, see Funnel, B. E. 1996. *The Role of the Bacterial Membrane in Chromosome Replication and Partition*. Chapman and Hall, New York. Sequestration of the origin also plays a role in regulating transcription in the oriC region, including the transcription of *dnaA*. For a discussion of this point, see Bogan, J. A., and C. E. Helmstetter. 1997. DNA sequestration and transcription in the oriC region of *Escherichia coli*. *Molec. Microbiol.* **26**:889–896.

12. Funnell, B. E. 1996. *The Role of the Bacterial Membrane in Chromosome Replication and Partition*. Chapman and Hall, New York.

13. Lohman, T. M., and K. P. Bjornson. 1996. Mechanisms of helicase-catalyzed DNA unwinding. *Ann. Rev. Biochem.* **65**:169–214.

14. Scherzinger, E., G. Ziegelin, M. Bárcena, J. M. Carazo, and E. Lanka. 1997. The RepA protein of plasmid RSF1010 is a replicative DNA helicase. *J. Biol. Chem.* **272**:30228–30236.

15. Helicases separate the strands of the double helix by breaking the hydrogen bonds between the bases. The energy required to do so is provided by nucleotide triphosphates, usually ATP. Helicases are oligomers, typically dimers or hexamers. There is a central hole in the hexamer through which one or both strands (it is not known which) passes. DnaB is a triangle-shaped hexamer and consists of a trimer of dimers. It moves in the 5' to 3' direction along the lagging strand template. There is more than one helicase in *E. coli*. However, the major helicase for bacterial DNA replication is DnaB. For a further discussion of helicases, see Lohman, T. M., and K. P. Bjornson. 1996. Mechanisms of helicase-catalyzed DNA unwinding. *Ann. Rev. Biochem.* **65**:169–214.

16. The realization that DNA copied off one one template is made in short fragments came from experiments where bacteriophage T4 that was infecting *E. coli* was labeled with tritiated thymidine for various lengths of time (2 to 60 sec). The DNA was denatured with base, and the intermediates were separated from the template DNA by centrifugation in a sucrose gradient. It was found that when the DNA was labeled for a short period of time, substantial radioactivity was recovered in small pieces of DNA about 1500 nucleotides long. As the labeling time was increased, the radioactivity in the smaller pieces remained constant but accumulated in high-milecular-weight DNA. Okazaki fragments are made during the synthesis of all DNA molecules in viruses, prokaryotes, and eukaryotes.

17. The RNA polymerase that synthesizes most cellular RNA is required for the initiation of DNA replication. However, it is not clear whether it is required to synthesize the leading strand RNA primer or whether it is required to separate the DNA strands at *oriC* (so that DnaA can bind) by transcribing *oriC*. If the latter were the case, then primase may synthesize the leading strand primer as well as the lagging strand primer.

18. Kuempel, P. L., A. J. Pelletier, and T. M. Hill. 1989. Tus and the terminators: The arrest of replication in prokaryotes. *Cell* **59**:581–583.

19. Khatri, G. S., T. MacAllister, P. R. Sista, and D. Bastia. 1989. The replication terminator protein of *E. coli* is a DNA sequence-specific contra-helicase. *Cell* **59**:667–674.

20. Manna, A. C., K. S. Pai, D. E. Bussiere, C. Davies, S. W. White, and D. Bastia. 1996. Helicase–contrahelicase interaction and the mechanism of termination of DNA replication. *Cell* **87**:881–891.

21. *B. subtilis* also has 6 Ter sites (TerI through TerVI), and they are polar as in *E. coli*. The protein in *B. subtilis* that binds to the Ter sequences is called RTP (replication terminator protein). The *B. subtilis* Ter sites and RTP seem unrelated in sequences to the *E. coli* terminators or Tus.

22. Rothfield, L. I. 1994. Bacterial chromosome segregation. *Cell* 77:963–966.

23. *E. coli* mutants that fail to unlink daughter chromosomes are called Par⁻ mutants. They are temperature sensitive for growth, and at 42°C the DNA is a large mass in the center of an elongated cell. The *par* genes code for different proteins including proteins involved in DNA replication: *dnaG* is required for synthesis of the RNA primer. *parA* and *parD* are the same as *gyrB* and *gyrA*, which encode two polypeptide subunits of DNA gyrase, B and A, respectively. *parE* and *parC* encode subunits of topoisomerase IV, which are highly homologous to the B and A subunits in DNA gyrase. They both form similar heterotetrameric dimers (A$_2$B$_2$). Topoisomerase IV is required to unlink the daughter chromosomes.

Mutants defective in chromosome segregation rather than in detachment of daughter chromosomes have a different phenotype and are not *par* gene mutants. They produce anucleate cells and sometimes multinucleate filaments with abnormally arranged nucleoids. The genes involved in segregation of daughter chromosomes include the *minD* (*E. coli*), *muk* (*E. coli*), *spoOJ* (*B. subtilis*), and *spoIIIE* (*B. subtilis*) genes. Consider the Muk mutants. (Muk is derived from the Japanese word mukaku, which means anucleate.) When grown at 22°C *mukB*⁻ forms normal nucleated rods with a small percent of anucleate cells (5%). At higher temperatures (e.g., 42°C) the cells die, and anucleate cells and multinucleate filaments with abnormal patterns of nucleoid segregation accumulate (clumps and isolated nucleoids). When examining two paired daughter cells, it was seen that some of the anucleate cells were paired with cells that appeared to have more than one nucleoid, reflecting the fact that both daughter chromosomes remained with one of the daughter cells upon division. It has been suggested that MukB might be a motor that moves the daughter chromosomes to opposite cell poles. Another example of segregation-defective mutants are *minD* mutants. The *minD*⁻ mutants produce anucleate rods of normal size, as well as nucleated rods. (The *min* genes are also involved in regulating the site of septum formation, as discussed in Section 2.2.5.) A third example is the *spoOJ* gene in *Bacillus subtilis*. Mutations in *spoOJ* result in a small proportion (1.4%) of anucleate cells. The predicted amino acid sequence of SpoOJ indicates that it may be related in function to ParB, a DNA-binding protein used for plasmid partitioning. SpoIIIE is a DNA translocase that threads the chromosome through the septum into the forespore. It is discussed in Section 18.13.1. Mutations in *spoIIIE* are not defective in chromosome partitioning during vegetative growth.

24. Kato, J-I, Y. Nishimura, R. Imamura, H. Niki, S. Hiraga, and H. Suzuki. 1990. New topoisomerase essential for chromosome segregation in *E. coli*. *Cell* 63:393–404.

25. Kato, J-I, Y. Nishimura, M. Yamada, H. Suzuki, and Y. Hirota. 1988. Gene organization in the region containing a new gene involved in chromosome partition in *Escherichia coli*. *J. Bacteriol.* 170:3967–3977.

26. Cell division and DNA metabolism are coupled. That is to say, cell division is inhibited whenever DNA replication or partitioning is blocked. This is advantageous in that it lowers the frequency of formation of anucleate cells. When cell division is inhibited, long cells or filaments form. If chromosome separation or segregation is impaired, then the nucleoids are abnormally placed in the elongated cells or filaments. There are two mechanisms responsible for inhibiting cell division. One of these requires SulA. Blocking DNA replication results in the synthesis of SulA as part of the SOS response. (See Section 19.1.3 for a discussion of the SOS response and SulA.) SulA inhibits the formation of the FtsZ septal ring; hence cell division is inhibited. There is also a SulA-independent block in cell division whenever DNA replication or segregation are inhibited.

27. Ward, D., and A. Newton. 1997. Requirement of topoisomerase IV *parC* and *parE* genes for cell cycle progression and developmental regulation in *Caulobacter crescentus*. *Molec. Microbiol.* 26:897–910.

28. Kuempel, P. L., J. M. Henson, L. Dirks, M. Tecklenburg, and D. F. Lim. 1991. *dif*, A redA- independent recombination site in the terminus region of the chromosome of *Escherichia coli*. *The New Biologist* 3:799–811.

29. Blakely, G., G. May, R. McCulloch, L. K. Arciszewska, M. Burke, S. T. Lovett, and D. J. Sherratt. 1993. Two related recombinases are required for site-specific recombination at *dif* and *cer* in *E. coli* K12. *Cell* 75:351–361.

30. Webb, C. D., A. Teleman, S. Gordon, A. Straight, and A. Belmont. 1997. Bipolar localization of the replication origins of chromosomes in vegetative and sporulating cells of *B. subtilis*. *Cell* 88:667–674.

31. Errington, J. 1998. Dramatic new view of bacterial chromosome segregation. *ASM News* 64:210–217.

32. Wheeler, R. T., and L. Shapiro. 1997. Bacterial chromosome segregation: Is there a mitotic apparatus? *Cell* **88**:577–579.

33. Mohl, D. A., and J. W. Gober. Cell cycle-dependent polar localization of chromosome partitioning proteins in *Caulobacter crescentus*. *Cell* **88**:675–684.

34. ParA and ParB are proteins required for the partitioning of certain low-copy plasmids such as F factor, R1, RK2, or the prophage form of P1. ParB is a DNA-binding protein that binds to the *parS* sequence in the plasmid DNA. It is thought that ParA and ParB form a complex that is somehow involved in partitioning of plasmids. Some bacteria possess homologues to the *parA* and *parB* genes, and as discussed in Section 10.1.4, there is good evidence that ParA and ParB function in chromosome partitioning in *C. crescentus*.

35. Immunofluorescent microscopy to visualize ParA and ParB was performed with synchronized cultures and the appropriate antibodies. The bacteria were fixed with glutaraldehyde and formaldehyde and then washed and treated with lysozyme. After further treatment with methanol and acetone, they were incubated with antibody. The antibody was visualized with secondary antibody, which was goat–antirabbit IgG conjugated to either fluorescein or biotin. The biotin-conjugated antibody was visualized with streptavidin-conjugated Texas red which binds to avidin.

36. Wake, R. G., and J. Errington. 1995. Chromosome partitioning in bacteria. *Ann. Rev. Gen.* **29**:41–67.

37. Bone, E. J., J. A. Todd, D. J. Ellar, M. G. Sargent, and A. W. Wyke. 1985. Membrane particles from *Escherichia coli* and *Bacillus subtilis*, containing penicillin-binding proteins and enriched for chromosomal-origin DNA. *J. Bacteriol.* **164**:192–200.

38. Jacob, F., S. Brenner, and F. Cuzin. 1963. On the regulation of DNA replication in bacteria. *Cold Spring Harbor Symp. Quant. Biol.* **28**:329–348.

39. Hiraga, S. 1992. Chromosome and plasmid partition in *Escherichia coli*. *Ann. Rev. Biochem.* **61**:283–306.

40. Niki, H., R. Imamura, M. Kitaoka, K. Yamanaka, T. Ogura, and S. Hiraga. 1992. *E. coli* MukB protein involved in chromosome partition forms a homodimer with a rod-and-hinge structure having DNA binding and ATP/GTP binding activities. *The EMBO Journal* **11**:5101–5109.

41. Yamanaka, K., T. Ogura, H. Niki, and S. Hiraga. 1996. Identification of two new genes, *mukE* and *mukF*, involved in chromosome partitioning in *Escherichia coli*. *Mol. Gen. Gen.* **250**:241–251.

42. Erickson, H. P. 1997. FtsZ, a tubulin homologue in prokaryote cell division. *Trends in Cell Biol.* **7**:362–367.

43. Jaffé, A., R. D'Ari, and S. Hiraga. 1988. Minicell-forming mutants of *Escherichia coli*: Production of minicells and anucleate rods *J. Bacteriol.* **170**:3094–3101.

44. Frameshift mutations are mutations due to the insertion or removal of one or more (but not three) bases from DNA. Acridine dyes such as ethidium bromide and proflavine cause frameshift mutations. They insert between the bases, thus increasing the distance between bases and preventing them from aligning correctly. As a consequence the two strands of DNA slip with respect to one another. This leads to the insertion or deletion of a base.

45. DNA polymerase III has 10 different protein subunits coded for by separate genes. One of these, the $\alpha$ subunit, is encoded by the *polC* (*dnaE*) gene and is responsible for polymerization. The subunit responsible for the $3'$ exonuclease activity is the $\varepsilon$ subunit encoded by the *dnaQ* (*mutD*) gene. These two subunits, along with the $\phi$ subunit, comprise the core of the polymerase. The $\tau$ subunit dimerizes the core. The $\beta$ subunit is required for optimal processivity. The other subunits (i.e., $\gamma, \delta, \delta', \chi, \psi$) form a clamp that keeps the polymerase on the template. The subunits that are not part of the core are sometimes called accessory proteins. The core plus the accessory proteins is referred to as the holoenzyme.

46. Modrich, P. 1991. Mechanisms and biological effects of mismatch repair. *Ann. Rev. Genet.* **25**:229–253.

47. Kolodner, R., 1996. Biochemistry and genetics of eukaryotic mismatch repair. *Genes Dev.* **10**:1433–1442.

48. Modrich, P. 1995. Mismatch repair, genetic stability and tumor avoidance. *Philos. Trans. R. Soc. Lond. B* **347**:89–95.

49. Modrich, P. 1994. Mismatch repair, genetic stability and cancer. *Science* **266**:1959–1960.

50. Bacteria have one RNA polymerase that synthesizes ribosomal, messenger, and transfer RNA. The enzyme that synthesizes the RNA in Okazaki fragments is called primase. Eukaryotes have 3 nuclear RNA polymerases and a mitochondrial RNA polymerase.

51. McClure, W. R. 1985. Mechanism and control of transcription initiation in prokaryotes. *Ann. Rev. Biochem.* **54**:171–204.

52. Platt, T. 1986. Transcription termination and the regujlation of gene expression. *Ann. Rev. Biochem.* 55:339–372.

53. Henkin, T. M. 1996. Control of transcription termination in prokaryotes. *Ann. Rev. Genet.* 30:35–37.

54. Ishihama, A. 1997. Promoter selectivity control of RNA polymerase, pp. 53–70. In: *Mechanisms of Transcription.* F. Eckstein and D. M. J. Lilley (Eds.). Springer-Verlag. Berlin.

55. Steitz, T. A., D. H. Ohlendorf, D. B. McKay, W. F. Anderson, and B. W. Matthews. 1982. Structural similarity in the DNA-binding domains of catabolite gene activator and *cro* repressor proteins. *Proc. Natl. Acad. Sci.* 79:3097–3100.

56. The eukaryotic cytosolic ribosome is 80S and has two subunits, 40S and 60S. The 40S subunit contains an 18S rRNA. The 60S subunit contains a 28S, 5S, and a 5.8S rRNA. The rRNA transcript in eukaryotes is 45S. It is processed in the nucleus to produce the 18S, 28S, and 5.8S rRNAs that are present in the ribosomes. The 5S rRNA is on a separate transcript.

57. Björk, G. R., J. U. Ericson, C. E. D. Gustafsson, T. G. Hagervall, Y. H. Jöhnsson, and P. M. Wilkström. 1987. Transfer RNA modification. *Ann. Rev. Biochem.* 56:263–287.

58. Hershey, J. W. B. 1987. Protein synthesis. In: *Escherichia coli and Salmonella typhimurium: Cellular and Molecular Biology.* Vol. 1, pp. 613–647. F. C. Neidhardt, J. L. Ingraham, K. B. Low, M. Magasanik, M. Schaechter, and H. E. Umbarger (Eds.). American Society for Microbiology, Washington, D.C.

59. Protein synthesis by cytosolic eukaryotic ribosomes differs in some respects from protein synthesis by bacterial ribosomes. (1) The ribosomal subunits are 40S and 60S. (2) The first amino acid is methionine rather than formyl methionine. (But still, a special tRNA[Met] is used for initiation.) (3) There are at least 9 initiation factors, rather than 3. These are: eIF3 and eI4C, which bind to the 40S subunit and stimulate further steps in initiation; the cap protein (or CBPI), which binds to the 5′ end of the mRNA and aids in binding the mRNA to the 40S subunit; eIF2, which aids in binding of Met–tRNA[Met] to the 40S subunit; eIF4a, eIF4B, and eIF4F, which help to locate the start AUG; eIF5, which facilitates the dissociation of the other initiation factors from the 40S subunit so that it can combine with the 60S subunit to form the 80S ribosome; and eIF6, which stimulates the dissociation of the 80S ribosome not engaged in protein synthesis to 40S and 60S subunits. The elongation and translocation phases are analogous to what occurs in bacterial ribosomes. The cytosolic eukaryotic ribosomes use 3 elongation factors, eEF1α, eEF1βγ, and eEF2, which are analogous to EF-Tu, EF-Ts, and EF- G. Termination is also similar. Whereas in bacteria there are 3 release factors that are specific for the 3 termination codons, protein synthesis by cytosolic eukaryotic ribosomes uses 1 release factor, eRF, which recognizes 3 different termination codons. Eukaryotic mRNA differs from prokaryotic mRNA in not being polycistronic and in not having a special ribosome binding analogous to the Shine–Dalgarno sequence upstream of the initiation codon. Eukaryotes use the first AUG at the 5′ end of the mRNA.

60. Arnez, J. G., and D. Moras. 1997. Structural and functional considerations of the aminoacylation reaction. *TIBS* 22:211–216.

61. Although there are 61 codons for the 20 amino acids, there are fewer than 61 tRNAs. This is because some tRNAs can read more than one codon for a particular amino acid. In order to understand the following discussion, bear in mind that the anticodon has opposite polarity to the codon so that the third nucleotide in the codon always pairs with the first nucleotide in the anticodon. Crick realized that the first two nucleotides in the codon form strong pairs with their partners in the anticodon and in fact are responsible for most of the codon specificity. That is to say, the third nucleotide in the codon frequently does not change the specificity of the codon. Therefore, codons with the same nucleotide in the first two positions but differing in the third nucleotide can often code for the same amino acid. This actually depends upon the base at the first position in the anticodon. If the first base in the anticodon is either C or A, then the base in the third position of the codon is always G or U, respectively, and there is only one codon per amino acid and therefore only one tRNA per codon. However, if the base in the first position in the anticodon is I (inosinate), G, or U, then the third base in the codon forms a loose pair or "wobble" with its partner in the anticodon. This allows different codons for a single amino acid to be read by the same tRNA. For example, inosinate, which has hypoxanthine as its base, can be found in the first position of the anticodon for some amino acids. Inosinate can hydrogen bond weakly to U, C, or A. This means that the third base in the codon can be U, C, or A and all three codons will bind to the same anticodon, i.e., to the same tRNA. Likewise, U in the first position in the anticodon can pair with either A or G in the third position of the codon, so that in this case two codons can be read by the same tRNA. Similarly, G in the first position of the anticodon can pair with C or U. Thus the number of tRNAs can be fewer than the number of codons. At least 32 different tRNAs are required to read the 61 codons. It has been

suggested that the advantage of introducing a "wobble" in the third position is that it is makes the bonding between the codon and anticodon less tight and increases the rate of dissociation of the tRNAs from the ribosome during protein synthesis.

62. Ibba, M., A. W. Curnow, and D. Söll. 1997. Aminoacyl–tRNA synthesis: Divergent routes to a common goal. *TIBS* 22:39–42.

63. Becker, H. D., J. Reinbolt, R. Kreutzer, R. Giegé, and D. Kern. 1997. Existence of two distinct aspartyl–tRNA synthetases in *Thermus thermophilus*. Structural and biochemical properties of the two enzymes. *Biochem.* 36:8785–8797.

64. Many organisms possess fewer than 20 different aminoacyl–tRNA synthetases to activate the 20 standard amino acids. That is to say, in a few instances the same aminoacyl–tRNA synthetase charges two different tRNAs with the same amino acid. For example, it has been found that *Bacillus subtilis*, other gram-positive bacteria, the gram-negative *Rhizobium meliloti*, some archaebacteria, chloroplasts, and mitochondria, use the same enzyme, GluRS, to charge tRNA$^{Gln}$ (has anticodon for Gln codon) and tRNA$^{Glu}$ (has anticodon for Glu codon) with glutamate. Then an amidotransferase recognizes the Glu–tRNA$^{Gln}$ and ω amidates the glutamate to form Gln–tRNA$^{Gln}$, which places the Gln at the Gln codon. A similar situation occurs in the halophilic archaebacterium *Haloferax volcanii* with Asn–tRNA$^{Asn}$, where both tRNA$^{Asp}$ and tRNA$^{Asn}$ are charged by the same enzyme with aspartate. The aspartate is then converted to asparagine on tRNA$^{Asn}$, which places Asn at the Asn codon. Note the similarity to the fate of methionine, which is charged onto tRNA$^{fMet}$. A special transformylase recognizes the Met–tRNA$^{fMet}$ and formylates the Met to fMet, which is used to initiation translation. In all these cases the tRNA is charged with the wrong amino acid, which is subsequently modified to the correct amino acid. There is also an example of two different aminacyl synthetases activating the same amino acid and charging the same tRNA. The thermophilic bacterium *Thermus thermophilus* has two aspartyl–tRNA synthetases (AspRS1 and AspRS2) that charge tRNA$^{Asp}$. They are encoded by separate genes.

65. Arnez, J. G., and D. Moras. 1997. Structural and functional considerations of the aminoacylation reaction. *TIBS* 22:211–216.

66. The amino acids are attached to the 3′-terminal CCA of tRNAs. Some synthetases attach the amino acid directly to the 3′ hydroxyl, whereas other synthetases attach the amino acid to the 2′-hydroxyl, and then a transesterification reaction moves the amino acid to the

3′-hydroxyl. A description of the structural and functional features of the aminoacyl–tRNA synthetases is reviewed in Ref. 65.

67. Noller, H. F., V. Hoffarth, and L. Zimniak. 1992. Unusual resistance of peptidyl transferase to protein extraction procedures. *Science* 256:1416–1419.

68. Certain RNA molecules are known to have enzymatic activity. They are called *ribozymes*. Two well-studied examples are RNase P and self-splicing group I introns. The self-splicing intron is an endonuclease that cuts itself out of RNA precursor molecules and joins the exons together to form the mature RNA. Another ribozyme is RNase P, which is actually a nucleoprotein. However, the catalytic activity resides in the RNA. RNase P acts on tRNA precursors. Transfer RNA is made from longer precursors, from which nucleotides are removed from the ends. RNase P is an RNA molecule that acts as an endonuclease and removes nucleotides from the 5′ end of tRNA molecules. Self-splicing introns and RNase P are widespread in prokaryotes and eukaryotes.

69. The primary structure of proteins refers to the sequence of amino acids. The secondary structure refers to regular arrangements of the polypeptide chain, e.g., an α helix in which the polypeptide backbone is wound around the long axis of the molecule in a helix and the R groups of the amino acids point out. The tertiary structure refers to the folding of the polypeptide chain. The quaternary structure refers to the spatial arrangement of folded polypeptide subunits in a multimeric protein.

70. Ewalt, K. L., J. P. Hendrick, W. A. Houry, and F. U. Hartle. 1997. In vivo observation of polypeptide flux through the bacterial chaperonin system. *Cell* 90:491–500.

71. GroEL and GroES are sometimes referred to as the GroE system. *E. coli* requires both for growth at all temperatures, although as discussed in Chapter 19, their synthesis is greatly stimulated by heat shock. Possibly the excess GroEL at higher temperatures protects unstable proteins or aids in refolding partially denatured proteins when normal temperatures are restored. GroEL (also called Hsp60) is a double-ring structure ("double doughnut") with 14 identical subunits. It binds to the protein that is to be folded, and the proteins are believed to be folded in the cavity of the "double doughnut." ATP is also necessary. ATP binds to GroEL, producing a conformational change and increases the affinity of GroEL for the protein substrate. The hydrolysis of ATP lowers the affinity and aids in the release of the folded protein. The role of GroES (also called Hsp10), which is a single ring of

severn subunits, is to aid in the release of the folded protein from GroEL One might view the GroEL cavity within which the protein folds as a "safe haven" where a protein might fold spontaneously without interacting with another protein to form aggregates.

72. Gaitanaris, G. A., A. Vysokanov, S-C. Hung, M. E. Gottesman, and A. Gragerov. 1994. Successive action of *Escherichia coli* chaperones *in vivo*. *Molec. Microbiol.* **14**:861–869.

73. Following are the methods used by Gaitanaris et al. to show that DnaK and DnaJ associate with nascent polypeptides on ribosomes and that GroEL transiently associates with completed polypeptides after their release from ribosomes. Cell lysates of *E. coli* were centrifuged through a sucrose gradient. The ribosome fractions were isolated and the proteins were separated by gel electrophoresis and then transferred (blotted) onto nitrocellulose filters. The filters were treated with anti-DnaK, anti-DnaJ, and anti-GroEL antibodies, and bound antibody was detected using alkaline phosphatase–coupled secondary antibodies. Both DnaK and DnaJ were associated with the ribosomal fractions isolated from the sucrose density gradients, but GroEL was not. When cells were briefly treated with puromycin to release nascent polypeptides from ribosomes, DnaK and DnaJ were released from the ribosomes. Newly synthesized polypeptides were associated with GroEL. This was determined by pulse-labeling cells with $^{35}$S-methionine, chasing with unlabeled methionine for various times, and separating cell fractions on a sucrose gradient. The proteins were separated by gel electrophoresis and blotted onto nitrocellulose, and GroEL was detected using anti-GroEL antibody as described above. Radioactive proteins were transiently associated with fractions containing GroEL. However, none of the GroEL-associated proteins was in the ribosomal fractions, indicating that they became associated with GroEL after their release from the ribosomes (Gaitanaris, G. A., A. Vysokanov, S-C. Hung, M.E. Gottesman, and A. Gragerov. 1994. Successive action of *Escherichia coli* chaperones *in vivo*. *Molec. Microbiol.* **14**:861–869).

74. Fedorov, A. N., and T. O. Baldwin. 1997. Cotranslational protein folding. *J. Biol. Chem.* **272**:32715–32718.

75. Stoller, G., K. P. Rücknagel, K. H. Nierhaus, F. X. Schmid, G. Fischer, and J-U Rahfeld. 1995. A ribosome-associated peptidyl-prolyl *cis/trans* isomerase identified as the trigger factor. *The EMBO J.* **14**:4939–4948.

# 11

# Cell Wall and Capsule Biosynthesis

Studying cell wall and capsule synthesis in bacteria is instructive in showing how logistic problems of extracellular biosynthesis are solved. For example, the subunits of the cell wall and capsule polymers are synthesized as water-soluble precursors in the cytosol. How do the subunits traverse the lipid barrier in the cell membrane to the sites of polymer assembly? A second problem concerns the final stages of peptidoglycan synthesis. During peptidoglycan synthesis the newly polymerized glycan chains become cross-linked by peptide bonds on the outside cell surface. What is the source of energy for making the peptide cross-links at a site where there is no ATP? This chapter considers these and other aspects of the biosynthesis of peptidoglycan and lipopolysaccharide, as well as capsular and other extracellular polysaccharides. The biosynthesis of cytosolic polysaccharides is also considered.

## 11.1 Peptidoglycan

### 11.1.2 Structure

Peptidoglycan is a heteropolymer of glycan chains cross-linked by amino acids (Section 1.2.3). The peptidoglycan is a huge molecule, since it surrounds the entire cell and appears to be covalently bonded throughout. A schematic

drawing of how this might look in gram-negative bacteria is shown in Fig. 11.1 (see also Fig. 1.8). Peptidoglycan confers strength to the cell wall, and if one were enzymatically to destroy the integrity of the peptidoglycan (with lysozyme) or prevent its synthesis (with antibiotics), then the cell is likely to swell through the weak areas and lyse as a result of the internal turgor pressures.

## The chemical composition of peptidoglycan

Peptidoglycan is made of glycan strands of alternating residues of N-acetylmuramic acid and N-acetylglucosamine linked by $\beta$-1,4 glycosidic bonds between the C1 of N-acetylmuramic acid and the C4 of N-acetylglucosamine (Fig. 11.2). N-acetylmuramic acid is a modified form of N-acetylglucosamine in which a lactyl group has been attached to the C3 carbon. Attached to each N-acetylmuramic acid is a tetrapeptide. The tetrapeptide is L-alanyl-$\gamma$-D-glutamyl-L-$R_3$-D-alanine. The amino acid in position 3 varies with the species of bacterium. Gram-negative bacteria generally have *meso*-diaminopimelic acid. (However, some spirochetes contain ornithine instead of *meso*-diaminopimelic acid.) In contrast, there is much more variability in the amino acids in position 3 in gram-positive bacteria (Fig. 11.2).

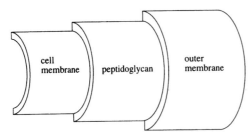

**Fig. 11.1** The topological relationship of the peptidoglycan to the cell membrane and the rest of the cell wall. In gram-negative bacteria such as *E. coli*, the peptidoglycan is a thin layer sandwiched between the inner and outer membranes. In gram-positive bacteria there is no outer membrane, and the peptidoglycan is a thick layer usually covalently bonded to other molecules, e.g., teichoic acids.

## Cross-linking

The tetrapeptide chains are cross-linked to each other by peptide bonds (Fig. 11.3A,B). There is a great deal of variability in the composition of the cross-links between the different groups of bacteria. In fact, the amino acid composition and location of the cross-links have been used for taxonomic purposes. In most instances the peptide bridge is from the carboxyl in the terminal D-alanine in one tetrapeptide to an amino group in the amino acid in the L-R$_3$ position in another tetrapeptide. In some bacteria the cross-linking is direct, as, for example, between D-alanine and diaminopimelic acid in gram-negative bacteria and many *Bacillus* species. However, in most gram-positive bacteria there is a bridge of one or more amino acids. Some examples are a bridge of five glycine residues in *Staphylococcus aureus*, three L-alanines and one L-threonine in *Micrococcus roseus*, three glycines and two L-serines in *Staphylococcus epidermidis*, and so on. Sometimes the bridge is from the terminal D-alanine to the α-carboxyl of D-glutamic acid of another tetrapeptide. Since this is a connection between two carboxyl groups, a bridge of amino acids containing a diamino acid is necessary.

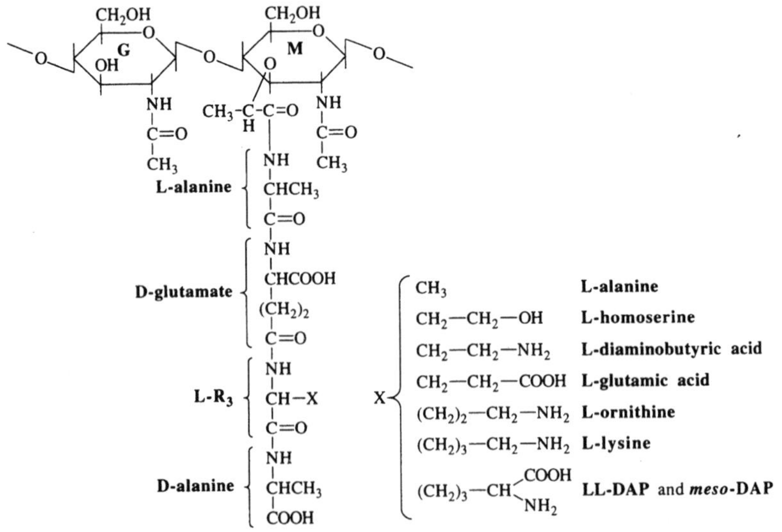

**Fig. 11.2** The disaccharide–peptide subunit in peptidoglycan. The glycan backbone consists of alternating residues of N-acetylglucosamine (G) and N-acetylmuramic acid (M) linked β-1,4. The carboxyl group of the lactyl moiety in N-acetylmuramic acid is substituted with a tetrapeptide. The amino acids in the tetrapeptide are usually L-alanine, D-glutamic, L-R$_3$ (residue 3), which is an amino acid that varies with the species, and D-alanine. The peptide linkages are all α except for that between D-glutamic and the amino acid in postion 3, which is γ linked. X can be any of one a large number of side chains, some examples of which are shown. The α-carboxyl of glutamic acid can be free, an amide, or substituted, e.g., by glycine.

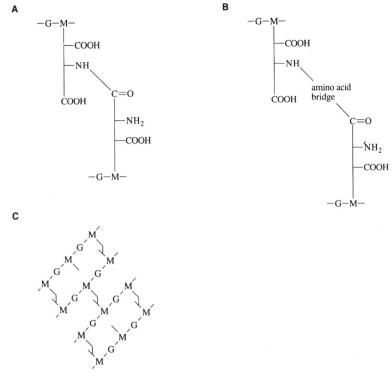

**Fig. 11.3** Peptidoglycan cross-linking. (A) Direct crosslink between diaminopimelate and D-alanine as it occurs in gram-negative bacteria. (B) An amino acid bridge between two tetrapeptides as in many gram-positive bacteria. Sometimes the bridge is between the C-terminal D-alanine and the α-carboxyl of D-glutamic acid. When this occurs, there must be a diamino acid in the bridge. (C) Schematic drawing of crosslinked peptidoglycan. Tetrapeptides are indicated by straight lines. The lengths of the glycan chain in some bacteria have been measured to be about 20–100 sugar residues, and apparently vary in length within the same bacterium. The degree of crosslinking depends upon the bacterium. In *E. coli* about 50% of the peptide chains are crosslinked. This is considered a relatively uncrosslinked peptidoglycan. In some bacteria, about 90% of the chains are crosslinked. Abbreviations: G, N-acetylglucosamine; M, N-acetylmuramic acid.

## 11.1.3 Synthesis

Peptidoglycan is made in several stages: (1) The precursors to the peptidoglycan are UDP derivatives of the amino sugars that are made in the cytosol; (2) the amino sugars are then transferred to a lipid carrier in the membrane, which carries the amino sugars across the membrane; (3) the peptidoglycan is polymerized on the outer surface of the membrane; and (4) a transpeptidation reaction cross-links the peptidoglycan. (For reviews, see Refs. 1 and 2.)

## Synthesis of the UDP derivatives: UDP-N-acetylglucosamine and UDP-N-acetylmuramyl-pentapeptide

The two amino sugars that are precursors to the peptidoglycan are N-acetylglucosamine and N- acetylmuramic acid. Both amino sugars are made from fructose-6-phosphate (Fig. 11.4). In step 1, glutamine donates an amino group to fructose-6-phosphate, converting it to glucosamine-6-phosphate. Then in step 2, a transacylase transfers an acetyl group from acetyl–CoA to the amino group

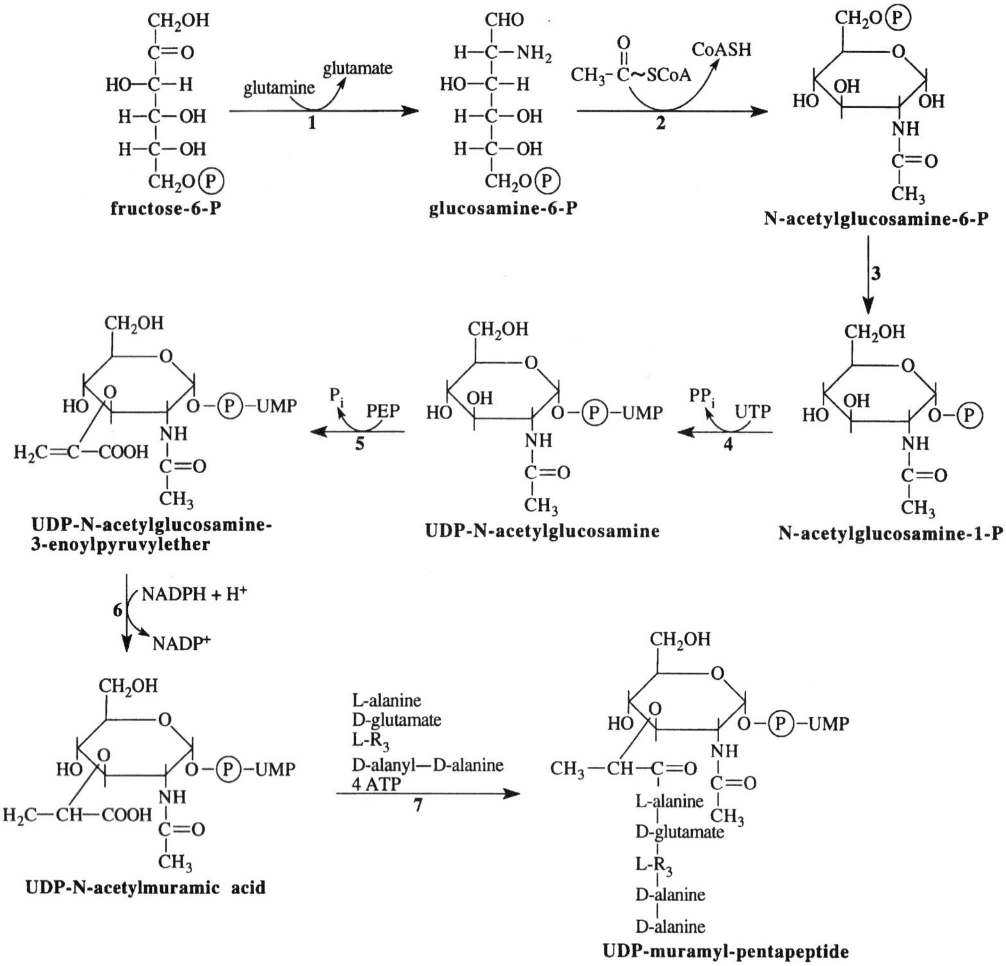

**Fig. 11.4** Synthesis of N-acetylglucosamine and N-acetylmuramyl-pentapeptide. Enzymes: 1, glutamine: fructose-6-phosphate aminotransferase; 2, glucosamine phosphate transacetylase; 3, N-acetylglucosamine phosphomutase; 4, UDP–N-acetylglucosamine pyrophosphorylase; 5, enoylpyruvate transferase; 6, UDP–N acetylenolpyruvoylglucosamine reductase. Steps 7: The UDP–N-acetylmuramic acid is converted to the pentapeptide derivative by the sequential additions of L-alanine, D-glutamate, L-R$_3$, and D-alanine–D-alanine by separate enzymes.

on glucosamine-6-phosphate to make *N*-acetylglucosamine-6-phosphate, which is isomerized in step 3 to *N*- acetylglucosamine-1-phosphate. The latter attacks UTP, displacing pyrophosphate in step 4 to form UDP-*N*-acetylglucosamine. The reaction is driven to completion by a pyrophosphatase. Some of the UDP-*N*-acetylglucosamine (UDP-GlcNAc) is used as the precursor to the *N*-acetylglucosamine in peptidoglycan, and some is converted to UDP-*N*-acetylmuramic acid (UDP-MurNAc). The UDP- GlcNAc is converted to UDP-MurNAc by the addition of a lactyl group to the sugar in step 5. In this

reaction, the C3 OH of the sugar displaces the phosphate from the α carbon of phosphoenolpyruvate, forming the enol pyruvate ether derivative of the UDP-*N*-acetylmuramic acid. Then, in step 6, the enol derivative is reduced to the lactyl moiety by NADPH. The UDP-MurNAc is converted into UDP-MurNAc-pentapeptide by the sequential addition of five amino acids, L-alanine, D-glutamate, L-R$_3$ (residue 3), and the dipeptide D-alanyl–D-alanine in step 7. Each reaction is catalyzed by a separate enzyme and requires ATP to activate the carboxyl group of the amino acid. The activated carboxyl is probably an

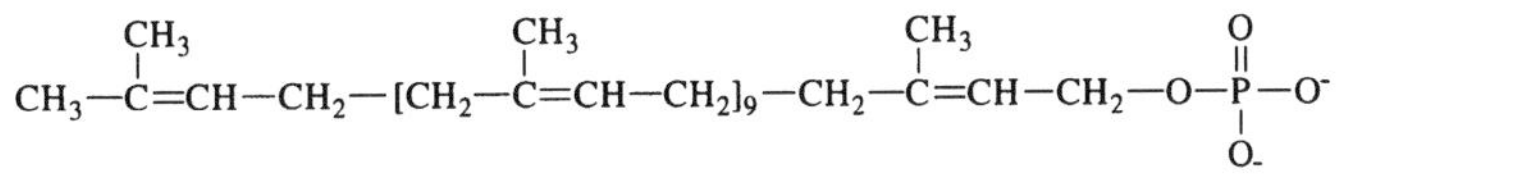

$$CH_3-\overset{\overset{\displaystyle CH_3}{|}}{C}=CH-CH_2-[CH_2-\overset{\overset{\displaystyle CH_3}{|}}{C}=CH-CH_2]_9-CH_2-\overset{\overset{\displaystyle CH_3}{|}}{C}=CH-CH_2-O-\overset{\overset{\displaystyle O}{\|}}{\underset{\underset{\displaystyle O_-}{|}}{P}}-O^-$$

**Fig. 11.5** The structure of undecaprenyl phosphate. Undecaprenyl phosphate, is a $C_{55}$ isoprenoid phosphate that carries precursors to peptidoglycan, lipopolysaccharide, and teichoic acids through the cell membrane.

acyl-phosphate, which is attacked by the incoming amino group displacing the phosphate. (See Fig. 7.7.) The products are ADP and inorganic phosphate. (The fifth D-alanine is removed during the cross-linking reaction described later.) [The synthesis of the peptide subunit in the archaeal pseudomurein (Fig. 1.13) takes place via UDP-peptide intermediates, which also appear to form peptide bonds via acyl-phosphates.[3] The D-alanine–D-alanine is made by separate enzymes. The first of these is a racemase, which converts L-alanine to D-alanine. Then an ATP-dependent D-alanyl–D-alanyl synthetase makes D-alanyl–D-alanine from two D-alanines. The racemase and synthetase are inhibited by the antibiotic D-cycloserine. The MurNAc-pentapeptide is transferred to the lipid carrier in the membrane, as described next.

### Reactions in the membrane

The lipid carrier is called *undecaprenyl phosphate* or *bactoprenol*. Undecaprenyl phosphate is a $C_{55}$ isoprenoid phosphate whose structure is shown in Fig. 11.5. Undecaprenyl phosphate not only serves as a carrier for peptidoglycan precursors, but also serves as a carrier for the precursors of other cell wall polymers (e.g., lipopolysaccharide and teichoic acids). Interestingly, eukaryotes also use an isoprenoid phosphate, dolichol phosphate, to carry oligosaccharide subunits across the endoplasmic reticulum (ER) membrane to be attached to glycoproteins in the ER lumen. Dolichol phosphate is larger than undecaprenyl phosphate but has the same structure. The nucleotide sugars diffuse to the membrane where undecaprenyl phosphate (lipid-P) attacks the UDP-MurNAc-pentapeptide, displacing UMP (Fig. 11.6, step 1). The product is lipid-PP-MurNAc-pentapeptide. The GlcNAc is then transferred from UDP-GlcNAc to the MurNAc on the lipid carrier (step 2). This occurs when the C4 hydroxyl in

the MurNAc attacks the C1 carbon in UDP-GlcNAc, displacing the UDP. The product is the disaccharide precursor to the peptidoglycan, lipid-PP-MurNAc(pentapeptide)-GlcNAc.

The disaccharide-lipid moves to the other side of the membrane (step 3). Exactly how the disaccharide-lipid moves through the membrane is not understood, but it has been speculated that it is aided by unidentified proteins. On the outside surface of the membrane, the lipid-disaccharide is transferred to the growing end of the acceptor glycan chain (step 4). Step 4 is a transglycosylation where the C4 hydroxyl of the incoming GlcNAc attacks the C1 of the MurNAc in the glycan, displacing the lipid-PP from the growing glycan chain. This reaction is catalyzed by a membrane-bound enzyme called transglycosylase. Notice that the growing glycan chain remains anchored via the lipid carrier to the membrane at the site of the transglycosylase. The lipid-PP released from the growing glycan chain is hydrolyzed by a membrane-bound pyrophosphatase to lipid-P and $P_i$ (step 5). This reaction is very important because it helps to drive the transglycosylation reaction to completion, since the hydrolysis of the phosphodiester results in the release of substantial energy. This hydrolysis also regenerates the lipid-P, which is necessary for continued growth of the peptidoglycan as well as other cell wall polymers (e.g., lipopolysaccharide and teichoic acids). The hydrolysis is inhibited by the antibiotic bacitracin.

### Making the peptide cross-link

The problem of providing the energy to make the peptide cross-link outside the cell membrane is solved by using a reaction called transpeptidation. In the transpeptidation reaction, an $:NH_2$ group from the diamino acid in the 3 position (e.g., DAP) attacks the carbonyl carbon in the peptide bond, holding the two

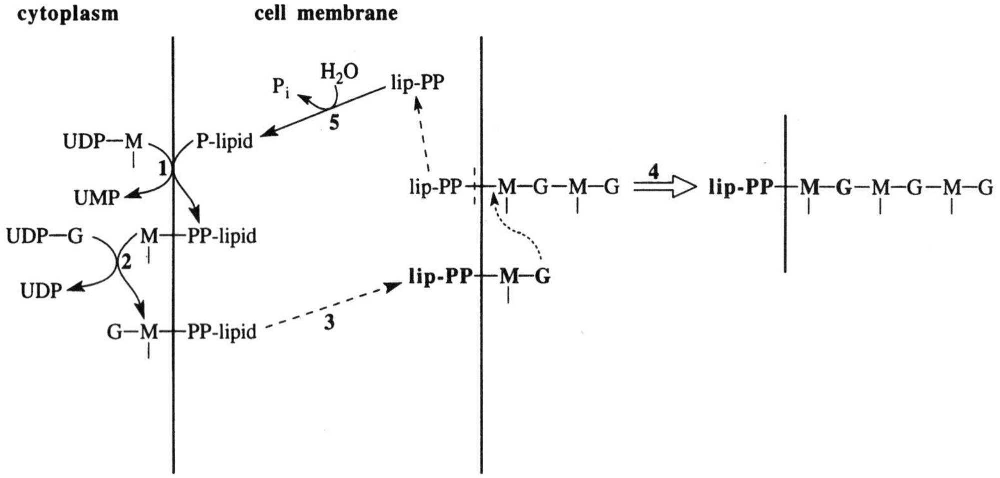

**Fig. 11.6** Extension of the glycan chain during peptidoglycan synthesis. *Step 1*. The MurNac pentapeptide (M–) is transferred to the phospholipid carrier (undecaprenyl phosphate) on the cytoplasmic side of the cell membrane. *Step 2*. The GlcNAc (G) is transferred to the MurNAc pentapeptide to form the disaccharide–PP–lipid precursor. *Step 3*. The disaccharide–PP–lipid precursor moves to the external face of the membrane. *Step 4*. A transglycosylase transfers the incoming disaccharide to the growing glycan, displacing the lipid–PP from the growing chain. Thus the growing chain remains anchored to the membrane by the lipid carrier at the site of the transglycosylase. *Step 5*. The lipid–PP released from the growing chain is hydrolyzed to lipid–P by a membrane-bound pyrophosphatase. Note that the glycan chain grows at the reducing end; i.e., displacements occur on the C1 of muramic acid.

D-alanine residues together in the pentapeptide, and displaces the terminal D-alanine. The result is a new peptide bond (Fig. 11.7). The transpeptidation reaction is inhibited by penicillin.

### Penicillin-binding proteins (PBPs)

Bacteria that have peptidoglycan have in their membranes proteins that bind penicillin that are called penicillin-binding proteins, or PBPs. For example, *E. coli* has at least 7 such proteins. These are involved in catalyzing the late steps of peptidoglycan synthesis. The high-molecular-weight PBPs from *E. coli* (1a, 1b, 2, 3) have been shown to catalyze the transglycosylation and transpeptidation steps; that is, they are bifunctional enzymes. [The low-molecular-weight PBPs from *E. coli* (4, 5, 6) appear to be dispensible.] One of the PBPs (i.e., PBP3) is specifically required for peptidoglycan synthesis in the septum. (See the discussion of cell division in Section 2.2.5.)

## 11.2 Lipopolysaccharide

### 11.2.1 Structure

Lipopolysaccharide (LPS) is a complex polymer of polysaccharide and lipid in the outer membrane of gram-negative bacteria. The outer membrane is described in more detail in Section 1.2.3. The best-characterized lipopolysaccharides are those of *E. coli* and *Salmonella typhimurium* (Fig. 11.8).[4] The lipolysaccharides of other bacteria are very similar.

As seen in Fig. 11.8, lipopolysaccharide is composed of three parts: (1) a hydrophobic region called lipid A, which is composed of a backbone of two glucosamine residues linked β 1,6 and esterified via the hydroxyl groups to fatty acids; (2) a core polysaccharide region, whose composition is similar in all the Enterobacteriaceae and which is connected to lipid A via 3-deoxy-D-manno-octulosonate

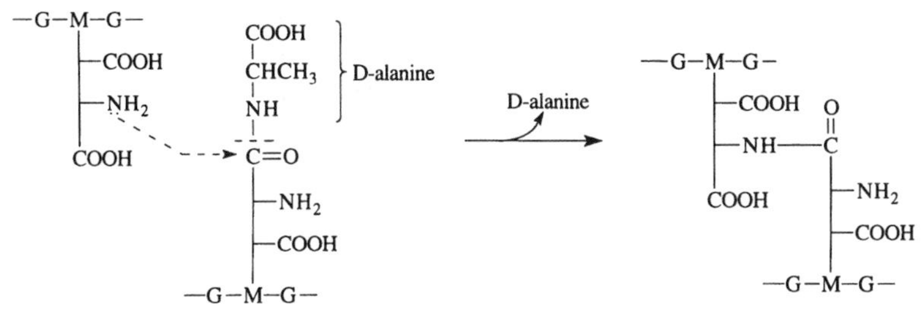

**Fig. 11.7** The transpeptidation reaction. This is a nucleophilic displacement where the nucleophilic nitrogen attacks the carbonyl, displacing the terminal D-alanine The reaction is inhibited by penicillin. Terminal D-alanine residues in chains not participating in crosslinking are removed by a D-alanine carboxypeptidase.

(KDO)[5]; and (3) a distal polysaccharide region connected to the core. The distal polysaccharide region is sometimes called the O-antigen or the repeat oligosaccharide. It is made up of repeating units of four to six sugars that vary considerably in composition between different strains of bacteria. The repeating oligosaccharide may have as many as 30 units. The structure of KDO is shown in Fig. 11.9. The arrangement of the LPS in the outer envelope is shown in Fig. 11.10, which shows that the LPS is anchored in the outer envelope by the hydrophobic lipid A region while the repeating oligosaccharide protrudes into the medium.

### The fatty acids in lipid A

Four identical fatty acids are attached directly to the glucosamine in lipid A from *E. coli* and *Salmonella typhimurium*. These are a $C_{14}$ hydroxy fatty acid, $\beta$-hydroxy-myristic acid (3-hydroxytetradecanoic acid), linked via ester bonds to the 3' hydroxyls of the glucosamine, and via amide bonds to the nitrogen of the glucosamine. These fatty acids appear to be found uniquely in lipid A, and

their presence in a bacterium implies the presence of a lipopolysaccharide-containing lipid A. Esterified to the hydroxyls of two of the $\beta$-hydroxymyristic acids are long-chain saturated fatty acids. In *E. coli*, these are lauric acid ($C_{12}$) and myristic acid ($C_{14}$). The fatty acids attached to the glucosamine anchor the lipopolysaccharide into the outer membrane.

### 11.2.2 Synthesis of the lipopolysaccharide

#### Lipid A

The lipid A portion of the lipopolysaccharide is synthesized from UDP-GlcNAc, which, as described in Fig. 11.4, is made from fructose-6-phosphate.[6] A model for lipid A synthesis in *E. coli* proposed by Raetz is shown in Fig. 11.11.[7] Lipid A is synthesized in the cytoplasmic membrane, but the early steps are catalyzed by three cytoplasmic enzymes. The first enzyme, UDP-GlcNAc acyltransferase, transfers $\beta$-OH myristic acid from an ACP derivative to C3 of UDP-GlcNAc to form the monoacyl derivative.

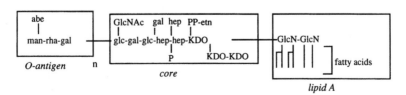

**Fig. 11.8** The LPS of *Salmonella typhimurium*. Abbreviations: abe, abequose; man, mannose; rha, rhamnose; gal, galactose; GlcNAc, N-acetylglucosamine; glc, glucose; hep, heptose (L-glycero-D-mannoheptose); KDO, 3-deoxy-D-mannooctulosonic acid; etn, ethanolamine; P, phosphate; GlcN, glucosamine.

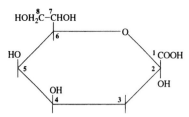

Fig. 11.9 The structure of KDO. The linkage between KDO residues has been reported to be a glycosidic linkage between a C2 hydroxyl in one KDO with either the C4 or C5 hydroxyl in a second KDO. The linkage to lipid A has been suggested to be from the C2 hydroxyl of KDO to the C6 hydroxyl of lipid A.

(Recall that fatty acids are synthesized as ACP derivatives, Chapter 9.) (This is actually a branch point in the metabolism of UDP-GlcNAc because some of the UDP-GlcNAc is incorporated into peptidoglycan. It reacts with PEP to form UDP-MurNAc, and it also condenses with UDP-MurNAc to form UDP-MurNAc-GlcNAc.) Then a deacetylase removes the acetate from the nitrogen on C2. Interestingly, the gene for the deacetylase is *envA*, which had previously been known to be required for cell separation after septum formation. (See Section 2.2.5.) After deacetylation a second molecule of $\beta$-OH myristic is transferred from the ACP derivative by a third enzyme (an N-acyl transferase) to the nitrogen to form the 2,3-diacyl derivative. Some of the latter loses UMP and is converted to 2,3- diacylglucosamine-1-P (also called lipid X), which condenses with UDP-2,3-diacylglucosamine to form the disaccharide linked in $\beta$ 1,6 linkage. A phosphate is added from ATP to form the 1,4 diphosphate derivative, and this is then modified by the addition of KDO from CMP derivatives and

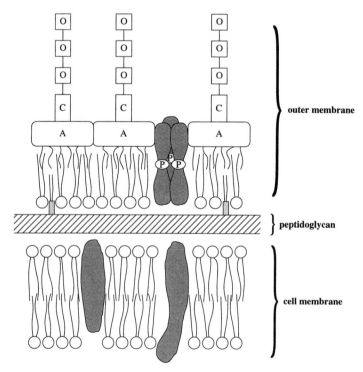

Fig. 11.10 The arrangement of the LPS in the outer membrane of some gram-negative bacteria. Abbreviations: A, lipid A; C, core; O, oligosaccharide; P, porin. The outer envelope in the enteric gram-negative bacteria is asymmetric, with the lipopolysaccharide confined to the outer leaflet of the lipid bilayer. However, some bacteria, for example, penicillin-sensitive strains of *Neisseria* and *Treponema*, also have phospholipid in the outer leaflet. Phospholipids in the outer membrane make the bacterium more sensitive to hydrophobic antibiotics. Refer to Section 1.2.3 for a more complete discussion of the outer membrane.

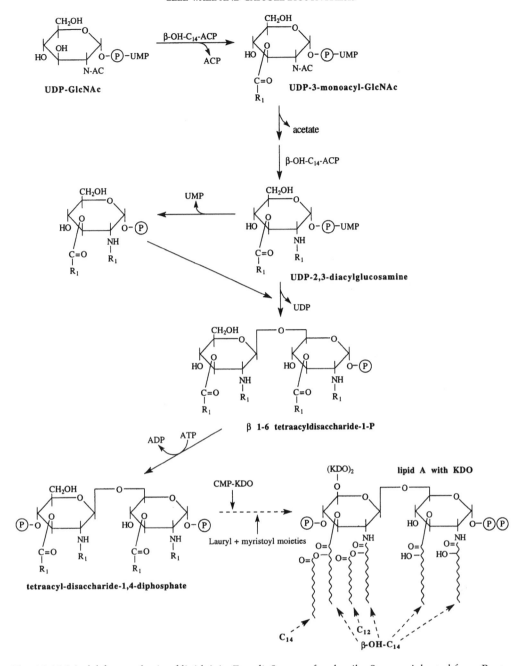

**Fig. 11.11** Model for synthesis of lipid A in *E. coli*. See text for details. *Source:* Adapted from Raetz, C. R. H. 1987. Structure and biosynthesis of lipid A in *Escherichia coli*, Vol. 1, pp. 498–503. In: *Escherichia coli* and *Salmonella typhimurium*: Cellular and Molecular Biology, F. C. Neidhardt, J. L. Ingraham, K. B. Low, B. Magasanik, M. Schaechter, and H . E. Umbarger (Eds.). American Society for Microbiology, Washington, D.C.

the esterification of fatty acids (lauryl and myristoyl) to the OH of the $\beta$-hydroxymyristic moieties. The fatty acids are donated by ACP derivatives.

## Core

The core in the *Enterobacteriaceae* contains of an inner region consisting of KDO, heptose, ethanolamine, and phosphate, and an outer region that consists of hexoses. The biosynthesis of the inner region is not fully understood. The outer core region grows as hexose units are donated one at a time from nucleoside diphosphate derivatives to the nonreducing end of the growing glycan chain attached to the KDO. The addition of each sugar is catalyzed by a specific glycosyl transferase, which is membrane bound. The core-lipid A portion of the LPS is translocated across the cell membrane to the periplasmic surface, where the LPS is completed by attachment of the O-antigen (see Fig. 11.13).[8]

## O-Antigen

The O-antigen region is synthesized by a mechanism that is different from that of core synthesis (Fig. 11.12). Whereas the core is synthesized via the addition of sugars one at a time to the growing end of the glycan chain, the O-antigen is synthesized as a separate polymer on a lipid carrier and then transferred as a unit to the core. The lipid carrier is undecaprenyl phosphate (i.e., the same molecule that carries the peptidoglycan precursors across the membrane). First, the repeat unit of the O-antigen is synthesized on the lipid carrier. This is done by a series of consecutive reactions in which a sugar moiety is transferred from a nucleoside diphosphate carrier to the nonreducing end of the growing repeat unit (Fig. 11.12, steps 1–4). Each of these reactions is catalyzed by a different enzyme. Then the repeat unit is transferred as a block to the growing oligosaccharide chain (step 5). The lipid-PP of the acceptor oligosaccharide chain is displaced and enters the lipid pyrophosphate pool, where it is hydrolyzed to lipid-P by a bacitracin-sensitive enzyme. Finally, the completed oligosaccharide chain is transferred

to the lipid A-core (step 6), displacing the lipid pyrophosphate.

## How the lipopolysaccharide might be assembled

Figure 11.13 depicts a model for how the LPS might be assembled. The O-antigen tetrasaccharide subunit is probably synthesized on the lipid carrier on the cytoplasmic side of the membrane and then moves to the periplasmic side, where it is added to the growing O-antigen anchored to the membrane by its lipid carrier. The lipid carrier on the growing oligosaccharide is displaced. The core-lipid A region may also be assembled on the cytoplasmic surface and translocated to the periplasmic surface. The final completion of the lipopolysaccharide would take place on the periplasmic surface, where the O-antigen is transferred to the core-lipid A, displacing the lipid-PP. The lipid-P is regenerated via a phosphatase and enters a common pool of lipid-P also used in the biosynthesis of peptidoglycan. It is not known how the lipopolysaccharide moves through the periplasm into the outer envelope.

# 11.3 Extracellular Polysaccharide Synthesis and Export in Gram-Negative Bacteria

## 11.3.1 Overview

The student should refer to Section 1.2.2 for a description of extracellular polysaccharides synthesized by bacteria and the biological roles they play, and to recent reviews.[9–12] The extracellular polysaccharides are critical for cell survival in the natural habitat, although mutants lacking them can survive in the laboratory. They are either closely associated with the cell wall (capsular polysaccharides) or exist as an amorphous layer loosely attached to the cell (slime polysaccharides). The attachments of the extracellular polysaccharides to the bacterial surface varies according to the polysaccharide. Slime polysaccharides are not attached at all and are readily released

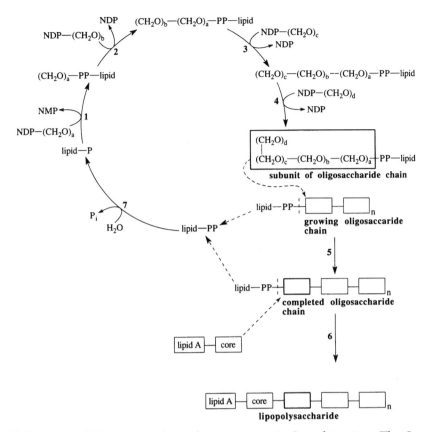

**Fig. 11.12** Synthesis of O-antigen and attachment to core in *S. typhimurium*. The O-antigen is polymerized on the same phospholipid carrier that is used for peptidoglycan synthesis. Each sugar is added from a nucleotide diphosphate derivative using a specific sugar transferase (reactions 1–4). The nucleotide that is used depends upon the sugar and the specific glycosyl transferase. For example, in *S. typhimurium* the nucleotide precursors are: UDP–galactose, TDP–rhamnose, GDP–mannose, and CDP–abequose, in that order. The first sugar that is added (e.g., galactose in *S. typhimurium*) is transferred as a phosphorylated derivative to the lipid carrier. Hence the NMP, i.e., UMP in the case of *E. coli* and *S. typhimurium*, is released. When the O-antigen tetrasaccharide subunit is finished, it is transferred to the growing O-antigen chain by an enzyme called O-antigen polymerase (reaction 5). The completed O-antigen is transferred from its lipid carrier to core-lipid A by an enzyme called O-antigen:lipopolysaccharide ligase (reaction 6). The lipid-PP that is displaced is hydrolyzed by a bacitracin-sensitive phosphatase (reaction 7). All the reactions take place in the cell membrane. There exist immunoelectron microscopy data to suggest that the ligase reaction and perhaps the polymerase reaction take place on the periplasmic surface of the cell membrane. Presumably, the lipid carrier ferries the subunits across the cell membrane. It is not known how the LPS crosses the periplasm to enter the outer envelope. See Mulford, C. A., and M. J. Osborn. 1983. An intermediate step in translocation of lipopolysaccharide to the outer membrane of *Salmonella typhimurium*. *Proc. Natl. Acad. Sci. USA* 80:1159–1163.

from the cell surface. Capsular polysaccharides in gram-negative bacteria are generally attached via a hydrophobic molecule that anchors them to the lipid portion of the outer membrane. The hydrophobic anchor might be lipid A or phosphatidic acid. Lipopolysaccharides, which are discussed in Section 11.2, are attached to the outer membrane via lipid A (Figure 11.10). Figure 11.14 summarizes the modes of attachments.

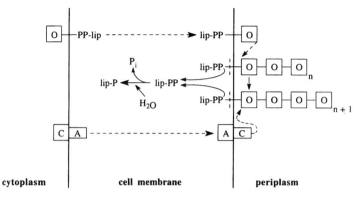

**Fig. 11.13** A model for LPS assembly. The O-antigen subunit is synthesized on the cytoplasmic surface and then moves on the lipid carrier to the periplasmic surface of the membrane. A polymerase then transfers the O-antigen subunit to the growing O-antigen chain. The core-lipid A is also synthesized on the cytoplasmic surface and is translocated to the periplasmic surface. The O-antigen is transferred to the core to complete the LPS. The entire LPS is translocated to the outer membrane. Abbreviations: O, O-antigen; C, core; lip, lipid carrier; A, lipid A.

## Escherichia coli capsules

*E. coli* is an example of a gram-negative bacterium that may possess an extracellular capsule surrounding its outer membrane. The capsule masks the antigenic determinants of the O polysaccharides in the LPS. An easy way to detect the presence of the capsule is to heat the cells. Heating removes the capsule and makes the cells agglutinable with anti-O antigen antibody. The capsular antigens are called K antigens (from the German "Kapselantigene"), and there are at least 80 types that differ in their antigenicity and composition. The K antigens are acidic polysaccharides and are divided into two groups, group I and group II. Group I polysacharides have a high molecular weight (>100,000) and contain either glucuronic or galacturonic acids. Group II polysaccharides have a somewhat lower molecular weight (<50,000) and consist of other constituents in addition to, or instead of, glucuronic acid, including N-acetylneuraminic acid (NeuNAc), also called sialic acid, 2-keto-3-deoxymanno-octonic acid (KDO), N-acetylmannosamine, or phosphate. The repeating unit varies from two to six monosaccharides, depending upon the polysaccharide. Many appear to be substituted at their reducing end with lipid A (group I) or KDO-phosphatidic acid (group II polysaccharides), presumably to anchor them in the outer membrane

(Fig. 11.14). (Another extracellular polysaccharide that is anchored to the outer membrane via lipid A is LPS. See Section 11.2.) Interestingly, other bacteria may synthesize similar or identical capsular polysaccharides. For example, *Neisseria meningitidis* makes a capsular polysaccharide that is identical to the K1 capsule (a group II polysaccharide) of *E. coli*. *Haemophilus influenzae* also makes a capsule quite similar to some of the K polysaccharides of *E. coli*. Recently, group II capsular polysaccharides have been subdivided into groups II and III.[13]

## Summary of synthesis of extracellular polysaccharides

Most extracellular polysaccharides are synthesized from intracellular nucleoside diphosphate-sugar precursors and must be transported through the cell membrane to the outside of the cell. The nucleoside diphosphate sugars are synthesized from sugar-1-P and nucleoside triphosphates, as described in Section 11.1.2 for the synthesis of N-acetylglucosamine. If the extracellular polysaccharides are synthesized by gram-negative bacteria, then they must be transported through the cell membrane, periplasm, and outer membrane. As we shall see, not all polysaccharides are synthesized and exported the same way.

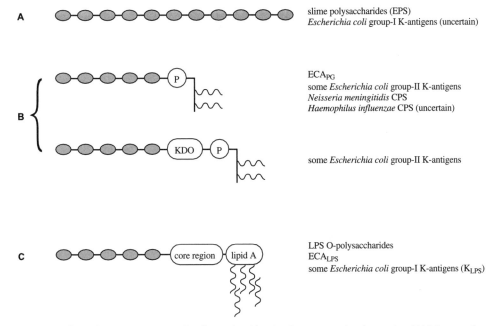

**Fig. 11.14** Cell-surface association of polysaccharides in Gram-negative bacteria. (A) No membrane anchor. (B) Attachment through diacylglycerolphosphate. (C) Attachment through lipid A. KDO, 3-deoxy-D-*manno*-octulosonic acid; P, phosphate. *Source:* Whitfield C., and M. A. Valvano. 1993. Biosynthesis and expression of cell-surface polysaccharides in gram-negative bacteria. *Adv. in Microbial Physiology* 35:125–246. Academic Press, New York.

## Steps in synthesis and assembly of extracellular polysaccharides in gram-negative bacteria

The synthesis of some extracellular polysaccharides takes place via undecaprenol intermediates in a pathway similar to the synthesis of the oligosaccharide of lipopolysaccharide, and are translocated through the membrane in an undecaprenol-dependent manner. However, other polysaccharides are not synthesized using lipid intermediates and are translocated through the membrane in an entirely different way, using specific transporters.

## 11.3.2 Polysaccharide synthesis using undecaprenol-diphosphate intermediates

Several bacteria synthesize certain exopolysaccharides via undecaprenol intermediates similar to the pathway for the biosynthesis of the oligosaccharide repeat unit in lipopolysaccharide and the disaccharide repeat unit in peptidoglycan (Figs. 11.6 and 11.12).

Polysaccharides that are synthesized via this pathway include the group I K antigens of *E. coli*, the extracellular polysaccharide xanthan made by *Xanthomonas campestris*, and the capsular polysaccharide of *Klebsiella (Aerobacter) aerogenes*. As an example, we will consider the biosynthesis of the *K. aerogenes* capsular polysaccharide.

## K. aerogenes capsular polysaccharide

The capsule made by *K. aerogenes* is a polymer of repeating tetrasaccharides composed of galactose, mannose, and glucuronic acid in a molar ratio of 2:1:1. The glucuronic is attached as a branch at each mannose residue. Its synthesis resembles the synthesis of the oligosaccharide in lipopolysaccharide and is depicted in Fig. 11.15.[14] The tetrasaccharide repeating unit is synthesized on undecaprenol diphosphate from nucleotide diphosphate sugar intermediates. Each repeating unit is then added as a block to the growing oligosaccharide attached to undecaprenol diphosphate.

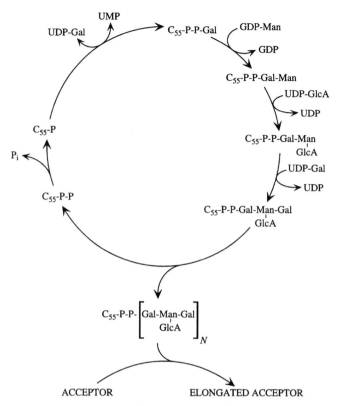

**Fig. 11.15** Synthesis of capsular polysaccharide in *Aerobacter*. Each monosaccharide is donated to the tetrasaccharide repeat unit from a UDP derivative. This takes place on the inner surface of the cell membrane. Polymerization takes place while the sugars are attached to undecaprenol diphosphate. The tetrasaccharide repeat units are transferred from the lipid carrier to the growing chain, presumably on the outer surface of the cell membrane. See Figs. 10.12 and 10.13 for a model as to how this might occur. UDP, uridine diphosphate; Gal, galactose; Man, mannose; GlcUA, glucuronic acid, lipid-PP, undecaprenol diphosphate.

### 11.3.3 Synthesis of E. coli group II K1 antigen: A pathway that may use an undecaprenol monophosphate intermediate

#### Chemical nature and biological role of the K1 capsule

The *E. coli* K1 capsule (a group II antigen), *Neisseria meningitidis* group B capsular polysaccharide, and capsular polymers of *Pasteurella haemolytica* and *Moraxella nonliquefaciens* are homopolymers of sialic acid approximately 200 residues long. (For a review, see Ref. 15.) The mechanism of growth of the polysialic acid capsule of *E. coli* has been of interest for a long time. This is in part because it is a virulence factor for strains of *E. coli* that cause neonatal meningitis,

septicaemia, and urinary tract infections in children. Polysialic acid capsules are virulence factors in part because they resemble polysaccharides on host tissue and consequently are poorly immunogenic. Thus they make the bacteria more resistant to the immune response. Additionally, capsules containing sialic acid make the cells more resistant to complement-mediated killing apparently because they inhibit the activation of complement by the alternative system.

#### Biosynthesis of the K1 capsule

The mechanism of polysialic acid chain elongation in *E. coli K1* has been well studied both biochemically and genetically. (See Refs. 16 and 17 and references therein.) The polymer grows by the stepwise addition of single sialic acid residues, apparently from CMP-sialic acid. This can be demonstrated by adding CMP-[$^{14}$C]-sialic acid to incubation mixtures containing washed membranes. The enzyme that catalyzes the reaction is called *sialytransferase,* which is loosely attached to the cell membrane. The cell extracts make undecaprenol monophosphate-linked sialic acid residues by transferring sialic acid from CMP-sialic acid to undecaprenol phosphate, and transfer the sialic acid to endogenous acceptors in the membrane (reviewed in Ref. 8.) However, doubts have been raised whether the lipid-linked sialyl residues are an obligate intermediate in the elongation reaction.[14] Assuming that they are, then one model is that the sialic acid is transferred from CMP to undecaprenol phosphate to form sialyl monophosphorylundecaprenol, which transfers the sialic acid to the nonreducing end of the growing polysialic chain. It is not known whether the growing polymer is attached to undecaprenol. Neither is the identity of the initial receptor for the first sialic acid known, and is simply referred to as the endogenous receptor. The endogeneous receptor does not contain sialic acid.[18] The completed polysaccharide has phospholipid attached to its reducing end, which presumably anchors the polysaccharide to the outer membrane. It is not known at what stage of the biosynthesis of the polysialyl polymer the phospholipid is attached.

## 11.3.4 Pathways not involving undecaprenol derivatives

The synthesis of some extracellular polysaccharides, including some of the group II K antigen capsules, such as K5 in *E. coli,* alginate synthesized by *Azotobacter vinelandii* and *Pseudomonas aeruginosa,* and cellulose synthesized by *Acetobacter xylinum,* does not involve undecaprenol intermediates. In several instances, the pathways of polymerization are not known. Where characterized, it is found that the polysaccharides are synthesized from nucleotide diphosphate precursors added to the growing oligosaccharide chain. For example, alginate, which is a linear

copolymer of D-mannuronic and D-guluronic acid, is synthesized by brown algae from GDP-mannuronic and GDP-guluronic acid. Another example is cellulose, which is polymerized from UDP-glucose by a membrane-bound cellulose synthetase in *Acetobacter xylinum.*

## 11.3.5 Export of polysaccharides

### Undecaprenol phosphate-based translocation

Because polysaccharides are polymerized from nucleotide derivatives of the sugars that are made in the cytoplasm, one can assume that at least the first steps in polymerization take place on the cytoplasmic surface of the cell membrane. Because of this, there must be a mechanism to transport newly synthesized polysaccharide across the cytoplasmic membrane. One of these postulates that the role of undecaprenol phosphate is to serve as part of a transmemebrane-assembly process, which synthesizes and moves completed polysaccharides to the periplasmic surface of the cell membrane. This is depicted in Fig. 11.16(a).

### Translocation using ABC transporters

There is evidence for a specific transport mechanism that does not require undecaprenol. This evidence has been derived from examining the genes (*kps* cluster) for biosynthesis and export of group II capsular polysaccharides (including K1 and K5) in *E. coli* and is depicted in Fig. 11.16(b). (See Ref. 13 for a review.) Mutations in a particular region of the *kps* gene cluster (region 3) required for biosynthesis of the polysaccharide result in the accumulation of the polysaccharide in the cytoplasm. The deduced amino acid sequences of the protein products of these genes (KpsM and KpsT) indicate that they belong to the family of ABC transporters. Accordingly, they are referred to as the KpsMT transporter and probably form a protein channel through which the polysaccharides are translocated either during or after biosynthesis. (See Section 16.3.3 for a discussion of ABC transporters.) Similar genes have been found in the

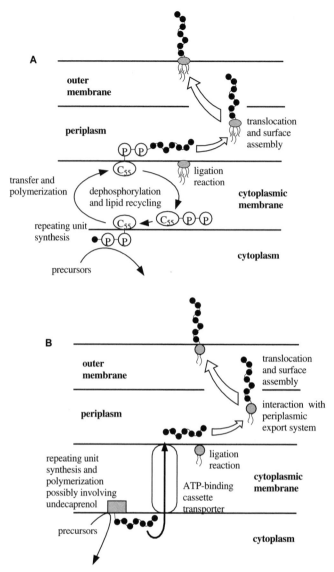

**Fig. 11.16** Models for synthesis and export of polysaccharides in gram-negative bacteria. (A) *rfe*-independent O-polysaccharide biosynthesis in Salmonella enterica. (B) Group II capsular polysaccharide biosynthesis in *Escherichia coli*. C55, undecaprenol; P, phosphate. Source: Whitfield C., and M. A. Valvano. 1993. Biosynthesis and expression of cell-surface polysaccharides in gram-negative bacteria. *Adv. In Microbial Physiology* 35:135–246. Academic Press, New York.

biosynthetic gene cluster for extracellular polysaccharide synthesis in other bacteria, including *Haemophilus, Neisseria Rhizobium, Agrobacterium,* and may be widespread. Thus it can be concluded that many extracellular polysaccharides are exported through the cell membrane in protein channels using energy derived at least in part from the hydrolysis of ATP. (A $\Delta p$ is also required.) How these polysaccharides reach the

outer membrane surface is not clear. Various possible mechanisms are reviewed in Refs. 8 and 13.

## 11.4 Levan and Dextran Synthesis

Certain bacteria (e.g., the lactic acid bacteria) synthesize extracellular polymers of fructose, called *levans*, or glucose, called *dextrans*, from sucrose. The synthesis is catalyzed by extracellular enzymes called *levansucrase* or *dextransucrase*. The reaction is:

$$y \text{ sucrose } + (\text{fructose})_n \rightarrow (\text{fructose})_{n+y}$$
$$+ \, y \text{ glucose} \qquad (11.1)$$

$$y \text{ sucrose } + (\text{glucose})_n \rightarrow (\text{glucose})_{n+y}$$
$$+ \, y \text{ fructose} \qquad (11.2)$$

In both reactions a monosaccharide is transferred from the disaccharide to the reducing end of the growing oligosaccharide chain. The polysaccharide chain remains attached to the enzyme during its elongation. The dextran is a glucose polymer (glucan) in which the glucose residues are attached via $\alpha 1$–6 linkages. A different enzyme makes a glucan with $\alpha 1$–3 linkages called *mutan*. *Streptococcus mutans,* the bacterium that is a principal cause of tooth decay, forms dextrans, mutans, and fructans. The glucans enable the bacteria to adhere to the surface of the teeth.

## 11.5 Glycogen Synthesis

Certain bacteria synthesize glycogen, a polymer of glucose, as an intracellular carbon and energy reserve. Glycogen is a branched polysaccharide. The linear portion consists of glucose residues connected by $\alpha 1$–4 linkages, and the branches are attached via $\alpha 1$–6 linkages. The oligosaccharides are extended by the addition of glucose units donated by ADP-glucose in a reaction catalyzed by *glycogen synthetase*. A *branching enzyme* transfers 6–8 glucose segments from the linear portion of the oligosaccharide to form an $\alpha 1$–6 linked branch. The synthesis of nucleoside diphosphate derivatives of sugars from sugar-1-P and nucleoside triphosphates is described in Section 11.1.2.

## 11.6 Summary

There are two major problems that must be solved in order to synthesize the cell wall. One is how to move the precursors through the cell membrane, and the second is how to make peptide bonds outside the cell membrane far from the cellular ATP pools. Bacteria employ

a lipid carrier, undecaprenyl phosphate, to carry the cell wall precursors through the membrane. Interestingly, eukaryotes employ a similar compound, dolichol phosphate, to move oligosaccharide precursors through the ER membrane to synthesize glycoproteins. The peptidoglycan peptide cross-link forms as a result of a transpeptidation reaction. During the transpeptidation reaction an amino group from the diamino acid (e.g., DAP) displaces a terminal D-alanine. This is an exchange of a peptide bond for one that was made in the cytosol at the expense of ATP. ATP also provides the energy to make the glycosidic linkages in the cell wall polymers because ATP drives the synthesis of the sugar-PP-lipid intermediates. The formation of the glycosidic linkage is a straightforward displacement of the lipid pyrophosphate from the C1 carbon of the *N*-acetylmuramic acid by the OH on the incoming C4 carbon of *N*-acetylglucosamine. The reaction is driven to completion by the subsequent hydrolysis of the lipid pyrophosphate to the lipid phosphate and inorganic phosphate.

Lipopolysaccharide synthesis can be thought of as occurring in four stages: (1) synthesis of the lipid A portion; (2) synthesis of the core region by adding one sugar at a time to lipid A from nucleoside diphosphate precursors; (3) synthesis of the complete repeat oligosaccharide; and (4) attachment of the repeat oligosaccharide to the core. All these events are associated with the membrane. It appears that the lipid A and core portions are synthesized on the cytoplasmic side of the membrane and then translocated in an unknown manner to the periplasmic surface. The oligosaccharide (O-antigen) subunits are assembled on undecaprenyl pyrophosphate and transferred to a growing oligosaccharide chain. When the oligosaccharide is complete, it is attached to the core. How the lipopolysaccharide enters the outer envelope is not known.

Capsular and extracellular polysaccharides are synthesized and exported in at least two ways. One pathway is similar to the pathway for the synthesis of the oligosaccharide in lipopolysaccharide and involves undecaprenol intermediates. A second pathway appears more complex and appears to involve

ABC-type transporters to move the polysaccharide through the cell membrane.

## Study Questions

1. O-Antigen synthesis requires a lipid carrier, but core synthesis does not. Offer a plausible explanation for the difference.

2. During the synthesis of the pentapeptide in the peptidoglycan precursor, an ATP is expended to make the peptide bond as each amino acid is added to the growing pentapeptide. The products are ADP and $P_i$. Write a plausible mechanism by which ATP is used to provide the energy to make a peptide bond between two amino acids. (Note: m-RNA and ribosomes are not involved. The addition of each amino acid is catalyzed by a separate enzyme specific for that amino acid.)

3. Peptidoglycan and lipid A share a common pathway early in their syntheses. Outline the early stages of both pathways up to the branch point.

4. Carriers of subunit moieties play important roles in biosynthesis. Usually, the carriers are involved in more than one pathway. What carrier is common to both peptidoglycan and lipopolysaccharide synthesis?

5. Important enzymes in cell wall peptidoglycan synthesis are membrane-bound transglycosidases that transfer carbohydrate subunits to the growing polymer. What ensures that the growing end of the polymer remains at the site of the transglycosidase?

6. Show how ATP drives the synthesis of UDP-GlcNAc from glucose. Focus on reactions in which phosphoryl and nucleotide groups are transferred. You must show how ATP drives the synthesis of UTP. There are two phosphate groups in MurNAc(pentapeptide)-PP-lipid. Show how one of them is derived from ATP. What eventually happens to this phosphate?

7. Covalent bond formation in the cytoplasm is driven by high-energy molecules such as ATP or other nucleoside triphosphates. Peptidoglycan synthesis offers some examples of how covalent bonds can be formed outside the cytoplasm without access to a source of high-energy molecules such as nucleoside triphosphates, which are cytoplasmic. Two examples are the transpeptidation reaction, which results in the synthesis of a peptide bond, and the transglycosidase reaction, which results in the synthesis of a glycosidic bond. These reactions are displacement reactions in which a diamino acid (e.g., diaminopimelic acid) displaces D-alanine on the pentapeptide (transpeptidation), and the N-acetylglucosamine moiety of the incoming lipid–disaccharide pentapeptide displaces the lipid pyrophosphate in the growing glycan chain (transglycosidase). If one carefully examines the pathway for peptidoglycan synthesis, one can see that nucleoside triphosphates are indeed used in the cytoplasm in reactions that ultimately provide the energy for the transpeptidation and transglycosidase reactions. Write these cytoplasmic reactions and show how they drive the transpeptidation and transglycosidase reactions.

## NOTES AND REFERENCES

1. van Heijenhoort. 1996. Murein synthesis. In: *Escherichia coli* and *Salmonella*: Cellular and Molecular Biology. Vol. 1, pp. 1025–1034. F. C. Neidhardt (Ed.). American Society for Microbiology, Washington, D.C.

2. J.-V. Höltje. 1998. Growth of the stress-bearing and shape-maintaining murein sacculus of *Escherichia coli*. Microbiol. and Molec. Biol. Rev. **62**:181–203.

3. Hartmann, E., and H. Konig. 1994. A novel pathway of peptide biosynthesis found in methanogenic Archaea. *Arch. Microbiol.* **162**:430–432.

4. Raetz, C. R. 1996. Bacterial lipopolysaccharide: A remarkable family of bioactive macroamphiphiles. In: *Escherichia coli and Salmonella: Cellular and Molecular Biology*. Vol. 1, pp. 1035–1063. F. C. Neidhardt (Ed.). American Society for Microbiolgy, Washington, D.C.

5. This used to be called 2-keto-3-deoxyoctonate.

6. Rick, P. D. 1987. Lipopolysaccharide biosynthesis. In: *Escherichia coli and Salmonella typhimurium: Cellular and Molecular Biology*. Vol. 1. pp. 648–662. Neidhardt, F. C., J. L. Ingraham, K. B. Low, B. Magasanik, M. Schaecter, and H. E. Umbarger (Eds.). American Society for Microbiology, Washington, D.C.

7. Raetz, C. R. H. 1987. Structure and biosynthesis of lipid A. In: *Esherichia coli and Salmonella typhimurium: Cellular and Molecular Biology.* Vol. 1. pp. 498–503. Neidhardt, F. C., J. L. Ingrahm, K. B. Low, B. Magasanik, M. Schaechter, and H. E. Umbarger (Eds.). American Society for Microbiology, Washington, D.C.

8. Mulford, C. A., and M. J. Osborn. 1983. An intermediate step in translocation of lipopolysaccharide to the outer membrane of *Salmonella typhimurium. Proc. Natl. Acad. Sci. USA* 80:1159–1163.

9. Whitfield, C., and M. A. Valvano. 1993. Biosynthesis and expression of cell-surface polysaccharides in gram-negative bacteria, pp. 135–246. In: *Adv. Microbial. Phys.* Vol. 35. A. H. Rose (Ed.). Academic Press, New York.

10. Rick, P. D., and R. P. Silver. 1996. Enterobacterial common antigen and capsular polysaccharides, pp. 104–122. In: *Escherichia coli and Salmonella: Cellular and Molecular Biology,* Vol. 1. F. C. Neidhardt (Ed.). American Society for Microbiology, Washington, D.C.

11. Roberts, E. S. 1995. Bacterial polysaccharides in sickness and in health. *Microbiology* 141:2023–2031.

12 Roberts, I. S. 1996. The biochemistry and genetics of capsular polysaccharide production in bacteria. *Ann. Rev. Microbiol.* 50:285–315.

13. Russo, T. A., S. Wenderoth, U. B. Carlino, J. M. Merrick and A. J. Lesse. 1998. Identification, genomic organization, and analysis of the group III capsular polysaccharide genes *kpsD, kpsM, kpsT,* and *kpsE* from an extraintestinal isolate of *Escherichia coli* (CP9, 04/K54/H5). *J. Bacteriol.* 180:338–349.

14. Troy, F. A., F. E. Frerman, and E. C. Heath. 1971. The biosynthesis of capsular polysaccharide in *Aerobacter aerogenes. J. Biol. Chem.* 246:118–133.

15. Bliss, J. M., and R. P. Silver. 1996. Coating the surface: A model for expression of capsular polysialic acid in *Escherichia coli* K1. *Molec. Microbiol.* 21:221–231.

16. Steenbergen, S. M., and E. R. Vimr. 1990. Mechanism of polysialic acid chain elongation in *Escherichia coli* K1. *Molec. Microbiol.* 4:603–611.

17. Cieslewixz, M., and E. Vimr. 1997. Reduced polysialic acid capsule expression in *Escherichia coli* K1 muntants with chromosomal defects in *kpsF. Molec. Microbiol.* 26:237–249.

18. Rohr, T. E., and F. A. Troy. 1980. Structure and biosynthesis of surface polymers containing polysialic acid in *Escherichia coli. J. Biol. Chem.* 255:2332–2342.

# 12

# Inorganic Metabolism

Inorganic molecules such as derivatives of sulfur, nitrogen, and iron are used by prokaryotes in a variety of metabolic ways related to energy metabolism and biosynthesis.

1. There are *assimilatory pathways* in which inorganic nitrogen and sulfur are incorporated into sulfur- and/or nitrogen-containing organic material (e.g., amino acids and nucleotides). Most prokaryotes can do this. In addition, many prokaryotes can also utilize nitrogen gas as a source of nitrogen, a process called *nitrogen fixation*.

2. There are *dissimilatory pathways* in which inorganic compounds are used instead of oxygen as electron acceptors, a process called *anaerobic respiration*. The reduced products are excreted into the environment. During anaerobic respiration a $\Delta p$ is created in the same way as during aerobic respiration (see Section 4.6). For example, many facultative anaerobes can use nitrate as an electron acceptor, reducing it to ammonia or nitrogen gas. Several obligate anaerobes use sulfate as an electron acceptor, reducing it to hydrogen sulfide. There also exist bacteria that can use $Fe^{3+}$ or $Mn^{4+}$ as an electron acceptor during anaerobic growth.[1-3] The latter organisms are responsible for most of the reduction of iron and manganese that takes place in sedimentary organic matter (e.g., in lake sediments). However, only a few of the iron and manganese

reducers have been isolated, and little is known about their physiology.

3. There are *oxidative pathways* in which inorganic compounds such as $H_2$, $NH_3$, $NO_2^-$, $S°$, $H_2S$, and $Fe^{2+}$, rather than organic compounds, are oxidized as a source of electrons and energy. Organisms that derive their energy and electrons for biosynthesis in this way are called *chemolithotrophs*. (If they use $CO_2$ as the sole or major source of carbon, then they are called *chemolithoautotrophs*.)

## 12.1 Assimilation of Nitrate and Sulfate

Many bacteria can grow in media in which the only sources of nitrogen and sulfur are inorganic nitrate salts and sulfate salts. The nitrate is reduced to ammonia, and the ammonia is incorporated into the amino acids glutamine and glutamate using the GS/GOGAT system (see Section 9.3.1). Glutamate and glutamine are the sources of amino groups for the other nitrogen-containing organic compounds. The sulfate is reduced to $H_2S$, which is immediately incorporated into the amino acid cysteine using the O-acetylserine pathway described in Section 9.3.1. Cysteine, in turn, is the source of sulfur for other organic molecules [e.g., methionine, coenzyme A (CoASH), and acyl carrier protein (ACPSH)].

## Nitrate assimilation

Nitrate can serve as the source of cellular nitrogen for plants, fungi, and many bacteria. Although some bacteria (e.g., *Klebsiella pneumoniae,*) assimilate nitrate during aerobic growth, certain other bacteria (e.g., *E. coli*, and *Salmonella typhimurium,*) assimilate nitrate only during anaerobic growth. (See ref. 4 for a review of nitrate transport and reduction.) The nitrate is first reduced to ammonia. Since the oxidation state for the nitrogen in nitrate ($NO_3^-$) is +5 and for the nitrogen in ammonia ($NH_3$) it is −3, eight electrons must be transferred to nitrate in order to reduce it to ammonia. The enzymes involved in the assimilatory reduction of nitrate to ammonia are *cytoplasmic nitrate reductase,* which reduces nitrate to nitrite, and *cytoplasmic nitrite reductase*, which reduces nitrite to ammonia. The electron donors for the nitrate reductase can be NADH (via a separate NADH oxidoreductase), ferredoxin, or flavodoxin, depending upon the bacterium, as explained in Note 5. The electron donor for the major nitrite reductase in *E. coli* is NADH. The electron transport pathway is shown in Fig. 12.1 as proceeding in two-electron transfer steps. The ammonia that is formed is incorporated into glutamine via *glutamine synthetase* (GS). The glutamine then serves as the amino donor for purine, pyrimidine, and amino sugar biosynthesis (Sections 9.2.2, 9.2.3, and 11.1.3), as well as for the synthesis of glutamate (glutamate synthetase). Glutamate is the amino donor for amino acid biosynthesis via the transamination reactions described in Section 9.3.1.

## Sulfate assimilation

Many bacteria can use sulfate ($SO_4^{2-}$) as their principal source of sulfur. The sulfate is first reduced to sulfide ($H_2S$ and $HS^-$) and then incorporated into cysteine. (See Note 6 for an explanation why $H_2S$ and $HS^-$ are the major species of sulfide in the cytoplasm.) Since the oxidation level of sulfur in $SO_4^{2-}$ is +6 and in $S^{2-}$ is −2, a total of eight electrons are required to reduce sulfate to sulfide. The first step in the reduction is the formation of adenosine-5′-phosphosulfate (APS) catalyzed by the enzyme ATP sulfurylase (Fig. 12.2, reaction 1). Here sulfate acts as a nucleophile and displaces pyrophosphate ($PP_i$). (See Section 7.1.1 for a discussion of nucleophilic displacements.) The pyrophosphate is subsequently hydrolyzed to inorganic phosphate, thus driving the synthesis of APS to completion. There is a sound thermodynamic reason for making the AMP derivative of sulfate prior to its reduction. Attaching the sulfate to AMP raises the reduction potential, making APS a better electron acceptor than free sulfate. (The $E_0'$ for the reduction potential of sulfate to sulfite is very low, i.e., −520 mV, such that its reduction

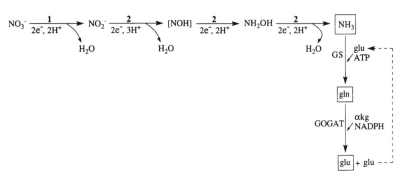

**Fig. 12.1** Assimilatory nitrate reduction. This pathway is present in all bacteria that reduce nitrate to ammonia, which is then incorporated into cell material. The enzymes are found in the cytosol and are not coupled to ATP formation. Nitrate is reduced via two-electron steps to nitrite, nitroxyl, hydoxylamine, and ammonia. The ammonia is incorporated into organic carbon via glutamine synthetase (GS) and the GOGAT enzyme, or via glutamate dehydrogenase. (See Section 9.3.1.) Enzymes: 1, nitrate reductase; 2, nitrite reductase. Abbreviations: glu, glutamate; α-kg, α-ketoglutarate.

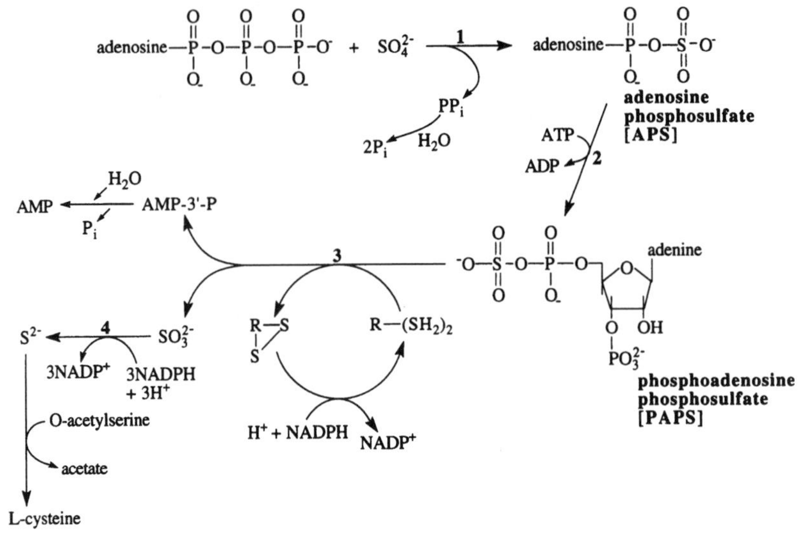

Fig. 12.2 Assimilatory sulfate reduction. Enzymes: 1, ATP sulfurylase; 2, APS phosphokinase; 3, PAPS reductase; 4, sulfite reductase; 5, O-acetylserine sulfhydrylase. R(SH)$_2$ is reduced thioredoxin.

even by H$_2$, $E'_0 = -420$ mV, is endergonic.) The second reaction is the phosphorylation of APS to form adenosine-3'-phosphate-5'-phosphosulfate (PAPS), catalyzed by APS kinase (reaction 2). The reaction is an attack by the 3' hydroxyl of the ribose of APS on the terminal phosphate of ATP, displacing ADP. The PAPS is then reduced to sulfite with the release of AMP-3'-phosphate (reaction 3). The reductant is a sulfhydryl protein called thioredoxin, which in turn accepts electrons from NADPH. (Thioredoxin is used in other metabolic pathways as a reductant. For example, recall that thioredoxin also reduces the nucleoside diphosphates to the deoxynucleotides (Section 9.2.5). The sulfite is then reduced by NADPH to hydrogen sulfide (H$_2$S) (reaction 4). Hydrogen sulfide is very toxic and does not accumulate. The sulfide enzymatically displaces acetate from O-acetylserine to form cysteine (see Section 9.3.1). The AMP-3'-phosphate is hydrolyzed to AMP and P$_i$, thus helping to drive the overall reaction to completion. Note that three ATPs are used to reduce sulfate to sulfide, two to make the PAPS derivative, and a third to phosphorylate the AMP released from AMP-3'- phosphate to ADP (reaction 3). (The ADP is then converted to ATP via respiratory phosphorylation or substrate-level phosphorylation.)

## 12.2 Dissimilation of Nitrate and Sulfate

In the *dissimilatory* pathways the nitrate and sulfate are used as electron acceptors during anaerobic respiration. The reduced products are excreted rather than being incorporated into cell material. Whereas many facultative anaerobes are capable of dissimilatory nitrate reduction and employ this pathway when oxygen is not available, the use of sulfate as an electron acceptor is restricted to obligately anaerobic bacteria called *sulfate reducers*.

### 12.2.1 Dissimilatory nitrate reduction

Dissimilatory nitrate reduction (nitrate respiration) is usually facultative and occurs as a substitute for aerobic respiration when the oxygen levels become very low. It takes place in membranes, and a $\Delta p$ is usually made. The products of nitrate respiration can be nitrite, ammonia, or nitrogen gas.

### Denitrification

When the nitrate (or nitrite) is reduced to nitrogen gas (or nitric oxide gas, NO, or nitrous oxide gas, N$_2$O), the process is called *denitrification*, which can be an ecologically important drain of nitrogen from the soil. Denitrification occurs in the soil when conditions become anaerobic (e.g., in water-logged soil), and also during composting and sludge digestion. The denitrifiers reduce the available nitrate in the soil, but they are also important in anaerobic niches for the breakdown of undesirable biodegradable materials, plant materials, complex organic compounds, and so forth.[7] The breakdown of organic materials by denitrifiers is usually more complete than via a fermentative process. Many bacteria denitrify, but the most commonly isolated denitrifiers are *Alcaligenes* and *Pseudomonas* species, and to a lesser extent *Paracoccus denitrificans*. The electron transport pathway for denitrification by *P. denitrificans* is summarized in Section 4.7.2.

## 12.2.2 Dissimilatory sulfate reduction

### A general description of the sulfate reducers

The sulfate reducers are heterotrophic strict anaerobes that grow in anaerobic muds, mostly in anaerobic parts of fresh water, and in sea water.[8] They carry out anaerobic respiration, during which sulfate is reduced to $H_2S$. Some can be grown autotrophically with $H_2$ as the source of electrons and $SO_4^{2-}$ as the electron acceptor. Formerly the sulfate reducers were believed to use as carbon sources only a limited variety of compounds (e.g., formate, lactate, pyruvate, malate, fumarate, ethanol, and a few other simple compounds). Recently, however, it has been realized that, depending upon the species, many other carbon sources can be used, including straight-chain alkanes and a variety of aromatic compounds. Sulfate reducers comprise a very diverse group of organisms that include both gram-positive and gram-negative bacteria, as well as archaea. An example of the latter is *Archaeoglobus*, a hyperthermophile isolated from sediments near hydrothermal vents. Gram-positive spore-forming sulfate reducers belong to the genus *Desulfotomaculum*, which is very diverse. The most prominent of the gram-negative sulfate reducers belong to the genus *Desulfovibrio*, which is also phylogenetically diverse.

Traditionally, the sulfate reducers are divided into two physiological groups, I and II.

Those in group I cannot oxidize acetyl–CoA to $CO_2$ and therefore excrete acetate when growing on certain carbon sources (e.g., lactate or ethanol). The group I genera include *Desulfovibrio* and most *Desulfotomaculum* species. *Group II organisms can oxidize acetyl–CoA to $CO_2$*. Group II sulfate reducers are found in several genera, including *Desulfotomaculum* and *Desulfobacter*.

There exist two pathways for oxidizing acetyl–CoA anaerobically to $CO_2$; a modified citric acid cycle and the acetyl–CoA pathway. *Desulfobacter* has a modified citric acid cycle that resembles that found in aerobes except that: (1) Instead of a citrate synthase there is an ATP–citrate lyase; and (2) the $NAD^+$-linked $\alpha$-ketoglutarate dehydrogenase is replaced by a ferredoxin-dependent enzyme. The pathway is called the reductive tricarboxylic acid pathway, although it can be used in the oxidative direction (Section 13.1.9). Other sulfate reducers (e.g., *Desulfobacterium autotrophicum*, *Desulfotomaculum acetooxidans*, and the archaeon *Archaeoglobus fulgidus*) oxidize acetyl–CoA to $CO_2$ using the acetyl-CoA pathway described in Section 13.1.3. Autotrophic $CO_2$ fixation by facultatively autotrophic sulfate reducers using these pathways in the reductive direction is also described in Sections 13.1. 3 and 13.1.4. It should also be pointed out that many sulfate reducers are known to be able to ferment pyruvate to acetate, or to acetate and propionate in the absence of sulfate. This is described in Section 14.8.

### The path of electrons to sulfate in Desulfovibrio

*Desulfovibrio* carries out an anaerobic respiration during which electrons flow in the cell membrane to sulfate as the terminal electron acceptor, reducing it to $H_2S$. Electron flow is coupled to the generation of a $\Delta p$, which is used for ATP synthesis via respiratory phosphorylation. As stated above, these electrons may come from the oxidation of organic compounds (e.g., lactate). Since dissimilatory sulfate reduction takes place in membranes, involves cytochromes, and generates a $\Delta p$, it is very different from the assimilatory pathway, which is a soluble pathway and does not generate a $\Delta p$ or ATP. A pathway of electron transport has been proposed for the genus *Desulfovibrio* (Fig. 12.3). It is called the hydrogen cycling model. Lactate is oxidized to pyruvate in the cytoplasm, yielding two electrons (Fig. 12.3, reaction 1). The oxidation of lactate to pyruvate is catalyzed by a membrane-bound lactate dehydrogenase, which is probably a flavoprotein. The pyruvate is then oxidized to acetyl–CoA and $CO_2$ by pyruvate–ferredoxin oxidoreductase, an enzyme found in other anaerobes (reaction 2). (See Section 7.3.2 for a description of the pyruvate–ferredoxin oxidoreductase reaction.) The acetyl–CoA is used to generate ATP via a substrate-level phosphorylation, using two enzymes common in bacteria, phosphotransacetylase and acetate kinase (reactions 3 and 4). (See Section 7.3.2 for a description of the phosphotransacetylase and acetate kinase reactions.) Since each lactate that is oxidized to acetyl–CoA yields four electrons, two lactates must be oxidized to provide the eight electrons to reduce one sulfate to sulfide. The model proposes that the electrons are transferred from lactate dehydrogenase and pyruvate–ferredoxin oxidoreductase to a cytoplasmic hydrogenase and then to $H^+$,

producing $H_2$ (reaction 5), and the $H_2$ diffuses out of the cell into the periplasm.

In the periplasm, the $H_2$ is oxidized by a periplasmic hydrogenase, and the electrons are transferred to cytochrome $c_3$ (reaction 6). From cyt $c_3$ the electrons travel through a series of membrane-bound electron carriers to APS reductase and sulfite reductase in the cytosol (reactions 7, 9, 10). A pyrophosphatase pulls the sulfurylation of ATP to completion (reaction 8). Note that according to the scheme proposed in Fig. 12.3, the inward flow of electrons across the membrane leaves the protons from the hydrogen on the outside, thus generating a $\Delta p$. An examination of the scheme reveals that the $\Delta p$ is necessary for growth. The two ATPs made via substrate-level phosphorylation from the two moles of acetyl–CoA are used up in reducing the $SO_4^{2-}$. This follows because, after reduction to sulfite, AMP is produced, and it requires the energy equivalent of two ATPs to make one ATP from one AMP. Thus, without using the $\Delta p$, there would be no ATP left over for growth. Some strains of *Desulfovibrio* can grow on $CO_2$ and acetate as the sole sources of carbon, and the $\Delta p$ produced by the redox reaction between $H_2$ and $SO_4^{2-}$ is the sole source of energy.

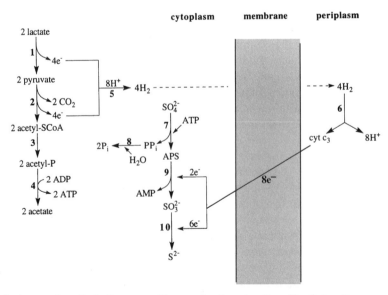

**Fig. 12.3** Pathway for dissimilatory sulfate reduction in *Desulfovibrio*. Enzymes: 1, lactate dehydrogenase; 2, pyruvate–ferredoxin oxidoreductase; 3, phosphotransacetylase; 4, acetate kinase; 5, cytoplasmic hydrogenase; 6, periplasmic hydrogenase; 7, ATP sulfurylase; 8, pyrophosphatase; 9, APS reductase; 10 sulfite reductase.

There are also facultatively autotrophic strains of sulfate reducers that grow on $CO_2$, $H_2$, and $SO_4^{2-}$, (e.g., *Desulfobacterium autotrophicum*). All these strains derive their ATP from the $\Delta p$ created during sulfate reduction. It should be pointed out that serious reservations have been expressed as to whether free $H_2$ is actually an electron carrier during lactate oxidation by sulfate in *Desulfovibrio* (i.e., whether the hydrogen cycling model is generally valid). (See Note 9.) Clearly, there is still much to be learned regarding how various species of *Desulfovibrio* couple electron transport to the generation of a $\Delta p$.

## 12.3 Nitrogen Fixation

From an ecological point of view, one of the most important metabolic processes carried out by prokaryotes is nitrogen fixation (i.e., the reduction of $N_2$ to $NH_3$). As far as is known, eukaryotes have not evolved this capability. Since fixed nitrogen is usually limiting for plant growth, the ability of prokaryotes to fix nitrogen is necessary to maintain the food chain. The ammonia that is produced via nitrogen fixation is incorporated into cell material using glutamine synthetase and glutamate synthase (Section 9.3.1). It used to be thought that nitrogen fixation was restricted to a few bacteria such as *Azotobacter*, *Rhizobium*, *Clostridium*, and the cyanobacteria. It is now realized that nitrogen fixation is a capability widespread among many different families of bacteria and also occurs in archaea. The enzyme responsible for nitrogen fixation, nitrogenase, is very similar in the different bacteria, and the nitrogen fixation genes have homologous regions. This has led to the suggestion that the nitrogenase gene may have been transferred laterally between different groups of bacteria. Organisms that fix nitrogen encompass a wide range of physiological types and include aerobes, anaerobes, facultative anaerobes, autotrophs, heterotrophs, and phototrophs. Nitrogen fixation takes place when $N_2$ is the only or the major source of nitrogen because the genes for nitrogen fixation are repressed by exogenously supplied sources of fixed nitrogen (e.g., ammonia.) The signal transduction pathway responsible for repression is discussed in Chapter 18. It is a remarkable fact that biological nitrogen reduction takes place at all. The nitrogen molecule is so stable that very high pressures and temperatures in the presence of inorganic catalysts are necessary to make it reactive in nonbiological systems. Industrially, nitrogen gas is reduced to ammonia using the Haber process, which requires 200 atm and 800°C. Yet prokaryotes carry out the reduction at atmospheric pressures and ordinary temperatures.

### 12.3.1 The nitrogen-fixing systems

The biological nitrogen fixing systems are listed below[10–13]:

1. *Rhizobium*, *Bradyrhizobium*, and *Azorhizobium* in symbiotic relationships with leguminous plants (soybeans, clover, alfalfa, string beans, peas, i.e., plants that bear seeds in pods). The bacteria infect the roots of the plants and stimulate the production of *root nodules*, within which the bacteria fix nitrogen. The plant responds by feeding the bacteria organic nutrients made during photosynthesis.

2. *Nonleguminous plants in symbiotic relationships with nitrogen-fixing bacteria*. For example, the water fern *Azolla* makes small pores in its fronds within which a nitrogen-fixing cyanobacterium, *Anaebaena azollae*, lives. The *Azolla–Anabaena* symbiotic system is used to enrich rice paddies with fixed nitrogen. Another example is the alder tree, which has nitrogen-fixing nodules containing *Frankia*, a bacterium resembling the streptomycetes. A third example is *Azospirillum lipoferum*, which is a $N_2$-fixing rhizosphere bacterium that is found around the roots of tropical grasses.

3. *Many free-living soil and aquatic prokaryotes*. As indicated in the introduction, many different prokaryotes fix nitrogen. They include *Azotobacter*, *Clostridium*, certain species of *Desulfovibrio*, the photosynthetic bacteria, and various cyanobacteria. Nitrogen fixation is not confined to the (eu)bacteria. Recently, some methanogens have been reported to be nitrogen fixers.[14]

Nitrogen fixation is sensitive to oxygen. The enzyme that fixes nitrogen, *nitrogenase*, is inhibited by oxygen. Thus for many prokaryotes nitrogen fixation takes place only under anaerobic or microaerophilic conditions. However, some prokaryotes can fix nitrogen while growing in air. As described later, they have evolved systems to protect the nitrogenase from oxygen.

## 12.3.2 The nitrogen-fixation pathway

### Nitrogenase

As mentioned, the enzyme that reduces nitrogen gas is called *nitrogenase*.[15, 16] The major nitrogenase in nitrogen-fixing organisms is a molybdenum-containing enzyme that consists of two multimeric proteins. One of these is usually called the molybdenum–iron protein (MoFe protein). It is also known as dinitrogenase, or component I. The second protein is called the iron protein (Fe protein) and is also known as dinitrogenase reductase, or component II. Both of the proteins contain FeS centers.

The MoFe protein is a tetramer ($\alpha_2\beta_2$) of four polypeptides. When it is extracted with certain solvents, a cofactor called the iron–molybdenum cluster (FeMoco) is removed. The cofactor contains approximately one-half of the iron and labile sulfide of the protein. Thus the MoFe protein contains FeMoco plus additional FeS centers. (See Note 17 for more information about the MoFe protein.)

The Fe protein is a dimer ($\gamma_2$) of two identical polypeptides. The dimer contains a single $Fe_4S_4$ cluster, which is responsible for reducing FeMoco during nitrogen fixation.

Twenty-one genes have been identified to be necessary for the expression and regulation of the nitrogenase enzyme system in *Klebsiella pneumoniae*. They are called the *nif* genes. The presence of sufficient $NH_4^+$ represses the synthesis of nitrogenase. Other nitrogen sources (nitrates, amino acids, urea) also suppress the synthesis of nitrogenase, probably by producing ammonia. The *K. pneumoniae* system has been used as a model for the regulation of expression of the nitrogen-fixing genes (Chapter 18).

### The nitrogenase reaction

The nitrogenase reaction is a series of reductions during which 0.5 mole of $N_2$ and 4 moles of $H^+$ are reduced to 1 mole of $NH_3$ and 0.5 mole of $H_2$. (Fig. 12.4). Nitrogenase reaction:

$$4e^- + 0.5N_2 + 4H^+ + 8ATP$$
$$\longrightarrow NH_3 + 0.5H_2 + 8ADP + 8P_i$$

Since the oxidation state of $N_2$ is 0 and the oxidation state of the nitrogen in $NH_3$ is $-3$, this requires three electrons per nitrogen atom. A fourth electron is transferred to a proton to reduce it to hydrogen gas. The electrons are transferred one at a time in an ATP-dependent reaction from the $Fe_4S_4$ cluster in the Fe protein to the MoFe cluster in the MoFe protein and from there to $N_2$. The details of the electron transport pathway through the

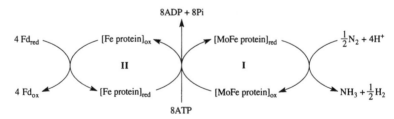

Fig. 12.4 The nitrogenase reaction. The enzyme system consists of two components. Component I is called the molybdenum–iron protein (MoFe protein) or dinitrogenase. Component II is called the iron protein (Fe protein) or dinitrogenase reductase. Both of the proteins contain FeS centers. A low potential reductant, either ferrodoxin or flavodoxin, reduces component II, which transfers the electrons to component I. Component I reduces $N_2$. There is always some $H_2$ produced. ATP is required despite the fact that the overall reduction of $N_2$ by ferredoxin or flavodoxin is an exothermic reaction.

proteins and the role of ATP are not well understood. (See Note 18 for a more complete discussion of electron flow in the MoFe protein and the role of ATP.) However, it is clear that the hydrolysis of at least two ATP molecules is required per electron transferred between the two proteins. Therefore, about 16 moles of ATP are necessary to convert one mole of nitrogen gas to two moles of ammonia. This is a great deal of energy (about 800 kJ). Recall that only two moles of ATP are generated during the fermentation of one mole of glucose to lactic acid, and 38 moles of ATP are produced during the complete oxidation of one mole of glucose to carbon dioxide and water. Therefore, an organism growing on nitrogen gas must consume a large fraction of the ATP that it produces in order to reduce the nitrogen to ammonia. Not surprisingly, bacteria do not fix nitrogen gas if an alternative source of nitrogen is present. As stated previously, this is because the nitrogen fixation genes are repressed when nitrogen sources other than $N_2$ are available.

During nitrogen reduction, protons are also reduced to hydrogen gas. The production of hydrogen gas appears to be wasteful of electrons and ATP. Indeed, some bacteria (e.g., *Azotobacter*) are very good at scavenging the hydrogen gas with a hydrogenase. The hydrogen gas is used to generate electrons for the nitrogenase.

## Other nitrogenases

Although the nitrogenase described above is certainly the major one, other nitrogenases have recently been discovered. For example, *Azotobacter vinelandii* can synthesize three nitrogenases, encoded by three different genes. The three nitrogenases have different metal requirements. The molybdenum-containing nitrogenase (nitrogenase I) is made when the organism is grown in media containing molybdenum. Nitrogenase II is a vanadium-containing nitrogenase that is synthesized when the cells are grown in media lacking molybdenum but containing vanadium. Instead of FeMoco, the nitrogenase contains a vanadium cofactor called FeVaco. When both molybdenum and vanadium are lacking in the media, *Azotobacter* makes a third nitrogenase, called nitrogenase III, which requires only iron.

## The source of electrons for nitrogen reduction

The nitrogenases are reduced by ferredoxins (FeS proteins) or flavodoxins (flavoproteins) in most known systems. These electron carriers have midpoint potentials sufficiently low to reduce nitrogenase (-400 to -500 mV). The source of electrons for the ferredoxins and flavodoxins varies with the metabolism of the organism. For example, during heterotrophic anaerobic growth the oxidation of pyruvate to acetyl–CoA and carbon dioxide generates reduced ferredoxin (pyruvate:ferredoxin oxidoreductase, Section 7.3.2) or flavodoxin (pyurvate:flavodoxin oxidoreductase) that donates electrons to nitrogenase. (*Clostridium pasteurianum* uses the ferredoxin enzyme, whereas *Klebsiella pneumoniae* uses the flavodoxin enzyme.) The path of electrons to nitrogenase in aerobic and phototrophic bacteria and in cyanobacteria is not as well understood, but is thought to involve ferredoxin or flavodoxin as the immediate electron donor. The ferredoxin or flavodoxin might be reduced by NAD(P)H (or some other electron carrier) generated during metabolism, for example, during the oxidation of carbohydrate. However, a source of energy (i.e., the protonmotive force) would be necessary to drive the reduction of ferredoxin and flavodoxin by NAD(P)H (reversed electron transport, Sections 3.7.1 and 4.5) because the midpoint potential of the $NAD(P)^+/NAD(P)H$ couple is $-320$ mV as compared to $-400$ to $-500$ mV for the ferredoxins and flavodoxins. Alternatively, the cyanobacteria and green sulfur bacteria could use light energy to reduce the electron donor for nitrogenase during photosynthetic noncyclic electron flow. For example, it has been suggested that photosystem I and some electron donor other than water might reduce nitrogenase in the heterocyst, and similarly that a light-generated reductant might drive the reduction of nitrogenase in the green sulfur bacteria. Recall that these two photosystems have reaction centers that produce a reductant at a sufficiently low potential to reduce ferredoxin (Sections 5.3 and 5.4).

## Protecting the nitrogenase from oxygen

All nitrogenases are rapidly inactivated by oxygen in vitro. Some nitrogen-fixing microorganisms are strict anaerobes (e.g., the clostridia or nitrogen-fixing sulfate reducers). Others are facultative anaerobes (i.e., can grow aerobically or anaerobically). These fix nitrogen only when they are living anaerobically (e.g., the purple photosynthetic bacteria, and *Klebsiella* spp). Some microorganisms are microaerophilic (i.e., can grow only in low levels of oxygen). Some of these are nitrogen fixers and can grow on nitrogen gas under microaerophilic conditions. But what about strict aerobes or the cyanobacteria that produce oxygen in the light? Various strategies have evolved to protect the nitrogenase in those microorganisms that fix nitrogen in air.[19, 20]

The *Azotobacter* species have a very active respiratory system that is suggested to utilize oxygen rapidly enough to lower the intracellular concentrations in the vicinity of the nitrogenase. These organisms are also able to protect their nitrogenase from inactivation by associating it with protective proteins. For example, the nitrogenase of *Azotobacter* can be isolated as an air-tolerant complex with a redox protein called the Shethna, FeS II, or protective protein.

The rhizobia in root nodules exist in plant vesicles in the inner cortex of the nodule as modified cells called *bacteroids*. Using oxygen-sensitive microelectrodes, it can be shown that the oxygen concentrations in the inner cortex are much lower than in the surrounding tissue. The oxygen levels in the vicinity of the bacteroid-containing cells are controlled by a boundary of densely packed plant cells between the inner and outer cortex. The control of oxygen access to the inner cortex is achieved by regulating the intercellular spaces, which are either air filled or contain variable amounts of water, within the boundary layer. However, the bacteroids are dependent upon oxygen for respiration, and within the nodule a plant protein called leghemoglobin binds oxygen and delivers it to the bacteroids.

It is not understood how the unicellular cyanobacteria or the nonheterocystous (see next) filamentous cyanobacteria protect their nitrogenase from photosynthetically produced or atmospheric oxygen. In fact, most of these strains fix nitrogen only when grown under anoxic or micro-oxic conditions. Anoxic conditions are maintained in the light only when photosynthetic oxygen production is experimentally inhibited. Micro-oxic conditions (also called microaerobic) refer to growth conditions when atmospheric oxygen is absent but photosynthetically produced oxygen is present. It is presumed that intracellular oxygen levels under these conditions are significantly lower than atmospheric levels, hence the term *micro-oxic*. Whether this is indeed the case, especially under conditions of high illumination, has not been shown. However, there are some unicelulular and nonheterocystous filamentous cyanobacteria that do fix nitrogen under aerobic conditions, that is, at oxygen concentrations that are approximately equal to atmospheric. How they protect their nitrogenase from oxygen is not known but has generated much interest. These include the unicellular *Gloethece* (formerly called *Gloeocapsa*), certain strains of the unicellular *Synechococcus*, and a marine filamentous cyanobacterium called *Trichodesmum*. Nitrogen fixation by unicellular and filamentous nonheterocystous cyanobacteria, including protection of the nitrogenase, have been reviewed by Bergman et al.[21] Several filamentous cyanobacteria protect their nitrogenase from oxygen by fixing nitrogen in special cells called heterocysts, which allows these organisms to fix nitrogen under aerobic growth conditions. This is discussed next.

## Heterocysts

Filamentous cyanobacteria such as *Anabaena* and *Nostoc* protect their nitrogenase by differentiating approximately 5 to 10% of their vegetative cells into special nitrogen-fixing cells called heterocysts in the absence of combined nitrogen[22] (Fig. 12.5). Heterocysts differ from vegetative cells in:

1. Possessing the nitrogenase enzymes

2. Having only photosystem 1 (PS1), and therefore not producing oxygen

3. Not fixing $CO_2$

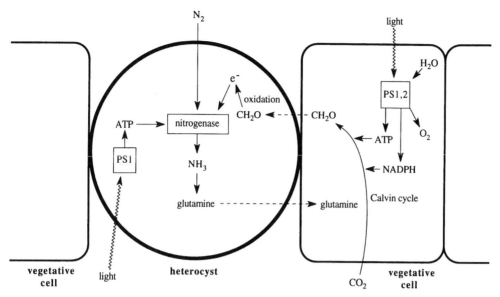

**Fig. 12.5** Heterocyst interactions with vegetative cells in *Anabaena*. The heterocyst reduces dinitrogen to ammonia, which is then incorporated into glutamine via glutamine synthetase. The glutamine then enters the vegetative cells, where it serves as a source of fixed nitrogen for growth. The vegetative cells fix carbon dioxide into carbohydrate using the Calvin cycle. Some of the carbohydrate enters the heterocyst, where it serves as a source of carbon and NADPH. The NADPH reduces the nitrogenase via ferredoxin:NADP oxidoreductase and ferredoxin. ATP is made via cyclic photophosphorylation in the heterocyst using PSI. Since PSII is lacking in the heterocyst, oxygen is not produced there.

4. Being surrounded by a thick cell wall consisting of glycolipid and polysaccharide that is believed to serve as a permeability barrier to atmospheric oxygen

5. Not dividing

The ATP made during cyclic photophosphorylation by photosystem I is used to fix the nitrogen. But where do the heterocysts get the reducing power? What happens is that the heterocysts fix $N_2$ and feed reduced nitrogen to the rest of the filament (Fig. 12.5). In turn, the heterocyst is fed carbohydrate made from $CO_2$ by the vegetative cells. The heterocyst oxidizes the carbohydrate, reducing ferredoxin, which, in turn, reduces the nitrogenase. Therefore, this is a complex situation where two different cell types in the filament are feeding each other.

Since the heterocysts do not have photosystem 2, they do not produce oxygen. However, there is still the problem of protecting the nitrogenase from atmospheric oxygen. It has been suggested that the crystalline glycolipid and polysaccharide cell wall may present a diffusion barrier to oxygen. Heterocysts also have a high rate of respiration, which presum-

ably also contributes to a low internal oxygen environment.

## 12.4 Lithotrophy

While most organisms derive energy from oxidizing organic nutrients (chemoorganotrophs) or from the absorption of light (phototrophs), there exist many prokaryotes that derive energy from the oxidation of inorganic compounds such as $H_2$, $CO$, $NH_3$, $NO_2^-$, $H_2S$, $S^o$, $S_2O_3^{2-}$, or $Fe^{2+}$. This type of metabolism is called *lithotrophy*, and the organisms are called *lithotrophs*.[23]

### 12.4.1 The lithotrophs

The lithotrophs are physiologically diverse and exist among several different groups of bacteria and archaea. Many of the lithotrophs are aerobes (i.e., they carry out electron transport with oxygen as the terminal electron acceptor). However, some are facultative

**Table 12.1** Chemoautotrophs

| Bacterial group | Typical species | Electron donor | Electron acceptor | Carbon source | Product |
|---|---|---|---|---|---|
| Hydrogen-oxidizing | *Alcaligenes eutrophus* | $H_2$ | $O_2$ | $CO_2$ | $H_2O$ |
| Carbon-monoxide oxidizing (carboxydobacteria) | *Pseudomonas carboxydovorans* | CO | $O_2$ | $CO_2$ | $CO_2$ |
| Ammonium-oxidizing | *Nitrosomonas europaea* | $NH_4^+$ | $O_2$ | $CO_2$ | $NO_2^-$ |
| Nitrite-oxidizing | *Nitrobacter winogradskyi* | $NO_2^-$ | $O_2$ | $CO_2$ | $NO_3^-$ |
| Sulfur-oxidizing | *Thiobacillus thiooxidans* | $S, S_2O_3^{2-}$ | $O_2$ | $CO_2$ | $SO_4^{2-}$ |
| Iron-oxidizing | *Thiobacillus ferrooxidans* | $Fe^{2+}$ | $O_2$ | $CO_2$ | $Fe^{3+}$ |
| Methanogenic | *Methanobacterium thermautotrophicum* | $H_2$ | $CO_2$ | $CO_2$ | $CH_4$ |
| Acetogenic | *Acetobacterium woodii* | $H_2$ | $CO_2$ | $CO_2$ | $CH_3COOH$ |

*Source:* Schlegel, H. G., and H. W. Jannasch. 1992. Prokaryotes and their habitats, PP. 75–125. In: *The Prokaryotes,* Vol. I. A. Balows, H. G. Trüper, M. Dworkin, W. Harder and K.-H. Schleifer (eds.). Springer-Verlag, Berlin.

anaerobes using nitrate or nitrite as the electron acceptor when oxygen is unavailable, and a few are obligate anaerobes that use sulfate or $CO_2$ as the electron acceptor. Most of the lithotrophs are autotrophs (i.e., their sole or major source of carbon is $CO_2$). They are called *chemoautotrophs* or *chemolithoautotrophs*. The lithotrophs vary with regard to the autotrophic $CO_2$ fixation pathway they use (Chapter 13). Other lithotrophs are facultatively heterotrophic. The facultative heterotrophs include all the bacterial hydrogen oxidizers, some sulfur-oxidizing thiobacilli, and some thermophilic iron-oxidizing bacteria. Some representative species are listed in Table 12.1.

Table 12.2 is a list of redox potentials of the inorganic substrates at pH 7. Theoretically, all these organisms, with the possible exception of the hydrogen oxidizers and the CO oxidizers, must carry out reversed electron transport to generate NAD(P)H for biosynthesis because the electron donor is more electropositive than the NAD$^+$/NADH couple ($E_0' = -0.32$ V). Reversed electron flow is driven by the $\Delta p$ (Section 4.5). Because of the relatively small $\Delta E_h$ between the inorganic electron donor and oxygen, and the need to reverse electron transport, the energy yields, and therefore the cell yields, are relatively small compared to growth on organic substrates (Fig. 12.6).

## Aerobic hydrogen-oxidizing bacteria and carboxydobacteria

The hydrogen-oxidizing bacteria are usually facultative and can live either autotrophically or heterotrophically. However, some always require organic carbon for growth and are called chemolithoheterotrophs. The hydrogen oxidizers can be found in aerobic or anaerobic environments where $H_2$ is available. The hydrogen gas itself is produced as a byproduct of nitrogen fixation (e.g., in the rhizosphere of nitrogen-fixing plants and in cyanobacterial blooms). Hydrogen gas is also produced

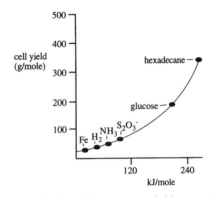

**Fig. 12.6** Cell yields versus available energy in inorganic and organic electron sources. *Source:* Adapted from Brock, T. D., and M. T. Madigan. 1991. *Biology of Microorganisms.* Reprinted by permission of Prentice-Hall, Englewood Cliffs, New Jersey.

**Table 12.2** Redox potentials of inorganic compounds[a]

| Compound | $E_0'$ (mV) |
|---|---|
| $CO_2/CO$ | $-540$ |
| $SO_4^{2-}/HSO_3^-$ | $-516^b$ |
| $H+/H_2$ | $-414^b$ |
| $S°/HS^-$ | $-270^b$ |
| $HSO_3^-/HS^-$ | $-116^b$ |
| $NO_3^-/NO_2^-$ | $+420$ |
| $NO_2^-/NH_3$ | $+440$ |
| $FE^{3+}/Fe^{2+}$ | $+772^b$ |
| $O_2/H_2O$ | $+818^b$ |

[a] For comparison, the standard potential at pH 7 for $NAD^+/NADH$ is $-320V$.
[b] Data from Thauer, R. K., Jungermann, K., and K. Decker. 1977. Energy conservation in chemotrophic anaerobic bacteria. *Bacteriol. Rev.* **41**:100–180.

in anaerobic environments via fermentations, where some of it escapes into the aerobic atmosphere. (However, most of the $H_2$ produced anaerobically is utilized by the sulfate reducers and methanogens.) Among the hydrogen-oxidizing bacteria are some that can also grow on carbon monoxide (CO) as the sole source of energy and carbon using oxygen, or in some cases nitrate (denitrifiers) as the electron acceptor. They are called *carboxydobacteria*. The hydrogen-oxidizing bacteria and the carboxydobacteria are represented by several genera, including representatives from *Pseudomonas, Arthrobacter, Bacillus,* and *Rhizobium*. Anaerobic hydrogen oxidizers include some sulfate-reducing bacteria and some archaea, including the methanogens when growing autotrophically on $CO_2$, and certain sulfur-dependent archaea that use elemental sulfur as the electron acceptor, reducing it to hydrogen sulfide.

### Ammonia-oxidizing bacteria

Bacteria that oxidize ammonia as a source of energy are called nitrifiers.[24] There are at least five genera of nitrifiers. They are *Nitrosomonas, Nitrosococcus, Nitrosospira, Nitrosolobus,* and *Nitrosovibrio*. All the nitrifiers are aerobic obligate chemolithoautotrophs that assimilate $CO_2$ via the Calvin cycle. Ammonia that is produced in the anaerobic niches by deamination of amino acids, urea, or uric acid, or via dissimilatory nitrate

reduction diffuses into the aerobic environment, where it is oxidized by the nitrifiers. The nitrifying bacteria are often found at the aerobic/anaerobic interfaces, where they capture the ammonia as it diffuses from the anaerobic environments, as well as in the more highly aerobic parts of the soil and water.

*Nitrosomonas* oxidizes ammonia to nitrite. Along with *Nitrobacter,* which oxidizes nitrite to nitrate, it is responsible for a major portion of the conversion of ammonia to nitrate, a process called *nitrification*. The first oxidation is the oxidation of ammonia to hydroxylamine catalyzed by *ammonia monooxygenase* (AMO). The reaction is:

$$2H^+ + NH_3 + 2e^- + O_2 \longrightarrow NH_2OH + H_2O$$

In this reaction, one oxygen atom is incorporated into hydroxylamine and the other is reduced to water. The oxidation of hydroxylamine to nitrite is catalyzed by the enzyme *hydroxylamine oxidoreductase* (HAO):

$$NH_2OH + H_2O \longrightarrow HONO + 4e^- + 4H^+$$

Of the four moles of electrons removed from one mole of hydroxylamine, two are used in the ammonia monooxygenase reaction, approximately 1.7 are transferred to oxygen via cytochrome oxidase, and about 0.3 moles are used to generate NAD(P)H via reversed electron transport.

$$2H^+ + 0.5O^2 + 2e^- \longrightarrow H_2O$$

A proposed electron transport scheme showing the topological arrangement of the electron carriers is shown in Fig. 12.7. The electron transport scheme shown in Fig. 12.7 is speculation based upon the known location of the enzymes and their redox potentials. The scheme proposes that ammonia oxidation to hydroxylamine takes place in the cytoplasm, although it is not known whether this occurs in the cytoplasm or the periplasm. (Ammonia monooxygenase is in the cell membrane, but it is unknown whether the substrate binding site is exposed to the cytoplasm or the periplasm.)

Assuming cytoplasmic oxidation, the hydroxylamine diffuses across the cell membrane to the periplasm, where it is oxidized to nitrite by hydroxylamine oxidoreductase, a periplasmic enzyme. During this oxidation

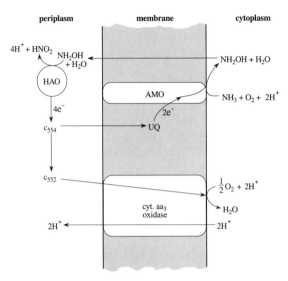

**Fig. 12.7** Model for the electron transport scheme in *Nitrosomonas*. Ammonia is oxidized to hydroxylamine by the enzyme ammonia monooxygenase (AMO). Although the oxidation of ammonia is depicted as occurring in the cytoplasm, it has not been ruled out that the oxidation takes place in the periplasm. The reaction requires two electrons to reduce one of the oxygen atoms to water. The hydroxylamine diffuses across the membrane to the periplasm, where it is oxidized to nitrite by a complex cytochrome called hydroxylamine oxidoreductase (HAO). The electrons are passed to a periplasmic cytochrome c and from there to ubiquinone in the membrane. Electrons travel from ubiquinone in two branches. One branch passes two electrons to the ammonia monooxygenase, and the second branch leads to oxygen via cytochromes c and $aa_3$. All four electrons end up in water. *Source:* Adapted from Hooper, A. B. 1989. Biochemistry of the nitrifying lithoautotrophic bacteria, pp. 239–265. In: *Autotrophic Bacteria,* H. G. Schlegel and B. Bowien (Eds.). Springer-Verlag, Berlin.

four electrons are transferred to periplasmic cytochrome $c_{554}$. The four electrons are transferred from cytochrome $c_{554}$ to ubiquinone in the membrane. It is suggested that two electrons travel from ubiquinone to the ammonia monooxygenase enzyme and two electrons to cytochrome $aa_3$ via membrane cytochrome $c_{553}$ and periplasmic cytochrome $c_{552}$. Cytochrome $aa_3$ is presumed to act as a proton pump. The $\Delta p$ that is created as a result of the oxidation of ammonia and hydroxyamine and the reduction of oxygen is the sole source of energy for these bacteria.

*Nitrite-oxidizing bacteria*

The nitrite oxidizers are *Nitrobacter, Nitrococcus, Nitrospina,* and *Nitrospira.* They are aerobic obligate chemolithoautotrophs, with the exception of *Nitrobacter,* which is a facultative autotroph (i.e., it can also be grown

heterotrophically). The details of the electron transport scheme have not been fully elucidated. However, a proposed model is shown in Fig. 12.8. Electrons travel from nitrite to oxygen via a periplasmic cytochrome c. There is a thermodynamic problem here. Cytochrome c exists at a midpoint potential ($E_{m',7}$) of +270 mV, whereas the midpoint potential of the nitrate/nitrite couple is more electropositive at +420 mV. Because electrons do not spontaneously flow toward the lower redox potential, energy must be provided to drive the electrons over the 150 mV difference in order for nitrite to reduce cytochrome c. Nichols and Ferguson suggested that the membrane potential drives the electrons from nitrite to cytochrome c.[25] It was proposed that nitrite oxidation takes place on the cytoplasmic surface of the membrane, and that the electrons flow across the membrane to cytochrome c

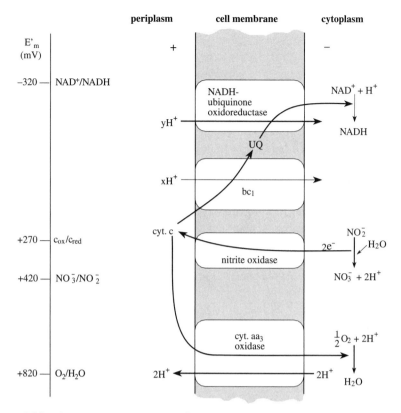

**Fig. 12.8** A model for electron transport in *Nitrobacter*. The electrons travel from nitrite via membrane-bound nitrite oxidase to cytochrome c, which is at a more negative potential. It is proposed that nitrite oxidation takes place on the cytoplasmic side of the membrane and that the membrane potential drives the electrons transmembrane to cytochrome c. From cytochrome c the electrons diverge. Most travel to oxygen at a more positive potential, and a $\Delta p$ is created. The coupling site is the cytochrome $aa_3$ oxidase, which is a proton pump. Other electrons travel to $NAD^+$, which is at a more negative potential. The $\Delta p$ drives the electrons in reverse flow to $NAD^+$. This is accomplished by coupling electron transport with the return of protons down the proton potential to the cytoplasmic side. The scheme presumes the presence of a $bc_1$ complex as well as a reversible NADH dehydrogenase complex. *Source:* Adapted from Nicholls, D. G., and S. J. Ferguson. 1992. *Bioenergetics 2*. Academic Press, London.

located on the periplasmic side, which is typically 170 mV more positive than the cytoplasmic side (Fig. 12.8). In this way, the membrane potential lowers the potential difference between the nitrate/nitrite couple and the $c_{ox}/c_{red}$ couple by approximately 170 mV. The student will recognize this model as *reversed electron transport* driven by the membrane potential (which is consumed in the process). The role of the membrane potential in driving electron transport from nitrite to cytochrome c is consistent with the observation that experimental procedures that lower the membrane potential (e.g., incubation with proton ionophores)

decrease electron transfer from nitrite to oxygen in inverted membrane vesicles from *Nitrobacter*. The electrons then flow from cytochrome c back across the membrane to oxygen through cytochrome $aa_3$ oxidase driven by a favorable midpoint potential difference of about +440 mV. A $\Delta p$ is created by the outward pumping of protons by the cytochrome $aa_3$ oxidase. Electron flow is also reversed from cytochrome c to $NAD^+$ through two coupling sites: a ubiquinone–cytochrome c oxidoreductase (probably a $bc_1$ complex), and a reversible NADH–ubiquinone oxidoreductase. The model proposes that reversed electron flow to $NAD^+$ is coupled to the *influx* of protons (i.e., it is driven by the $\Delta p$).

## Sulfur-oxidizing prokaryotes

The sulfur-oxidizing prokaryotes include the photosynthetic sulfur oxidizers (Chapter 5) and the nonphotosynthetic sulfur oxidizers. (See Refs. 26–28 for reviews.) It is the latter that concern us here. Almost all the known ones are gram-negative bacteria. Sulfur oxidizers comprise a physiologically heterogeneous group. They may be obligate autotrophs (grow only on $CO_2$ as the carbon source) or facultative heterotrophs (can also use organic carbon as the source of carbon). Some are neutrophiles (grow best around pH 7), whereas others are acidophiles (grow best between pH 1 and 5). Sulfur-oxidizing acidophiles can be isolated from sulfur and coal mines that produce sulfuric acid. An example is *Thiobacillus thiooxidans*, which can grow at a pH of 1, although the optimum is between 2 and 3. They can be mesophilic (growth temperature optimum between 25 and 40°C) or thermophilic (growth temperature optimum above 55°C). Some sulfur bacteria (e.g., *Beggiatoa)* can be grown only mixotrophically (i.e., using $H_2S$ as the energy source and organic carbon as the source of carbon). The nonphotosynthetic sulfur prokaryotes include the bacteria *Beggiatoa* and *Thiothrix*, bacteria belonging to the genus *Thiobacillus* and *Paracoccus,* and an archaeon belonging to the genus *Sulfolobus*, a thermophilic acidophile that grows in sulfur acid springs, where the temperature can be 90°C and the pH can be between 1 and 5. (See the discussion of *Sulfolobus* and other archaea in Section 1.1.1.) The thiobacilli are the most prominent of the sulfur oxidizers, and their sulfur oxidation pathways are the most well studied.

Sulfur compounds commonly used as sources of energy and electrons include hydrogen sulfide ($H_2S$), elemental sulfur ($S°$), and thiosulfate ($S_2O_3^{2-}$), all of which can be oxidized to sulfate. (Laboratory studies often utilize thiosulfate or tetrathionate because they are stable in air. Tetrathionate has an additional advantage because, unlike thiosulfate, it is stable at acidic pH values.) Although most sulfur bacteria are aerobes, a few can

be grown anaerobically using nitrate as the electron acceptor. The sulfur bacteria can be found in nature growing near sources of sulfur [e.g., sulfur deposits, sulfide ores, hot sulfur springs, sulfur mines, and coal mines that are sites of iron pyrite ($FeS_2$) deposits]. Sulfide oxidizers can sometimes be found in large accumulations in thin layers between the aerobic and anaerobic environments. The reason they are sometimes confined to the aerobic/anaerobic interface is that $H_2S$, which is produced anaerobically by sulfate reducers, is rapidly oxidized by oxygen. Thus the sulfide-oxidizing bacteria grow where oxygen levels are relatively low in order to compete with the rapid chemical oxidation of sulfide.

There has been much confusion in the literature regarding the intermediate stages of sulfur oxidation. Probably more than one pathway exist. Two pathways that are receiving much research attention are described next.

*Paracoccus versutus* is a neutrophilic facultative lithoautotroph. In addition to growth on thiosulfate (but not polythionates), it can be grown on various organic carbon sources. (As reviewed in Ref. 28, this organism was formerly called *Thiobacillus versutus.*) Oxidation of thiosulfate takes place in the periplasm on a multienzyme complex (Fig. 12.9). No free intermediates are released, and both atoms of thiosulfate are oxidized to sulfate. Protons are released in the periplasm during the oxidations and the electrons are transferred *electrogenically* from the periplasmic side of the membrane to the cytoplasmic side through cytochrome $c_{552}$ and then cytochrome $aa_3$ oxidase to oxygen. (See Section 3.2.1 for a discussion of electrogenic movement of electrons.) A proton current is maintained by the release of protons in the periplasm during the oxidations and their consumption in the cytoplasm during the reduction of oxygen. The expected $H^+/O$ ratio ($H^+$ produced in the periplasm or translocated from the cytoplasm to the periplasm per oxygen atom reduced) can be obtained from Fig. 12.9 (and confirmed experimentally). The ratio is 2.5 for thiosulfate oxidation. ATP is synthesized by a proton-translocating ATP synthase driven by the $\Delta p$. If one assumes that the $H^+/ATP$ ratio for the ATP synthase is three (a consensus value), then

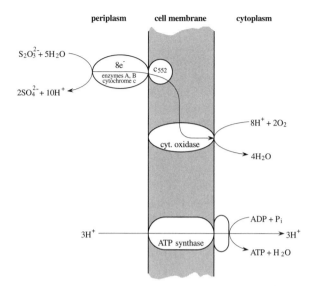

**Fig. 12.9** Model for thiosulfate oxidation in *Thiobacillus versutus*. Thiosulfate is oxidized to sulfate in the periplasm by a multienzyme complex consisting of enzyme A, enzyme B, and cytochromes c. The electrons are electrogenically transferred to oxygen via a membrane-bound cytochrome $c_{552}$ and a cytochrome oxidase (cytochrome aa$_3$). A $\Delta p$ is created by the release of protons in the periplasm via the oxidations, the consumption of protons in the cytoplasm during oxygen reduction, and electrogenic flow of electrons across the membrane to oxygen. A proton-translocating ATP synthase makes ATP. *Source:* Adapted from Kelly, D. P. 1989. Physiology and biochemistry of unicellular sulfur bacteria, pp. 193–217. In: *Autotrophic Bacteria*. H. G. Schlegel and B. Bowien (Eds.). Springer-Verlag, Berlin.

the maximum P/O ratio would be 2.5/3 or 0.83 for thiosulfate oxidation. (See Section 4.5.2 for a discussion of the relationship between the size of the proton current and the upper limit of ATP that can be made.)

*Thiobacillus tepidarius* is a neutrophilic thermophilic lithoautotroph that has been isolated from hot springs. It can be grown on hydrogen sulfide, thiosulfate, trithionate, and tetrathionate as electron donors. The pathway of oxidation of thiosulfate to sulfate (called the polythionate pathway) begins in the periplasm, but the bulk of the oxidations take place in the cytoplasm. Two molecules of thiosulfate are oxidized to tetrathionate in the periplasm (Fig. 12.10). A periplasmic cytochrome c accepts the two electrons generated from the thiosulfate oxidation and transfers these electrons to a membrane-bound cytochrome c. The tetrathionate is believed to be transported into the cell, where it is oxidized to four sulfates via the intermediate sulfite (generating 14 elec-

trons). The 14 electrons generated during the tetrathionate oxidations in the cytoplasm are transferred to a proton-translocating ubiquinone–cytochrome b complex in the cell membrane, which reduces membrane-bound cytochrome c. Cytochrome c transfers all the electrons to cytochrome oxidase (probably cytochrome o), which reduces oxygen. Because the oxidations of tetrathionate and sulfite are cytoplasmic rather than periplasmic, the proton current is due to proton translocation by the ubiquinone–cytochrome b complex (coupled to the cytoplasmic oxidations), rather than to periplasmic oxidations, as is the case for *T. versutus*.

Some thiobacilli can incorporate a substrate-level phosphorylation when oxidizing sulfur. The substrate-level phosphorylation occurs at the level of sulfite. The sulfite reacts with adenosine monophosphate (AMP) to form adenosine phosphosulfate (APS), in a reaction catalyzed by APS reductase. The APS is oxidized to sulfate, producing ADP

## INORGANIC METABOLISM

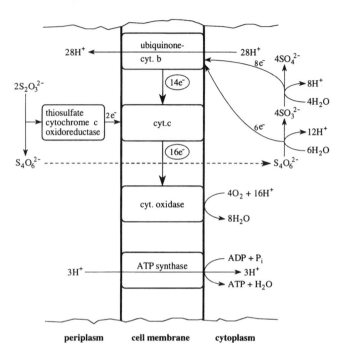

**Fig. 12.10** A model for thiosulfate oxidation in *Thiobacillus tepidarius*. Thiosulfate is oxidized to tetrathionate in the periplasm by a thiosulfate–cytochrome c oxidoreductase. It is hypothesized that the tetrathionate is transported into the cell in symport with protons where it is oxidized to sulfate. The electrons travel to oxygen via a proton-translocating ubiquinone–cytochrome b system (probably a bc₁ complex) and cytochrome oxidase, which appears to be a cytochrome o. *Source:* Adapted from Kelly, D. P. 1989. Physiology and biochemistry of unicellular sulfur bacteria, pp. 193–217. In: *Autotrophic Bacteria.* H. G. Schlegel and B. Bowien (Eds.). Springer-Verlag, Berlin; and Smith, D. W., and W. R. Strohl, 1991. Sulfur-oxidizing bacteria, pp. 121–146. In: *Variations in Autotrophic Life,* J. M. Shively and L. L. Barton (Eds.). Academic Press, New York.

or ATP, using either ADP sulfurylase or ATP sulfurylase:

$$SO_3^{2-} + AMP \xrightarrow{\text{APS reductase}} APS + 2e^-$$

$$APS + P_i \xrightarrow{\text{ADP sulfurylase}} ADP + SO_4^{2-}$$

$$APS + PP_i \xrightarrow{\text{ATP sulfurylase}} ATP + SO_4^{2-}$$

Recall that ATP sulfurylase and APS reductase are used by dissimilatory sulfate reducers to reduce sulfate to sulfite in reactions that consume ATP (Section 12.2.2). By running these reactions in the oxidative direction, sulfur oxidizers can use the energy in pyrophosphate to make ATP.

Figure 12.11 summarizes the inorganic sulfur oxidation pathways. Elemental sulfur exists as an octet ring of insoluble sulfur ($S_8^\circ$). It is first activated by reduced glutathione (GSH) to form a linear polysulfide (G-S-S₈-H). Sulfide ($S^{2-}$) also reacts with GSH and is oxidized to linear polysulfide. The sulfur atoms are removed from the polysulfide one at a time during the oxidation to sulfite ($SO_3^{2-}$).

Another elemental sulfur oxidation pathway has been reported for some thiobacilli, in which S° is oxygenated by a sulfur oxygenase:

$$S^\circ + O_2 + H_2O \xrightarrow{\text{sulfur oxygenase}} H_2SO_3$$

However, the oxygenase reaction cannot account for sulfur oxidation under anaerobic conditions when the oxidant provided is nitrate (*T. denitrificans*) or $Fe^{3+}$ (*T. ferrooxidans*). Furthermore, the oxygenase reaction

330

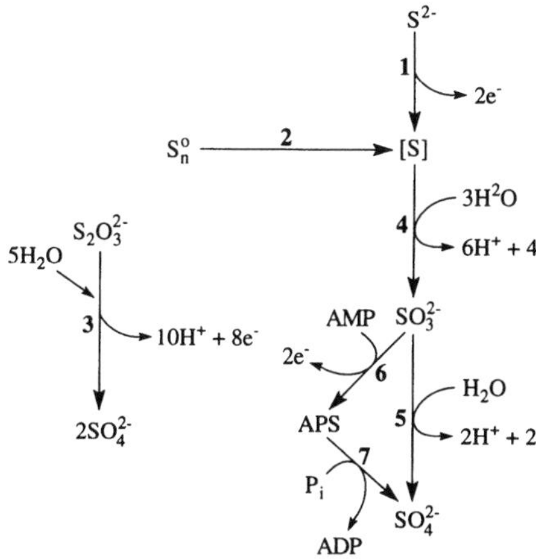

**Fig. 12.11** A summary of sulfur oxidation pathways. 1, oxidation of sulfide to linear polysulfide [S]; 2, conversion of elemental sulfur to linear polysulfide; 3, thiosulfate multienzyme complex; 4, sulfur oxidase; 5, sulfite oxidase; 6, APS reductase; 7, ADP–sulfurylase. *Source:* Adapted from Gottschalk, G. 1985. *Bacterial Metabolism,* Springer-Verlag, Berlin.

cannot conserve energy for the cell, since the electrons do not enter the respiratory chain. (Even though the periplasm might be acidified with $H_2SO_3$ and thus generate a $\Delta pH$, this by itself cannot generate net ATP in a growing cell because protons entering via the ATP synthase must not accumulate in the cytoplasm. The extrusion of protons from the cells or the utilization of protons to form water requires electron transport.)

## Iron-oxidizing bacteria

A few bacteria derive energy from the aerobic oxidation of ferrous ion to ferric ion.[29] Most of these are also acidophilic sulfur oxidizers (i.e., they oxidize sulfide to sulfuric acid). The acidophilic iron oxidizers can be found growing at the sites of geological deposits of iron sulfide minerals [e.g., pyrite ($FeS_2$) and chalcopyrite ($CuFeS_2$)], where water and oxygen are also present. The iron sulfide minerals are uncovered during mining operations, and the presence of acid mine water at these sites is due to the growth of the iron sulfide oxidizers. Since $Fe^{2+}$ is rapidly oxidized chemically by oxygen to $Fe^{3+}$ at neutral pH but only slowly at acid pH, the acidic environment is conducive to growth of the iron oxidizers. An example is *Thiobacillus*

*ferrooxidans,* which is an autotroph able to use either ferrous ion or inorganic sulfur compounds as a source of energy and electrons. *Thiobacillus ferrooxidans* can be grown on ferrous ion (e.g., $FeSO_4$) at optimal external pH values between 1.8 and 2.4. It maintains an internal pH of around 6.5 and therefore a $\Delta pH$ of about 4.5. The membrane potential ($\Delta\Psi$) at low pH is reversed, and energy in the $\Delta p$ is due entirely to the $\Delta pH$ component (Section 3.3). (As discussed in Section 15.1.3, the inversion of the $\Delta\Psi$ is necessary for maintenance of the $\Delta pH$ and may be due to electrogenic influx of $K^+$.)

### 1. Growth on ferrous sulfate
Figure 12.12 illustrates a model for the electron transport pathway in *T. ferrooxidans* growing on the aerobic oxidation of ferrous sulfate. The pathway proposes that extracellular $Fe^{2+}$ is oxidized to $Fe^{3+}$ by a $Fe^{3+}$ complex in the outer membrane. The electrons travel to a periplasmic cytochrome c. There is also a copper protein called rusticyanin in the periplasm, which is thought to be part of the electron transport scheme. From cytochrome c the electrons flow across the membrane via cytochrome oxidase (cytochrome $a_1$) to oxygen, generating a $\Delta\Psi$, inside negative. (The inversion of the $\Delta\Psi$ may be due to electrogenic influx of $K^+$, as discussed in Section 15.1.3.) The oxidation of ferrous ion drives the consumption of protons in the cytoplasm during oxygen reduction. The $\Delta pH$ is maintained because the rate of respiration and proton consumption matches the rate at which protons enter the cell through the ATPase or through leakage. From an energetic point of view, one would not expect any proton pumping by the cytochrome oxidase. The reason for this is that the difference in midpoint potentials between the $Fe^{3+}/Fe^{2+}$ couple (pH 2) and the $O_2/H_2O$ (presumed to be at pH 6.5) is very small, perhaps only 0.08 V or less.[30] (See Note 31 for a more complete discussion.) Under these circumstances, a two-electron transfer would generate at the most only 0.16 eV. A simple calculation reveals that this would not be enough energy to pump protons against the pH gradient. At an external pH of 2 and an internal pH of 6.5, the $\Delta pH$ is 4.5. This is

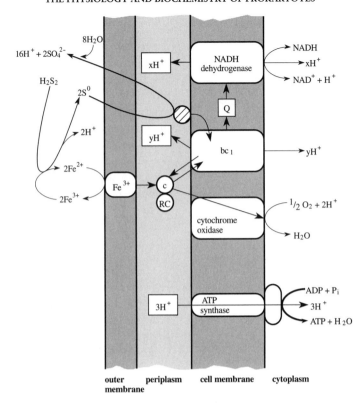

**Fig. 12.12** Electron transfer in *Thiobacillus ferrooxidans*. Ferrous ion is oxidized to $Fe^{3+}$ by a $Fe^{3+}$ complex in the outer membrane, and the electrons flow through periplasmic cytochrome c and a copper protein, rusticyanin (RC), to a membrane cytochrome oxidase of the $a_1$ type. The $\Delta p$ is maintained by the inward flow of electrons (contributing towards a negative $\Delta \Psi$) and the consumption of cytoplasmic protons during oxygen reduction, maintaining a $\Delta pH$. The consumption of two protons per two electrons in the cytoplasm and the uptake of three protons through the ATP synthase indicate that a maximum of $\frac{2}{3}$ of an ATP can be made per oxidation of two ferrous ions. When the bacteria are oxidizing iron pyrite ($FeS_2$), the $Fe^{3+}$ that is produced outside the cell envelope is chemically reduced to $Fe^{2+}$ by $S^{2-}$. The resultant $S^{\circ}$ is oxidized to $SO_4^{2-}$, and the electrons pass through a proton-translocating $bc_1$ complex to periplasmic cytochrome c and thence to cytochrome oxidase. The figure also illustrates $\Delta p$-driven reverse electron transport from $S^{\circ}$ and from $Fe^{2+}$ to $NAD^+$ through coupling sites that bring protons into the cell.

equivalent to 0.06(4.5) or 0.27 V at 30°C (eq. 3.10). Thus each proton would have to be energized by approximately 0.27 eV in order to be pumped out of the cell, even in the presence of a small $\Delta \Psi$, inside positive, which may be on the order of +0.01 to +0.02 V.

ATP is synthesized by a membrane $H^+$-translocating ATP synthase driven by a $\Delta p$ of approximately −250 mV. Electrons from $Fe^{2+}$ must also move toward a lower redox potential in order to generate NAD(P)H. Figure 12.12 also illustrates how reversed electron flow from $Fe^{2+}$ to $NAD^+$ might occur through a $bc_1$ complex and quinone. Reversed

electron flow is probably driven by the $\Delta p$ as protons enter the cell down the $\Delta p$ gradient through the $bc_1$ complex and the NADH dehydrogenase complex.

### 2. Growth on pyrite

Pyrite is a crystalline ore of iron that is usually written as ferrous disulfide ($FeS_2$). It is actually a stable crystal of discrete $^-S–S^-$ disulfide ions ($S_2^{2-}$) and ferrous ion ($Fe^{2+}$). The ferrous ion and disulfide are oxidized to ferric ion and sulfuric acid, respectively, according to the following overall reaction:

$$4FeS_2 + 2H_2O + 15O_2$$
$$\longrightarrow 4Fe^{3+} + 8SO_4^{2-} + 4H^+$$

However, the oxidation is not straightforward.[32] The oxidation of $FeS_2$ is the result of several redox reactions (Fig. 12.12). When the bacteria are growing on iron pyrite, the $Fe^{2+}$ is oxidized extracellularly by a complex of $Fe^{3+}$ in the outer membrane (Fig. 12.12). The outer membrane iron complex then transfers the electrons to the periplasmic cytochrome c, which transfers the electron to cytochrome oxidase. The $Fe^{3+}$ that is formed extracellularly is recycled to $Fe^{2+}$ by disulfide from the iron pyrite, according to the following reaction, which occurs spontaneously:

$$H_2S_2 + 2Fe^{3+} \longrightarrow 2S^o + 2Fe^{2+} + 2H^+$$

The elemental sulfur that is produced is oxidized by the bacteria to sulfate, with the electrons traveling through a proton-translocating $bc_1$ complex to the periplasmic cytochrome c and from there to cytochrome oxidase.

## 12.4.2 A review of the energetic considerations for lithotrophic growth

### ATP synthesis

As discussed in Section 3.7.1, the energy from respiratory oxidation–reduction reactions is first converted into a $\Delta p$, which is then used to drive ATP synthesis via the ATP synthase. For illustrative purposes, we will consider growth on nitrite in an aerated culture as a source of energy and electrons. The $E'_m$ for $NO_3^-/NO_2^-$ is +0.42 V. The $E'_m$ for $O_2/H_2O$ is +0.82 V. The difference in potential is therefore +0.82–0.42 or +0.40 V. Recall that in the respiratory chain, each coupling site is associated with pairs of electrons traveling over a midpoint potential difference of approximately 0.2 V or more. The conclusion is that there is sufficient energy for a coupling site between nitrite and oxygen. Another way of examining this question is to consider the amount of energy required to synthesize an ATP. The energy required to synthesize ATP will vary with the concentrations of ATP, ADP, and $P_i$. However, we will assume a value of 0.4 to 0.5 eV,

which is a reasonable estimate. Therefore, two electrons traveling over a redox gradient of 0.40 V should give 0.80 eV of energy, which is sufficient for the synthesis of an ATP. (See Note 33.)

### NADH reduction requires reversed electron transport

In order to grow, all cells must be able to make NAD(P)H, because this is a major source of electrons for reductions that occur during biosynthesis. For example, the Calvin cycle, which reduces $CO_2$ to the level of carbohydrate, requires two NADH per $CO_2$. When the cells are growing on $NO_2^-$ and $CO_2$, what can they use as a source of electrons to reduce $NAD^+$? Clearly not $CO_2$, since that is already the highest oxidized form of carbon. So we are left with $NO_2^-$. However, a comparison of the electrode potentials points to a problem. The $E'_m$ for $NO_3^-/NO_2^-$ is +0.42 V, and the $E'_m$ for $NAD^+/NADH$ is −0.32 V. Electrons flow spontaneously only to the more electropositive acceptor. Thus, in order to make the electrons flow from $NO_2^-$ to $NAD^+$, the flow must be reversed, and this requires energy. How much energy is required? The potential difference ($\Delta E'_0$) is −0.32–0.42 or −0.7 V. Thus each electron would have to be energized by 0.7 eV. The source of energy is the $\Delta p$. The inward flow of protons through the coupling sites down the $\Delta p$ gradient drives electron transport in reverse (see Section 3.7.1).

## 12.5 Summary

When it comes to inorganic metabolism, prokaryotes are versatile creatures. They can reduce oxidized forms of sulfur and nitrogen, as well as nitrogen gas, for incorporation into cell material, and they can oxidize a variety of inorganic substrates, trapping the energy released as a $\Delta p$. They can also carry out anaerobic respiration using oxidized forms of iron, manganese, nitrogen, and sulfur as electron acceptors. One can add $CO_2$, which is used by the methanogens, to this list of electron acceptors.

There is both an assimilatory and dissimilatory route for nitrate reduction. The assimilatory route is catalyzed by cytosolic enzymes that reduce nitrate to ammonia. The ammonia is then incorporated into amino acids using the GOGAT enzyme system. All bacteria that grow on nitrate as a source of cell nitrogen must have the assimilatory pathway.

The dissimilatory route differs in being membraneous and in producing a $\Delta p$ that can be used for ATP synthesis. A common dissimilatory route reduces nitrate to $N_2$. Nitrate dissimilation to $N_2$ is also called denitrification. Organisms that carry out denitrification are widespread. However, the enzymes for denitrification are oxygen sensitive and furthermore are formed only under anaerobic conditions or when oxygen tensions are very low. These bacteria are generally facultative anaerobes and will carry out an aerobic respiration when oxygen is available. Several bacteria are not denitrifiers, but can carry out an anaerobic respiration in which nitrate is reduced to nitrite by a membrane-bound nitrate reductase, creating a $\Delta p$.

Most bacteria can use sulfate as a source of cell sulfur. They employ a cytosolic assimilatory pathway in which the sulfate is activated by ATP to form adenosine-3′-phosphate-5′-phosphosulphate (PAPS). The formation of PAPS raises the $E_h$ of sulfate so that it is a better electron acceptor. The reduction of PAPS yields sulfite, which is reduced to $H_2S$ via a soluble sulfite reductase. The $H_2S$ does not accumulate but is immediately incorporated into O-acetylserine to form L-cysteine. The L-cysteine donates the sulfur to the other sulfur-containing compounds.

Certain strict anaerobes can use sulfate as an electron acceptor for anaerobic respiration and reduce it to $H_2S$. These are the sulfate reducers. Organic carbon is oxidized, and the electrons are transferred to protons via a hydrogenase to form $H_2$. It is suggested that the $H_2$ diffuses into the periplasm, where it is oxidized by a cytochrome c that transfers the electrons to membrane carriers. Electrons travel across the membrane to the inner surface, creating a membrane potential. The protons resulting from the periplasmic oxidation of $H_2$ remain on the outer surface or in the periplasm, contributing to the proton gradient. Dissimilatory sulfate reduction differs from assimilatory reduction in that APS is reduced rather than PAPS, the electron carriers are membraneous, and a $\Delta p$ is created.

Nitrogen fixation is carried out by a diversity of prokaryotes, including both bacteria and archeae. These include cyanobacteria, photosynthetic bacteria, strict heterotrophic anaerobes (*Clostridium pasteuranium*, *Desulfovibrio vulgaris*), and several obligate and facultative aerobes (e.g., rhizobia, *Azotobacter*, *Klebsiella*, and methanogens). The reduction of $N_2$ is an ATP-dependent process catalyzed by the enzyme nitrogenase. Hydrogen gas is always a byproduct of nitrogen fixation. The nitrogenase is oxygen sensitive and therefore must be protected from oxygen, either in specialized cells (heterocysts) in the case of cyanobacteria, or leguminous nodules, or perhaps by an unusually high respiratory rate (*Azotobacter*), by binding to protective proteins, or by growth in an anaerobic environment (e.g., photosynthetic bacteria).

The oxidation of inorganic substances by oxygen is the source of energy for lithotrophs. Many lithotrophs are aerobic autotrophs. Electron transport takes place in the membrane, and a $\Delta p$ is created. The $\Delta p$ is used to drive ATP synthesis and reversed electron flow so that NADH can be generated for biosynthesis.

Iron-oxidizing bacteria such as *Thiobacillus ferrooxidans* live in acid environments (around pH 2) where iron sulfide minerals and oxygen are available. The bacteria maintain a $\Delta pH$ of around 4.5 units, by taking up cytoplasmic protons during the reduction of oxygen to water. The $\Delta pH$ drives the synthesis of ATP via a membrane ATP synthase.

Lithotrophic activities are of immense ecological significance because they are necessary for the recycling of inorganic nutrients through the biosphere. For example, consider the nitrogen cycle. All organisms use ammonia and nitrate, or amino acids, as the source of nitrogen. These are called "fixed" forms of nitrogen. Approximately 97% of all the nitrogen incorporated in living tissue comes from fixed nitrogen. However, the majority of the nitrogen on this planet is in the form of $N_2$ and unavailable to most organisms. Furthermore, due to the denitrification activities of bacteria living anaerobically, there is a constant drain of fixed nitrogen from the biosphere. That is to say, living systems eventually oxidize reduced forms of nitrogen to nitrate, which is then reduced to nitrogen gas by the denitrifying bacteria. Therefore, we all depend upon the prokaryotes that reduce $N_2$ for our supply of fixed nitrogen. The sulfate- reducing bacteria are responsible for much of the $H_2S$ produced and feed sulfide to the photosynthetic sulfur bacteria and aerobic sulfur-oxidizing bacteria. The combined activities of the sulfate reducers and the sulfide oxidizers accounts for much of the elemental sulfur deposits. Hydrogen sulfide can also have deleterious effects. Since $H_2S$ is toxic, it can occasionally produce harmful effects on fish, waterfowl, and even plants when the soil becomes anaerobic. The latter can occur in rice paddies. The $H_2S$ can also cause corrosion of metal pipes in anaerobic soils and waters.

## Study Questions

1. What are the major distinguishing features between assimilatory and dissimilatory nitrate reduction and assimilatory and dissimilatory sulfate reduction?

2. During assimilatory sulfate reduction O-acetylserine reacts with $H_2S$ to form cysteine. The O-acetylserine is formed from serine and acetyl–CoA. Write a reaction showing the chemical structures suggesting how O-acetylserine might be formed from serine and acetyl–CoA.

3. Clostridia can reduce nitrogen gas to ammonia using electrons derived from pyruvate (generated from glucose using the EMP pathway). The immediate reductant for the nitrogenase is reduced ferredoxin. How is reduced ferredoxin generated from pyruvate?

4. What is the minimum number of $Fe^{2+}$ that must be oxidized by oxygen to $Fe^{3+}$ in order to reduce one $NAD^+$? Solve this problem by focusing on electron volts.

5. The $\Delta pH$ drives the synthesis of ATP in the acidophilic iron-oxidizing bacteria. How do they maintain the $\Delta pH$?

6. What are some of the mechanisms used by nitrogen fixers to protect the nitrogenase from oxygen?

## NOTES AND REFERENCES

1. Lovley, D. R. 1991. Dissimilatory Fe(III) and Mn(IV) reduction. *Microbiol. Rev.* 55:259–287.

2. Lovley, D. R. 1993. Dissimilatory metal reduction. *Ann. Rev. Microbiol.* 47:263–290.

3. Nealson, K. H., and D. Saffarini. 1994. Iron and manganese in anaerobic respiration: Environmental significance, physiology, and regulation. *Ann. Rev. Microbiol.* 48:311–343.

4. Lin, J. T., and V. Stewart. 1998. Nitrate assimilation by bacteria. *Adv. Microbial Physiol.* 39:1–30.

5. In *Klebsiella pneumoniae* the electron donor for nitrate reductase appears to be an NADH-dependent oxidoreductase. Thus the electrons travel from NADH to the oxidoreductase to the nitrate reductase. Other bacterial nitrate reductases use reduced ferredoxin or flavodoxin. The major cytoplasmic nitrite reductase in *E. coli* uses NADH as the electron donor. For example, see the following references: Lin, J. T., B. S. Goldman, and V. Stewart. 1994. The *nas* FEDCBA operon for nitrate and nitrite assimilation in *Klebsiella pneumoniae* M5al. *J. Bacteriol.* 176:2551–2559. Gangeswaran, R., D. J. Lowe, and R. R. Eady. 1993. Purification and characterization of the assimilatory nitrate reductase of *Azotobacter vinelandii. Biochem. J.* 289:335–342.

6. Most of the reduced intracellular sulfur is $H_2S$ and $HS^-$, rather than $S^{2-}$. This is because the $pK_1$ of $H_2S$ is 7.04, the $pK_2$ is 11.96, and the cytoplasmic pH is usually close to 7. For example, neutrophilic bacteria have a cytoplasmic pH of approximately 7.5.

7. Casella, S., and W. J. Payne. 1996. Potential of denitrifiers for soil environment protection. *FEMS Microbiol. Lett.* 140:1–8.

8. Hansen, T. 1994. Metabolism of sulfate-reducing prokaryotes. *Antonie van Leeuwenhoek* 66:165– 185.

9. Thauer has argued that free $H_2$ may not be the electron carrier for lactate oxidation by sulfate. Even at very low $H_2$ partial pressures the reduction of $H^+$ by lactate is thermodynamically unfavorable. Furthermore, it has been demonstrated that $H_2$ does not affect lactate oxidation

by sulfate by *Desulfovibrio vulgaris*, and *Desulfovibrio sapovorans* growing on lactate plus sulfate does not contain hydrogenase. Reviewed in Thauer, R. K. 1989. Energy metabolism of sulfate-reducing bacteria, pp. 397–413. In: *Autotrophic Bacteria*. H. G. Schlegel and B. Bowien (Eds.). Science Tech Publishers, Madison, WI.

10. Reviewed by Postgate, J. 1987. *Nitrogen Fixation*. Edward Arnold, London.

11. Reviewed in Stacey, G., R. H. Burris, and H. J. Evans (Eds.). 1992. *Biological Nitrogen Fixation*. Chapman and Hall, New York, London.

12. Reviewed in Dilworth, J. J., and A. R. Glenn (Eds.). 1991. *Biology and Biochemistry of Nitrogen Fixation*. Elsevier. Amsterdam, New York, Oxford, Tokyo.

13. Peuppke, S. G. 1996. The genetic and biochemical basis for nodulation of legumes by rhizobia. *Crit. Rev. Biotech.* 16:1–51.

14. Lobo, A. L., and S. H. Zinder. 1992. Nitrogen fixation by methanogenic bacteria, pp. 191–211. In: *Biological Nitrogen Fixation*. Stacey, G., Burris, H. R., and H. J. Evans (Eds.). Chapman and Hall, New York.

15. Dean, Dennis R., J. T. Bolin, and L. Zheng. 1993. Nitrogenase metalloclusters: Structures, organization, and synthesis. *J. Bacteriol.* 175:6737–6744.

16. Peters, J. W., K. Fisher, and D. R. Dean. 1995. Nitrogenase structure and function: A biochemical–genetic perspective. *Ann. Rev. Microbiol.* 49:335–366.

17. The MoFe protein has two pairs of metalloclusters. One pair is iron–sulfur clusters (Fe$_8$S$_{7-8}$) known as P clusters, and the other pair is iron–sulfur–molybdate clusters (Fe$_7$S$_9$Mo-homocitrate) known as FeMo cofactors.

18. There are two types of metalloclusters in the MoFe protein. These are a FeMo cofactor and a P cluster. The P cluster consists of two Fe$_4$S$_4$ clusters. Each $\alpha\beta$ dimer has a FeMo cofactor and a P cluster, making two copies of each type of cluster per tetramer. One model of electron flow proposes that the electron travels from the Fe$_4$S$_4$ cluster in the Fe protein in an ATP-dependent reaction to a P cluster in the MoFe protein. From the P cluster the electron moves to the FeMo cofactor, and from there to N$_2$. The precise role of ATP in electron transfer is not known, although it has been shown to bind as the Mg$^{2+}$ chelate to the Fe protein. It is presumed that binding and/or hydrolysis of MgATP causes conformational changes in the nitrogenase that facilitate electron transfer. For example, conformational changes in the proteins might alter the environment and redox potential of the metal clusters, or bring the Fe$_4$S$_4$ cluster in the Fe protein and the P cluster in the FeMo protein into closer proximity. Specifically, it has been proposed that binding of ATP to the Fe protein lowers the redox potential of the Fe$_4$S$_4$ cluster and also leads to association (complex formation) between the Fe protein and the MoFe protein. Association of the two proteins causes ATP hydrolysis and electron transfer. One model proposes that complex formation and hydrolysis of ATP is accompanied by a configurational change in the MoFe protein, which moves a P cluster close to a Fe$_4$S$_4$ cluster in the Fe protein, thus facilitating electron transfer. The two proteins dissociate after electron transfer. Proposals as to how the MoFe protein might reduce N$_2$ are derived from model systems. The chemical reactivity of N$_2$ is increased when it binds to a transition metal, especially in a complex with other ligands, such as sulfur. This has been studied in model systems where chemically synthesized metal complexes containing molybdenum bound to N$_2$ are reduced with artificial reductants. It is thought that the N$_2$ becomes activated when it binds to the molybdenum in the FeMo cofactor in nitrogenase and, while bound, is reduced to ammonia.

19. Eady, R. R. 1992. The dinitrogen-fixing bacteria, pp. 535–553. In: *The Prokaryotes*. Vol. I. Balows, A., H. G. Truper, M. Dworkin, W. Harder, and K.-H. Schleifer (Eds.). Springer Verlag, Berlin.

20. Yates, M. G., and F. O. Cambell. 1989. The role of oxygen and hydrogen in nitrogen fixation, pp. 383–416. In: *SGM Symposium*, Vol. 42, *The Nitrogen and Sulphur Cycles*, Cole, J. A., and S. Ferguson (Eds.). Cambridge University Press, Cambridge.

21. Bergman, B., J. R. Gallon, A. N. Rai, and L. J. Stal. 1997. N$_2$ fixation by non-heterocystous cyanobacteria. *FEMS Microbiol. Rev.* 19:139–185.

22. Haselkorn, R., and W. J. Buikema. 1992. Nitrogen fixation in cyanobacteria, pp. 166–190. In: *Biological Nitrogen Fixation*. Stacey, G., R. H. Burris, and H. J. Evans (Eds.). H. J. Chapman and Hall, New York.

23. Kelly, D. P. 1990. Energetics of chemolithotrophs, pp. 479–503. In: *The Bacteria*, Vol. XII. Krulwich, T. A. (Ed.). Academic Press, New York.

24. Hooper, A. B. 1989. Biochemistry of the nitrifiying lithoautotrophic bacteria, pp. 239–265, In: *Autotrophic Bacteria*. Schlegel, H. G., and B. Bowien (Eds.). Springer-Verlag, Berlin.

25. Nicholls, D. G., and S. J. Ferguson. 1992. *Bioenergetics* 2. Academic Press, London.

26. Kelly, D. P. 1989. Physiology and biochemistry of unicellular sulfur bacteria, pp. 193–217. In: *Autotrophic Bacteria*. Schlegel, H. G., and B. Bowien (Eds.). Springer-Verlag, Berlin.

27. Kelly, D. P., W-P Lu, and R. K. Poole. 1993. Cytochromes in *Thiobacillus tepidarius* and the respiratory chain involved in the oxidation of thiosulphate and tetrathionate. *Arch. Microbiol.* 160:87–95.

28. Friedrich, C. G. 1998. Physiology and genetics of sulfur-oxidizing bacteria, pp. 235–289. In: Adv. Microbiol. Physiol. R. K. Poole (Ed.). Academic Press, New York.

29. Ingledew, W. J. 1990. Acidophiles, pp. 33–54. In: *Microbiology of Extreme Environments*. Edwards, C. (Ed.). McGraw–Hill Publishing Company, New York.

30. Ingledew, W. J. 1982. *Thiobacillus ferrooxidans:* The bioenergetics of an acidophilic chemolithotroph. *Biochem. Biophys. Acta* 683:89–117.

31. The amount of energy available from the oxidation of $Fe^{+2}$ by $O_2$ depends upon whether the oxygen is reduced on the periplasmic side of the membrane or on the cytoplasmic side. The $E_{m'2}$ of $Fe^{3+}/Fe^{2+}$ is usually stated to be about 0.77–0.78 V. The $E_{m,7}$ of $O_2/H_2O$ is 0.82 V. The $E_{m'2}$ of $O_2/H_2O$ is 1.12 V. If the oxygen were reduced in the cytoplasm, then 0.82–0.77 or 0.05 eV would be available per electron. If the oxygen were reduced in the periplasm, then 1.12–0.77 or 0.35 eV would be available per electron. However, as pointed out by Nichols and Ferguson, periplasmic reduction of oxygen with consumption of periplasmic protons would not be helpful because a $\Delta p$ would not develop. (There would be no electrogenic electron flow and no differential consumption of protons between periplasm and cytoplasm.) Bringing protons to the periplasm from the cytoplasm in order to reduce oxygen would require energy to overcome the proton concentration gradient between periplasm and cytoplasm.

32. Ehrlich, H. L., J. W. Ingledew, and J. C. Salerno. 1991. Iron- and manganese-oxidizing bacteria, pp. 147–170. In: *Variations in Autotrophic Life*. Shively, J. M., and L. L. Barton (Eds.). Academic Press, New York.

33. According to the chemiosmotic theory, the actual number of ATPs made will depend upon the ratio of protons translocated per electron to the number translocated by the ATP synthase, [i.e., $(H^+/e^-)/H^+/ATP)$] (see Section 4.5.2).

# 13

# C₁ Metabolism

Many prokaryotes can grow on $C_1$ compounds as their sole source of carbon. Some common $C_1$ compounds that support growth are:

1. Carbon dioxide ($CO_2$)

2. Methane ($CH_4$))

3. Methanol ($CH_3OH$)

4. Methylamine ($CH_3NH_2$)

A few strictly anaerobic prokaryotes that use carbon dioxide as a source of cell carbon also use it as an electron acceptor, reducing it to methane, deriving ATP from the process. These are the methanogens, which are placed in one of the two prokaryotic domains (i.e., the Archaea) (Section 1.1.1).

## 13.1 Carbon Dioxide Fixation Systems

Prokaryotes that use $CO_2$ as the sole or major carbon source are called *autotrophs,* and the pathway of $CO_2$ assimilation is called autotrophic $CO_2$ fixation.[1] There are three major autotrophic $CO_2$ fixation pathways in prokaryotes. These are the Calvin cycle (also known as the Calvin–Benson–Bassham or CBB cycle), the acetyl–CoA pathway, and the reductive tricarboxylic acid path-

way. The Calvin cycle is the most prominent of the autotrophic $CO_2$ fixation systems and is found in photosynthetic eukaryotes, most photosynthetic bacteria, cyanobacteria, and chemoautotrophs. It does not occur in the archaea or in certain obligately anaerobic or microaerophilic bacteria. Instead, the acetyl–CoA pathway and the reductive carboxylic acid pathway are found in these organisms. The reductive carboxylic acid pathway occurs in the green photosynthetic bacteria *Chlorobium* (anaerobes), in *Hydrogenobacter* (microaerophilic), in *Desulfobacter* (anaerobes), and in the archaeons *Sulfolobus* and *Thermoproteus*. The acetyl–CoA pathway is more widespread and is found in methanogenic archaea, some sulfate-reducing bacteria, and in facultative heterotrophs that synthesize acetic acid from $CO_2$ during fermentation. The latter bacteria are called *acetogens*.

### 13.1.1 The Calvin cycle

The Calvin cycle is a pathway for making phosphoglyceraldehyde completely from $CO_2$.[2] The pathway operates in the chloroplasts of plants and algae. However, in bacteria where there are no similar organelles, the Calvin cycle is found in the cytosol.

## Summing the reactions of the Calvin cycle

During the Calvin cycle, three $CO_2$ molecules are reduced to phosphoglyceraldehyde, which is at the oxidation level of glyceraldehyde ($C_3H_6O_3$). Twelve electrons are required, and these are provided by six NAD(P)H. [Recall that each NAD(P)H carries a hydride ion (one proton and two electrons) (Section 8.3)]. However, NAD(P)H is not a sufficiently strong reductant to reduce $CO_2$ without an additional source of energy. Each reduction thus requires an ATP. Therefore, six ATPs are required for the six reductions. However, the overall reaction, shown below, indicates that nine ATPs are required. The three extra ATPs are used to recycle an intermediate, ribulose-1,5-bisphosphate (RuBP), that is catalytically required for the pathway.

$$3CO_2 + 9ATP + 6NAD(P)H + 6H^+ + 5H_2O$$
$$\longrightarrow \text{phosphoglyceraldehyde} + 9ADP$$
$$+ 8P_i + 6NAD(P)^+$$

## The Calvin cycle can be divided into two stages

The Calvin cycle is a complicated pathway, but it will be familiar because it resembles the pentose phosphate pathway in certain key reactions (Section 8.4). It is convenient to think of the Calvin cycle as occurring in two stages:

1. Stage 1 is a reductive carboxylation of RuBP to form phosphoglyceraldehyde (PGALD).

2. Stage 2 consists of sugar rearrangements, regenerating RuBP using some of the PGALD produced in Stage 1.

## Stage 1

The first reaction is the carboxylation of RuBP, forming two moles of 3-phosphoglycerate. The enzyme is called RuBP carboxylase or ribulose-1,5-bisphosphate carboxylase ("rubisco"). The reaction (and its probable mechanism) is shown in Fig. 13.1. An enolate anion probably forms that is carboxylated on the C2. Hydrolysis of the intermediate yields two moles of 3-phosphoglycerate (PGA).

The PGA is then reduced to PGALD via 1,3-bisphosphoglycerate (BPGA). These reactions also take place in glycolysis. The reactions of Stage 1 can be summarized as follows:

$$3\,CO_2 + 3\,RuBP + 3\,H_2O \longrightarrow 6\,PGA$$
$$6\,PGA + 6\,ATP \longrightarrow 6\,BPGA + 6\,ADP$$
$$6\,BPGA + 6\,NAD(P)H + 6\,H^+ \longrightarrow 6\,PGALD$$
$$+ 6NAD(P)^+ + 6\,P_i$$

$$\overline{\begin{array}{l}3\,CO_2 + 3\,RuBP + 3\,H_2O + 6\,ATP + 6\,NAD(P)H \\ + 6\,H^+ \longrightarrow 6\,PGALD + 6\,ADP + 6\,P_i \\ + 6\,NAD(P)^+\end{array}}$$

Stage 1 is summarized as reactions 1 and 2 in Fig. 13.2.

## Stage 2

The whole point of the sugar rearrangements of Stage 2 is to regenerate three RuBPs from 5 of the six phosphoglyceraldehydes made in Stage 1. For reference, refer to Fig. 13.2. (The Calvin cycle is summarized without chemical structures in the discussion below. See the section on the relationshiop of the Calvin cycle to glycolysis and the pentose phosphate pathway.) Since RuBP has five carbons and PGALD has three carbons, a two-carbon fragment must be transferred to phosphoglyceraldehyde. That is to say, a transketolase reaction is required. The $C_2$ donor is fructose-6-phosphate, which is made from PGALD (reactions 3, 4, and 5, Fig. 13.2). The synthesis of fructose-6-phosphate from PGALD uses glycolytic enzymes and fructose-1,6-bisphosphate phosphatase. One molecule of PGALD first isomerizes to dihydroxyacetone phosphate, which condenses with a second molecule of PGALD to form fructose-1,6-bisphosphate. The fructose-1,6-bisphosphate is then dephosphorylated to fructose-6-phosphate by the phosphatase (reaction 5). The fructose-6-phosphate then donates a two-carbon fragment in a transketolase reaction to PGALD, forming erythrose-4-phosphate and xylulose-5-phosphate (reaction 6).

The xylulose-5-phosphate is isomerized to ribulose-5-phosphate (reaction 10). The erythrose-4-phosphate is condensed with the second dihydroxyacetone phosphate

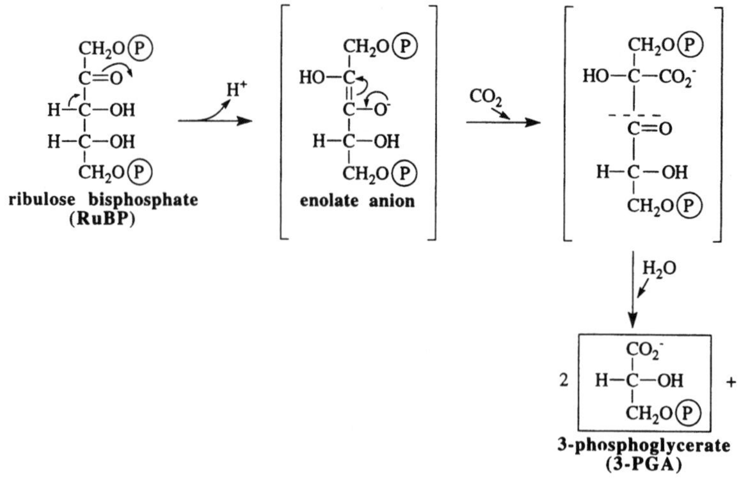

**Fig. 13.1** Carboxylation of RuBP: The "rubisco" reaction. One possible mechanism is that the carbonyl group in RuBP attracts electrons, resulting in the dissociation of a hydrogen from the carbon on the *cis* HCOH group. Electrons then shift to form the enolate anion, which becomes carboxylated at the C2. Hydrolysis of the carboxylated intermediate yields two 3-PGA molecules.

to form a seven-carbon ketose diphosphate, sedoheptulose-1,7-bisphosphate (reaction 7). The formation of sedoheptulose-1,7-bisphosphate is analogous to the condensation of dihydroxyacetone phosphate with PGALD to form fructose-1,6-bisphosphate, but is catalyzed by a different aldolase, sedoheptulose-1,7-bisphosphate aldolase. (However, as discussed later, the bacterial sedoheptulose-1,7-bisphosphate aldolase and the fructose-1,6-bisphosphate aldolase may actually be a single bifunctional enzyme.) A phosphatase hydrolytically removes the phosphate from the $C_1$ to form sedoheptulose-7-phosphate (reaction 8). This is an irreversible reaction and prevents the pentose phosphates that are formed from sedoheptulose-7-phosphate in the next step from being converted back to phosphoglyceraldehyde. The sedoheptulose-7-phosphate is a two-carbon donor in a second transketolase reaction using the last of the phosphoglyceraldehyde as an acceptor (reaction 9). The products are the pentose phosphates, ribose-5-phosphate and xylulose-5-phosphate, both of which are isomerized to ribulose-5-phosphate (reactions 10 and 11). The ribulose-5-phosphate is then phosphorylated to ribulose-1,5-bisphosphate by a ribulose-5-phosphate kinase, which is unique to the Calvin cycle (reaction 12). Thus the

pentose isomerase reactions and the transketolase reactions are the same in both the Calvin cycle and the pentose phosphate pathway. The reactions of Stage 2 are summarized as follows:

2 PGALD $\longrightarrow$ 2 DHAP

PGALD + DHAP $\longrightarrow$ FBP

FBP + $H_2O$ $\longrightarrow$ F6P + $P_i$

F6P + PGALD $\longrightarrow$ erythrose-4-P

  + xylulose-5-P

erythrose-4-P + DHAP $\longrightarrow$

  sedoheptulose-1,7-bisphosphate

sedoheptulose-1,7-bisphosphate + $H_2O$ $\longrightarrow$

  sedoheptulose-7-P + $P_i$

sedoheptulose-7-P + PGALD $\longrightarrow$

  ribose-5-P + xylulose-5-P

2 xylulose-5-P $\longrightarrow$ 2 ribulose-5-P

ribose-5-P $\longrightarrow$ ribulose-5-P

3 ribulose-5-P + 3 ATP $\longrightarrow$

  3 ribulose-1,5-bisphosphate + 3 ADP

_____

5 PGALD + 2 $H_2O$ + 3 ATP $\longrightarrow$

  3 ribulose-1,5-bisphosphate + 2$P_i$ + 3 ADP

# C₁ METABOLISM

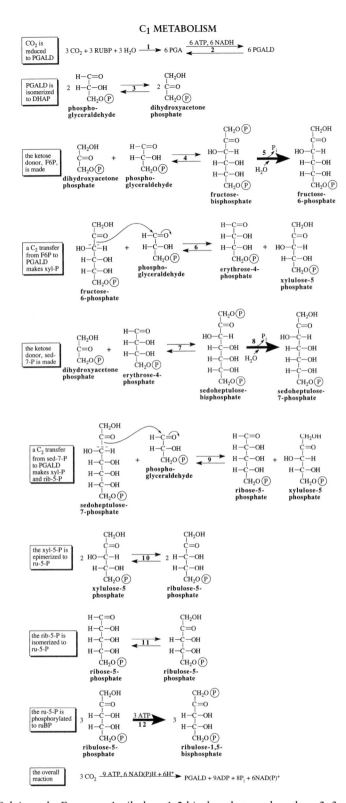

**Fig. 13.2** The Calvin cycle. Enzymes: 1, ribulose-1,5-bisphosphate carboxylase; 2, 3-phosphoglycerate kinase and triosephosphate dehydrogenase; 3, triosephosphate isomerase; 4, fructose-1,6-bisphosphate aldolase; 5, fructose-1,6-bisphosphate phosphatase; 6 and 9, transketolase; 7, sedoheptulose-1,7-bisphosphate aldolase; 8, sedoheptulose-1,7-bisphosphatase; 10, phosphopentose epimerase; 11, ribose phosphate isomerase; 12, phosphoribulokinase.

Summing Stages 1 and 2 yields

$$3CO_2 + 9ATP + 6NAD(P)H + 6H^+ + 5H_2O$$
$$\longrightarrow PGALD + 9ADP + 8P_i + 6NAD(P)^+$$

## The carbon balance

Complex pathways can be seen in simpler perspective by examining the carbon balance. The carbon balance for the Calvin cycle is:

$$3\,C_1 + 3\,C_5 \longrightarrow 6\,C_3$$
$$2\,C_3 \longrightarrow C_6$$
$$C_6 + C_3 \longrightarrow C_4 + C_5$$
$$C_4 + C_3 \longrightarrow C_7$$
$$\underline{C_7 + C_3 \longrightarrow 2\,C_5}$$
$$3\,C_1 \longrightarrow C_3$$

The carbon balance for the Calvin cycle illustrates that the pathway produces one $C_3$ from three $C_1$ molecules.

## The relationship of the Calvin cycle to glycolysis and the pentose phosphate pathway

Most of the reactions of the Calvin cycle also take place in glycolysis and the pentose phosphate pathway. In Fig. 13.3, reactions 1–12 are the Calvin cycle. Reaction 1 is the carboxylation of ribulose-1,5-bisphosphate to form 3-phosphoglycerate. This reaction is unique to the Calvin cycle. The 3-phosphoglycerate enters the glycolytic pathway. Reactions 2 through 6 are reactions that take place during the reversal of glycolysis, whereby 3-phosphoglycerate is transformed into fructose-6-phosphate (see Section 8.1). The fructose-6-phosphate enters the pentose phosphate pathway at the level of the sugar rearrangement reactions (reaction 7) (Section 8.5). Reaction 7 is the transketolase reaction, which forms erythrose-4-phosphate and xylulose-5-phosphate. Now the Calvin cycle diverges from the pentose phosphate pathway. Whereas the pentose phosphate pathway synthesizes sedoheptulose-7-phosphate from erythrose-4-phosphate via the reversible transaldolase reaction (TA), the Calvin cycle uses the aldolase (reaction 7) and the irreversible phosphatase (reaction 8) to synthesize sedoheptulose-7-phosphate. This has important consequences regarding the directionality

of the Calvin cycle. Because the phosphatase (reaction 8) is an irreversible reaction, the Calvin cycle proceeds only in the direction of pentose phosphates, whereas the pentose phosphate pathway is reversible from sedoheptulose-7-phosphate. (An irreversible Calvin cycle makes physiological sense, since the sole purpose of the Calvin cycle is to regenerate ribulose-1,5-bisphosphate for the initial carboxylation reaction.) Once the sedoheptulose-7-phosphate is formed, the synthesis of ribulose-5-phosphate (reactions 10 and 11) takes place via the same reactions as in the pentose phosphate pathway. Reaction 12 of the Calvin cycle is the phosphorylation of ribulose-5-phosphate to form ribulose-1,5-bisphosphate (RuMP kinase), a second reaction that is unique to the Calvin cycle. The sedoheptulose-1,7-bisphosphate aldolase and the sedoheptulose-1,7-bisphosphate phosphatase also occur in the ribulose monophosphate pathway, which is a formaldehyde-fixing pathway that uses Calvin cycle reactions (Section 13.2.1).[3]

## 13.1.2 The acetyl–CoA pathway

Bacteria that use the acetyl–CoA pathway include methanogens, acetogenic bacteria, and most autotrophic sulfate-reducing bacteria.[4] (See Note 5 for a definition of the acetogenic bacteria.) This section presents a general summary of the acetyl–CoA pathway without describing the individual reactions. The first part explains how acetyl–CoA is made from $CO_2$ and $H_2$, the second part explains how the acetyl–CoA is incorporated into cell material, and the third part summarizes how the acetyl–CoA pathway is used by methanogens for methanogenesis. Sections 13.1.3 and 13.1.4 describe the individual reactions of the acetyl–CoA pathway in acetogens and in methanogens.

### Making acetyl–CoA from $CO_2$ using the acetyl–CoA pathway

The acetyl–CoA pathway can be viewed as occurring in four steps. The first step is a series of reactions that result in the reduction of $CO_2$ to a methyl group [$CH_3$]. The methyl group is bound to tetrahydrofolic acid (THF) in bacteria and to tetrahydromethanopterin (THMP) in archaea. The methyl group

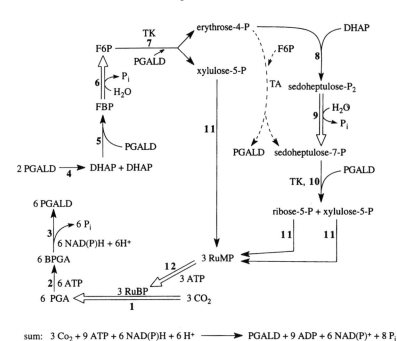

**Fig. 13.3** Relationships between the Calvin cycle, glycolysis, and the pentose phosphate pathway. Reactions 1–12 are the Calvin cycle. Note that the only reactions unique to the Calvin cycle are reactions 1, 8, 9, and 12. Reaction 1 is catalyzed by ribulose-1,5-bisphosphate carboxylase. Reactions 2–5 are glycolytic reactions. Reaction 6 is catalyzed by fructose-1,6-bisphosphatase. Reaction 7 is the transketolase reaction also found in the pentose phosphate pathway. Reactions 8 and 9 are the sedoheptulose–bisphosphate aldolase and the sedoheptulose-1,7-bisphosphatase reactions. Reaction 10 is the transketolase reaction also found in the pentose phosphate pathway. Reactions 11 are the pentose epimerase and isomerase reactions also present in the pentose phosphate pathway. Reaction 12 is the RuMP kinase reaction. The pentose phosphate pathway differs from the Calvin cycle only in that the pentose phosphate pathway synthesizes sedoheptulose-7-phosphate via a reversible transaldolase reaction (dotted lines), whereas the Calvin cycle synthesizes sedoheptulose-7-phosphate via an aldolase and phosphatase reaction (reactions 8 and 9). Since the phosphatase is irreversible, the Calvin cycle irreversibly converts phosphoglyceraldehyde to pentose phosphates.

is then transferred to the enzyme *carbon monoxide dehydrogenase*. Carbon monoxide dehydrogenase also catalyzes the reduction of a second molecule of carbon dioxide to a bound carbonyl group [CO], which will become the carboxyl group of acetate. During autotrophic growth, the electrons are provided by hydrogen gas via a hydrogenase:

$$CO_2 + 3H_2 + H^+ \longrightarrow [CH_3] + 2H_2O \quad (1)$$

$$CO_2 + H_2 \longrightarrow [CO] + H_2O \quad (2)$$

The bound [CH₃] and [CO] are then condensed by carbon monoxide dehydrogenase to form bound acetyl [CH₃CO]:

$$[CO] + [CH_3] \longrightarrow [CH_3CO] \quad (3)$$

The bound acetyl reacts with bound CoASH,

[CoAS], to form acetyl–CoA, which is released from the enzyme:

$$[CH_3CO] + [CoAS] \longrightarrow CH_3COSCoA \quad (4)$$

The acetyl–CoA is converted to acetate via phosphotransacetylase and acetate kinase (generating an ATP) and excreted, or it is assimilated into cell material. As mentioned, many anaerobic bacteria can grow autotrophically with the acetyl–CoA pathway using H₂ as the source of electrons. These bacteria must generate ATP from the Δ*p* generated during the reduction of the CO₂ by H₂. The Δ*p* is required for net ATP synthesis because the ATP generated during the conversion of acetyl–CoA to acetate is balanced by the ATP used for the formation of formyl–THF (Fig. 13.4).

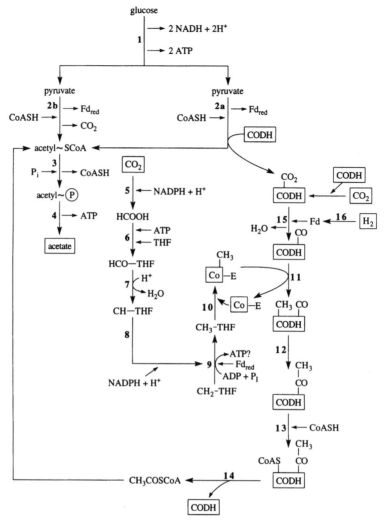

**Fig. 13.4** The acetyl–CoA pathway in *Clostridium thermoaceticum*. During heterotrophic growth, the carboxyl group in some pyruvate molecules is directly transferred to CODH without being released as free $CO_2$. The evidence for this is that pyruvate is required in cell-free extracts for the synthesis of acetic acid from $CH_3$–THF even in the presence of $CO_2$. Furthermore, radioisotope experiments confirm that the carboxyl in acetate is derived from the carboxyl in pyruvate without going through free $CO_2$. However, the cells can be grown on $CO_2$ and $H_2$. Under these circumstances, the CODH reduces $CO_2$ to CO–CODH using hydrogenase and electrons from $H_2$ (reactions 15 and 16). Enzymes: 1, glycolytic; 2, pyruvate:ferredoxin oxidoreductase; 3, phosphotransacetylase; 4, acetate kinase; 5, formate dehydrogenase; 6, formyltetrahydrofolate (HCO–THF) synthetase; 7, methenyltetrahydrofolate (CH–THF) cyclohydrolase; 8, methylenetetrahydrofolate (CH$_2$–THF) dehydrogenase; 9, methylenetetrahydrofolate reductase; 10, methyltransferase; 11, corrinoid enzyme; 12, 13, 14, 15, carbon monoxide dehydogenase (CODH); 16, hydrogenase.

## Incorporating acetyl–CoA into cell material

For acetogens growing autotrophically, part of the acetyl–CoA that is produced must be incorporated into cell material. The glyoxylate cycle, which is generally associated with aerobic metabolism, is not present in these organ-isms, and therefore a different pathway must operate. As shown below, the acetyl–CoA is reductively carboxylated to pyruvate via *pyruvate synthase*, a ferredoxin-linked enzyme. The pyruvate can be phosphorylated to phosphoenolpyruvate using PEP synthetase. The phosphoenolpyruvate is used for biosynthesis in the usual way. A more detailed discussion of acetyl–CoA assimilation in methanogens is given in Section 13.1.4.

$$\text{acetyl–CoA} + \text{Fd(red)} + \text{CO}_2 \xrightarrow{\text{pyruvate synthase}}$$

$$\text{pyruvate} + \text{Fd(ox)} + \text{CoASH}$$

$$\text{Pyruvate} + \text{ATP} + \text{H}_2\text{O} \xrightarrow{\text{PEP synthetase}}$$

$$\text{PEP} + \text{AMP} + \text{P}_i$$

## 13.1.3 The acetyl–CoA pathway in Clostridium thermoaceticum

The acetyl–CoA pathway was first investigated in *C. thermoaceticum*, an acetogenic bacterium that converts one mole of glucose to three moles of acetate.[6] Under these circumstances, the production of acetate is used as an electron sink during fermentation rather than for autotrophic growth. The pathway is illustrated in Fig. 13.4. Glucose is converted to pyruvate via the Embden–Meyerhof–Parnas pathway (reactions 1). Two of the acetates are synthesized from the decarboxylation of pyruvate using pyruvate–ferredoxin oxidoreductase, phosphotransacetylase, and acetate kinase (reactions 2–4). The third acetate is synthesized from CO$_2$ using the acetyl–CoA pathway. In the acetyl–CoA pathway, one of the carbon dioxides that is removed from pyruvate is not set free but instead becomes bound to the enzyme *carbon monoxide dehydrogenase* (CODH), where it will be used for acetate synthesis (reaction 2a). This bound CO$_2$ will eventually become the carbonyl group in acetyl–CoA. The second pyuvate is decarboxylated to release free CO$_2$ (reaction 2b).

In reactions 5–9, the free CO$_2$ is reduced to bound methyl. The free CO$_2$ first becomes reduced to formate (HCOOH) by formate dehydrogenase (reaction 5). The formate is then attached to tetrahydrofolic acid (THF) in an ATP-dependent reaction to make formyl–THF (HCO–THF) (reaction 6). (See Section 9.2.4 for a discussion of tetrahydrofolate reactions.) The formyl–THF is then dehydrated to form methenyl–THF (CH–THF) (reaction 7). The methenyl–THF is reduced by NADPH to methylene–THF (CH$_2$–THF) (reaction 8). The methylene–THF is reduced by ferredoxin

to methyl–THF (CH$_3$–THF) (reaction 9), a reaction thought to be coupled to ATP formation by a mechanism involving $\Delta p$. The methyl group is then transferred to a corrinoid enzyme, [Co]-E, which transfers the methyl group to the CO–CODH (reactions 10 and 11).[7] The CODH makes the acetyl moiety, [CH$_3$CO] (reaction 12). The CODH also has a binding site for CoASH and synthesizes acetyl–CoA from the bound intermediates (reactions 13 and 14). The CODH can also bind free CO$_2$ and reduce it to the level of carbon monoxide (reaction 15).

### Autotrophic growth

Following the entry of CO$_2$ at reactions 5 and 15 (Fig. 13.4), it can be seen that the acetyl–CoA pathway allows autotrophic growth on CO$_2$ and H$_2$. Carbon dioxide is reduced to [CH$_3$] and [CO] (reactions 5–10 and reaction 15). The [CH$_3$] and [CO] combine with CoASH to form acetyl–CoA (reactions 11–14). During acetogenesis an ATP can be made via substrate-level phosphorylation from acetyl–CoA (reactions 3 and 4). However, because an ATP is required to incorporate formate (reaction 6), there can be no net ATP synthesis via substrate-level phosphorylation alone. How do organisms using the acetyl–CoA pathway during autotrophic growth make net ATP? It is necessary to postulate that ATP is made by a chemiosmotic mechanism coupled to the reduction of CO$_2$ to acetate. For example, there may be electrogenic electron flow following periplasmic oxidation of H$_2$, or perhaps a quinone loop or a proton pump may be operating. (Mechanisms for generating a $\Delta p$ are reviewed in Sections 4.6 and 4.7.4.) It appears that homoacetogenic clostridia do indeed create a $\Delta p$ during the reduction of CO$_2$ to acetate and that the $\Delta p$ drives the synthesis of ATP via a H$^+$–ATP synthase.[8] But not all acetogenic bacteria use a proton circuit. *Acetobacterium woodii* couples the reduction of CO$_2$ to acetyl–CoA to a primary sodium ion pump.[9] The sodium ion translocating step has not been identified but is thought to be either the reduction of methylene–THF to methyl–THF (reaction 9) or some later step in the synthesis of acetyl–CoA because these are highly exergonic reactions. *A. woodii* also has a sodium ion–dependent ATP synthase. Thus these bacteria rely on a sodium ion current to make ATP when growing on $CO_2$ and $H_2$. A similar situation exists in *Propionigenium modestum*, where a $Na^+$-translocating methylmalonyl–CoA decarboxylase generates a $Na^+$ electrochemical potential that drives the synthesis of ATP (Section 3.8.1).

## Carbon monoxide dehydrogenase

It is clear that carbon monoxide dehydrogenase is a crucial enzyme for acetogenesis, for autotrophic growth in anaerobes, and for methanogenesis from acetate. Because of the importance of this enzyme, some of the evidence for its participation in synthesizing and degrading acetyl–CoA is described next.

The CODH catalyzes the breakage and formation of the carbon–carbon bond between the methyl group [$CH_3$] and the carbonyl group [CO] in the acetyl moiety. Evidence for this is the following exchange reaction carried out by CODH[10]:

$$^{14}CH_3–CODH + CH_3COSCoA$$

$$\longrightarrow CH_3–CODH + {}^{14}CH_3COSCoA$$

For the above reaction, the CODH was methylated with $CH_3I$.

The CODH catalyzes the breakage and formation of the thioester bond between CoASH and the carbonyl group of the acetyl moiety. Evidence for this is the following reaction carried out by CODH.[11] The asterisk refers to radioactive CoASH.

$$CH_3COSCoA + {}^*CoASH$$

$$\longrightarrow CH_3COS^*CoA + CoASH$$

The following exchange has also been demonstrated to be catalyzed by CO dehydrogenase[12]:

$$[1\text{-}^{14}C]acetyl–CoA + {}^{12}CO$$

$$\longrightarrow {}^{14}CO + [1\text{-}^{12}C]acetyl–CoA$$

The three exchange reactions described above mean that the CODH is capable of disassembling acetyl–CoA into its components, [$CH_3$], [CO], and [CoASH], and resynthesizing the molecule. It is therefore sufficient for making acetyl–CoA from the bound components.

## 13.1.4 The acetyl–CoA pathway in methanogens

### Acetyl–CoA synthesis from $CO_2$ and $H_2$

Among the archaea is a group called methanogens that can grow autotrophically on $CO_2$ and $H_2$ (Section 1.2). During autotrophic growth, the $CO_2$ is first incorporated into acetyl–CoA in a pathway very similar to the acetyl–CoA pathway that exists in the bacteria (e.g., in *C. thermoaceticum*).[13–16] In other words, one molecule of $CO_2$ is reduced to [CO] bound to CODH. A second molecule of $CO_2$ is reduced to bound methyl, [$CH_3$], which is transferred to the CODH. Then the CODH makes acetyl–CoA from the bound CO, $CH_3$, and CoASH. There are, however, differences between the acetyl–CoA pathway found in the methanogenic archaea and that in the bacteria. The differences have to do with the cofactors and the utilization of formate. For reference, the structures of the cofactors are drawn in Fig. 13.5. The differences between the two pathways are:

1. The carriers for the formyl and more reduced $C_1$ moieties in the methanogens is not tetrahydrofolic acid (THF) but rather tetrahydromethanopterin (THMP), a molecule resembling THF (see Fig. 13.5 for a comparison of the structures).

2. The methanogens do not reduce free $CO_2$ to formate. Nor do they incorporate formate into THMP to form formyl–THMP. What the methanogens do instead is fix $CO_2$ onto a $C_1$ carrier called methanofuran (MFR) and reduce it to formyl–MFR. The formyl group is then transferred to THMP. (Recall that the acetogenic bacteria reduce free $CO_2$ to formate and then incorporate the formate into formyl–THF.)

Figure 13.6 is a diagram showing how the acetyl–CoA pathway is thought to operate in methanogens. (The same pathway is shown in Fig. 13.7 but with the chemical structures drawn.) The $CO_2$ is first condensed with methanofuran (MFR) and reduced to formyl–MFR (reaction 1). The formyl group is then transferred to tetrahydromethanopterin

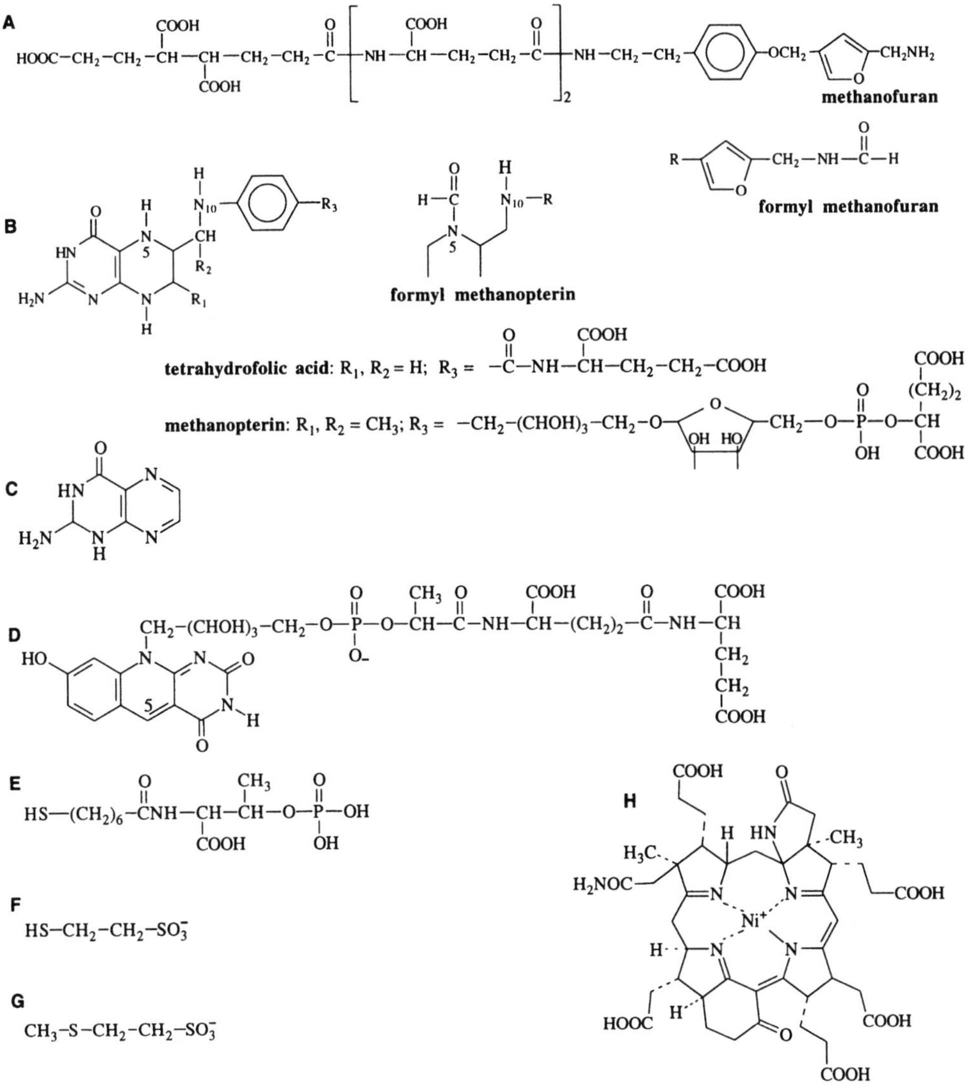

**A** ... methanofuran

formyl methanofuran

**B** ... formyl methanopterin

tetrahydrofolic acid: $R_1, R_2$ = H; $R_3$ = $-C-NH-CH-CH_2-CH_2-COOH$

methanopterin: $R_1, R_2$ = $CH_3$; $R_3$ = $-CH_2-(CHOH)_3-CH_2-O-$ ...

**C** Pterin

**D** ...

**E** ...

**F** $HS-CH_2-CH_2-SO_3^-$

**G** $CH_3-S-CH_2-CH_2-SO_3^-$

**H** ...

Fig. 13.5 Structures of coenzymes in the acetyl–CoA pathway and in methanogenesis. (A) Methanofuran (4-[N-(4,5,7-tricarboxyheptanoyl-*n*-L-glutamyl-*n*-L-glutamyl)-*p*-(*β*-aminoethyl)phenoxymethyl]-2-(aminomethyl)furan). The $CO_2$ becomes attached to the amino group on the furan and is reduced to the oxidation level of formic acid (formyl-methanofuran). (B) Tetrahydrofolic acid and methanopterin. They are both derivatives of pterin shown in (C). In formyl-methanopterin the formyl group is bound to $N_5$, whereas in tetrahydrofolic acid it is bound to $N_{10}$. (C) Pterin. (D) Oxidized coenzyme $F_{420}$. This is a 5-deazaflavin and carries two electrons but only one hydrogen (on the N1 nitrogen). Compare this structure to flavins that have a nitrogen at position 5 and therefore carry two hydrogens (Section 4.2.1). (E) Factor B or HS-HTP (7-mercaptoheptanoylthreonine phosphate). (F) Coenzyme M (2-mercaptoethanesulfonic acid. (G) Methyl–coenzyme M. (H) Coenzyme $F_{430}$. This prosthetic group is a nickel–tetrapyrrole.

(THMP) to form formyl–THMP (reaction 2). Then a dehydration produces methenyl–THMP (reaction 3). The methenyl–THMP is reduced to methylene–THMP (reaction 4), which is reduced to methyl–THMP (reaction 5). The pathway diverges at the methyl level. One branch synthesizes acetyl–CoA in reactions quite similar to the synthesis of acetyl–CoA in the bacteria (Fig. 13.4), and the other branch synthesizes methane (Section 13.1.5). For the synthesis of acetyl–CoA, the methyl group is transferred to a corrinoid

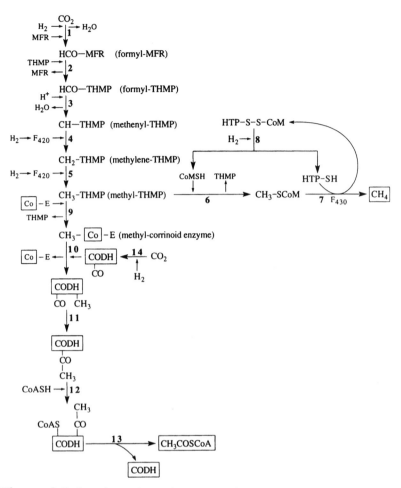

**Fig. 13.6** The acetyl–CoA pathway in methanogens. There are routes of entry of $CO_2$. In one sequence, $CO_2$ is bound to carbon monoxide dehydrogenase and is reduced to the oxidation level of carbon monoxide [CO] (reaction 14). The electron flow is from $H_2$ via a hydrogenase to ferredoxin to CODH. In a second sequence, $CO_2$ is attached to methanofuran (MFR) and becomes reduced to the level of formic acid (formyl–MFR) (reaction 1). The acceptor of electrons from $H_2$ in this reaction is not known. The formyl group is then transferred to a second carrier, tetrahydromethanopterin (THMP), to form formyl–THMP (reaction 2). The formyl–THMP is reduced to methyl–THMP ($CH_3$–THMP) (reactions 3–5). The $CH_3$–THMP donates the methyl group to a corrinoid enzyme for the synthesis of acetyl–CoA (reactions 9–13), or to CoMSH for the synthesis of methane (reactions 6–8). Enzymes: 1, formyl–MFR dehydrogenase (an iron–sulfur protein); 2, formyl–MFR:$H_4$MPT formyltransferase; 3, N5,N10-methenyl-$H_4$MPT cyclohydrolase; 4, N5,N10-methylene-$H_4$MPT dehydrogenase; 5, N5,N10-methylene-$H_4$MPT reductase; 6, N5-methyl-$H_4$MPT:CoMSH methyltransferase; 7, methyl-S–CoM reductase; 8, heterodisulfide reductase; 9, methyltransferase; 10–13, carbon monoxide dehydrogenase. THMP is also written as $H_4$MPT.

enzyme (Figs. 13.6 and 13.7, reaction 9). A second molecule of $CO_2$ is reduced by carbon monoxide dehydrogenase (CODH) to bound carbon monoxide [CO] (reaction 14). The CODH then catalyzes the synthesis of acetyl–CoA from [$CH_3$], [CO], and CoASH (reactions 11–13).

### 13.1.5 Methanogenesis from $CO_2$ and $H_2$

In addition to incorporating the [$CH_3$] into acetyl–CoA as described above, the methanogens can also reduce it to methane ($CH_4$) in a series of reactions not found

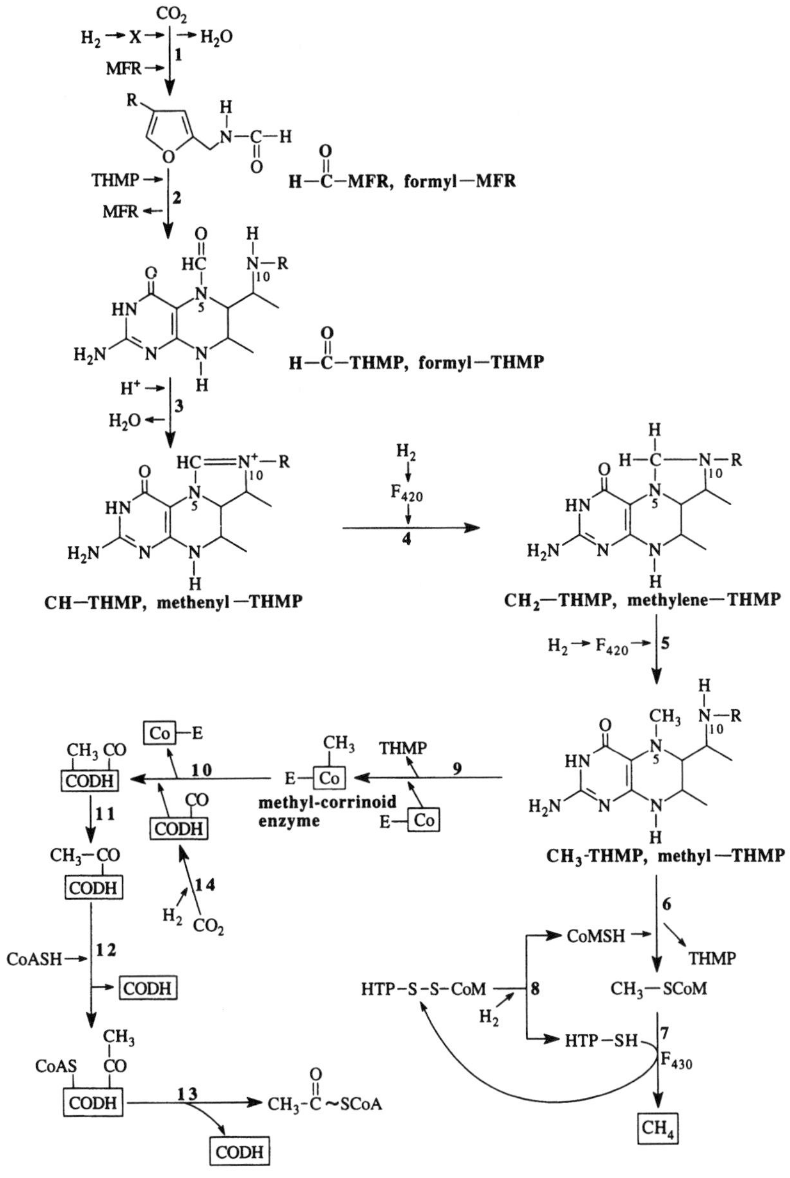

**Fig. 13.7** The acetyl–CoA pathway in methanogens. See the legend to Fig. 13.6.

among the bacteria. For methane synthesis the methyl group is transferred from CH$_3$–THMP to CoMSH to form CH$_3$.–SCoM (Figs. 13.6 and 13.7, reaction 6). The terminal reduction is catalyzed by the *methylreductase system* (reactions 7 and 8).[17] The methylreductase system has two components. One component is a methylreductase that reduces CH$_3$–SCoM to CH$_4$ and CoM–S–S–HTP. The electron donor for the methylreductase is HTP–SH.[18] A nickel-containing tetrapyrrole, F$_{430}$, is an electron carrier that is part of the methyl

reductase. The second component is an FAD-containing heterodisulfide reductase that reduces CoM–S–S–HTP to CoMSH and HTP–SH. The crystal structure of methylreductase has been published and a model for how it works has been proposed.[19]

### Energy conservation during methanogenesis

The production of methane yields ATP and is the only means of ATP formation for the methanogens. Probably, energy is

349

conserved for ATP synthesis at two sites in the pathway of methanogenesis. It appears that electron transfer to CoM–S–S–HTP (Fig. 13.7, reaction 8) occurs in the membranes and is accompanied by the generation of a $\Delta p$, which drives ATP synthesis via a $H^+$-translocating ATP synthase. (See Note 20 for experimental evidence.) The mechanism of the generation of the $\Delta p$ is not known, but could include $H_2$ oxidation on the outer membrane surface, depositing protons on the outside, followed by electrogenic movement of electrons across the membrane to CoM–S–S–HTP on the inside surface. This is a tactic often used by bacteria to generate a $\Delta p$ coupled to periplasmic oxidations. For example, recall the periplasmic oxidation of $H_2$ by *Wolinella succinogenes* coupled to the cytoplasmic reduction of fumarate described in Section 4.7.4, or the reduction of sulfate by *Desulfovibrio* described in Section 12.2.2. There may also be a second energy-coupling site. The transfer of the methyl group from $CH_3$–THMP to CoMSH (Fig. 13.7, reaction 6) is an exergonic reaction accompanied by the uptake of $Na^+$ into inverted vesicles, creating a primary sodium motive force (inside positive). How might the sodium motive force be used? The sodium motive force probably drives ATP synthesis in whole cells, since it has been demonstrated to energize the synthesis of ATP in washed inverted vesicles made from *Methanosarcina mazei Gö1*. There appear to be two different ATP synthases present, one coupled to protons and the other coupled to sodium ions.[21] It has been speculated that the sodium motive force created during methyl transfer to CoMSH may also be used to drive the formation of formyl–MFR from free $CO_2$ using $H_2$ as the reductant (Fig. 13.7, reaction 1). The latter reaction has a standard free energy change of approximately +8 kJ/mol.

### Unique coenzymes in the Archaea

As discussed in Section 1.1.1, there are several unique aspects to the biochemistry of the Archaea. For example, there are several coenzymes in archaea that are not found in bacteria or eucarya. These are represented by five coenzymes used in methanogenesis. They are: methanofuran (formerly called

CDR, carbon dioxide reduction enzyme), methanopterin, coenzyme M (CoSM), $F_{430}$, and HS–HTP, also known as factor B or coenzyme B (Fig. 13.5). Coenzyme M, $F_{430}$, and HS–HTP are found only in methanogens. The other coenzymes occur in other archaea. The electron carrier, $F_{420}$, also exists in bacteria and eucarya.

### 13.1.6 Methanogenesis from acetate

Species of methanogens within the genera *Methanosarcina* and *Methanotrhix* can use acetate as a source of carbon and energy.[22] In fact, acetate accounts for approximately two-thirds of the biologically produced methane. (See Note 23.) Those methanogens that convert acetate to methane and $CO_2$ carry out a dismutation of acetate in which the carbonyl group is oxidized to $CO_2$, and the electrons are used to reduce the methyl group to methane according to the following overall reaction:

$$CH_3CO_2^- + H^+ \longrightarrow CH_4 + CO_2$$

The acetate is first converted to acetyl–CoA. *Methanosarcina* uses acetate kinase and phsphotransacetylase to make acetyl–CoA from acetate.

$$\text{acetate} + \text{ATP} \xrightarrow{\text{acetate kinase}} \text{acetyl-phosphate}$$
$$+ \text{ADP}$$

$$\text{acetyl-phosphate} + \text{CoA} \xrightarrow{\text{phosphotransacetylase}}$$

$$\text{acetyl–CoA} + \text{inorganic phosphate}$$

*Methanotrhix* uses acetyl–CoA synthetase to make acetyl–CoA.

$$\text{acetate} + \text{ATP} \xrightarrow{\text{acetyl–CoA synthetase}}$$

$$\text{acetyl-AMP} + \text{pyrophosphate}$$

$$\text{acetyl–AMP} + \text{CoA} \longrightarrow \text{acety–CoA} + \text{AMP}$$

Reactions similar to those of the acetyl–CoA pathway in methanogens convert acetyl–CoA to methane and $CO_2$, deriving ATP from the process. Examine Fig. 12.6, beginning with acetyl–CoA. In reaction 12, acetyl–CoA combines with carbon monoxide dehydrogenase (CODH), which catalyzes the breakage of the C–S bond and the release of CoA. In reaction 11, the CODH catalyzes the breakage of the C–C, bond forming the carbonyl [CO] and methyl [CH$_3$] groups. Then the carbonyl group is oxidized to CO$_2$ (reaction 13) and the methyl group is transferred to a corrinoid enzyme (reaction 10). (Note that the electrons removed from the carbonyl group are not released in hydrogen gas, as implied in reaction 13, but are used to reduce the methyl group to methane.) The methyl group is transferred to tetrahydromethanopterin (THMP) (reaction 9).[24] [Acetate-grown cells of *Methanosarcina* species also contain large amounts of a derivative of THMP, tetrahydrosarcinapterin (H$_4$SPT), which has also been reported to serve as a methyl group carrier during methanogenesis from acetate.[25]] The methyl group is then transferred to coenzyme M (CoMSH) (reaction 6) and reduced to methane using the electrons derived from the oxidation of the carbonyl group (reaction 7). The electron donors for the reduction of the methyl group are HTP–SH and CH$_3$–S–CoM. Each of these contribute one electron from their respective sulfur atoms, and the oxidized product is the heterodisulfide, HTP–S–S–CoM. The reduction of HTP–S–S–CoM to CoMSH and HTP–SH (reaction 8) is coupled to the oxidation of the carbonyl group (reaction 13). The electron transport pathway that links the oxidation of the carbonyl group with the reduction of HTP–S–S–CoM is thought to generate ATP, presumably via a $\Delta p$.[19]

Methanogenesis from acetate is summarized below:

$$CH_3CO\text{-}CoA \longrightarrow [CH_3] + [CO] + CoA \quad (7)$$

$$[CO] + H_2O \longrightarrow CO_2 + 2[H] \quad (8)$$

$$[CH_3] + 2[H] + ADP + P_i \longrightarrow CH_4 + ATP \quad (9)$$

It was reported that ATP can also be made during the oxidation of [CO] to CO$_2$ and H$_2$.[26]

### 13.1.7 Incorporation of acetyl–CoA into cell carbon by methanogens

The glyoxylate cycle, one of the means of incorporating net acetyl–CoA into cell material, is not present in prokaryotes that use the acetyl–CoA pathway. How do they grow on the acetyl–CoA? The acetyl–CoA must be converted to phosphoenolpyruvate, which feeds into an incomplete citric acid pathway and into gluconeogenesis. The enzymes to make phosphoenolpyruvate from acetyl–CoA are widespread in anaerobes. A suggested pathway for acetyl–CoA incorporation by methanogens is discussed below.

Figure 13.8 shows a proposed pathway for acetyl–CoA incorporation by methanogens.[27] The first reaction is the carboxylation of acetyl–CoA to form pyruvate, a reaction catalyzed by *pyruvate synthase*. The pyruvate is then phosphorylated to form phosphoenolpyruvate using *PEP synthetase*. The phosphoenolpyruvate has two fates. Some of it enters the gluconeogenic pathway, and some is carboxylated to oxaloacetate via *PEP carboxylase*. In *M. thermoautotrophicum* the oxaloacetate is reduced to $\alpha$-ketoglutarate via an incomplete reductive citric acid pathway. That is to say, the reactions between citrate and $\alpha$-ketoglutarate do not take place. *M. thermoautotrophicum* lacks citrate synthase and so must make its $\alpha$-ketoglutarate this way. However, another methanogen, *M. barkeri*, does have citrate synthase and synthesizes $\alpha$-ketoglutarate in an incomplete oxidative pathway via citrate and isocitrate. Thus no methanogen seems to have a complete citric acid pathway.[28]

### 13.1.8 Using the acetyl–CoA pathway to oxidize acetate to CO$_2$ anaerobically

Several anaerobes can oxidize acetate to CO$_2$ without the involvement of the citric acid cycle. They do this by operating the acetyl–CoA pathway in reverse. These include some of the sulfate reducers that grow heterotrophically on acetate and oxidize it to CO$_2$.[29] To do this, they start with reactions 4 and 3 to make acetyl–CoA (Fig. 13.4) $\beta$ and then reverse reactions 14 through 11. This produces CH$_3$–[Co-E] and CO–[CODH]. In reactions 15, CO–[CODH] is oxidized to CO$_2$. The CH$_3$ is oxidized to CO$_2$ via reactions 10 through 5. In this way, both carbons in acetate are converted to CO$_2$. A $\Delta p$ is made during electron transport. (However, it may be that the enzymes involved in the oxidative acetyl–CoA pathway are not identical to those of the reductive acetyl–CoA pathway that

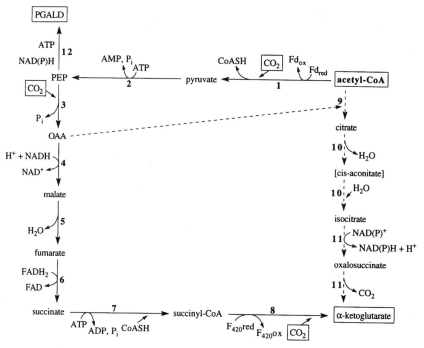

**Fig. 13.8** Assimilation of acetyl–CoA into cell material in methanogens. The acetyl–CoA is carboxylated to form pyruvate, which is then phosphorylated to make phosphoenolpyruvate (PEP). In *M. thermoautotrophicum,* the PEP is carboxylated to oxaloacetate (OAA), which is reduced to α-ketoglutarate, or it is reduced to phosphoglyceraldehyde. *M. barkeri* synthesizes α-ketoglutarate by a different pathway. That is to say, the oxaloacetate condenses with another acetyl–CoA to form citrate, and the citrate is then oxidized to α-ketoglutarate via aconitate, isocitrate, and oxalosuccinate (dotted lines). Enzymes: 1, pyruvate synthase; 2, PEP synthetase; 3, PEP carboxylase; 4, malate dehydrogenase; 5, fumarase; 6, fumarate reductase; 7, succinyl–CoA synthetase; 8, α-ketoglutarate synthase; 9, citrate synthase; 10, aconitase; 11, isocitrate dehydrogenase; 12, enolase, mutase, phosphoglycerate kinase, triosephosphate dehydrogenase.

operates in autotrophs.) Some of the group II sulfate reducers (complete oxidizers) (e.g., *Desulfotomaculum acetoxidans*) use this pathway. Other group II sulfate reducers use the reductive tricarboxylic acid pathway to oxidize acetate (Section 13.1.6). Archaea sulfate reducers (e.g., *Archaeoglobus fulgidus*) may also use the acetyl–CoA pathway to oxidize acetate to $CO_2$ anaerobically.

### 13.1.9 The reductive tricarboxylic acid pathway (reductive citric acid cycle)

Bacteria that use this pathway are strict anaerobes belonging to the genera *Desulfobacter* and *Chlorobium*, and the aerobic *Hydrogenobacter*.[30] (See Note 31.) The pathway is also present in the Archaea and is thought

to have evolved earlier than the Calvin cycle as an autotrophic carbon dioxide fixation pathway.[32] The overall reaction is the synthesis of one mole of oxaloacetate from four moles of carbon dioxide. In the reductive tricarboxylic acid pathway, phosphoenolpyruvate (PEP) is carboxylated to form oxaloacetate (OAA) using the enzyme *PEP carboxylase* (Fig. 13.9, reaction 1). The oxaloacetate is reduced to succinate, which is derivatized to succinyl–CoA. The succinyl–CoA is carboxylated to α-ketoglutarate (the precursor to glutamate). This is a reductive carboxylation requiring reduced ferredoxin, and carried out by *α-ketoglutarate synthase*. The α-ketoglutarate is carboxylated to isocitrate, which is converted to citrate. Therefore, this is really a reversal of the citric acid pathway, substituting *fumarate reductase* for succinate

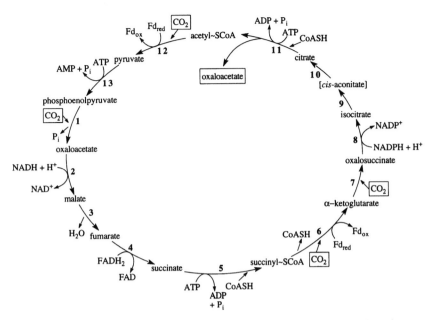

**Fig. 13.9** The reductive tricarboxylic acid pathway. Enzymes: 1, PEP carboxylase; 2, malate dehydrogenase; 3, fumarase; 4, fumarate reductase; 5, succinyl–CoA synthetase (succinate thiokinase); 6, α-ketoglutarate synthase; 7, 8 isocitrate dehydrogenase; 9, 10 aconitase; 11, ATP–citrate lyase; 12, pyruvate synthase; 13, PEP synthetase.

dehydrogenase and *α-ketoglutarate synthase* for α-ketoglutarate dehydrogenase. Thus far, one mole of phosphoenolpyruvate (C$_3$) has been converted to one mole of citrate (C$_6$). The citrate is split by a special enzyme, called *ATP-dependent citrate lyase*, to acetyl–CoA (C$_2$) and oxaloacetate (C$_4$). (The ATP-dependent citrate lyase is found in eukaryotic cells but is rare in prokaryotes. See Note 33. Most prokaryotes use the citrate synthase reaction, which proceeds only in the direction of citrate.) The oxaloacetate can be used for growth. The acetyl–CoA is used to regenerate the phosphoenolpyruvate as follows. The acetyl–CoA (C$_2$) is reductively carboxylated to pyruvate (C$_3$) using reduced ferredoxin and the enzyme *pyruvate synthase*. There is a thermodynamic reason why reduced ferredoxin and not NADH is the electron donor. NADH is not a sufficiently strong reductant to reduce acetyl–CoA and CO$_2$ to pyruvate. Reduced ferredoxin has a potential more electronegative than that of NADH and is therefore a stronger reductant. The pyruvate is then converted to phosphoenolpyruvate using

the enzyme *PEP synthetase*, thus regenerating the phosphoenolpyruvate. This discussion and Section 13.1.2 emphasize the importance of pyruvate synthase and PEP synthetase for anaerobic growth on acetate.

Note that to reverse the citric acid cycle, three new enzymes are required:

1. *Fumarate reductase* replaces succinate dehydrogenase. This commonly occurs in other prokaryotes that synthesize succinyl–CoA from oxaloacetate under anaerobic conditions.

2. *α-Ketoglutarate synthase* replaces the NAD$^+$-linked α-ketoglutarate dehydrogenase.

3. *ATP-dependent citrate lyase* replaces citrate synthase.

In addition, pyruvate synthase replaces pyruvate dehydrogenase, and PEP synthetase replaces pyruvate kinase.

Summary of the reductive tricarboxylic acid pathway:

$$CO_2 + PEP \longrightarrow OAA + P_i$$

$$OAA + NADH + H^+ + FADH_2 + Fd(red) + ATP + CO_2 \longrightarrow \alpha kg + NAD^+ + FAD + Fd(ox) + ADP + P_i$$

$$\alpha kg + NADPH + H^+ + CO_2 \longrightarrow citrate$$

$$citrate + ATP + CoASH \longrightarrow OAA + acetyl-CoA + ADP + P_i$$

$$acetyl-CoA + Fd(red) + CO_2 \longrightarrow pyruvate + Fd(ox)$$

$$pyruvate + ATP \longrightarrow PEP + AMP + P_i$$

---

$$4CO_2 + 2NAD(P)H + 2H^+ + 2Fd(red) + FADH_2 + 3ATP \longrightarrow OAA + 2NAD(P)^+ + FAD + 2Fd(ox) + 2ADP + AMP + 4P_i$$

### Anaerobic acetate oxidation by reversal of the reductive tricarboxylic acid pathway

As mentioned previously, some group II sulfate reducers reverse the reductive tricarboxylic acid pathway to oxidize acetate to $CO_2$.[28] An example is *Desulfobacter postgatei*. As shown in Fig. 13.9, by reversing reactions 11 through 6, acetyl–CoA is oxidized to $CO_2$. The oxaloacetate used in reaction 11 is regenerated in reactions 5 through 2. Acetyl–CoA is made from acetate by transferring the CoA from succinyl–CoA. Hence, reaction 5 is replaced by a CoA transferase.

### 13.2 Growth on $C_1$ Compounds Other than $CO_2$: The Methylotrophs

Many aerobic bacteria can grow on compounds other than $CO_2$ that do not have carbon–carbon bonds. These bacteria are called *methylotrophs*.[34, 35] Compounds used for methylotrophic growth include single carbon compounds such as methane ($CH_4$), methanol ($CH_3OH$), formaldehyde (HCHO), formate (HCOOH), and methylamine ($CH_3NH_2$), as well as multicarbon compounds without C–C bonds such as trimethylamine [$(CH_3)_3N$], dimethyl ether [$(CH_3)_2O$], and dimethyl carbonate ($CH_3OCOOCH_3$). These compounds appear in the natural habitat as a result of fermentations and the breakdown of plant and animal products and pesticides. (See Note 36 for a description of where these compounds are found.)

The methylotrophs are divided according to whether they can also grow on multicarbon compounds. Those that *cannot* are called *obligate* methylotrophs. The obligate methylotrophs that grow on methanol or methylamine *but not on methane* are aerobic gram-negative bacteria that belong to two genera, *Methylophilus* and *Methylobacillus*. (Some strains of *Methylophilus* can use glucose as a sole carbon and energy source.[37]) The obligate methylotrophs that grow on methane or methanol are called *methanotrophs*. (See Note 38.) They are gram-negative and fall into five genera, *Methylomonas*, *Methylococcus*, *Methylobacter*, *Methylosinus*, and *Methylocystis*. All the methanotrophs form extensive intracellular membranes and resting cells, either cysts or exospores. The intracellular membranes are postulated to be involved in methane oxidation, since they are not present in methylotrophs that grow on methanol but not on methane.

Those methyltrophs that grow on *either* $C_1$ compounds (methanol or methylamine) or multicarbon compounds are called *facultative* methylotrophs. The facultative methylotrophs are found in many genera and consist of both gram-positive and gram-negative bacteria. They include species belonging to the genera *Bacillus*, *Acetobacter*, *Mycobacterium*, *Arthrobacter*, *Mycobacterium*, *Hyphomicrobium*, *Methylobacterium*, and *Nocardia*. Some species of *Mycobacterium* can grow on methane as well as on methanol or multicarbon compounds.

Methylotrophs assimilate the $C_1$ carbon source via either the ribulose monophosphate (RuMP) pathway or the serine pathway. The RuMP pathway assimilates formaldehyde into cell material, whereas the serine pathway assimilates carbon dioxide *and* formaldehyde. (See Note 39.) There also exist bacteria that grow on methanol and oxidize it to $CO_2$, which is assimilated via the ribulose bisphosphate (RuBP) pathway (Calvin cycle). These have been called "pseudomethylotrophs" or autotrophic methylotrophs.

## 13.2.1 Growth on methane

Methane is oxidized to $CO_2$ in a series of four reactions. The first oxidation is to methanol using a mixed-function oxidase called methane monooxygenase. There are two different methane monooxygenases: one in the membrane, and one soluble. All methanotrophs have the membrane-bound enzyme. It is not yet known whether the soluble monooxygenase, which is the better characterized enzyme, is present in all methanotrophs.

$$CH_4 + NADH + H^+ + O_2$$
$$\longrightarrow CH_3OH + NAD^+ + H_2O$$

In the above reaction, one atom of oxygen is incorporated into methanol, and the other atom is reduced to water using NADH as the reductant. The second reaction is the oxidation of methanol to formaldehyde catalyzed by methanol dehydrogenase, which has as its prosthetic group a quinone called pyrroloquinoline quinone (PQQ) (Fig. 13.10).

$$CH_3OH + PQQ \longrightarrow HCHO + PQQH_2$$

In the third reaction the formaldehyde is oxidized to formate by formaldehyde dehydrogenase:

$$HCHO + NAD^+ + H_2O$$
$$\longrightarrow HCOOH + NADH + H^+$$

And the formate is oxidized to $CO_2$ by formate dehydrogenase:

$$HCOOH + NAD^+ \longrightarrow CO_2 + NADH + H^+$$

Because the formaldehyde and formate dehydrogenases are soluble enzymes that use $NAD^+$, they are probably located in the cytoplasm. Of the two NADHs produced, one is reutilized in the monooxygenase reaction, and the second is fed into the respiratory chain in

the usual way. However, methanol dehydrogenase is a periplasmic enzyme in gram-negative bacteria. It is thought that the methanol diffuses into the periplasm, where it is oxidized by the dehydrogenase. The electrons are transferred to periplasmic cytochromes c that in turn transfer the electrons to a membrane cytochrome oxidase, which probably pumps protons out of the cell during inward flow of electrons to oxygen. A $\Delta p$ is established by the inward flow of electrons, the outward pumping of protons by the cyt aa$_3$ oxidase, and the release of protons in the periplasm and consumption in the cytoplasm. (See Section 4.7.2 for a description of electron transport and the generation of a $\Delta p$ during methanol oxidation in *P. denitrificans*.) A small number of gram-positive bacteria are also methylotrophic. However, the biochemistry of methanol oxidation may not be the same as in gram-negative bacteria.

Some of the formaldehyde produced during methane oxidation is incorporated into cell material rather than oxidized to $CO_2$. Depending upon the particular bacterium, one of two formaldehyde fixation pathways is used. The two pathways are the serine pathway and the ribulose monophosphate cycle.

## The serine pathway

The serine pathway produces acetyl–CoA from formaldehyde and carbon dioxide. The pathway is shown in Fig. 13.11. The formaldehyde is incorporated into glycine to form serine in a reaction catalyzed by serine hydroxymethylase (reaction 1). In this reaction, the formaldehyde is first attached to tetrahydrofolic acid to form methylene–THF. (Tetrahydrofolic reactions are described in Section 9.2.4.) Methylene–THF then donates the C1

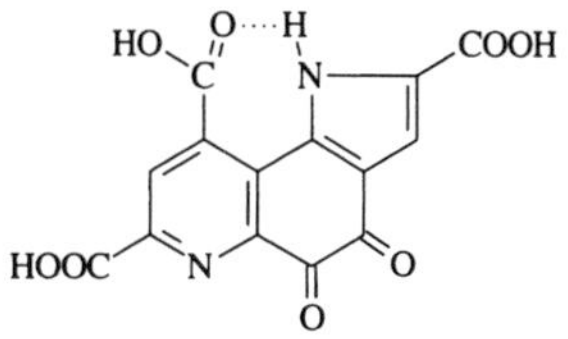

**Fig. 13.10** Structure of pyrroloquinoline–quinone (PQQ). *Source:* From Gottschalk, G. 1986. *Bacterial Metabolism.* Springer-Verlag, Berlin.

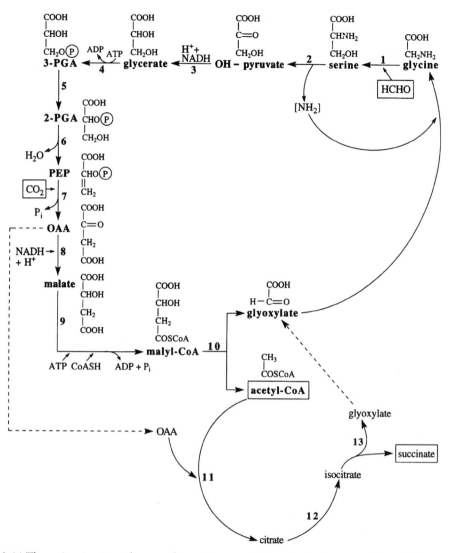

**Fig. 13.11** The serine–isocitrate lyase pathway. Enzymes: (1), serine hydroxymethylase; (2), transaminase; (3), hydroxypyruvate reductase; (4), glycerate kinase; (5), phosphoglycerate mutase; (6), enolase; (7), PEP carboxylase; (8), malate dehydrogenase; (9), malyl–CoA synthetase; (10), malyl–CoA lyase; (11), citrate synthase; (12), *cis*-aconitase; (13), isocitrate lyase.

unit to glycine to form serine, regenerating THF. The serine is then converted to hydroxypyruvate via a transaminase that aminates glyoxylate, regenerating glycine (reaction 2). Hydroxypyruvate is reduced to glycerate (reaction 3), which is phosphorylated to 3-phosphoglycerate (reaction 4). The 3-phosphoglycerate is then converted to 2-phosphoglycerate (reaction 5), which is dehydrated to phosphoenolpyruvate (reaction 6). The phosphoenolpyruvate is carboxylated to oxaloacetate (reaction 7). The oxaloacetate is

reduced to malate (reaction 8). Malyl–CoA synthetase then converts malate to malyl–CoA (reaction 9), which is split by malyl–CoA lyase to acetyl–CoA and glyoxylate (reaction 10). Thus the glyoxylate is regenerated, and the product is acetyl–CoA. So far, the following has happened:

$$HCHO + CO_2 + 2\ NADH + 2\ H^+$$
$$+ 2\ ATP + CoASH \longrightarrow$$
$$acetyl\text{–}CoA + 2\ NAD^+ + 2\ ADP + 2\ P_i$$

But how is acetyl–CoA incorporated into cell material? What happens in *some* methylotrophs is that the serine pathway goes around *a second time* to generate a *second* oxaloacetate. Then the second oxaloacetate condenses with the acetyl–CoA to form citrate (reaction 11). The citrate isomerizes to isocitrate (reaction 12). The isocitrate is cleaved by isocitrate lyase to form succinate and glyoxylate (reaction 13). The succinate can be assimilated into cell material via oxaloacetate and PEP (Section 8.12). The second glyoxylate can be used for the second round of the serine pathway, which produces the second oxaloacetate. Therefore, the serine pathway and the isocitrate lyase pathway (called the serine–isocitrate lyase pathway) can be be described by the following overall reaction:

$$2 \text{ HCHO} + 2 \text{ CO}_2 + 3 \text{ NADH} + 3 \text{ H}^+$$
$$+ 3 \text{ ATP} \longrightarrow$$
$$\text{succinate} + 3 \text{ NAD}^+ + 3 \text{ ADP}$$
$$+ 3 \text{ P}_i + 2 \text{ H}_2\text{O}$$

Alternatively, the succinate can be converted to the second oxaloacetate via fumarate and malate, and the second glyoxylate can be converted via the serine pathway to 3-phosphoglycerate, which is assimilated into cell material.

However, only a few methylotrophs have isocitrate lyase, and it has not been established how the acetyl–CoA is converted to glyoxylate or otherwise assimilated in those strains lacking isocitrate lyase. The reactions are summed below:

$$2 \text{ CH}_2\text{O} + 2 \text{ glycine} \longrightarrow 2 \text{ serine} \tag{1}$$

$$2 \text{ serine} + 2\text{glyoxylate} \longrightarrow 2 \text{ glycine}$$
$$+ 2 \text{ hydroxypyruvate} \tag{2}$$

$$2 \text{ hydroxypyruvate} + 2 \text{ NADH} + 2\text{H}^+$$
$$\longrightarrow 2 \text{ glycerate} + 2\text{NAD}^+ \tag{3}$$

$$2 \text{ glycerate} + 2 \text{ ATP} \longrightarrow 2 \text{ 3-PGA}$$
$$+ 2 \text{ ADP} + 2 \text{ P}_i \tag{4}$$

$$2 \text{ 3-PGA} \longrightarrow 2 \text{ 2-PGA} \tag{5}$$

$$2 \text{ 2-PGA} \longrightarrow 2 \text{ PEP} + 2\text{H}_2\text{O} \tag{6}$$

$$2 \text{ PEP} + 2 \text{ CO}_2 \longrightarrow 2 \text{ OAA} + 2\text{P}_i \tag{7}$$

$$\text{OAA} + \text{NADH} + \text{H}^+ \longrightarrow \text{malate}$$
$$+ \text{NAD}^+ \tag{8}$$

$$\text{malate} + \text{ATP} + \text{CoASH} \longrightarrow \text{malyl–CoA}$$
$$+ \text{ADP} + \text{P}_i \tag{9}$$

$$\text{malyl–CoA} \longrightarrow \text{glyoxylate} + \text{acetyl–CoA} \tag{10}$$

$$\text{acetyl–CoA} + \text{OAA} \longrightarrow \text{citrate} \tag{11}$$

$$\text{citrate} \longrightarrow \text{isocitrate} \tag{12}$$

$$\text{isocitrate} \longrightarrow \text{glyoxylate} + \text{succinate} \tag{13}$$

$$2 \text{ CH}_2\text{O} + 2 \text{ CO}_2 + 3 \text{ NADH} + 3\text{H}^+ + 3 \text{ ATP}$$
$$\longrightarrow \text{succinate} + 3 \text{ NAD}^+ + 3 \text{ ADP}$$
$$+ 3 \text{ P}_i + 2 \text{ H}_2\text{O}$$

## *The ribulose monophosphate cycle*

There are several methanotrophs that use the *ribulose monophosphate* (RuMP) pathway instead of the serine pathway for formaldehyde assimilation.[40] The pathway is shown in Fig. 13.12. It is convenient to divide the RuMP pathway into three stages. Stage 1 begins with condensation of formaldehyde with ribulose-5-phosphate to form hexulose-6-phosphate (reaction 1). The reaction is catalyzed by hexulose phosphate synthase. The hexulose phosphate is then isomerized to fructose-6-phosphate by hexulose phosphate isomerase (reaction 2). In order to synthesize a three-carbon compound (e.g., dihydroxyacetone phosphate or pyruvate) from formaldehyde (a one-carbon compound), Stage 1 must be repeated three times.

In Stage 2, one of the three fructose-6-phosphates is split into two C3 compounds. There are two pathways for the cleavage, depending upon the organism. In one pathway, fructose-6-phosphate is phosphorylated to fructose-1,6-bisphosphate, which is cleaved via the fructose-1,6-bisphosphate aldolase to phosphoglyceraldehyde and dihydroxyacetone phosphate. This is called the FBPA pathway, for fructose bisphosphate aldolase. If this route is taken, then the net product of the pathway is dihydroxyacetone phosphate. In the second pathway, fructose-6-phosphate is isomerized to glucose-6-phosphate, which is oxidatively cleaved to phosphoglyceraldehyde

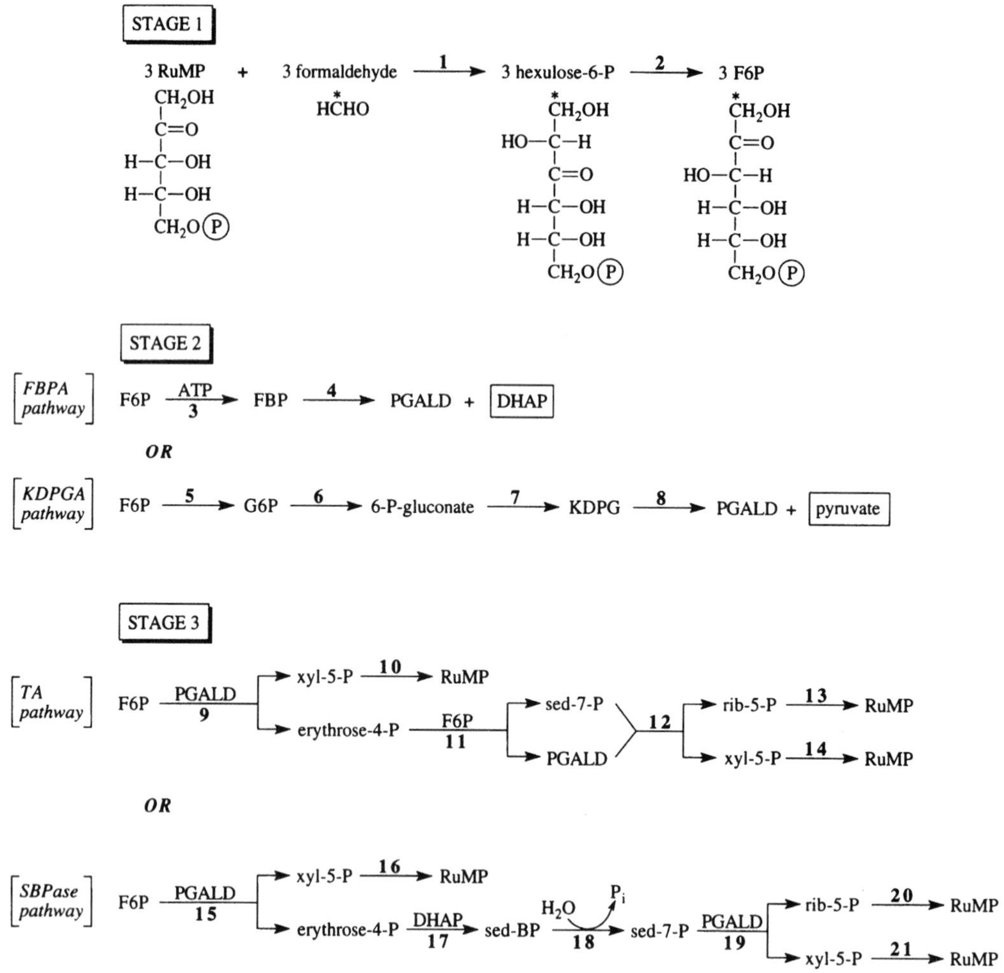

Fig. 13.12 The ribulose–monophosphate cycle. Enzymes: 1, hexulose-6-phosphate synthetase; 2, hexulose-6-phosphate isomerase; 3, phosphofructokinase; 4, fructose-1,6-bisphosphate (FBP) aldolase; 5, glucose phosphate isomerase; 6, glucose-6-phosphate dehydrogenase; 7, 6-phosphogluconate dehydratase; 8, 2-keto-3-deoxy-6-phosphogluconate (KDPG) aldolase; 9, 12, 15, 19, transketolase; 10, 14, 16, 21, ribulose-5-phosphate epimerase; 11, transaldolase; 13, 20 ribose-5-phosphate isomerase; 17, sedoheptulose-1,7-bisphosphate aldolase; 18, sedoheptulose-1,7-bisphosphatase. Abbreviations: RuMP, ribulose-5-phosphate; F6P, fructose-6-phosphate; FBP, fructose-1,6-bisphosphate; PGALD, phosphoglyceraldehyde; DHAP, dihydroxyacetone phosphate; KDPG, 2-keto-3-deoxy-6-phosphogluconate; xyl-5-P, xylulose-5-phosphate; rib-5-P, ribose-5-phosphate; sed-BP, sedoheptulose-1,7-bisphosphate.

and pyruvate using the enzymes of the Entner–Doudoroff pathway (Section 8.6). This is called the KDPGA pathway, for the 2-keto-3-deoxy-6-phosphogluconate aldolase. If this route is taken, then the net product of the pathway is pyruvate. Both pathways also produce phosphoglyceraldehyde.

Stage 3 is a sugar rearrangement stage during which the phosphoglyceraldehyde produced in Stage 2 and two fructose-6-phosphates produced in Stage 1 are used to regenerate the three ribulose-5-phosphates. In some bacteria the rearrangements take place using the pentose phosphate cycle enzymes (called the TA pathway, for transaldolase), whereas in other bacteria the rearrangements use the enzymes of the closely related Calvin cycle (called the SBPase pathway, for sedoheptulose-1,7-bisphosphate aldolase). Taking into consideration the alternative pathways in Stages 2 and 3, there are four different variations of the RuMP cycle that might occur in the methylotrophs. The obligate methanotrophs (*Methylococcus* and *Methylomonas*) and obligate methylotrophs (*Methylophilus*, *Methylobacillus*) that have been examined all use the KDPGA mode of cleavage and the TA pathway of sugar arrangement. The facultative methylotrophs thus far examined use the FBPA mode of cleavage. Some use the SBPase sugar rearrangment pathway, and some use the TA pathway. Use of the KDPG aldolase and the SBPase pathways in combination has not yet been found.

## 13.3 Summary

There are three characterized $CO_2$-fixation pathways used for autotrophic growth: the Calvin cycle, the acetyl–CoA pathway, and the reductive tricarboxylic acid pathway. The Calvin cycle is the only one in the aerobic biosphere. It is present in green plants, algae, cyanobacteria, chemoautotrophs, and most photosynthetic bacteria. The Calvin cycle uses the transketolase and isomerization reactions of the pentose phosphate pathway and several glycolytic reactions to reduce $CO_2$ to phosphoglyceraldehyde. It bypasses the transaldolase reaction and synthesizes sedoheptulose-7-P via an aldolase and an irreversible phosphatase and proceeds only in the direction of pentose phosphate. The pathway has two unique enzymes, RuMP kinase and RuBP carboxylase.

The acetyl–CoA pathway is widespread among anaerobes and occurs in both the archaea and the bacteria. However, the archaea use different coenzymes to carry the C1 units. The pathway is used not only reductively for autotrophic growth but can also be used for anaerobic oxidation of acetate to $CO_2$ (e.g., by certain group II sulfate reducers). An analogous pathway is used by methanogens to form $CH_4$ and $CO_2$ from acetate. A key enzyme in the acetyl–CoA pathway is carbon monoxide dehydrogenase (CODH), which is capable of reducing $CO_2$ to the level of carbon monoxide and catalyzing its condensation with bound methyl and CoA to form acetyl–CoA. The bound methyl is made by reducing a second molecule of $CO_2$ via a separate series of enzymatic reactions and transferring the methyl group to the CODH. The reduction of $CO_2$ to bound methyl also takes place during methanogenesis. In methanogenesis, the bound methyl is further reduced to methane, an energy-yielding reaction, rather than transferred to CODH. Growth on acetyl–CoA requires the synthesis of phosphoenolpyruvate, because the glyoxylate pathway is not present in these organisms. It involves a ferredoxin-linked carboxylation of acetyl–CoA to pyruvate catalyzed by pyruvate synthase, followed by the phosphorylation of pyruvate to phosphoenolpyruvate via the PEP synthetase. The incorporation of phosphoenolpyruvate into cell material takes place using reactions common to heterotrophs.

The third $CO_2$ fixation pathway has been found among the photosynthetic green sulfur bacteria, *Hydrogenobacter*, and *Desulfobacter* species. The pathway is a reductive citric acid pathway and synthesizes oxaloacetate from four moles of $CO_2$. There are two carboxylation reactions. One carboxylation is common, even among heterotrophs. It is the carboxylation of phosphoenolpyruvate to form oxaloacetate catalyzed by PEP carboxylase. The oxaloacetate is reduced to succinyl–CoA. The second carboxylation is found only in some strict anaerobes. It is the ferredoxin-linked carboxylation of succinyl–CoA to form $\alpha$-ketoglutarate catalyzed by $\alpha$-ketoglutarate synthase. The $\alpha$-ketoglutarate is converted to citrate via a reversal of the reactions of the citric acid cycle, and the citrate is cleaved to oxaloacetate and acetyl–CoA via an ATP-dependent citrate lyase. The bacteria can incorporate the oxaloacetate into cell carbon. The phosphoenolpyruvate is regenerated from acetyl–CoA via its carboxylation to pyruvate using pyruvate synthase and phosphorylation of the latter to phosphoenolpyruvate using PEP synthetase. Therefore, the bacteria are synthesizing oxaloacetate by carboxylating phosphoenolpyruvate, and regenerating the phosphoenolpyruvate via a reductive citric acid pathway. The pathway can also operate in the oxidative direction and oxidize acetate to $CO_2$. This occurs in some group II sulfate reducers.

There are many bacteria that grow aerobically on $C_1$ compounds such as methane or methanol. They oxidize the $C_1$ compounds to $CO_2$, deriving ATP from the respiratory pathway in the usual way, and incorporate the rest of the $C_1$ at the level of formaldehyde, into cell carbon. Bacteria that do this are called methylotrophs. A subclass of methylotrophs are those bacteria that are able to grow on methane. These are called methanotrophs. Two pathways for formaldehyde incorporation have been found in the bacteria. They are the serine–isocitrate lyase pathway, which incorporates both formaldehyde and carbon dioxide, and the ribulose monophosphate cycle, which incorporates only formaldehyde.

The use of $C_1$ compounds is an important part of the carbon cycle. The methane produced by methanogens escapes into the aerobic atmosphere and is transformed back into carbon dioxide by aerobic methane oxidizers. The carbon dioxide is reduced to organic carbon in the aerobic environments by both photosynthetic eukaryotes. and both photosynthetic and chemoautotrophic prokaryotes. It is also reduced to organic carbon anaerobically by photosynthetic prokaryotes, acetogens, and methanogens. Methanogenesis occurs in a variety of anaerobic environments, including swamps and marshes, the rumen, anaerobic microenvironments in the soil, lake muds, rice paddies, and the intestine of termites. Indeed, close to 70% of the atmospheric methane is produced by methanogens. This is approximately $10^8$ tons of methane per year. (The other 30% or so originates from abiogenic sources such as biomass burning and coal mines.) When one considers not only the vast amounts of methane produced by the methanogens, but also the fact that they are responsible for cycling much of the carbon into a gaseous form for reutilization by the methane oxidizers and eventually the $CO_2$-fixing organisms, it is clear that they play a critical life-supporting role in the biosphere.

## Study Questions

1. Contrast the number of ATPs required to fix three moles of $CO_2$ into phosphoglyceraldehyde using the Calvin cycle, the acetyl–CoA pathway, and the reductive carboxylic acid pathway.

2. Describe the role of carbon monoxide dehydrogenase in methane and carbon dioxide production from acetate, and in acetyl–CoA synthesis from $CO_2$.

3. Why can anaerobes carboxylate acetyl–CoA and succinyl–CoA but aerobes cannot?

## NOTES AND REFERENCES

1. Autotrophic bacteria are those that use carbon dioxide as the sole source of carbon, except perhaps for some vitamins that may be required for growth. The autotrophs include the photoautotrophs and the chemoautotrophs. The prefix indicates the source of energy and the suffix the source of carbon. The *photo*autotrophs use light as their source of energy, and they include the plants, algae, and photosynthetic bacteria. The *chemo*autotrophs use inorganic chemicals as their source of energy (e.g., hydrogen gas, ammonia, nitrite, ferrous ion, and inorganic sulfur). Thus far, the only known chemoautotrophs are bacteria.

2. The properties of the Calvin cycle enzymes and their regulation are reviewed in Bowien, B. 1989. Molecular biology of carbon dioxide assimilation in aerobic chemolithotrophs, pp. 437–460. In: *Autotrophic Bacteria*. Schlegel, H. G., and B. Bowien (Eds.). Science Tech Publishers, Madison, WI, and Springer Verlag, Berlin.

3. There is little information available about the bacterial aldolases and phosphatases that function in the Calvin cycle. However, it appears that the sedoheptulose-1,7-bisphosphatase and the fructose-1,6-bisphosphatase are a single bifunctional enzyme, capable of using either fructose-1,6-bisphosphate or sedoheptulose-1,7-bisphosphate as the substrate. In addition, the fructose-1,6-bisphosphate aldolase and the sedoheptulose-1,7-bisphosphate aldolase may also be a single bifunctional enzyme.

4. Reviewed by Wood, H. G., and L. G. Ljungdahl. 1991. Autotrophic character of the acetogenic bacteria, pp. 201–250. In: *Variations in Autotrophic Life*. Shively, J. M., and L. L. Barton (Eds.). Academic Press, New York.

5. The "acetogenic bacteria" are defined as anaerobic bacteria that synthesize acetic acid solely from $CO_2$ and secrete the acetic acid into the media. Many of the acetogenic bacteria are facultative heterotrophs that use the acetyl–CoA pathway to reduce $CO_2$ to acetic acid as

an electron sink, but these can also be grown autotrophically on $CO_2$ and $H_2$.

6. The early history is reviewed in Wood, H. G. 1985. Then and Now. *Ann. Rev. Biochem.* 54:1–41.

7. Corrinoid enzymes are proteins containing vitamin B$_{12}$ derivatives as the prosthetic group. Vitamin B$_{12}$ is a cobalt-containing coenzyme.

8. Reviewed by Wood, H. G., and L. G. Ljungdahl. 1991. Autotrophic character of the acetogenic bacteria, pp. 201–250. In: *Variations in Autotrophic Life.* Shively, J. M., and L. L. Barton (Eds.). Academic Press, New York.

9. Heise, R., V. Muller, and G. Gottschalk. 1993. Acetogenesis and ATP synthesis in *Acetobacterium woodii* are coupled via a transmembrane primary sodium ion gradient. *FEMS Microbiol. Lett.* 112:261– 268.

10. Lu, W.-P., S. R. Harder, and S. W. Ragsdale. 1990. Controlled potential enzymology of methyl transfer reactions involved in acetyl–CoA synthesis by CO dehydrogenase and the corrinoid/iron–sulfur protein from *Clostridium thermoaceticum. J. Biol. Chem.* 265: 3124–3133.

11. Pezacka, E., and H. G. Wood. 1986. The autotrophic pathway of acetogenic bacteria: Role of CO dehydrogenase disulfide reductase. *J. Biol. Chem.* 261:1609–1615.

12. Ragsdale, S. W., and H. G. Wood. 1985. Acetate biosynthesis by acetogenic bacteria: Evidence that carbon monoxide dehydrogenase is the condensing enzyme that catalyzes the final steps of the synthesis. *J. Biol. Chem.* 260:3970–3977.

13. Fuchs, G. 1990. Alternatives to the Calvin cycle and the Krebs cycle in anaerobic bacteria: Pathways with carbonylation chemistry, pp. 13–20. In: *The Molecular Basis of Bacterial Metabolism.* Hauska, G., and R. Thauer (Eds.). Springer-Verlag, Berlin.

14. Reviewed in, Jetten, M. S. M., A. J. M. Stams, and A. J. B. Zehnder. 1992. Methanogenesis from acetate: A Comparison of the acetate metabolism in *Methanothix soehngenii* and *Methanosacrcia* spp. *FEMS Microbiol. Rev.* 88:181–198.

15. Wolfe, R. S. 1990. Novel coenzymes of archaebacteria, pp. 1–12. In. *The Molecular Basis of Bacterial Metabolism.* Hauska, G., and R. Thauer (Eds.). Springer-Verlag, Berlin.

16. Weiss, D. S., and R. K. Thauer, 1993. Methanogenesis and the unity of biochemistry. *Cell* 72:819–822

17. Olson, K. D., L. Chmurkowska-Cichowlas, C. W. McMahon, and R. S. Wolfe, 1992.

Structural modifications and kinetic studies of the substrates involved in the final step of methane formation in *Methanobacterium thermautotrophicum, J. Bacteriol.* 174:1007–1012.

18. HTP–SH is also known as factor B or coenzyme B. It is 7-mercaptoheptanolythreonine phosphate.

19. Ermler, U., W. Grabarse, S. Shima, M. Goubeaud, and R. K. Thauer. 1997. Crystal structure of methyl–coenzyme M reductase: The key enzyme of biological methane formation. *Science* 278:1457–1462.

20. There are several reasons for suggesting that a $\Delta p$ is generated during electron transfer to the mixed disulfide: (1) *M. barkeri* creates a $\Delta p$ during catabolism of acetate. (2) The membranes of *M. thermophila* and *M. barkeri* contain electron carriers, including cytochrome b and FeS proteins. (3) Over 50% of the heterodisulfide reductase is in the membranes.

21. Becher, B., and V. Müller. 1994. $\Delta\mu_{Na}^{+}$ drives the synthesis of ATP via an $\Delta\mu_{Na}^{+}$-translocating F$_1$F$_0$–ATP synthase in membrane vesicles of the archaeon *Methanosarcina mazei* Gol. *J. Bacteriol.* 176:2543–2550.

22. Ferry, James, G. 1992. Methane from acetate. *J. Bacteriol.* 174:5489–5495.

23. The methanogens are paramount in recycling organic carbon into gaseous forms of carbon in anaerobic habitats (Section 13.4). Most of the methane that enters the aerobic atomosphere is oxidized to $CO_2$ by the aerobic methanotrophs (Section 12.2).

24. Fischer, R., and R.K. Thauer. 1989. Methyl-tetrahydromethanopterin as an intermediate in methogenesis from acetate in *Methanosarcina barkeri. Arch. Microbiol.* 151:459–465.

25. Grahame, D. A. 1991. Catalysis of actyl–CoA clevage and tetrahydrosarcinapterin methylation by a carbon monoxide dehydrogenase-corrinoid enzyme complex. *J. Biol. Chem.* 266:22227–22233.

26. Bott, M., B. Eikmanns, and R. K. Thauer. 1986. Coupling of carbon monoxide oxidation to $CO_2$ and $H_2$ with the phosphorylation of ADP in acetate-grown *Methanosarcia barkeri. Eur. J. Biochem.* 159:393–398.

27. Stupperich, E., and G. Fuchs. 1984. Autotrophic synthesis of activated acetic acid from two $CO_2$ in *Methanobacterium thermautotrophicum.* I. Properties of the in vitro system. *Arch. Microbiol.* 139:8–13.

28. Reviewed in Danson, M. J. 1988. *Archaebacteria:* The comparative enzymology of their central metabolic pathways, pp. 165–231. In:

*Advances in Microbial Physiology,* Vol. 29. Rose, A. H., and D. W. Tempest (Eds.). Academic Press, New York.

29. Hansen, T. A. 1993. Carbon metabolism of sulfate-reducing bacteria, pp. 21–40. In: *The Sulfate-Reducing Bacteria: Contemporary Perspectives.* Odom, J. M., and R. Singleton, Jr. (Eds.). Springer-Verlag, New York.

30. Reviewed by Amesz, J. 1991. Green photosynthetic bacteria and heliobacteria. pp. 99–119. In: *Variations in Autotrophic Life.* Shively, J. M., and L. L. Barton (Eds.). Academic Press, New York.

31. *Desulfobacter* is a group II sulfate-reducing bacterium that grows on fatty acids, especilly acetate, using sulfate as an electron acceptor and producing sulfide (Section 11.2.2). *Chlorobium* is a green sulfur photoautotroph (Section 5.3). *Hydrogenobacter* is an aerobic obligately chemolithotrophic hydrogen oxidizer.

32. Reviewen in Schonheit, P., and T. Schafer. 1995. Metabolism of hyperthermophiles. *World J. Microbiol. And Biotech.* 11:26–57.

33. In eukaryotic cells the ATP-dependent citrate lyase is used to generate acetyl–CoA from citrrate. The acetyl–CoA is then used as a precursor to fatty acids and steroids in the cytosol.

34. Reviewed in *Methane and Methanol Utilizers.* 1992. Colin Murrell, J., and H. Dalton. Plenum Press, New York and London.

35. Reviewed in, Lindstrom, M. E. 1992. The aerobic Methylotrophic bacteria, pp. 431–445. In: *The Prokaryotes,* Vol. 1. Second edition. Balows, A., H. G. Truper, M. Dworkin, W. Harder, and K.-H. Schleifer (Eds.). Springer-Verlag, Berlin.

36. Methanol appears in the environment as a result of the microbial breakdown of plant products with methoxy groups (e.g., pectins and lignins). Formate is a fermentation end product excreted by fermenting bacteria. Methylamines can result from the breakdown products of some pesticides and certain other compounds, including carnitine and lecithin derivatives. Carnitine is a trimethylamine derivative that is present in many organisms, and in all animal tissues, especially muscle. It functions as a carrier for fatty acids across the mitochondrial membrane into the mitochondria, where the fatty acids are oxidized. Lecithin is phosphatidylocholine, which is a phospholipid containing a trimethylamine group (Section 9.1.3)

37. Green, P. N. 1992. Taxonomy of methylotrophic bacteria, pp. 23–84. In: *Methane and Methanol Utilizers.* Colin Murrell, J., and H. Dalton (Eds.). Plenum Press, New York and London.

38. The methanotrophs are responsible for oxidizing about one-half of the methane produced by methanogens. The rest of the methane escapes to the atmosphere.

39. Most of the nonmethanotrophs that are obligate methylotrophs use the RuMP pathway. Within the methanotrophs, some use the RuMP pathway and some the serine pathway.

40. Dijkhuizen, L., P. R. Levering, and G. E. De Vries. 1990. The physiology and biochemistry of aerobic methanol-utilizing gram-negative and gram-positive bacteria, pp. 149–181. In: *Methane and Methanol Utilizers. Biotechnology Handbooks.* Vol. 5. Colin Murrell, J., and H. Dalton (Eds.). Plenum Press, New York, London.

# 14

# Fermentations

Growing in the numerous anaerobic niches in the biosphere (muds, sewage, swamps, and so on) are prokaryotes that can grow indefinitely in the complete absence of oxygen, a capability that is rare among eukaryotes. (Eukaryotes require oxygen to synthesize unsaturated fatty acids and sterols.) Anaerobically growing prokaryotes reoxidize NADH and other reduced electron carriers either by anaerobic respiration (e.g., using nitrate, sulfate, or fumarate as the electron acceptor) (Chapters 4 and 11), or they carry out a fermentation.

A *fermentation* is defined as a pathway in which NADH (or some other reduced electron acceptor that is generated by oxidation reactions in the pathway) is reoxidized by metabolites produced by the pathway. The redox reactions occur in the cytosol rather than in the membranes, and ATP is produced via substrate-level phosphorylation.

It should not be concluded that fermentation occurs only among prokaryotes. Eukaryotic microorganisms can live fermentatively (e.g., yeast), although as mentioned, oxygen is usually necessary unless the medium is supplemented with sterols and unsaturated fatty acids. Furthermore, certain animal cells are capable of fermentation, such as, for example, muscle cells and human red blood cells.

Fermentations are named after the major end products they generate. For example, yeast carry out an ethanol fermentation, and muscle cells and red blood cells carry out a lactic acid fermentation. There are many different types of fermentations carried out by microorganisms. However, the carbohydrate fermentations can be grouped into six main classes: *lactic, ethanol, butyric, mixed acid, propionic,* and *homoacetic.* A homoacetic fermentation of glucose is described in Section 13.1.3.

## 14.1 Oxygen Toxicity

Many anaerobic prokaryotes (strict anaerobes) are killed by even small traces of oxygen. Strict anaerobes are killed by oxygen because toxic products of oxygen reduction accumulate in the cell. The toxic products of oxygen are produced when single electrons are added to oxygen sequentially.[1] These toxic products are hydroxyl radical (OH·), superoxide radical ($O_2^-$), and hydrogen peroxide ($H_2O_2$). A more complete discussion of oxygen toxicity can be found in Section 19.4.

The superoxide radical forms because oxygen is reduced by single electron steps:

$$O_2 + e^- \longrightarrow O_2^-$$

A small amount of superoxide radical is always released from the enzyme when oxygen is reduced by electron carriers such as flavoproteins or cytochromes. This is because the electrons are transferred to oxygen one at a time. The hydroxyl radical and hydrogen peroxide are derived from the superoxide radical.

Aerobic and aerotolerant organisms do not accumulate superoxide radicals because they have an enzyme called *superoxide dismutase* that is missing in strict anaerobes. The superoxide dismutase catalyzes the following reaction:

$$O_2^- + O_2^- + 2H^+ \longrightarrow H_2O_2 + O_2$$

Notice that one superoxide radical transfers its extra electron to the second radical, which is reduced to hydrogen peroxide. Strict anaerobes also lack the enzyme that converts hydrogen peroxide to water and oxygen. That enzyme is catalase:

$$H_2O_2 + H_2O_2 \longrightarrow 2H_2O + O_2$$

Catalase catalyzes the transfer of two electrons from one hydrogen peroxide molecule to the second, oxidizing the first to oxygen and reducing the second to two molecules of water. Table 14.1 shows the distribution of catalase and superoxide dismutase in aerobes and anaerobes.

If the $H_2O_2$ is not disposed of, then it can oxidize transition metals such as free iron(II) in the *Fenton reaction* and form the free hydroxyl radical, OH· (see Note 2 for a definition of transition metals):

$$Fe^{2+} + H^+ + H_2O_2 \longrightarrow Fe^{3+} + OH· + H_2O$$

## 14.2 Energy Conservation by Anaerobic Bacteria

An examination of mechanisms of energy conservation and ATP production in anaerobic bacteria reveals a variety of methods that were described in previous chapters. Some of these are reviewed in Fig. 14.1. Most fermenting bacteria make most or all of their ATP via substrate-level phosphorylations, and they create a $\Delta p$ (needed for solute transport, motility, and so on) by reversing the membrane-bound ATPase. Many anaerobic bacteria also generate a $\Delta p$ by reducing fumarate. This is an anaerobic respiration that can occur during fermentative metabolism when fumarate is produced. During fumarate reduction, NADH dehydrogenases donate electrons to fumarate via menaquinone and a $\Delta p$ is created, perhaps via a quinone loop. (See Section 4.6.1 for a discussion of quinone loops.) The production of a $\Delta p$ by fumarate respiration in fermenting bacteria probably spares ATP that would normally be hydrolyzed to maintain the $\Delta p$. Some anaerobes (e.g., *Wolinella succinogenes*) carry out a periplasmic oxidation of electron donors, such as $H_2$ or formate in the case of *W. succinogenes*, in order to create a $\Delta p$. (See Section 4.7.4.) There are several other means available to anaerobic bacteria for generating a protonmotive force or a sodium motive force. Some anaerobes and facultative anaerobes are capable of creating an electrochemical ion gradient by symport of organic acids

Table 14.1 The distribution of catalase and superoxide dismutase

| Bacterium | Superoxide dismutase | Catalase |
| --- | --- | --- |
| Aerobes or facultative anaerobes | | |
|   *Escherichia coli* | + | + |
|   *Pseudomonas species* | + | + |
|   *Deinococcus radiodurans* | + | + |
| Aerotolerant bacteria | | |
|   *Butyribacterium rettgeri* | + | − |
|   *Streptococcus faecalis* | + | − |
|   *Steptococcus lactis* | + | − |
| Strict anaerobes | | |
|   Clostridium pasteurianum | − | − |
|   Clostridium acetobutylicum | − | − |

*Source*: Stainer, R. Y., J. L. Ingraham, M. L. Wheelis, and P. R. Rainter. 1986. *The Microbial World*. (Reprinted by permission of Prentice Hall, Englewood Cliffs, NJ.)

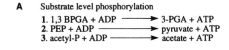

**A**   Substrate level phosphorylation

1. 1,3 BPGA + ADP $\longrightarrow$ 3-PGA + ATP
2. PEP + ADP $\longrightarrow$ pyruvate + ATP
3. acetyl-P + ADP $\longrightarrow$ acetate + ATP

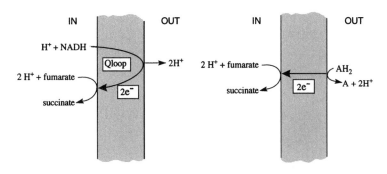

**B**   Fumarate respiration

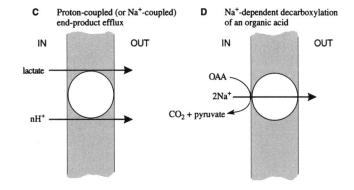

**C**   Proton-coupled (or Na$^+$-coupled) end-product efflux

**D**   Na$^+$-dependent decarboxylation of an organic acid

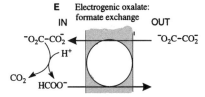

**E**   Electrogenic oxalate: formate exchange

**Fig. 14.1** Energy conservation in anaerobic bacteria. (A) Substrate-level phosphorylation: 1, the PGA kinase reaction; 2, the pyruvate kinase reaction; 3, the acetate kinase reaction. (B) Fumarate respiration. When the electron donor is NADH, a Q loop probably operates to translocate protons out of the cell. When the electron donor is periplasmic, proton translocation is not necessary. (C) Efflux of an organic acid coupled to protons or sodium ions, e.g., the coupled efflux of protons and lactate by the lactate bacteria. (D) Decarboxylation of an organic acid coupled to Na$^+$ efflux, e.g., the decarboxylation of oxaloacetate by *Klebsiella*. (E) Electrogenic oxalate:formate exchange in *Oxalobacter*.

out of the cell with protons or sodium ions. The organic acids are produced during fermentation, and the energy to create the electrochemical gradient is due to the concentration gradient of the excreted organic acid (high inside). This has been demonstrated for lactate excretion (proton symport) by the lactate bacteria, and for succinate ex-

cretion (sodium ion symport) by a rumen bacterium, *Selenomonas ruminantium*. (See Section 3.8.3 for a discussion of lactate and succinate excretion in symport with protons and sodium ions.) *Klebsiella pneumoniae* is capable of generating an electrochemical sodium ion gradient by using the energy released from the decarboxylation of oxaloacetate to pyruvate (Section 3.8.1). The decarboxylase pumps $Na^+$ out of the cell. The sodium potential that is created is used to drive the uptake of oxaloacetate, which is used as a carbon and energy source. *Oxalobacter formigenes* creates a proton potential by catalyzing an electrogenic anion exchange coupled to the decarboxylation of oxalate to formate (Section 3.8.2).

## 14.3 Electron Sinks

A major problem that must be addressed during fermentation is what to do with the electrons removed during oxidations. For example, how is the NADH reoxidized? Respiring organisms do not have this problem, since the electrons travel to an exogeneous electron acceptor (e.g., oxygen or nitrate). Since fermentations usually occur in the absence of an exogenously supplied electron acceptor, the fermentation pathways themselves must produce the electron acceptors for the electrons produced during the oxidations. The electron acceptors are called "electron sinks" because they dispose of the electrons removed during the oxidations, and the reduced products are excreted into the medium. Consequently, fermentations are characterized by the excretion of large quantities of reduced organic compounds such as alcohols, organic acids, and solvents. Frequently, hydrogen gas is also produced, since protons are used as electron acceptors by hydrogenases. Generally, one can view fermentations in the following way, where B is the electron sink:

$$AH_2 + NAD^+ + P_i + ADP$$
$$\longrightarrow B + NADH + H^+ + ATP$$
$$B + NADH + H^+$$
$$\longrightarrow BH_2 \text{ (excreted)} + NAD^+$$
$$NADH + H^+ \xrightarrow{\text{hydrogenase}} H_2 + NAD^+$$

For example, there is the lactate fermentation:

$$\text{glucose} + 2 \text{ ADP} + 2 \text{ P}_i \longrightarrow$$
$$2 \text{ lactate} + 2 \text{ ATP}$$

or an ethanol fermentation:

$$\text{glucose} + 2 \text{ NAD}^+ + 2 \text{ ADP} + 2 \text{ P}_i \longrightarrow$$
$$2 \text{ pyruvate} + 2 \text{ NADH} + 2 \text{ H}^+ + 2 \text{ ATP}$$
$$2 \text{ pyruvate} \longrightarrow 2 \text{ acetaldehyde} + 2 \text{ CO}_2$$
$$2 \text{ acetaldehyde} + 2\text{NADH} + 2\text{H}^+ \longrightarrow$$
$$2 \text{ ethanol} + 2\text{NAD}^+$$

---

$$\text{glucose} + 2 \text{ ADP} + 2 \text{ P}_i \longrightarrow$$
$$2 \text{ ethanol} + 2 \text{ CO}_2 + 2 \text{ ATP}$$

## 14.4 The Anaerobic Food Chain

The fermentation of amino acids, carbohydrates, purines, pyrimidines, and so on to organic acids and alcohols (acetate, ethanol, butanol, propionate, succinate, butyrate, and so on) by prokaryotes, and the conversion of these fermentation end poducts to $CO_2$, $CH_4$, and $H_2$ by the combined action of several different types of bacteria, is called the anaerobic food chain (Fig. 14.2). It takes place in anaerobic environments such as in muds, the bottom of lakes, and sewage treatment plants. This is an important part of the carbon cycle and serves to regenerate gaseous carbon (i.e., carbon dioxide and methane), which is reutilized by other microorganisms and plants throughout the biosphere. As illustrated in Fig. 14.2, the process can be viewed as occurring in three stages. Fermenters produce organic acids, alcohols, hydrogen gas, and carbon dioxide. These fermentation end products are oxidized to $CO_2$, $H_2$, and acetate by organisms that have been only partially identified. Finally, the methanogens grow on the acetate, $H_2$ and $CO_2$, converting these to $CH_4$ and $CO_2$.

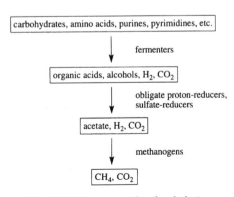

Fig. 14.2 The anaerobic food chain.

## 14.4.1 Interspecies hydrogen transfer

Some anaerobic bacteria use protons as the major or sole electron sink. They include the *obligate proton-reducing acetogens* that oxidize butyrate, propionate, ethanol, and other compounds to acetate, $H_2$, and $CO_2$. The physiology of these organisms is not well understood. Probably the electrons travel from the organic substrate to an intermediate electron carrier such as $NAD^+$ to hydrogenase to $H^+$. Some of these oxidations are:

$$CH_3CH_2CH_2COO^- + 2H_2O \longrightarrow$$
$$2CH_3COO^- + H^+ + 2H_2,$$
$$\Delta G'_0 = +48.1 \text{ kJ}$$

$$CH_3CH_2COO^- + 3H_2O \longrightarrow CH_3COO^-$$
$$+ HCO_3^- + H^+ + 3H_2,$$
$$\Delta G'_0 = +76.1 \text{ kJ}$$

$$CH_3CH_2OH + H_2O \longrightarrow CH_3COO^-$$
$$+ H^+ + 2H_2,$$
$$\Delta G'_0 = +9.6 \text{ kJ}$$

Notice that the above reactions are thermodynamically unfavorable under standard conditions at pH 7. In fact, the proton reducers live symbiotically with $H_2$ utilizers that keep the $H_2$ levels low and pull the reaction towards $H_2$ production. The $H_2$ utilizers are methanogens and sulfate reducers. (Under conditions of sufficient sulfate, the sulfate reducers predominate over methanogens. The sulfate reducers also oxidize ethanol, lactate, and other organic acids to acetate.) The symbiotic relationship between the obligate proton reducers and the hydrogen utilizers is called a *syntrophic* association.[3] It is also called *interspecies hydrogen transfer*. It should be pointed out that many other fermenting bacteria besides the obligate proton reducers have dehydrogenases that transfer electrons from NADH to protons and can use protons as an electron sink when grown in the presence of hydrogen gas utilizers. An example is *Ruminococcus albus*, which is discussed in Section 14.12. Hydrogenasess are also coupled to the thermodynamically favored oxidation of reduced ferredoxin, as in pyruvate:ferredoxin oxidoreductase found in the clostridia and sulfate reducers (Section 7.3.2).

## 14.5 How to Balance a Fermentation

A written fermentation is said to be balanced when the hydrogens produced during the oxidations equal the hydrogens transferred to the fermentation end products. Only under these conditions is all the NADH and reduced ferredoxin recycled to the oxidized forms. It is important to know whether a fermentation is balanced, because if it is not, then the overall written reaction is incorrect. There are two methods used to balance fermentations, the O/R method and the available hydrogen method.

### The O/R method

This is a bookkeeping method to keep track of the hydrogens. An oxidation–reduction (O/R) balance is computed, as described later. One arbitrarily designates the O/R value for formaldehyde ($CH_2O$) and multiples thereof, $(CH_2O)_n$, as zero, and uses that formula as a standard with which to compare the reduction level of other molecules. The following are the steps involved in determining the O/R value of any molecule:

1. Add or subtract water to the molecule in question to make the C/O ratio 1. This will allow a comparison to $(CH_2O)_n$. For example, the formula for ethanol is $C_2H_6O$. The C/O ratio is 2. One water must be added so that the C/O ratio is 1. This changes $C_2H_6O$ to $C_2H_8O_2$. Acetic acid is $C_2H_4O_2$. Since the C/O ratio is already one, nothing further need be done. The formula for carbon dioxide is $CO_2$. In order to make the C/O ratio one, one must subtract $H_2O$ (i.e., [$CO_2$–$H_2O$]). The result is $C(-2H)O$.

2. Now compare the number of hydrogens in the modified formula to $(CH_2O)_n$, which has the same number of carbons as in the modified formula. For ethanol, $C_2H_8O_2$ is compared to $(CH_2O)_2$. There are 8H in $C_2H_8O_2$ but only 4H in $(CH_2O)_2$. Thus ethanol has an additional 4H. For carbon dioxide, $C(-2H)O$, there are $-4H$ compared

to $CH_2O$. For acetic acid, $C_2H_4O_2$, there are the same number of hydrogens as in $(CH_2O)_2$.

3. Add $-1$ for each additional 2H and $+1$ for a decrease in 2H. Thus the O/R for ethanol is $-2$, for $CO_2$ it is $+2$, and for acetic acid it is 0.

Since both the oxidized and reduced fermentation end products originate from the substrate, the sum of the O/R of the products equals the O/R of the substrate if the fermentation is balanced. For example, the O/R for glucose is 0. When one mole of glucose is fermented to two moles of ethanol and two moles of carbon dioxide, the O/R of the products is $(-2 \times 2) + (+2 \times 2) = 0$. Often one simply takes the ratio $(+/-)$ of the O/R of the products when a carbohydrate is fermented. For a balanced fermentation, $+/-$ should be 1.

### The available hydrogen method

Like the O/R method, this procedure is merely one of bookkeeping. It has nothing to do with the chemistry of the reactions. According to this method, one "oxidizes" the molecule to $CO_2$ using water to obtain the "available hydrogen." For example:

$$C_6H_{12}O_6 + 6H_2O \longrightarrow 24H + 6CO_2$$

Thus glucose has 24 available H. The available H in all of the products must add up to the available H in the starting material. In Table 14.2, the concentration of products is given per 100 moles of glucose used. The available H in the glucose is $24 \times 100 = 2400$. The available H in the products adds up to 2,242. Thus the balance is $2,400/2,242 = 1.07$.

## 14.6 Propionate Fermentation Using the Acrylate Pathway

The genus *Clostridium* comprises a heterogeneous group of bacteria consisting of gram-positive, anaerobic, spore-forming bacteria that cannot use sulfate as a terminal electron acceptor. They can be isolated from anaerobic regions (or areas of low oxygen levels) in soil. The clostridia ferment organic nutrients to products that can include alcohols, organic acids, hydrogen gas, and carbon dioxide. *C.*

*propionicum* oxidizes three moles of lactate to two moles of propionate, 1 mole of acetate, and 1 mole of carbon dioxide, and produces 1 mole of ATP. The pathway is called the *acrylate pathway* because one of the intermediates is acrylyl–CoA. The bacteria derive ATP via a substrate-level phosphorylation during the conversion of acetyl-P to acetate catalyzed by acetate kinase. Since only one acetate is made per three lactates used, the pathway yields one-third of an ATP per lactate. Growth yields are proportional to the amount of ATP produced, and it is to be expected that the growth yields for these organisms are very low. (The molar growth yield for ATP is 10.5 g cells per mole of ATP synthesized; Section 2.2.6.)

### 14.6.1 The fermentation pathway of C. propionicum

A molecule of lactate is oxidized to pyruvate yielding 2[H] (Fig. 14.3, reaction 1). The pyruvate is then oxidized to acetyl–CoA and $CO_2$, yielding 2[H] again (reaction 2). The acetyl–CoA is converted to acetate and ATP via acetyl-P (reactions 3 and 4). During the oxidations, 4[H] are produced that must be reutilized. The electron acceptor is created from a second and third molecule of lactate (actually lactyl–CoA). The lactate acquires a CoA from propionyl–CoA (reaction 5). The lactyl–CoA is dehydrated to yield the unsaturated molecule, acrylyl–CoA (reaction 6). Each acrylyl–CoA is then reduced to propionyl–CoA using up the 4[H] (reaction 7). The fermentation is thus balanced. The propionate is produced during the CoA transfer step (reaction 5). This is catalyzed by *CoA transferase*, an enzyme that occurs in many anaerobes.

What can we learn from this pathway? The bacteria use a standard method for making ATP under anaerobic conditions. They decarboxylate pyruvate to acetyl–CoA, and then using a *phosphotransacetylase* (to make acetyl-P) and an *acetate kinase*, they make ATP and acetate. These reactions are widespread among fermenting bacteria. The production of acetate is presumed to be associated always with the synthesis of two moles of ATP per mole of acetate if the bacteria

**Table 14.2** Balancing an acetone-butanol fermentation

| Substrate and products | Yield (mol/100 mol substrate) | Carbon (mol) | O/R value | O/R value (mol/100mol) | Available H | Available H (mol/100mol) |
|---|---|---|---|---|---|---|
| Glucose | 100 | 600 | 0 | — | 24 | 2,400 |
| Butyrate | 4 | 16 | −2 | −8 | 20 | 80 |
| Acetate | 14 | 28 | 0 | — | 8 | 112 |
| CO$_2$ | 221 | 221 | +2 | +442 | 0 | — |
| H$_2$ | 135 | — | −1 | −135 | 2 | 270 |
| Ethanol | 7 | 14 | −2 | −14 | 12 | 84 |
| Butanol | 56 | 224 | −4 | −224 | 24 | 1,344 |
| Acetone | 22 | 66 | −2 | −44 | 16 | 352 |
| Total | | 569 | | −425, +442 | | 2,242 |

*Source:* Gottschalk, G. 1986. *Bacterial Metabolism.* Springer-Verlag, Berlin.

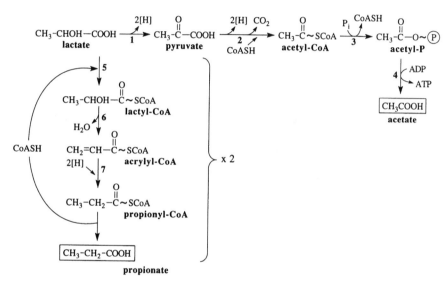

**Fig. 14.3** Propionate fermentation via the acrylate pathway. Enzymes: 1, lactate dehydrogenase; 2, pyruvate–ferredoxin oxidoreductase; 3, phosphotransacetylase; 4, acetate kinase; 5, CoA transferase; 7, a dehydrogenase. Reaction 6 is not sufficiently characterized.

are growing on glucose and using the EMP pathway. One ATP is produced from acetyl–CoA, and the second ATP is produced during the production of the pyruvate in the EMP pathway. Another common reaction among fermenting bacteria is the transfer of coenzyme A from one organic molecule to another, a reaction catalyzed by *CoA transferase.* The other way of attaching a coenzyme A molecule to a carboxyl group is to transfer an AMP or a phosphate to the carboxyl group from ATP making an acyl phosphate or an acyl–AMP, and then to displace the AMP or phosphate

with CoASH. (Recall the activation of fatty acids prior to their degradation; Section 9.1.1.) However, fermenting organisms must conserve ATP. The CoA transferase reaction is one way this can be done.

## 14.7 Propionate Fermentation Using the Succinate–Propionate Pathway

Many bacteria produce propionic acid as a product of fermentation using a pathway different from the acrylate pathway. The other pathway is called the *succinate–propionate pathway*, which yields more ATP than the acrylate pathway per mole of propionate formed. One of the organisms that utilizes this pathway, *Propionibacterium*, ferments lactate as well as hexoses to a mixture of propionate, acetate, and $CO_2$. *Propionibacterium* is a gram-positive anaerobic, nonsporulating, nonmotile, pleomorphic rod that is part of the normal flora in the rumen of herbivores, on human skin, and in dairy products (e.g., cheese). *Propionibacterium* is used in the fermentation process that produces Swiss cheese. The characteristic sharp flavor of this cheese is due to the propionate, and the holes in the cheese are due to the carbon dioxide produced.

The pathway illustrated in Fig. 14.4 shows that three molecules of lactate are oxidized to pyruvate (reaction 1). This yields six electrons. Then one pyruvate is oxidized to acetate, $CO_2$, and ATP, yielding two more electrons (reaction 2). We now have eight electrons to use up. The other two pyruvates are carboxylated to yield two molecules of oxaloacetate (reaction 5). The reaction is catalyzed by *methyl malonyl–CoA transcarboxylase*. The two oxaloacetates are reduced to two malates, consuming a total of four electrons (reaction 6). The two malates are dehydrated to two fumarates (reaction 7). The two fumarates are reduced via fumarate reductase to two succi-

nates, consuming four electrons (reaction 8). The latter reduction is coupled to the generation of a $\Delta p$. The fermentation is now balanced. The two succinates are converted to two molecules of succinyl–CoA via a CoA transferase (reaction 9). The two molecules of succinyl–CoA are isomerized to two molecules of methylmalonyl–CoA in an unusual reaction in which COSCoA moves from the $\alpha$-carbon to the $\beta$-carbon in succinyl–CoA to form methylmalonyl–CoA (reaction 10). The reaction can be viewed as an exchange between adjacent carbons of a H for a COSCoA. The enzyme that carries out this reaction is *methylmalonyl–CoA racemase*, an enzyme that requires vitamin $B_{12}$ as a cofactor. The two molecules of methylmalonyl–CoA donate the carboxyl groups to pyruvate via the transcarboxylase and in turn become propionyl–CoA (reaction 5). The propionyl–CoA donates the CoA to succinate via the CoA transferase and becomes propionate (reaction 9). Notice the important role of transcarboxylases and CoA transferases. These enzymes allow the attachment of $CO_2$ and CoA to molecules without the need for ATP.

### The fumarate reductase as a coupling site

When comparing Figs. 14.3 and 14.4, we learn that in metabolism there is sometimes

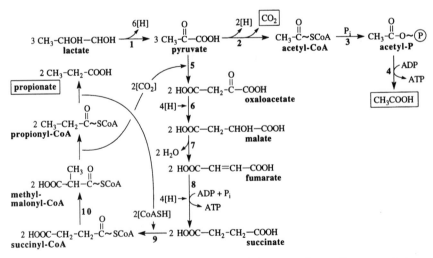

Fig. 14.4 Propionate fermentation by the succinate–propionate pathway. Enzymes: 1, lactate dehydrogenase (a flavoprotein); 2, pyruvate dehydrogenase (an $NAD^+$ enzyme); 3, phosphotransacetylase; 4, acetate kinase; 5, methylmalonyl–CoA–pyruvate transcarboxylase; 6, malate dehydrogenase; 7, fumarase; 8, fumarate reductase; 9, CoA transferase; 10, methylmalonyl–CoA racemase.

more than one route to take from point A to point B. *Propionibacterium* and other bacteria that use the succinate–propionate pathway use a circuitous route, but one that sends electrons through an energy-coupling site via the membrane-bound fumarate reductase. Electron flow to fumarate requires a quinone, and presumably the $\Delta p$ is generated via a redox loop involving the quinone (Section 4.6). The use of fumarate as an electron acceptor during anaerobic growth is widespread among bacteria. (See the discussion of the mixed acid fermentation in Section 14.10.) The electron donors besides lactate include NADH, $H_2$, formate, and glycerol-3-phosphate. The $\Delta p$ that is established can be used for ATP synthesis, for solute uptake, or to spare ATP that might be hydrolyzed to maintain the $\Delta p$.

## The transcarboxylase reaction spares an ATP

One way to carboxylate pyruvate is to use pyruvate carboxylase and $CO_2$ (Section 8.8). However, this requires an ATP. *Propionibacterium* has a transcarboxylase called *methylmalonyl–CoA–pyruvate transcarboxylase* that transfers a carboxyl group from methylmalonyl–CoA to pyruvate; hence an ATP is not used (Fig. 14.4, reaction 5). By using the transcarboxylase, the bacteria save energy by substituting one covalent bond for another. However, not all fermenting bacteria that produce propionate from pyruvate use the methylmalonyl–CoA–pyruvate transcarboxylase. For example, *Veillonella alcalescens* and *Propionigenum modestum* use a sodium-dependent decarboxylase to remove the carboxyl group from methylmalonyl–CoA while generating an electrochemical potential (Section 3.8). The sodium-dependent decarboxylase pumps sodium ions out of the cell, generating a sodium ion potential, which can be used as a source of electrochemical energy (e.g., for solute uptake or ATP synthesis).

## 14.7.1 The PEP carboxytransphosphorylase of propionibacteria and its physiological significance

### The reaction

Propionibacteria can produce succinate as well as propionate as an end product of fermentation when growing on carbon sources such as glucose that enter the glycolytic pathway. This means that they must have an enzyme to carboxylate a $C_3$ intermediate to form the $C_4$ product. The $C_3$ intermediate that is carboxylated is phosphoenolpyruvate (an intermediate in glycolysis). The phosphoenolpyruvate is carboxylated to oxaloacetate, which is then reduced to succinate via reactions 6, 7, and 8 shown in Fig. 14.4. The enzyme that catalyzes the carboxylation of phosphoenolpyruvate is called PEP carboxytransphosphorylase, and it catalyzes the following reaction (see Note 4 for a description of other $C_3$ carboxylases):

$$PEP + CO_2 + P_i \longrightarrow oxaloacetate + PP_i$$

During the carboxylation, a phosphoryl group is transferred from PEP to inorganic phosphate to form pyrophosphate. Pyrophosphate is a high-energy compound and propionibacteria have enzymes that phosphorylate fructose-6-phosphate to fructose-1,6-bisphosphate and serine to phosphoserine using pyrophosphate as the phosphoryl donor.

### Physiological significance

The carboxylation of phosphoenolpyruvate or pyruvate to oxaloacetate and the reduction of the oxaloacetate to succinate is a widespread pathway among fermenting bacteria. (For example, see the mixed acid fermentation, Section 14.10.) These reactions were also discussed in the context of the reductive citric acid pathway (Section 8.10). The pathway from PEP or pyruvate to succinate is extremely important for anaerobes and serves three purposes: (1) The fumarate is an electron sink, enabling NADH to be reoxidized; (2) the fumarate reductase is a coupling site (i.e., a $\Delta p$ is generated); and (3) the succinate can be converted to succinyl–CoA, which is required for the biosynthesis of tetrapyrroles, lysine, diaminopimelc acid, and methionine. With respect to fumarate acting as an electron sink, Gest has suggested that the carboxylation of the $C_3$ glycolytic intermediate and the reductive pathway from oxaloacetate to succinate may have evolved when the earth's atmosphere was still anaerobic, and served the purpose of balancing fermentations, thus sparing one of the two pyruvates derived from glucose for biosynthesis.[5] The reactions between oxaloacetate and succinate may have later become part of the oxidative citric acid cycle during the evolution of aerobic metabolism.

## 14.8 Acetate Fermentation (Acetogenesis)

As discussed in Section 13.1.3 when describing the acetogenic bacterium *Clostridium thermoaceticum*, some bacteria use $CO_2$ as an electron sink and reduce it to acetate as a fermentation end product using the acetyl–CoA pathway. This is called acetogenesis. Another acetogenic bacterium is the sulfate reducer *Desulfotomaculum thermobenzoicum*, when growing on pyruvate in the absence of sulfate.[6] (Another sulfate reducer, *Desulfobulbus propionicus*, ferments pyruvate to a mixture of acetate and propionate using the succinate–propionate pathway.) The acetogenic pathway is shown in Fig. 14.5. Four pyruvate molecules are oxidatively decarboxylated to 4 acetyl–CoA molecules, producing eight electrons (reaction 1). The acetyl–CoA is converted to acetyl phosphate via phosphotransacetylase (reaction 2). The acetyl phosphate is converted to acetate and ATP via the acetate kinase (reaction 3). Six of the electrons are used to reduce $CO_2$ to bound methyl, $[CH_3]$ (reaction 4). This requires an ATP in the acetyl–CoA pathway to attach the formic acid to the THF (see Section 13.1.3). The remaining two electrons are used to reduce a second molecule of $CO_2$ to bound carbon monoxide, $[CO]$ (reaction 5). The $[CH_3]$, $[CO]$, and CoASH combine to yield acetyl–CoA (reaction 6). The fourth acetyl–CoA is converted to acetate with the formation of an ATP (reactions 7 and 8).

## 14.9 Lactate Fermentation

The lactic acid bacteria are a heterogeneous group of aerotolerant anaerobes that ferment glucose to lactate as the sole or major product of fermentation. They include the genera *Lactobacillus*, *Sporolactobacillus*, *Streptococcus*, *Leuconostoc*, *Pediococcus*, and *Bifidobacterium*. Lactic acid bacteria are found living on the skin of animals, in the gastrointestinal tract, and in other places (e.g., mouth and throat). Some genera live in vegetation and in dairy products. Several lactic acid bacteria are medically and commercially important organisms. These include the genus *Streptococcus*, several of which are pathogenic. The lactic acid bacteria are also important in various food fermentations (e.g., the manufacture of butter, cheese, yogurt, pickles, and sauerkraut) Although they can live in the presence of air, they metabolize glucose only fermentatively and derive most or all of their ATP from substrate-level phosphorylation. Under certain growth conditions they may transport lactate out of the cell in electrogenic symport with $H^+$, creating a $\Delta\Psi$ (Section 3.8.3). There are two major types of lactate fermentations: *homofermentative* and *heterofermentative*. The former uses the Embden–Meyerhof–Parnas pathway (glycolysis), and the latter uses the pentose phosphate pathway. A third pathway, called the *bifidum pathway* is found in *Bifidobacterium bifidum*.

### Homofermentative lactate fermentation

The homofermentative pathway produces primarily lactate. The bacteria use the glycolytic pathway to oxidize glucose to pyruvate. This

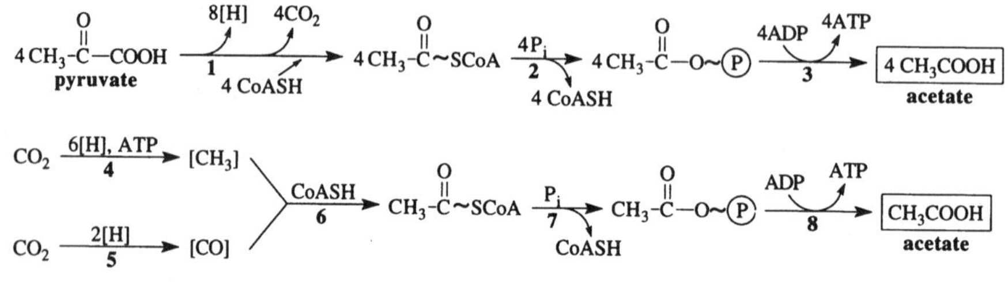

Fig. 14.5 Acetogenesis from pyruvate by *Desulfotomaculum thermobenzoicum*. Enzymes: 1, pyruvate dehydrogenase; 2, 7, phosphotransacetylase; 3, 8, acetate kinase; 4, enzymes of the acetyl–CoA pathway; 5, 6, carbon monoxide dehydrogenase.

nets them two ATPs per mole of glucose via the oxidation of phosphoglyceraldehyde. The NADHs produced during this oxidative step are used to reduce the pyruvate, forming lactate. The overall reaction is:

$$\text{glucose} + 2\text{ADP} + 2\text{P}_i \longrightarrow 2 \text{ lactate} + 2\text{ATP}$$

Whenever lactate is produced during fermentations, it is always the result of the reduction of pyruvate. It can be assumed that one ATP is made per mole of lactate produced, since the ATP yield per pyruvate is one.

## Heterofermentative lactate fermentation

The heterofermentative lactate fermentation produces lactate using the decarboxylation and isomerase reactions of the pentose phosphate pathway (Fig. 14.6). The glucose is oxidized to ribulose-5-phosphate using glucose-6-phosphate dehydrogenase and 6-phosphogluconate dehydrogenase (reactions 1–3). Four [H] are produced in the form of two NADHs. The ribulose-5-phosphate is isomerized to xylulose-5-phosphate using the epimerase (reaction 4). Then an interesting reaction occurs during which the xylulose-5-phosphate is cleaved with the aid of inorganic phosphate to form phosphoglyceraldehyde and acetyl phosphate (reaction 5). The enzyme that catalyzes this reaction is called *phosphoketolase* and requires thiamine pyrophosphate (TPP) as a cofactor. The phosphoglyceraldehyde is oxidized to pyruvate via reactions also found in the glycolytic pathway, yielding an ATP and a third NADH, and the pyruvate is reduced to lactate by one of the 3 NADHs (reactions 9–14). The acetyl phosphate produced in the phosphoketolase reaction is reduced to ethanol using the second and third NADH (reactions 6–8). Thus the fermentation is balanced. The overall reaction is:

$$\text{glucose} + \text{ADP} + \text{P}_i \longrightarrow \text{ethanol} + \text{lactate}$$
$$+ \text{CO}_2 + \text{ATP}$$

Note that the heterofermentative pathway produces only one ATP per glucose, in contrast to the homofermentative pathway, which produces two ATPs per glucose.

## Bifidum pathway

The bifidum pathway ferments 2 glucose to two lactates and three acetates with the production of 2.5 ATPs per glucose. The overall reaction is:

$$2 \text{ glucose} + 5 \text{ ADP} + 5 \text{ P}_i \longrightarrow 3 \text{ acetate}$$
$$+ 2 \text{ lactate} + 5 \text{ ATP}$$

The ATP yields are therefore greater than for the homo- or heterofermentative pathways. The pathway uses reactions of the pentose phosphate pathway (Section 8.5) and the homofermentative pathway. Two glucose molecules are converted to two fructose-6-P, requiring two ATPs. One of the fructose-6-P molecules is cleaved by fructose-6-phosphate phosphoketolase to erythrose-4-P and acetyl-P. The acetyl-P is converted to acetate via acetate kinase with the formation of an ATP. The erythrose-4-P reacts with the second fructose-6-P in a transaldolase reaction to form glyceraldehyde-3-P and sedoheptulose-7-P. These then react in a transketolase reaction to form xylulose-5-P and ribose-5-P. The latter isomerizes to form a second xylulose-5-P. The two xylulose-5-P are cleaved by xylulose-5-phosphate phosphoketolase to two glyceraldehyde-3-P and two acetyl-P. The two glyceraldehyde-3-P are converted to two lactate, with the production of four ATP molecules using reactions of the homofermentative pathway, and the two acetyl-P are converted to two acetates with the production of two more ATP molecules. Thus seven ATPs are produced per two glucose molecules fermented, but since two ATPs were used to make the two fructose-6-P molecules, the net gain in ATP per glucose is 5/2 or 2.5. Note that since glucose-6-P is not oxidized to 6-phosphogluconate, the acetyl-P can serve as a phosphoryl donor for ATP synthesis rather than be reduced to ethanol.

## Synthesis of acetyl–CoA or acetyl-P from pyruvate by lactic acid bacteria

Not all the pyruvate produced by lactic acid bacteria needs be converted to lactate. Depending upon the species, the lactic acid bacteria may have one of three enzymes or enzyme complexes for decarboxylating pyruvate

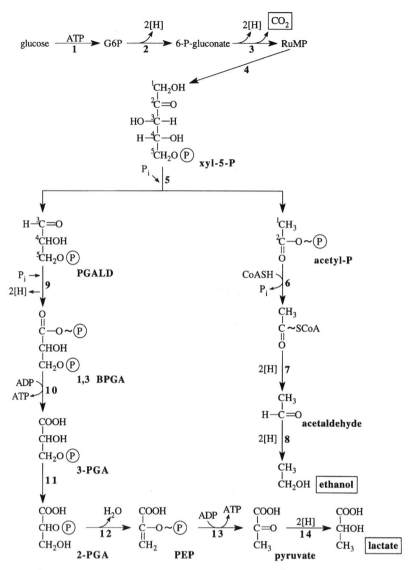

**Fig. 14.6** Heterofermentative lactate fermentation. Enzymes: 1, hexokinase; 2, glucose-6-phosphate dehydrogenase; 3, 6-phosphogluconate dehydrogenase; 4, ribulose-5-phophate epimerase; 5, phosphoketolase; 6, phosphotransacetylase; 7, acetaldehyde dehydrogenase; 8, alcohol dehydrogenase; 9, PGALD dehydrogenase; 10, PGA kinase; 11, phosphoglycerate mutase; 12, enolase; 13, pyruvate kinase; 14, lactate dehydrogenase.

(Section 7.3.2). Streptococci have pyruvate dehydrogenase, usually found in aerobically respiring bacteria; several lactic acid bacteria are known to have pyruvate formate lyase, an enzyme also found in the Enterobacteriaceae; *L. plantarum* and *L. delbruckii* use the flavoproteins pyruvate oxidase and lactate oxidase in coupled reactions that convert two pyruvates to one acetyl-P and one lactate:

$$\text{pyruvate} + P_i + \text{FAD} \xrightarrow{\text{pyruvate oxidase}} \text{acetyl-P}$$

$$+ CO_2 + FADH_2$$

$$\text{pyruvate} + FADH_2 \xrightarrow{\text{lactate oxidase}} \text{lactate} + \text{FAD}$$

## 14.10 Mixed Acid and Butanediol Fermentation

The enteric bacteria are facultative anaerobes. In the absence of oxygen several physiological changes take place as part of the adaptation to anaerobic growth. These changes are:

1. Terminal reductases replace the oxidases in the electron transport chain.

2. The citric acid cycle is modified to become a reductive pathway. $\alpha$-Ketoglutarate dehydrogenase and succinate dehydrogenase are missing or occur at low levels, the latter being replaced by fumarate reductase.

3. Pyruvate–formate lyase is substituted for pyruvate dehydrogenase. This means that the cells oxidize pyruvate to acety-l-CoA and formate, rather than to acetyl–CoA, $CO_2$, and NADH.

4. They carry out a mixed acid or butanediol fermentation.

The mixed acid and butanediol fermentations are similar in that both produce a mixture of organic acids, $CO_2$, $H_2$, and ethanol. The butanediol fermentation is distinguished by producing large amounts of 2,3-butanediol, acetoin, more $CO_2$ and ethanol, and less

acid. The mixed acid fermenters belong to the genera *Escherichia, Salmonella*, and *Shigella*. All three can be pathogenic and cause intestinal infections such as dysentery, typhoid fever (*Salmonella typhi*), or food poisoning. Butanediol fermenters are *Serratia, Erwinia*, and *Enterobacter*.

## Mixed acid fermentation

The products of the mixed acid fermentation are succinate, lactate, acetate, ethanol, formate, carbon dioxide, and hydrogen gas (Fig. 14.7). Each of these is made from one phosphoenolpyruvate and $CO_2$ or one pyruvate. For example, the formation of succinate is due to a carboxylation of phosphoenolpyruvate to oxaloacetate followed by two reductions to form succinate (reactions 10–13). All the other products are formed from pyruvate. The formation of lactate from pyruvate is simply a reduction (reaction 4). Pyruvate is also

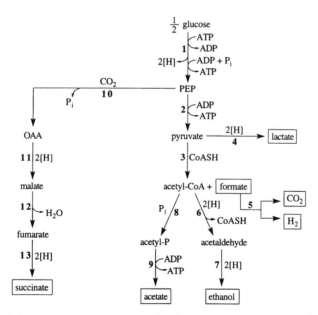

Fig. 14.7 Mixed acid fermentation. Enzymes: 1, glycolytic enzymes; 2, pyruvate kinase; 3, pyruvate–formate lyase; 4, lactate dehydrogenase; 5, formate–hydrogen lyase; 6, acetaldehyde dehydrogenase; 7, alcohol dehydrogenase; 8, phosphotransacetylase; 9, acetate kinase; 10, PEP carboxylase; 11, malate dehydrogenase; 12, fumarase; 13, fumarate reductase. Note the ATP yields: per succinate, approximately 1; per ethanol, 1; per acetate, 2; per formate, 1; per $CO_2$ and $H_2$, 1; per lactate, 1. Energy equivalent to approximately 1 ATP is conserved per succinate formed because the fumarate reductase reaction takes place in the cell membrane and generates a $\Delta p$. Note also the reducing equivalents used in the production of the end products: per succinate, 4; per ethanol, 4; per acetate, 0; per lactate, 2; per formate, 0. The number of reducing equivalents used must equal the number produced during glycolysis. Therefore, only certain ratios of end products are compatible with a balanced fermentation.

decarboxylated to acetyl–CoA and formate using the pyruvate–formate lyase (reaction 3). The acetyl–CoA can be reduced to ethanol (reactions 6 and 7) or converted to acetate and ATP via acetyl-P (reactions 8 and 9). The formate can be oxidized to $CO_2$ and $H_2$ by the enzyme system formate–hydrogen lyase (reaction 5). This system actually consists of two enzymes: formate dehydrogenase and an associated hydrogenase. The formate dehydrogenase oxidizes the formate to $CO_2$ and reduces the hydrogenase, which transfers the electrons to two protons to form hydrogen gas. *Shigella* and *Erwinia* do not contain formate–hydrogen lyase and therefore do not produce gas. (See Note 7.) Each of the pathways following phospho-enolpyruvate or pyruvate can be viewed as a metabolic branch that accepts different amounts of reducing equivalent (i.e., 0, 2[H], or 4[H]), depending upon the pathway. The reducing equivalents in the different branches are: succinate (4), ethanol (4), lactate (2),

acetate (0), and formate (0). During glycolysis, 2 [H] are produced for each phospho-enolpyruvate or pyruvate formed. Therefore, in order to balance the fermentation, 2 [H] must be used for each phosphoenolpyruvate or pyruvate formed. The pathways that utilize 4 [H] per phosphoenolpyruvate or pyruvate formed spare the second phosphoenolpyruvate or pyruvate for biosynthesis. However, they may do this at the expense of an ATP. For example, the reduction of acetyl–CoA to ethanol uses 4 H, but this is at the expense of an ATP that can be formed when acetyl–CoA is converted to acetate. In this context, the production of succinate is particularly valuable. The pathway utilizes 4 H and also includes a coupling site (fumarate reductase).

## Butanediol fermentation

The butanediol fermentation is characterized by the production of 2,3-butanediol and acetoin (Fig. 14.8). Glucose is oxidized via the

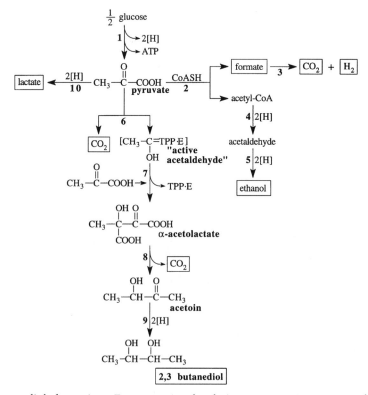

**Fig. 14.8** Butanediol formation. Enzymes: 1, glycolytic enzymes; 2, pyruvate–formate lyase; 3, formate–hydrogen lyase; 4, acetaldehyde dehydrogenase; 5, alcohol dehydrogenase; 6, 7, $\alpha$-acetolactate synthase; 8, $\alpha$-acetolactate decarboxylase; 9, 2,3-butanediol dehydrogenase; 10, lactate dehydrogenase.

glycolytic pathway to pyruvate (reactions 1). There are three fates for the pyruvate. Some of it is reduced to lactate (reaction 10), some is converted to acetyl–CoA and formate (reaction '2'), and some is used for the synthesis of 2,3-butanediol (reactions 6–9). The formate is converted to $CO_2$ and $H_2$ (reaction 3), and the acetyl–CoA is reduced to ethanol (reactions 4 and 5). The first free intermediate in the butanediol pathway is $\alpha$-acetolactate. This is formed by the enzyme $\alpha$-acetolactate synthase, which decarboxylates pyruvate in a thiamine pyrophosphate (TPP)-dependent reaction to enzyme-bound "active acetaldehyde" (reaction 6). The active acetaldehyde is transferred by the $\alpha$-acetolactate synthase to pyruvate to form $\alpha$-acetolactate (reaction 7). The $\alpha$-acetolactate, a $\beta$-keto carboxylic acid, is decarboxylated to acetoin (reaction 8). The acetoin is reduced by NADH to 2,3-butanediol (reaction 9). The production of butanediol is favored under slightly acidic conditions, and is a way for the bacteria to limit the decrease in external pH caused by the synthesis of organic acids from pyruvate.

### How thiamine pyrophosphate catalyzes the decarboxylation of $\alpha$-keto carboxylic acids

The decarboxylation of $\alpha$-keto carboxylic acids presents a problem because there is no electron- attracting carbonyl group $\beta$ to the C–C bond to withdraw electrons, as there is in $\beta$-keto carboxylic acids. (See Section 8.10.2 for a description of the decarboxylation of $\beta$-keto carboxylic acids.) The problem is solved by using thiamine pyrophosphate. A proposed mechanism is illustrated in Fig. 14.9. A proton dissociates from the thiamine pyrophosphate to form a dipolar ion, which is stabilized by the positive charge on the nitrogen atom. The negative center of the dipolar ion adds to the keto group of the $\alpha$-keto carboxylic acid. Then the electron attracting $=N^+-$ group pulls electrons away from the C–C bond, facilitating the decarboxylation. The product is an $\alpha$-ketol called "active aldehyde." The flow of electrons is reversed, and the "active aldehyde" then attacks a positive center on another molecule. The product varies, depending upon the enzyme. As shown

in Fig. 14.9, $\alpha$-acetolactate synthase catalyzes the condensation of "active acetaldehyde" with pyruvate to form $\alpha$-acetolactate, and pyruvate decarboxylase catalyzes the addition of a proton to form acetaldehyde. Pyruvate dehydrogenase and $\alpha$-ketoglutarate dehydrogenase catalyze the condensation with lipoic acid to form the acyl-lipoate derivative.

## 14.11 Butyrate Fermentation

Butyrate fermentations are carried out by the butyric acid clostridia. The clostridia are a heterogeneous group of anaerobic spore-forming bacteria that can be isolated from anaerobic muds, sewage, feces, or other anaerobic environments. They all are classified in the genus *Clostridium*. Some are saccharolytic (i.e., they ferment carbohydrates) and/or proteolytic. The proteolytic clostridia are important in the anaerobic decomposition of proteins, called "putrefaction." Other clostridia are more specialized and will ferment only a few substrates (e.g., ethanol, acetate, certain purines, or certain amino acids). The butyric acid clostridia ferment carbohydrates to butyric acid. The fermentation products also include hydrogen gas, carbon dioxide, and small amounts of acetate. The bacteria first oxidize the glucose to two moles of pyruvate using the Embden–Meyerhof–Parnas pathway (Fig. 14.10). This produces two NADH, plus two ATP molecules. The pyruvate is then decarboxylated to acetyl–CoA, $CO_2$, and $H_2$ using pyruvate–ferredoxin oxidoreductase and hydrogenase (reaction 1). The acetyl–CoA is condensed to form acetoacetyl–CoA (reaction 2), which is reduced to $\beta$-hydroxybutyryl–CoA using one of the two NADHs (reaction 3). The $\beta$-hydroxybutyryl–CoA is reduced to butyryl–CoA using the second NADH (reactions 4 and 5). The CoASH is displaced by inorganic phosphate (reaction 6), and the butyryl phosphate donates the phosphoryl group to ADP to form ATP and butyrate (reaction 7). This pathway therefore utilizes three substrate-level phosphorylations, the phosphoryl donors being 1,3-bisphosphoglycerate, phosphoenolpyruvate, and butyryl-P. Note the role of the hydrogenase (reaction 9) in the

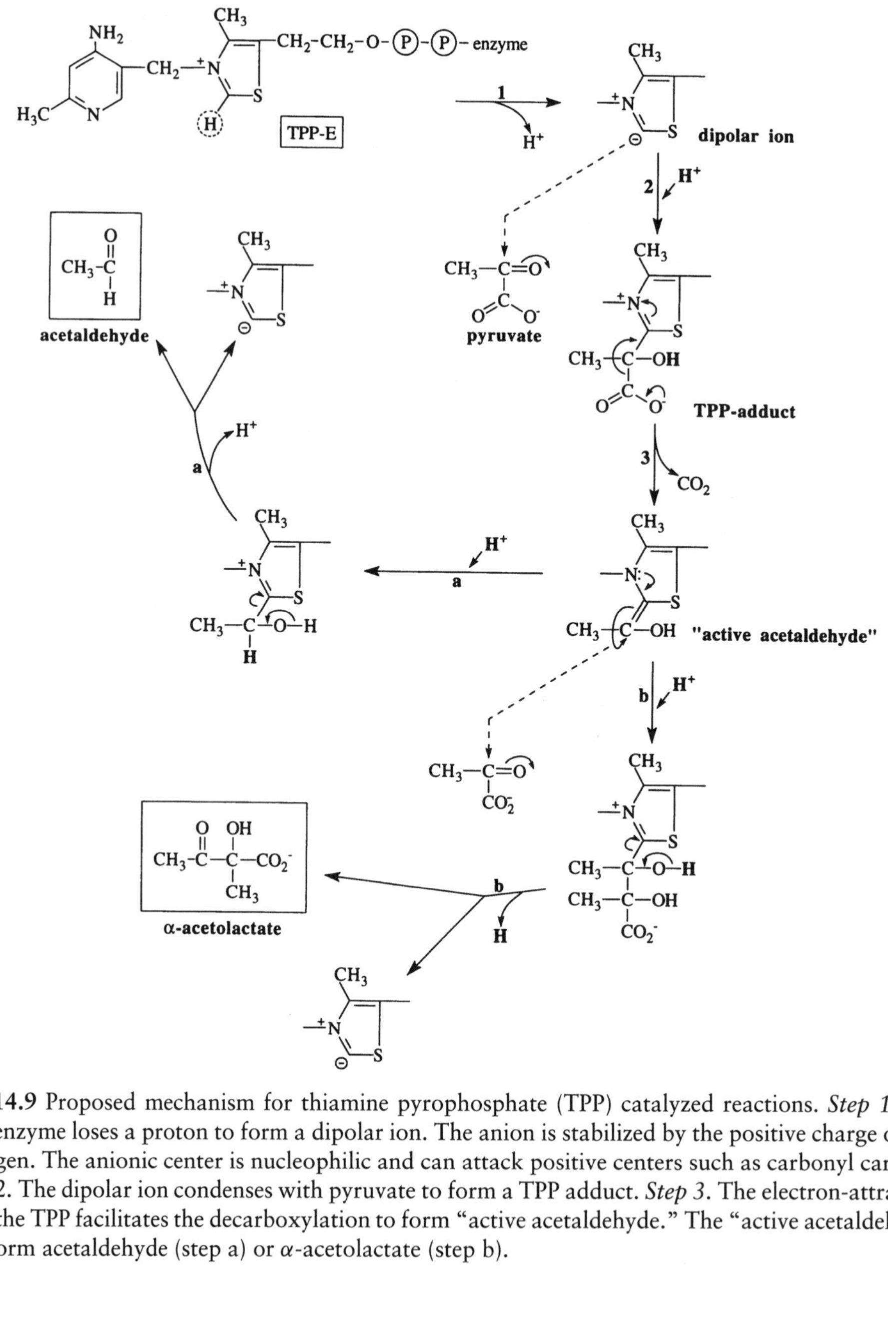

**Fig. 14.9** Proposed mechanism for thiamine pyrophosphate (TPP) catalyzed reactions. *Step 1.* The TPP enzyme loses a proton to form a dipolar ion. The anion is stabilized by the positive charge on the nitrogen. The anionic center is nucleophilic and can attack positive centers such as carbonyl carbons. *Step 2.* The dipolar ion condenses with pyruvate to form a TPP adduct. *Step 3.* The electron-attracting N in the TPP facilitates the decarboxylation to form "active acetaldehyde." The "active acetaldehyde" can form acetaldehyde (step a) or α-acetolactate (step b).

hydrogen sink. (For a discussion of acetyl–CoA condensations, see Section 8.11.1.)

### 14.11.1 Butyrate and butanol-acetone fermentation in C. acetobutylicum

Some clostridia initially make butyrate during fermentation, and when the butyrate accumulates and the pH drops (due to the butyric acid), the fermentation switches to a butanol–acetone fermentation. As we shall see, the butyrate is actually taken up by the cells and converted to butanol and acetone. The accumulation of butyric acid in media of low pH can be toxic because the undissociated form of the acid is lipophilic, and can enter the cell acting as an uncoupler and also resulting in a decrease in the $\Delta$pH. (The p$K$ of butyric acid is 4.82.) Butanol and acetone production by the clostridia was at one time the second largest

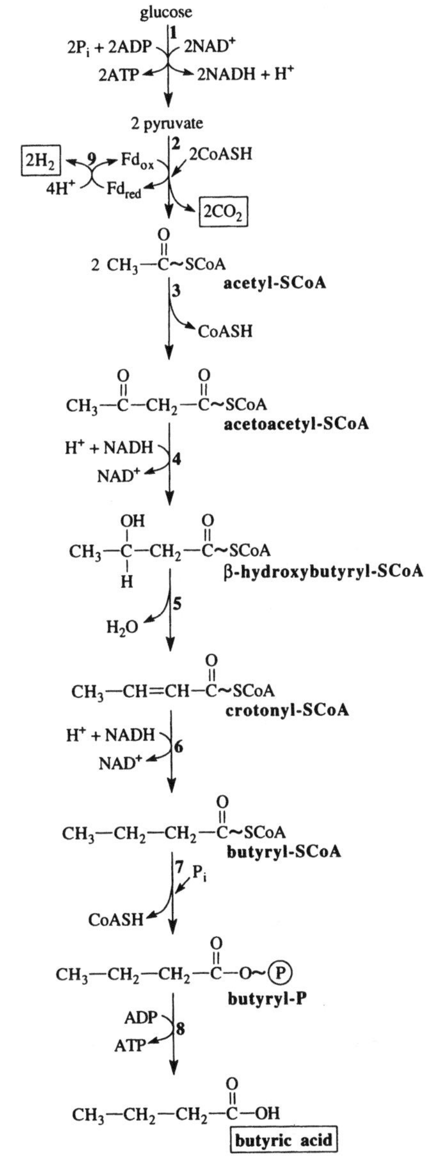

Fig. 14.10 The butyrate fermentation. The glucose is degraded via glycolysis to pyruvate, which is then oxidatively decarboxylated to acetyl–CoA. Two molecules of acetyl–CoA condense to form acetoacetyl–CoA, which is reduced to butyryl–CoA. A phosphotransacetylase makes butyryl-P, and a kinase produces butyrate and ATP. Two ATPs are produced in the glycolytic pathway per glucose and one ATP from butyryl-P. Note the production of hydrogen gas as an electron sink. This actually allows the production of the third ATP from butyryl-P, rather than reduce the butyryl–SCoA to butanol in order to balance the fermentation. Enzymes: 1, glycolysis; 2, pyruvate–ferredoxin oxidoreductase; 3, acetyl–CoA acetyltransferase (thio-

industrial fermentation process, second only to ethanol. The solvents are now synthesized chemically.

As an example of the butyrate fermentation and the shift to the butanol–acetone fermentation, we will consider the fermentation of carbohydrates carried out by *C. acetobutylicum*.[8] During exponential growth the bacteria produce butyrate, acetate, $H_2$, and $CO_2$. This is called the acidogenic phase. When the culture enters stationary phase, the acids are taken up by the cells, concomitant with the fermentation of the carbohydrate, and converted to butanol, acetone, and ethanol. This is called the solventogenic phase. Pentoses are also fermented, and these are converted to fructose-6-phosphate and phosphoglyceraldehyde via the pentose phosphate pathway (Fig. 14.11). The pyruvate formed from the sugars is oxidatively decarboxylated to acetyl–CoA and $CO_2$ via the pyruvate–ferredoxin oxidoreductase (reaction 1). The acetyl–CoA has two fates. Some of it is converted to butyrate and ATP as described in Fig. 14.10 (reactions 2–7). Some acetyl–CoA is also converted to acetate via acetyl-P in a reaction that yields an additional ATP (reactions 8 and 9). The ability to send electrons to hydrogen via the NADH:ferredoxin oxidoreductase (reaction 16) allows the bacteria to produce more acetate and hence more ATP, rather than reduce acetyl–CoA to butyrate. Notice that twice as much ATP is generated per acetate produced as opposed to butyrate. However, reduction of protons by NADH is not favored energetically and is limited by increasing concentrations of hydrogen gas. The NADH oxidoreductase can be pulled in the direction of hydrogen gas by other bacteria that utilize hydrogen in a process called interspecies hydrogen transfer, explained in Section 14.4.1. In the solventogenic phase, butyrate and acetate are taken up by the cells and reduced to butanol and ethanol (dotted lines). The acids are converted to their CoA derivatives by accepting a CoA from acetoacetyl–CoA (reaction 10). The reaction is catalyzed by CoA transferase. (Recall that

lase); 4, β-hydroxybutyryl–CoA dehydrogenase; 5, crotonase; 6, butyryl–CoA dehydrogenase; 7, phosphotransbutyrylase; 8, butyrate kinase; 9, hydrogenase.

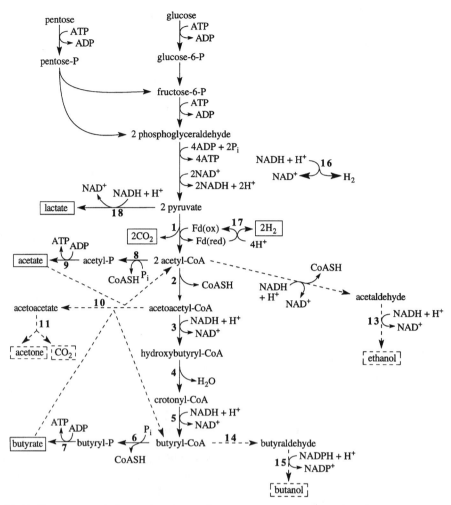

Fig. 14.11 Butyrate and butanol–acetone fermentation in *C. acetobutylicum*. Carbohydrates are oxidized to acetyl–SCoA. Pentose phosphates are converted to fructose-6-phosphate and phosphoglyceraldehyde via the pentose phosphate reactions. Glucose is oxidized to pyruvate via the Embden–Meyerhof–Parnas pathway. The pyruvate is oxidized to acetyl–SCoA. In the butyric acid fermentation the acetyl–SCoA is converted to acetate and butyrate (solid lines). When the acetate and butyrate levels rise, they are taken up by the cells and converted to butanol and ethanol while carbohydrates continue to be fermented (dotted lines). During the butanol–acetone fermentation, the acetoacetyl–SCoA donates the CoASH to butyrate and acetate and becomes acetoacetate, which is decarboxylated to acetone. Enzymes: 1, pyruvate–ferredoxin oxidoreductase; 2, acetyl–CoA acetyltransferase (thiolase); 3, hydroxybutyryl–CoA dehydrogenase; 4, crotonase; 5, butyryl–CoA dehydrogenase; 6, phosphotransbutyrylase; 7, butyrate kinase; 8, phosphotransacetylase; 9, acetate kinase; 10, acetoacetyl–SCoA:acetate/butyrate:CoA transferase; 11, acetoacetate decarboxylase; 12, acetaldehyde dehydrogenase; 13, ethanol dehydrogenase; 14, butyraldehyde dehydrogenase; 15, butanol dehydrogenase; 16, NADH–ferredoxin oxidoreductase and hydrogenase; 17, hydrogenase; 18, lactate dehydrogenase. *Source:* Adapted from Jones, D. T. and D. R. Woods. 1986. Acetone–butanol fermentation revisited. *Microbiol. Rev.* 50:484–524.

CoA transferase is also used in the acrylate pathway for propionate fermentation; Section 14.6.) The acetoacetate that is formed is decarboxylated to acetone and $CO_2$ (reaction 11). The acetyl–CoA is reduced to ethanol (reactions 12 and 13), and the butyryl–CoA is reduced to butanol (reactions 14 and 15).

The molar ratios of the fermentation end products in clostridial fermentations will vary according to the strain.[9] For example,

*C. acetobutylicum* is an important solvent-producing strain, and when grown at $pH_0$ below 5 produces butanol and acetone in the molar ratio of 2:1 with small amounts of isopropanol, whereas *C. sporogenes* and *C. pasteurianum* produce very little solvent. *Clostridium butyricum* forms butyrate and acetate in a ratio of about 2:1, whereas *C. perfringens* produces these acids in a ratio of 1:2, with significant amounts of ethanol and lactate.

## 14.12 Ruminococcus albus

*Ruminococcus albus* is a rumen bacterium that can ferment glucose to ethanol, acetate, $H_2$, and $CO_2$ using the glycolytic pathway (Fig. 14.12). It is a good fermentation to examine because it illustrates how growth of bacteria in mixed populations can influence fermentation end products. The fermentation is easy to understand. The glucose is first oxidized to two moles of pyruvate, yielding two moles of NADH. Then each mole of pyruvate is oxidized to acetyl–CoA, $CO_2$, and $H_2$ using the pyruvate–ferredoxin oxidoreductase and hydrogenase. One of the acetyl–CoA molecules is converted to acetate allowing an ATP to be made. The second acetyl–CoA is reduced to ethanol using the two NADH. Thus the fermentation is balanced. The overall reaction is the conversion of one mole of glucose to one mole of ethanol, one mole of acetate, two moles of hydrogen, and two moles of carbon dioxide, yielding three ATPs. Something different happens when the bacterium is grown in a co-culture with a methanogen. Growth with a methanogen shifts the fermentation in the direction of acetate with the concomitant production of more ATP. This is explained in the following way: *R. albus* has a hydrogenase that transfers electrons from NADH to $H^+$ to produce hydrogen gas. When hydrogen accumulates in the medium, the hydrogenase does not oxidize NADH, because the equilibrium favors $NAD^+$ reduction. The NADH therefore reduces acetyl–CoA to ethanol. However, the methanogen utilizes the hydrogen

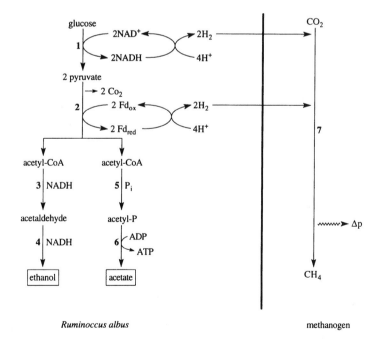

**Fig. 14.12** Fermentation of glucose by *R. albus*. *R. albus* ferments glucose to a mixture of ethanol, acetate, $CO_2$, and $H_2$. Methanogens draw off the $H_2$, thus stimulating electron flow to $H_2$. The result is a shift in the fermentation end products toward acetate accompanied by a greater ATP yield. The production of $H_2$ by one species and its utilization by another is called interspecies hydrogen transfer. The methanogens can also utilize the acetate produced by *R. albus*.

gas for methane production and keeps the hydrogen levels very low. In the presence of the methanogen, the NADH in *R. albus* reduces protons to hydrogen gas instead of reducing acetyl–CoA. The result is that *R. albus* converts the acetyl–CoA to acetate. This is also an advantage to the methanogen, since methanogens can also use acetate as a carbon and energy source. The transfer of hydrogen gas from one species to another is called *interspecies hydrogen transfer* and is an example of nutritional synergism found among mixed populations of bacteria. (See the discussion of interspecies hydrogen transfer in Section 14.4.1.)

## 14.13 Summary

Fermentations are cytosolic oxidation–reduction pathways wherein the electron acceptor is an organic compound, usually generated in the pathway.

The source of ATP in fermentative pathways is substrate-level phosphorylation. For sugar fermentations, these are the phosphoglycerate kinase, pyruvate kinase, acetate kinase, and butyrate kinase reactions. In other words, ATP is made from bisphosphoglycerate, phosphoenolpyruvate, acetyl-P, and butyryl-P. Butyryl-P is a phosphoryl donor during butyrate fermentations.

However, other means of conserving energy are available for fermenting bacteria. For example, a $\Delta p$ can be created during electron flow to fumarate, the fumarate being generated during fermentation. Other means of creating a proton potential or sodium potential exist in certain groups of bacteria. These include efflux of carboxylic acids in symport with protons or sodium ions, decarboxylases that function as sodium ion pumps, and oxalate:formate exchange.

Besides making ATP, fermenting bacteria must have someplace to unload the electrons removed during oxidation reactions. Of course, the reason for this is that they must regenerate the $NAD^+$, oxidized ferredoxin, and FAD to continue the fermentation. The electron acceptors are sometimes referred to as "electron sinks." During a fermentation the electron sinks are created from the carbon source. In fact, all the electron sinks for the

major carbohydrate fermentations are either pyruvate itself or are synthesized from pyruvate or phosphoenolpyruvate plus $CO_2$. Protons can also be used as electron sinks and many fermenting bacteria have hydrogenases that reduce protons to hydrogen gas.

The excreted end products of fermentations, including hydrogen gas, are used by other anaerobic bacteria so that an anaerobic food chain develops. At the bottom of the food chain are the methanogens that convert hydrogen gas, carbon dioxide, and acetate to $CO_2$ and $CH_4$. These are recycled to organic carbon in the biosphere as they are used by autotrophic and methanotrophic organisms as a source of carbon.

## Study Questions

1. Write a fermentation balance using both the O/R and the available hydrogen method for the following:

   $C_6H_{12}O_6$ (glucose) $+ H_2O \longrightarrow$
   
   $C_2H_4O_2$ (acetate) $+ C_2H_6O$ (ethanol)
   
   $+ 2H_2 + 2CO_2$

   If the EMP pathway is used, what is the yield of ATP?

2. *C. propionicum* and *Propionibacterium* both carry out the following reaction:

   3 lactate $\longrightarrow$ acetate $+ 2$ propionate $+ CO_2$

   *C. propionicum* nets one ATP, but *Propionibacterium* derives more. How might *Propionibacterium* make more ATP from the same overall reaction?

3. Write an ethanol fermentation using the Entner–Doudoroff pathway. Contrast the ATP yields with an ethanol fermentation using the EMP pathway.

4. Consider the following fermentation data for *Selenomonas ruminantium* (in mmol)[10] of products formed per mmol of glucose.

   | | |
   |---|---|
   | Lactate | 0.31 |
   | Acetate | 0.70 |
   | Propionate | 0.36 |
   | Succinate | 0.61 |

What is the fermentation balance using the O/R method and the available hydrogen method? What percentage of the glucose carbon is recovered in end products?

5. Consider the following data for fermentation products made by *Selenomonas rumanantium* in pure culture and in coculture with *Methanobacterium rumanantium* (in moles per 100 moles of glucose).

| Product | Selenomonas | Selenomonas + Methanobacterium |
|---|---|---|
| Lactate | 156 | 68 |
| Acetate | 46 | 99 |
| Propionate | 27 | 20 |
| Formate | 4 | 0 |
| Methane | 0 | 51 |
| $CO_2$ | 42 | 48 |

*Source:* Chen, M., and M. J. Wolin. 1977. Influence of $CH_4$ production by *Methanobacterium ruminantium* on the fermentation of glucose and lactate by *Selenomonas ruminantium*. *App. Env. Microbiol.* 34:756–759.

Offer an explanation accounting for the shift from lactate to acetate in the coculture compared to the pure culture. How might you expect this to affect the growth yields of *Selenomonas*?

## NOTES AND REFERENCES

1. Henle, E. S., and S. Linn. 1997. Formation, prevention, and repair of DNA damage by iron/hydrogen peroxide. *J. Biol. Chem.* 272:19095–19098.

2. Transition elements are defined as those elements in which electrons are added to an outer electronic shell before one of the inner shells is complete. There are nine of them. They are all metals, and they include iron, which is the one that occurs in highest amounts in living systems.

3. Shink, B. 1997. Energetics of syntrophic cooperation in methanogenic degradation. *Microbiol. and Mol. Biol. Rev.* 61:262–280.

4. Other enzymes besides PEP carboxytransphosphorylase that carboxylate $C_3$ glycolytic intermediates to oxaloacetate are PEP carboxylase and pyruvate carboxylase (Section 8.8) and PEP carboxykinase (Section 8.12). The latter enzyme usually operates in the direction of PEP synthesis, but in some anaerobes (e.g., *Bacteroides*) it functions to synthesize oxaloacetate.

5. Gest, H. 1983. Evolutionary roots of anoxygenic photosynthetic energy conversion, pp. 215–234. In: *The Phototrophic Bacteria: Anaerobic Life in the Light.* Studies in Microbiology, Vol. 4. Ormerod, J. G. (Ed.). Blackwell Scientific Publications, Oxford.

6. Tasaki, M., Y. Kamagata, K. Nakamura, K. Okamura, and K. Minami. 1993. Acetogenesis from pyruvate by *Desulfotomaculum thermobenzoicum* and differences in pyruvate metabolism among three sulfate-reducing bacteria in the absence of sulfate. *FEMS Microbiol. Lett.* 106:259–264.

7. Gas production is generally observed as a bubble in an inverted vial placed in the fermentation tube. The bubble is due to $H_2$, since $CO_2$ is very soluble in water.

8. Jones, D. T., and D. R. Woods. 1986. Acetone-butanol fermentation revisited. *Microbiol. Rev.* 50:484–524.

9. Hamilton, W. A. 1988. Energy transduction in anaerobic bacteria, pp. 83–149. In: *Bacterial Energy Transduction.* Anthony, C. (Ed.). Academic Press, New York.

10. Michel, T. A., and J. M. Macy. 1990. Generation of a membrane potential by sodium-dependent succinate efflux in *Selenomonas ruminantium*. *J. Bacteriol.* 172:1430–1435.

# 15

# Homeostasis

Homeostasis refers to the ability of living organisms to maintain an approximately constant internal environment despite changes in the external milieu. As applies to the bacterial cell, this means (among other things) the ability to maintain a steady intracellular pH and a constant osmotic differential across the cell membrane, despite fluctuations in external pH and osmolarity. The regulatory mechanisms underlying homeostasis are generally unknown, although they are a subject of active research.

## 15.1 Maintaining a $\Delta pH$

### 15.1.1 Neutrophiles, acidophiles, and alkaliphiles

Bacteria can be found growing in habitats that vary in pH from approximately pH 1 to 2 in acid springs to as high as pH 11 in soda lakes and alkaline soils. However, regardless of the external pH, the internal pH is usually maintained within 1–2 units of neutrality, which is necessary to maintain viability (Table 15.1).[1-6] Thus bacteria maintain a pH gradient ($\Delta pH$) across the cell membrane. For example, bacteria that grow optimally between pH 1 and 4 (i.e., *acidophiles*) have an internal pH of about 6.5–7. That is to say, they maintain a $\Delta pH$ of greater than 2.5 units, inside alkaline. Those that grow optimally between pH 6 and

8 (i.e., *neutrophiles)* have an internal pH of around 7.5–8.0 and can maintain a $\Delta pH$ of 0.5–1.5 units, inside alkaline. Those that grow best at pH 9 or above (often in the range of pH 10–12) (i.e., *alkaliphiles*) have an internal pH of 8.4–9. Therefore, alkaliphiles maintain a negative $\Delta pH$ of about 1.5–2 units, inside acid. (This lowers the $\Delta p$. For a discussion of this point, see Section 3.10.)

### 15.1.2 Demonstrating pH homeostasis

One method to demonstrate pH homeostasis is to perturb the cytoplasmic pH by changing the external pH, and then study the recovery process. For example, when *E. coli* is exposed to a rapid change in the external pH (e.g., of 1.5–2.5 units), the internal pH initially changes in the direction of the external pH. However, a recovery soon occurs as the internal pH bounces back to its initial value (Fig. 15.1). In other bacteria, pH homeostasis may be so rapid that a temporary change in the internal pH is not even measurable.

Table 14.1 pH homeostasis in bacteria

| Bacteria | $pH_{out}$ | $pH_{in}$ | $\Delta pH$ (in–out) |
|---|---|---|---|
| Neutrophile | 6–8 | 7.5–8 | + |
| Acidophile | 1–4 | 6.5–7 | + |
| Alkaliphile | 9–12 | 8.4–9 | − |

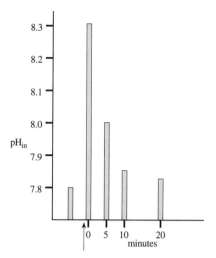

**Fig. 15.1** pH homeostasis in *E. coli*. Growing *E. coli* cells were shifted from pH 7.2 to 8.3 (arrow). The difference in pH (pH$_{in}$-pH$_{out}$) was measured using a weak acid or a weak base as described in Chapter 3, and the pH$_{in}$ was calculated. Immediately after shifting the cells to pH 8.3, the cytoplasmic pH rose to 8.3. However, within a few minutes the cytoplasmic pH was restored to its approximate original value. The mechanism by which *E. coli* acidifies the cytoplasm to maintain pH homeostasis is uncertain. *Source:* Data from Zilberstein, D., V. Agmon, S. Schuldiner, and E. Padan. 1984. *Escherichia coli* intracellular pH, membrane potential, and cell growth. *J. Bacteriol.* **158**:246–252.

### 15.1.3 The mechanism of pH homeostasis

#### Proton pumping

One would expect that many factors influence the intracellular pH. These include the buffering capacity of the cytoplasm and metabolic reactions that produce acids and bases. However, it is generally believed that the regulation of intracellular pH is, to a large extent, a consequence of controlling the flow of protons across the cell membrane.

The reason for believing this is that the ΔpH is dissipated when the proton pumps are inhibited or when proton ionophores are added to the medium. (Recall that proton ionophores equilibrate protons across the cell membrane.) The idea is that, when the cytoplasm becomes

too acid, protons are pumped out. This must be done electroneutrally (e.g., by bringing K$^+$ into the cell). When the cytoplasm becomes too basic, protons are brought in via exchange with outgoing K$^+$ or Na$^+$ (Fig. 15.2). This implies the existence of feedback mechanisms where the intracellular pH can signal proton pumps and antiporters. Regulation at this level is not understood. It should be pointed out that even the influx of protons depends upon the outgoing proton pumps. This is because the antiporters that bring in protons in exchange for sodium or potassium ions are driven by the Δp, either by the ΔpH, outside acid, and/or by the membrane potential. The latter is the case when the inflow of protons is electrogenic [i.e., when the ratio of H$^+$ to Na$^+$ (or K$^+$) on the antiporter exceeds one]. Therefore, inhibition of the proton pumps should lead to a situation where pH$_{in}$ = pH$_{out}$ (i.e., a collapse of the ΔpH), and this can be tested. The two major classes of proton pumps are those coupled to respiration and the proton translocating ATPase. The former can be inhibited by respiratory poisons such as cyanide, and the latter by inhibitors of the ATPase (e.g., DCCD) or by mutation. One can also short-circuit the proton flow by using proton ionophores. The ionophores will dissipate the proton potential, thus neutralizing the pumps.[7] These experiments show that in the absence of proton pumping the ΔpH falls as the protons tend to equilibrate across the cell membrane.

#### The role of K$^+$ in maintaining a ΔpH

A problem in using the proton pumps to move protons out of the cell is that the pumps are electrogenic, and the number of protons that can be pumped out of the cell is limited by the membrane potential that develops. (See the discussion of membrane capacitance; Section 3.2.2.) Thus, in order for proton pumping to raise the intracellular pH, the excess membrane potential must be dissipated either by the influx of cations or the efflux of anions. In neutrophilic bacteria, K$^+$ influx dissipates the membrane potential, allowing more protons to be pumped out of

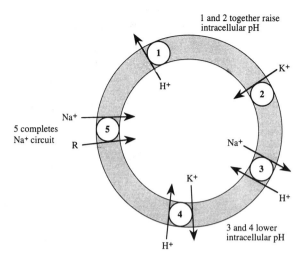

**Fig. 15.2** Mechanisms of pH homeostasis. (1) Primary proton pumps create a membrane potential. (2) The uptake of $K^+$ dissipates the membrane potential, allowing extrusion of protons via the pumps and an alkalinization of the cytoplasm. The bulk solution is kept electrically neutral because of the counterions that had neutralized the $K^+$. Therefore, a $\Delta\Psi$ is changed into a $\Delta pH$, inside alkaline. (3) and (4) Cation/proton antiporters pump $Na^+$ and $K^+$ out and bring in $H^+$. These are suggested to be the major mechanism for acidifying the cytoplasm. (5) Sodium ion uptake systems complete the sodium circuit.

the cell. This means that $K^+$ influx is required to raise the intracellular pH.

This is seen in Fig. 15.3, where the addition of $K^+$ to *E. coli* cells suspended in media of low pH caused an immediate increase in intracellular pH, which stabilized at approximately pH 7.6. There is much more uncertainty as to how neutrophiles might *decrease* their intracellular pH. Potassium ions and sodium ions have been suggested to play a role in lowering the intracellular pH by bringing protons back into the cell via the $H^+/K^+$ and $H^+/Na^+$ antiporters (Fig. 15.2). However, the evidence for this is not as strong in neutrophiles as it is in alkaliphiles.

### In alkaliphiles the $Na^+/H^+$ antiporter acidifies the cytoplasm

See Ref. 8 for a review of bioenergetics in alkaliphiles, including pH homeostasis. Since alkaliphiles live in a very basic medium, their main problem is keeping a cytoplasmic pH more acid than the external pH, perhaps by as much as 2 units. In other words, they must always be bringing protons into the cell. In contrast to research with neutrophiles, a strong case can be made for the acidification

of the cytoplasm of alkaliphiles by $Na^+/H^+$ antiporters[9] (Fig. 15.2). When alkaliphiles are placed in a medium without $Na^+$ at pH 10 or 10.5, the internal pH quickly rises to the value of the external pH. However, when $Na^+$ is present in the external medium, the internal pH does not rise upon shifting to the more basic medium. (See Note 10 for more detail.) Furthermore, mutants of alkaliphiles that cannot grow at pH values above 9 are defective in $Na^+/H^+$ antiporter activity. The antiporter is electrogenic ($H^+ > Na^+$) and driven by the $\Delta\Psi$, which is generated by the primary proton pumps of the respiratory chain. The sodium ion circuit is completed when sodium ion enters the cell via $Na^+$/solute symporters that are also driven by the $\Delta\Psi$. The use of $Na^+$/solute symporters has the advantage that solute transport is driven by the sodium potential rather than the proton potential, the latter being low because of the inverted $\Delta pH$.

### pH homeostasis in acidophiles

Acidophiles differ from other bacteria in that the external pH is several units *lower* than the cytoplasmic pH.[6] The maintenance of the large

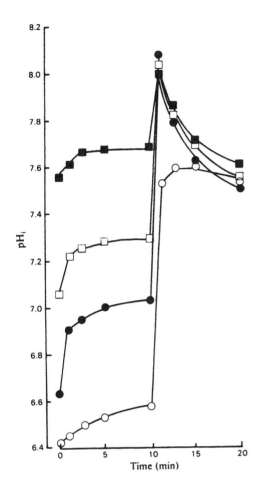

**Fig. 15.3** Uptake of $K^+$ in *E. coli* is associated with pH homeostasis. Washed *E. coli* cells were suspended in buffer without $K^+$ at pH 5.3 (open circles), 6.8 (closed circles), 7.15 (open squares), or 7.6 (closed squares). The cytoplasmic pH ($pH_i$) was determined using the distribution of a weak acid. Glucose was added shortly after 0 min and $K^+$ was added at 10 min. Potassium uptake was complete within 10 min after its addition. Immediately upon the addition of $K^+$ the intracellular pH, which had been lowered by the extracellular pH, rose and stabilized at approximately pH 7.6. The rise in the cytoplasmic pH was due to pumping protons out of the cell in response to the depolarization of the membrane by the influx of potassium ions. The mechanism for acidification of the cytoplasm, i.e., recovery from overshoot of the $pH_{in}$, is unknown. *Source:* From Kroll, R. G., and I. R. Booth. 1983. The relationship between intracellular pH, the pH gradient and potassium transport in *Escherichia coli. Biochem. J.* **216**:709–716.

$\Delta pH$ requires an *inverted* $\Delta\Psi$ at low $pH_0$; otherwise proton efflux would be limited by a positive $\Delta\Psi$ as well as a low $pH_o$, and proton influx would be promoted by a negative $\Delta\Psi$ as well as a high $pH_{in}$. Accordingly, acidophilic bacteria have small membrane potentials, which can be inside positive at acidic $pH_0$. For example, the membrane potential of *Thiobacillus ferrooxidans* is +10 mV at $pH_0$ 2, and the membrane potential of *Bacillus acidocaldarius* is +20 to +30 mV at $pH_0$ 2.5. However, acidophiles pump protons out of the cell during electron transport, generating a membrane potential that is outside positive like other bacteria. How is the membrane potential inverted? It has been suggested that the maintenance of the inverted membrane potential in acidophiles is due to an *inward* flux of $K^+$ greater than an outward flux of protons. This might be due to the electrogenic influx of $K^+$ catalyzed by an ATP-dependent $K^+$ pump known to exist. Thus the method of maintaining a $\Delta pH$ in acidophiles and neutrophiles may be similar in relying on $K^+$ influx to depolarize the membrane. It has also been suggested that the *efflux* of $K^+$ in acidophiles may be slowed when $pH_{in}$ falls, thus limiting the entry of protons against a positive $\Delta\Psi$.

Although the protons are not being pumped against a membrane potential when the external pH is acid, they are being pumped against a proton gradient. The $\Delta pH$ in acidophiles can be 4 to 5 units, which is equivalent to 240–300 mV (i.e., 60 $\Delta pH$). This is a large concentration gradient against which to pump protons. The energy to pump the protons is derived from aerobic respiration. However, iron-oxidizing acidophilic bacteria do not generate sufficient energy during electron transport to pump protons out of the cell because the $\Delta E_h$ between $Fe^{3+}/Fe^{2+}$ and $O_2/H_2O$ is very small (less than 100 mV). They appear to regulate their $\Delta pH$ by consuming cytoplasmic protons during respiration, rather than by proton pumping. This is discussed in Section 12.4.1.

## 15.2 Osmotic Pressure and Osmotic Potential

### 15.2.1 Osmotic pressure

When two solutions are separated by a membrane that allows the passage of water but not of solute, then the water will diffuse from the less concentrated to the more concentrated side in order to equalize the water concentration. What is happening is that the concentration of water (actually the thermodynamic activity of water) in the less concentrated solution is higher than in the concentrated solution. Thus the water is simply following its concentration gradient. The diffusion of water into the more concentrated solution is called *osmosis*. If the water were diffusing from a side with pure water, then the pressure that would have to be applied to stop the osmotic flow of water is called the *osmotic pressure*, which is given the symbol $\Pi$. If a solution is sufficiently dilute so that one can discount the interactions between the solute molecules, then $\Pi = RTC_s$, where $R$ is the gas constant, $T$ is temperature in degrees kelvin, and $C_s$ is the molar concentration of solute particles. It is important to point out that $C_s$ represents the concentration of independent particles that contribute to the osmotic pressure. For example, $C_s$ is equal to the sum of the concentration of ions produced when a salt completely ionizes in solution. The osmolarity (OsM) of a solution is equal to $\Pi/RT$, and the units are molarity. $\Pi/RT$ is determined experimentally. At higher solute concentrations one must take into account the interactions between solute molecules, and $\Pi/RT$ is not equal to $C_s$ but rather to the sum of the *effective* molar concentrations of the solutes. Often concentrations are expressed as molality (moles of solute per kilogram of solvent) rather than molarity, and the units of osmotic pressure are given as *osmolality*.

### 15.2.2 Osmotic potential

For a review of osmotic potential and its regulation, see Ref. 11. Csonka has pointed out that the term osmotic *potential* is more useful than osmotic pressure in that it

emphasizes that water flows from solutions of a high osmotic potential to solutions of a low osmotic potential.[12] [The osmotic potential is numerically equal to the osmotic pressure but has a negative sign (eq. 15.3).] The osmotic potential, $\pi$, is a function of the activity ($a$) of the solvent. For water, this would be:

$$\pi = (RT/V_w) \ln a_w \qquad (15.1)$$

where $R$ is the universal gas constant, $T$ is the absolute temperature, $V_w$ is the partial molal volume of water, and $a_w$ is the activity of water. The activity of pure water is defined as one, making the osmotic potential zero. Solutes tend to lower the activity of water, therefore making the osmotic potential of solutions negative. This is because in an ideal solution (i.e., one where the interactions between solute and solvent molecules are independent of concentration) the activity of the solvent is equal to its mole fraction. For example, for water:

$$a_w = n_w/(n_w + n_s) \qquad (15.2)$$

where $n_w$ equals the number of water molecules and $n_s$ equals the number of solute molecules. For dilute solutions, the osmotic potential is related to the molar concentration (moles of solute per liter), $C_s$, as:

$$\pi = -RTC_s \qquad (15.3)$$

Equation 15.3[13] points out that solutions with higher concentrations of solutes have more negative osmotic potentials. Thus water flows into these solutions.

### 15.2.3 Turgor pressure and its importance for growth

Because the cytoplasm of most bacterial cells is much more concentrated in particles than is the medium in which the cells are suspended, the cytoplasm has a more negative osmotic potential (a more positive osmotic pressure) than the medium, and water flows into the cell. The incoming water expands the cell membrane, which exerts a pressure directed outward against the cell wall. The pressure exerted against the cell wall is called the *turgor pressure*. The turgor pressure is equal to the difference in osmotic pressure between the medium and the cytoplasm.

$$P = \Delta\Pi = \Pi_{in} - \Pi_{out}$$
$$= RT(OsM_{in} - OsM_{out}) \quad (15.4)$$

where $P$ is the turgor pressure, and OsM is the total concentration (in units of molarity) of osmotically active solutes in the cells and in the medium.

The turgor pressures in gram-positive bacteria are about 15–20 atm and in gram-negative bacteria between 0.8 and 5 atm.[14–16] This, of course, is the reason that bacterial cell walls must be so strong. In bacteria, the tensile strength of the cell wall is due to the peptidoglycan. It is important to realize that the turgor pressure provides the force that expands the cell wall and is necessary for the growth of the wall and cell division.[17] In fact, sudden decreases in turgor pressure brought on by increasing the osmolarity of the suspending medium result in a cessation of growth accompanied by the inhibition of a variety of physiological activities (e.g., nutrient uptake and DNA synthesis). Therefore, the physiological significance of osmotic homeostasis is that it maintains an internal turgor pressure necessary for growth.

Bacteria have the capability of adjusting their internal osmolarity to changes in external osmolarity in order to maintain cell turgor. How bacteria detect differences in external osmolarity and transfer the appropriate signals to the adaptive cellular machinery is largely unknown. The problems associated with analyzing the signaling are discussed in Section 15.2.4. Before we address these problems, we will examine the evidence for osmotic regulation and identify the molecules primarily responsible for maintaining the osmotic differential between the cytoplasm and the external medium.

## 15.2.4 Adaptation to high osmolarity

When cells are placed in media of high osmolarity, they increase the intracellular concentrations of certain solutes called *osmolytes,* thus ensuring that the internal osmolarity is always higher than the external, and that cell turgor is therefore maintained. The osmolytes used by bacteria are sometimes called *compatible solutes* to reflect their relative nontoxicity. Some compatible solutes are not synthesized by the cells but are accumulated intracellularly from the medium via transport. Organic compatible solutes that are accumulated via transport but are not synthesized are called *osmoprotectants.* One of the most important compatible solutes in bacteria is $K^+$, whose intracellular concentration is sufficiently high to be a major contributor to the internal osmolarity and hence turgor. (See Note 18.) Bacteria that live in high-osmolarity madia have proportionally higher intracellular concentrations of K+ because of uptake. Therefore, $K^+$ is important for two major aspects of homeostasis: maintenance of cell turgor and maintenance of a $\Delta$pH (Section 15.1.3). Bacteria employ other compatible solutes in addition to $K^+$. The situation with regards to compatible solutes can be summarized as follows.

Solutes that increase intracellularly in many different bacteria in response to high external osmolarity include $K^+$, the amino acids glutamate, glutamine, and proline, the quarternary amine betaine (also called trimethylglycine or glycine betaine), and certain sugars (e.g., trehalose, which is a disaccharide of glucose). Very few chemoheterotrophic bacteria can synthesize betaine de novo, and they probably find it in the environment. It is, however, synthesized by cyanobacteria and phototrophic bacteria and is probably abundant in saline environments where these phototrophs are growing. Betaine is also synthesized by several halophilic and halotolerant archaebacteria. Pathways for the biosynthesis of organic compatible solutes have been reviewed.[16]

### Osmotic homeostasis in halobacteria

As described in Section 1.1.1, the halobacteria (archaeons) live in waters where the NaCl concentrations are from 3 to 5 M. In order to prevent water from exiting the cell via osmosis and also to maintain a positive internal turgor pressure, the cytoplasm is kept quite salty. However, the cation is $K^+$ (KCl), rather than $Na^+$ (NaCl). Potassium ion uptake systems maintain internal $K^+$ concentrations on the order of 3 M. The $Na^+$ is exported from the cytoplasm.[19] Of course there must be adaptations in the cytoplasm in order for the proteins to cope with such a high ionic strength. It is well known that ionic interactions are weakened in the presence of high concentrations of salt, and this can have a profound effect on the tertiary and quaternary structure of proteins leading, to unfolding and dissociation of subunits. In fact, however, the proteins and structures in extreme halophiles *require* high salt (at least 1 M) for stability and activity. Even the cell envelope of halobacteria disintegrates when the salt concentration is lowered. (See Ref. 20 for a review of the effect of salt on halophilic macromolecules and Ref. 21 for a general review of the biology of the halophilic bacteria.)

## Osmotic homeostasis in E. coli

It is possible to observe osmotic homeostasis by rapidly increasing the external osmolarity of the medium and identifying the intracellular compatible solutes that maintain turgor. As an example of osmotic homeostasis, some experiments performed with *E. coli* will be considered.[22] When *E. coli* is shifted to

a medium of high osmolarity, a series of adjustments take place (Fig. 15.4). First, there is an influx of $K^+$ via the potassium uptake systems discussed in Chapter 16. These are thought to respond to a decrease in turgor pressure.[23] However, there are two problems that must be solved by the cell in order that there be an influx of $K^+$. Steps must be taken to (1) preserve electrical neutrality in the cytoplasm and (2) prevent depolarization of the cell membrane, which would lead to a drop in the $\Delta p$. In fact, *E. coli* handles these problems in two ways. There is a transient alkalinization of the cytoplasm as protons are pumped out to prevent depolarization of the membrane and to aid in maintaining cytoplasmic neutrality during the early rapid uptake of $K^+$. However, the cytoplasm reacidifies, and the the main reason for cytoplasmic neutrality is the rapid accumulation during reacidification of glutamic acid, which ionizes providing counterions to the increased levels of $K^+$.[24, 17] The accumulation of glutamic acid is due either to increased synthesis or decreased utilization, or a combination of the two.

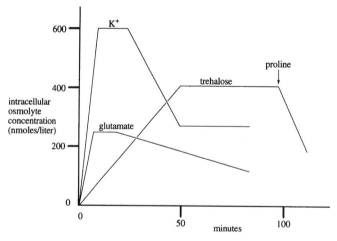

Fig. 15.4 Changes in intracellular osmolyte concentrations when *E. coli* is shifted to a medium of higher osmolarity. Growing cells were shifted to a medium containing 0.5 M NaCl at time zero. The intracellular osmolyte concentrations are the differences between the concentrations in cells subjected to osmotic upshock and control cells that were not. Upon a shift to a higher osmolarity medium the cells accumulated potassium ion and synthesized glutamate. The potassium glutamate was replaced by newly synthesized trehalose. The addition of proline to the medium (arrow) caused the displacement of trehalose by proline. Proline was also capable of inducing an early efflux of $K^+$ if added earlier. *Source:* Data adapted from Dinnbier et al. 1988. Transient accumulation of potassium glutamate and its replacement by trehalose during adaptation of growing cells of *Escherichia coli* K-12 to elevated sodium chloride concentrations. *Arch. Microbiol.* **150**:348–357.

Thus the initial major compatible solute is potassium glutamate. However, after several minutes the potassium glutamate is replaced by the newly synthesized sugar, trehalose, which then becomes the major compatible solute. This is due to the excretion of the $K^+$ and the catabolism or excretion of glutamate as well as the synthesis of trehalose. If proline or betaine are present in the medium, then *E. coli* transports these into the cell and replaces the trehalose or the excess $K^+$, indicating that *E. coli* preferentially use proline and betaine as osmotic stabilizers rather than trehalose or $K^+$. (See Note 25.) (The $K^+$ is excreted, and the trehalose is catabolized.) The preferential use of some osmoprotectants over others is common among bacteria. In many bacteria betaine is a preferred osmoprotectant that is readily accumulated from media of high osmolarity and can even suppress the uptake of other osmoprotectants, as well as causing the excretion of $K^+$ and the catabolism of trehalose, thus replacing these osmolytes.

## Effect of osmolarity on transcription and on activities of enzymes

As part of adaptation to a change in media osmolarity, bacteria synthesize new enzymes or transporters that are responsible for the biosynthesis of compatible solutes or the transport of these into the cells. Increased transcription of some of the genes activated by an osmotic upshift is due to increased levels of the transcription factor $\sigma^s$ (sigma S). The transcription factor $\sigma^s$ is a subunit of RNA polymerase that is responsible for recognizing $\sigma^s$-dependent promoters that are activated during osmotic upshifts as well as during starvation. (See Section 2.2.2 and Ref. 26 for a review.) The increase in $\sigma^s$ is regulated by medium osmolarity at the level of translation and protein turnover. A few examples of the effect of osmolarity on transcription and enzyme acitivities are:

1. *E. coli* activates enzymes for the synthesis of periplasmic oligosaccharides when shifted to low- osmolarity media. This is explained in Section 15.2.3.

2. *Staphylococcus aureus* activates a preexisting proline uptake system when shifted to high-osmolarity media.[27] Proline is an osmoprotectant.

3. *E. coli* and *S. typhimurium* increase the transcription of *proU*, an operon that codes for a proline (and betaine) transport system when shifted to high-osmolarity media.

4. Another set of genes whose transcription is increased in high-osmolarity media is the *kdp* operon in *Escherichia*, which codes for a high-affinity $K^+$ transport system. (See Section 18.11 for a discussion of the regulation of the *kdpABC* operon.) As mentioned, $K^+$ is a major compatible osmolyte in most bacteria.

One can ask how bacteria sense changes in the external osmolarity and transmit the appropriate signals to the genome or to certain enzymes. As indicated earlier, the answers are largely unkown. Perhaps the best studied signaling system that responds to changes in external osmolarity includes the genes for the OmpF and OmpC porins in *Escherichia* and *Salmonella* (see Section 1.2.3). The total amounts of OmpF and OmpC are fairly constant, but the ratio changes with the osmolarity and temperature of the medium. In *high-osmolarity* media the transcription of the *ompF* gene, which codes for the *larger OmpF channel* is *repressed*, and the transcription of the *ompC* gene, which codes for the *smaller OmpC channel* is *increased*. The result is a switch to a smaller porin channel in high-osmolarity media. Why bacteria should switch from one porin to the other is not clear, but it probably has nothing to do with osmotic homeostasis. The smaller OmpC channel is probably an advantage in the intestinal tract, where these bacteria live, because of the presence of toxic molecules (e.g., bile salts).[28] The argument is that, in the intestinal tract, the osmotic pressures are higher, thus favoring the smaller OmpC channels, whereas in ponds and streams where the bacteria are also found, the lower osmotic pressures favor the larger OmpF channels, which may allow more efficient uptake of nutrients. Consistent with this hypothesis is that higher temperatures, expected in the intestines of animals as opposed to habitats outside the body, also repress the transcription of *ompF*.

Regardless of the physiological significance of the switch in porins, the system is of interest to us here because it is regulated in some way by osmotic pressure, and the signaling pathway from the membrane to the genome is being dissected experimentally.[29] The signaling pathway, which involves a membrane sensor protein called EnvZ, is reviewed in Section 18.10.1.

### 15.2.5 Adaptation to low-osmolarity media

Thus far we have been considering the adaptation of bacteria to *high-osmolarity* media, which tends to suck water out of the cytoplasm and lower the turgor pressure. As discussed above, bacteria respond to high osmolarity by raising the intracellular concentrations of compatible solutes. Many bacteria have means for adjusting to low-osmolarity media, thus limiting cell turgor pressure. Not very much is known about the details. One response of bacteria to low-osmolarity media is to decrease the concentration of cytoplasmic osmolytes. This might occur via specific excretion of osmolytes or their catabolism. For example, *E. coli* excretes $K^+$ via special transporters that may respond to cell turgor.[30]

### Osmotic homeostasis in the periplasm

The periplasm is reportedly filled with a gel whose volume is still a matter of controversy (Section 1.2.4). (See Note 31.) It has been suggested that gram-negative bacteria adapt to low-osmolarity media by raising the osmolarity of the periplasm so that the cytoplasm never actually "sees" the low-osmolarity external medium. In this way swelling of the cell membrane with concomitant compression of the periplasm is minimized, as well as the turgor pressure across the cell membrane. (The option of lowering the osmolarity of the cytoplasm by excreting solutes is limited. The cytoplasm must maintain a minimum concentration of salts and other solutes, approximately 300 mosM, to support growth.) In fact, it has been reported that the periplasm remains as a separate compartment under all conditions of external osmolarity, with an osmolarity apparently isoosmotic with that of the cytoplasm.[32] However, since the outer membrane of *Escherichia* and *Salmonella*, and by inference other gram-negative bacteria, is permeable to small molecules of molecular weight less than 600 because of the nonspecific diffusion channels formed by the porins, one can ask how bacteria maintain an osmolarity in the periplasm higher than that of the external medium. One possibility is that gram-negative bacteria synthesize and/or accumulate periplasmic osmolytes when grown in low-osmolarity media. Indeed, many gram-negative bacteria besides E. coli, including those belonging to the genera *Salmonella, Pseudomonas, Agrobacterium, Acetobacter, Klebsiella, Enterobacter, Bradyrhizobium, Brucella, Xanthomonas, Alcaligenes,* and *Rhizobium*, synthesize periplasmic β-glucan oligosaccharides (called membrane-derived oligosaccharides, or MDO, in E. coli) when grown in media of *low* osmolarity.[33–35] It has been shown that *E. coli, Agrobacterium, Rhizobium,* and *Bradyrhizobium* increase the synthesis of periplasmic β-glucans when grown in low-osmolarity media. The enzymes that synthesize the oligosaccharides are constitutive; therefore, their *activities* are increased when the cells are grown in low-osmolarity media.[30] The increase in the synthesis of the periplasmic oligosaccharides when the bacteria are grown in media of low osmolarity has led to the suggestion that the role of the oligosaccharides is to raise the osmolarity in the periplasm. The glucans generally have anionic groups because they are substituted with phosphorylated compounds (i.e., phosphoglycerol, phosphoethanoamine, and phosphocholine, and sometimes succinic acid). The anionic oligosaccharides would be expected to be very effective in raising the osmolarity of the periplasm because cations would accumulate in the periplasm in response to the negatively charged nonpermeable oligosaccharides. (See Note 36.) However, *E. coli* mutants unable to synthesize the MDOs show no growth defects in low-osmolarity media.[37] If it is necessary to maintain a high periplasmic osmolarity when the external osmolarity is decreased, then there must exist alternative mechanisms besides the synthesis of MDOs.

## 15.2.6 Conceptual problems

It has been suggested that the periplasm is maintained isoosmotic with respect to the cytoplasm. If indeed this is the case, then this has important consequences for our understanding of how the cell walls of gram-negative bacteria are able to resist high turgor pressures. It is generally believed that the turgor pressure is exerted across the cell membrane, not the outer membrane, and that the overlying peptidoglycan acts as a strong retainer against which the cell membrane is pressed. If the periplasm were isoosmotic with the cytoplasm, then the turgor pressure would not be across the cell membrane and against the peptidoglycan, but rather against the outer membrane. However, the outer membrane is not built to withstand high turgor pressures. If indeed the turgor pressure were exerted against the outer membrane, then one must suppose that the peptidoglycan reinforces the outer membrane by being tightly bonded to it at numerous sites. Perhaps the lipoprotein molecules that are covalently bonded to the peptidoglycan are anchored sufficently to the outer membrane via hydrophobic bonding to provide the needed stability (Section 1.2.3).[38] Another conceptual difficulty regarding a periplasm isoosmotic with the cytoplasm is that it is not clear how turgor pressure would be sensed by cell membrane proteins if there were no differential pressure across the cell membrane. (This is discussed in Section 15.2.5.) Clearly, much more needs to be learned in order to come to a better understanding of turgor pressure and osmotic regulation in the periplasm.

## 15.2.7 What is the nature of the signal sensed by the osmosensors?

The signals to which the putative osmosensors respond are not understood.[9] Indeed, the different osmosensors may not even respond to the same type of signal. Several possibilities for osmosensor responses have been discussed, including membrane proteins that are sensitive to: (1) pressure against the peptidoglycan sacculus; (2) membrane stretch; (3) changes in the concentrations of intracellular solutes (i.e., $K^+$); and (4) water activity. It is important to know whether the periplasm is truly isoosmotic with the cytoplasm, as discussed above, for under these conditions there should be no pressure differential across the cell membrane, and mechanisms (1) and (2) above would not apply.

It should also be understood that after the cells adapt to an osmotic upshift and begin to grow again, the osmotic differential between the cytoplasm and medium is presumably restored, and, therefore, systems still activated under these circumstances cannot be responding to a change in turgor pressure or related events such as membrane stretching. They could, however, be responding to increased concentrations of specific solutes or some other parameters not dependent upon a changed turgor pressure. For example, when E. coli is shifted to a high-osmolarity medium, it continues to repress the ompF gene and stimulate the ompC gene, even though adaptation to the high-osmolarity medium has taken place. Similarly, the proU genes that specify proline and glycine betaine uptake systems in E. coli and S. typhimurium remain induced in high-osmolarity media, also implying that the sensor is not responding to a changed osmotic pressure differential on both sides of the cell membrane. On the other hand, the kdp genes for potassium ion, uptake are quickly but only transiently induced by an upshift in osmolarity. This fact, plus the finding that only nonpermeable solutes induce transcription of the kdp operon, suggests that the osmosensor for these genes may indeed detect turgor pressure or something closely related, such as membrane stretch. (See Section 18.11 for a discussion of the kdp gene products.)

## 15.3 Summary

Bacteria must maintain a fairly constant internal pH and adjust their osmolarity in response to the external osmolarity. Very little is known about the details of how this is done. In broad outline, however, the internal pH is adjusted by using proton pumps to extrude protons and antiporters that bring protons into the cell in exchange for sodium or potassium ions. What is not clear is precisely how the activities of the pumps and antiporters are regulated by the external pH, although it is reasonable to suggest that they respond to the internal pH by some sort of feedback mechanism. The acidophilic iron-oxidizing bacteria represent a special problem in that there is very little energy obtainable by $Fe^{2+}$ oxidation to pump protons out of the cell. They maintain a $\Delta pH$ by consuming cytoplasmic protons during oxygen respiration.

The reason that bacteria must maintain an osmotic differential across the cell membrane is that the resulting turgor pressure is essential for growth of the cell wall and for cell division. A variety of physiological activities come to a halt when the turgor pressure is suddenly decreased (e.g., nutrient uptake and DNA synthesis). When the external osmotic pressure is increased (causing decreased turgor), bacteria respond by increasing the concentration of internal osmolytes to raise the internal osmotic pressure and turgor. An important molecule in this regard is potassium ion, which increases, in some cases transiently, in response to an increase in external osmolarity. A counterion must also increase, and in *E. coli* this is glutamate. In *E. coli*, the $K^+$ is subsequently replaced by trehalose. If proline or betaine are in the medium, then they replace the $K^+$ and trehalose.

It is clear that transcriptional changes occur when bacteria are shifted into media of different osmolarities. For example, when *E. coli* is grown in high-osmolarity media, the transcription of several genes is increased. These include *proU*, the operon that codes for the uptake of two osmoprotectants, proline and betaine, and *kdp*, the operon that codes for the proteins required for the uptake of $K^+$. Also, when *E. coli* and *S. typhimurium* are grown in high external osmolarity, the gene for the OmpC porin is activated, whereas the gene for the OmpF porin is repressed. This leads to more OmpC and less OmpF. A protein in the cell membrane called EnvZ has a periplasmic domain and is thought to be an osmosensor (Chapter 18).

There is also osmoregulation of the periplasm in gram-negative bacteria. Gram-negative bacteria adapt to media of low osmolarity by synthesizing membrane-derived oligosaccharides (MDO) in the periplasm. The idea is that the multiple-charged anionic MDO molecules accumulate cations in the periplasm, thus raising its osmolarity. However, mutants of *E. coli* defective in MDO synthesis show no growth defects in low-osmolarity media. It must be concluded that other mechanisms of osmoregulation of the periplasm must be present and that osmoregulation of the periplasm is not well understood at this time.

## Study Questions

1. $Na^+/H^+$ and $K^+/H^+$ antiporters are an important way to lower the intracellular pH. What is the evidence for this in alkaliphiles?

2. No one knows for sure how acidophiles establish an inverted membrane potential. How might it affect pH homeostasis? What experiments would support the hypothesis that an electrogenic $K^+$ pump was responsible for the inverted membrane potential in acidophiles?

3. Upon shifting *E. coli* to a higher osmolarity, there is a transient uptake of $K^+$. What is the role of $K^+$ in this regard? What keeps the cytoplasm neutral? How is the membrane potential maintained? Is there a complication with the $\Delta pH$? Describe the role of $K^+$ in pH homeostasis.

4. What is meant by turgor pressure? What is the evidence that turgor pressure is necessary for cell growth?

5. What is meant by periplasmic osmotic homeostasis? What might be the role of MDOs?

## NOTES AND REFERENCES

1. Padan, E., D. Zilberstein, and S. Schuldner. 1981. pH homeostasis in bacteria. *Biochim. Biophys. Acta* 650:151–166.

2. Bakker, E. P. 1990. The role of alkali-cation transport in energy coupling of neutrophilic and acidophilic bacteria: An assessment of methods and concepts. *FEMS Microbiol. Rev.* 75:319–334.

3. Krulwich, T. A., A. A. Guffanti, and D. Seto-Young. 1990. pH homeostasis and bioenergetic work in alkaliphiles. *FEMS Microbiol. Rev.* **75**:271–278.

4. Matin, A. 1990. Keeping a neutral cytoplasm; the bioenergetics of obligate acidophiles. *FEMS Microbiol. Rev.* **75**:307–318.

5. Krulwich, T. A., and D. M. Ivey. 1990. Bioenergetics in extreme environments, pp. 417–447. *The Bacteria*, Vol. XII. Krulwich, T. A. (Ed.). Academic Press, New York.

6. Booth, I. R. 1985. Regulation of cytoplasmic pH in bacteria. *Microbiol. Rev.* **49**:359–378.

7. Harold, F. M., E. Pavlasova, and J. R. Baarda. 1970. A transmembrane pH gradient in *Streptococcus faecalis*: Origin, and dissipation by proton conductors and N,N′-dicyclohexylcarbodiimide. *Biochim. Biophys. Acta.* **196**:235–244.

8. Ivey, D.M., M. Ito, R. Gilmour, J. Zemsky, A. A. Guffanti, M.G. Sturr, D.B. Hicks, and T. A. Krulwich. 1998. Alkaliphile bioenergetics, pp. 181–210. In: *Expremophiles: Microbial Life in Extreme Environments*. K. Horikoshi and W. D. Grant (Eds.). John Wiley and Sons, New York.

9. Reviewed in, Krulwich, T. A., A. A. Guffanti, and D. Seto-Young. 1990. pH homeostasis and bioenergetic work in alkaliphiles. *FEMS Microbiol. Rev.* **75**:271–278.

10. Careful measurements have been made of the $\Delta$pH of the facultative alkaliphile *Bacillus firmus* to show that the bacteria are capable of maintaining a $\Delta$pH of up to two units as the external pH is raised. When pH$_{out}$ was 7.5, pH$_{in}$ was also 7.5. When the external pH was increased from 7.5 to 10.5, the cytoplasmic pH increased only to 8.2. At an external pH of 11.4, the cytoplasmic pH was only 9.5. Growth rates did not begin to slow until the external pH was greater than 10.6. For a review, see: Ivey, D. M., M. Ito, R. Gilmour, J. Zemsky, A. A. Guffanti, M. G. Sturr, D. B. Hicks, and T. A. Krulwich. 1998. Alkaliphile bioenergetics, pp. 181–210. In: *Expremophiles: Microbial Life in Extreme Environments*. K. Horikoshi and W. D. Grant (Eds.). John Wiley and Sons, New York.

11. Csonka, L. N., and A. D. Hanson. 1991. Prokaryotic osmoregulation: Genetics and physiology. *Ann. Rev. Microbiol.* **45**:569–606.

12. Csonka, L. N. 1989. Physiological and genetic responses of bacteria to osmotic stress. *Microbiol. Rev.* **53**:121–147.

13. A derivation of eq. (15.3) can be found in the review article by Csonka. See Ref. 12.

14. Ingraham, J. L. 1987. Effect of temperature, pH, water activity and pressure on growth, pp. 1543–1554. In: Neidhardt, F. C, J. L. Ingraham, K. B. Low, B. Magasanik, M. Schaechter, and H. E. Umbarger (Eds.). *Escherichia coli and Salmonella typhimurium: Cellular and Molecular Biology.* American Society for Microbiology, Washington, D.C.

15. Koch, A. L., and M. F. S. Pinette. 1987. Nephelometric determination of tugor pressure in growing gram-negative bacteria. *J. Bacteriol.* **169**:3654–3668.

16. Walsby, A. E. 1986. The pressure relationships of halophilic and non-halophilic prokaryotic cells determined by using gas vesicles as pressure probes. *FEMS Microbiol. Rev.* **39**:45–49.

17. Koch, A. L. 1991. Effective growth by the simplest means: The bacterial way. *ASM News* **57**:633–637.

18. Glutamate and K$^+$ are present in high concentrations in most (eu)bacteria. The former is the major organic anion, and the latter is the major inorganic cation. Glutamate is generally synthesized rather than accumulated from the medium, using either glutamate dehydrogenase or glutamate synthase, discussed in Section 9.3.1. In *E. coli*, potassium ion is taken up by four different transport systems, including the Kdp and Trk systems discussed in Section 16.3.3. Potassium ion efflux is catalyzed by two systems, the KefB and KefC transporters.

19. Galinski, E. A. 1995. Osmoadaption in bacteria. *Adv. Microbial Phys.* **37**:273–328.

20. Lanyi, J. K. 1974. Salt-dependent properties of proteins form extremely halophilic bacteria. *Bacteriol. Rev.* **38**:272–290.

21. *The Biology of Halophilic Bacteria.* 1992. R. H. Vreeland and L. I. Hochstein (Eds.). CRC Press, Ann Arbor.

22. Dinnbier, U., E. Limpinsel, R. Schmid, and E. P. Bakker. 1988. Transient accumulation of potassium glutamate and its replacement by trehalose during adaption of growing cells of *Escherichia coli* K-12 to elevated sodium chloride concentrations. *Arch. Microbiol.* **150**:348–357.

23. Epstein, W. 1986. Osmoregulation by potassium transport in *Escherichia coli*. *FEMS Microbiol. Rev.* **39**:73–78.

24. McLaggan, D., T. M. Logan, D. G. Lynn, and W. Epstein. 1990. Involvement of $\gamma$-glutamyl peptides in osmoadaptation of *Escherichia coli*. *J. Bacteriol.* **172**:3631–3636.

25. Betaine is synthesized by oxidizing the hydroxyl group in choline to a carboxyl group:

$(CH_3)_3N^+-CH_2CH_2OH$ (choline)

$\longrightarrow (CH_3)_3N^+-CH_2COOH$ (betaine)

Betaine and glycinebetaine (N,N,N-trimethyl-glycine) are different names for the same molecule.

26. Hengge-Aronis, R. 1996. Back to log phase: $\sigma^s$ as a global regulator in the osmotic control of gene expression in *Escherichia coli*. *Molec. Microbiol.* 21:887–893.

27. Townsend, D. E., and B. J. Wilkinson. 1992. Proline transport in *Staphylococcus aureus:* A high-affinity system and a low-affinity system involved in osmoregulation. *J. Bacteriol.* 174:2702–2710.

28. Nikaido, H., and M. Vaara. 1985. Molecular basis of bacterial outer membrane permeability. *Microbiol. Rev.* 49:1–32.

29. Stock, J. B., A. J. Ninfa, and A. M. Stock. 1989. Protein phosphorylation and regulation of adaptive responses in bacteria. *Microbiol. Rev.* 53:450–490.

30. Bakker, E. P., I. R. Booth, U. Dinnbier, W. Epstein, and A. Gajewska. 1987. Evidence for multiple potassium export systems in *Escherichia coli*. *J. Bacteriol.* 169:3743–3749.

31. The volume of the periplasm in *Escherichia* and *Salmonella* has been reported to be as high as 20–40% of the total cell volume by some investigators and as low as 5% of the total cell volume by others. One must conclude that there is some uncertainty regarding the size of the periplasm and that, furthermore, the periplasm of different types of bacteria may vary significantly in physical dimensions.

32. Stock, J. B., B. Rauch, and S. Roseman. 1977. Periplasmic space in *Salmonella typhimurium* and *Escherichia coli*. *J. Biol. Chem.* 252:7850–7861.

33. Miller, K. J., E. P. Kennedy, and V. N. Reinhold. 1986. Osmotic adaptation by gram-negative bacteria: Possible role for periplasmic oligosaccharides. *Science* 231:48–51.

34. Weissborn, A. C., M. K. Rumley, and E. P. Kennedy. 1992. Isolation and characterization of *Escherichia coli* mutants blocked in production of membrane-derived oligosaccharides. *J. Bacteriol.* 174:4856–4859.

35. Kennedy, E. P. 1996. Membrane-derived oligosaccharides (periplasmic beta-d-glucans) of *Escherichia coli*. pp. 1064–1071. In: *Escherichia coli and Salmonella: Cellular and Molecular Biology*. F. C. Neidhardt (Ed.). ADM Press, Washington, D.C.

36. Whenever a solution (1) of a nonpermeant ion is separated from another solution (2) (e.g., by a membrane permeable to other ions), there will be an unequal distribution of ions at equilibrium. The compartment with the nonpermeant ion will contain a higher concentration of permeant ions and hence a higher osmotic pressure (or lower osmotic potential). For the case of the periplasm and the outer membrane, suppose the nondiffusible oligosaccharide is $R^-$ and the diffusible ions are $K^+$ and $Cl^-$. Then $K^+$ will passively accumulate in the periplasm as the counterion to $R^-$. At equilibrium, the concentration of %$K^+$ will be greatest on the same side of the membrane as $R^-$ (the periplasm), whereas the concentration of $Cl^-$ will be greater on the other side (the medium). Because the $K^+$ can diffuse across the outer membrane and $R^-$ cannot, a voltage potential, outside positive, develops across the membrane. This is the Donnan potential, which is really a $K^+$ diffusion potential. The following equations explain the relationships between the diffusible and nondiffusible ions. At equilibrium,

$$[K^+]_1 = [R^-]_1 + [Cl^-]_1 \qquad (15.5)$$

$$[K^+]_2 = [Cl^-]_2 \qquad (15.6)$$

Equations (15.5) and (15.6) describe electrical neutrality in the two solutions.

Because almost all the $K^+$ that diffuses across the membrane must be coupled with $Cl^-$ or else the diffusion potentials would prevent further cation (or anion) flow, the diffusion kinetics can be described by a second-order rate equation:

$$k[K^+]_1[Cl^-]_1 = k[K^+]_2[Cl^-]_2$$

where $k$ is the rate constant for diffusion through the membrane. $k$ cancels out, and therefore:

$$[K^+]_1[Cl]_1 = [K^+]_2[Cl]_2 \qquad (15.7)$$

But, eq. (15.6) states that $[K^+]_2 = [Cl^-]_2$; therefore:

$$[K^+]_1[Cl^-]_1 = [K^+]_2^2 \qquad (15.8)$$

But, eq. (15.5) states that $[Cl^-]_1 = [K^+]_1 - [R^-]_1$; therefore:

$$[K^+]_1([K^+]_1 - [R^-]_1) = [K^+]_2^2 \qquad (15.9)$$

and rearranging:

$$[K^+]_1^2 - [K^+]_1[R^-]_1 = [K^+]_2^2$$

Thus the concentration of $K^+$ on the side without $R^-$ (side 2) is less than the concentration on the side with $R^-$ (side 1) by $[K^+]_1[R^-]_1$. Note that if R⁻ is multivalent, then even more $K^+$ will accumulate, because each negative charge on R⁻ is balanced by a $K^+$.

37. Kenendy, E. P. 1982. Osmotic regulation and the biosynthesis of membrane-derived oligosaccharides in *Escherichia coli. Proc. Natl. Acad. Sci. USA* **79**:1091–1095.

38. The difficulty in reconciling a periplasm isoosmotic with the cytoplasm with the fragility of the outer membrane was suggested to me by Arthur Koch.

# 16

# Solute Transport

Bacterial cell membranes consist in large part of a phospholipid matrix that acts as a permeability barrier blocking the diffusion of water soluble molecules into and out of cells (see Section 1.2.5). This has the advantage of allowing the bacterium to maintain an internal environment different from the external, and one conducive to growth. For example, metabolites can be maintained at an intracellular concentration that is orders of magnitude higher than the extracellular extracellular. This has two important consequences: (1) the promotion of more rapid enzymatic reactions, and (2) the retention of metabolic intermediates within the cell. The lipid barrier also minimizes the passive diffusion of ions, including protons, and thus functions to maintain the electrochemical proton and sodium ion gradients that are important for driving ATP synthesis, solute transport, and other membrane activities. Since the phospholipid presents a permeability barrier, virtually everything that is not lipid soluble enters and leaves the cell through integral membrane proteins that have various names, including transporters, carriers, porters, or permeases.

The amino acid sequences of a few transporters, deduced from nucleotide sequences, imply that they form multiple transmembrane loops that fold in the membrane, perhaps making an internal channel through which the solute is passed. The solute might move along amino acid side chains that face into the channel. For example, the lactose/proton symporter from *E. coli* probably forms 12 transmembrane loops.[1] The importance of transporters is easy to demonstrate. Mutants that lack a transporter for a required nutrient will grow very poorly in low concentrations of the nutrient. In the natural environment it would not be expected that such mutants would survive. But they can be isolated during a screen for slow growers and maintained as laboratory cultures. Although not discussed in this chapter, transport through the outer membrane of gram-negative bacteria is an important area of research. The role of porins and of TonB in outer membrane transport is discussed in Sections 1.2.3 and 1.2.4, respectively.

## 16.1 Reconstitution into Proteoliposomes

*Proteoliposomes* are artificial membrane vesicles of protein and phospholipid that are of enormous value in studying solute transport, and it is instructive to describe how they are made and some of their properties. There are several methods to prepare proteoliposomes, but they are all similar and rely upon the fact that membrane proteins solubilized in detergent will integrate into phospholipid bilayers when the detergent is removed by dilution or dialysis.[2] One way to prepare phospholipid bilayers is first to disperse the phospholipids in water, where they spontaneously aggregate to form spherical vesicles called *liposomes* consisting of concentric layers of phospholipid (Fig. 16.1). The liposomes are then subjected to high-frequency sound waves (sonic oscillation), which breaks them into smaller vesicles surrounded by a single phospholipid bilayer resembling the lipid bilayer found in natural membranes. Then the protein, which has been solubilized in detergent, is mixed with the sonicated phospholipids in the presence of detergent and buffer, and the suspension is diluted into buffer, which lowers the concentration of detergent. The protein leaves the detergent and becomes incorporated into the phospholipid bilayer. The protein and lipid membrane vesicles that form are called proteoliposomes. One can "load" the proteoliposomes with ATP or other substrates by including these in the dilution buffer. When the proteoliposomes are incubated with solute, they catalyze uptake of the solute into the vesicles, provided the appropriate transporter has been incorporated. The vesicles can be reisolated (e.g., by centrifugation or filtration), and the amount of solute taken into the vesicles can be quantified. Some examples of transporters that have been reconstituted into proteoliposomes and used to demonstrate catalyzed transport are the lactose permease from *E. coli*, the oxalate/formate antiporter from *Oxalobacter formigenes*, the $Na^+/H^+$ antiporter from *E. coli*, and the histidine permease from

*Salmonella typhimurium*.[3–6] (Proteoliposomes are also used to study electron transport. For example, they have been important in experiments that establish that certain cytochrome oxidases are proton pumps.)

## 16.2 Kinetics of Solute Uptake

### 16.2.1 Transporter-mediated uptake

The existence of transporters can be revealed by the kinetics of solute uptake. If one were to add a solute (e.g., an amino acid or sugar) at different concentrations to a bacterial suspension, and plot the initial rate of uptake into the cell as a function of the external concentration of solute, a curve such as that shown in Fig. 16.2 would be generated. Notice that the curve is a hyberbola that approaches a maximum rate. The kinetics for transporter-mediated solute uptake can be rationalized by assuming that the only significant route of entry for the solute is on a limited number of transporters. That is to say, the solute does not passively leak into the cell to any significant extent. Therefore, the rate at which the solute enters is directly proportional to the fraction of transporters that are occupied with solute. As the external concentration of solute increases, a progressively larger fraction of the transporters bind solute, and the rate of transport increases to a maximum rate ($V_{max}$), when there are no unloaded transporters.

The concentration of solute that produces one-half the maximum initial rate of transport

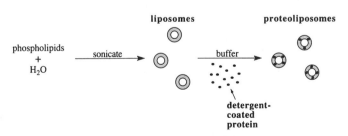

Fig. 16.1 Preparation of proteoliposomes. Phospholipids are dispersed in water and sonicated. Small vesicles each surrounded by a lipid bilayer form. The vesicles are mixed with detergent-solubilized protein and diluted into buffer. Proteoliposomes form with the protein incorporated into the bilayer. The proteoliposomes can be loaded with substrate or ATP by including these in the dilution buffer.

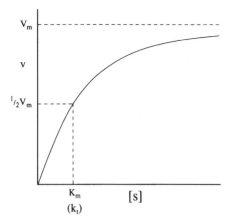

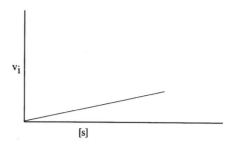

**Fig. 16.3** The initial rate of uptake of solute in the absence of a transporter. In the absence of a transporter the rate of solute entry is relatively slow and does not approach a maximum even at very high concentrations of [S], but is proportional to the concentration gradient.

**Fig. 16.2** Kinetics of transport. Bacteria are incubated with different concentrations of solute (S), and the initial rate of solute uptake is measured for each solute concentration. The rate ($v$) approaches a maximum ($V_m$) as the fraction of transporter molecules bound to solute reaches a maximum. The concentration of solute that gives $\frac{1}{2}V_m$, the $K_m$ (sometimes called $k_t$) is characteristic for each transporter.

## 16.3 Energy-Dependent Transport

A transporter simply facilitates the entry and exit of the solute across the membrane. At equilibrium, it does not bias the transport in any direction and therefore cannot, on its own, cause the accumulation of solute against a concentration gradient. However, we know that many transport mechanisms catalyze the accumulation of solutes into the cell. The internal concentration when the steady state is reached can be several orders of magnitude higher than the external concentration. Of course, this requires energy. The source of energy can be *chemical*, *light*, or *electrochemical*.

is called the $K_m$ or sometimes the $K_t$. It is frequently assumed to be a measure of the affinity of the solute for the transporter and is called the affinity constant. However, it will be referred to here simply as the solute concentration that gives $\frac{1}{2}V_{max}$. The value of the $K_m$ is characteristic of the transporter and can range from less than one micromolar to several hundred micromolar. The kinetics shown in Fig. 16.2 are described by the Michaelis–Menten equation for enzyme catalysis, which is explained in Section 6.2.1.

Bacterial transport systems are now divided into two categories, *primary transport* and *secondary transport*. Primary transport systems are driven by an energy-producing (exergonic) metabolic event. Primary transport systems include: proton-translocation-driven by ATP, light, or oxidation–reduction reactions (Sections 3.7.1, 3.7.2, 3.8.4, 4.5), light-driven chloride transport (Section 3.9), sodium ion transporting decarboxylases (Section 3.8.1), the uptake of inorganic or organic solutes driven by ATP hydrolysis (described later in this chapter), and the uptake of sugars driven by phosphoryl group transfer from phosphoenolpyruvate, called the phosphotransferase (PTS) system (described later in this chapter). During transport by the phosphotransferase system the sugar accumulates inside the cell

### 16.2.2 Uptake in the absence of a transporter

What if there were no transporter and the nutrient simply diffused into the cell? Slow-growing mutants have been isolated that have no functional transporter for a particular nutrient. The kinetics of uptake in these mutants reflects what one would expect in the absence of a transporter and is shown in Fig. 16.3. Note that the rates of uptake are low and are proportional to the concentration gradient, with no saturation at very high external concentrations of solute.

as the phosphorylated derivative. Secondary transport systems are driven by electrochemical gradients (e.g., proton and sodium ion gradients). During secondary transport the solute moves "down" an electrochemical ion gradient, usually of protons or sodium ions. (See Note 7.)

Secondary transport is coupled to primary transport by the proton in the following way. Primary transport:

$$H_{in}^+ + energy \longrightarrow H_{out}^+$$

Secondary transport:

$$H_{out}^+ + S_{out} \longrightarrow H_{in}^+ + S_{in}$$

The distinction between primary and secondary transport is an important one because it emphasizes a central feature of the chemiosmosis theory, that is, that the coupling between energy-yielding reactions and energy-requiring reactions in the membrane is via ion currents. In the example shown above, the coupling between the energy-dependent uptake of S and the primary energy-yielding reaction is via the proton current, which is most commonly used.

"Active transport" refers to primary transport, during which the solute is not chemically modified (e.g., the ATP-dependent uptake of histidine). Active transport is therefore a subclass of primary transport. (It should be pointed out that, prior to the chemiosmotic theory, the distinction between primary and secondary transport was not made, and an older definition of active transport was *any* transport that results in a concentration gradient where the solute is accumulated in a chemically unmodified form. However, the current definition of active transport restricts its usage to primary transport in which the solute does not chemically change.)

Secondary transport is catalyzed by uniporters, symporters, and antiporters that use electrochemical ion gradients to accumulate solutes. Examples are illustrated in Fig. 16.4. *Uniport* refers to solute translocation in the absence of a coupling ion. For example, the uptake of $K^+$ down its electrochemical gradient is uniport. *Symport* refers to solute uptake in which two solutes are carried on the carrier in the same direction. An example of this

would be the uptake of a solute coupled to the uptake of one or more protons or sodium ions. *Antiport* refers to the coupled movement of two solutes in opposite directions. For example, an exchange of $H^+$ for $Na^+$ on the same carrier is antiport.

### 16.3.1 Secondary transport

#### A general description

The way that solute transport is coupled to ion currents is that the transporter functions only when it transports both the ion and the solute in one direction (symport) or the solute in one direction *and* the ion in the other (antiport) (Fig. 16.4).[8] However, there are also transporters that move just an ion along its electrochemical gradient (uniport). In all these cases, the transporter is part of the electrical circuit. This is analogous to saying that, in order for an electrical current to make a motor turn, the current must flow through the motor. Most bacteria employ both proton and sodium symporters but mainly the former. (See Note 9 for more information.) (Sodium ion symporters are common in bacteria for amino acid transporters.) However, certain bacteria (e.g., alkaliphiles, halophiles, and marine bacteria) rely more heavily on sodium symporters. The reason for this becomes clear when one considers their ecological niches. The alkaliphiles live in environments with pH values in the range of 9–11. Because their cytoplasmic pH is below 8.5, the $\Delta pH$ ($pH_{in}$-$pH_{out}$) is negative rather than positive. This decreases the $\Delta p$. These organisms therefore depend upon sodium ion and the $\Delta \Psi$ for solute transport. Some of these bacteria also have flagella motors powered by a sodium potential rather than a proton potential.[10, 11] Halophilic bacteria require high external NaCl concentrations (3–5 M) in order to grow. They use predominantly $Na^+$/solute symporters. Marine bacteria also live in high concentrations of $Na^+$ (close to 0.5 M) and rely heavily on $Na^+$/solute symporters.[12] The use of ionophores can aid in identifying the ion current that is responsible for secondary transport. This is discussed in Section 3.4. It might be added that in contrast to transporters driven by ATP described in Section 16.3.3, the

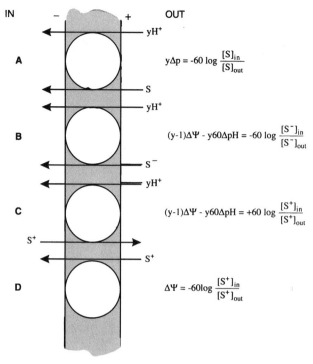

**Fig. 16.4** Some examples of solute transport driven by the $\Delta\Psi$ and the $\Delta pH$. (A) Symport of protons with an uncharged solute. (B) Symport of protons with a monovalent anion. (C) Antiport of protons with a monovalent cation. (D) Uniport of a monovalent cation. Another possibility, not shown, is symport of protons with cations. The ratio of protons to solute is given by $y$, which is assumed to have a value of 1 or greater.

characterized transporters driven by the $\Delta p$ (e.g., the Lac permease) are relatively simple in composition, consisting of a single protein that traverses the membrane in several loops.

## Energetics of transport

The free energy in joules required to move a mole of solute from outside a cell to inside the cell is:

$$\Delta G = RT \ln([S]_{in}/[S]_{out}) \quad J/mol \quad (16.1)$$

$$= 60 \log_{10}([S_{in}]/[S]_{out}) \quad mV \text{ at } 30°C$$

where $[S]_{in}$ is the concentration inside the cell and $[S]_{out}$ is the concentration outside the cell. For example, if $[S]_{in}/[S]_{out} = 10^3$, then $\Delta G = 17.4$ kJ/mol at 30°C.[13] This means that at least 17.4 kJ of work must be applied against the concentration gradient (at 30°C) to move one mole of S into the cell when the concentration ratio, in/out, is $10^3$, and remains as such. This would be the case in a steady-state situation

where $S_{in}$ is used as fast as it is brought into the cell. Note that 17.4 kJ does *not* refer to the energy required to move $10^3$ moles of S to one side of the membrane to establish a ratio of $10^3/1$. Equation (16.1) can also be expressed as an electrical potential in volts, since $(RT/F)(2.303) = 0.06$ V, or 60 mV, at 30°C. Figure 16.4 illustrates some examples of solute transport coupled to an electrochemical gradient. These are discussed next.

## Some examples of solute transport coupled to an electrochemical gradient

### 1. Symport of an uncharged solute with protons (Fig. 16.4A)

Many solutes are transported by symport. Assume that the stoichiometric ratio of H⁺/S is $y$. At equilibrium, the force accompanying the diffusion of the solute down its concentration gradient (out of the cell) is $-60 \log_{10}([S_{in}]/[S]_{out})$ mV, and is balanced by the $\Delta p$

drawing the solute in the opposite direction (into the cell) so that at 30°C:

$$y(\Delta\Psi - 60\ \Delta pH) = y\ \Delta p$$

$$= -60\ \log_{10}([S]_{in}/[S]_{out}) \quad (16.2)$$

[Alternatively, one can write the total driving force, which in this case would be equal to $y\ \Delta p + 60\ \log_{10}([S]_{in}/[S]_{out})$, and set this equal to 0 (equilibrium). Rearrangement of the equation then yields eq. 16.2. A similar procedure produces eqs. 16.3, 16.4, and 16.5.] For example, if the ratio ($y$) of protons to solute transported is 1:1 and the concentration gradient $[S]_{in}/[S]_{out}$ is $10^3$, then the minimum $\Delta p$ that is required to maintain that concentration gradient would be $-180$ mV. Notice that $[S]_{in}/[S]_{out}$ is an exponential function (logarithmic function) of $y\ \Delta p$. For example, when $y$ is increased to 2, the maximum concentration gradient attained at equilibrium is squared. Theoretically, very large concentration gradients can be maintained by the $\Delta p$.

*2. Symport of a monovalent anion with protons (Fig. 16.4B)*

The $\Delta p$ can also drive the uptake of anions. However, whether or not the $\Delta\Psi$ is part of the driving force depends upon whether a net charge is transported. For example, for a monovalent anion, the ratio $H^+/R^-$ must be greater than one in order that the $\Delta\Psi$ be part of the driving force. The relative contributions of the $\Delta\Psi$ and the $\Delta pH$ to the driving force are as follows:

$$(y-1)\ \Delta\Psi - y60\ \Delta pH$$

$$= -60\ \log_{10}([S^-]_{in}/[S^-]_{out}) \quad (16.3)$$

Equation 16.3 is the same as eq. 3.25, whose derivation can be found in Section 3.8.3. (See also eq. 3.23 in Section 3.8.3 for a multivalent anion.)

*3. Antiport of a monovalent cation with $H^+$ (Fig. 16.4C)*

An example is the proton:sodium ion antiporter, which is widespread among the bacteria. It is used for pumping sodium ions out of the cell. The equation is similar to eq. 16.3, except that the sign is changed for the expression for the

electrochemical potential of $S^+$ because it is moving out of the cell; that is, $[S^+]_{in}/[S^+]_{out}$ is $<1$.

$$(y-1)\ \Delta\Psi - y60\ \Delta pH$$

$$= +60\ \log_{10}([S^+]_{in}/[S^+]_{out}) \quad (16.4)$$

*4. Electrogenic uniport of a cation (Fig. 16.4D)*

For some transporters, the membrane potential alone can provide the driving force for the uptake of cations (e.g., for $K^+$ uptake). The relationship between $\Delta\Psi$ and the uptake of cations is

$$\Delta\Psi = -60\ \log_{10}([S^+]_{in}/[S^+]_{out}) \quad (16.5)$$

For a divalent cation, the driving force would be $2\ \Delta\Psi$, since there is twice as much charge per ion. Equation 16.5 is one form of the Nernst equation.

### 16.3.2 Evidence for solute/proton or solute/sodium symport

One way to demonstrate coupling of transport to proton or sodium ion influx is to measure the alkalinization of the medium (decrease in protons) or a decrease in the extracellular $Na^+$ concentration when bacteria or membrane vesicles are incubated with the appropriate solutes.[14] The changes in the external pH or $Na^+$ concentration occur as a result of symport with the solute and can be measured with $H^+$ or $Na^+$-selective electrodes. It is necessary to ensure that $H^+$ or $Na^+$ influx is electroneutral and a membrane potential, which would impede further influx of cations, does not develop. Thus the experiments are done in the presence of a permeant anion (e.g., $CNS^-$) that serves as a counterion to protons or sodium ions, or $K^+$ plus valinomycin whose efflux can exchange for the incoming protons or sodium ions. (This is explained in Note 15.) The experiments may be done so that the ion influx driven by the solute concentration (i.e., symport) is demonstrated rather than accumulation of the solute against a concentration gradient. An example is the experiment reported by West in 1970.[16] West suspended *E. coli* in dilute buffer and added the sugar lactose, which promoted an increase in the external pH (Fig. 16.5). This

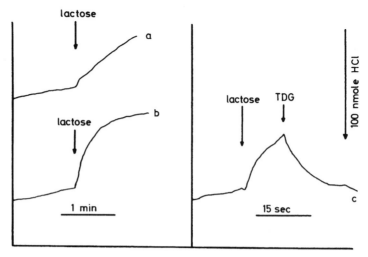

**Fig. 16.5** Lactose-dependent proton uptake. Lactose was added to a suspension of *Escherichia coli* without (a) and with (b) CNS⁻ and the pH was measured with a pH meter. Lactose caused the immediate uptake of protons. Uptake was stimulated by CNS⁻, which prevented a membrane potential, inside positive, from developing. Lactose uptake was inhibited by thiodigalactoside (TDG), which is an inhibitor of the lac permease (c). These data reflect symport between lactose and protons. *Source:* From West, I. C. 1970. Lactose transport coupled to proton movements in *Escherichia coli.* *Biochem. Biophys. Res. Commun.* **41**:655–661.

implied that the lactose permease catalyzed a symport of lactose with protons. Similar experiments have since been performed by other investigators, demonstrating that many sugars and amino acids enter bacteria in symport with either protons or sodium ions.

### 16.3.3 Primary transport driven by ATP

Transport systems exist that are driven by ATP (or some phosphorylated derivative in equilibrium with ATP). These are: (1) H⁺ transport [i.e., the ATP synthase (ATPase)]; (2) K⁺ transport in *E. coli*; (3) some transport systems in gram-positive bacteria; and (4) transport systems in gram-negative bacteria that use periplasmic binding proteins. (The student should review the discussion of the periplasm in Section 1.2.4.) The latter systems rely on binding proteins that combine with sugars and amino acids in the periplasm and transfer these to the actual transporters in the cell membrane. They are distinguished from all other transport systems in that they are not functional in cells that have been osmotically shocked, a treatment that makes the outer envelope permeable and

causes the release of periplasmic proteins. (The procedure for osmotic shock is described in Note 17.) Shock-sensitive transport systems are also called *periplasmic permeases*.

### Shock-sensitive transport systems

Shock-sensitive transport systems are characteristic of gram-negative bacteria and are responsible for the transport of a wide range of solutes, including sugars, amino acids, and ions.[18–22] They usually consist of a transporter, which is a membrane complex consisting of four subunits (two of which are identical) and a periplasmic solute binding protein (Fig. 16.6). In this way they are more complex than transporters driven by the Δp, which consist of a single protein that crosses the membrane in several loops. Solute transport via the shock-sensitive systems can be visualized as occurring in several sequential steps, as illustrated in Fig. 16.6. This model is based upon the histidine permease system, which is described in Note 23.[24] It should be pointed out that the histidine permease system is an example of a large superfamily of proteins that bind ATP and catalyze its hydrolysis to drive import (solute uptake) or export (proteins, peptides, polysaccharides).[25] They are called ABC

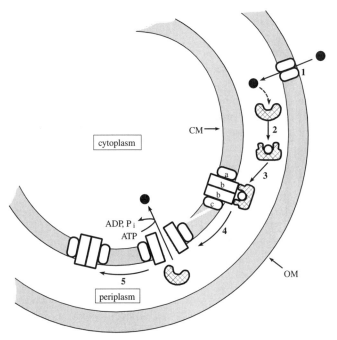

**Fig. 16.6** Model for periplasmic transport. (1) The solute enters the periplasm through a pore in the outer membrane. (2) Inside the periplasm the solute binds to a binding protein, which undergoes a conformational change when it binds the solute. (3) The binding protein carrying the solute binds to the transporter (a, b, c) located in the inner (cell) membrane. (4) The transporter is thought to undergo a conformational change that may result in the opening of a pore through which the solute diffuses to the cytoplasm. (5) The putative pore closes again when the binding protein is released from the permease. The transporter has a binding site for ATP, which has been demonstrated to be hydrolyzed during histidine transport, and presumably during the transport of other solutes. An alternative hypothesis speculates that the binding protein triggers conformational changes in the permease that make a binding site available for the solute. The solute is then passed from one binding site to another on the permease until it is released inside the cell, rather than diffusing through a pore.

(ATP-binding cassette) transporters because they all have a protein with a homologous sequence of amino acids (about 200) that contains an ATP-binding motif. ABC transporters are not confined to gram-negative bacteria. They occur in gram-positive bacteria (where an equivalent of the periplasmic binding protein is part of the cell membrane) and in eukaryotic cells.[26] See the discussion of ABC transporters in Section 17.4.1.

**Step 1.** The solute enters the periplasm through an outer membrane pore (e.g., through a porin). (Consult Section 1.2.3 for a description of the outer membrane and of porins.)

**Step 2.** The solute binds to a specific periplasmic binding protein to form a solute:binding protein complex. The bind-

ing protein undergoes a conformational change that allows it to bind productively to the transporter in the membrane.

**Step 3.** The liganded binding protein binds to the transporter in the cell membrane, delivering the solute to the transporter. The membrane-bound transporter is a complex consisting of four proteins, two of which are identical.

**Step 4.** The transporter complex translocates the solute across the cell membrane, perhaps through a channel that opens transiently. ATP hydrolysis, catalyzed by the transporter complex, occurs at this step. The ATP-binding domains of the transporter are exposed to the cytoplasm.

**Step 5.** The transporter complex returns to its unstimulated state.

In some cases the membrane-bound transporter (also called permease) interacts with more than one type of binding protein. For example, the histidine permease, besides transporting histidine also transports arginine via the arginine–ornithine binding protein. Several different binding proteins for branched-chain amino acids use the same membrane transporter. The shock-sensitive transport systems are characterized by very high efficiencies. They are capable of maintaining concentration gradients of approximately $10^5$, and $K_m$ values for uptake are in the 0.1–1 $\mu$M range. This means that they can scavenge very low concentrations of solute and accumulate these to relatively high internal concentrations.

## Evidence that ATP is the source of energy for histidine uptake

Direct evidence that ATP is the source of energy for the transport of histidine via a periplasmic binding protein was obtained by Bishop et al.[27] They extracted the proteins, including the histidine carrier proteins, from the membranes of *E. coli* and incorporated these proteins into proteoliposomes. (See Section 16.1 for a description of proteoliposomes.) The proteoliposome vesicles were sometimes loaded with ATP by including the ATP in the dilution step. When the reconstituted proteoliposomes were incubated with the histidine binding protein (HisJ) and histidine, they transported histidine only when ATP was present (Fig. 16.7). The transport of histidine was not affected by the ionophores valinomycin (and $K^+$), nigericin, or the proton ionophore FCCP. (Ionophores and their physiological activities are described in Section 3.4.) These experiments demonstrated that histidine uptake via its periplasmic binding protein was driven by ATP and not an electrochemical potential.

## ATP-driven $K^+$ influx

Potassium ion is the principal cation in bacteria, and it not only plays a role in osmotic and pH homeostasis (Chapter 15) but is also a cofactor for many enzymes and ribosomes. Bacteria accumulate $K^+$ to a level several orders of magnitude higher than the external concentrations. Most of what we know about $K^+$ transport is derived from studies of *E. coli*.[28, 29] There are two transport systems. The major route for $K^+$ uptake occurs via the TrK system, which is always present (constitutive) and operates at a high rate, but relatively low affinity (high $K_m$, i.e., 2 mM) for $K^+$. This transporter requires both a $\Delta$p and ATP in order to function, and the reasons for the dual requirement are not clear. (It has been speculated that the energy source is actually the $\Delta p$ and that ATP acts as a positive regulator.[30]) When *E. coli* is

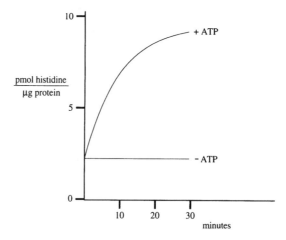

**Fig. 16.7** Transport of histidine by reconstituted proteoliposomes. The histidine carrier proteins plus other membrane proteins were solubilized in detergent and incorporated into proteoliposomes in the presence or absence of ATP. Incubation mixtures contained the histidine-binding protein. Histidine was taken up by the proteoliposomes that were loaded with ATP. *Source:* Adapted from Bishop et al., 1989.

grown in media where the $K^+$ concentrations are very low, the cells synthesize a second $K^+$ transport system called the Kdp system, which serves to scavenge $K^+$. However, the induction depends upon the osmolarity of the medium. The threshold concentration of $K^+$ that prevents induction is higher when the osmolarity is higher. This means that the Kdp system can be induced by raising the osmotic pressure of the medium without altering the external $K^+$ concentrations as long as the external $K^+$ concentration is not too high. This reflects the important role for $K^+$ as an intracellular osmolyte.[31] (See Section 15.2.4 for a discussion of $K^+$ as an osmolyte.) The Kdp system has a very low $K_m$ for $K^+$ (2 $\mu$M) and uses ATP as the energy source. There are three structural proteins in the Kdp system, KdpA, KdpB, and KdpC, all of which are located in the cell membrane. These genes are encoded in the *kdpABC* operon. KdpB is a transmembrane protein (predicted from the nucleotide sequences suggesting several membrane spanning loops of hydrophobic domains) and catalyzes the ATP-dependent uptake of $K^+$ (Fig. 16.8). It may form a channel through which $K^+$ moves across the membrane.[32] KdpA is exposed to the periplasm and is believed to bind to $K^+$, delivering it to KdpB. The reason for believing this is that mutations in the KdpA protein produce changes in the $K_m$ for $K^+$ uptake, suggesting that it has the substrate binding site. The function of KdpC is not known. There are two proteins that are required for the transcription of the *kdpABC* operon. These are the KdpD protein, located in the cytoplasmic membrane, and the KdpE protein, located in the cytosol. These two proteins are part of a two-component signaling system thought to sense turgor pressure and transfer the signal to the *kdpABC* promoter.[33–37] (See Section 18.11 for a discussion of this point.)

### 16.3.4 The phosphotransferase (PTS) system

This transport system differs from those driven by ATP or electrochemical gradients in that it catalyzes the accumulation of carbo hydrates as the *phosphorylated derivatives*

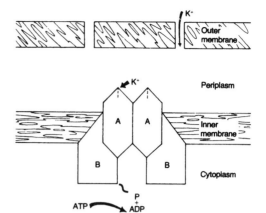

Fig. 16.8 The Kdp system for $K^+$ uptake. $K^+$ is thought to bind to KdpA in the periplasm and then be transported through the membrane via KdpB. The energy for $K^+$ transport is derived from ATP hydrolysis. The function of KdpC, which is not shown, is unknown. *Source:* From Epstein, W., and L. Laimins. 1980. Potassium transport in *Escherichia coli*: Diverse systems with common control by osmotic forces. *TIBS* 5:21–23.

instead of as the free sugar.[38–40] The phosphoryl donor in these transport systems is phosphoenolpyruvate (PEP), an intermediate in glycolysis (Section 8.1). Because the carbohydrate is modified (by phosphorylation), this type of transport is not referred to as active transport but rather as group translocation. The phosphotransferase (PTS) system is characteristic of anaerobic and facultatively anaerobic bacteria, but seems to be lacking in many aerobes. It also does not occur in eukaryotes. The overall reaction of the PTS system for glucose transport is

$$PEP + \text{carbohydrate}_{(out)}$$
$$\longrightarrow \text{pyruvic acid} + \text{carbohydrate-P}_{(in)}$$

During the reaction, PEP donates a phosphoryl group to the carbohydrate, which accumulates inside the cell as the phosphorylated derivative. The phosphoryl group is transferred to the carbohydrate via a consecutive series of reactions, beginning with PEP as the initial donor. This is illustrated in Fig. 16.9 and described as individual steps:

**Step 1.** In the soluble part of the cell, the phosphoryl group is transferred from phosphoenolpyruvate to enzyme I (EI).

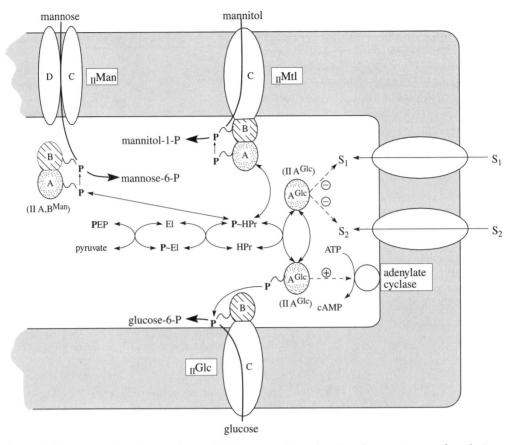

**Fig. 16.9** The sugar phosphotransferase (PTS) system. The phosphoryl group is transferred via a series of proteins to the sugar. The sugar-specific carrier proteins in the membrane, $_{II}$Man, $_{II}$Mtl, $_{II}$Glc, accept the phosphoryl group from P-HPr and transfer it to the sugar-forming mannose-6-P, mannitol-1-P, and glucose-6-P. Enzyme II can be a single protein with three domains, A, B, and C, as in the mannitol system, or separate proteins, as in the glucose and mannose systems. The mannose carrier consists of two proteins, C and D. Note that the phosphoryl group travels from HPr to IIA to IIB to the sugar, which is translocated by IIC into the cell in an unknown manner. At some stage during translocation the sugar becomes phosphorylated. However, phosphorylation of the sugar need not take place during translocation per se. That is to say, the sugar may be phosphorylated on the inside surface of the cell membrane prior to its release into the cytoplasm. Also shown is the stimulation of adenylate cyclase by P-EIIA-$^{Glc}$ (formerly called P-III$^{Glc}$) and the inhibition of non-PTS sugar carriers ($S_1$ and $S_2$ carriers) by EIIA$^{Glc}$. *Source:* Adapted from Postma, P. W., J. W. Lengeler, and G. R. Jacobon, 1993. Phosphenolpyruvate:carbohydrate phosphotransferase systems of bacteria. *Microbiol. Rev.* 57:543–594.

$$PEP + EI \longrightarrow EI\text{-}P + pyruvate$$

**Step 2.** Enzyme I transfers the phosphoryl group to a small cytoplasmic protein called HPr.

$$EI\text{-}P + HPr \longrightarrow HPr\text{-}P + EI$$

Enzyme I and HPr are common to all the PTS carbohydrate uptake systems.

**Step 3.** The HPr then transfers the phosphoryl group to a carbohydrate-specific permease complex in the membrane called enzyme II, which transfers the phosphoryl group to the carbohydrate during carbohydrate uptake into the cell.

$$HPr\text{-}P + EII \longrightarrow HPr + EII - P$$

$$EII\text{-}P + [CH_2O]_{out} \longrightarrow EII + [CH_2O\text{-}P]_{in}$$

Enzyme II has three domains, A, B, and C. The phosphoryl group is transferred from HPr to domain A, then to domain B, and finally to the carbohydrate in a reaction that requires domain C, which is

always an integral membrane protein. This is summarized below:

$$HPr\text{-}P + EIIA \longrightarrow EIIA\text{-}P + HPr$$

$$EIIA\text{-}P + EIIB \longrightarrow EIIA + EIIB - P$$

$$EIIB\text{-}P + carbohydrate_{(out)}$$

$$\xrightarrow{EIIC} EIIB + carbohydrate\text{-}P_{(in)}$$

How EIIC brings the carbohydrate into the cell is not understood.

Different carbohydrate uptake systems differ with respect to the number of separate proteins that constitute "enzyme II." It can be from one to four, one of which (IIC), is always membrane bound and catalyzes the transport of the carbohydrate into the cell. For example, enzyme II for mannitol uptake (II$^{Mtl}$ in Fig. 16.9) is a single membrane-bound protein with three domains, A, B, C. Domain C is in the membrane, whereas domains A and B project into the cytoplasm. However, in some transport systems the domains A, B, and C are on separate enzyme II proteins. Thus enzyme II for glucose consists of two proteins, IIA and IIBC (Fig. 16.9). In this case, IIBC is membrane bound, whereas IIA (formerly called III$^{Glc}$) is cytoplasmic. In a third case (EII$^{Man}$), EIIAB exist as a single cytplasmic protein, whereas there are two membrane-bound EII proteins, IIC and IID. In the cellobiose PTS system in *E. coli*, the enzymes II are three proteins (i.e., IIA and IIB in the cytoplasm and IIC in the membrane).

## Catabolite regulation by the PTS system in enteric bacteria

It has been known for many years that when bacteria using the PTS system for the transport of glucose are presented with a choice of glucose and another carbon source, they will preferentially utilize the glucose and delay the use of the other carbon source until the glucose

is depleted.[41] This is called glucose repression, or catabolite repression, and is responsible for diauxic growth, described in Section 2.2.4. The phenomenon is not restricted to glucose, since many PTS carbohydrates are used in preference over other carbon sources. (See Note 42.) (See Section 18.13.1 for a discussion of a different catabolite repression system in *E. coli* that is due to the Cra system, and Section 18.13.2 for a discussion of catabolite repression in gram-positive bacteria.)

## The model

A widely accepted model for regulation by the PTS system in enteric bacteria is illustrated in Fig. 16.9. The model postulates the following:

1. IIA$^{Glc}$ inhibits several enzymes required for carbohydrate metabolism, including certain non-PTS sugar transporters such as the lactose and melibiose transporter, the MalK protein, which is essential for the maltose transport system, and glycerol kinase. P-IIA$^{Glc}$ is dephosphorylated to IIA$^{Glc}$ by IICB$^{Glc}$ during glucose transport. Therefore, according to the model, glucose transport into the cell inhibits the above-mentioned enzymes. The inhibition by glucose of transport and metabolism of non-PTS carbohydrates is called *inducer exclusion*.

2. P-IIA$^{Glc}$ stimulates a membrane-bound enzyme called adenylate cyclase, which makes cyclic-AMP (c-AMP), which in turn stimulates the transcription of many genes that code for catabolic enzymes. It is actually a complex of c-AMP and the C-AMP receptor protein (CRP), which regulates transcription. (The CRP protein is also called the catabolite activator protein, or CAP.) During glucose uptake by the PTS system, P-IIA$^{Glc}$ is dephosphorylated; hence adenylate cyclase is no longer stimulated, and transcription of c-AMP-dependent genes is inhibited. However, there is always a basal level of c-AMP synthesized regardless of the carbon source. This is necessary because the transcription of genes required for the metabolism of many PTS sugars also requires c-AMP.

3. The model also explains how PTS carbohydrates in addition to glucose can also depress the entry of non-PTS sugars and inhibit the expression of c-AMP-dependent

genes. The uptake of PTS carbohydrates would be expected to draw phosphoryl groups away from P-HPr toward the sugars. This should decrease the phosphorylation state of IIA$^{Glc}$, which is phosphorylated by P-HPr. Because the phosphorylation of IIA$^{Glc}$ by P-HPr is reversible, phosphate would be expected to flow from P-IIA$^{Glc}$ to the PTS carbohydrates. The subsequent increase in IIA$^{Glc}$ would be expected to inhibit the enzymes required for uptake and metabolism of non-PTS sugars, and the decrease in P-IIA$^{Glc}$ would be expected to inhibit the adenylate cyclase.

### Rationale for the model

1. The reason for believing that the PTS system is involved in the utilization of glucose-repressed carbon sources was the original finding that mutants lacking HPr or enzyme I are unable to grow on glucose-repressed carbon sources. In E. coli, these carbon sources include lactose, maltose, melibiose, glycerol, rhamnose, xylose, and citric acid cycle intermediates. The reason for this is that they cannot phosphorylate IIA$^{Glc}$, which stimulates the production of c-AMP by activating the adenylate cyclase. The requirement for EI and Hpr for growth on non-PTS sugars is not restricted to E. coli or other enterics, although it is less well studied in other bacteria.

2. Because the addition of glucose to wild-type cells has the same repressive effect on growth using non-PTS sugars and citric acid cycle intermediates as does mutations in HPr or enzyme I, it was suggested that P-IIA$^{Glc}$ is required for growth on these substances. The reasoning is that: (1) A defect in enzyme I or HPr will result in the inability to phosphorylate IIA$^{Glc}$; and (2) the addition of glucose to wild-type cells should result in the dephosphorylation of P-IIA$^{Glc}$.

3. The involvement of adenylate cyclase was implicated when it was discovered that the addition of c-AMP, the product of adenylate cyclase, could overcome the Hpr mutant phenotype with respect to growth. Since c-AMP was known to be required for the transcription of several genes, a model was postulated that stipulated a stimulation by P-

IIA$^{Glc}$ of the enzyme that synthesizes c-AMP (adenylate cyclase) (Fig. 16.9). Mutations in enzyme I or HPr, or the addition of glucose to wild-type cells, should lower the amounts of P-IIA$^{Glc}$ and thus cause a decrease in the levels of c-AMP. Of course, it is possible that the inhibition of the adenylate cyclase is due to an increase in amounts of IIA$^{Glc}$ rather than a decrease in levels of P-IIA$^{Glc}$. However, mutations that result in lowered IIA$^{Glc}$ activity (crr mutants) do not relieve the inhibition of adenylate cylase by glucose. Also it was reported several years ago that the addition of PTS carbohydrates to E. coli cells made permeable with toluene inhibited adenylate cyclase activity, which according to the current model can be interpreted as due to the lowering of the levels of P-IIA$^{Glc}$ by the PTS carbohydrates.[43, 44] However, it should be emphasized that the evidence for stimulation of adenylate cyclase by P-IIA$^{Glc}$ is thus far indirect, and based primarily on genetic evidence. What is lacking is direct evidence that P-EII$^{Glc}$ binds to the adenylate cyclase and/or stimulates its activity.

4. Mutants defective in HPr are also defective in the *uptake* of certain non-PTS sugars. The model explains this by postulating that IIA$^{Glc}$ inhibits the carriers for these sugars. The evidence for this is abundant. Uptake in HPr mutants is restored by a mutation in the gene for IIA$^{Glc}$, which results in lowered activity of IIA$^{Glc}$ (crr mutants). Also IIA$^{Glc}$ is capable of binding to the lactose carrier (permease) and of inhibiting the transport of the lactose analog (TMG) by liposomes made with the lactose permease.[45, 46]

## 16.4 How to Determine the Source of Energy for Transport

Methods are available to distinguish whether bacterial transport is driven by the electrochemical proton potential ($\Delta p$) or by ATP.

### The ATP synthase must be inactivated

In order to determine whether the source of energy is ATP hydrolysis or the $\Delta p$, it is necessary to isolate these sources of energy from each other and to perturb them independently (i.e., to increase and decrease the ATP and $\Delta p$ levels independently of each other). Since bacteria use ATP synthase to interconvert the proton potential and ATP, one cannot vary the ATP levels and $\Delta p$ independent of each other while the ATP synthase is functioning. To circumvent this problem, investigators either use mutants defective in the ATP synthase ("unc" mutants) or add inhibitors of the ATP synthase (e.g., N,N'-dicyclohexylcarbodiimide, DCCD). Once this is done, the ATP *or* the $\Delta p$ can be decreased or increased independent of each other.

### Perturbing the intracellular levels of ATP

How are intracellular ATP levels manipulated? Once the ATP synthase is inactivated, bacteria must rely completely on substrate-level phosphorylation, to synthesize ATP, and the problem becomes one of interfering with substrate-level phosphorylation. In order to lower the levels of substrate-level phosphorylation one can: (1) starve the cells of endogeneous energy reserves whose metabolism produces ATP via substrate-level phosphorylation; and (2) add an inhibitor of substrate-level phosphorylation (e.g., arsenate). (See Note 47.) Conversely, one can *increase* the production of ATP from substrate-level phosphorylation in starved cells by feeding them an energy source (e.g., glucose). However, one must be careful. Glucose can also stimulate respiration, and respiration can be a source of $\Delta p$. Therefore, to stimulate substrate-level phosphorylation with glucose without encouraging respiration, respiration should be prevented by using a respiratory inhibitor (e.g., cyanide) or by incubating the cells anaerobically.

### Perturbing the $\Delta p$

How is the $\Delta p$ manipulated? One can use ionophores that collapse the proton potential. (However, see Note 48. ) (See Section 3.4 for a discussion of ionophores.) It is also possible to *increase* electrochemical potentials in de-energized cells (e.g., starved cells, or in vesicles). For example, if there is a high $K^+$ concentration in the cells, the addition of valinomycin will produce a potassium diffusion potential. (See Fig. 3.5.) Also, the addition of certain substrates that feed electrons directly into the electron transport chain (e.g., D-lactate or succinate) will produce a proton potential without increasing substrate-level phosphorylation. A summary of some of these methods is presented in Table 16.1. However, caution must be observed when interpreting data derived from the use of inhibitors on whole cells because of indirect effects of the inhibitors.[21] Also it is not possible to distinguish between ATP and other high-energy molecules in equilibrium with ATP (e.g., PEP) when studying whole cells. Ideally, one should incorporate the carrier into proteoliposomes and directly test potential energy sources, as was described in Section 16.3.3 for the histidine permease.

## 16.5 A Summary of Bacterial Transport Systems

A summary of bacterial transport systems is illustrated in Fig. 16.10. These include primary transport systems (driven by chemical energy) and secondary transport systems (driven by electrochemical ion gradients). Note the role of the proton circuit for secondary transport. Most transport is in symport with protons. Although several symporters use $Na^+$ instead

Table 16.1 Summary of data that can distingush between energy sources for transport using whole cells

| | $\Delta p$ | ATP |
|---|---|---|
| Stimulated by: | | |
| glucose | no[a] | yes |
| D-lactate | yes | no |
| Inhibited by: | | |
| DNP[b] | yes | no[c] |
| arsenate | no | yes[d] |

[a] Respiration must be inhibited.
[b] One can use any combination of ionophores that abolishes both the $\Delta pH$ and the $\Delta \Psi$.
[c] The ATP synthase must be inhibited.
[d] Arsenate prevents ATP formation via substrate level phosphorylation. This is the only source of ATP when the ATP synthase is not operating.

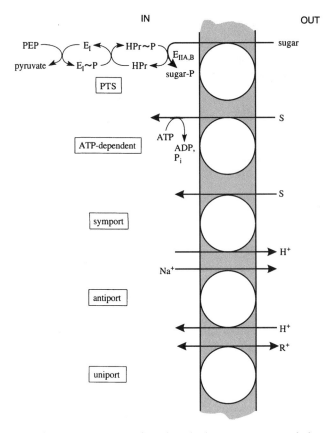

**Fig. 16.10** A summary of transport systems found in the bacteria. Any single bacterium uses diverse transport systems to take up sugars, amino acids, ions, vitamins, organic acids, etc. Many of these systems are powered by electrochemical energy (ion gradients). They catalyze symport, antiport, and uniport. There are also transport systems that use chemical energy instead of electrochemical energy. An example of the latter is the PTS system, which is specific for carbohydrates, is driven by phosphoenolpyruvate, and accumulates the sugar as the phosphorylated derivative. Other chemically driven transport systems use ATP or a high-energy molecule in equilibrium with ATP. These include shock-sensitive transport systems (periplasmic permeases) that are found in gram-negative bacteria and other chemically driven uptake systems found in both gram-negative and gram-positive bacteria.

of $H^+$, the return of the $Na^+$ to the extracellar space requires a proton circuit in most bacteria.

## 16.6 Summary

Bacteria employ several different types of energy-dependent transport systems. A single bacterium may have representatives of all the transport systems, except for the shock-sensitive systems that are present only in gram-negative bacteria. Transport systems may be classified as being either primary or secondary, depending on how they are coupled to the energy source. The difference between the two is that primary transport systems are directly coupled to an energy-generating reaction (e.g., ATP or PEP hydrolysis), whereas secondary transport systems are energized by an existing electrochemical gradient (e.g., a proton or sodium ion gradient) itself having been produced by energy-yielding reactions such as ATP hydrolysis or redox reactions. That is to say, secondary transport is indirectly coupled to an energy-yielding chemical reaction by ion currents. (This is one of the predictions of the chemiosmotic theory.) Active transport is defined as primary transport in which the solute accumulates in a chemically unmodified form.

ATP hydrolysis is used to drive some primary transport systems. These include osmotic shock-sensitive (periplasmic permeases) systems present in gram-negative bacteria that transport a range of solutes similar to those systems driven by electrochemical gradients. They consist of a periplasmic binding protein plus three or four membrane proteins. Other primary transport systems utilizing ATP include the ATP synthase (translocates protons) and the Kdp $K^+$-ATPase in *E. coli* ($K^+$ influx).

The transport of carbohydrates via the sugar–phosphotransferase (PTS) system driven by phosphoenolpyruvate is also a primary transport system. It is widespread in bacteria, being found in anaerobic and facultative anaerobes, but not in strict aerobes. However, the PTS system is absent in eukaryotes. The sugar is phosphorylated during transport, and therefore this is not active transport. Sometimes this type of transport is called group translocation. Transport by the PTS system results in the inhibition of transport of non-PTS sugars and the inhibition of adenylate cyclase, leading to catabolite repression by glucose. This can be explained in terms of the regulatory roles of $A^{Glc}$ ($III^{Glc}$) and $P$-$A^{Glc}$ ($P$-$III^{Glc}$).

There exist other catabolite repression systems besides the one mediated by $IIA^{Glc}$. Enteric bacteria have a c-AMP-independent catabolite repression system mediated by FruR, which is also called Cra. FruR is also an activator of certain genes. Thus FruR represses enzymes required for sugar catabolism and activates enzymes required for gluconeogenesis. Glycolytic intermediates such as fructose-1,6-bisphosphate remove FruR from the FruR-regulated operons, resulting in repression of gluconeogenic genes and stimulation of glycolytic genes. Gram-positive bacteria have a catabolite-repression system mediated by HPr(ser-P), which combines with the protein CcpA to repress catabolite-repressible genes. Hpr(ser-P) is a phosphorylated form of HPr made by a special ATP-dependent kinase that is stimulated by fructose-1,6-bisphosphate. All these catabolite-repression systems involve the repression of genes by carbohydrates. There also exist catabolite-repression systems where genes are repressed when cells are grown on organic acids, rather than carbohydrates. As discussed in Section 2.2.4, several obligately aerobic bacteria (e.g., *Rhizobium*) will grow first on organic acids when given a mixture of glucose and the organic acid. Clearly there must be yet other systems of catabolite repression mediated by intracellular signals remaining to be discovered.

Secondary transport systems require ion symport, antiport, or uniport and make use of electrochemical ion gradients to transport a diverse range of solutes including sugars, amino acids, and ions. The coupling ions are either protons or sodium ions, depending upon the transporter. This type of transport is found in bacteria, fungi, plants, protozoa, and higher eukaryotes. It is the simplest of the transport systems and consists of a single membrane-spanning protein.

The type of transport system used by a bacterium depends upon the organism, not the solute being transported. For example, *Pseudomonas* (a strict aerobe) actively transports glucose using symport with $H^+$, whereas *E. coli* (a facultative anaerobe) transports the same sugar using the PTS system. Furthermore, a single bacterium can use several different types of transport for the same class of compound. For example, *E. coli* uses the PTS system for some sugars but uses $H^+$ symport, $Na^+$ symport, or ATP for others. Harold has speculated regarding the diversity of transport systems in the bacteria, pointing out that whereas transport systems coupled to electrochemical gradients have the advantage of being simple in composition, they are limited by the $\Delta p$ with respect to the concentration gradients that can be attained.[49] Furthermore, they operate close to equilibrium and can theoretically be reversed if the $\Delta p$ suffers a transient decrease (e.g., during starvation). On the other hand, transport systems powered by ATP or PEP are structurally more complex, but have the advantage of being driven unidirectionally by the relatively large free energies available in ATP and PEP.

### Study Questions

1. Suppose the transport of X was driven by symport with $H^+$ in a 1:1 ratio. Assume X is not charged, the $\Delta\Psi$ is $-120$ mV, and the $\Delta$pH is 1. What is the expected $X_{in}/X_{out}$?

What would be the answer if X were $X^-$? If X were $X^-$ and the $\Delta pH$ were 2?

*ans.* 1,000, 10, 100.

2. What is the explanation of the curious fact that mutants in the PTS system that are defective in $E_1$ or HPr are also defective in the transport of some sugars that do not use the PTS system?

3. What are proteoliposomes, and how are they prepared?

4. Solute transport might be driven by the $\Delta p$ or by ATP. Describe some experiments that could distinguish between the two sources of energy. (Hint: You will have to manipulate the ATP and $\Delta p$ separately, and you might want to use proteoliposomes for some of your experiments.)

5. What is the procedure to induce "osmotic shock"? What is the cellular location of the proteins released by osmotic shock? What are the functions of some of the proteins released by osmotic shock?

## NOTES AND REFERENCES

1. Brooker, R. J. 1990. The lactose permease of *Escherichia coli. Res. Microbiol.* 141:309–316.

2. Racker, E., B. Violand, S. O'Neal, M. Alfonzo, and J. Telford. 1979. Reconstitution, a way of biochemical research; some new approaches to membrane-bound enzymes. *Arch. Biochem. Biophys.* 198:470–477.

3. Vitanen et al. 1986. Purification, reconstitution, and characterization of the *lac* permease of *Escherichia coli.* In: *Methods in Enzymology*, Vol. 125. Fleischer, S., and B. Fleischer (Eds.). Academic Press, New York.

4. Maloney, P. C., V. Anantharam, and M. J. Allison. 1992. Measurement of the substrate dissociation constant of a solubilized membrane carrier. *J. Biol. Chem.* 267:10531–10536.

5. Taglich, D., E. Padan, and S. Schuldiner. 1991. Overproduction and purification of a functional $Na^+/H^+$ antiporter coded by nhaA (ant) from *Escherichia coli. J. Biol. Chem.* 266:11289–11294.

6. Bishop, L., R. Agbayani, Jr., S. V. Ambudkar, P. C. Maloney, and G. F-L. Ames. 1989. Reconstitution of a bacterial periplasmic permease in proteoliposomes and demonstraton of ATP hydrolysis concomitant with transport. *Proc. Natl. Acad. Sci. USA* 86:6953–6957.

7. The proton gradient is created by a primary transport system. Some bacteria can create sodium gradients with a primary transport system, but most create a sodium gradient by converting a proton gradient into a sodium ion gradient using a $H^+/Na^+$ antiporter.

8. Wright, J. K., R. Seckler, and P. Overath. 1986. Molecular aspects of sugar:ion cotransport. *Ann. Rev. Biochem.* 55:225–248.

9. Bacteria, yeast, and plants use primarily the proton as the coupling ion for symport, whereas animals rely on the sodium ion. This may be because the cell membranes of most bacteria, fungi, and plants create a proton potential by pumping protons out of the cell via a proton-translocating ATPase, whereas the cell membranes of animal cells create a sodium potential by pumping sodium ions out of the cell via a $Na^+,K^+$-ATPase, which exchanges $Na^+$ for $K^+$.

10. Hirota, N., and Y. Imae. 1983. $Na^+$-driven flagellar motors of an alkalophilic *Bacillus* strain YN-1. *J. Biol. Chem.* 258:10577–10581.

11. Imae, Y., and T. Atsumi. 1989. $Na^+$-driven bacterial flagellar motors. *J. Bioenerg. Biomemb.* 21:705–716.

12. Maloy, S. R. 1990. Sodium-coupled cotransport, pp. 203–224. In: *The Bacteria*, Vol. XII. Krulwich, T. A. (Ed.). Academic Press, New York.

13. $R = 8.3144$ J deg$^{-1}$ mol$^{-1}$; $T = K = 273.16 + °C$; to convert natural logarithms to $\log_{10}$ multiply by 2.303.

14. Wilson, D. M., T. Tsuchiya, and T. H. Wilson. 1986. Methods for the study of the melibiose carrier of *Escherichia coli.* In: *Methods in Enzymology*. Fleischer, S., and B. Fleischer (Eds.). 125:377–387.

15. The negative charge on $SCN^-$ is delocalized over the three atoms of the molecule, and this allows it to penetrate the lipid bilayer.

16. West, I. C. 1970. Lactose transport coupled to proton movements in *Escherichia coli. Biochem. Biophys. Res. Commun.* 41:655–661.

17. Gram-negative cells are shocked in the following way. They are first suspended in a hypertonic solution of Tris buffer, EDTA, and sucrose. Such treatment removes much of the divalent metal cations that are holding the lipopolysaccharide together along with the lipopolysaccharide, and plasmolyzes the cells. Then the cells are rapidly diluted into water or dilute $MgCl_2$ to neutralize the EDTA. This results in the release of the periplasmic proteins. The treatment inhibits cellular functions that depend on periplasmic binding proteins (e.g., ATP-dependent transport of sugars and amino

acids). Other cellular functions, including other transport systems, are retained.

18. Ames. G. F.-L. 1988. Structure and mechanism of bacterial periplasmic transport systems. *J. Bioenerg. Biomembr.* **20**:1–18.

19. Ames, G. F.-L. 1990. Energetics of periplasmic transport systems, pp. 225–245. In: *The Bacteria*, Vol. 12. Krulwich, T. A. (Ed.). Academic Press, New York.

20. Ames, G. F.-L. 1986. Bacterial periplasmic transport systems: Structure, mechanism, and evolution. *Ann. Rev. Biochem.* **55**:397–425.

21. Ames, G. F.-L, and A. K. Joshi. 1990. Energy coupling in bacterial periplasmic permeases. *J. Bacteriol.* **172**:4133–4137.

22. Furlong, C. E. 1987. Osmotic-shock-sensitive transport systems, pp. 768–796. In: *Escherichia coli and Salmonella typhimurium: Cellular and Molecular Biology*. Vol. 1. Neidhardt, F. C., J. L. Ingraham, K. B. Low, B. Magasanik, M. Schaechter, and H. E. Umbarger (Eds.). ASM Press. Washington, D.C.

23. In thes specific case of the histidine permease system, the transporter consists of two hydrophobic proteins, HisQ and HisM, which span the membrane, and two identical hydrophilic proteins, HisP. The HisP protein may be a peripheral membrane protein bound to the inner surface of the membrane, or it may span the membrane, being separated from the hydrophobic lipids by the HisQ and HisM proteins. The HisP protein binds ATP and is responsible for ATP hydrolysis. The periplasmic histidine binding protein is called HisJ.

24. Ames, G. F.-L., and H. Lecar. 1992. ATP-dependent bacterial transporters and cystic fibrosis: Analogy between channels and transporters. *FASEB* **6**:2660–2666.

25. Fath, M. J., and R. Kolter. 1993. ABC transporters: Bacterial exporters. *Microbiol. Rev.* **57**:995–1017.

26. Boos, W., and J. M. Lucht. 1996. Periplasmic binding protein-dependent ABC transporters. In: *Escherichia coli* and *Salmonella: Cellular and Molecular Biology*, pp. 1175–1209, Vol. 1. F. C. Neidhardt et al. (Eds.) ASM Press, Washington, D.C.

27. Bishop, L., R. Agbayani, Jr., S. V. Ambudkar, P. C. Maloney, and G. F.-L. Ames. 1989. Reconstitution of a bacterial periplasmic permease in proteoliposomes and demonstration of ATP hydrolysis concomitant with transport. *Proc. Natl. Acad. Sci. USA* **86**:6953–6957.

28. Epstein, W., and L. Laimins. 1980. Potassium transport in *Escherichia coli*: Diverse systems with common control by osmotic forces. *TIBS* **5**:21–23.

29. Rosen, B. 1987. ATP-coupled solute transport systems. pp. 760–767. In: *Escherichia coli and Salmonella typhimurium: Cellular and Molecular Biology*. Vol. 1. Neidhardt, F. C., J. L. Ingraham, K. B. Low, B. Magasanik, M. Schaechter, and H. E. Umbarger (Eds.). ASM Press. Washington, D.C.

30. Stewart, L. M. D., E. P. Bakker, and I. R. Booth. 1985. Energy coupling to $K^+$ uptake via the Trk system in *Escherichia coli*: The role of ATP. *J. Gen. Microbiol.* **131**:77–85.

31. Epstein, W. 1986. Osmoregulation by potassium transport in *Escherichia coli*. *FEMS Microbiol. Rev.* **39**:73–78.

32. Buurman, E. T., K-T. Kim, and W. Epstein. 1955. Genetic evidence for two sequentially occupied $K^+$ binding sites in the Kdp transport ATPase. *J. Biol. Chem.* **270**:6678–6685.

33. Walderhaug, M. O., J. W. Polarek, P. Voelkner, J. M. Daniel, J. E. Hesse, K. Altendorf, and W. Epstein. 1992. KdpD and KdpE, proteins that control expression of the *kdpABC* operon, are members of the two-component sensor–effector class of regulators. *J. Bacteriol.* **174**:2152–2159.

34. Voelkner, P., W. Puppe, and K. Altendorf. 1993. Characterization of the KdpD protein, the sensor kinase of the $K^+$-translocating Kdp system of *Escherichia coli*. *Eur. J. Biochem.* **217**:1019–1026.

35. Nakashima, K., H. Sugiura, H. Momoi, and T. Mizuno. 1992. Phosphotransfer signal transduction between two regulatory factors involved in the osmoregulated kdp operon in *Escherichia coli*. *Molec. Microbiol.* **6**:1777–1784.

36. Sugiura, A., K. Hirokawa, K. Nakashima, and T. Mizuno. 1994. Signal-sensing mechanisms of the putative osmosensor KdpD in *Escherichia coli*. *Molec. Microbiol.* **14**:929–938.

37. Sugiura, A., K. Nakashima, K. T. Tanaka, and T. Mizuno. 1992. Clarification of the structural and functional features of the osmoregulated kdp operon of *Escherichia coli*. *Molec. Microbiol.* **6**:1769–1776.

38. Saier, M. J., Jr., and A. M. Chin. 1990. Energetics of the bacterial phosphotransferase system in sugar transport and the regulation of carbon metabolism, pp. 273–299. In: *The Bacteria*, Vol. XII. Krulwich, T. A. (Ed.). Academic Press, New York.

39. Meadow, N. D., D. K. Fox, and S. Roseman. 1990. The bacterial phosphoenolpyru-

vate:glycose phosphotransferase system. *Ann. Rev. Biochem.* 59:497–542.

40. Postma, P. W., J. W. Lengeler, and G. R. Jacobson. 1993. Phosphoenolpyruvate:carbohydrate phosphotransferase systems of bacteria. *Microbiol. Rev.* 57:543–594.

41. Saier, M. H., Jr. 1989. Protein phosphorylation and allosteric control of inducer exclusion and catabolite repression by the bacterial phosphoenolphyruvate:sugar phosphotransferase system. *Microbiol. Rev.* 53:109–120.

42. Inhibition of transport and metabolism of non-PTS carbohydrates such as lactose, melibiose, maltose, and glycerol (the class I PTS carbohydrates) by PTS carbohydrates is enhanced in mutants of *E. coli* that have less EI activity (leaky *ptsI* strains).

43. Harwood, J. P., C. Gazdar, C. Prasad, A. Peterkofsky, S. J. Curtis, and W. Epstein. 1976. Involvement of the glucose enzymes II of the sugar phosphotransferase system in the regulation of adenylate cyclase by glucose in *Escherichia coli. J. Biol. Chem.* 251:2462–2468.

44. Peterkofsky, A., and C. Gazdar. 1975. Interaction of enzyme I of the phosphoenolpyruvate:sugar phosphotransferase system with adenylate cyclase of *Escherichia coli. Proc. Natl. Acad. Sci. USA* 72:2920–2924.

45. Nelson, S. O., J. K. Wright, and P. W. Postma. 1983. The mechanism of inducer exclusion. Direct interaction between purified III$^{Glc}$ of the phosphoenolpyruvate:sugar phosphotransferase system and the lactose carrier of *Escherichia coli. EMBO J.* 2:715–720.

46. Osumi, T., and M. H. Saier, Jr. 1982. Regulation of lactose permease activity by the phosphoenolpyruvate:sugar phosphotransferase system: Evidence for direct binding of the glucose-specific enzyme III to the lactose permease. *Proc. Natl. Acad. Sci. USA* 79:1457–1461.

47. Arsenate can substitute for inorganic phosphate in the synthesis of 1,3-diphosphoglycerate. The acyl-arsenate is quickly chemically hydrolyzed to the carboxylic acid and ATP is not made.

48. As long as the ATP synthase is inhibited, the ATP levels should not be affected. However, if the ATP synthase is functioning, then collapsing the $\Delta p$ will shift the equilibrium of the ATP synthase in the direction of ATP hydrolysis.

49. Harold, F. M. 1986. *The Vital Force: A Study of Bioenergetics.* W. H. Freeman and Co., New York.

# 17

# Protein Export and Secretion

Many proteins synthesized by cytosolic ribosomes are destined for export to various cellular locations (cell membrane, periplasm, outer envelope, cell wall) or secretion into the medium. The traffic in exporting these proteins to their correct locations is considerable. For example, cell membranes alone contain approximately 300 different proteins. In gram-negative bacteria there may be 100 various proteins in the periplasm. The outer envelope of gram-negative bacteria is the site of perhaps 50 different proteins. In addition, both gram-negative and gram-positive bacteria also export proteins that are part of surface layers (glycocalyx) and appendages (flagella, fibrils, pili), as well as extracellular hydrolytic enzymes (e.g., proteases, lipases, nucleases, and saccharidases). Pathogenic bacteria may also secrete toxins that adversely affect host cells in addition to hydrolytic enzymes that degrade host connective tissue, which presumably facilitates the spread of the bacteria. The translocation process into or through the membranes is referred to as *protein export*. If the protein is exported to the extracellular medium or to the bacterial cell surface, the process is called *secretion*. (The extracellular transport of nonproteinaceous compounds, e.g., end products of fermentation, is called excretion.)

What is it about the structure of a protein that determines whether it will be translocated and to where? What mechanisms are responsible for the translocation of exportable proteins through the cell membrane which is generally nonpermeable to proteins? What is the source of energy for protein translocation? These are important areas of research not only in prokaryotes but in eukaryotes as well, where proteins synthesized on cytosolic ribosomes are exported to different cell compartments such as mitochondria and chloroplasts, or are secreted out of the cell via the endoplasmic reticulum. By far, most research on bacterial protein translocation through the cell membrane has been with *Escherichia coli*, which has served as a model system primarily because of the ease of genetic manipulation. Indeed, virtually all the proteins involved in *E. coli* protein translocation through the cell membrane have been purified and the genes have been cloned. This has allowed the formulation of a model that describes the general features by which most proteins are translocated through the cell membrane. It is called the *Sec system*, or "general secretory pathway" (GSP).

## 17.1 The Sec System

### 17.1.1 The components

Before describing the model for protein export using the Sec system (Section 17.1.2), the components of the Sec system will be introduced.[1-7] These are (1) a *leader peptide*, (2) a *chaperone protein*, and (3) a membrane-bound *translocase* (SecY/E/G). Other membrane proteins that are involved in translocation (SecD, SecF, and yajC) are described in Refs. 6–10. (See Note 11.)

## The leader peptide

Most proteins that are translocated by bacteria are synthesized with a leader peptide (also called leader sequence or signal sequence) at the amino terminal end that is removed during or after translocation. The leader peptide is necessary for attachment and insertion of the protein into the cell membrane. The protein with the leader peptide is called a precursor protein or a *preprotein*. If it is secreted to the outside, it is sometimes called a *presecretory* protein. There is ample evidence that the leader peptide is necessary for the initial stages of translocation through the membrane.

If the leader peptide is altered by amino acid substitutions, or deleted, then translocation is impaired or does not occur at all. Leader peptides have three regions. They are: (1) a basic region (positively charged near neutral pH) at the extreme N-terminal end that attaches to the membrane, perhaps to negatively charged phospholipids; (2) a central hydrophobic region (core) that inserts into the membrane; and (3) a C-terminal region that contains a recognition site for a peptidase that removes the leader peptide during or after translocation (Fig. 17.1). The primary amino acid sequences of different leader peptides can vary significantly. Although the leader peptide is important for the initial stages of translocation through the membrane, it does not determine the final destination of the protein. Two reasons for believing this are:

1. There are no obvious differences in amino acid sequences between leader peptides of periplasmic and outer membrane proteins.

**a** Met Lys Ala Thr Lys|Leu Val Leu Gly Ala Val Ile Leu Gly Ser Thr Leu Leu Ala Gly△Cys

**b** (Met) Met Ile Thr Leu Arg Lys|Leu Pro Leu Ala Val Ala Val Ala Ala Gly Val Met Ser Ala Gln Ala Met Ala△Val

**c** Met Lys Ile Lys Thr Gly Ala Arg|Ile Leu Ala Leu Ser Ala Leu Thr Thr Met Met Phe Ser Ala Ser Ala Leu Ala△Lys

**d** Met Ser Ile Gln His Phe Arg|Val Ala Leu Ile Pro Phe Phe Ala Ala Phe Cys Leu Pro Val Phe Ala△His

**e** Met Lys Thr Lys|Leu Val Leu Gly Ala Val Ile Leu Thr Ala Gly Leu Ser Gly Ala Ala△Glu

**f** Met Lys Lys Ser Leu Val Leu Lys|Ala Ser Val Ala Val Ala Thr Leu Val Pro Met Leu Ser Phe Ala△Ala

**g** Met Lys Lys|Leu Leu Phe Ala Ile Pro Leu Val Val Pro Phe Tyr Ser His Ser△Ala

**Fig. 17.1** Leader peptides of exported proteins in *E. coli*. The leader sequence consists of a basic amino terminal end that has positively charged lysine and arginine residues, followed by a hydrophobic region. The boundary between the hydrophilic basic region and the hydrophobic region is denoted with a vertical line. The cleavage site for the leader peptide peptidase is at the carboxy terminal end (triangle): a, lipoprotein, located in outer membrane; b, phage lambda receptor, located in outer membrane; c, maltose binding protein located in periplasm; d, β-lactamase, located in periplasm; e, arabinose binding protein, located in periplasm; f, fd phage major coat protein, located in cell membrane; g, fd phage minor coat protein, located in cell membrane. Note that although the leader peptides share similar features with respect to charged and hydrophobic regions, the amino acid sequences are not the same. Nor are there any differences that can distinguish between leader sequences of outer membrane proteins, periplasmic proteins, and cell membrane proteins. *Source:* After Osborn, M. J., and H. C. P. Wu. 1980. Proteins of the outer membrane of gram-negative bacteria. *Ann. Rev. Microbiol.* 34:369–422. Reproduced, with permission, from the *Annual Review of Microbiology*, Vol. 34, 1980, by Annual Reviews, Inc.

One would expect to find such differences if the leader peptide specified the final location of the protein.

2. Exchange of leader peptide between proteins destined for two different compartments using recombinant DNA technology does not influence their final destination.

## Chaperone proteins

When exportable proteins are synthesized, the nascent preprotein binds to a soluble chaperone protein upon leaving the ribosome. The chaperone protein brings the preprotein to the membrane and also prevents the preprotein from assuming a tightly folded configuration or aggregating into a complex that cannot be translocated. The signal peptide retards the folding of the preprotein, giving the chaperone protein time to bind. A major chaperone protein in *E. coli* is SecB. SecB binds to the mature domain (not the leader sequence) of the preprotein and prevents premature folding in the cytoplasm and aggregation. SecB protein recognizes the unfolded topology of the proteins rather than specific amino acid sequences. SecB protein also binds to SecA at the membrane site of translocation and delivers the preprotein to SecA, which is a peripheral component of the translocase. Many proteins translocate using the Sec system but without requiring SecB. These use other chaperone proteins. See Section 19.1.3 for a discussion of chaperone proteins and their functions.

## Sec A and the SecY/E/G translocase

The preprotein is transferred from SecB to SecA, which is attached to the membrane-bound translocase, SecY/E/G.[12] It is not known how the translocase moves proteins through the membrane. For this discussion, it will be assumed that the translocase forms a hydrophilic channel for proteins. The formation of a hydrophilic channel is merely speculation at this time, but can be rationalized because many of the proteins secreted into the external medium do not have significantly hydrophobic regions that would facilitate their movement through the lipid bilayer. The translocase consists of an integral membrane protein complex. The purified complex

is composed of three different polypeptides, SecY, SecE, and SecG. The SecY protein spans the membrane several times, and it may form a channel. Although SecG does not appear to be absolutely necessary for protein translocation, it had a large stimulatory effect on protein translocation when incorporated into proteoliposomes along with SecY and SecE, and antibodies against SecG inhibited protein translocation in everted membrane vesicles.[13]

## 17.1.2 A model for protein secretion

Protein export can be divided into several stages that are illustrated in Fig. 17.2. Before the protein is exported, it must bind to a chaperone protein, which brings the protein to the membrane-bound translocase. As will be described, one of the chaperone proteins is SecB, which brings many preproteins to the membrane for export through the cytoplasmic membrane. As reviewed in Ref. 2 and discussed in Note 23, there may actually be a cascade of chaperones that ultimately deliver the protein to SecB.

**Step 1.** The preprotein–SecB protein complex binds to the SecA–translocase complex in the membrane.

**Step 2.** The preprotein binds to SecA. The SecB protein may be released at this step.

**Step 3.** SecA binds ATP. Another model postulates that SecA binds ATP first, and this induces a conformational change in SecA that facilitates its binding to the preprotein. A more complete discussion of the role of ATP can be found in Note 14.

**Step 4.** The amino terminus of the leader peptide leaves SecA and enters the membrane. The positively charged amino terminus binds to the negatively charged phospholipid head groups or perhaps to the SecY/E/G complex and remains attached to the cytoplasmic side of the membrane. The hydrophobic region of the leader sequence spontaneously inserts into the membrane, forming a loop.

**Step 5.** The N-terminus of the leader peptide remains on the cytoplasmic side of the

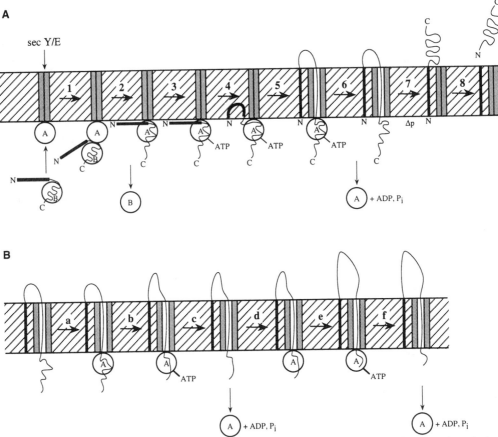

**Fig. 17.2** A model for protein translocation in *E. coli*. (A) 1. The preprotein binds, either during or immediately after its synthesis, to SecB, forming a preprotein/SecB complex that moves to the SecA/SecY/E membrane complex (the translocase) and binds to SecA. 2. The preprotein is transferred to SecA and SecB is released. 3. SecA binds ATP. 4. The leader peptide leaves SecA and is inserted into the lipid bilayer as a loop. 5. The carboxy terminus of the leader peptide "flips" to the periplasmic side, and the preprotein enters the translocase channel. A short segment of the preprotein is translocated. 6. ATP is hydrolyzed and SecA is released into the cytosol. 7. Translocation continues, but now driven by the $\Delta p$ alone. 8. The leader peptide is cleaved, releasing the protein into the periplasm. (B) In the absence of a $\Delta p$ in vitro there can be several rounds of SecA binding and release, promoting ATP-dependent translocation. a, binding SecA; b, binding of ATP to SecA accompanied by the translocation of a small segment; c, hydrolysis of ATP and release of SecA; d, rebinding of SecA as the cycle continues. Symbols: A, SecA; B, SecB.

membrane, possibly attached to the negatively charged phospholipid head groups, while the carboxy-terminal end "flips" into the lipid bilayer as the preprotein enters the translocase (SecY/E/G). It is envisaged that a channel opens up within the translocase to accommodate the protein. During these initial stages, a small segment of the preprotein (perhaps 20 to 30 amino acids) is translocated into the membrane. The precise mechanism by which all this occurs is

not known. Limited translocation occurs in vitro using nonhydrolyzable analogs of ATP, indicating that it is the binding of ATP rather than its hydrolysis that provides the energy for translocation.[15, 16]

**Step 6.** ATP hydrolysis takes place, and SecA is released from the preprotein and the membrane.

**Step 7.** The rest of the protein is translocated, driven by the $\Delta p$. In agreement with this, translocation in respiring cells is immediately inhibited by uncouplers that dissipate the $\Delta p$ while the intracellular ATP levels do not change. Furthermore, the direction of translocation was reversed when the polarity of the $\Delta p$ was reversed in proteoliposomes, translocating an outer membrane protein (proOmpA).[17] However, it may also be that there occur cycles of ATP-driven SecA binding and translocation followed by periods of $\Delta p$-driven translocation after SecA is released. (See the section on recycling of SecA later.)

Step 8. The leader sequence is cleaved by a leader peptidase, and the translocated protein is released into the periplasm. The leader peptide must be removed from periplasmic and outer membrane proteins in order for them to leave the membrane surface, but it need not be removed for translocation per se to take place. According to this model, the initiation of translocation requires the hydrolysis of one ATP. The rest of translocation is driven by $\Delta p$ and does not require ATP. However, see the discussion of the recycling of SecA next.

## Recycling of SecA

The initial stages of translocation, which consist of the insertion of the leader peptide into the membrane and a limited amount of translocation (steps 1–4), require SecA and ATP binding. ATP hydrolysis releases SecA, and the remainder of translocation is driven by the $\Delta p$. However, experiments have demonstrated that, in the absence of a $\Delta p$, ATP can drive translocation to completion in vitro. (See Note 18.) This is because SecA can rebind to the portion of the protein not yet translocated and promote successive rounds of ATP-dependent translocation (Fig. 17.2B, steps a, b, c, d). For some proteins, it may be necessary to increase the SecA concentrations in order to demonstrate complete translocation in the absence of a $\Delta p$.[19] The extent of the in vitro requirement for the $\Delta p$ apparently varies with the preprotein that is used.[20] Although SecA and ATP can be shown to drive translocation to completion in certain in vitro experiments in which there is no $\Delta p$, the evidence favors

the conclusion that the major driving force for proton translocation in vivo is the $\Delta p$ and not ATP. (See the description of step 7, above.) That is not to say that SecA binding and rebinding may not be important beyond the initiation stages of translocation. Cycles of SecA rebinding and dissociation may, in fact, occur in vivo, promoting limited translocation in conjunction with translocation driven by the $\Delta p$. One possible role for cycles of SecA binding and ATPase-dependent dissociation from the preprotein as it is being translocated might be to unfold untranslocated regions that have assumed a tight tertiary configuration that cannot be threaded through the translocase. One might expect such foldings to occur, since protein secretion need not be coupled to translation, and the untranslocated regions of the protein that are still in the cytosol may fold in a way not suitable for translocation.

## Cotranslational translocation

Polysomes translating presecretory proteins associate with membranes, and it seems that some proteins might be translocated while they are being translated, although in E. coli most proteins are exported posttranslationally. The idea is that the ribosome translating a messenger RNA for a preprotein is closely associated with a membrane translocase so that the polypeptide is threaded through the translocase during translation. Chaperones might be necessary if the nascent polypeptide entered the cytosol before engaging the translocase. Cotranslational translocation may involve the signal recognition particle (SRP), which is described later (Section 17.3).

## Protein export independent of the Sec proteins

It would be misleading to imply that all proteins translocated by E. coli require the Sec proteins, although most do. The insertion of certain integral membrane proteins is not affected by mutations in the sec genes. These include the phage M13 procoat protein. Sec-independent proteins like the M13 protein may also have leader sequences, even though they do not require SecA. Thus, although a major route for protein translocation is via the Sec pathway, another pathway(s) obviously exists.

## 17.2 The Translocation of Membrane-Bound Proteins

Thus far the description of protein translocation has been confined to those proteins that ultimately reside in either the periplasm or the outer membrane. The export of several of the inner membrane proteins is also Sec dependent and occurs in a similar fashion. The question is, what keeps these proteins from being translocated through the membrane into the periplasm? The answer lies in internal hydrophobic regions of the protein that stop translocation and anchor the protein into the membrane because they bind to the lipid. Sometimes it is the signal sequence itself that anchors the protein in the membrane (Fig. 17.3A,a). These signal sequences differ from those discussed above in not being recognized by the signal peptidase and are therefore not cleaved. Other times it is an internal hydrophobic region called the "signal-membrane anchor" or "stop-transfer" sequence, which anchors the protein after the signal sequence has been cleaved (Fig. 17.3A, b). The "stop-transfer" regions have a stretch of 15 or more hydrophobic amino acids, which vary from one protein to another. Proteins that span the membrane several times do so via a series of alternating uncleaved signal sequences and "stop-transfer" signals (Fig. 17.3A, c). (These include the membrane- bound solute transporters discussed in Chapter 16.) A model for the insertion of proteins that span the membrane multiple times is shown in Fig. 17.3B. A noncleavable signal sequence initiates insertion into the membrane. At some point, translocation stops as an internal hydrophobic "stop-transfer" signal enters the translocase. According to the model, the translocase opens laterally when the "stop signal" enters, allowing the diffusion of the "stop- transfer" signal into the lipid bilayer. Then the channel closes. A downstream secondary signal sequence reinitiates translocation, and the process is repeated until the entire protein is translocated into the membrane. This is not a well-understood process.

## 17.3 The E. coli SRP

The components required for the insertion of inner membrane proteins in E. coli are not as well understood as in the case of translocation of proteins into the periplasm. Almost all proteins translocated into the periplasm depend upon the Sec system as described earlier. However, it has been suggested that in E. coli the placement of some inner membrane proteins, especially those that do not have extensive periplasmic loops, may involve a "signal recognition particle" or SRP.[21] E. coli contains a ribonucleoprotein particle called SRP that consists of a 4.5S RNA and a 48-kDa protein subunit. A membrane receptor for the SRP (i.e., FtsY) is also present. Mutants lacking either SRP or FtsY are not viable, although there is no profound effect on the export of proteins to the periplasm or outer membrane. This is consistent with the conclusion that SRP is primarily involved with the assembly of certain inner membrane proteins. (Reviewed in Ref. 22 and discussed in Note 23.) Both SRP and FtsY are homologous to SRP and its receptor found in eukaryotes. Precisely how the E. coli SRP functions in protein translocation is not understood at the present time, but it has been speculated that the E. coli SRP functions in a manner similar to the functioning of SRP in eukaryotes. A brief account of the role of SRP in eukaryotes is given in Section 17.3.1.

### 17.3.1. SRP-dependent protein translocation across the endoplasmic reticulum (ER) membrane in eukaryotes: Comparison to prokaryotes

A comparison of protein translocation in eukaryotes and prokaryotes (Bacteria and Archaea) is given in Refs. 7 and 24. The student is encouraged to read these reviews for a more detailed account of the similarities and differences between the secretory pathways in eukaryotes and prokaryotes. In eukaryotic cells proteins that are destined for locations

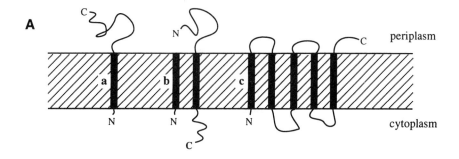

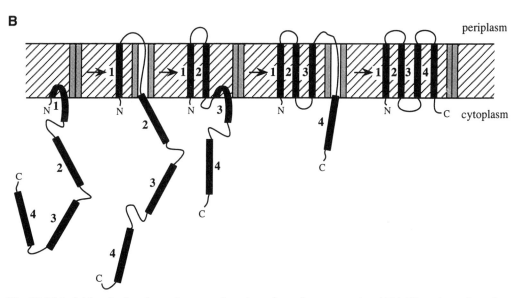

**Fig. 17.3** Model for the Sec-dependent translocation of membrane proteins. (A) (a) Protein anchored to the membrane by its leader peptide. (b) Protein anchored to the membrane by a "stop-transfer signal." The leader sequence has been proteolytically removed and is shown in the membrane. (c) Protein anchored to membrane by alternating leader peptides and stop-transfer signals. (B) Insertion of a protein with four hydrophobic domains. Domain 1 inserts into the membrane as a signal peptide. This opens the translocase channel and and initiates translocation. When domain 2 (stop-transfer signal) enters the translocase, the putative channel opens laterally, allowing domain 2 to diffuse laterally into the lipid matrix. The channel then closes. Domain 3 (secondary signal sequence) reinitiates translocation by inserting into the lipid bilayer and reopening the channel. When domain 4 (stop-transfer signal) enters the translocase, it stops translocation and moves laterally out of the translocase channel into the lipid matrix. *Source:* Adapted from Pugsley, A. P. 1993. The complete general secretory pathway in gram-negative bacteria. *Microbiol. Rev.* 57:50–108.

other than the cytosol, nucleus, mitochondria, or chloroplasts are synthesized on ribosomes attached to the endoplasmic reticulum (ER). The proteins are secreted into the lumen of the ER as they are being translated (i.e., secretion and translation are coupled). After being secreted into the lumen of the ER, the proteins move to the Golgi complex and from there to various destinations such as the cell membrane, secretory vesicles, or lysosomes. It has been known since the 1970s that proteins destined to be translocated across the ER membrane have a *signal sequence* at their N-terminal ends that is later removed by a *signal peptidase* on the luminal side of the ER. This is analogous to the leader peptide found on the presecretory proteins in prokaryotes. (Some secreted or membrane proteins in eukaryotes, e.g., hen egg-white ovalbumin, have a signal sequence that is internal rather than at the N-terminal end.)

Eukaryotes have a 16S ribonucleoprotein particle called the "signal recognition particle" (SRP) that binds to the signal sequence as the protein is being synthesized on cytosolic ribosomes and also binds to the ribosome directly. The binding of SRP stops translation. The SRP–ribosome–nascent polypeptide complex then diffuses to the ER, where the SRP binds to a receptor (docking protein) and delivers the ribosome–nascent polypeptide complex to the translocase in the ER. Once the SRP is removed from the ribosome, translation resumes and the newly synthesized polypeptide is translocated through the translocase as it is being translated. The translocase is a heterotrimeric machine called the Sec61 complex, which is similar in some respects to the bacterial SecY/E/G. Meanwhile the SRP is released into the cytosol and can guide another ribosome–nascent polypeptide complex to the ER. Eukaryotes also have a post-translational secretory pathway that is SRP independent. It has been speculated that in eukaryotes only if the SRP fails to bind to the nascent polypeptide do the post-translational pathways come into play. This can happen because the affinity of SRP for the signal sequence decreases as the polypeptide gets longer, and thus there is only a short period of time when SRP can bind effectively. As in bacteria, the post-translational pathway appears to involve the use of chaperones analogous to SecB that prevent the presecretory protein from folding into a conformation that prevents translocation.

## 17.4 Extracellular Protein Secretion

Proteins that are secreted into the medium include enzymes such as proteases, nucleases, lipases, carbohydrases, toxins, and other virulence factors.[25–27] Some proteins whose secretion has been studied include immunoglobulin A proteases secreted by gram-negative pathogenic bacteria belonging to the genera *Neisseria, Haemophilus,* and *Serratia,* cholera and pertussis toxin secreted by *Vibrio cholerae* and *Bordetella pertussis,* respectively, pilin proteins, and hemolysins (toxins that lyse red blood cells), secreted by various gram-negative bacteria. Protein secretion has been studied in gram-positive bacteria as well. These include penicillinase production and the secretion of various proteases in *Bacillus* species. In some cases, a particular sequence of amino acids at the carboxy end seems sufficient for translocation through the outer envelope (self-promoted secretion). However, in most instances a special secretion machinery seems to be necessary. There are four major secretion machineries: Two are *sec*-independent (Type I and Type III), and two are *sec*-dependent (Type II and Type IV). There is also a fifth secretion system, which has been called Type V.[41] This last system is not well characterized, but consists of approximately 10 proteins that form a pore through which DNA or protein can be transferred through membranes from cell to cell. It is responsible for conjugal transfer of plasmids between gram-negative bacteria, for the transfer of oncogenic T-DNA by *Agrobacterium tumefaciens* into plant cells (Section 18.19.6), and the secretion of *Bordetella pertussis* toxin. This secretion system is reviewed in Ref. 28. See Fig. 17.4 for an overview of some of the secretion systems.

### 17.4.1 Sec-independent protein secretion

There are two *sec*-independent secretion pathways. They are the Type I and the Type III pathways. The Type I pathway is described next, and the Type III pathway in Section 17.4.4. Examples of proteins secreted by the Type I system include haemolysin by *E. coli,* proteases by *Erwinia chrysanthemi* and *Pseudomonas aeruginosa,* and leucotoxin by *Pasteurella haemolytica.* These proteins do not have leader sequences and do not require any of the *sec* genes for secretion. The proteins are secreted directly from the cytoplasm to the outside of the cell. As described next, the machinery to secrete the proteins consists of two inner (cytoplasmic) membrane proteins and an outer membrane protein.

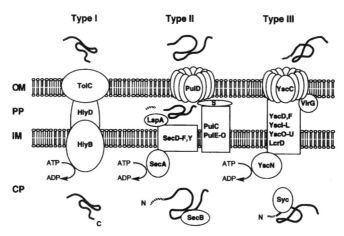

**Fig. 17.4** Overview of Types I, II, and III secretion systems. Type I is represented by hemolysin secretion by *E. coli*. Type II is represented by the pullunase secretion system by *Klebsiella oxytoca*. Type III is represented by Yop secretion by *Yersinia*. OM, outer membrane; PP, periplasm; IM, inner membrane; CP, cytoplasm. SecB and Syc are cytoplasmic chaperone proteins. LspA is a periplasmic peptidase that cleaves off the amino-terminal signal sequence. PulD and YscC (Types II and III outer membrane proteins) are homologous. *Source:* From Hueck, C. J. 1998. Type III protein secretion systems in bacterial pathogens of animals and plants. *Microbiol. Molec. Biol. Rev.* **62**:379–433.

Genes required for secretion have been identified by mutant analysis and are called *secretory genes*. The secretory genes are usually contiguous with the structural gene for the protein that is secreted. The proteins coded for by the secretory genes form a secretory apparatus. Since the carboxyl terminus of the secreted protein is required for secretion, it would appear that the secretion apparatus recognizes this region of the protein. Presumably, this is done by the ABC transporter, described next. There are three secretory proteins. Two are inner membrane proteins, and one is an outer membrane protein. One of the inner membrane proteins is an ATP-binding protein that belongs to the superfamily of proteins called the *ABC exporters* (ATP-binding cassette), which also catalyze solute uptake and extracellular polysaccharide export. (See Sections 11.3.5 and 16.3.3). It is sometimes referred to as the *ABC transporter*. The fact that the ABC transporter has an ATP binding site indicates that it drives the secretion of proteins. However, the mechanism of translocation is not understood. The other two proteins are called auxiliary proteins. One of these, the membrane fusion protein (MFP), is anchored in the inner membrane and is believed to span the periplasm. The other auxiliary protein is

an outer membrane protein (OMP). The cellular location of the secretory proteins suggest that they may form a channel through the inner and outer membrane, allowing proteins to move directly from the cytoplasm to the external medium without entering the periplasm. The three transporter proteins in the different Type I systems have different names, but they are homologous proteins, and the secretion systems can substitute for one another in mutants.[29, 30] (There are actually families of Type I systems. Within a family there is sequence similarity between the components of the secretion apparatus, and they can substitute for one another. When components from different families are interchanged, there is a significant drop in protein secretion. See Ref. 31 for a review.) Two illustrative examples of ABC transporters follow (Fig. 17.4). In *E. coli,* the structural gene for α-hemolysin is *hlyA*. It is located next to *hlyB* and *hlyD*, the genes for the ABC transporter and the auxiliary inner membrane secretory protein, respectively. The gene for the auxiliary outer membrane secretory protein, *TolC*, is not linked to these genes. This is an exception to the rule that all three secretory genes are linked to the gene for the secreted protein, but TolC is a multifunctional protein and is also used for the transport of other molecules. In *E. chrysanthemi,* there are four contiguous genes for four homologous secreted proteases, and these are located next to one set of the secretory genes, which are *prtD* (the gene for the ABC transporter), *prtE* (the gene for the auxiliary inner membrane protein), and *prtF* (the gene for the auxiliary outer membrane protein).

## 17.4.2 Sec-dependent protein secretion

The Type II pathways appear to be the *major route* by which proteins are exported in gram-negative bacteria. Type II pathways secrete proteins in two steps (Fig. 17.4). Step 1 uses the GSP (general secretory pathway) to secrete the protein into the periplasm. (The GSP pathway is the *sec*-dependent pathway described in Section 17.1.) The leader sequence is proteolytically removed on the external face of the cell membrane, and the proteins enter the periplasm for the second step of secretion to the extracellular milieu. The translocation machinery and mechanism of the second step are not understood. It has been speculated that the secreted proteins may move in vesicles from the periplasm to the outer membrane, where they fuse with channels in the outer membrane made by the Type II secretory apparatus. Even if this were true, it does not address the problem of how the secreted proteins are selected for inclusion into the postulated vesicles. About 12 to 14 proteins are required for passage through the outer membrane, which is suggestive of a complex translocation machinery. These proteins may be located in both the inner and outer membranes.

There are several Type II systems, and each can secrete more than one kind of protein. For example, a secretion machinery called the Out system exists in *Erwinia chrysanthemi* and *E. carotovora,* causative agents of soft-rot disease in plants. The enzymes secreted by the Out system are pectate lyase, exo-poly-α-D-galacturonosidase, pectin methylesterase, and cellulase. Mutants in the *out* genes have a defect in the secretion of cellulases and pectinases and accumulate these enzymes in the periplasm. Other GSP systems exist in other bacteria and are responsible for the secretion of different proteins. For example, the *pul* genes are required for the secretion of pullulanase by *Klebsiella pneumoniae,* and the *eps* genes are required for the secretion of cholera toxin and protease by *Vibrio cholerae.*[32] (See Note 33 for a description of the pullulanases.) The *eps* genes are also required for correct assembly of the proteins in the outer membrane.

A resident Type II system need not secrete extracellular proteins of related bacteria when the structural genes are introduced, even though there may be homology between the GSP proteins from the donor and recipient bacteria. For example, there is homology in the proteins between the *out* gene products from *Erwinia* and the *pul* gene products from *Klebsiella.* Yet when the pectate lyase gene from *Erwinia* was expressed in *Klebsiella,* the enzyme was not secreted. There can even be specificity between Out systems from two different *Erwinia* strains. For example, the *E. chrysanthemi* Out system did not secrete an extracellular pectate lyase encoded by a gene from a different species of *Erwinia,* even though both species of *Erwinia* use the Out system. The conclusion is that, when the enzyme structural gene is expressed heterogenerically, the protein products are not secreted. This is because each GSP system distinguishes not only between different secreted proteins, but also recognizes "self" proteins. The basis for this distinction is not understood. However, if the GSP system *as well as its cognate protein* are transferred to a different genus of bacterium, then protein secretion can occur. For example, when the structural gene for pullulanase as well as the adjoining secretion genes are transferred from *Klebsiella pneumoniae* to *E. coli,* pullulanase is synthesized and secreted.[34]

There is a relationship between the Type II pathway and the assembly of Type IV pili that are present on several pathogenic bacteria including *Pseudomonas aeruginosa, Neisseriae gonorrhoeae,* and some pathogenic *E. coli.* (The pili are used to adhere to host tissue.) For example, consider *Ps. aeruginosa.*[35] Although pilus assembly and protein secretion via the Type II pathway operate independently of each other, they do use proteins that are homologs of each other. For example, protein export uses four cell membrane proteins (XcpT, U, V, and W) that are homologs of the major subunit of Type IV pili (i.e., PilA). There are also accessory genes for pilus biogenesis that are similar to genes required for protein secretion. In addition, both pilus assembly and protein export require the same leader peptidase (PilD, XcpA) (i.e., encoded by the same gene), which removes the leader peptide from the N-terminus of the pilin subunits as well as from the four homologs of the export machinery. Thus there is clearly a relationship between the mechanism of pilus biogenesis and protein secretion. Presumably, the proteins of the Type II pathway interact with each other during protein secretion in a way that resembles the formation of pili, hence the use of homologous proteins. The shape of this putative structure involved in secretion is completely unknown, but it is not believed to resemble a pilus. It has also been reported that the pilin subunit PilA can be detected in cells of *Ps. aeruginosa* as a complex with one of the export protein homologs; that is, XcpT and mutants defective in PilA production are less efficient in protein secretion.[35] What this means in terms of the structure or function of the Type II secretion apparatus is a matter of speculation, but it does point to an interesting relationship between pilus production and protein secretion via the Type II system.

### 17.4.3 A sec-dependent protein secretion that does not appear to be Type II; the Type IV pathway

The proteins secreted by this system use the GSP pathway to move through the cell membrane (inner membrane), but do not require a separate secretion apparatus to move through the outer membrane. They have been called *autotransporters*. The protein perhaps forms a pore in the outer membrane through which it passes, and then is released into the extracellular fluid via autoproteolytic cleavage. As an example of self-promoted protein secretion that does not appear to require a secretion apparatus, we will consider the secretion of immunoglobulin A protease by the causative agent of gonorrhoeae (i.e., *Neisseria gonorrhoeae*). The protease is synthesized as a precursor protein with an N-terminal leader sequence typical for exported proteins. The leader sequence functions with the *sec* machinery and is removed by the peptidase after translocation through the inner membrane. Up to this stage, the secretion is similar to the secretion of proteins via the Type II system. However, the immunoglobulin A protease has information in its sequence that directs it through the outer envelope to the outside. At the carboxy end of the protein, there is a helper domain that aids the protein in traversing the outer membrane, where it anchors the protein. The protein is released into the medium by proteolytic cleavage (autocatalytic) at the C-terminal end, leaving the helper peptide embedded in the membrane. Some other proteins are secreted in a similar fashion, although the proteolytic cleavage may be catalyzed by a separate enzyme. How the secreted proteins cross the periplasm and insert into the outer envelope is not known. The helper domain can promote translocation of hybrid proteins across the outer membrane in *E. coli*, which would not normally release the passenger protein into the medium. It would thus appear that the helper region may be sufficient for translocation through the outer envelope without the need for an accessory secretion system that has been identified thus far.

### 17.4.4 Type III protein secretion

The Type III protein secretion systems are GSP-independent systems that form an apparatus used for the *secretion of virulence proteins* in some gram-negative animal pathogenic bacteria such as *Yersinia, Salmonella,* and *Shigella,* and several plant pathogens.[36, 37] (For reviews, see Refs. 38–41.) The secretion apparatus consists of approximately 20 proteins (mostly inner membrane proteins) and is quite similar among unrelated bacteria, although the proteins that are secreted differ considerably. The secreted proteins are called Ipas in *Shigella,* Sips in *Salmonella,* Harpins and Pops in plant pathogens, and Yops in *Yersinia.* (See Ref. 41 for a review.) Similar secretion systems exist for flagellum secretion and assembly in *E. coli, Salmonella,* and *B. subtilis.* (In other words, most of the inner membrane proteins are homologs to the flagellar biosynthetic proteins.) Furthermore, an outer membrane protein of the Type III secretion system is homologous to PulD, which is an outer membrane protein that is part of the Type II secretion system. (See Fig. 17.4). The Type III systems have the following characteristics (Fig. 17.4).

1. There is no leader sequence typical of the Sec pathway. Furthermore, as with the Type I system, there is no processing of the N-terminal end during secretion. (See Note 42.)

2. They all require accessory proteins for secretion.

3. Proteins are secreted through both the inner and outer membranes, and generally into the cytosol of target cells.

4. Secretion is initiated upon reception of an activation signal. In at least some instances, the activation signal occurs when the bacterium adheres to the target cell.

### Yersinia

Some of the secreted virulence proteins have cytoxic effects on the target cells. As an example of the latter, we will consider the Yops proteins (*Yersinia* outer proteins), which are approximately 12 proteins secreted by bacteria of the genus *Yersinia*. (See Refs. 43 and 44 for reviews.) There are three species of *Yersinia* that cause disease. (See Note 45 for a description of these diseases.) *Yersinia pestis* is the causative agent of plague. Both *Y. enterocolitica* and *Y. pseudotuberculosis* can cause severe gastroenteritis, local abscess formation, and peritonitis in humans. *Yersinia* are resistant to phagocytosis by macrophages and polymorphonuclear leukocytes (PMNs), and this enables them to establish a systemic infection. (As explained later, they resist phagocytosis and killing by injecting Yops proteins into target cells.) They multiply mainly outside of host cells, but are also capable of invading host cells such as epithelial cells.

### 1. Injection of cytotoxic Yops proteins by Yersinia into target cells

Some of the Yops proteins form a secretion and delivery apparatus that injects other, cytotoxic, Yops proteins into target eukaryotic cells to which the bacterium is adhering. (A similar situation holds for virulence proteins secreted by *S. typhimurium* via the Type III system. These proteins also enter the host cell.[41]) Four Yops proteins that enter the target cells are YopE, YopH, YpkA, and YopM. YopE is cytotoxic to host cells, and like the other Yops proteins, must be injected by adhering bacteria in order for host cell damage to occur. YopE disrupts the actin microfilaments in the target cells and in this way prevents uptake of the bacteria by phagocytic cells as well as collapsing the cytoskeleton. YopH is a protein tyrosine phosphatase. It dephosphorylates macrophage proteins, and perhaps disrupts phosphate-dependent signal transduction necessary for normal phagocytosis. YopH also prevents the production of toxic forms of oxygen that are produced by macrophages upon phagocytosis of pathogens[46]; YpkA is a serine/threonine kinase and perhaps also disrupts signal transduction. The intracellular action of YopM is unknown. Upon adhering to the surface of a host cell, *Yersinia* injects Yops into the cytoplasm of the host cell, producing a cytotoxic effect In order that the transfer of Yops take place, the proteins must cross the inner and outer bacterial membranes as well as the cell membrane of the target cell. A model for how this might occur is shown in Fig. 17.5. The secretion apparatus consists of a channel through the inner and outer envelope of the bacterium and a microinjection device by which the Yop proteins are injected directly into the target cell. Note that in the model YopN is postulated to be a stop valve that prevents the secretion of Yops proteins into the extracellular medium. YopN has been postulated to be a surface sensor that upon contact with the target cell undergoes a conformational change, allowing secretion to take place. It has been known that $Ca^{2+}$ chelation in vitro results in the secretion of large amounts of Yops proteins into the medium, regardless of the presence of eukaryotic cells. Additionally, mutants that are unable to secrete YopN secrete the other Yops proteins regardless of the presence of $Ca^{2+}$. These observations have led to the suggestion that $Ca^{2+}$ chelation results in the removal of YopN, and this leads to deregulated release of Yops.[43]

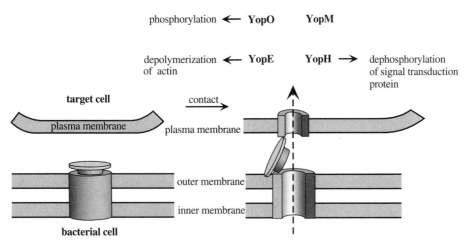

phosphorylation ← **YopO**     **YopM**

depolymerization ← **YopE**     **YopH** → dephosphorylation
of actin                                of signal transduction
                                        protein

**Fig. 17.5** Model for secretion and translocation into target cell of Yops proteins; the Type III system. The protein secretion apparatus in *Yersinia* is called Ysc and is suggested to form a channel that crosses the inner and outer membranes. The Ysc apparatus is very complicated and consists of approximately 20 proteins, some of which are inner membrane proteins and some of which are outer membrane proteins. When there is no contact between *Yersinia* and the target cell, then there is a plug, called YopN, lying across the opening of the Ysc channel. YopN binds to a receptor on the surface of the target cell and the Ysc channel is opened. At the same time the bacterium makes an injection device consisting of YopB and YopD that spans the host cell plasma membrane. Yop proteins are then injected into the host cell. The injected Yop proteins are believed to have several toxic effects, including the depolymerization of actin and inactivation of host-cell signaling systems. *Source:* Adapted from Cornelis, G. R., and H. Wolf-Watz. 1997. The *Yersinia* Yop virulon: A bacterial system for subverting eukaryotic cells. *Molec. Microbiol.* **23**:861–867.

## 17.5 Summary

Bacteria translocate (export) proteins into the cell membrane or secrete them through the membrane into the periplasm and outer envelope of gram-negative bacteria, the cell wall of gram-positive bacteria, and the extracellular medium. The translocation machinery and requirements for targeting to different destination sites are being studied in several bacteria, especially in *E. coli*.

The Sec system is necessary for the translocation of most proteins across the cell membrane. A preprotein is made with a leader sequence at the amino-terminal end. The role of the leader sequence is to initiate translocation by inserting into the lipid bilayer. The leader sequence also aids in preventing premature folding of the newly synthesized protein. The preprotein binds to a chaperone protein in the cytosol. An important chaperone protein is Sec B. Two roles of the chaperone protein are to prevent folding of the preprotein into a configuration that disallows translocation and to prevent the formation of protein aggregates. A third role for the chaperone protein is to deliver the preprotein to the translocase. The preprotein–chaperone protein complex binds to the translocase on the inner membrane surface, and the translocation of the preprotein into the membrane is initiated. The initial stages of translocation require ATP and SecA, but the remainder of translocation can be driven by the Δp. Once the preprotein has been translocated through the membrane, the leader sequence is removed by a peptidase on the outer surface of the cell membrane. A few proteins (e.g., some phage proteins) are translocated independently of the Sec system, but these are a small minority. *E. coli* also has a signal recognition particle (SRP) that may guide certain proteins destined to remain in the inner membrane.

Many proteins are secreted into the extracellular space. They must traverse the periplasm and outer envelope in gram-negative bacteria and the cell wall in gram-positive bacteria. Although a few proteins seem to be secreted without additional proteins other than the Sec proteins, most require a second secretion apparatus. Two major secretion systems for getting through the outer envelope in gram-negative bacteria are called Type I and Type II. Type I is independent of the Sec proteins, and the secreted proteins seem to travel from the cytoplasm to the exterior without stopping in the periplasm. The Type II occurs in two stages. Stage 1 uses the general secretion pathway (GSP) to secrete the protein into the periplasm. Stage 2 transports the protein from the periplasm through the outer envelope to the exterior. It is not understood how proteins destined for secretion cross the periplasm and outer envelope. A third system that uses the GSP pathway but is not Type II also exists. A very important protein secretion system that is responsible for the secretion of virulence proteins in certain gram-negative pathogens is called the Type III system. It is Sec independent.

## Study Questions

1. What is the experimental basis for believing that the leader peptide is necessary for translocation? What are two reasons for believing that the leader peptide is not necessary for determining the final destination of the protein?

2. What are the postulated roles for chaperone proteins in protein translocation?

3. What is the role of Sec A in protein translocation?

4. What are the functions of the different domains of the leader sequence (i.e., the basic N-terminal, middle hydrophobic, and carboxy terminus)?

5. What is the evidence, based upon mutational analysis, that there are two secretory systems for proteins secreted into the medium and that one is *sec* dependent and one is *sec* independent.

6. How does the Type III protein secretion system differ from Types I and II? How is it similar?

## NOTES AND REFERENCES

1. Driessen, A. J. M. 1992. Bacterial protein translocation: Kinetic and thermodynamic role of ATP and the protonmotive force. *TIBS* 17:219–223.

2. Ito, K. 1992. SecY and integral membrane components of the *Escherichia coli* protein translocation system. *Molec. Microbiol.* 6:2423–2428.

3. Wickner, W., A. J. M. Driessen, and F.-U. Hartl. 1991. The enzymology of protein translocation across the *Escherichia coli* plasma membrane. *Ann. Rev. Biochem.* 60:101–124.

4. Saier, Jr., M. H., P. K. Werner, and M. Muller. 1989. Insertion of proteins into bacterial membranes: Mechanism, characteristics, and comparisons with the eucaryotic process. *Microbiol. Rev.* 53:333–366.

5. Pugsley, A. P. 1993. The complete general secretory pathway in gram-negative bacteria. *Microbiol.* Rev. 57:50–108.

6. Murphy, C. K., and J. Beckwith. 1996. Export of proteins to the cell envelope in *Escherichia coli.* pp. 967–978. In: *Escherrichia coli* and *Salmonella*: Cellular and Molecular Biology. F. C. Neidhardt (Ed.). ASM Press. Washington, D.C.

7. Rapoport, T. A., B. Jungnickel, and U. Kutay. 1996. Protein transport across the eukaryotic endoplasmic reticulum and bacterial inner membranes. *Ann. Rev. Biochem.* 65:271–303.

8. Gardel, C., K. Johnson, A. Jacq, and J. Beckwith. 1987. SecD, a new gene involved in protein export in *Escherichia coli. J. Bacteriol.* 169:1286–1290.

9. Wickner, W., and M. R. Leonard. 1996. *Escherichia coli* preprotein translocase. *J. Biol. Chem.* 271:29514–29516.

10. Duong, F., and W. Wickner. 1997. The SecDFyajC domain of preprotein translocase controls preprotein movement by regulating SecA membrane cycling. *The EMBO Journal* 16:4871–4879.

11. In *E. coli* the SecYEG translocase can be isolated from the membrane in a complex with three other proteins, i.e., SecD, SecF, and yajC. There appears to be only 10% as much SecD-FyajC as SecYEG in *E. coli*, suggesting either that some SecYEG translocases operate without SecDFyajC or that SecDFyajC cycles between operating SecYEG translocases. Mutants lacking SecDF are defective in translocation and growth. Overexpression of *secDF* suppresses mutations in the leader peptide sequence. Anti-SecD antibodies have been reported to block translocation

in spheroplasts. Recent in vitro studies using isolated membranes with depleted, wild-type, or enriched DF levels indicate that SecDFyajC stabilizes the inserted form of SecA in the membrane and slows the movement of the preprotein (proOmpA). As a consequence, translocation intermediates accumulate in the membrane, and these are translocated using the driving force of the $\Delta p$. Thus SecDFyajC plays a regulatory role in protein secretion and facilitates the coordination between ATP and $\Delta p$ driven translocation, lowering the amount of ATP used for translocation. For a discussion of the role of SecD-FyajC in protein secretion, see Duong, F., and W. Wickner. 1997. The SecDFyajC domain of preprotein translocase controls preprotein movement by regulating SecA membrane cycling. *The EMBO J.* **16**:4871–4879.

12. Manting, E. H., C. Van der does, and A. J. M. Driessen. 1997. In vivo cross-linking of the SecA and SecY subunits of the *Escherichia coli* preprotein translocase. *J. Bacteriol.* **179**:5699–5704.

13. Nishiyama, K-I., S. Mizushima, and H. Tokuda. 1993. A novel membrane protein involved in protein translocation across the cytoplasmic membrane of *Escherichia coli. The EMBO J.* **12**:3409–3415.

14. The situation regarding SecA and ATP seems to be more complicated than indicated in the model. SecA has two binding sites for ATP and possibly two ATPase activities. One idea is that SecA binds ATP, which induces a conformational change that enhances its binding to the preprotein. A first round of hydrolysis (or release) of ATP might then induce a conformation change in SecA such that the preprotein is inserted into the membrane so that it interacts with SecY and SecE. A second round of ATP hydrolysis, termed "translocation ATPase," seems to be required for the translocation of the preprotein across the cytoplasmic membrane, although, as the model suggests, perhaps only the release of ATP and not its hydrolysis is required for translocation. See Ref. 7 for a discussion of these points.

15. Schiebel, A., A. J. M. Driessen, F.-U. Hartl, and W. Wickner. 1991. $\Delta\mu_{H^+}$ and ATP function at different steps of the catalytic cycle of preprotein translocase. *Cell* **64**: 927–939.

16. The in vitro system consists of inverted *Escherichia coli* inner membrane vesicles translocating the outer membrane protein proOmpA.

17. Driessen, A. J. M. 1992. Precursor protein translocation by the *Escherichia coli* translocase is directed by the protonmotive force. *EMBO J.* **11**:847–853.

18. There are several ways to separate effects of the $\Delta p$ from those of SecA and ATP in vitro. For example, membrane vesicles or proteoliposomes will not have a $\Delta p$ unless one is imposed. Different methods can be used to impose a $\Delta p$. For example, the stimulation of electron transport by incubating inner membrane vesicles with NADH has been used. Vesicles are prepared from *unc⁻* cells so that the $\Delta p$ and ATP are not interconvertible. A $\Delta p$ has also been created by light in proteoliposomes in which bacteriorhodopsin was incorporated, or by adding reduced cytochrome c to proteoliposomes in which cytochrome c oxidase has been incorporated. Proton ionophores can be used to dissipate the $\Delta p$. SecA can be inactivated with anti-SecA antibody, and ATP levels can be experimentally manipulated.

19. Yamada, H., S-ichi Matsuyama, H. Tokuda, and S. Mizushima. 1989. A high concentration of SecA allows proton motive force-independent translocation of a model secretory protein into *Escherichia coli* membrane vesicles. *J. Biol. Chem.* **264**:18577–18581.

20. Yamada, H., H.Tokuda, and S. Mizushima. 1989. Protonmotive force-dependent and -independent protein translocation revealed by an efficient in vitro assay system of *Escherichia coli. J. Biol. Chem.* **264**:1723–1728.

21. De Gier, J-W. L., Q. A. Valent, G. V. Heijne, and J. Juirink. 1997. The *E. coli* SRP: Preferences of a targeting factor. *FEBS Lett.* **408**:1–4.

22. Duong, F., J. Eichler, A. Price, M.R. Leonard, and W. Wickner. 1997. Biogenesis of the gram-negative bacterial envelope. 1997. *Cell* **91**:567–573.

23. Chaperones may aid in separating preproteins destined for secretion from cytosolic proteins. There are two chaperones that are bound to the ribosome. These are "trigger factor" and the SRP particle. A decision is made when at least some of the nascent proteins leave the ribosome to bind to either one of these two chaperones. If a leader sequence is not present, then the nascent protein binds preferentially to the "trigger factor." One model speculates that nascent proteins bound to the "trigger factor" are transferred to another chaperone called GroEL, which takes part in ensuring proper folding of the protein. DnaJ/K/GrpE chaperones may also take part in folding the nascent proteins, as discussed in Section 10.3.10. These folded proteins would not be secreted. A subset of nascent proteins that are destined to be integral cell membrane proteins bind to SRP via their leader sequence while still associated with the ribosome. Nascent proteins with more hydrophobic core regions in their leader sequence are more likely to bind to SRP. The preproteins are eventually delivered

to the membrane. It is not known whether the SRP-dependent pathway delivers cell membrane proteins co- or post-translationally. (The SRP-dependent pathway in eukaryotes operates co-translationally.) SRP-dependent cell membrane proteins include cytochrome oxidase d. Other cell membrane proteins, e.g., mannitol and maltose permease, do not require SRP for their assembly. SRP is primarily involved in the membrane assembly of proteins that do not have long periplasmic loops. SRP is required for viability. It is clear that SRP is not required for the export of proteins through the cell membrane, e.g., to the periplasm or the outer envelope. Preproteins destined for passage through the cell membrane to the periplasm or outer membrane become associated with SecB, which brings them to the SecA/SecYEG translocase. It seems that the SRP and SecB pathways might be parallel pathways, i.e., a SRP-targeting pathway that delivers mostly cell membrane proteins that do not have long periplasmic loops, and the SecB pathway that mostly delivers proteins destined for excretion. SecB is also required for the insertion into the membrane of membrane proteins with long periplasmic loops. The SecB-dependent pathway operates mostly post-translationally, whereas, as mentioned, the SRP-dependent pathway might operate co-translationally. Although the SRP-dependent pathway and the SecB-dependent pathway may operate in parallel under usual growth conditions, it has not been ruled out that under certain circumstances, e.g., overloading of the SecB-dependent pathway, SRP does not substitute for SecB in the secretion of proteins with less hydrophobic leader signals. Under this scenario both the SRP-targeting pathway and the SecB-targeting pathway can deliver preprotein to the translocase. The decision as to which pathway to use would depend upon the hydrophobicity of the leader sequence and the availability of SecB. (See the review by De Gier et al. for a discussion of this point as well as other aspects of the *E. coli* SRP. De Gier, J-W., Q. A. Valent, G. Von Heijne, and J. Luirink. 1997. The *E. coli* SRP: Preferences of a targeting factor. *FEBS Lett.* 408:1–4.)

Trigger factor is a peptidyl–prolyl *cis/trans* isomerase attached to the ribosomal 50S subunit. It catalyzes the *cis–trans* isomerization of the peptide bond that is on the N-terminal side of proline residues. This may aid in the folding of the nascent protein and perhaps in the delivery of the nascent protein to other chaperones such as GroEL. Normally the hydrogen attached to the substituted amino group in peptide bonds is opposite (*trans*) to the oxygen of the carbonyl group. This is the most stable configuration. However, when the nitrogen atom is contributed by proline, the peptide linkage is occasionally *cis* (in about 5% of cases) in native proteins. Probably a certain proportion of the prolyl bonds

must isomerize from *trans* to *cis* in order for proper folding to take place. For a review of prolyl isomerases see: Hunter, T. 1998. Prolyl isomerases and nuclear function. *Cell* 92:141–143. For a report of the *E. coli* trigger factor see: Stoller, G., P. K. Rücknagel, K. H. Nierhaus, F. X. Schmid, G. Fischer, and J-U. Rahfeld, 1995. A ribosome-associated peptidyl–prolyl *cis/trans* isomerase identified as the trigger factor. *The EMBO J.* 14, 4939–4948.

24. Pholschröder, M., W. A. Prinz, E. Hartmann, and J. Beckwith. 1997. Protein translocation in the three domains of life: Variations on a theme. *Cell* 91:563–566.

25. Lory, S. 1992. Determinants of extracellular protein secretion in gram-negative bacteria. *J. Bacteriol.* 174:3423–3428.

26. Salmond, G. P. C., and P. J. Reeves. 1993. Membrane traffic wardens and protein secretion in gram-negative bacteria. *TIBS* 18:7–12.

27. Wandersman, C. 1996. Secretion across the bacterial outer membrane. pp. 955–966. In: *Escherichia coli* and *Salmonella*: Cellular and Molecular Biology. F. C. Neidhardt (Ed.). ASM Press, Washington, D.C.

28. Winans, S. C., D. L. Burns, and P. J. Christie. 1996. Adaptation of a conjugal transfer system for the export of pathogenic macromolecules. *Trends in Microbiol.* 4:64–68.

29. For example, colicin was exported through the alpha-hemolysin export system (HlyBD system) in strains of *E. coli* defective in the colicin export system.[30]

30. Fath, M. J., R. C. Skvirsky, and R. Kolter. 1991. Functional complementation between bacterial MDR-like export systems: Colicin V, alpha-hemolysis, and *Erwinia* protease. *J. Bacteriol.* 173:7549–7556.

31. Binet, R., S. Letoffe, J. M. Ghigo, P. Delepelaire, and C. Wandersman. 1997. Protein secretion by gram-negative bacterial ABC exporters–a review. *Gene.* 192:7–11.

32. Sandkvist, M., L. O. Michel, L. P. Hough, V. M. Morales, M. Bagdasarian, M. Koomey, V. J. Dirita, and M. Bagasarian. 1997. General secretion pathway (eps) genes required for toxin secretion and outer membrane biogenesis in *Vibrio cholerae. J. Bacteriol.* 179:6994–7003.

33. Pullulan is a starch-like polysaccharide formed by the fungus *Pullularia pullulans*. Pullulanases are enzymes that catalyze the cleavage of certain $\alpha$-1,4 and/or $\alpha$-1,6 linkages in starch and related oligosaccharides such as pullulan. They have been isolated from bacteria, archaea, and fungi. Some pullulanases are used industrially as starch-debranching enzymes as an early step in the conversion of starch to saccharides.

Further saccharification is catalyzed by α- or β-amylases. There are four types of pullulanases, and these differ in the specificities of the bonds that are cleaved. The pullulanase from *Klebsiella pneumoniae* hydrolyzes α-1,6 linkages.

34. d'Enfert, C., A. Ryter, and A. P. Pugsley. 1987. Cloning and expression in *Escherichia coli* of the *Klebsiella pneumoniae* genes for production, surface localization and secretion of the lipoprotein pullulanase. *EMBO J.* 6:3531–3538.

35. Lu, H-M., S. T. Motley, and S. Lory. 1997. Interactions of the components of the general secretion pathway: Role of *Pseudomonas aeruginosa* type IV pilin subunits in complex formation and extracellular protein secretion. *Molec. Microbiol.* 25:247–259.

36. Rosqvist, R., S. Håkansson, Å. Forsberg, and H. Wolf-Watz. 1995. Functional conservation of the secretion and translocation machinery for virulence proteins of yersiniae, salmonellae, and shigellae. *EMBO J.* 14:4187–4195.

37. Kubori, T., Y. Matsushima, D. Nakamura, J. Uralil, M. Lara-Tejero, A. Sukhan, J. E. Galán, and S-I Aizawa. 1998. Supramolecular structure of the *Salmonella typhimurium* type III protein secretion system. *Science* 280:602–605.

38. Wattiau, P., S. Woestyn, and G. R. Cornelis. 1996. Customized secretion chaperones in pathogenic bacteria. *Mol. Microbiol.* 20:255–262.

39. Galán, J. E. 1996. Molecular genetic bases of *Salmonella* entry into host cells. *Mol. Microbiol.* 20:263–271.

40. Galán, J. E. 1996. Molecular genetic bases of *Salmonella* entry into host cells. *Mol. Microbiol.* 20:263–271.

41. Hueck, C. J. 1998. Type III protein secretion systems in bacterial pathogens of animals and plants. *Microbiol. and Mol. Biol. Rev.* 62:379–433.

42. There has recently been a suggestion that the signal for secretion by the type III secretion system is actually in the 5′ end of the mRNA that encodes the secreted protein. See Anderson, D. M., and O. Schneewind. 1997. A mRNA signal for the type III secretion of Yop proteins by *Yersinia enterocolitica*. *Science* 278:1140–1143.

43. Cornelis, G. R., and H. Wolf-Watz. 1997. The *Yersinia* Yop virulon: A bacterial system for subverting eukaryotic cells. *Mol. Microbiol* 23:861–867.

44. Silhavy, T. J. 1997. Death by lethal injection. *Science* 278:1085–1086.

45. *Yersinia enterocolitica* causes a severe enterocolitis, especially in young children. The symptoms include fever, diarrhea, and abdominal pain. The disease reservoir is other animals, e.g., farm animals, cats, and dogs. It is transmitted to humans via contaminated foods, e.g., milk, and undercooked meat. It can also be found in well water and lakes. The bacteria produce an enterotoxin that causes diarrhea. They adhere to and invade the intestinal epithelial cells. *Y. pseudotuberculosis* is primarily a pathogen of wild and domestic animals and is transmitted to humans by eating infected meat. The bacteria cause necrotic lesions in the liver, spleen, and lymph nodes. In humans the mesenteric lymph nodes will swell, accompanied by severe abdominal pain. This is usually accompanied by diarrhea and fever. *Yersinia pestis* causes plague. The bacteria live in wild rodents (rats, mice, ground squirrels, chipmunks, prairie dogs) and are transmitted to humans by fleas that have bitten these rodents and then bite humans. The wild rodents are relatively resistant to disease. After being introduced into the human the bacteria grow in the lymph nodes. The bacteria can multiply within the lymph cells. A high fever and swollen lymph nodes (buboes) develop, and the infection at this stage is called bubonic plague. If the infection is not treated, then the bacteria spread to the blood and cause a septicemia. Once in the blood the bacteria spread throughout the body to all the organs. Subcutaneous bleeding from ruptured blood cells produces black spots on the skin, and the disease at this stage has been called Black Death. If the lungs become infected, the infection is called pneumonic plague. Mortality from pneumonic plague approaches 100%. All three pathogenic species of *Yersinia* can prevent uptake by professional phagocytic cells such as macrophages. Hence they are found primarily extracellularly during an infection. Prevention of phagocytosis is due to the type III secretion system, as described in the text.

46. Bliska, J. B., and D. S. Black. 1995. Inhibition of the Fc receptor-mediated oxidative burst in macrphages by *Yersinia pseudotuberculosis* tyrosine phosphatase. *Infect. Immun.* 63:681–685.

# 18

# Adaptive and Developmental Changes

Bacteria may alter cell morphology, cell metabolism, gene transcription, and cell behavior in response to environmental fluctuations such as the availability of respiratory electron acceptors, the supply of carbon, nitrogen, or phosphate, changes in the osmolarity and temperature of the medium. In this way bacteria survive in and adapt to ever-changing environmental conditions. In addition, bacteria may undergo physiological or developmental responses to signaling molecules produced by other cells and engage in cooperative behavior. Underlying the adaptations and responses are sophisticated detection systems with which the bacterium continuously monitors the environment and transmits signals across the cell membrane to specific intracellular targets, which can be the transcriptional machinery, enzymes, or a cellular component such as the flagellum motor. This chapter describes some of these signaling circuits and physiological as well as behavioral responses.

There are many signaling pathways that bacteria use to transmit environmental and intercellular signals to a cell target, be it cellular proteins (e.g., flagellar motors) or transcription factors. In most cases the signal causes the cells to activate or inactivate a cytoplasmic transcription factor. The activation (or inactivation) of transcription factors as a result of the signaling pathway usually occurs either by covalent modification (e.g.,

phosphorylation or dephosphorylation) of the transcription factor, or conformational change upon binding a co-inducer. Two-component systems, which will be described first, work by phosphorylating or dephosphorylating a transcription factor. Other systems (e.g., the quorum-sensing systems that use acylated homoserine lactones as signals) act by binding the signal to the transcription factor, and in that way the signal serves as a co-inducer. In some systems that respond to oxygen, the activity of the transcription factor is regulated by oxygen. The FNR system is an example of the latter. Other systems (e.g., glucose repression via the phosphotransferase system) act by lowering the amounts of a co-inducer (cyclic AMP) for gene transcription, hence gene repression results. Thus there are several different mechanisms of signal transduction.

The signaling systems that will be described include two-component signaling systems, the FNR system, acylated homoserine lactone systems, the formate regulon system, and catabolite repression signaling systems in both gram-positive and gram-negative bacteria. It will be shown how the various signaling systems underlie adaptive physiological changes, chemotaxis, aerotaxis, phototaxis, bioluminescence by *Vibrio*, the production of virulence factors by pathogenic bacteria, the initiation of *Bacillus* sporulation, myxobacteria intercellular communication and multicellular development, and cell-cycle-dependent DNA replication and transcription in *Caulobacter*. All of these, with the exception of the *Caulobacter* system, have in common particular responses by the bacteria to fluctuations in the environment (e.g., to nutrient or oxygen supply) or to signals from other cells.

## 18.1 An Introduction to Two-Component Signaling Systems

In several systems, a signal transduction pathway exists, called a two-component system, that includes a *histidine kinase* protein that receives a signal and transmits it to a partner *response regulator* protein.[1] The response regulator protein in turn transmits the signal to the target. The signal is transmitted between the histidine kinase and the response regulator via *phosphorelay*. Specifically, the histidine kinase autophosphorylates (using ATP as the phosphoryl donor) at a histidine residue in the carboxyl-terminal region (comprising approximatelty 240 amino acids) called the *transmitter domain* in response to a stimulus, and then transfers the phosphoryl group to an aspartate residue in the amino-terminal region (comprising about 120 amino acids) of the partner response regulator protein called the *receiver domain*. This activates the response regulator, which transmits the signal to its target (Section 18.1.2 and Fig. 18.1). Most of the known phosphorylated response regulators stimulate or repress the transcription of specific genes. (Exceptions include P-CheB and P-CheY, which affect the chemotaxis machinery; Section 18.4). See Note 2 for a further discussion of domains in sensor kinases and response regulators.

The signaling pathway also includes a phosphatase that dephosphorylates the response regulator, returning it to the nonstimulated state, where it once again can respond to the signal. The phosphatase may be the histidine kinase itself, the response regulator, or a separate protein. Histidine kinases may reside in the membrane or in the cytoplasm, although they are often in the membrane, whereas the response regulators are in the cytoplasm. Even though signaling systems that consist of a

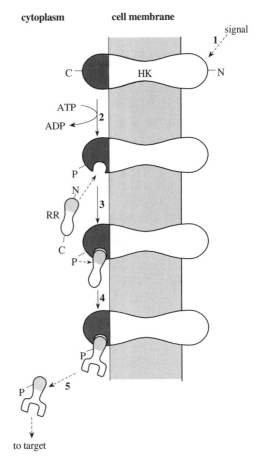

Fig. 18.1 Two-component regulatory systems. 1. A transmembrane histidine kinase (HK) is activated by a signal at its N-terminal domain. 2. The activated protein autophosphorylates in the C-terminal domain. 3. The response regulator protein (RR) binds to the C-terminal end of the histidine kinase, and the phosphoryl group is transferred from the histidine kinase to the response regulator, thus activating the latter. 4. The activated response regulator leaves the histidine kinase and stimulates its target. Shaded and stippled areas of the histidine kinase and response regulator represent conserved amino acid sequences typical for the respective class of protein. The change in shape of the proteins represents a presumed conformational change. In some systems the histidine kinase is cytoplasmic and detect signals within the cytoplasm.

histidine kinase and a response regulator protein are called "two-component" systems, the number of proteins in the signal transduction pathway is often more than two. Two-component signaling systems involving histidine kinases also occur in fungi, including yeast, plants, slime molds, and presumably other eukaryotes.[3,4]

It must be emphasized that the histidine kinase need not be the first protein in the signal transduction pathway to respond to the signal. In other words, it need not be the sensor. In many systems, signals first interact with protein(s) other than the histidine kinase, and the stimulus is relayed to the histidine kinase. For example, in the *E. coli* chemotaxis system described in Section 18.4, the transmembrane proteins (called chemoreceptors or MCP proteins) are sensor proteins that respond to chemoeffectors, and as a consequence, change the activity of a cytoplasmic histidine kinase (CheA). Another example where the initial receiver of the signal is not the histidine kinase occurs in the PHO regulon control system, described in Section 18.9, which is repressed by inorganic phosphate. (A regulon is a set of noncontiguous genes or operons controlled by the same transcription regulator.) The proteins that initially bind inorganic phosphate are in the phosphate transport system (Pts), which is believed to bind inorganic phosphate and then stimulate the enzymatic activity of the membrane-bound PhoR histidine kinase. A third (and more complicated) example occurs in the Ntr regulon, which is repressed by ammonia (Section 18.8). The ammonia levels determine the concentrations of glutamine and $\alpha$-ketoglutarate via the enzymes glutamine synthetase and glutamate synthase, respectively. The $\alpha$-ketoglutarate and glutamine in turn influence the activity of a bifunctional enzyme, uridylyl transferase–uridylyl-removing (UT-UR) enzyme, that modifies a signal transduction protein, $P_{II}$, which in turn regulates the activity of a cytoplasmic histidine kinase ($N_{II}$). In this case the histidine kinase is indeed far removed from the initial signal, ammonia.

Two-component signaling systems have been discovered in many bacteria, both gram negative and gram positive, and have been implicated in a wide range of physiological responses.[5-7] These include nitrogen assimilation, outer membrane porin synthesis, chemotaxis in *E. coli* and *S. typhimurium*, nitrogen fixation in *Klebsiella* and *Rhizobium*, sporulation in *Bacillus*,

myxobacteria development, oxygen regulation of gene expression in *E. coli*, the uptake of carboxylic acids in *Rhizobium* and *Salmonella*, and the production of virulence factors by *Salmonella* and *Bordetella*. It is clearly a widespread and important signal transduction system that enables bacteria to adapt to changes in the external milieu. As mentioned, there is also evidence to suggest that similar systems occur in eukaryotes.[8,9]

### 18.1.1 Components

The components of the "two-component" systems include:

1. A *histidine kinase* (HK), which is sometimes called a *sensor/kinase*. The histidine kinase receives a signal and autophosphorylates at a histidine residue:

$$HK + ATP \xrightarrow{\text{signal}} HK\ P + ADP$$

2. A *partner response regulator* (RR). The partner response regulator is phosphorylated at an aspartate residue by HK-P and sends a signal to its target (e.g., the genome or the flagella motor):

$$RR + HK\ P \longrightarrow RR\ P\ (\text{active}) + HK$$

(However, not all signals result in increased synthesis of RR-P. See Section 18.1.2.)

3. A *phosphatase*. The phosphatase inactivates the RR-P. The phosphatatse may be the histidine kinase, the response regulator, or a separate protein.

$$RR\ P + H_2O \longrightarrow RR + P_i$$

As mentioned, some signals stimulate phosphatase activity and thus act as inhibitors or repressors rather than stimulators or inducers. We will see some examples of this later.

### 18.1.2 Signal transduction in the two-component systems

Figure 18.1 illustrates a simplified model that summarizes signal transduction in two-component systems. In this model the signal stimulates phosphorylation of the response regulator protein. (The model will be modified later to accommodate differences between signaling systems, such as the inclusion of a separate sensor protein that receives the signal and transmits it to the histidine kinase, and systems where the signal actually results in less phosphorylation of the regulatory protein and hence a suppression of a particular response rather than an activation.) In the example shown in Fig. 18.1, the histidine kinase (HK) is depicted as a transmembrane protein composed of three domains: an N-terminal domain that is presumed to be at the outer surface of the cell membrane and to bind to a signaling ligand in the periplasm; a hydrophobic domain that is transmembrane; and a conserved C-terminal domain that is cytoplasmic. Some histidine kinases (e.g., CheA and NR$_{II}$) are cytoplasmic proteins (Sections 18.4.4 and 18.8.1). Signal transduction as depicted in Fig. 18.1 can be conveniently thought of as occurring in three steps:

**Step 1.** In response to a stimulus at the N-terminal domain, the histidine kinase autophosphorylates at the C-terminal domain. The phosphoryl donor is ATP.

**Step 2.** The phosphoryl group is transferred from the histidine kinase to its partner response regulator protein. All the response regulator proteins are related in having conserved amino acid sequences in the N-terminal domain (usually) that may bind to the conserved C-terminal region of the histidine kinases.

**Step 3.** The response regulator becomes activated on being phosphorylated and changes the activity of its target. The effect is usually the stimulation or repression of gene transcription. As we shall see, some response regulator proteins (e.g., NarL) do both depending upon the target gene (Section 18.3). Other response regulators may have targets other than the genome. For example, in chemotaxis, the phosphorylated derivatives of the response regulators CheY and CheB change the direction of the flagella motors and the extent of methylation of the chemoreceptor proteins in the

membrane, respectively (Section 18.4.4). The response regulator proteins differ at their C-terminal domains, and this probably confers specificity with regard to their targets and activities.

Not all two-component systems respond to the signal by increasing the phosphorylation of the response regulator protein. Histidine kinases are sometimes bifunctional enzymes that can act either as kinases *or* as phosphatases when stimulated, depending upon the particular histidine kinase. When stimulation of the histidine kinase promotes phosphatase activity, this results in the *dephosphorylation* of the response regulator and hence a *repression* of transcription. (In the absence of the signal, the kinase phosphorylates the response regulator and transcription is stimulated.) An example where the signal results in stimulating the phosphatase activity of the histidine kinase, and therefore represses gene transcription, is the repression of the Ntr regulon by ammonia. In the presence of excess ammonia, the histidine kinase, NR$_{II}$, acts as a phosphatase rather than as a kinase and inactivates the response regulator, NR$_{I}$, which in its phosphorylated form is a positive transcription factor (Section 18.8). Another example may be the repression of the PHO regulon by inorganic phosphate. The histidine kinase, PhoR, appears to respond to excess inorganic phosphate by dephosphorylating the response regulator, PhoB (Section 18.9). However, as explained in Sections 18.8 and 18.9, NR$_{II}$ and PhoR do not themselves bind the ammonia or phosphate, respectively. Furthermore, in the case of chemotaxis, the positive chemoeffectors *suppress* histidine kinase activity rather than stimulate the activity. This results in less RR-P (i.e., CheY-P) formed, and consequently the cells swim smoothly toward the attractant. The reason for this is that CheY-P causes the cells to tumble and swim randomly. (See Section 18.4.)

## Amino acid sequences define histidine kinases and response regulator proteins

The histidine kinases are defined by a conserved sequence of about 200 amino acids at the C-terminal end. The C-terminal domain is the site of the conserved histidine residue that becomes phosphorylated in response to a stimulus (Fig. 18.2). As indicated in Fig. 18.1, most of the known histidine kinases are transmembrane. The C-terminal end is in the cytoplasm, where it interacts with the response regulator protein. The N-terminal end may be exposed on the extracellular membrane surface (the periplasm in gram-negative bacteria). As stated earlier, the amino acid sequence at the N-terminal domain varies with the different histidine kinases, presumably because they respond to different stimuli. The response to the stimulus within the N-terminal region causes the enzyme to autophosphorylate the conserved histidine residue in the cytoplasmic domain (Fig. 18.1).

The *response regulator proteins* are defined by a conserved amino terminal domain of about 100 amino acids (Fig. 18.3). The conserved amino terminal end of the response regulator protein is thought to interact with the conserved carboxy end of the histidine kinase and become phosphorylated at an aspartate residue. The phosphate is eventually removed by a phosphatase, which may be the histidine kinase, the response regulator protein, or perhaps a third protein. Since the carboxy ends of the histidine kinases and the amino ends of the different response regulator proteins are conserved, it is theoretically possible that a histidine kinase in one signaling system may activate a response regulator of

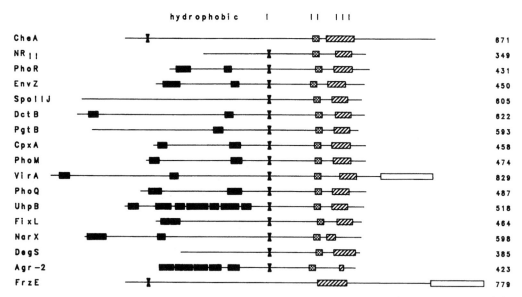

**Fig. 18.2** Structures of the histidine protein kinases. The histidine protein kinases are part of the two-component regulatory systems. They autophosphorylate at a histidine residue and then transfer the phosphate to a response regulator protein. Most of the histidine kinases are believed to be transmembrane proteins with an extracytoplasmic amino terminus that responds to stimuli. Exceptions are CheA, NR$_{II}$, and FrzE, which are cytoplasmic. Hydrophobic domains that presumably span the membrane are indicated by the black boxes at the amino terminal end. The cytoplasmic domain that includes the phosphorylated histidine residue is indicated by a filled X (domain I). Regions II and III (stippled and hatched-filled boxes) represent regions in the carboxy terminal domain where certain amino acids appear with high frequency at specific locations in the sequence. These are called conserved regions. For example, when the amino acid sequences are lined up for comparison, position 43 may be a glycine in all the proteins (totally conserved), whereas position 44 may be arginine in, e.g., 60% of the proteins, and so on (partially conserved). The empty boxes at the extreme carboxy ends are homologous to response regulator domains at their amino-terminal ends. *Source:* From Stock, J. B., A. J. Ninfa, and A. M. Stock. 1989. Protein phosphorylation and regulation of adaptive responses in bacteria. *Microbiol. Rev.* 53:450–490.

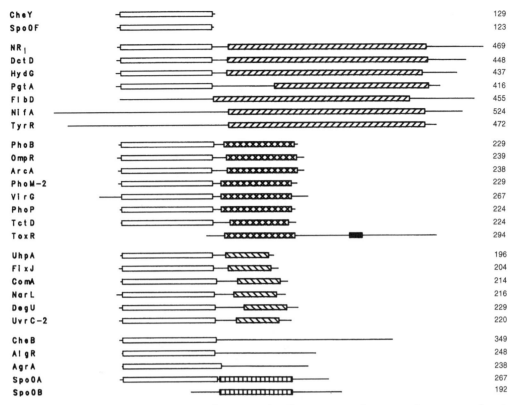

**Fig. 18.3** Structures of the response-regulator proteins. Response-regulator proteins are cytoplasmic proteins that are phosphorylated, or presumed to be phosphorylated, by the histidine kinase proteins. The phosphorylated regulator proteins transmit the signal to the genome or to some other cellular machinery, e.g., the flagellar motor. The amino-terminal region of the response-regulator protein has conserved amino acids (open boxes). Approximately 20–30% of the amino acids are identical at corresponding positions when the sequences are aligned. This is the region thought to interact with the carboxy domain of the histidine kinase, and to become phosphorylated. Other conserved regions are indicated by cross-hatched boxes. These are in the carboxy-terminal domain, which is thought to interact with target molecules. For example, $NR_I$, DctD, NifA share a homologous carboxy-terminal domain that is thought to interact with one of the *E. coli* sigma factors, sigma 54. $NR_I$ and NifA have a common carboxy-terminal region that is thought to bind to DNA. One of the proteins, ToxR, spans the membrane, and the hydrophobic region is indicated by the filled box. The numbers to the right are the lengths of the protein in amino acid residues predicted from the nucleotide sequences. *Source:* From Stock, J. B., A. J. Ninfa, and A. M. Stock. 1989. Protein phosphorylation and regulation of adaptive responses in bacteria. *Microbiol. Rev.* 53:450–490.

a different system. This possible interaction between different signaling systems would be an example of what has been termed "cross-regulation," which is the regulation of a response regulator by a signal that comes from some source other than the cognate histidine kinase.[10] Cross-regulation is discussed in Sections 18.9.1 and 18.12.

## 18.2 Responses by Facultative Anaerobes to Anaerobiosis

A shift from an aerobic to an anaerobic atmosphere results in extensive changes in the metabolism of facultative anaerobes due to the repression of genes required for aerobic growth and the induction of genes necessary for anaerobic growth.[11] These adaptive responses, which have stimulated much interest, are described next. Most of the discussion focuses on what has been learned from studying *E. coli* and related bacteria. However, similar systems exist in other bacteria. The metabolic changes will be discussed first, followed by a description of the regulation of the relevant genes. Many genes that are responsive to anaerobiosis are globally regulated by two systems: the Arc system (two-component), and the FNR system (not two-component), which will be discussed first.[12, 13] This will be followed by a description of the regulation of the formate hydrogen–lyase system, which also responds to anaerobiosis, but which is regulated by formate and a transcription activator, FhlA. The RegB/RegA system will also be described. This is a a two-component system that stimulates the transcription of genes for the light-harvesting complex and photoreaction center of certain photosynthetic bacteria under anaerobic incubation conditions. A third system (NarL/NarP/NarX/NarQ), one that regulates the response to nitrate and nitrite as electron acceptors under anaerobic conditions, is discussed in Section 18.3.

### 18.2.1 Metabolic changes accompanying shift to anaerobiosis

During aerobic respiration, the citric acid pathway is cyclic and oxidative (Fig. 18.4a). However, in the absence of oxygen several important changes take place (Fig. 18.4b). These include the replacement of fumarase A and succinate dehydrogenase by fumarase B and fumarate reductase, respectively, and the repression of the synthesis of $\alpha$-ketoglutarate dehydrogenase. The result of these enzymatic changes is the conversion of the oxidative citric acid cycle into a reductive noncyclic pathway. (Not all bacteria have a reductive citric acid pathway when respiring anaerobically. For example, *Pseudomonas stutzeri* oxidizes glucose completely to $CO_2$ when growing anerobically using nitrate as an electron acceptor, and appears to have an oxidative citric acid cycle.[14] Similarly, *Paracoccus denitrificans* has a complete citric acid cycle when growing anaerobically using nitrate as the electron acceptor.[15]) Additionally, when

*E. coli* is growing anaerobically, acetyl–CoA is no longer made by pyruvate dehydrogenase but rather by pyruvate–formate lyase. This is an advantage under anaerobic conditions because it decreases the amount of NADH that must be reoxidized. The acetyl–CoA thus formed is converted to acetate and ethanol, and the formate is converted to hydrogen gas and carbon dioxide via formate hydrogen-lyase. In *E. coli* there is also a decrease in synthesis of other enzymes used during aerobic growth, such as those of the glyoxylate cycle and fatty acid oxidation.

Depending upon the presence of particular electron acceptors, major changes also take place in the respiratory pathway.[16] Facultative anaerobes such as *E. coli* carry out aerobic respiration in the presence of oxygen, but in the absence of oxygen they carry out anaerobic respiration using either nitrate, fumarate, or some other electron acceptor (e.g., TMAO or DMSO). (TMAO and DMSO are formed naturally in nature and presumably exist in the intestine, where *E. coli* grows.) There is a hierarchy of electron acceptors that are used, oxygen being the most preferred with nitrate second, followed by fumarate and the other electron acceptors. The hierarchy parallels the maximum work that can be done when electrons travel over the electrode potential gradient to the terminal acceptor. The work is proportional to the difference in electrode potential between the electron acceptor and donor. For example, the maximum work that can be done is greatest when oxygen is the electron acceptor ($E'_m = +0.82$ V), less when nitrate is the electron acceptor ($E'_m = +0.42$ V), and least when fumarate is the electron acceptor ($E'_m = +0.03$ V). As described in Section 4.4, the electrons are passed to these terminal electron acceptors from reduced quinone. For example, ubiquinone ($E'_m = +0.1$ V) transfers electrons from the reductant to either the cytochrome oxidase or the nitrate reductase module. Menaquinone ($E'_m = -0.074$ V), rather than ubiquinone, transfers electrons to the fumarate reductase complex. (Fumarate reductase is at too low a potential to accept electrons from ubiquinone.) *E. coli* determines which electron acceptors will be used, in part by regulating the transcription of genes coding for the reductases. Thus, in the

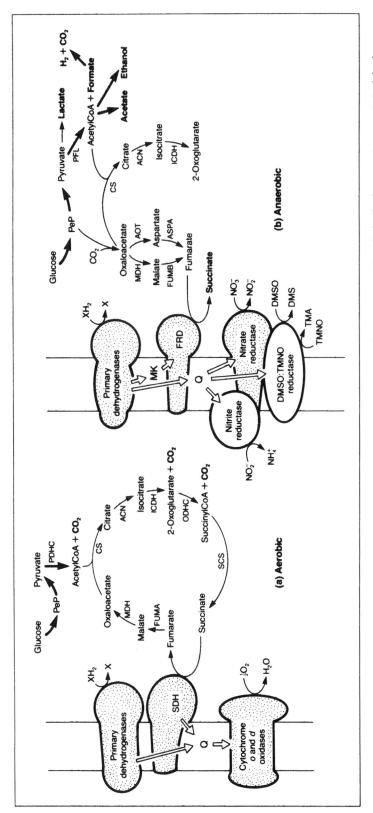

**Fig. 18.4** A comparison of respiratory systems and carbon metabolism in *E. coli*. (a) An oxidative citric acid cycle feeds electrons into succinic dehydrogenase (SDH) and NADH dehydrogenase (a primary dehydrogenase). (b) The citric acid pathway is noncyclic and reductive because the α-ketoglutarate dehydrogenase activity is absent (or severely diminished). The electron transport chain has been altered so that electrons flow to acceptors other than oxygen. When nitrate is present, nitrate reductase (NR) is synthesized. Nitrate represses the synthesis of fumarate reductase (FRD) and DMSO:TMNO reductase. In the absence of an electron acceptor for anaerobic respiration, the major energy-yielding pathways are fermentative. *Source:* From Spiro, S., and J. R. Guest. 1991. Adaptive responses to oxygen limitation in *Escherichia coli. TIBS* **16**:310–314.

presence of oxygen, the genes for nitrate reductase, fumarate reductase, and the other reductases are repressed, and therefore only aerobic respiration can take place. (However, see the discussion in Section 4.7.2 regarding denitrifying enzymes not sensitive to oxygen in some facultative anaerobes.) In the absence of oxygen but in the presence of nitrate, nitrate reductase genes are induced (by nitrate), but the genes for fumarate reductase and the other reductases are repressed by nitrate. Therefore, nitrate respiration takes place. In the absence of both oxygen and nitrate there is no longer any represssion of fumarate reductase, and so fumarate respiraton takes place. Changes also take place within the aerobic respiratory pathway. When oxygen levels are high, *E. coli* uses cytochrome o, encoded by the *cyo* operon, as the terminal oxidase. When oxygen becomes limiting (and during stationary phase), cytochrome o is replaced by cytochrome d, encoded by the *cyd* operon. One advantage to this is that cytochrome d has a higher affinity for oxygen ($K_m = 0.23$–$0.38$ $\mu$M) than has cytochrome o ($K_m = 1.4$–$2.9$ $\mu$M).

Thus oxygen represses the synthesis of the anaerobic reductases, ensuring that oxygen is used as the electron acceptor in air. Nitrate induces the synthesis of nitrate reductase and represses the synthesis of the other reductases, ensuring that nitrate is used as the electron acceptor in the presence of nitrate anaerobically. In the absence of any exogenously supplied electron acceptor, *E. coli* relies on fermentation as its major source of ATP.

## 18.2.2 Regulatory systems

Four systems for the regulation of gene expression by oxygen or nitrate in *E. coli* will be described. Two of these are activated by anaerobiosis (ArcA/B and FNR), the third by nitrate and nitrite (NarL/NarP/NarX/NarQ system), and the fourth is activated by anaerobiosis and formate, and repressed by nitrate (the FhlA regulon). Photosynthetic bacteria have a regulatory system that will also be described called the RegA/B system, which is activated under anaerobic conditions

and controls the induction of photosynthetic genes. A common way to test the regulation of gene transcription is to construct a fusion between the promoter of the gene of interest and the gene for $\beta$-galactosidase (*lacZ*). Then strains harboring the fusion are tested under different conditions to see how the test conditions affect the production of $\beta$-galactosidase. (See Note 25 for a more detailed explanation of *lacZ* fusions.) The four systems in *E. coli* for the regulation of gene expression by oxygen or nitrate are:

1. The Arc (aerobic respiratory control) system (two-component). This represses under anaerobic conditions the transcription of several genes that are expressed only during aerobic growth. In addition, it stimulates the transcription of a much smaller number of genes that are expressed during microaerophilic or anaerobic conditions. The Arc system functions during aerobic and anaerobic growth.

2. The FNR (fumarate nitrate reductase) system (not a two-component system). This stimulates the transcription of many genes that are required for fermentation and anaerobic respiraton, and represses the transcription of some genes that function only during aerobic growth. More than 75 genes are regulated by FNR in *E. coli*. The FNR system is active only during anaerobic growth.

3. The NarL/NarP/NarX/NarQ system, also called the Nar system (two component). This stimulates the transcription of the nitrate reductase and other genes required for nitrate and nitrite metabolism, and represses the transcription of the other terminal reductase genes. The system is active only under anaerobic growth conditons in the presence of nitrate or nitrite.

4. The FhlA regulon, also called the formate regulon, consists of genes repressed by oxygen and nitrate and induced by formate. The regulon includes several genes, including those for formate hydrogen–lyase, which is necessary to convert the fermentation end product formate to $H_2$ and $CO_2$.

The Arc and FNR are global regulatory systems sensitive to oxygen which regulate the transcription of genes in different pathways. These two sytems are discussed next, followed by a description of two other aerobic/anaerobic regulatory systems that are more specific regarding the target pathways, that is, the formate hydrogen-lyase system in *E. coli* required under anaerobic (and acidic) conditions for the transformation of formate to carbon dioxide and hydrogen gas, and the RegB/RegA system for the transcription under anaerobic conditions of genes in *Rhodobacter capsulatus* required for photosynthesis. The Nar system is described in Section 18.3.

## The Arc system

The Arc system consists of a membrane-associated histidine kinase, ArcB, and a partner response regulator, ArcA. Mutations in the *arc* genes cause derepression of several genes and repression of a relatively short list of others, indicating that phosphorylated ArcA is both a repressor and activator of gene transcription under anaerobic conditions. By analyzing these mutants, it has been possible to conclude that the Arc system is responsible for:

1. Anaerobic repression of genes for:

   Citric acid cycle enzymes

   Glyoxylate cycle enzymes

   Several dehydrogenases for aerobic growth (e.g., pyruvate dehydrogenase)

   Fatty acid oxidation enzymes

   Cytochrome o oxidase

2. Anaerobic induction of the gene for:

   Pyruvate formate–lyase[17, 18]

3. Induction in low oxygen of the genes for:

   Cytochrome d oxidase and cobalamin synthesis[19] and (along with FNR), the pyruvate formate–lyase gene.[17]

Interestingly, ArcA-P, but not ArcB, may also activate the transcription of the mating system genes, as explained in Note 20.

It appears that the ArcB protein monitors the level of oxygen indirectly and that when oxygen levels are sufficiently low, the ArcB protein autophosphorylates using ATP as the phosphoryl donor, and then transfers the phosphoryl group to ArcA (Fig. 18.5).[21, 22] How does ArcB detect changes in the levels of oxygen? Apparently, it is not oxygen itself that is the signaling molecule. This conclusion has been reached from two lines of evidence. The level of expression of the *sdh* operon (succinate dehydrogenase), which is under ArcA/ArcB control, varies with the midpoint potential of the electron acceptor. (The expression of the *sdh* operon is complex and controlled by factors in addition to ArcA/ArcB. See Note 23 for further comments.) The level of expression is highest with oxygen, lower with nitrate, and lowest with fumarate. This parallels the midpoint potential of the electron acceptor, which is most positive for oxygen, less positive for nitrate, and least positive for fumarate. The ArcB protein may respond to a reduced carbon compound in the cell, such as the reduced form of an electron transport carrier (e.g., flavoprotein, quinone, cytochrome), or NADH, or some metabolic intermediate that might accumulate anaerobically, rather than to fumarate, nitrate, and oxygen per se. One way to test this hypothesis is to delete the cytochrome o and d genes and measure the expression of genes under the control of the Arc system. Deletion of the cytochrome oxidase genes inhibits electron transport at the terminal step, and would be expected to have extensive metabolic consequences, including, for example, an increase in the ratio of reduced to oxidized forms of electron carriers, and the accumulation of metabolites that are formed in the absence of an exogenous terminal electron acceptor. If this is the case, then deletion of the cytochrome oxidase genes would be expected to mimic the absence of oxygen. This experiment was done using *cyo–lacZ* and *cyd–lacZ* fusions as probes to monitor the expression of the *cyo* and *cyd* genes.[24] (Gene fusions are explained in Note 25.) When both the *cyo* and *cyd* genes were deleted and the cells were grown in air, *cyo–lacZ* expression was lowered and *cyd–lacZ* expression increased, as if oxygen were absent. (See Note 26 for a description of the control experiments.)

One interpretation of these experiments is that the reduced form of an electron transport

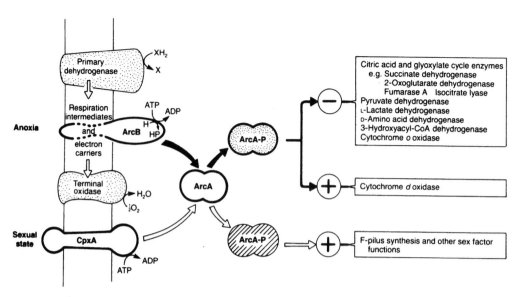

**Fig. 18.5** The ArcA/ArcB regulatory system in *E. coli*. ArcB is a membrane protein activated by anoxia, perhaps by a reduced form of an electron carrier. The model postulates that the activated form of ArcB becomes phosphorylated and then phosphorylates ArcA, which then becomes a repressor of aerobically expressed enzymes and an inducer of cytochrome d oxidase. Recently it was reported that ArcA-P also activates the genes for cobalamin synthesis (*cob* genes) in *Salmonella typhimurium*. See Andersson, D. I. 1992. Involvement of the Arc system in redox regulation of the cob operon in *Salmonella typhimurium*. *Molecular Microbiology* 6:1491–1494. ArcA also responds to CpxA, a membrane-bound sensor protein that is necessary for the synthesis of the F-pilus and other sex factor functions. *Source*: From, Spiro, S., and J. R. Guest. 1991. Adaptive responses to oxygen limitation in *Escherichia coli*. *TIBS* **16**:310–314.

carrier (e.g., a quinone) may signal ArcB, which in turn autophosphorylates and then phosphorylates ArcA (Fig. 18.5). The control of the phosphorylating activity of ArcB also appears to involve cellular metabolites that increase when oxygen is lacking, since pyruvate, acetate, D-lactate, or NADH increase the rate and level of phosphorylation of ArcB in vitro, apparently by inhibiting dephosphorylation. The expected result would be the transfer of the phosphoryl group to ArcA. These and other aspects of the regulation of the ArcA/ArcB system are discussed by Iuchi et al.[27] and Lynch and Lin.[28]

## The FNR system

When *E. coli* is shifted from aerobic to anaerobic growth, a number of genes are induced, while others are repressed (Fig. 18.4). The *fnr* gene codes for a protein (FNR) that under anaerobic conditions acts as a positive regulator of transcription for many genes that are expressed only during anaerobic growth,

and a repressor for certain genes that are not expressed during anaerobic growth (Fig. 18.6).[29, 30] Genes whose expression requires FNR include those coding for the anaerobic respiratory enzymes fumarate reductase (*frdABCD*) and nitrate reductase (*narGHJI*), as well as several other enzymes, including pyruvate formate lyase (*pfl* genes), formate dehydrogenase-N, (the respiratory formate dehydrogenase, *fdnGHI*), aspartase (*aspA*), anaerobic fumarase B (*fumB*), and glycerol-3-phosphate dehydrogenase (*glpA*). Mutations in the *fnr* gene result in an inability to grow anaerobically on fumarate or nitrate as electron acceptors. FNR is also a negative regulator for several aerobically expressed genes, including those for cytochrome o oxidase (*cyoABCDE*), cytochrome d oxidase (*cydAB*), succinate dehydrogenase (*sdhCDAB*), and superoxide disumutase (*sodA*). Many of the genes regulated by FNR are also regulated by the ArcA/ArcB system and the Nar system.

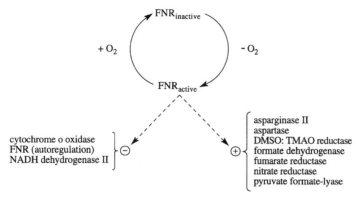

**Fig. 18.6** Fnr is a transcription regulator protein during anaerobic growth. In the absence of oxygen FNR becomes activated to become an inducer for many anaerobically expressed genes and a repressor for certain aerobically expressed genes. It has been speculated that Fnr might become activated by the reduction of bound ferric ion, causing a conformational change in the protein. This is not an example of a two-component regulatory system.

There is no evidence that FNR is phosphorylated or is part of a two-component regulatory system, and it is discussed here in the context of a protein involved in aerobic/anaerobic regulation of gene expression.

Studies have shown that the FNR protein in *E. coli* is present in comparable amounts in both aerobically and anerobically growing cells, and it is believed to be largely in an inactive state during aerobic growth.[31, 32] Active FNR is a homodimer of an iron–sulfur protein with an oxygen-labile [4Fe–4S] cluster in each monomer. Upon exposure to oxygen, the [4Fe–4S] cluster is oxidized and can even be lost from the protein. (See Note 33 for a further explanation.) When this happens, a fraction of the FNR loses its iron clusters and becomes an apoprotein.[34–36] Both the protein bearing an oxidized cluster and the apoprotein are inactive. They bind with low affinity to DNA and do not stimulate transcription. The oxidation of the iron–sulfur clusters as well as their loss are reversible, and under anaerobic conditions the Fe–S clusters are restored.

FNR expression need not be regulated exactly the same in all bacteria. For example, whereas in *E. coli* the transcription of *fnr* seems to be similar under both aerobic and anaerobic growth conditions and the amounts of active FNR increase anaerobically solely at the post-translational level, transcription of *fnr* is strongly stimulated in *Bacillus subtilis* during oxygen limitation. The anaerobic activation of transcription of *fnr* in *B. subtilis* requires *resDE*, a two-component system[37] (*resD* is a response regulator gene, and *resE* is a histidine kinase gene.) The *activity* of FNR in *B. subtilis* may also be regulated by anaerobiosis in a fashion similar to its regulation in *E. coli*.

An interesting parallel exists between the FNR protein and another transcriptional regulator, the cAMP receptor protein, CRP (cyclic AMP receptor protein) (also called the catabolite activator protein, CAP). The CRP protein is a positive transcriptional regulator for catabolite-sensitive genes (i.e., those genes repressed by glucose). CRP, in response to binding to cAMP, binds to specific sites on the promoter region of the target gene to activate transcription. A comparison between the nucleotide-derived amino acid sequences of FNR and CRP reveals that the structure of FNR is very similar to CRP, including the presence of a DNA-binding domain that allows the dimer to bind, and a nucleotide-binding domain (although numerous attempts to show specific binding of nucleotides to FNR have failed). This has led to the suggestion that FNR and CRP are examples of a family of proteins that regulate transcription at the promoter region of target genes. Indeed, there have been several reports of proteins resembling FNR in bacteria other than the enterics, and these FNR-like proteins take part in the regulation of a variety of metabolic activities.[38-40]

In summary, ArcA and FNR are activated by anaerobiosis and regulate the enzymological changes that accompany the shift from aerobic to anaerobic growth. For example, activated ArcA (ArcA-P) primarily represses several aerobically expressed genes, including pyruvate dehydrogenase, succinate dehydrogenase, $\alpha$-ketoglutarate dehydrogenase, and cytochrome o oxidase (Fig. 18.5). It is also a positive regulator for some genes expressed when oxygen is low or lacking, including the cytochrome d oxidase gene and the pyruvate–formate lyase gene. Similarly, activated FNR represses several genes normally expressed during aerobic growth, including the genes for cytochrome o oxidase, succinate dehydrognase, and pyruvate dehydrogenase, and is required for the induction of many genes during anaerobic growth, including pyruvate–formate lyase, fumarate reductase, and nitrate reductase (compare Fig. 18.5 with Fig. 18.6. Also, see Note 23.) It is important to point out that in several cases FNR and ArcA regulate the same gene. For example, they are both positive regulators for the pyruvate–formate lyase gene. Sometimes they have opposite effects on the same gene. For example, ArcA is a positive regulator for the cytochrome d oxidase gene, whereas FNR is a negative regulator for that gene. Multiple controls of the same gene are a common theme in bacterial physiology. However, a note of caution must be introduced when drawing conclusions from physiological studies of mutants regarding co-regulation by ArcA and FNR. This is because FNR has been reported to stimulate anaerobic *arcA* expression when such was monitored using *arcA–lacZ* fusions.[40]

## Regulation of the formate hydrogen–lyase pathway

When enterobacteria such as *E. coli* are grown anaerobically, formic acid is converted to $H_2$ and $CO_2$ via the enzyme complex formate hydrogen–lyase (Figs. 14.7 and 18.4b). Formate hydrogen–lyase consists of two enzymes, formate dehydrogenase H and hydrogenase 3, whose genes are repressed by oxygen and nitrate and induced by formate. Induction also requires an acidic pH in the external medium, which probably means that the cells make formate hydrogen–lyase to prevent the medium from becoming too acidic due to the production of formic acid from pyruvate (the pyruvate formate–lyase reaction). (*E. coli* actually makes three distinct formate dehydrogenases and three different hydrogenases. See Note 41 for a discussion of this point.) The genes for formate hydrogen–lyase are part of a regulon that has two names: the FhlA regulon and the formate regulon. The regulon is controlled by a transcription factor called FhlA and by formate, which is a coactivator of FhlA. (See Ref. 42 for a review.) (See Note 43 for a further description of the regulon.) Under aerobic conditions, formate levels are kept low, and because of this the regulon is not induced. Two different mechanisms keep the formate levels low under aerobic conditions. The synthesis of formate from pyruvate is prevented under aerobic conditions because the *pfl* gene, which codes for pyruvate formate–lyase, which synthesizes acetyl–CoA and formate from pyruvate and CoASH, requires active FNR as a positive regulator. Since oxygen prevents activation of FNR, oxygen represses the *pfl* gene, thus preventing formate synthesis. Oxygen and nitrate also lower the levels of formate in the cell because they induce the synthesis of respiratory formate dehydrogenases (FDH-O and FDH-N), which oxidize formate to $CO_2$.

## 18.3 Response to Nitrate and Nitrite: The Nar Regulatory System

### Backround information

*E. coli* can be grown anaerobically using nitrate ($NO_3^-$) as the electron acceptor. The process is called nitrate respiration and is reviewed in Section 12.2.1. In fact, when *E. coli* is given a choice of electron acceptors under anaerobic conditions, one of them being nitrate, it will utilize the nitrate. This may be advantageous, because nitrate has a more positive redox potential than the alternative electron acceptors, and therefore more energy is potentially available from electron transport when nitrate is the electron acceptor. The reason that nitrate is preferentially used as an electron acceptor during anaerobic respiration is that it induces the synthesis of a membrane-bound nitrate reductase and represses the synthesis of the other reductases (e.g., fumarate reductase, DMSO/TMAO reductase, and formate-dependent nitrite reductase, cytochrome $c_{552}$). The formate-dependent nitrite reductase operon (nrf, nitrite reduction by formate) encodes a periplasmic nitrite reductase that catalyzes nitrite reduction using formate as the electron donor. A $\Delta p$ is produced.[44] (See Note 45, which states the various genes in these operons.)

## Pathway of nitrate reduction

The nitrate ($NO_3^-$) is first reduced to nitrite ($NO_2^-$), which in turn is reduced to ammonia ($NH_4^+$). Part of the ammonia is assimilated into cellular nitrogenous compounds, and part is excreted into the medium. Thus under anaerobic conditions E. coli can use nitrate as both an electron acceptor during anaerobic respiration and as a source of cellular nitrogen.

## The enzymes involved

The two major enzymes for the reduction of nitrate to ammonia during nitrate respiration in E. coli are a membrane-bound nitrate reductase, which reduces nitrate to nitrite (a two-electron transfer), and a cytoplasmic NADH-linked nitrite reductase, which reduces nitrite to ammonia (a six-electron transfer). Nitrate reductase is also called respiratory nitrate reductase and is encoded by the narG operon. It translocates one proton per electron from the cytoplasm to the periplasm during quinol oxidation and is therefore a coupling site. This is discussed in Section 4.7.1 and reviewed in Ref. 46. The role of the cytoplasmic NADH-linked nitrite reductase is to remove nitrite (which is toxic) produced from nitrate and to regenerate $NAD^+$.

As mentioned earlier, E. coli also makes a periplasmic nitrite reductase involved in anaerobic respiration, cytochome $c_{552}$, encoded by the nrfA gene. It is also called respiratory nitrite reductase and is a coupling site, but the mechanism is not understood. The enzyme is induced by nitrite and repressed by nitrate. For a review of nitrate and nitrite respiration, see Ref. 47.

## The nar system and gene regulation: An overview

The responses to both nitrate and nitrite are regulated by a two-component signaling system that consists of two membrane-bound sensor kinases (NarX and NarQ) and two cytoplasmic response regulator proteins (NarL and NarP), rather than one sensor kinase and one response regulator protein.[48-51] (See Note 52 for how the nar genes were discovered.) The biochemical evidence for this is summarized later. The result of this regulation is that in the presence of excess nitrate there is preferential synthesis of enzymes that use nitrate as an electron acceptor (i.e., nitrate reductase), and in the presence of excess nitrite there is preferential synthesis of enzymes that use nitrite as an electron acceptor (i.e., respiratory nitrite reductase, cytochrome $c_{522}$). Both nitrate and nitrite induce the formation of a cytoplasmic nitrite reductase, and this enzyme probably serves to prevent the accumulation of toxic nitrite in the cytoplasm. The reason why nitrate stimulates transcription of the respiratory nitrate reductase gene (and inhibits transcription of the respiratory nitrite reductase gene) is that nitrate causes increased amounts of NarL-P, and NarL-P stimulates the expression of the former (narG) and inhibits expression of the latter (nrfA). (See Fig. 18.7.) Nitrite does not stimulate expression of the respiratory nitrate reductase gene but does stimulate expression of the respiratory nitrite reductase gene, and the reason for this is that nitrite lowers the amounts of NarL-P (hence depressing trasncription of narG) and increases the amounts of NarP-P. The higher concentrations of NarP-P stimulate expression of the respiratory nitrite reductase gene (nrfA). (See Fig. 18.7). The lower concentrations of NarL-P also stimulate expression of nrfA, and this has been explained by an activation site for the nrfA operon, which has a high affinity for NarL-P.[49]

## The roles of NarX and NarQ

### 1. NarX
In the presence of nitrate NarX phosphorylates NarL. However, very importantly, in the presence of nitrite NarX dephosphorylates NarL-P (Fig. 18.7A, reaction 1). This accounts

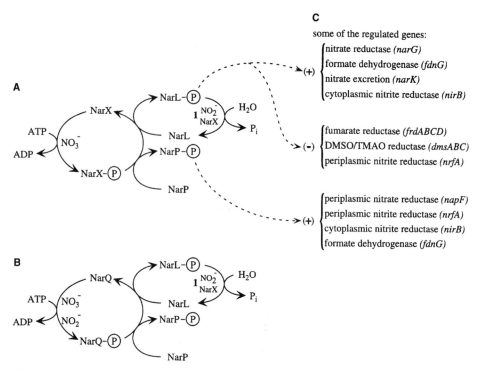

**Fig. 18.7** A model for nitrate/nitrite control of anaerobic gene expression. The membrane-bound histidine kinase sensor proteins are NarQ and NarX. The cytoplasmic response regulator proteins are NarL and NarP. Nitrate stimulates the phosphorylation of NarL and NarP via NarX and NarQ. Nitrite stimulates the phosphorylation of NarL and NarP via NarQ, but causes the dephosphorylation of NarL-P by stimulating the phosphatase activity of NarX. The result is that nitrate increases the amounts of NarL-P and nitrite lowers the amounts of NarL-P. Some of the genes regulated by the response regulator proteins are listed. In the presence of nitrate the higher levels of NarL-P induce the *narG* operon, which encodes respiratory nitrate reductase, and repress the *nrfA* operon, which encodes respiratory nitrite reductase. The *nrfA* operon is induced, however, in the presence of nitrite. This is due to two effects: (1) Nitrite stimulates the phosphorylation of NarP via NarQ, and NarP-P stimulates transcription of the *nrfA* operon. (2) The lower concentrations of NarL-P caused by nitrite actually induce the *nrfA* operon. This has been explained by a second site for the *nrfA* operon (an activation site) that has a high affinity for NarL-P, which stimulates transcription when bound to this site. See Darwin, A. J., K. L. Tyson, J. W. Busby, and V. Stewart. 1997. Differential regulation by the homologous response regulators NarL and NarP of *Escherichia coli* K12 depends on the DNA binding site arrangement. *Molec. Microbiol.* 25:583–595.

for why there is less NarL-P in the presence of excess nitrite. As a consequence, the nitrate reductase operon (*narG*) is stimulated by nitrate but not by nitrite, and the periplasmic nitrite reductase gene (*nrfA*) is repressed by nitrate but not by nitrite. The levels of NarP-P, on the other hand, are the same for both nitrate and nitrite, and therefore the differential effects of nitrate and nitrite on gene transcription depend upon NarL-P and not NarP-P. There are several other examples of a signaling molecule lowering the levels of

a phosphorylated response regulator protein by stimulating its dephosphorylation. Another example of a signal that does this is $P_{II}$, which represses the *glnALG* operon by stimulating the dephosphorylation of $NR_I$-P by $NR_{II}$ (Fig. 18.11).

### 2. NarQ

In contrast to the differential effect of nitrate and nitrite on NarX activity with respect to NarL, the response of NarQ to either nitrate or nitrite is the same: phosphorylation of

NarL and NarP (Fig. 18.7B). In fact, nitrate regulation mediated by NarL is essentially normal in null mutants of either *narX* or *narQ*. Furthermore, NarX or NarQ work equally well in regulating NarP-mediated gene transcription in response to nitrate or nitrite. Therefore, it is the NarX/NarL couple that is responsible for the differential response to nitrate and nitrite.

### The role of IHF

Integration host factor (IHF) is known to activate and repress genes. See the discussion of the nucleoid in Section 1.2.6 for a description of IHF and its role in DNA bending and the regulation of gene activity, as well as Sections 10.2.5, 18.8.2, and 18.10.1 for more backround information on IHF. The transcription of the nitrate reductase gene (*narG*) and the nitrite extrusion protein gene (*narK*), which catalyzes the electrogenic excretion of nitrite, are also stimulated by integration host factor.[53] This indicates that IHF brings a loop of DNA to which NARL-P is attached close to the promoter of these genes. (See Fig. 10.26.)

### Some biochemical and genetic evidence

There exist biochemical data in support of the model that NarX and NarQ are kinases that autophosphorylate and transfer the phosphoryl group to the response regulator proteins. It has been demonstrated with purified proteins that NarX and NarQ can autophosphorylate using ATP as the phosphoryl donor and transfer the phosphoryl group to NarL.[54, 55] NarX and NarQ can also negatively regulate the NarL protein by dephosphorylating NarL-P, although NarX is more active than NarQ in this respect.[56] NarX and NarQ have two transmembrane regions near the N-terminal end and a region in between exposed to the periplasm. Mutations in the periplasmic region of NarX or NarQ result in mutants that either have lost the ability to respond to nitrate and nitrite or behave as if nitrate or nitrite were present even in their absence.[57, 52] The mutants were analyzed using *narG–lacZ* and *frdA–lacZ* fusions. (See Note 25 for a description of gene fusions.) From these data it has

been concluded that nitrate and nitrite are detected in the periplasmic region, and that the signal is transduced across the cell membrane to the response regulator proteins. Both NarL and NarP have been demonstrated to bind to specific sites in target operons.[49]

## 18.4 Response to Attractants and Repellents: Chemotaxis

*Chemotaxis* refers to the ability of bacteria to move along a concentration gradient toward a chemical attractant (positive chemotaxis), or away from a chemical repellent (negative chemotaxis).[58–61] The attractants and repellents are called *chemoeffectors*. Most research concerning chemotaxis has employed flagellated, swimming bacteria, and the discussion that follows is restricted to these. The student should refer to Section 1.2.1 for a description of bacterial flagella.

### 18.4.1 Bacteria measure changes in concentration over time

One might think that bacteria swimming along a chemical concentration gradient are detecting the spatial gradient itself, but this is not the case. Calculations indicate that, because of the small size of bacteria, the difference in concentration of a chemoeffector between both ends of the cell would be too small to be measured accurately.[62] What the bacterium actually measures is the absolute concentration of chemoeffector and then compares it to the concentration previously measured. In other words, *bacteria measure changes in concentration over time*. They actually "remember" the previous concentration. If the bacterium finds that it is in a higher concentration of an attractant or a lower concentration of repellent than at a previous time, it continues to move in that direction. If the bacterium finds that it is in a lower concentration of attractant or a higher concentration of repellent than at a previous time, it moves in a *randomly* different direction. The propensity to swim randomly is fundamental to chemotaxis, as will be made clear in Section 18.4.2.

Some bacteria swim randomly because they periodically tumble, whereas other bacteria swim randomly without tumbling. Chemotaxis in bacteria that tumble is understood best, and will be discussed first. Random swimming without tumbling is discussed in Section 18.4.8.

## 18.4.2 Tumbling

*E. coli* and many other bacteria swim along a path for a few seconds and then tumble. Very importantly, when they recover from the tumble, they swim in a *randomly* different direction. If nothing affected the frequency of tumbles and swimming, then bacteria would just swim in random directions. This is the situation when they are not responding to chemoeffectors. However, *attractants decrease the frequency of tumbles, whereas repellents increase the frequency of tumbles.*[63, 64] This is illustrated in Fig. 18.8. If a bacterium moves into an area where the concentration of chemoattractant is higher than at a previous moment, the cell detects the higher concentration (because of increased binding to chemoreceptors), tumbles less frequently, and continues to move in that direction. Likewise, if the cell swims into an area where the concentration of repellent is higher, then tumbling increases and the cell changes its direction and moves away from the repellent. But why does the cell tumble?

*E. coli* and *S. typhimurium* are peritrichously flagellated bacteria in which 6 to 8 flagella protruding randomly from the cell surface are wrapped around each other in a helical bundle that extends to the rear of the swimming cell (Fig. 18.9). The flagella are usually in the conformation of a left-handed helix. When the flagella rotate counterclockwise (viewed from the tip of the flagellum toward the cell), they remain in a bundle as a helical wave moves from the proximal to the distal portion of the flagellum. This pushes the bacterium forward. However, when the flagella rotate clockwise, they unwind (fly apart), and there is no longer coordinated movement. The result is tumbling. Restoration of the original direction of rotation causes the flagella to come together again and the bacterium to resume swimming,

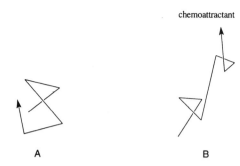

chemoattractant

A                                        B

**Fig. 18.8** Chemoeffectors bias random swimming. (A) The swimming pattern is smooth swimming for a short period of time interspersed with brief periods of tumbling. After tumbling, the bacterium resumes swimming in a randomly different direction. (B) A chemoattractant decreases the frequency of tumbling, thus prolonging the periods of smooth swimming. The result is that the bacterium swims in the direction of higher concentrations of chemoattractant. A chemorepellent increases the frequency of tumbling, causing the bacterium to swim away from the repellent.

but in a randomly different direction. Thus the repellents cause the flagellar motor to reverse and the cells to tumble, whereas attractants decrease the frequency of reversal and promote smooth swimming. The tumbling and smooth swimming responses to changes in chemoeffector concentrations can be seen by observing the swimming of individual cells in the microscope. It is also possible to monitor the change in direction of flagellar rotation by tethering the bacteria to a glass slide by a flagellum. One way to do this is to use an antibody to the flagellum that sticks to both the flagellum and the glass slide.[65] The cells are grown in minimal media with glucose as the carbon source. Under these conditions, the average number of flagella per cell is reduced, and cells tethered by a single flagellum can be observed. When the flagellar motor rotates, the cell rotates because the flagellum is fixed to the slide. It is then possible to infer that the addition of attractants causes the flagella to rotate counterclockwise and the addition of repellents causes clockwise rotation.[66]

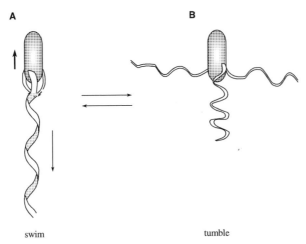

swim            tumble

**Fig. 18.9** Flagella fly apart during tumbling. In *E. coli* the flagella are peritrichously located and form a trailing bundle while the cell is swimming. (A) The filaments are in a left-handed helix. When the motors rotate counterclockwise, a helical wave travels proximal to distal, pushing the cell forward. (B) When the motors rotate clockwise, the filaments undergo a transition to a right-handed wave form, causing them to fly apart and the cells to tumble. *Source:* Adapted from MacNab, R. M. 1987. Motility and chemotaxis, pp. 732–759. In *Escherichia coli and Salmonella typhimurium*. Neidhardt, F. C., et al. (Eds.). American Society for Microbiology, Washington, D.C.

## 18.4.3 Adaptation

An important aspect of the chemotactic response is *adaptation*. Bacteria adapt (i.e., become desensitized) to a chemoeffector so that after a short period of time (seconds to minutes), they no longer respond to it at that concentration, but may respond to a higher concentration. In fact, one could say that adaptation is the way that the bacterium "remembers" the previous concentration of chemoeffector. Why is an adaptation circuit built into the signaling pathway? Adaptation makes good sense. When a bacterium adapts to a chemoattractant, it resumes tumbling at the nonstimulated frequency and will stay in the area, but if it encounters an increasing concentration gradient of attractant, tumbling will again be suppressed. Consequently, the bacterium will swim toward and remain in the area of the higher concentration of chemoattractant. Another way of saying this is that adaptation allows the bacterium to respond to local gradients that it may encounter while randomly swimming. Adaptation to repellents is also easily rationalized. If a bacterium did not adapt to the repellent, it would continue to tumble at the stimulated rate even when

swimming toward the lower concentration of repellent, and hence be trapped in the area of repellent.

### 18.4.4 Proteins required for chemotaxis

Chemotaxis employs complex regulatory circuits involving both cytoplasmic and membrane proteins. A description of the proteins and their functions will be given first, followed by an explanation of the regulatory circuits.

### Cytoplasmic proteins involved in chemotaxis

Mutants of *E. coli* and *Salmonella typhimurium* have been isolated that fail to show chemotaxis. These mutants have resulted in the identification of six genes required for chemotaxis. The genes are *cheA*, *cheB*, *cheR*, *cheW*, *cheY*, and *cheZ*. Deletions of any one of these genes prevents chemotaxis without affecting motility. The Che proteins are cytoplasmic proteins that are part of a signal transduction pathway between the attractant or repellent and the flagellar motor switch.[67] (See Note 68 for a further discussion.) CheA belongs to the class of signaling proteins called histidine kinases. It has conserved amino acid sequences at the carboxy-terminal domain similar to the other histidine kinases, is phosphorylated at a histidine residue, and transfers the phosphoryl group to a response regulator protein (Fig. 18.1).[69] CheY and CheB are *response regulator proteins*. They have aminoterminal domains similar to the other response regulators and undergo phosphorylation and dephosphorylation.

## Membrane proteins involved in chemotaxis

*Chemoreceptor proteins* for the chemoeffectors are built into the bacterial cell membrane. The chemoreceptor proteins are also called *receptor–transducer proteins*, or simply *transducer proteins*. The chemoreceptor proteins are thought to loop across the membrane with a periplasmic domain separating two hydrophobic membrane-spanning regions that connect to cytoplasmic domains. This is similar to the configuration illustrated in Fig. 17.3b for a protein anchored to the membrane by a stop-transfer signal and a leader peptide, which for the case of the chemoreceptor proteins protein would be uncleaved. The cytoplasmic domain interacts with cytoplasmic components such as the CheW and CheA proteins, whereas the periplasmic domain interacts with chemoeffectors or periplasmic proteins to which the chemoeffectors are bound. The cytoplasmic domain of the chemoreceptor proteins also contains four or five glutamate residues that become methylated and demethylated as part of the adaptation response described in Section 18.4.5. Therefore, the chemoreceptor proteins are also called *methyl-accepting chemotaxis proteins* (MCPs). E. coli has four different MCPs. These are the *Tsr* protein (taxis to serine and away from repellents), *Tar* (taxis to aspartate and maltose and away from repellents), *Trg* (taxis to ribose, glucose, and galactose), and *Tap* (taxis to dipeptides). *Salmonella* does not have *Tap*, but it does have *Tcp*, which senses citrate. Some chemoeffectors, primarily certain sugars, first bind to periplasmic binding proteins, also called primary chemotaxis receptors, and the

sugar-bound periplasmic binding proteins interact with specific receptor–transducers in the membrane. Under these circumstances, the receptor–transducer (MCP) acts as a *secondary* chemoreceptor rather than a primary chemoreceptor. These are the same periplasmic binding proteins that bind sugars and interact with membrane-bound transporters in the shock-sensitive permease systems described in Section 16.3.3. They therefore function in both chemotaxis and solute transport. However, the majority of periplasmic binding proteins in *E. coli* function only in transport, not in chemotaxis. (See Note 70 for a more complete discussion of the MCP proteins.)

## 18.4.5 A model for chemotaxis

### Stimulation

The cytoplasmic domains of the receptor–transducer proteins (Tar, Tsr, Trg, or Tap) control the autophosphorylation activity of CheA in an activity that requires CheW (Fig. 18.10):

$$CheA + ATP \longrightarrow CheA\text{-}P + ADP$$

CheA-P is a protein kinase that phosphorylates CheY:

$$CheA\text{-}P + CheY \longrightarrow CheA + CheY\text{-}P$$

CheY-P has an autophosphatase activity that is considerably enhanced by CheZ[71]:

$$CheY\text{-}P + H_2O \xrightarrow{\text{CheZ}} CheY + P_i$$

In the nonstimulated state there is a steady-state level of CheY-P that is determined by the rate of phosphorylation of CheY and the rate of dephosphorylation of CheY-P.

CheY-P is a response regulator protein. It binds to the flagellar motor switch, FliM, FliG, FliN, (primarily FliM), and initiates a reversal of the motor so that it rotates in a CW (clockwise) direction, causing the cell to tumble.[72] In the nonstimulated cell, there is a certain intermediate level of CheY-P that supports normal run–tumble behavior.

When an attractant or repellent stimulate a chemoreceptor protein, a signal is sent to the

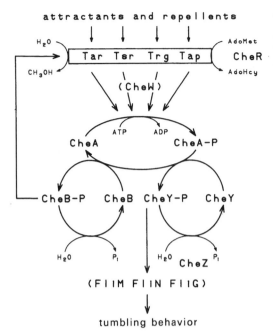

**Fig. 18.10** A model for signal transduction during chemotaxis in *E. coli* and *S. typhimurium*. Attractants bind to the receptor–transducer proteins (Tar, Tsr, etc.) in the membrane. The receptors also mediate the response to certain repellents (leucine, indole, acetate, $Co^{2+}$, and $Ni^{2+}$). This activates a CheW-dependent change in the rate of autophosphorylation of CheA and the proteins that are phosphorylated by CheA (CheB and CheY). Attractants reduce the rate of autophosphorylation, and repellents increase the rate of autophosphorylation. The phosphoryl group is transferred from CheA-P to CheY and CheB. CheY-P interacts with the switch proteins (FliM, FliN, FliG) to cause clockwise rotation of the motor and tumbling. Thus repellents increase tumbling because they increase the level of CheY-P, and attractants reduce tumbling because they reduce the level of CheY-P. CheB-P is a methylesterase whose activity results in demethylation of the receptor–transducer proteins. As CheB-P goes up, methylation goes down. Thus repellents reduce the level of methylation because they increase the level of CheB-P and attractants increase methylation because they reduce the levels of CheB-P. CheB-P autodephosphorylates. CheR is a methyltransferase that methylates the receptor–transducer proteins. The more highly methylated receptor–transducer protein does not transmit the chemoattracant signal to CheA. Hence adaptation to a chemoattractant is due to

cytoplasmic region of the receptor–transducer protein. The signal is presumably a conformational change in the periplasmic region of the chemoreceptor protein that is propagated to the cytoplasmic region of the chemoreceptor protein, which interacts with CheW/ChA. An attractant *decreases* the rate of autophosphorylation of CheA, and a repellent *increases* the rate of autophosphorylation. Thus attractants promote CCW (counterclockwise) rotation and smooth swimming, whereas repellents promote CW rotation and tumbling. (Chemotaxis in the gram-positive *Bacillus subtilis* differs from that in the enterics in several ways, including the fact that the attractant causes an *increase* in CheY phosphorylation and CheY-P causes the flagellar motor to rotate CCW, resulting in smooth swimming.[73])

## A model for adaptation

CheA-P also phosphorylates a second response regulator protein, CheB:

$$\text{CheA-P} + \text{CheB} \longrightarrow \text{CheA} + \text{CheB-P}$$

CheB-P is the active form of a methylesterase that removes methyl groups from glutamate residues in the receptor–transducer proteins. There is also a methyltransferase, CheR, that adds methyl groups to these residues. As a result of the stimulation of the chemoreceptor protein by the chemoattractant, the levels of CheB-P also decrease because there is less of the phosphoryl donor, CheA-P. When autophosphorylation of CheA and hence phosphorylation of CheB is inhibited by

increased methylation of the receptor–transducer proteins. Adaptation to a repellent is due to undermethylation of the receptor–transducer protein. CheZ is thought to be a phosphatase that removes the phosphate from CheY-P. The relative activities of CheA-P and CheZ determine the level of CheY-P, hence the frequency of tumbling. Theoretically, changes in the activity of CheZ could also alter the frequency of tumbling. *Source:* From Stock, J. B., A. J. Ninfa, and A. M. Stock. 1989. Protein phosphorylation and regulation of adaptive responses in bacteria. *Microbiol. Rev.* 53:450–490.

binding of the attractant to the chemoreceptor protein, the removal of methyl groups is slowed relative to the addition of methyl groups and *all* the chemoreceptor proteins *become more highly methylated.* (There are four or five glutamate residues that can be methylated.) The more highly methylated chemoreceptor protein stimulates tumbling, hence adaptation to that concentration of attractant. There is ample evidence, both in vivo and in vitro, that the methylated forms of the chemoreceptor proteins stimulate the activity of CheA, hence the phosphoryation of CheY, CW rotation, and tumbling.[74, 75] (See Note 76.) (Another way of saying this is that methylation of the chemoreceptor protein cancels the attractant-induced signal.)

As a result of the methylation, the prestimulus levels of CheY-P and CheB-P are restored, and the cell resumes the prestimulus run–tumble behavior, despite the presence of bound attractant. As long as the attractant is bound to the receptor–transducer protein, it remains highly methylated, and the cell is said to be *adapted* to that particular concentration of attractant. However, the *other* (nonoccupied) chemoreceptor proteins quickly lose their extra methyl groups so that the cell is adapted only to the specific attractant. If the chemoreceptor protein is not fully saturated with bound attractant, then the cells will respond to the attractant when it swims into areas of higher concentration of attractant. However, as the chemoreceptor protein binds more chemoattractant, the average number of methyl groups per transducer is increased two- to fourfold, thus balancing the increased binding of attractant and resulting in adaptation to the higher concentration of chemoattractant. (See Note 77.) When the chemoattractant is removed, the overmethylated chemoreceptor protein stimulates CheA autophosphorylation. This results in increased CheY-P and consequent tumbling.[78] However, the methylesterase CheB-P also increases and returns the chemoreceptor protein to the methylation level of the prestimulus state and normal run–tumble behavior. Adaptation to a repellent differs from adaptation to an attractant in that during repellent adaptation the activated CheB-P *demethylates* the chemoreceptor protein that responds to the repellent.

In summary, then, binding of the attractant to the chemoreceptor inhibits the kinase activity of CheA, which leads to a fall in the levels of CheY-P, hence smooth swimming. However, binding of the attractant to the chemoreceptor also promotes the methylation of the chemoreceptor. The methylated chemoreceptor stimulates the kinase activity of CheA, which leads to a rise in the levels of CheY-P and tumbling; thus adaptation takes place.

### Some supporting evidence in favor of the model

The phenotypes of chemotaxis mutants support the model. This is summarized in Table 18.1. For example, *cheW*, *cheA*, and *cheY* mutants do not tumble, whereas *cheB* and *cheZ* mutants tumble constantly. This is in agreement with the model, since it stipulates that CheW and CheA promote the phosphorylation of CheY, and CheY-P causes the motor to reverse and tumbling to take place. CheB competes with CheY for the phosphoryl group on CheA-P and would therefore be expected to promote smooth swimming. Hence a mutation in *cheB* should make the cells tumble. CheZ dephosphorylates CheY, and therefore a mutation in *cheZ* should result in high levels of CheY-P and therefore increased tumbling. There are also biochemical data to support the model. When an attractant binds to the extracellular domain of the chemoreceptor protein, several events occur: (1) the rate of phosphorylation of CheY decreases; (2) the rate of demethylation is reduced; and (3) the rate of methylation is increased. There is also a great deal of evidence indicating that the methylated form of the chemoreceptors stimulate CheyY phosphorylation and tumbling.[73, 74, 76]

### An adaptation mechanism that does not rely on methylation

If cells are experimentally exposed to a large and abrupt change in the concentration of a stimulus, such as the addition of an attractant, then adaptation, which can take minutes, is certainly correlated with an increased level of receptor methylation, as described earlier. On the other hand, it has been pointed out that the methylation-dependent adaptation system may be too slow for adaptation

**Table 18.1** Effects of chemotaxis mutants

| Gene | Mutant phenotype | Rationale for mutant phenotype |
|------|------------------|-------------------------------|
| $cheY^-$ | Smooth swimming | CheY-P is required for tumbling |
| $cheA^-$, $cheW^-$ | Smooth swimming | CheA-P and CheW phosphorylate CheY |
| $cheB^-$ | Tumbling | CheB competes with CheY for the phosphoryl group from CheA-P |
| $cheZ^-$ | Tumbling | CheZ dephosphorylates Chy-P |

by cells swimming in gradients and in fact a methylation-independent system may exist.[79] During chemotaxis along an attractant concentration gradient the cells encounter very small changes in stimulus concentrations (as opposed to the experiments where large concentrations of attractant are added to a cell suspension and the behavior of the cells is monitored) and adaptation takes only about 1 second. It has been argued that the increase in methylation of the MCP proteins is too small in 1 second to account for adaptation to attractants in gradients. Furthermore, $cheR^- cheB^-$ double mutants that show no changes in methylation are nevertheless chemotactic in swarm plates and also show adaptation to the addition of very small concentrations of attractant when CWW rotation of tethered cells is measured.[79–82] (See Note 83 for a description of chemotaxis assays and how to measure responses to attractants and repellents using tethered cells.) Apparently there exists a methylation-independent system that allows cells to respond chemotactically to small concentrations of stimulus and to be chemotactic in gradients. Furthermore, chemotaxis to sugars transported by the PTS system does not involve MCP proteins; hence adaptation cannot be via methylation of these proteins. (See Section 18.4.7.)

## 18.4.6 Mechanism of repellent action

Repellents for *E. coli* and *S. typhimurium* include indole, glycerol, ethylene glycol, phenol (for *S. typhymurium* only), organic acids (e.g., formic, acetic, benzoic, salicylic), alcohols, $Co^{2+}$ and $Ni^{2+}$ (for *E. coli* only), and hydrophobic amino acids (leucine, isoleucine, tryptophan, valine). In general, the mechanism of action of organic repellents is not well understood, except that the response requires the receptor–transducer (MCP) proteins. Yet there is no direct evidence for specific binding between an organic repellent and a specific MCP. In some cases (e.g., glycerol, ethylene glycol, and aliphatic alcohols) there is not even specificity regarding which receptor–transducer protein is required. This has been learned by using mutants lacking one or more of the receptor–transducer proteins and showing that any of the remaining receptor–transducer proteins can mediate the repellent response. However, in other instances there is specificity. For example, the response to leucine, isoleucine, and valine requires the Tsr protein, and the response to the cations $Co^{2+}$ and $Ni^{2+}$ requires Tar. Because much higher concentrations of organic repellents (in the millimolar range) are usually required than for attractants (in the micromolar range), it has been suggested that the organic repellents (which all have some hydrophobic character as well as polar groups) do not actually bind to the receptor–transducer proteins but act indirectly by perturbing the membrane at the site of the receptor–transducer proteins. For example, they might make the membrane more fluid, which could result in the alteration of receptor–transducer activity. Recently, a study was carried out to investigate the action of repellents on the membrane fluidity of *E. coli*.[84] It was concluded that one could *not* account for repellent activity by changes in membrane fluidity, suggesting that repellents may indeed bind to receptor–transducer proteins, however, with low affinity, and in some cases low specificity. In summary, the response to repellents *does* involve the receptor–transducer proteins (MCP proteins), but the mechanism of how they affect the activities of the receptor–transducer proteins is not understood. However, there is more known about the repellent activity of weak organic acids. These are believed to affect the activities of the receptor–transducer proteins Tsr and Tap by lowering the intracellular pH.

### 18.4.7 Chemotaxis that does not use MCPs: The phosphotransferase system (PTS) is involved in chemotaxis toward PTS sugars

In addition to functioning in sugar uptake, the PTS system is also important for chemotaxis toward PTS sugars.[85] (See Section 16.3.4 for a description of the PTS system.) Some of the evidence for this is that:

1. Mutants of *E. coli* defective in EI and HPr are usually defective in chemotaxis toward PTS sugars.

2. EII mutants defective in chemotaxis toward the sugar specifically recognized by EII also exist.

#### A comparison of PTS-mediated chemotaxis with MCP-dependent chemotaxis

*E. coli* mutants lacking CheB, CheR, or the MCP proteins display normal chemotaxis to PTS sugars. This means that methylation-dependent adaptation does not occur in the PTS system and there must be another adaptation mechanism. However, CheA, ChW, and CheY *are* required for chemotaxis toward PTS sugars.[86] One model postulates that EI-P can cause the phosphorylation of CheA as well as the incoming PTS sugar. Thus, as the sugar becomes phosphorylated during its uptake, the levels of CheA-P fall, causing lower levels of CheY-P and hence smooth swimming. (Reviewed in Ref. 87.)

### 18.4.8 Chemotaxis that is not identical with the model proposed for the enteric bacteria

Chemotaxis in many bacteria differs in several respects from chemotaxis in *E. coli*. For example, some bacteria do not reverse the rotation of their flagella and do not tumble.

Other bacteria do reverse the direction of rotation of their flagella, but this causes them to back up, not tumble. Additionally, some bacteria must metabolize the chemoattractant in order to generate the chemoattractant signal; that is, the initial signal does not come from the binding of the chemoattractant to an MCP-like protein. How they adapt is not known. Of course, as discussed earlier, even *E. coli* does not always use the MCP proteins, such as when it responds to PTS sugars. Several bacteria have multiple *che* genes, especially *cheY*. It is not known why this is the case. The following is a description of chemotaxis in bacteria other than *E. coli*, where it has been studied in detail.

#### Bacteria that do not tumble

Not all bacteria tumble, yet they are capable of chemotaxis. For example, *Rhodobacter sphaeroides* and *Rhizobium* (*Sinorhizobium*) *meliloti* rotate their flagella only in a clockwise direction and do not tumble. (See *Rhodobacter sphaeroides* and *Rhizobium meliloti* later.) Yet they swim randomly and exhibit chemotaxis.

In some bacteria with polar flagella (flagella at one pole or both poles, either single or in bundles) the reversal of the flagellar motor causes the cells to reverse the direction of swimming rather than tumble. An example is *Caulobacter* or certain photosynthetic bacteria that change direction by reversing the direction of flagellar rotation. Because they do not back up in an absolutely straight line, and/or do not swim forward in a straight line, the direction of forward swimming is random to the original forward direction.

#### Differences in the signaling pathways

Since chemotaxis is an evolutionary adaptation to the bacterium's environment, one might suspect that there would be some differences in the chemotactic signaling pathways among the bacteria. This is the case. There exist bacteria that appear to have more than one chemotaxis signalling pathway, as suggested by two or more copies of some of the *che* genes. In addition, the presence of membrane-bound MCP proteins that are closely related to the enteric MCPs is not established in all bacteria, although similar cytoplasmic proteins have sometimes been detected. Indeed, the mechanism of adaptation is not known for several bacteria and in some cases certainly differs from the mechanism described for the enterics with respect to MCP proteins. Some bacteria whose swimming behavior and chemotaxis pathway are not exactly like that of *E. coli* are described next.

## Rhodobacter sphaeroides

*R. sphaeroides* is a photosynthetic bacterium that can live aerobically heterotrophically in the dark or anaerobically as a photo-heterotroph in the light. (See Section 5.1.2 on the purple nonsulfur phototrophs.) It has a single medially (subpolar) located flagellum that rotates in one direction only (clockwise), pushing the cell forward. The cells stop swimming intermittently, and when swimming resumes, the cells swim randomly in a different direction. It appears that while the bacterium is stopped, the flagellum is coiled against the cell body and rotates slowly, resulting in reorientation of the cell. (Reviewed in Ref. 87.) *R. sphaeroides* is attracted to weak organic acids, sugars and polyols, glutamate, ammonia, and certain cations (e.g., $K^+$ and $Rb^+$). As opposed to the enteric bacteria, metabolism of the chemoattractant appears necessary for chemotaxis, because if metabolism is blocked (e.g., by a specific inhibitor), then chemotaxis is also inhibited.

*R. sphaeroides* has homologs to all the *E. coli che* genes except *cheZ*.[88, 89] (See Note 90 for a description of how the *che* genes were detected in *R. sphaeroides*.) Also, although a typical enteric-like membrane-bound MCP appears to be absent, a *cytoplasmic* methyl-accepting chemotaxis protein called transducer-like protein (TlpA) with some similarity to the enteric MCPs does exist.[91] (See Note 92 for a further description of how this was determined.) Recall that metabolism is required for chemotaxis in *R. sphaeroides*. Perhaps TlpA responds to a cytoplasmic metabolic intermediate and transfers the signal to CheA. In contrast to *E. coli*, the *che* genes exist in multiple copies in *R. sphaeroides*. There are two copies of *cheA* and *cheW* and three copies of *cheY* genes.

This is because there are two independent chemosensory pathways in *R. sphaeroides* as opposed to one in *E. coli*. (Reviewed in Ref. 93.) The reason for believing that two independent chemosensory pathways exist is that mutants in which the originally discovered chemotaxis operon containing homologs to *cheA*, *cheW*, *cheR* and homologs to two *cheY* genes is deleted, the cells are still capable of chemotaxis as well as the wild type.[89] When the deletion mutant was mutagenized and chemotaxis and phototaxis minus mutants were screened, a second chemotaxis operon was discovered containing homologs to *cheY*, *cheA*, *cheR*, *cheB*, two homologs to *cheW*, and *tipC*, a transducerlike protein. Mutations of this second chemotaxis operon (*cheA⁻*) in a wild-type backround produced chemotaxis mutants, suggesting that it is part of an operative chemotaxis pathway. The question remains as to the role of the first chemotaxis operon, since its absence or presence does not seem markedly to affect chemotaxis. How *R. sphaeroides* senses carbohydrates that act as attractants is not understood. As mentioned, metabolism of the carbohydrate is necessary, suggesting that the chemotactic signal is generated during metabolism and not by binding of the sugar to an MCP-like protein.[94] (See Note 95.)

Since *R. sphaeroides* changes its direction as a result of stopping rather than tumbling, it is reasonable to suppose that an increased concentration of chemoattractant might decrease the frequency of stops (the equivalent of the suppression of tumbling), thus ensuring that the cells continue to swim in the direction of the chemoattractant. Likewise, a decreased concentration of chemoattractant might increase stopping frequency, and hence promote changes in the direction of swimming. However, it does not happen exactly this way. When anaerobically grown cells tethered to glass via antiflagellin antibody were examined, it was observed that an increase in attractant concentration did not cause a change in rotational behavior, but when the attractant concentration was *decreased*, the cells transiently stopped rotating. (Reviewed in Ref. 87.) It appears that cell accumulation in response to a chemoattractant gradient by *R. sphaeroides* is dependent upon sensing a *reduction* and not an increase in the concentration of chemoattractant. This is opposite to the situation in *E. coli*, which responds to an increase in attractant concentration.

Some metabolites, especially weak organic acids, cause an increase in the speed of swimming by *R. sphaeroides*, a response called *chemokinesis*.[96] Many compounds that cause chemokinesis also cause chemotaxis. However, the two phenomena are separable because: (1) most amino acids and sugars that are chemoattractants do not cause chemokinesis; and (2) metabolism is required for chemotaxis but not for chemokinesis. (It appears that transport across the cell membrane may be involved in chemokinetic signaling.) The increased rate of swimming in response to certain metabolites lasts for a long time (i.e., there is no rapid adaptation). It has been pointed out that in the absence of adaptation, chemokinesis should result in the spreading of the population rather than its accumulation.[97]

### Rhizobium meliloti

*Rhizobium meliloti* (also called *Sinorhizobium meliloti*) is an aerobic bacterium that can live symbiotically with leguminous plants and fix nitrogen. (See Section 12.3.1.) The bacteria enter the plants via plant rootlets to which they are attracted by root exudates that include amino acids, carbohydrates, and flavones (cyclized isoprenoid compounds made from phenylalanine and malonyl–CoA). Chemotaxis studies with *R. meliloti* employ L-amino acids and D-mannitol as attractants. *R. meliloti* has 5–10 peritrichously located flagella that form a bundle when the cell swims. Swimming consists of straight runs interupted by very quick turns. The cells do not seem to stop before turning as, for example, does *R. sphaeroides*. It has been suggested that the flagella never actually stop rotating. When the cell is swimming in one direction, the flagella rotate as a single bundle. The cell turns when the flagellar motors rotate at different rates and the bundle flies apart. (Reviewed in Ref. 87.) Cells tethered to a microscope slide by a single flagellum (the other flagella were mechanically sheared off the cell) rotated only

clockwise with very brief (< 0.1 sec), intermittent stops.[98]

*R. meliloti* has homologs to the enteric *che* genes. Present in *R. meliloti* are *cheY* (2 copies), *cheA*, *cheW*, *cheR*, and *cheB*.[99] (There is no *cheZ*.) Mutations in the *che* genes make the cells defective in chemotaxis. *R. meliloti* possesses a *tlpA* gene, which, as in *R. sphaeroides*, encodes a cytoplasmic MCP-like protein. Other MCP-like proteins have been detected, as reviewed in Ref. 87. As noted previously, *R. meliloti* flagella turn in only one direction (clockwise), and the cell does not seem to stop momentarily before it turns in a new direction. *R. meliloti* responds to attractants by increasing its swimming speed (chemokinesis). This is equivalent to decreasing the frequency of tumbles in *E. coli*. It has been speculated that the rate of flagellar rotation is determined by CheY-P, which perhaps slows the flagellar motor. Thus the presence of the attractant might inhibit the autophosphorylation of CheA, which would decrease the levels of CheY-P and increase swimming speed.[87] This would result in chemotaxis toward the attractant if the cells were also capable of adaptation.

### Azospirillum brasilense

There have been reports that *Azospirillum brasilense* lacks a methylation-dependent pathway for chemotaxis toward organic compounds.[100, 101] In *Azospirillum brasilense*, it appears that the strongest chemoattractants are electron donors to the aerobic redox chain. This, plus the fact that chemoattraction toward oxygen is so strong in this organism that it masks responses to other chemoattractants, has led to the suggestion that the signaling system involves the respiratory chain. It is reasonable to suggest that the signaling system in these cases (and in phototaxis and aerotaxis discussed next) may involve the redox level of one of the electron carriers or the protonmotive force.

## 18.5 Photoresponses

As discussed next, tactic responses to light as well as to electron acceptors such as oxygen and nitrate also exist among the prokaryotes. The student should consult the review by Hellingwerf, Hoff, and Crielaard, which discusses various photosensors, including a newly discovered family of photosensors called photoactive yellow protein (PYP), which appears to be widespread among the bacteria.[102] It seems that in several of these cases the signalling pathway to the flagellum motor uses the Che proteins that function in the chemotaxis signaling pathway. (See Section 18.4.4.) Thus one might view various sensory signaling pathways as differering in the initial sensory receptors but converging at the level of the Che proteins (e.g., CheA). The signaling pathways for responses other than chemotaxis are generally not well characterized in detail at present. Photoresponses in halobacteria and photosynthetic bacteria will be discussed, followed by a descripiton of the signaling pathway for responses to terminal electron acceptors.

## 18.5.1 Halobacteria

The student should review the description of the halophilic archaea in Sections 1.1.1 and 3.8.4, and the discussion of the role of light and retinal pigments in establishing a $\Delta p$ and in $Cl^-$ transport in Sections 3.8.4 and 3.9.

### Swimming behavior

Halobacteria have a single flagellum consisting of 5–10 filaments in a right-handed helix at one pole of the cell. (Reviewed in Ref. 103.) (Actually, the cells are monopolarly flagellated only during exponential growth. In the stationary phase they are mostly bipoplarly flagellated.) The available evidence suggests that all the filaments are attached to the same flagellum motor.[103] When the flagellum motor turns clockwise (CW), the cells swim forward, and when the motor turns counterclockwise (CCW), the swimming direction is reversed. (The flagellar filaments remain together even when the direction of rotation is reversed.) There is a brief period between CW and CCW rotation where the flagellum is not rotating and the bacterium changes its orientation, perhaps due to Brownian motion. When swimming resumes, it is in a direction random to the original direction.

### Photoresponse

The cells are capable of both positive and negative photoresponses depending upon the wavelength of light. The bacteria are attracted toward orange/red light (500–600 nm), which is where bacteriorhodopsin and halorhodopsin absorb, and are repelled by UV/blue light, which can damage DNA. Attracting light retards the process of switching; hence the bacterium swims in the direction of the light. Repulsive light increases the switching frequency; hence the bacteria swim away from repulsive light. This is not really phototaxis, in that the bacteria do not perceive a gradient of light and are simply responding to changes in intensity. In this way, it is similar to chemotaxis.

### Signal transfer to flagellar motor

The photoreceptors are rhodopsins similar to bacteriorhodopsin.[104–106] As a consequence of absorbing light, there is a conformational change in the photoreceptor, which results in a signal (probably a conformational change) to a closely associated membrane-bound MCP-like protein called Htr. It has been suggested that the Htr then signals proteins homologous to the Che proteins, which then alter the frequency of flagellar reversal. From this perspective the signaling pathway for photosensing is similar to chemotaxis except that in photosensing the sensor (Htr) responds to a light-activated photoreceptor, whereas in chemotaxis the sensor (MCP) responds to the binding of a chemoeffector or a periplasmic protein to which the chemoeffector is bound (Section 18.4.4). The signal in both cases would be transferred via phosphorelay to the flagellar motor via the Che proteins. Exactly how the absorption of light by the photoreceptor activates HtrI is not known, but see Note 107 for a more complete discussion of photosensing.

## 18.5.2 Photosynthetic bacteria

### Avoidance of the dark

Photosynthetic bacteria such as *Chromatium*, *Thiospirillum*, and *Rhodospirillum* that are swimming in cell suspension will reverse their direction of swimming when they swim out of a light area into a dark area, resulting in their accumulation in the light.[106] This is easily observed microscopically by shutting the condenser so that there is a small circle of light in an otherwise dark field. For example, when *Rhodospirillum rubrum*, a polarly flagellated photoheterotroph, swims from an area of light into an area of darkness, the polar flagellar bundles reverse their direction of rotation, and the cells back up into the light. Such behavior has been called *scotophobia* because it is an avoidance of the dark rather than a seeking of light. Experiments with a horizontal beam of light in a microscope field indicate that free-swimming photosynthetic bacteria do not respond to the direction of propagation of the light but rather to a decrease in intensity.[108] If the cells can reverse their direction of swimming by reversing the rotation of the flagella, then they back up whenever they start to swim across the light/dark boundary. They become trapped throughout the beam of light, not simply near its source, indicating that they are not responding to the direction from which the the light is coming but rather to the decrease in intensity as they cross into the dark. Not all photosynthetic bacteria can reverse the direction of flagella rotation and therefore do not back up when crossing a light/dark boundary. An example is *R. sphaeroides*. In this case the flagellum rotates in only one direction, and therefore the photoresponse when the light intensity is reduced is an increase in the frequency of stopping. When the cells stop swimming, their orientation changes, leading to a change in swimming direction when flagellar rotation resumes. Interestingly, when a narrow beam of light exists horizontally across a microscope field, this results in the accumulation of bacteria in the dark, rather than in the light.[108] This is because the cells do not reverse when crossing the light/dark boundary but stop and swim off in a randomly different direction, which more often than not brings them into the dark.

### Role of photosynthetic electron transport

Mutants in the photosynthetic reaction center do not respond to a decrease in light intensity, suggesting that the response involves either photosynthetic electron transport, the $\Delta p$, or both. (A decrease in light intensity in wild-type cells causes a decrease in $\Delta p$ and in photosynthetic electron transport.) This conclusion is supported by the finding that inhibitors of electron transport prevent the photoresponses even though the cells swim normally.[109] The evidence gathered for *R. sphaeroides* suggests that it is photosynthetic electron transport (probably the redox level of one of the carriers) rather than the $\Delta p$ that signals a step down in the light intensity.[109] Some of the experimental evidence for this is derived from the use of the proton ionophore FCCP, which stimulates electron transport while it collapses the $\Delta p$. (See Section 3.4.1 for an explanation of uncouplers.) The experiments were done with cells tethered to glass by antiflagellin antibody. Under these conditions the cells rotated, and the frequency of stops was measured. The tethered cells showed no response to the addition of FCCP, although the $\Delta \Psi$ decreased to the same extent as caused by a step down in light intensity. This indicates that the cells respond to a decrease in photosynthetic electron transport rather than a decrease in the $\Delta p$ upon a step down in light intensity. Perhaps what happens is that when the rate of electron transport is decreased, a signal is sent to an Aer-like protein, which results in an increase in CheY-P that stops the flagellar motor. This, however, is speculation. (Aer proteins are MCP homologs required for response to oxygen gradients. See Section 18.6. Also, see Section 18.4.8 on *Rhizobium meliloti*.)

### Phototaxis

Individual cells of *Rhodospirillum centenum* in liquid culture show the scotophobic response described earlier for other photosynthetic bacteria. Cells grown in liquid media have a single polar flagellum and change the direction of swimming by reversing the rotation of the flagellum. *R. centenum* colonies move and are capable of phototaxis in gradients of light.[110] This is a true phototaxis in that the colonies move along a light gradient toward the source of light down an intensity gradient when a converging beam of light is used. When two attracant beams of light at 90° to each other are used, then the colonies move in a direction that is 45° between the beams, suggesting that they are integrating the signal from the light beams. The colonies move toward a source of infrared light and away from a source of visible light. The cells have lateral flagella when grown on agar and cooperatively move on the solid agar. The movement is called *swarming*. Mutants in the photoresponses have been isolated.[111] Some of the mutants that are not capable of photoresponses have defects in the synthesis of reaction center components or electron carriers that function in photosynthetic electron transport. This indicates that the photoresponses are in some way related to changes in photosynthetic electron transport as the cells move along the light gradient. Other mutants have defects in the *che* genes, indicating that the signal cascade proceeds from the photoreceptor through the Che proteins to the flagellar motors. Still other mutants are normal in their chemotaxis responses and in photosynthetic growth, and appear to have defects in the perception of phototactic light or in the signal transduction pathway upstream of the Che proteins. (See Note 112.)

## 18.6 Aerotaxis

Many aerobic bacteria, including *E. coli* and *Salmonella typhimurium*, swim toward a higher concentration of oxygen. The response to oxygen requires electron transport as well as a portion of the chemotaxis pathway, including CheW, CheA, and CheY.[113, 114] This was concluded using mutants lacking the terminal cytochrome oxidases o and d or the *che* genes. When wild-type *E. coli* is grown anaerobically in the presence of nitrate so that cytochrome o is absent and nitrate reductase is present, the cells are attracted to nitrate but not to oxygen. This indicates

that electron transport to oxygen generates the oxygen signal and that electron transport to nitrate generates the nitrate signal. When both cytochrome o and nitrate reductase are present, then the cells are attracted to both oxygen and nitrate in a competetive manner. Inhibitors of electron transport prevent taxis toward the electron acceptors. The primary signal might be a change in the $\Delta p$ or in the redox state of one of the electron carriers.

A protein called Aer (cytoplasmic but probably membrane associated) has been referred to as an MCP homolog and seems to be required as mutants in *aer* show reduced responses to oxygen gradients.[87] One can suppose that Aer senses the change in the redox state of a component of the electron transport chain and transmits the signal to CheA. For example, suppose that increased levels of oxidation of the electron carriers signaled Aer to lower the levels of CheA-P. This would result in lower amounts of CheY-P and hence smooth swimming and accumulation of cells in areas where the oxygen concentration was higher. Aer noncovalently binds FAD, and perhaps the redox state of the FAD alters the configuration and thus the activity of Aer.

## 18.7 Effect of Oxygen and Light on the Expression of Photosynthetic Genes in the Purple Photosynthetic Bacterium Rhodobacter capsulatus

*Rhodobacter capsulatus* is a purple nonsulfur photosynthetic bacterium that can grow heterotrophically in the dark, deriving its energy from aerobic respiration, or photoheterotrophically in the light without oxygen The synthesis of the photosynthetic apparatus is under the control of light and oxygen, and regulation is at the level of transcription. There are at least three regulatory systems that control transcription of the photosynthetic genes. One of these is a repressor system (the CrtJ system) activated by oxygen. A second is an inducer system (the RegB/RegA system) activated by anaerobiosis. And the third is an inducer system (the HvrA system) activated by low light.

### 18.7.1 Reponse to oxygen

The student should review the description of the light-harvesting complex in photosynthetic purple bacteria described in Chapter 5. The photosynthetic genes in the purple photosynthetic bacteria such as *R. capsulatus* exist in several operons collectively known as the *photosynthetic gene cluster*. These genes code for proteins required for the biosynthesis of the photosynthetic pigments (bacteriochlorophyll and carotenoids) as well as proteins that make up part of the reaction center and light-harvesting complexes.

### Aerobic repression

Oxygen represses the synthesis of photosynthetic pigments ensuring that photosynthesis occurs only under anoxic conditions. (The regulation of photosynthetic genes is reviewed in Ref. 115.) One of the genes in the photosynthetic gene cluster encodes an aerobic transcription repressor of genes for bacteriochlorophyll, carotenoid, as well as the genes for synthesis of light harvesting II complex (*puc* genes). The aerobic repressor is called CrtJ (or Pps), which is activated in the presence of oxygen. It is not known how CrtJ senses oxygen. Perhaps some functional group, for example, a sulfhydry group, or a metal cofactor, becomes oxidized, and this causes the protein to undergo a conformational change that allows it to bind tightly to target DNA. The student should review the discussion of another redox-sensitive transcription regulator, the FNR protein, which is inactivated by oxygen rather than being activated. See Section 18.2.2 and Ref. 32.

### Anaerobic induction (RegB/RegA)

The photosynthetic bacterium *Rhodobacter capsulatus* uses a two-component system (RegB/RegA) to stimulate the transcription of genes for its light-harvesting complex and photoreaction center under anaerobic incubation conditions.[116] Under anaerobic conditions a signal is sent to RegB, causing it to autophosphorylate. The signaling pathway is not understood. RegB-P then phosphorylates RegA.[116] DNA binding studies indicate that RegA-P is a transcriptional activator.[117] It has

also been reported that RegB is required in the photosynthetic bacterium *R. sphaeroides* for the expression of genes required for $CO_2$ fixation. Thus RegB/RegA may be a global regulator for many genes that serve a function in anaerobically growing *R. capsulatus* and *R. sphaeroides* that is analogous to the ArcB/ArcA system in *E. coli*. However, it is not understood how RegB senses anaerobiosis. (See the discussion in Section 18.2.2 of how ArcB might sense anaerobiosis. Recall that ArcB also autophosphorylates under anaerobic conditions. See Fig. 18.5.) In addition, homologs of RegA and RegB have been found in the nonphotosynthetic bacterium *Rhizobium meliloti* (cited in Ref. 117).

### Response to light

Photosynthetic pigment synthesis is increased anaerobically when the light intensity is decreased. This has the advantage of allowing the cells to capture more light energy despite the lower light intensities. *R. capsulatus* has a gene, called the *hvrA* gene, whose product (HvrA) is required for the stimulation of transcription of some genes in the photosynthetic gene cluster by low light intensity. This conclusion is supported by the finding that mutations in *hvrA* cause a loss in low-intensity light stimulation of transcription of target operons. It has been shown that HvrA binds to target promoters in the photosynthetic gene cluster, and it has been demonstrated that high light somehow reduces the intracellular levels of HvrA. However, not all the low-light-inducible genes require HvrA for transcription, and thus there is clearly another light-sensitive signaling system. In addition to transcriptional regulation by light, there appears to be post-transcriptional regulation.[118] This was reported for the synthesis of the peptides of the light-harvesting complex I (B800–850 peptides), which increase around fourfold in parallel with the number of B800–850–bacteriochlorophyll complexes when the light intensity is diminished. It was found, however, that low-light-grown cells had only $\frac{1}{4}$ the mRNA for the B800–850 peptides. We therefore have a situation where more protein is present under conditions of less mRNA. Perhaps there is decreased turnover of the protein under low light conditions. This is discussed in Ref. 118.

## 18.8 Response to Nitrogen Supply: The Ntr Regulon

Bacteria use inorganic and organic nitrogenous compounds as a source of nitrogen for growth. The inorganic nitrogen is in the form of ammonia, nitrate, or dinitrogen gas, and the organic nitrogen is in the form of amino acids, urea, and other nitrogenous compounds. Ultimately, all these nitrogenous compounds are either reduced or catabolized by the bacteria to ammonia, which is assimilated via the glutamine synthetase reaction. When the ammonia supply is adequate, then ammonia is the preferred source of nitrogen, and it represses genes required for the assimilation of other nitrogenous compounds. However, when ammonia in the growth medium becomes limiting, then genes required for ammonia production from external nitrogen sources are induced. Furthermore, under limiting ammonia conditions glutamine synthetase becomes more active, and the transcription of the gene for glutamine synthetase is also stimulated, resulting in more glutamine synthetase to scavenge the small amounts of ammonia for the synthesis of glutamine and glutamate. (The glutamic dehydrogenase, glutamine synthetase, and glutamine synthase reactions and their biological role are discussed in Section 9.3.1 and reviewed later.)

The genes that are regulated by the ammonia supply belong to the Ntr regulon.[119-121] As mentioned previously, a *regulon* is a set of noncontiguous operons or genes controlled by a common regulator, and for the *Ntr regulon*, the regulator is a response regulator protein called NR$_I$. The Ntr regulon consists of many genes, including the genes for glutamine synthetase, NR$_{II}$ (a histidine kinase, which is also called NtrB), and NR$_I$ (a response regulator, which is also called NtrC), all of which are in an operon called *glnALG* in *Escherichia coli* and *glnABC* in *Salmonella typhimurium*. Other genes in the Ntr regulon are the genes for the nitrogenase system in *Klebsiella* (*nif* genes), the genes for the degradation of urea

in *Klebsiella*, the genes for periplasmic binding proteins required for amino acid transport [e.g., for glutamine (*glnH*)], arginine, lysine, and ornithine (*argT*), and histidine (*hisY*) in *Salmonella*, the genes for histidine degradation (*hut*) in *Klebsiella*, and the genes for proline degradation (*put*) in *Salmonella*.[122] It should not be concluded automatically that if a gene system is under Ntr regulation in one genus of bacterium, it is also under Ntr regulation in all bacteria. This is clearly not the case. For example, the *hut* genes are not regulated by ammonia levels in *Salmonella typhimurium*, although they are so regulated in *Klebsiella pneumoniae*. Along the same lines, it should be emphasized that the following discussion regarding the Ntr system applies only to the enteric bacteria, where the regulation of nitrogen metabolism has been studied in detail. There is much less known about the regulation of nitrogen metabolism in other prokaryotes, and one should be careful about extrapolating from the enterics to other prokaryotes. For example, it appears that the Ntr system is not present in gram-positive bacteria. (Reviewed in Ref. 121.)

One can consider glutamine as the intracellular signal that informs the cell whether the ammonia levels are adequate. To understand this, let us review the biochemistry of ammonia assimilation (Section 9.3.1). When ammonia levels are high in the growth medium, the ammonia is incorporated into glutamate by the NADPH-dependent reductive amination of $\alpha$-ketoglutarate catalyzed by the enzyme glutamate dehydrogenase. Ammonia is also incorporated into glutamine in an ATP-dependent reaction with glutamate catalyzed by glutamine synthetase (GS). These reactions are crucial because glutamate supplies nitrogen to approximately 85% of the nitrogen-containing molecules in the cell, and glutamine supplies the remaining 15%. However, when the supply of ammonia is limiting (e.g., when cells derive ammonia from organic nitrogen compounds in the growth medium, or from the reduction of nitrate or dinitrogen gas), the glutamate dehydrogenase reaction does not significantly contribute to glutamate synthesis (discussed in Section 9.3.1). Instead, glutamate is made by transferring an amino group from the amide nitrogen of glutamine to

$\alpha$-ketoglutarate in a reaction catalyzed by glutamate synthase. The result of the glutamate synthase reaction is two molecules of glutamate, one of which is recycled back to glutamine using glutamine synthetase. Therefore, when the cells are limited for ammonia, glutamine is the immediate source of nitrogen for glutamate. As described below, glutamine levels parallel the changes in concentrations of extracellular ammonia, and the cell, using glutamine as an indicator of nitrogen supply, adjusts the transcription of the Ntr regulon and the activity of glutamine synthetase accordingly.

### 18.8.1 A model for the regulation of the Ntr regulon

This is a complex signaling pathway. (See Ref. 123 for a review.) It will be explained at two levels, with one giving a more detailed explanation than the other.

#### The less detailed explanation

Transcription of the Ntr regulon is inhibited by excess ammonia, and the inhibition is relieved when the ammonia concentration falls to below 1 mM. What happens is that excess ammonia results in an increase in the concentration of the active form of a protein, $P_{II}$ (also called GlnB), which *indirectly* represses transcription of the Ntr regulon by inactivating a positive transcription regulator protein ($NR_I$) via dephosphorylation. When ammonia concentrations decrease, the amount of active $P_{II}$ also decreases, and transcription of the Ntr regulon is stimulated because the response regulator is phosphorylated.

The levels of active $P_{II}$ are controlled by ammonia in the following way. When the ammonia concentration is increased, the intracellular concentration of glutamine rises. It is actually the glutamine that causes active $P_{II}$ levels, to increase and (as a consequence) transcription to be repressed. This is because glutamine stimulates an enzyme that converts $P_{II}$–UMP to $P_{II}$, the active form of $P_{II}$. [The glutamine stimulates a bifunctional enzyme called uridylyl transfer/removal enzyme (UT/UR) that in the presence of glutamine removes UMP from $P_{II}$–UMP. The enzyme is also called

GlnD.] As stated, the increase in $P_{II}$ represses the Ntr regulon.

Just the opposite happens when the ammonia levels fall. It happens in the following way: When the ammonia concentration is shifted from a high concentration to below 1 mM, the intracellular concentration of glutamine falls. [This is because the level of glutamine synthetase, the enzyme that synthesizes glutamine from ammonia, glutamate, and ATP, is low in cells grown in the presence of high ammonia. Additionally, glutamate dehydrogenase, the enzyme that synthesizes glutamate from $\alpha$-ketoglutarate, ammonia, and NADPH, has a high $K_m$ for ammonia (approximately 1 mM) and is therefore relatively inactive when the ammonia concentrations fall below 1 mM (Section 9.3.1, Fig. 9.16). The consequence of all this is that when the ammonia concentrations sharply decrease, there occurs a drop in the intracellular concentration of glutamine.] When the glutamine levels drop, the concentration of $\alpha$-ketoglutarate increases because it is the glutamine that donates an amino group to $\alpha$-ketoglutarate via the glutamate synthase reaction when the ammonia concentrations are low (Fig. 9.16). The increased amounts of $\alpha$-ketoglutarate and decreased amounts of glutamine lower the concentration $P_{II}$ by stimulating its conversion to $P_{II}$–UMP. [The $\alpha$-ketoglutarate stimulates the transferase activity of UT/UR, which adds UMP to $P_{II}$, converting it to $P_{II}$–UMP and glutamine inhibits this activity. Thus the combination of increased $\alpha$-ketoglutarate and decreased glutamine stimulates the enzyme.] The decrease in $P_{II}$ stimulates transcription of the Ntr regulon.

As stated earlier, $P_{II}$ influences transcription indirectly, as shown in Fig. 18.11. The reason why increasing the amount of $P_{II}$ inhibits transcription of the Ntr regulon is that $P_{II}$ activates the phosphatase activity of the histidine kinase, $NR_{II}$, and the phosphatase activity of $NR_{II}$ inactivates the response regulator (positive transcription factor), $NR_I$–P. When the the levels of $P_{II}$ decrease, $NR_{II}$ acts as a kinase rather than as a phosphatase and phosphorylates $NR_I$, which then activates transcription. Thus the sequence of events is as follows: (1) Low ammonia results

**A**

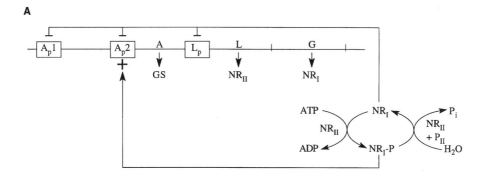

**B**

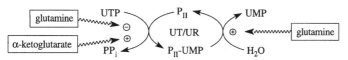

**Fig. 18.11** Model for transcriptional regulation of *gln ALG*. The protein products of the operon are glutamine synthetase (GS), a histidine protein kinase/phosphatase (NR$_{II}$), and a response regulator (NR$_I$). (A) Under ammonia-limiting conditions, NR$_{II}$ phosphorylates NR$_I$, forming NR$_I$-P. The increased levels of NR$_I$-P stimulate a large increase in transcription rate from Ap2, which uses sigma 54 RNA polymerase. As the levels of NR$_I$-P increase, other Ntr operons are also stimulated. Under conditions of excess ammonia, NR$_{II}$ is stimulated by P$_{II}$ to dephosphorylate NR$_I$-P. As a consequence, transcription from Ap$_2$ stops and sigma 70 RNA polymerase transcribes from promoters Ap1 and Lp at a low rate. Transcription from these promoters is further regulated by repression by NR$_I$. (B) The levels of P$_{II}$ are increased by glutamine (signaling high ammonia) and decreased by $\alpha$-ketoglutarate (signaling low ammonia). A bifunctional enzyme, uridylyl transferase/uridydyl removal (UT/UR), is stimulated by glutamine to remove UMP from P$_{II}$–UMP and stimulated by $\alpha$-ketoglutarate to add UMP to P$_{II}$. The enzyme adds a total of four UMP moieties to each of 4 subunits of P$_{II}$. Thus high glutamine (high ammonia) increases the levels of P$_{II}$, which stimulates the dephosphorylation of NR$_I$-P and represses transcription, whereas high $\alpha$-ketoglutarate (low ammonia) stimulates the addition of UMP to P$_{II}$, thus lowering the levels of P$_{II}$, and activates transcription.

in an increase in the concentrations of $\alpha$-ketoglutarate because its conversion to glutamine and glutamate is slowed; (2) the $\alpha$-ketoglutarate lowers the concentration of P$_{II}$ by stimulating its conversion to P$_{II}$–UMP; (3) when the P$_{II}$ levels fall, NR$_{II}$ no longer acts as a phosphatase but rather as a kinase and phosphorylates NR$_I$; and (4) NR$_I$-P stimulates transcription. When ammonia levels are high, then transcription is repressed because glutamine stimulates the conversion of P$_{II}$–UMP to P$_{II}$, and P$_{II}$ stimulates the phosphatase activity of NR$_{II}$, which results in a decrease in NR$_I$-P. These events are described in more detail next.

*The more detailed explanation*

The central operon involved is called the *glnALG* operon, which encodes three genes (Fig. 18.11A). They are: (1) *glnA*, the gene for glutamine synthetase; (2) *glnL*, the gene for NR$_{II}$, which is a bifunctional histidine kinase/phosphatase whose substrate is NR$_I$; and (3) *glnG*, the gene for NR$_I$, a response regulator. Under nitrogen excess (high ammonia), the operon is transcribed at a low level from the *glnAp$_1$* and *glnLp* promoters by sigma 70 RNA polymerase. The small amount of transcription from *glnAp1* is sufficient to guarantee the synthesis of enough glutamine synthetase to meet the cell's needs for glutamine when the ammonia concentrations are high. Part of the reason that only a small amount of transcription takes place from these promoters is that transcription from them is repressed by $NR_I$ (Fig. 18.11A).[124] Under nitrogen-limiting conditions, the *glnALG* operon is transcribed at a high frequency from the *glnAp₂* promoter by the sigma 54 RNA polymerase. (See Note 125 for a discussion of the sigma 54 polymerase and more on transcriptional regulation.) Transcription from the *glnp₂* promoter requires the phosphorylated form of the response regulator $NR_1$ (i.e., $NR_1$–P), reflecting the requirement of sigma 54 RNA polymerases for the binding of a transcription regulator upstream to the promoter in order to form an open complex. Since $NR_I$–P binds more than 100 bp upstream of the promote region, the DNA must bend in order for $NR_I$–P to make contact with the RNA polymerase (Fig. 18.12A). (See Note 126 for a model of how $NR_1$–P stimulates transcription.) The protein kinase, $NR_{II}$, is itself regulated by another protein, called $P_{II}$ (product of the *glnB* gene). When $P_{II}$ levels are high, $NR_{II}$ acts like a phosphatase rather than a kinase, and the levels of $NR_I$–P drop. Hence $P_{II}$ inhibits transcription of the *glnALG* operon. Under conditions of excess ammonia, $P_{II}$ is generated from $P_{II}$–UMP because glutamine stimulates the removal of UMP from $P_{II}$–UMP catalyzed by the bifunctional enzyme uridylyl transferase–uridylyl removal (UT/UR) (product of the *glnD* gene) (Fig. 18.11B). When ammonia levels fall, the $P_{II}$ is converted to $P_{II}$–UMP in a reaction catalyzed by UT/UR and stimulated by $\alpha$-ketoglutarate (and inhibited by glutamine). Under these circumstances, $NR_{II}$ acts as a kinase rather than as a phosphatase, and transcription is stimulated. In summary then:

1. High ammonia leads to high $P_{II}$ because glutamine stimulates the removal of UMP from $P_{II}$–UMP.

2. High $P_{II}$ leads to phosphatase activity of $NR_{II}$.

3. Phosphatase activity of $NR_{II}$ leads to dephosphorylation of $NR_I$–P.

4. Dephosphorylation of $NR_I$–P leads to a lowering of transcription of the *glnALG* operon, since $NR_1$–P is a positive transcription factor.

Thus high ammonia leads to the dephosphorylation of the response regulator, hence repression of the *glnALG* operon. As we shall see later, a similar situation exists for the regulation of the PHO operon. In this case a high concentration of inorganic phosphate leads to the dephosphorylation of the response regulator and a consequent repression of the PHO operon (Section 18.9).

A key enzyme in the signal transduction pathway is the bifunctional enzyme UT/UR that either adds or removes UMP from $P_{II}$. It can be considered a sensor protein that responds to the $\alpha$-ketoglutarate/glutamine ratio and modifies $P_{II}$, which in turn regulates the activity of the histidine kinase/phosphatase ($NR_{II}$). Alternatively, one might consider glutamine synthetase to be the sensor because it responds to the ammonia supply and determines the levels of glutamine and $\alpha$-ketoglutarate, which signal the UT/UR.

## Regulation of the other operons in the Ntr regulon

It has been estimated that the level of $NR_I$ in ammonia-repressed cells is about five molecules per cell, which when phosphorylated under conditions of ammonia starvation will activate transcription of *glnALG*. When the levels of $NR_I$–P become sufficiently high (around 70 molecules per cell) due to increased transcription of *glnALG*, it also activates transcription of the other operons in the Ntr regulon system.

The stimulation of some of the Ntr operons [e.g., *glnALG* and operons encoding proteins for the uptake of glutamine (*glnHPQ*) in *E. coli*, as well as histidine (*hisJQMP*) and the gene for arginine uptake (*argT*) in *S. typhimurium*) by $NR_I$–P is direct. However, in other instances, $NR_I$–P activates the transcription of a gene encoding a second positive regulator, which activates transcription of the target promoters. For example, in *Klebsiella pneumoniae*, $NR_1$–P activates the transcription of *nifAL*. (See Ref. 127 for a review.) The product of

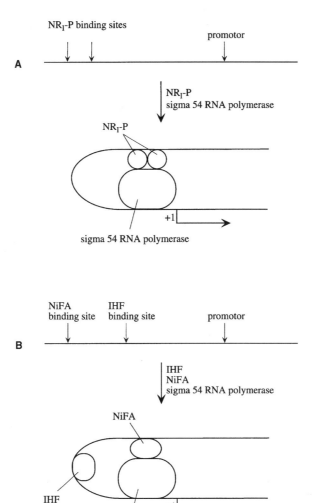

**Fig. 18.12** Activation by NR$_I$-P, and NifA plus IHF. Promoters read by sigma 54 polymerase require an activator that binds at an upstream site, sometimes over 100 bp away from the promoter. One model for how the activator contacts the polymerase is that the DNA bends. (A) Activation of *glnA* promoter by NR$_I$-P. Because the binding site for NR$_I$-P has an inverted repeat, each one capable of binding NR$_I$-P, it has been proposed that two NR$_I$-P molecules bind. (B) Activation of promoter by NifA. It is known that IHF is necessary for the transcriptional activation by NifA of most NifA-dependent genes. The model supposes that IHF binds at a site between the NifA binding site and the polymerase binding site and bends the DNA, bringing the activator to the RNA polymerase. See Magasanik, B. 1996. Regulation of nitrogen utilization, pp. 1344–1356. In *Escherichia coli and Salmonella, Cellular and Molecular Biology,* Vol. 1. F. C. Neidhardt et al. (Eds.). ASM Press, Washington, D.C.

*nifA* is a positive regulator of the *nif* genes (the nitrogen-fixation gene cluster in *K. pneumoniae*). NifA binds to upstream activator sequences (UAS) and makes contact with RNA polymerase, which is bound downstream. The contact between NifA and the polymerase is facilitated by a bending of the intervening DNA, as described later. (See Section 18.8.2, on the regulatory role of IHF, integration host factor, and Fig. 18.12B.) [In *K. pneumoniae*, the product of *nifL* inactivates the *nifA* product in response to oxygen and ammonia, thus making certain that the *nif* genes (except for *nifAL*) are expressed only during anaerobiosis and ammonia starvation.]

Transcription of the *nif* genes also relies on sigma 54. In *Klebsiella* species, but not *Salmonella typhimurium*, P–NR₁ also activates the transcription of *nac*, which encodes a positive regulator of the *hut* operons that encode enzymes for the catabolism of histidine as a source of nitrogen. The *hut* genes use a sigma 70 RNA polymerase (the major polymerase), rather than the sigma 54 polymerase.

### 18.8.2 Regulatory role of IHF

The student should review the discussion of the nucleoid in Section 1.2.6, where the DNA-binding protein IHF is described. See Ref. 128 for a review of IHF. IHF is a protein that is known to bend DNA. (Any bent DNA, including DNA to which IHF is bound in vitro, can be analyzed via gel electrophoresis because bent DNA fragments travel at different rates.) IHF activates certain genes and represses others. In some genes it binds at or near the promoter region and is suggested to interfere with the binding or activity of RNA polymerase. It also binds to specific locations on the DNA between the promotors and NifA-binding sites of genes activated by NifA (with the exception of one NifA activated gene, i.e., *nifB*). IHF is also required for the activation of one of the NR₁-dependent promoters (i.e., *glnHp2*), but not of the others. The model is that IHF

bends the DNA, bringing bound activator such as NifA or NR₁–P to the promoter where the activator stimulates transcription (Fig. 18.12B). (Reviewed in Ref. 123.) See Section 18.3 and 18.10.1 for other examples of IHF regulatory activity.

### 18.8.3 Effect of $P_{II}$–UMP and $P_{II}$ on glutamine synthetase activity

When ammonia levels are low, not only is transcription of the gene for glutamine synthetase stimulated, but the activity of the existing enzyme is also stimulated, ensuring that the low ammonia levels do not preclude incorporation of ammonia into glutamine and glutamic acid. The stimulation of enzyme activity is due to $P_{II}$–UMP, which is made when ammonia levels are low. However, the stimulation by $P_{II}$–UMP is indirect. To understand how $P_{II}$–UMP does this, we have to discuss how glutamine synthetase activity is regulated in the first place. When ammonia concentrations are high, glutamine synthetase activity is low because it has been adenylylated by an enzyme called adenylyl transferase (AT), which itself is activated by $P_{II}$, whose levels are high when ammonia levels are high (Fig. 18.13). (The adenylylated enzyme is further inactivated because it is susceptible to feedback inhibition by a variety of nitrogenous compounds that depend upon glutamine for their synthesis, e.g., glucosamine-6-phosphate, carbamoyl phosphate, CTP, AMP, tryptophan, and histidine.) However, adenylyl

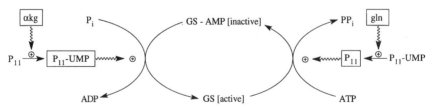

**Fig. 18.13** Adenylylation of glutamine synthetase regulates its activity. When glutamine (gln) levels are high (high ammonia), the enzyme is inactivated by adenylylation catalyzed by the bifunctional enzyme adenylyl transferasel/adenylyl removal (AT/AR), which is stimulated by $P_{II}$, itself produced from $P_{II}$–UMP in the presence of high glutamine. When the α-ketoglutarate (akg) levels are high (low ammonia), the glutamine synthetase is activated by deadenylylation via AR/AT stimulated by $P_{II}$(UMP), which is produced from $P_{II}$ in the presence of high α-ketoglutarate. Thus the ratio gln/akg regulates the activity of glutamine synthetase by changing the ratio $P_{II}$/$P_{II}$–UMP. A total of 12 AMP moieties can be added to each of 12 subunits of glutamine synthetase.

transferase is a bifunctional enzyme, and in the presence of $P_{II}$–UMP, it removes the AMP via a phosphorolysis reaction. In summary, then, low ammonia stimulates glutamine synthetase by increasing the levels of $P_{II}$–UMP (made by transferring UMP from UTP to $P_{II}$), which results in the deadenylylation of the enzyme, and high ammonia inhibits glutamine synthetase by increasing the levels of $P_{II}$ (made by removing UMP from $P_{II}$–UMP), which leads to adenylylation of the enzyme. Note that just as in the regulation of transcription, $P_{II}$ affects a separate enzyme that activates or inactivates a component. For example, in transcription, $P_{II}$ stimulates the phosphatase activity of $NR_{II}$, which results in the inactivation of the transcription regulator, $NR_I$. In the regulation of glutamine synthetase activity, $P_{II}$ activates AT, which in turn inactivates glutamine synthetase, whereas $P_{II}$–UMP activates the phosphorolysis, which in turn leads to activation of glutamine synthetase. $P_{II}$ thus turns out to be an important protein that signals the ammonia levels to both the transcription machinery and glutamine synthetase. Table 18.2 summarizes an overview of the regulation of transcription and glutamine synthetase activities by ammonia and $P_{II}$. While the post-translational control of glutamine synthetase by adenylylation occurs in *E. coli* and related bacteria, as well as in species of *Vibrio*, *Thiobacillus*, and *Streptomyces*, it does not occur in all bacteria (e.g., *Bacillus* and *Clostridium*). (Reviewed in Ref. 121.) This is another example of why one should use caution in extrapolating results from some bacteria to all the bacteria, especially with regards to the regulation of metabolism.

**Table 18.2** Regulation of transcription of *glnALG* and glutamine synthetase

| $NH_3$ | $P_{II}/P_{II}$–UMP | Transcription[a] | Glutamine[b] synthetase |
| --- | --- | --- | --- |
| High | High | Low | Inactive |
| Low | Low | High | Active |

[a]Indirectly repressed by $P_{II}$.
[b]Indirectly inactivated by $P_{II}$. Indirectly activated by $P_{II}$–UMP.

## 18.9 Response to Inorganic Phosphate Supply: The Pho Regulon

Bacteria have evolved a signaling system to induce the formation of phosphate assimilation pathways when the supply of phosphate becomes limiting. The following discussion pertains specifically to *E. coli* and related bacteria, although homologs of the histidine kinase (PhoR) and response regulator (PhoB) have been found in other bacteria, including *Bacillus*, *Pseudomonas*, *Rhizobium*, *Agrobacterium*, and *Synechococcus*.[129] When inorganic phosphate is in excess, it is transported into the cell by a low-affinity transporter called Pit. (See Ref. 130 for a comprehensive review of phosphate transport.) Pit consists of a single transmembrane protein that is driven by the $\Delta p$. (See the discussion of secondary transport in Section 16.3.1.) Under low inorganic phosphate conditions, *E. coli* stimulates the transcription of at least 38 genes (most of them in operons) involved in phosphate assimilation, including: genes encoding a periplasmic alkaline phosphatase that can generate phosphate from organic phosphate esters (*phoA*); an outer membrane porin channel for anions, including phosphate (*phoE*); a high-affinity inner membrane phosphate uptake system called the Pst system and a protein required for phosphate repression (*pstSCAB–phoU*); a histidine kinase and response regulator (*phoBR*), 14 genes for phosphonate[131] uptake and breakdown (*phnCDEFGHIJKLMNOP*); genes for glyceraldehyde-3-phosphate uptake; and a gene encoding a phosphodiesterase that hydrolyzes glycerophosphoryl diesters (deacylated phospholipids) (*ugpBAECQ*). All these genes (except for *pit*) are repressed by phosphate and are in the PHO regulon. The PHO regulon is controlled by PhoR, which appears to be a histidine kinase, and a response regulator, PhoB. The phosphorylated form of PhoB (i.e., PhoB-P) activates the transcription of the genes in the PHO regulon. The *pho* promoters use the sigma 70 RNA polymerase (encoded by *rpoD*). (See Note 132 for a description of how PhoR activates transcription.)

### 18.9.1 The signal transduction pathway

A model for the regulation of the PHO regulon in *E. coli* by $P_i$ is shown in Fig. 18.14.[133, 134] The components required for regulation are: (1) PstS, a periplasmic $P_i$ binding protein; (2) PstABC, integral membrane proteins required for $P_i$ uptake (PstB is the permease); (3) PhoU; (4) PhoR; and (5) PhoB. PhoR detects $P_i$. However, it is not known whether $P_i$ interacts with PhoR indirectly by binding to the PST system or directly by binding to PhoR. What is known is that under conditions of $P_i$ excess (about 4 $\mu$M) repression of the PHO regulon requires the PST system, PhoU, and a form of PhoR called PhoR$^R$. When $P_i$ becomes limiting, the PHO regulon is induced, and this requires a form of PhoR called PhoR$^A$ as well as PhoB, which is the response regulator. One model proposes that $P_i$ binds to PstS and that the $P_i$-bound PstS binds to PstABC, which is a member of the ABC transporter family. (See Note 135 for further comment on PST and its relationship to the ABC-type transporters.)

A "repressor complex" is postulated that consists of $P_i$ bound to PstS, PstABC, PhoU, and PhoR. PhoR is thought to be a histidine kinase/phosphatase bifunctional enzyme. (See Note 136.) In the "repressor complex" PhoR is suggested to be a monomer with phosphatase activity and to maintain PhoB in its dephosphorylated (inactive) state. (This would be analogous to the Ntr regulon, where $P_{II}$ stimulates the phosphatase activity of NR$_{II}$. Section 18.8.1.) The model further proposes that, when phosphate becomes limiting ($< 4\mu$M), PhoR is released from the "repressor complex" and functions as a histidine kinase, which is depicted in Fig. 18.14 as a dimer, although that has not been demonstrated. The dimer form of PhoR autophosphorylates and activates PhoB by transferring to it the phosphoryl group. In agreement with this model, Pst and PhoU are required for repression of the PHO regulon, but not for its activation. However, it must be added that the interaction of the Pst system and PhoU with PhoR to cause $P_i$ repression is not really understood.

In mutants that do not have PhoR, the Pho regulon is expressed constitutively; that is, the genes are not repressed by inorganic phosphate. This is because PhoB is always phosphorylated. This may be due to the postulated phenomenon of *cross-regulation*, which is a situation where a response regulator of one two-component system might be controlled by a different regulatory system. such as a related histidine kinase or by acetyl phosphate, whose levels may reflect growth conditions (Section 18.12.[134]) In this case the different regulatory systems that may phosphorylate PhoB are CreC (formerly called PhoM), which is induced by growth on glucose and is not regulated by phosphate, and the phosphoryl donor acetyl phosphate produced during growth on pyruvate.[134] The evidence for acetyl phosphate as a global signal for response regulator proteins is discussed in Section 18.12.

## 18.10 Response to Osmotic Pressure and Temperature: Regulation of Porin Synthesis

When growing in higher osmolarity or high temperature, *E. coli* increases the synthesis of the slightly smaller porin channel, OmpC, relative to the larger OmpF channel. For example, when *E. coli* is grown at 30°C and in the presence of 1% NaCl (which is the temperature and approximate osmolarity of the intestine), only OmpC is made.[137]

When growing at lower temperatures and at osmolarities that approximate the conditions in lakes and streams, OmpF is preferentially made. The changes in porin composition of the outer membrane do not change the intracellular osmotic pressure, and therefore are not part of a homeostatic response to osmotic pressure changes. One can rationalize the change in porins by assuming that the smaller OmpC channel is advantageous in the intestinal tract because it may retard the inflow of toxic substances such as bile salts, whereas in lower osmolarity environments, such as would be experienced outside the body, the larger OmpF channel might be an advantage to increase inward diffusion of dilute nutrients.

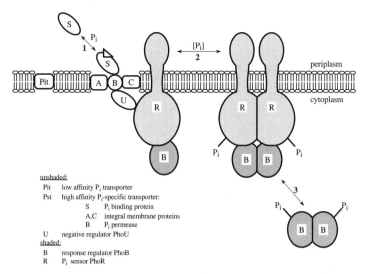

**Fig. 18.14** Model for the regulation of the PHO regulon. Inorganic phosphate in the periplasm binds to the phosphate binding protein, PstS. The $P_i$-PstS binds to the phosphate transporter, PstABC, in the cell membrane. A "repressor complex" forms between $P_i$-PstS, PstABC, PhoU, and PhoR. PhoR in the repressor complex acts as a phosphatase and maintains PhoB in the dephosphorylated state. When $P_i$ becomes limiting, PhoR is released from the repressor complex, autophosphorylates, and then phosphorylates PhoB. Phosphorylated PhoB is a positive transcription regulator for the PHO regulon. PhoR as a phosphatase is depicted as a monomer, and PhoR as a kinase is drawn as a dimer, although this has not been demonstrated. Unshaded: Pit, low-affinity $P_i$ transporter; Pst, high-affinity $P_i$-specific transporters: S, $P_i$ binding protein; A, C, integral membrane proteins; B, $P_i$ permease; U, negative regulator PhoU. Shaded: B, response regulator PhoB; R, $P_i$ sensor PhoR. *Source:* From Wanner, B. L. 1993. Gene regulation by phosphate in enteric bacteria. *J. Cellular Biochem.* **51**:47–54.

## 18.10.1 Porin synthesis is regulated by a two-component system

EnvZ is an inner membrane histidine protein kinase that has been postulated to function also as an osmotic sensor.[138] However, there is very little known concerning the signals to which EnvZ responds when it phosphorylates or dephosphorylates OmpR. EnvZ is a transmembrane protein, part of which is exposed to the periplasm (a loop near the N-terminal end) and part to the cytoplasm (the C-terminal domain). It is thought that the domain that senses osmolarity signals resides either in the periplasm or in the membrane. The response regulator is a cytoplasmic protein called OmpR. One model proposes that increased external osmolarity activates EnvZ so that it phosphorylates OmpR (Fig. 18.15). [However, mutations in EnvZ result in a high-osmolarity phenotype (constitutive OmpF⁻OmpC⁺ phenotype), indicating that perhaps EnvZ detects a signal in low-

osmolarity media rather than high-osmolarity media. The putative signal in low-osmolarity media would prevent the phosphorylation of OmpR by EnvZ.[139]] The model further proposes that high levels of P-OmpR repress *ompF* and stimulate *ompC* by binding to low-affinity sites upstream from the respective promoters, and that low levels of P-OmpR stimulate transcription of *ompF* by binding to high-affinity sites upstream from the *ompF* promoter. Thus, when the external osmotic pressure is raised, the levels of OmpR-P should increase and repress *ompF* while stimulating *ompC*. Several lines of evidence support the model[140–142]: (1) it has been demonstrated in vitro that EnvZ can accept a phosphoryl group from ATP and in turn transfer it to OmpR forming OmpR-P; (2) it is also known that when *E. coli* is shifted to a medium of high osmolarity, the levels of OmpR-P increase; (3) in a mutant of *E. coli* that has elevated levels of OmpR-P, *ompF* is repressed and *ompC* is expressed all the time (constitutively);

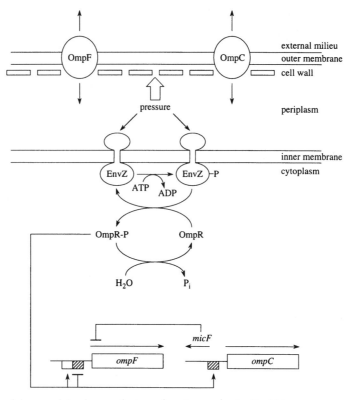

**Fig. 18.15** A model to explain the regulation of porin synthesis. EnvZ is a transmembrane histidine kinase/phosphatase whose substrate is the cytoplasmic response regulator OmpR. To explain how a shift to high-osmolarity media represses *ompF* and stimulates *ompC*, it is assumed that high osmotic pressures activate (not necessarily directly) the kinase activity of EnvZ. The resulting high levels of OmpR-P repress transcription of *ompF* and stimulate transcription of *ompC*. OmpR-P can effect these changes by binding to DNA sequences upstream of the respective promoters. A third gene, *micF*, is transcribed from the *ompC* promoter but in the opposite direction. Its transcription is also activated by OmpR-P. The gene *micF* codes for an RNA transcript that is complementary to the 5′ end of the *ompF* mRNA, i.e., the region where protein translation is initiated, and blocks translation. *MicF* RNA is therefore "antisense" RNA because it binds to the *ompF* RNA transcript that is translated into protein, i.e., the "sense" RNA. It may be that an additional control of *ompF* expression is the inhibition of translation of *ompF* mRNA by *micF* RNA. When the external osmotic pressures are lower, the concentrations of OmpR-P are too low to bind to the DNA sites for repression of *ompF* and stimulation of *ompC*. However, low levels of OmpR-P can stimulate the transcription of *ompF* because of high-affinity binding sites for OmpR-P upstream of the *ompF* promoter. *Source:* From Stock, J. B., A. J. Ninfa, and A. M. Stock. 1989. Protein phosphorylation and regulation of adaptive responses in bacteria. *Microbiol. Rev.* 53:450–490.

and (4) OmpR-P stimulates transcription from *ompF* and *ompC* promoters.

### Regulation of transcription by IHF and of translation by micF RNA

*ompf* and *ompC* are also repressed by integration host factor. IHF binds to nucleotide sequences in the promoter regions of the *ompF* and *ompC* genes in *Escherichia coli* and represses transcription. (See Section 18.8.2 for a discussion of IHF.) This was demonstrated in vitro for *ompF*, where it was also shown that OmpR reverses the inhibition, suggesting that IHF might somehow interfere with the activity of OmpR at the promoter site.[143]

There appears to be an additional mode of inhibition of *ompF* expression, particularly at levels of osmolarity that are not very high. Another regulatory gene, *micF*, is also stimulated by high concentrations of

OmpR-P and produces an RNA molecule that inhibits *ompF* mRNA translation by binding to the mRNA (Fig. 18.15).[144] Support for this conclusion derives from the observation that when multiple copies of *micF* are introduced on a high-copy-number plasmid, there is inhibition of in vivo synthesis of OmpF. Surprisingly, however, deletion of the *micF* gene does not result in any decrease in the synthesis of OmpF under high-osmolarity conditions (10–20% sucrose). One must therefore conclude that *micF* RNA plays a minor role compared to OmpR in osmoregulation in high-osmolarity media. However, *micF* does appear to play a major role in osmoregulation at low and medium levels of osmolarity (up to 6% sucrose), where one might expect a smaller activation of OmpR. This conclusion follows from the finding that a mutant lacking a functional gene for micRNA failed to reduce its levels of OmpF when the cells were grown in media containing 6% sucrose, whereas the wild-type cells did make less OmpF.[145] (MicF is an example of an antisense RNA. Other examples of antisense RNAs are known that function in regulating gene activity.[146])

## 18.11 Response to Potassium Ion: Regulation of the kdpFABC Operon

When *E. coli* is placed in a medium of high osmolarity, it responds by synthesizing a $K^+$ uptake system, called the KdpABC transporter, as long as the external $K^+$ concentrations are not too high. This is discussed in Section 16.3.3. The role of $K^+$ uptake in pH and osmotic homeostasis is discussed in Sections 15.1.3 and 15.2.2. There are two regulatory genes for the *kdpABC* operon, *kdpD* and *kdpE*, both of which are required for expression of the *kdpABC* genes. The model is that the KdpD protein, which is located in the inner (cell) membrane, senses a change in turgor pressure, and transmits the signal to KdpE, which activates the transcription of the *kdpABC* operon. (However, see Sections 15.2.4 and 15.2.5 for a discussion of whether the pressure differential due to an osmotic

gradient is across the cell membrane.) The KdpD and KdpE proteins are part of a two-component regulatory system, with KdpD being a sensory protein kinase that autophosphorylates, and KdpE being a response regulator protein that is phosphorylated by KdpD. The phosphorylated form of KdpD is thought to activate the *kdpABC* promoter.[147–149]

## 18.12 Acetyl Phosphate as a Possible Global Signal in Certain Two-Component Systems

It has been proposed that in the absence of the cognate HK proteins, that is, in mutants lacking specific histidine kinases, acetyl phosphate appears to be able to donate the phosphoryl group to the cognate response regulator protein, thus counteracting the effect of the mutation. (Reviewed in Ref. 150.) It must be emphasized that the extent to which acetyl phosphate does this in vivo, especially in wild-type cells, is not known. Nevertheless, it is an interesting phenomenon and potentially very significant. The following discussion pertains to the role that acetyl phosphate might play as a global regulator and the methods used to investigate this role.

### Changing the intracellular levels of acetyl phosphate to investigate its possible role

In order to study the role that acetyl phosphate plays in regulation in vivo, it is necessary that the investigator be able to change its intracellular concentrations. Mutants can be used for this purpose. Mutants lacking phosphotransacetylase (encoded by the *pta* gene) or acetate kinase (encoded by the *ackA* gene) have higher levels of acetyl phosphate, whereas mutants lacking both enzymes have lower levels of acetyl phosphate. To understand why this is so, we must first recall how *E. coli* makes acetyl phosphate. (See Section 7.3.2 for a review.) *Phosphotransacetylase* reversibly converts acetyl–CoA to acetyl phosphate. *Acetate kinase* makes acetyl phosphate directly from acetate and ATP. This is also a reversible reaction:

phosphotransacetylase: acetyl–CoASH + $P_i$

$\longrightarrow$ acetyl-P + CoASH

acetate kinase : acetate + ATP

$\longrightarrow$ acetyl-P + ADP

Mutants in *ackA* (acetate kinase gene) have increased levels of acetyl phosphate when growing on carbon sources such as glucose or pyruvate that yield acetyl–CoA (via pyruvate dehydrogenase), since they can form acetyl phosphate from acetyl–CoA using the phosphotransacetylase but cannot convert the acetyl phosphate to acetate without the acetate kinase. The same result might be achieved by growing *pta* (phosphotransacetylase) mutants on acetate, since these should be able to synthesize acetyl phosphate using the acetate kinase but not convert it to acetyl–CoA. On the other hand, mutants lacking *both pta* and *ackA* genes cannot make acetyl phosphate from pyruvate or acetate. Such mutants show the appropriate regulation expected if acetyl phosphate were able to phosphorylate response regulator proteins in mutants lacking the cognate histidine kinases.

The choice of carbon source also influences the levels of acetyl phosphate. Acetyl phosphate pools are lowest in cells grown on glycerol (< 0.04 mM), higher in cells grown on glucose (0.3 mM), and highest in cells grown on acetate (1.5 mM).[150]

Sometimes increasing the copy number of the gene that codes for acetate kinase (*ackA*) can indicate whether acetyl phosphate can phosphorylate the regulatory protein. For example, one of the genes of the Pho regulon, *phoA* (codes for alkaline phosphatase), could be expressed in the absence of the cognate HK proteins (mutants lacking both PhoR and CreC) by increasing the copy number of *ackA*.

## In vitro phosphorylation of a response regulator protein by acetyl phosphate

In some instances it has been possible to demonstrate in vitro phosphorylation of a response regulator protein by acetyl phosphate. It has been reported that $NR_I$ (the response regulator protein that regulates transcription of the Ntr regulon) could be phosphorylated by acetyl phosphate in vitro.[151]

Mutants lacking $NR_{II}$, the histidine kinase and phosphatase for the Ntr regulon, have residual regulation of expression of *glnA* (codes for glutamine synthetase), which is part of the Ntr regulon. This suggests the presence of an alternative donor of the phosporyl group for the response regulator, $NR_I$. The alternative donor may be acetyl phosphate.

## A regulatory role for OmpR and acetyl phosphate in flagellar biosynthesis in E. coli

In addition to regulating transcription of the porin genes (Section 18.10.1), OmpR also regulates the transcription of other genes, including flagellar genes. When *E. coli* is grown under stressful conditions (e.g., high osmolarity or elevated temperatures), flagellar synthesis is inhibited. The repression of the flagellar genes by high osmolarity is apparently due to a repressive effect of OmpR-P, which is known to regulate other genes in addition to OmpF and OmpC. Under conditions of high osmolarity, OmpR is phosphorylated by the sensor kinase EnvZ. However, acetyl phosphate can also phosphorylate OmpR, even in wild-type cells, and lead to the repression of flagellar genes.[152] In order to explain how this can be measured, it is necessary to introduce the master operon for flagellar genes and how the transcription of the operon can be monitored. The master operon for flagellar biosynthesis is *flhDC*, which codes for two DNA-binding proteins, FlhD and FlhC, that are required for transcription of all the genes in the flagellar regulon. The use of *flhDC'–lacZ* fusions can be used to monitor the activity of the *flhDC* promoter by doing β-galactosidase assays. (See Note 25 for a description of *lacZ* fusions.) It was shown that the expression of *flhDC* is substantially *increased* in *pta–ack* double mutants or *pta* mutants when compared to wild-type cells. These mutants can convert the carbon source to acetyl–CoA, but in the absence of phosphotransacetylase (encoded by *pta*) they cannot convert the acetyl–CoA to acetyl phosphate. Hence the conclusion that in wild-type cells acetyl phosphate represses the operon.

On the other hand, expression of the *flhDC* operon was *decreased* in *ack* mutants. The *ack* gene encodes acetate kinase, and in the absence of this enzyme the cells make acetyl phosphate from acetyl–CoA using the phosphotransacetylase but cannot metabolize the acetyl phosphate further in the absence of the acetate kinase. Hence the pool size of acetyl phosphate increases. All of this is consistent with the hypothesis that acetyl phosphate represses the *flhDC* operon.[152]

The involvement of OmpR in the repression of *flhDC'–lacZ* by acetyl phosphate was shown by using *ompR⁻* strains. *ompr⁻* mutants showed an enhanced expression of *flhDC'–lacZ*, indicating that OmpR is a negative regulator of *flhDC'–lacZ*, and this enhanced expression was not reduced in *ackA* mutants that have higher levels of acetyl phosphate, indicating that repression by acetyl phosphate requires OmpR. Acetyl phosphate has been shown to phosphorylate OmpR in vitro, and presumably it also does so in vivo. Since OmpR is phosphorylated by EnvZ when cells are grown at high osmolarities (see Section 18.10.1), it might be expected that *flhDC* expression would be decreased in bacteria grown at higher osmolarities, and this in fact is the case in wild-type cells but not in *ompR⁻* cells. Finally, it was demonstrated by a DNA mobility shift assay that phosphorylated OmpR binds to the *flhDC* promotor. The student should recall that in other instances where it was inferred that acetyl phosphate caused the phosphorylation of a response regulator in vivo, the effects were seen in mutants lacking the cognate histidine kinase. However, acetyl phosphate appears to phosphorylate OmpR even in cells that have the kinase, which is EnvZ.

It has been known that if *E. coli* is grown at elevated temperatures, it produces fewer flagella. This can possibly be explained by the repression of the flagellar operon by acetyl phosphate. The acetyl phosphate levels in cells grown at higher temperatures is elevated because the acetate kinase is less active. Therefore, the inhibition of flagellar synthesis at higher temperatures might be due to the increased levels of acetyl phosphate, which may phosphorylate OmpR, which may then repress the activity of *flhDC*.

### Acetyl phosphate can phosphorylate CheY

It has been suggested that acetyl phosphate can donate its phosphoryl group in vivo to CheY in the absence of the cognate histidine kinase (CheA). (See Section 18.4.5 for a discussion of the roles of the Che proteins in chemotaxis in *E. coli*.) The initial observation was that mutants of *E. coli* that do not have any of the cytoplasmic chemotaxis proteins except CheY swim smoothly except when incubated with acetate. Then they tumble. The response to acetate requires acetate kinase, the enzyme that synthesizes acetyl phosphate from acetate and ATP. This suggests that acetyl phosphate can phosphorylate CheY, and indeed this has been demonstrated in vitro. (Discussed in Ref. 154.) Acetate has no effect on chemotaxis to aspartate or serine in wild-type cells, indicating that the levels of CheY-P are not significantly influenced by acetyl phosphate in wild-type cells and are regulated entirely by CheA and CheZ in response to the appropriate chemoeffectors.[153] Thus the physiological role of acetate regarding CheY phosphorylation and chemotaxis in general is not clear at all. Perhaps it plays a subtle regulatory role in chemotaxis, somehow connecting chemotaxis to metabolism. Recently, it has been reported that acetyl–AMP can acetylate CheY with results on flagellar rotation similar to the phosphorylation of CheY, and therefore it appears that there are two pathways for stimulating CheY by acetate in *E. coli*.[154]

## 18.13 Response to Carbon Source: Catabolite Repression

Catabolite repression refers to the preferential use of one carbon source for growth over another when bacteria are grown in the presence of both carbon sources. It results in diauxic growth, as described in Section 2.2.4. There is more than one mechanism responsible for catabolite repression. One mechanism is discussed in Section 16.3.4, which describes catabolite repression by glucose as a consequence of glucose transport by the phosphotransferase system (PTS) in *E. coli* and related bacteria. The mechanism is that the uptake of glucose by the PTS system lowers the cAMP pools, and this slows the transcription of cAMP-dependent genes required to metabolize alternative carbon sources. Section 16.3.4 also points out that glucose uptake by the PTS system also inhibits permeases required for the uptake of other carbohydrate carbon sources (inducer exclusion). However, inducer exclusion by the PTS system and adjustment of cAMP pools are not the only means of catabolite repression. The discussions in Sections 18.13.1 and 18.3.2 describe two more catabolite repression systems. One is a system in *E. coli* that does not involve the PTS system or cAMP (i.e., the Cra system). A second is a different system, the CRE system, in the gram-positive bacterium *Bacillus subtilis*. The following discussion is restricted to catabolite repression by sugars. However, there exist many bacteria that preferentially use carboxylic compounds rather than carbohydrates as a source of carbon and will show diauxic growth with the appropriate mixtures. In these instances it is the carboxylic acid that represses the expression of genes required to metabolize the carbohydrate. Some of these systems have been recently reviewed.[155]

### 18.13.1 Catabolite repression in E. coli that does not involve cAMP or CRP: The Cra system

*E. coli* and *Salmonella* have an additional catabolite repressor system that does not involve cAMP. It has been called the FruR (fructose repressor) system and more recently the Cra (catabolite repressor/activator) system.[156, 157]

#### Cra is a transcriptional regulator

As described in Section 10.2.5, transcriptional regulators exist that induce certain genes and repress others. Cra is such a regulator. It represses genes coding for enzymes required for growth on sugars, that is, enzymes of the Embden–Meyerhoff–Parnas pathway (EMP

pathway or glycolytic pathway) and Entner–Doudoroff pathway (ED pathway), while inducing genes encoding enzymes required for growth on organic acids and amino acids, that is, enzymes of the citric acid cycle, glyoxylate cycle, and gluconeogenic enzymes. The reason for believing that Cra is an activator for genes required for growth on organic acids is that *cra* mutants cannot grow on lactate, pyruvate, and citric acid cycle intermediates and show lower activities of enzymes of the citric acid cycle, glyoxylate cycle, as well as gluconeogenic enzymes.[158] Recall that growth on these carbon sources requires the citric acid cycle, glyoxylate pathway (acetate), and gluconeogenesis (Sections 8.1.1, 8.8.3, and 8.12). The reason for believing that Cra is a repressor of genes required for growth on sugars is that mutations in *cra* lead to higher levels of enzymes required for the uptake and catabolism of sugars such as glucose and fructose. (For more backround information on Cra, see Note 159.)

#### A model for how Cra activates certain genes and represses other

The model proposes that Cra activates certain genes and inhibits others, depending on where it binds to the DNA in relationship to the RNA polymerase binding site. If it sits upstream of the binding site, then it is postulated to be an activator, and if it sits downstream of the binding site, then it is suggested to be a transcription inhibitor (Fig. 18.16). (See the discussion of the regulation of transcription in Section 10.2.5) This model is in fact supported by sequence analysis studies of Cra binding sites, which places these sites downstream of negatively controlled promoters and upstream of positively controlled promoters. (Reviewed in Ref. 160.)

#### Glucose and fructose remove Cra from DNA explaining catabolite repression

When glucose or fructose are in the growth medium, Cra no longer binds to the promoter regions of the target genes. Thus growth on the sugars can occur because the repressor (Cra) for the genes of the EMP and ED pathways is removed. However, the presence of glucose

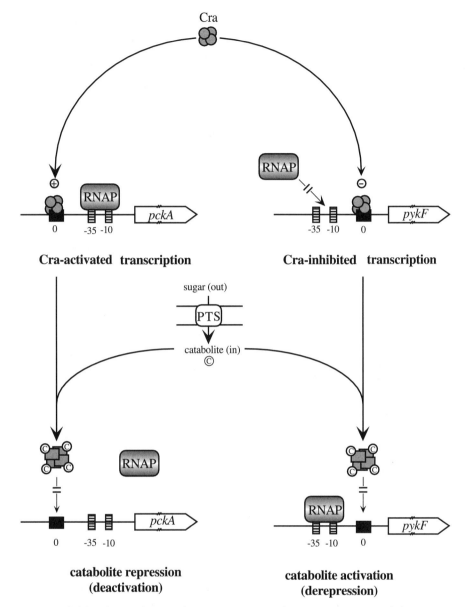

**Fig. 18.16** A model for the regulation of gene transcription by Cra. It is proposed that Cra activates gene transcription if it binds upstream of the RNA polymerase binding site, and that it inhibits gene transcription if it binds downstream of the RNA polymerase binding site. Catabolites derived from glucose or fructose are proposed to bind to Cra, preventing it from binding to DNA. Thus these catabolites would repress genes whose transcription is stimulated by Cra (gluconeogenic genes) and activate genes whose transcription is repressed by Cra (glycolytic genes). See text for details. *Source:* From Saier, Jr., M. H. 1996. Cyclic AMP-independent catabolite repression in bacteria. *FEMS Microbiol. Lett.* **138**:97–103.

or fructose prevents growth on organic acids such as acetate or succinate, or amino acids, because the activator (Cra) of genes that encode enzymes required for growth on these carbon sources is removed.

*A model for how glucose and fructose remove Cra from DNA*

Certain catabolites derived from glucose or fructose, that is, fructose-1-phosphate

(F1B) and fructose-1,6-bisphosphate (FBP) are thought to bind to Cra and remove it from the DNA, therefore diminishing the effect of Cra (Fig. 18.16). Thus these catabolites would repress genes activated by Cra (gluconeogenic genes) and activate genes inhibited by Cra (glycolytic genes). The evidence in favor of this model includes the finding that Cra does bind to Cra-regulated operons in vitro and fructose-1-phosphate or fructose-1,6-bisphosphate have been shown to displace Cra from DNA.

### A comparison of the Cra system and the cAMP system

The Cra system is similar to the cAMP system in that the sugar prevents catabolite-repressed genes from being activated. (See Section 16.3.4 and Fig. 16.9.) In one system the activator is cAMP–CRP, whose levels are lowered by glucose because the levels of the adenylyl cyclase activator, $IIA^{Glc}$-P, are lowered when incoming glucose is phosphorylated by $IIA^{Glc}$-P, whereas in the other system the activator of gluconeogenic genes is Cra, which is removed from the DNA by F1P or FBP, leading to catabolite repression of these genes. The Cra system differs from the cAMP system in that Cra is also a negative transcription regulator of genes required for sugar catabolism and provides for the activation of these genes by glucose and fructose because F1P and FBP remove Cra from these genes. Hundreds of genes are regulated by cAMP and Cra in the enteric bacteria.

### Repression of a nitrite reductase gene by Cra and catabolite activation

As discussed in Section 18.3, E. coli makes two different nitrite reductases under anaerobic conditions regulated by the Nar system. These are the formate-dependent periplasmic nitrite reductase encoded by the nrf operon and the NADH-linked cytoplasmic nitrite reductase encoded by the nir operon. Both nitrate and nitrite stimulate transcription of nir, whose product is required for detoxifying the nitrite formed from nitrate during nitrate respiration. Transcription of nrf is induced by nitrite but repressed by nitrate. In addition to regulation by the Nar system, these operons are induced

anaerobically by FNR. (See Section 18.2.2.) In addition to stimulation by FNR, nitrate, and nitrite, the nir operon is repressed by Cra and activated by catabolites.[161] (Transcription from the nrf operon is not regulated by Cra and is catabolite repressed in an unknown manner.) These experiments were done in E. coli strains carrying nir–lacZ and nrf–lacZ transcriptional fusions. Thus the nir and nrf promoters could be monitored by β-galactosidase assays. Transcription of the nir operon was repressed in poor medium but increased in rich medium containing glucose. Furthermore, in a $cra^-$ mutant transcription of the nir gene was increased (Fig. 18.17). Thus it was concluded that Cra inhibits transcription of the nir operon, and the inhibition is relieved by glucose. This is similar to the inhibition discussed earlier by Cra of the genes responsible for the uptake and catabolism of sugars such as glucose and fructose, and the release of the inhibition by fructose-1-phosphate (F1B) and fructose-1,6-bisphosphate (FBP) derived from glucose. (See the section on a model for how Cra activates certain genes and represses other and Fig. 18.16.) The binding of Cra to the nir promoter and the release of Cra from the DNA by fructose-1-phosphate was demonstrated in vitro.[161]

### 18.13.2 Catabolite repression in gram-positive bacteria via the CcpA system

A third system of catabolite repression has been discovered in low-GC (30–40%) gram-positive bacteria such as Bacillus, Staphylococcus, Streptococcus, Enterococcus, Lactococcus, and Lactobacillus. (These bacteria do not synthesize detectable cAMP or CRP.[162, 163]) It is called the CcpA system (catabolite control protein A system). CcpA is a transcription inhibitor of genes that are catabolite repressed by glucose. Recently a second transcription repressor, CcpB, has been identified in B. subtilis and proposed to act in parallel with CcpA via the same mechanism.[164]

### The model

A proposed model is drawn in Fig. 18.18. When cells are grown on glucose, the HPr

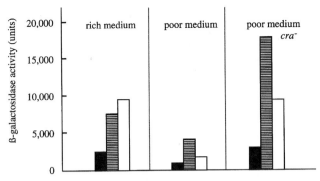

**Fig. 18.17** Repression of *nir* (NADH-dependent nitrite reductase) by Cra. *E. coli* was grown anaerobically in rich (LB plus glucose) or poor medium (minimal salts plus glycerol and fumarate) without nitrite (solid bars), with 2.5 mM nitrite (lined bars), or with 20 mM nitrate (open bars). The results with the wild-type strain are shown in (A) and (B), and the results using a *cra⁻* mutant are shown in (C). The expression of the *nir–lacZ* fusion was monitored by doing β-galactosidase assays. The expression of the nir operon was enhanced in rich medium or by a mutation in the *cra* gene. This is consistent with the hypothesis that Cra represses the operon and that a catabolite derived from glucose reverses the repression. See text for further details. *Source:* Adapted from Tyson, K., S. Busby, and J. Cole. 1997. Catabolite regulation of two *Escherichia coli* operons encoding nitrite reductases: Role of the Cra protein. *Arch. Microbiol.* **168**:240–244.

protein that is also part of the phosphotransferase (PTS) system becomes phosphorylated (Hpr-P) and binds to a transcription factor called CcpA. The Hpr-P–CcpA complex binds to a nucleotide sequence called CRE (catabolite responsive elements) (possibly as a ternary complex with fructose-1,6-bisphosphate), which is proximal to the promoter on genes susceptible to catabolite repression and inhibits transcription.[165] Thus the key event with respect to catabolite repression is that glycolytic intermediates stimulate the phosphorylation of Hpr and therefore the repression of CcpA-sensitive operons. This is explained next.

### The key regulatory event is the phosphorylation of Hpr stimulated by glycolytic intermediates

It is important to understand that the phosphorylation of Hpr is not due to the PEP-dependent kinase that functions in the PTS system. Rather, the phosphorylation is due to an ATP-dependent kinase. See Note 166 for a further explanation. Hpr is phosphorylated by a specific ATP-dependent kinase that is activated by the glycolytic intermediates fructose-1,6-bisphosphate and 2-phosphoglycerate, as well as by gluconate-6-P. Thus, when the cells

are growing on glucose and the levels of glycolytic intermediates rise, Hpr becomes phosphorylated and represses the transcription of genes having the CRE sequences. A phosphatase removes the phosphate from Hpr-P under starvation conditions.

The model also provides a role for inorganic phosphate ($P_i$). According to the model, inorganic phosphate would prevent catabolite repression by inhibiting the kinase and by stimulating a phosphatase that dephosphorylates HPr–(Ser-P).

### Genetic evidence for the model

Genetic evidence for the model is that mutations in CRE, or the genes coding for CcpA or CcpB, or mutations that lead to a failure to phosphorylate serine 46 in HPr, result in a failure of catabolite repression. (See Note 167 for a general description of how some of these mutants can be isolated.) Also the ATP-dependent kinase that phosphorylates HPr at serine 46 has been purified and shown to be stimulated by FBP and inhibited by inorganic phosphate ($P_i$). A $P_i$-activated phosphatase that dephosphorylates HPr–serine-P has been isolated from several gram-positive bacteria.

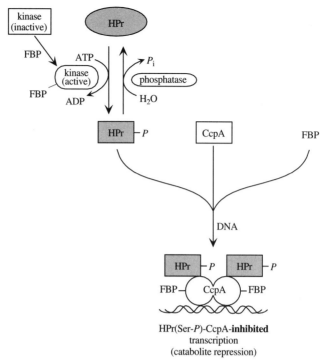

**Fig. 18.18** Catabolite repression in *B. subtilis*. An ATP-dependent Hpr kinase is activated by fructose-1,6-bisphosphate, FBP. Phosphorylated Hpr, Hpr-P, binds to a transcription regulator, CcpA, and possibly also to FBP to form a ternary complex. The ternary complex binds to the CRE region near the promoters of catabolite-repressable genes and represses transcription. *Source:* Adapted from Saier, Jr., M. H., S. Chauvaux, J. Deutscher, J. Reizer, and J-J Ye. 1995. Protein phosphorylation and regulation of carbon metabolism in gram-negative versus gram-positive bacteria. *TIBS* 20:267–269.

### Comparison to E. coli

A comparison of the catabolite repression mechanism of *E. coli* with that of *B. subtilis* shows that whereas catabolite repression in *E. coli* is due to a *decrease* in a *positive* regulator of catabolite repressable genes (i.e., cAMP–CRP or Cra), catabolite repression in *B. subtilis* is due to an *increase* in a *negative* regulator factor (i.e., CcpA–Hpr–Ser-P or CcpB–Hpr–Ser-P).

### 18.13.3 Inducer expulsion

Glucose prevents the accumulation of other sugars in some gram-positive bacteria, but the mechanism is not via the inhibition of sugar uptake by IIA$^{Glc}$, as is the case with *E. coli*. Two known mechanisms exist that lead to the equilibration of the sugar across the cell membrane, hence no intracellular accumulation. One of these is the uncoupling of

sugar transport from the proton motive force, and the second is the dephosphorylation of the incoming sugar-P. Consider the situation in *Lactobacillus brevis*, where Hpr(Ser-P) appears to bind to the lactose/proton symporter and uncouples sugar transport from proton symport.[155, 168] There are two phosphorylated forms of Her, one is active in the PTS system and the other in catabolite repression. See Section 18.13.2. The consequence of uncoupling the symporter from the proton gradient is that the symporter catalyzes facilitated diffusion rather than active transport. The result is that lactose equilibrates across the membrane and cannot be accumulated. In some gram-positive bacteria (streptococci, lactococci, and enterococci), but not others (bacilli, staphylococci), there exists a membrane-associated sugar-P phosphatase activated by HPr(Ser-P).[169] It is suggested that when the sugar-Ps are accumulated in the cell via the PTS system, they are dephosphorylated and then exit along their concentration gradient. This is called *inducer expulsion* and occurs upon the addition of glucose. The purified phosphatase has broad substrate specificity and dephosphorylates a long list of sugar-P molecules, including glucose-1-P, glucose-6-P, and fructose-6-P. Presumably only certain sugar-Ps are attacked because the enzyme's location on the membrane determines its accessibility to specific sugar-Ps being transported via sugar transporters into the cell.

## 18.14 Response to Carbon Source: Induction of a Permease for Dicarboxylic Acids in Rhizobium meliloti

*R. (Sinorhizobium) meliloti* is a nitrogen-fixing bacterium that lives symbiotically in the root nodules of leguminous plants. (See Section 18.4.8 for a discussion of chemotaxis in *R. meliloti*.) It grows most rapidly on C$_4$-dicarboxylic acids such as succinate, malate, and fumarate. *R. meliloti* has a well-characterized two-component system that regulates the production of a permease (transporter) that brings dicarboxylic acids into the cell. When *R. meliloti* is cultured with C$_4$-dicarboxylic acids, the gene for the dicarboxylic acid permease (*dctA*) is induced.

### The model

A transmembrane histidine kinase (sensor kinase) called DctB autophosphorylates at a histidine residue when the cells are incubated with dicarboxylic acids, and subsequently transfers the phosphoryl group to an aspartate residue in a cytoplasmic response regulator protein, DctD. DctD-P then activates the transcription of the dicarboxylic acid permease gene (*dctA*).

### Support for the model

In support of this model, autophosphorylation of the purified DctB protein has been demonstrated in vitro along with the transfer of the phosphoryl group to DctD and the binding of phosphorylated DctD to DNA.[170]

## The dicarboxylic acids themselves are not sufficient to activate the sensor kinase

The inclusion of dicarboxlic acids in the incubation mixture does not affect the autophosphorylation reaction, indicating that there may be another component (perhaps in the periplasm) that binds to the dicarboxylic acid and transmits the signal to the histidine kinase, or that the histidine kinase responds directly to the dicarboxyic acids only when integrated into the membrane.

## 18.15 Initiation of Sporulation in Bacillus subtilis: Response to Carbon Supply or Nitrogen Supply

When *Bacillus* is faced with limiting supplies of a carbon or nitrogen source, the cells sporulate, usually at the end of exponential growth and the beginning of stationary phase. The spore is a dormant stage in the life cycle and is resistant to environmental stresses such as heat, ultraviolet radiation, and toxic chemicals. They can remain dormant for hundreds of years but will germinate into growing cells (vegetative cells) when nutrient becomes available. *B. subtilis* makes the decision during a relatively brief period during the cell cycle whether to form a midcell septum and continue cell division, or to form a polar septum and sporulate.[171] The decision to sporulate is regulated by a phosphorelay signal transduction system described later.

*Bacillus* cells also communicate with each other via extracellular signaling peptides that aid them in determining when it is time to sporulate. (In addition, *Bacillus* also signal each other in order to become competent to take up exogenous DNA. See Section 18.16.) Thus *Bacillus* is an excellent system to study for investigating how bacteria couple extracellular signals to cell differentiation. In order to place the discussion of the phosphorelay system that regulates early events in sporulation into the context of sporulation, the morphological stages of sporulation will be described next, followed by a summary of the sporulation genes involved in the early stages of sporulation.

### 18.15.1 Stages in sporulation

#### Morphological description

The spore forms inside the progenitor cell as a consequence of an asymmetric septation and is called an endospore. The stages of endospore formation are shown in Fig. 18.19. Prior to sporulation the cells are said to be in Stage 0. As the cells get ready for sporulation, the chromosome condenses into an axial filament. This has been referred to as Stage I. Then a septum is laid down asymmetrically near one pole of the cell. Upon completion of the septum, the sporulating cell is said to be in Stage II. Once the septum is completed, the cell (now called a sporangium) is divided into two compartments: a mother cell compartment and a forespore compartment. After this has occurred, a copy of the chromosome is translocated into the forespore compartment. During Stage III the mother cell septum grows around the forespore (engulfment), pinching off the forespore. When engulfment is complete, the forespore in Stage III has two sets of membranes: an inner and an outer membrane. In Stage IV, a cortex, consisting of peptidoglycan, is synthesized

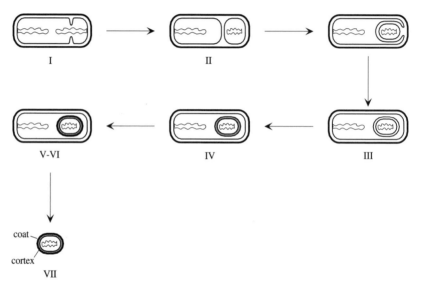

Fig. 18.19 Stages in sporulation of *Bacillus*. Vegetative cells are in Stage 0. The two chromosomes from the completed round of replication become aligned along the long axis of the cell in Stage I with their origins attached to opposite poles of the cell. (See the discussion in Section 10.1.5 regarding the attachment of the nucleoid to the cell poles.) At Stage II a septum has formed near one pole, dividing the cell into a mother cell compartment and a forespore, trapping approximately $\frac{1}{3}$ of the forespore chromosome in the forespore. DNA translocation across the forespore septum takes place, resulting in the forespore chromosome being moved entirely into the forespore. Once the septum forms, the cell is referred to as a sporangium. The mother cell engulfs the forespore, bringing the sporangium to Stage III where the forespore is set free in the cytoplasm of the mother cell as a protoplast. There is a space between the inner and outer membranes of the spore protoplast, and in Stage IV a cortex consisting of peptidoglycan is synthesized in that space. The cortex covers a peptidoglycan cell wall (the germ cell wall) that is laid down on the surface of the inner forespore membrane. A polypeptide multilayered coat is synthesized by the mother cell around the developing spore in Stage V. A proteinaceous exosporium is also made. The exosporium covers the spore coat either as a loosely fitting or more closely fitting layer, depending upon the species of *Bacillus*. The spore (also called an endospore) matures during Stage VI, developing resistance properties, and is released from the lysed sporangium in Stage VII. The entire process, including the release of the spore, takes between 8 and 10 h. Each flattened circle (wavy line) refers to a nucleoid.

in the space between the inner and outer forespore membrane. The cortex covers a layer of peptidoglycan called the germ cell wall, which is made on the surface of the inner forespore membrane. During Stage V a spore coat is synthesized surrounding the outer membrane. The coat is made of protein and is produced by the mother cell. The resistance properties of the spore develop during Stage VI, and the mature spore is released as a result of lysis of the mother cell during Stage VII. The entire process, including the release of the spore, takes around 8 to 10 h, and requires the activities of at least 113 sporulation-specific genes.[172]

## 18.15.2 A discussion of the roles of the genes involved in formation of the spore septum and chromosome partitioning into the forespore compartment

In certain ways the early stages of sporulation resemble cell division. For example, a septum is made separating the cell into two compartments and chromosome partitioning between the compartments takes place. These are poorly understood events with respect to cell division as well as sporulation, but it is useful to compare the two. One major difference between sporulation and cell division is that in cell division the chromosomes partition to opposite cell poles before the septum is made or invagination occurs, whereas partitioning of the chromosomes during sporulation is completed after septum formation. In this way, partitioning of the prespore DNA resembles DNA transfer between cells. (For a further discussion of the sporulation genes, see the article by Stragier and Losick.[173])

### FtsZ

The student should read Section 2.2.5 for a discussion of the role of FtsZ as well as other cell division genes in septum formation and cell division in *E. coli*. Prior to the formation of the asymmetric septum in Stage II, the cells form an FtsZ ring at both cell poles. The establishment of the FtsZ ring at the cell poles rather than in the cell center is an early event that distinguishes cell division from

sporulation. Very little is known about septum formation and the placement of the FtsZ rings during normal cell division as well as during sporulation.

### SpoA

SpoOA is a transcription factor that somehow induces the formation of the axial filament and the formation of the FtsZ rings at the poles. Mutations in *spoA* result in cells blocked at Stage 0 of sporulation. SpoOA is also directly or indirectly responsible for the expression of other sporulation genes. For example, SpoOA activates directly transcription of *spoIIA* and *spoIIG*, which encode the first prespore-specific sigma factor, $\sigma^F$, and the first mother-cell-specific sigma factor, $\sigma^E$, respectively. SpoOA also activates directly transcription of *spoIIE*, which encodes a protein that facilitates the formation of polar FtsZ rings.[174] Mutations in *spoIIE* reduce, but do not eliminate, polar septum formation under sporulation conditions. SpoIIE is also involved in the activation of $\sigma^F$ in the prespore compartment.

### $\sigma^H$

The subsequent formation of the septum at one of the two FtsZ rings requires the sporulation sigma factor $\sigma^H$, which presumably activates the transcription of a gene(s) whose product is required for formation of the septum. In the absence of $\sigma^H$ a polar septum does not form.

### $\sigma^E$

There is another sigma factor, $\sigma^E$, that is required to prevent septation at both poles, again suggesting the presence of a gene involved in septum formation, this one required to *prevent* septum formation at one of the poles. An interesting observation has been made with mutants lacking $\sigma^E$. These mutants form septa at both poles and transport a chromosome into both compartments, suggesting that in wild-type cells each chromosome is attached to an opposite cell pole. (See Section 10.1.5 for a description of experiments that demonstrate this.)

## SpoIIIE

The translocation of the chromosome into the forespore compartment after septum formation requires the product of the *spoIIIE* gene, which codes for a *DNA translocase* located at the leading edge of the growing septum that partitions the forespore compartment from the mother cell compartment. Septation actually bisects the prespore chromosome, trapping approximately 30% of the prespore chromosome, which includes the origin of replication, inside the forespore compartment, perhaps attached to the cell pole. The product of *spoIIIE* is required to move the remaining 70% through the septum into the forespore. The mechanism of transfer of 70% of the chromosome into the prespore compartment through the spore septum is believed to be similar to the transfer of plasmid DNA during conjugation between cells. (The prespore compartment and mother cell compartment are analogous to two separate cells.) This conclusion is supported by the finding that the carboxy-terminal domain of SpoIIIE has significant sequence similarity to DNA transfer proteins (Tra proteins) of conjugative plasmids of *Streptomyces*, and that mutations in this region block chromosome transfer into the prespore compartment.[175]

## SpoOJ

In the absence of SpoOJ, the 30% of the chromosome is not trapped in the forespore in *spoIIIE* mutants, suggesting that SpoOJ might have something to do with positioning the origin of replication (*oriC*) of the chromosome to the cell pole, perhaps by binding to sites near *oriC* and to proteins at the pole. Indeed, it has been recently demonstrated that SpOJ binds to sites near *oriC* of the *B. subtilis* chromosome (*parS* sites).[176] SpoOJ also functions during chromosome partitioning during vegetative growth. Mutations in *spoOJ* result in an approximately 100-fold increase in anucleate cells during vegetative growth. However, SpoOJ is not absolutely required for chromosome partitioning during vegetative cell division, since approximately 98% of the cells of a *spoOJ* mutant do partition the daughter chromosomes correctly. SpoOJ is homologous to ParB. (See the discussion of chromosome partitioning and the Par proteins in Section 10.1.5.).

## 18.15.3 Phosphorelay system for initiation of sporulation

The decision to sporulate rests upon integrating a variety of environmental and physiological signals that are transduced via various signaling pathways that result in the phosphorylation of SpoOA, which is a response regulator protein. SpoOA-P is responsible for activating the transcription of the sporulation genes. The phosphorelay system resembles the other two-component systems discussed in this chapter but differs in having intermediate phosphoryl donors between the histidine kinase and SpoOA.[177, 178]

### The model

The model shown in Fig. 18.20 stipulates that a histidine kinase receives a signal for sporulation and as a consequence autophosphorylates. It then transfers the phosphoryl group to an aspartate residue in a cytoplasmic response regulator protein called SpoOF. (The crystal structure of SpoOF has recently been reported.[179]) There are actually two independent histidine kinases that can phosphorylate SpoOF. One is KinA, which is cytoplasmic, and the other is KinB, which is membrane bound.[180] (See Note 181 for evidence of this.) It is not understood how the kinases are regulated. One might imagine that KinB responds to an external signal, whereas KinA responds to a cytoplasmic signal. Mutations in *kinB* prevent sporulation. However, *kinA* mutants will sporulate after a delay. SpoOF-P then transfers the phosphoryl group to a histidine residue in a phosphotransferase called SpoOB. SpoOB-P transfers the phosphoryl group to an aspartate residue in a second response regulator called SpoOA. SpoOA-P is directly or indirectly responsible for the transcription of other sporulation genes.

There are at least seven sporulation genes whose transcription is activated when SpoOA-P binds to their promoter regions. These genes include a gene that is responsible for determining the site of the forespore septum, four genes that are required to

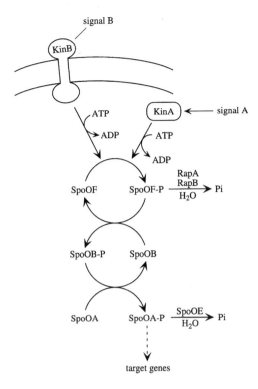

activate transcription of other sporulation genes in the forespore, and two genes that activate transcription of sporulation genes in the mother cell.[172, 173] SpoOA-P is also a repressor. It represses the *abrB* gene that encodes a DNA-binding protein that represses the transcription (during vegetative growth) of sporulation genes *spoOE*, *spoOH*, and *spoVG*.[182]

The phosphorelay system described for *B. subtilis* sporulation differs from most other two-component systems in that the phosphate is transferred between two response regulator proteins (i.e., SpoOF and SpoOA), via a phosphotransferase (i.e., SpoOB).

## Advantage to the phosphorelay

The phosphorelay system offers more sites for fine tuning or control than some of the more simple two-component systems. This might be necessary to ensure that *B. subtilis* does not stop multiplying and enter the sporulation pathway prematurely.

## Regulatory role of phosphatases

In many two-component systems the level of phosphorylation of the response regulatory protein is determined by the kinase and phosphatase activites of the histidine kinase. In other words, many histidine kinases are bifunctional enzymes and have both kinase and phosphatase activities. For example, see the discussion of NarX in Section 18.3, NR$_{II}$ in Section 18.8.1, and PhoR in Section 18.9.1. However, in the phosphorelay that controls sporulation the levels of the phosphorylated response regulator proteins are determined by separate phosphatases that respond to unknown signals. In fact, there exist several phosphatases that take part in adjusting the levels of the phosphorylated response proteins.[183] (Reviewed in Ref. 184.) Three of the phosphatases are SpoOE, RapA, and RapB. SpoOE dephosphorylates SpoOA-P but not SpoOF-P. RapA (SpoOL) and RapB (SpoOP) dephosphorylate SpoOF-P but not SpoOA-P. Regulation of the level of phosphorylated response regulators by separate phosphatases occurs in other phosphorelay systems. Recall that the level of CheY-P in the chemotaxis

**Fig. 18.20** Phosphorelay system during initiation of sporulation in *B. subtilis*. A histidine kinase autophosphorylates in response to a signal and then transfers the phosphoryl group to a response-regulator protein, SpoOF. There are two histidine kinases; KinA (cytoplasmic) and KinB (membrane bound). SpoOF-P then phosphorylates SpoOB, which in turn transfers the phosphoryl group to a response-regulator protein, SpoOA. SpoOA-P is a transcriptional activator for sporulation genes. (SpoOA-P also represses transcription of the negative regulator gene, *abrB*, that turns off transcription of genes normally expressed during the transition phase between exponential growth and sporulation.) There are at least three phosphatases that are important for regulating the level of SpoOA-P. They are RapA (SpoOL) and RapB (SpoOP), which dephosphorylate SpoOF-P, and SpoOE, which dephosphorylates SpoOA-P. For more information, see Stragier, P., and R. Losick. 1996. Molecular genetics of sporulation in *Bacillus subtilis*. *Ann. Rev. Gen.* 30:297–341; Perego, M., and J. A. Hoch. 1996. Protein aspartate phosphatases control the output of two-component signal transduction systems. *Trends in Genetics* 12:97–101.

system is regulated in part by a separate phosphatase (i.e., CheZ), in the enteric bacteria. See Section 18.4.5.

## How the RapA phosphatase is controlled

A model for how RapA phosphatase (response-regulator aspartate phosphatases) is regulated was published by Perego and Hoch.[185] (See Fig. 18.21.) The RapA phosphatase gene (rapA) is encoded in an operon with phrA (phosphatase regulator), which encodes a protein that regulates the synthesis or activity of RapA phosphatase. There exists a two-component signaling system consisting of a membrane-bound histidine kinase called ComP and a response regulator protein called ComA. (These are involved in positive signaling of the competence pathway and are more fully described in Section 18.16.) Upon receiving a signal, ComP phosphorylates ComA, which then activates the transcription of the rapA operon. RapA phosphatase is synthesized, and if its activity were not inhibited, sporulation would not take place because SpoOF-P would not accumulate. Thus the cells would enter the competence pathway rather than sporulate. Competence and sporulation occur about the same time during the culture period, and presumably the induction of rapA by the ComA/ComP system is to ensure that sporulation and the competence process do not take place in the same cell at the same time. However, the cells can inhibit the RapA phosphatase and initiate sporulation when it is time to do so. This requires PhrA, the other product of the rapA operon, to be secreted from the cell, where it is cleaved extracellularly to a pentapeptide (PhrA peptide) that is imported back into the cell through an oligopeptide transport system. (There is an oligopeptide permease, SpoOK, which is required for the response to the PhrA peptide, and it catalyzes the transport of PhrA peptide as well as other short peptides into the cell.[188])

Once inside the cell, PhrA peptide acts as an inhibitor of RapA phosphatase activity. (Genetic evidence for this is that if the phrA gene is deleted, then RapA phosphatase activity is very high, and sporulation is inhibited. It has also been demonstrated that the PhrA C-terminal pentapeptide inhibits RapA phosphatase activity in vitro.[186]) Thus PhrA

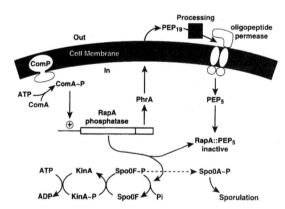

**Fig. 18.21** Regulation of RapA phosphatase by PhrA. An unknown signal for competence stimulates the ComP histidine kinase and causes phosphorylation of the ComA response regulator. The phosphorylated ComA then stimulates the transcription of the rapA operon, which encodes RapA phosphatase and PhrA. The RapA phosphatase dephosphorylates SpoOF-P so that sporulation does not occur while the cells are in the competence process. The phrA gene product is secreted to the cell surface, where it is processed to a pentapeptide (PEP$_5$) that re-enters the cell via the oligopeptide permease. PEP$_5$ inhibits RapA so that sporulation can take place. Certain steps in the secretion and/or processing or uptake of PhrA are probably regulated so that the inhibition of the RapA phosphatase does not take place while the cell is in the competence pathway. *Source:* From Perego, M. 1997. A peptide export–import control circuit modulating bacterial development regulates protein phosphatases of the phosphorelay. *Proc. Natl. Acad. Sci. USA* **94**:8612–8617.

peptide stimulates the sporulation pathway by lowering the amounts of active RapA phosphatase. It may be that control of initiation of sporulation involves, in part, regulation of the secretion and processing of PhrA and/or its uptake into the cell. The PhrA peptide does not accumulate in the extracellular medium supporting the growth of wild-type cells, and therefore it does not appear to be a quorum sensor. (See next.) In fact, it has been suggested that it is secreted to the cell wall, but not into the medium.[186]

## 18.16 Quorum Sensing by B. subtilis

Quorum sensing refers to the production of extracellular molecules that signal the population cell density. In gram-negative bacteria acylated homoserine lactones are frequently the extracellular signals produced by high-density populations, although lipids and amino acids may also be used as signals by some bacteria. (See Sections 18.18 and 18.19.) As discussed next, B. subtilis uses peptides to signal cell density. (As reviewed in Ref. 190, peptides are often used by gram-positive bacteria as intercellular signals.) It is well known that B. subtilis will sporulate more efficiently and is more competent (takes up exogenous DNA) at high cell densities. (Competence occurs in many different bacteria. See Ref. 187 for a review.) This is correlated with the production and accumulation in the media of extracellular peptides, which serve as cell density signals.[188] These peptides will be described, but first the competence regulatory pathway will be described, followed by a model to explain how the secreted peptides influence this pathway.

### 18.16.1 Regulation of competence

#### The model

In order for Bacillus subtilis to become competent to take up exogenous DNA, a transcription factor called ComK must be activated. ComK then activates genes that encode proteins required for the binding and uptake of exogenous DNA and its recombination into host DNA (Fig. 18.22). The activation of ComK requires ComS, which is a small polypeptide that is synthesized only at high cell densities because the gene that encodes it, comS, is transcribed only at high cell densities; so high cell densities result in the transcription of comS, whose product, ComS, activates ComK, which activates the transcription of genes required for competence. There is a way for Bacillus to sense when the cell density is sufficiently high so that comS should be transcribed. The high cell density signals that activate the transcription of comS are the peptides CSF and ComX, which are secreted by Bacillus and accumulate in the medium as the cell density increases. However, the peptides do not directly activate the transcription of comS. Instead, the peptides activate a two-component regulatory system that stimulates the transcription of comS. Therefore, there is a signal cascade leading to transcriptional activation of competence genes as follows:

1. Accumulation in the medium at high cell densities of CSF and ComX.

2. Stimulation of a two-component system by CSF and ComX.

3. Activation of transcription of comS by the two-component system.

4. Activation of ComK by ComS.

5. Activation of transcription of competence genes by ComK.

The details of the signal cascade are described next. For a review of the regulation of the initiation of competence, see Ref. 189. For a more general review of cell–cell communication in gram-positive bacteria, see Ref. 190.

#### Activation of the two-component regulatory system by CSF and ComX

The two-component signaling system activated by CSF and ComX consists of a membrane-bound histidine kinase called ComP, and a response regulator protein called ComA (Fig. 18.22). When stimulated by the extracellular peptide ComX, ComP autophosphorylates and then phosphorylates the

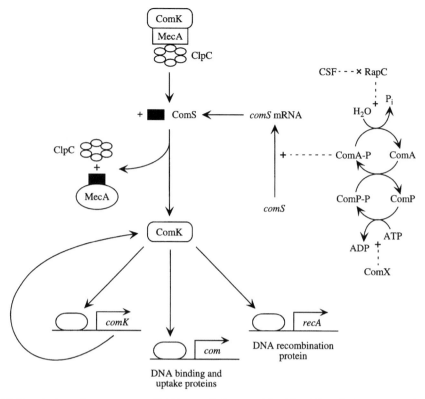

**Fig. 18.22** Regulation of competence. (A) Extracellular peptides (ComX and CSF) accumulate in the extracellular medium and stimulate the formation of ComS. ComS then activates ComK by displacing it from the ComK/MecA/ClpC complex. ComK stimulates the transcription of genes required for competence. (B) ComX activates the autophosphorylation of the histidine kinase ComP. ComP-P then phosphorylates the response regulator ComA. ComA-P stimulates the transcription of the *comS* gene. CSF inhibits the activity of the RapC phosphatase, thus promoting increased amounts of ComA-P. Thus the accumulation in the medium of CSF and ComX stimulates the expression of the competence gene, *comS*. *Source*: Adapted from Lazazzera, B. A., and A. D. Grossman. 1997. A regulatory switch involving a Clp ATPase. *BioEssays* **19**:455–458.

response regulator ComA. The amount of ComA-P is also regulated by CSF. CSF inhibits the activity of a phosphatase (RapC) that dephosphorylates ComA-P. ComA-P stimulates transcription of *comS*. (See Note 191 regarding *comS* and the *srfA* operon.) In this way, the accumulation in the medium of CSF and ComX stimulates the expression of the competence gene, *comS*. Notice that there are two ways in which the level of the response regulator protein ComA-P is increased. One signal (ComX) stimulates the kinase activity of ComP so that more ComA is phosphoryalted. A second signal (CSF) inhibits a phosphatase that removes the phosphate from ComA-P. It is not uncommon for the levels of a phosphorylated response regulator protein

to be regulated by both phosphorylation and dephosphorylation.

## High concentrations of CSF inhibit transcription of comS and stimulate sporulation

*Bacillus* is capable of switching from the competence pathway to the sporulation pathway. This is in part regulated by CSF. When CSF is present at relatively low concentrations (1–10 nM), it stimulates the transcription of *comS* by inhibiting the activity of the phosphatase that dephosphorylates ComA-P, and hence stimulates the competence pathway. However, when CSF is present at higher concentrations (20 nM–1 $\mu$M), it inhibits the transcription of *comS*. It is not known how CSF inhibits *comS* expression. One possibility is that CSF at high concentrations inhibits the histidine kinase (ComP), or perhaps activates a phosphatase that removes the phosphate from the response regulator ComA-P. The latter would be analogous to regulation of the Ntr regulon, where $P_{II}$ stimulates the phosphatase activity of $NR_{II}$. (See Section 18.8.1.) Interestingly, at these higher concentrations, CSF stimulates sporulation. It is unclear how high concentrations of CSF stimulate sporulation. Since CSF stimulates competence by inhibiting a phosphatase that dephosphorylates a response regulator protein, perhaps at high concentrations it stimulates sporulation in a similar manner.

### The activation of ComK by ComS

Recall that ComK must be activated by ComS before the former stimulates the transcription of the competetnce genes. However, the activation of ComK involves more than simply ComS (Fig. 18.22). ComK is kept in an inactive state in a complex with two other proteins, MecA and ClpC. The addition of ComS disrupts the complex and releases active ComK. ClpC belongs to the Clp ATPase family of chaperone proteins discussed in Section 19.1.3.

## 18.17 The Initiation of DNA Replication in Caulobacter and the Regulation of Transcription of Cell Cycle Genes

### 18.17.1 Life cycle

*Caulobacter* is a bacterium that undergoes an asymmetric cell division to produce a motile swarmer cell and a sessile stalked cell (Fig. 18.23). The swarmer cell differs morphologically from the stalked cell in having a polar flagellum and pili, and lacking a stalk. The two cells are also very different physiologically. Whereas the stalked cell synthesizes DNA and divides, the swarmer cell swims away and does not initiate DNA synthesis and cell division until it has shed its flagellum and grows a stalk at the same

pole. Thus the swarmer cell serves to disperse *Caulobacter*, whereas the stalked cell attaches to surfaces, grows, and reproduces. Organic nutrient is often concentrated at surfaces, and this is probably related to why *Caulobacter* attaches to surfaces.

### 18.17.2 CtrA is a response-regulator protein that represses the initiation of DNA replication and ftsZ transcription

As shown in Fig. 18.23, DNA replication takes place in the stalked cell but not in the swarmer cell. This is apparently due to a repressor of the initiation of DNA replication, CtrA (cell cycle transcription regulator A) that is present in the swarmer cell but not in the young stalked cell at the time DNA replication is initiated.[192–194] In order for DNA replication to be initiated, transcription must occur from a promoter within the origin of replication. CtrA represses this transcription; hence it prevents the inititation of DNA replication. CtrA also represses the transcription of *ftsZ* in swarmer cells, accounting, in part, for why there is no FtsZ in swarmer cells.[195, 196] (Recall that FtsZ is required for septation. Since swarmer cells do not divide, they do not need FtsZ See Section 2.2.5 for a discussion of the role of FtsZ in septation.)

### The levels of CtrA are regulated by cell-cycle-dependent proteolysis

As shown in Fig. 18.24, CtrA accumulates in the predivisional cell after DNA synthesis is initiated. The fact that it is present in the progeny swarmer cell but not in the progeny stalked cell is due to spatially regulated proteolysis in the stalked cell half of the predivisional cell, as well as during the swarmer–to–stalked cell transition. (This was shown by immunofluorescence microscopy.[194] See the discussion of FtsI in Section 2.2.5 and Note 66 in Chapter 2 for an explanation of immunofluorescent microscopy. See Note 197 for an explanation of how the cellular amounts of specific proteins can be measured.) This implies that there are cell cycle signals that activate proteolysis of CtrA at certain times (temporal

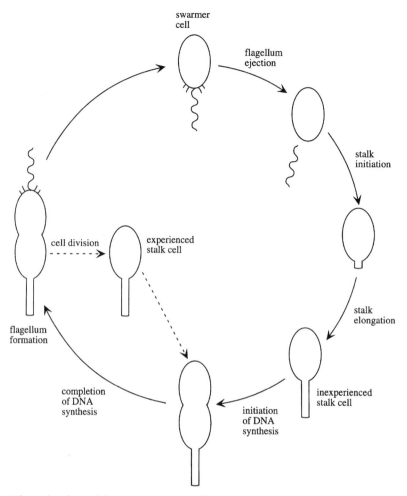

**Fig. 18.23** Life cycle of *Caulobacter crescentus*. The swarmer cell is a nongrowing cell specialized for dispersal. After a period of swimming it replaces its flagellum with a stalk and becomes a sessile growing and dividing cell. Cell division is asymmetric. Each division produces a motile swarmer cell and a stalked cell. The stalked cell continues to grow and divide. *Source:* Adapted from Dworkin, M. 1985. *Developmental Biology of the Bacteria.* The Benjamin/Cummings Publishing Company, Inc. Reading, MA.

control), and that mechanisms exist to ensure that proteolysis takes place only in certain regions of the cell (spatial control). The nature of these signals is not understood.

Proteolysis of CtrA allows DNA replication to be initiated and the *ftsZ* gene to be transcribed so that cell division can take place. After DNA replication has been initiated in the stalked cell, proteolysis ceases and CtrA accumulates again. The *ctrA* gene is transcribed in the stalked cell but not in the swarmer cell; hence the CtrA protein in the swarmer cell is inherited from the predivisonal cell. However, once the swarmer

cell becomes a stalked cell the *ctrA* gene is transcribed and CtrA accumulates.[198] As mentioned, when CtrA accumulates in the stalked cell, it represses the transcription of the *ftzZ* gene. However, this occurs only after sufficient FtsZ has accumulated for septation to occur in the stalked cell.[195] As septation progresses, the concentration of FtsZ declines due to proteolysis. Progeny swarmer cells do not have FtsZ, whereas progeny stalked cells do, indicating that proteolysis of FtsZ occurs only in the half of the cell destined to become the progeny swarmer cell. Notice that the fates of CtrA and FtsZ differ, depending upon in

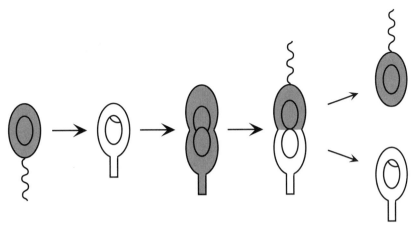

**Fig. 18.24** Relative levels of CtrA during the *Caulobacter* cell cycle are controlled by spatially regulated proteolysis. CtrA prevents the initiation of DNA replication and the transcription of *ftsZ*. It is present in swarmer cells and is degraded during the transition phase, when swarmer cells become stalked cells. The *ctrA* gene is transcribed in the stalked cell, but proteolysis in the half of the predivisional cell destined to be the progeny stalked cell ensures that only the swarmer cell is born with CtrA. The shaded portion represents CtrA. The circle within the cell represents DNA. *Source:* Adapted from Quon, K. C., B. Yang, I. J. Domian, L. Shapiro, and G. T. Marczynski. 1998. Negative control of bacterial DNA replication by a cell cycle regulatory protein that binds at the chromosome origin. *Proc. Natl. Acad. Sci. USA* **95**:120–125.

which part of the predivisional cell they lie. It appears that the region of the predivisional cell that is destined to become a swarmer cell degrades FtsZ but not CtrA, whereas the region of the predivisional cell destined to become a stalked cell degrades CtrA but not FtsZ.

CtrA is actually a global regulator, and, in addition to repressing transcription at the origin of replication and *ftsZ*, it activates transcription of genes required for the synthesis of flagella as well as the gene for a DNA methyltransferase required to fully methylate newly replicated DNA. Transcriptional regulators often activate transcription of some genes and inhibit transcription of other genes. See the discussion of Cra in Section 18.13.1 and the discussion of OmpR in Section 18.10.1.

### CtrA is a reponse regulator protein

CtrA is a response regulator that is activated by phosphorylation. Phosphorylation is cell cycle dependent, and occurs when the cells are midway through DNA replication. Neither the signal for phosphorylation nor the kinase (or

kinases) responsible for the phosphorylation of CtrA is known.

## 18.18 Intercellular Signaling in Myxobacteria

### 18.18.1 Life cycle

The myxobacteria are gram-negative gliding bacteria that can be isolated from soil, dung pellets, and decaying vegetation on the forest floor.[199] The cells glide on solid surfaces in populations called *swarms*. They are unique among the known prokaryotes in that they construct multicellular fruiting bodies. This can be induced in the laboratory by subjecting the cells to starvation on agar. Thus starvation is a signal for development, as it is in many microbial developmental systems. Figure 18.25 depicts the life cycle of a myxobacterium called *Myxococcus*.

The myxobacteria have two stages in the population life cycle, a vegetative stage and

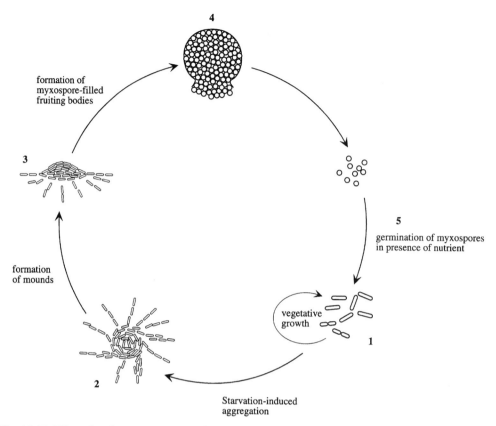

**Fig. 18.25** Life cycle of *Myxococcus xanthus*. 1. Vegetative growth. Cells grow in dense populations called swarms on solid surfaces. 2. Aggregation. When nutrients are depleted, cells glide into aggregation centers, each one consisting of many thousands of cells. 3. Mound formation. Each aggregation center becomes a mound of cells as bacteria continue to accumulate. 4. The mound develops into a fruiting body when the cells differentiate into resting cells called myxospores. Each myxospore is surrounded by a coat (capsule). 5. When nutrient becomes available once more, the myxospores germinate into vegetative cells, returning the population to a new growth phase. Once the cells deplete the supply of nutrients, they can aggregate and form myxospore-filled fruiting bodies within 24 h.

a developmental stage. The vegetative stage refers to the feeding and growing stage. During this period the cells in the swarms move in groups and feed on the source of nutrient, which in their natural habitat can be other bacteria that they lyse by secreting lytic enzymes. In the laboratory they can be grown on mixtures of peptides or amino acids. During the developmental stage the myxobacteria cells aggregate within the swarms into hundreds of aggregation centers consisting of approximately 100,000 cells each. The aggregation centers develop into multicellular fruiting bodies. Within the fruiting bodies the cells differentiate into resting cells called *myxospores*. Eventually, the fruiting bodies are transferred by nonspecific means, such as sticking to the the bristles of roaming insects, to new locations, where they germinate and produce a new swarm that enters the vegetative stage. It is important to realize that myxobacteria feed cooperatively; that is, they grow best on other bacteria or proteins such as casein when they are in swarms. This is because the digested products of the lytic enzymes secreted by the myxobacteria are shared. Thus the dispersal of fruiting bodies can be viewed as a means of ensuring that when the myxospores germinate, a feeding swarm will be produced, consisting of at least $10^5$ cells. In order to move as coherent groups of cells in the vegetative swarms while seeking nutrient, and also to move cooperatively to construct multicellular fruiting bodies, the myxobacteria have evolved systems of intercellular signaling. Intercellular signaling and gliding motility are discussed in Section 18.18.2, and the relationship of intercellular signaling to the construction of fruiting bodies and myxospores is discussed in Section 18.18.3.

### 18.18.2 Intercellular signaling and gliding motility

#### Adventurous motility (A motility) and social motility (S motility)

Gliding motility in *Myxococcus* is controlled by two sets of genes: system A genes (for adventurous) and system S genes (for social). Mutants in system A genes ($A^-S^+$) can glide only as groups; that is, system A genes are required for single cell motility. Mutants in system S genes ($A^+S^-$) show a motility pattern indicating that they glide as single cells (well-separated cells, some cell clusters); that is, system S genes are required for group motility. Social motility (the S system) requires that the cells synthesize type IV polar pili, and these are involved in a way that is not understood in S motility but not in A motility.[200–202] (See the discussion of type IV pili in Section 17.4.2.)

#### Intercellular signaling involved in A and S motility

Both single-cell and group motility are significantly stimulated when the cell density is increased, reflecting the fact that cell-to-cell communication stimulates gliding motility in both systems (Fig. 18.26). The signaling systems that stimulate single-cell and group motility are not understood, but appear to involve cell–cell contact in both the A and S motility systems.[203] For example, certain $A^-S^-$ cells will be stimulated to move when mixed with $A^+S^+$ but will not move if the wild-type and mutant cells are separated by a space on agar or are separated by a filter membrane. The signaling molecules have not been identified, but presumably they are bound to the cell surface or to fibrils or pili that extend from the cell surface. There are five different loci involved in A motility system

signaling: *cglB*, *cglC*, *cglD*, *cglE*, and *cglF* (*cgl* for conditional gliding). Mutants in these loci can be stimulated to move by either wild-type or by complementary mutants. For example, if $cglB^+ cglC^-$ is mixed with $cglB^- cglC^+$, they will both be stimulated to move, presumably because they each provide a signal that the other is lacking. This type of analysis has suggested that there are five signaling molecules for the A motility system. Very little is known about the signaling pathways.

A gene required for social motility called the *tgl* gene is necessary for the assembly of the polar-type IV pilus, which is required for S motility.[204] Tgl is a membrane-associated protein that can stimulate *tgl*⁻ cells, which have the pilus subunits, to assemble the subunits into pili upon contact.[205, 206] Nothing is known regarding how the stimulation between the cells occurs. It has been speculated that perhaps Tgl is somehow transferred from *tgl*⁺ cells to *tgl*⁻ cells when the cells come

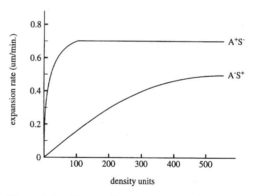

Fig. 18.26 Rate of swarm expansion as a function of initial cell density using $A^-S^+$ and $A^+S^-$ mutants. Microdroplets of cells at various density units (100 density units = $4 \times 10^8$ cells/ml) were placed on nutrient agar, and the rate of swarm expansion was measured. The swarms expanded linearly with time as the cells glided away from the center of the drop. Approximately 90% of swarm expansion was due to movement and 10% to growth, as determined by comparing swarm expansion using wild-type cells with swarm expansion using nonmotile mutants. *Source:* Data from Kaiser, D., and C. Crosby. 1983. Cell movement and its coordination in swarms of *Myxococcus xanthus*. *Cell Motil.* 3:227–245.

into contact. It is a matter of speculation as to how the transfer might occur, since Tgl is apparently attached to either the inner or outer membranes (or both) with a portion exposed to the periplasmic region; that is, it is not a cell-surface protein. Perhaps it is transferred via some process of membrane–vesicle fusion between cells.

## 18.18.3 Intercellular signaling and multicellular development

Early evidence for cell-to-cell signaling involved in development came from the isolation of *Myxococcus* mutants that were unable to form myxospores.[207] (The formation of myxospores is a convenient measure of the ability of the cells to aggregate and form fruiting bodies.) These cells had mutations in five genes: *asg*, *bsg*, *csg*, *dsg*, and *esg*, and were able to sporulate when mixed with wild-type cells. They would also sporulate when mixed with each other in complementary pairs, such as $bsg^+csg^-$ and $bsg^-csg^+$. The existance of the five classes of mutants that are capable of extracellular complementation indicates that five extracellular signals (A, B, C, D, E) exist for development. As described later, two signals, the A and C signals, have been isolated, and these will be discussed because they are the ones best characterized. A third signal, the E signal, may be a branched-chain fatty acid.[208]

The A signal is an early signal necessary for aggregation. Its signaling pathway will be discussed first. There is evidence that increased levels of ppGpp are also required for the expression of A-signal-dependent genes. See Section 2.2.2 for a discussion of ppGpp, and Note 209 for a further discussion.) A speculative model for the sequence of events involved in A signaling is presented next, followed by the evidence in support of the model.

### A model for A signaling

1. AsgA is suggested to be a cytoplasmic sensor-kinase that receives a starvation signal and phosphorylates a response regulator protein, AsgB.

2. AsgB-P activates genes responsible for the production of the A signal. It has a helix–turn–helix motif similar to many DNA binding proteins. (See the description of the helix–turn–helix motif in Section 10.2.5.)

3. The A signal is produced and stimulates a membrane-bound sensor-kinase called SasS to autophosphorylate.

4. SasS-P phosphorylates a cognate response regulator protein (unidentified) that induces the transcription of developmental genes.

Thus it is suggested that there are two two-component signaling systems: system 1 and system 2. System 1 responds to a starvation signal and stimulates the production of A signal. System 2 responds to the A signal and stimulates transcription of developmental genes.

### The chemical nature of the A signal

The A-signal activity (item 3 in the model) has been isolated from media in which wild-type cells have been subjected to starvation (shaking the cells in buffer) and shown to be a mixture of proteases, a specific set of amino acids, and peptides containing those amino acids.[210, 211] The model is that *Myxococcus* secretes proteases and proteins in response to starvation and that the proteases degrade the proteins to produce the amino acids that are the actual signaling molecules. In support of the model, it has been demonstrated that the amino acids isolated from the medium, as well as the mixture of proteases, restore development when added to *asg* mutants. It has been suggested that the A signal is a cell density signal during starvation.[212] When the concentration of amino acids exceeds 10 $\mu$M, which corresponds to what is produced at a cell density greater than $3 \times 10^8$/ml, the signal informs the population of cells that the cell density has reached the requisite number required to build fruiting bodies. From this point of view, the A signal is a quorum signal analogous to other quorum signals, including acylated homoserine lactones produced by a wide range of gram-negative bacteria, and peptide signals used by various bacteria, including *B. subtilis*, described in Sections 18.15 and 18.19.

## The roles of the asg *genes in the production of the A signal*

Signal A is produced when the cells are starved. The production of the A signal requires three genes important for transcriptional regulation. Two of the genes encode proteins that are thought to be part of a phosphorelay system similar to other two-component systems, that is, *asgA* (encoding a putative sensor-kinase) and *asgB* (encoding a putative response-regulator protein). The third gene, *asgC*, encodes a sigma factor. The nucleotide sequence of *asgA* suggests that its protein product, AsgA, has both histidine kinase and response-regulator domains; that is, it might be a bifunctional protein.[213] Interestingly, there is no membrane spanning domain, indicating that AsgA is a cytoplasmic protein, and therefore cannot directly sense an extracellular starvation signal. Perhaps it receives a signal from an unidentified membrane sensor, or alternatively, from a small cytoplasmic molecule whose levels might reflect nutrient supply. Cytoplasmic sensor kinases are known in other two-component systems. For example, recall that KinA is a cytoplasmic histidine kinase in the sporulation phosphorelay pathway (Section 18.15.3) and that $NR_{II}$ is a cytoplasmic histidine kinase for the Ntr regulon (Section 18.8). It has also been deduced from nucleotide analysis that AsgB is a DNA-binding protein that may be a transcription factor that alters gene expression, resulting in the production of the A signal, and that the *asgC* gene encodes the major sigma factor RpoD.[214, 215] One would expect that a mutation in the major sigma factor would result in a defect in growth. However, the *asgC* mutants do not show a general growth defect, indicating that the mutation does not affect transcriptional initiation in genes required for growth. One suggestion is that perhaps the mutation prevents the RpoD sigma factor from exchanging with a developmentally specific sigma factor required for transcription of a gene(s) required for production of the A signal.[215] Clearly, there is much to be learned

about the role of the *asg* genes in coupling the starvation signal to the synthesis of A signal.

## The SasS/SasR proteins are part of a two-component system that responds to the A signal

As discussed earlier, the Asg proteins appear to be involved in a signaling cascade induced by starvation that activates genes required to produce the A signal. Once the A signal is produced and accumulates in the external medium, how is it sensed by the cells? At least one pathway for sensing the A signal involves the product of the *sasS* gene.[216] According to the nucleotide sequence of *sasS*, SasS is a membrane-bound sensor–histidine kinase. The model is that SasS responds to the A signal by phosphorylating a cognate response regulator protein (SasR) that regulates the transcription of genes expressed early in development. In order to find *sasS*, it was necessary to use a developmental reporter gene, a gene that is expressed only during development. The reporter gene that was used was named gene *4521*. Gene *4521* is fused to Tn5 *lac* such that transcription from the *4521* promoter results in the synthesis of $\beta$-galactosidase, which can be measured. (See Note 25 for an explanation of how *lacZ* fusions are constructed and their use.) The expression of the reporter gene was shown to increase early in development and to require A signal. (The addition of the A signal rescues the expression of $\beta$-galactosidase in *asg* mutants.) However, expression of *4521* requires *both* starvation and the A signal; neither one is sufficient. Second site mutations in the *asg⁻* strains that expressed *4521* during growth as well as starvation were isolated, and these mapped to the *sasB* locus. Thus mutations in *sasB* expressed *4521* despite the absence of the A signal and regardless of whether the cells were starved. They are called *suppressor* mutants because they suppress the *asg⁻* phenotype. One of the mutant genes in the *sasB* locus, *sasS*, has a nucleotide sequence, suggesting that it encodes a transmembrane sensor–histidine kinase, and this is the reason for suggesting that SasS is a histidine kinase. The model is that SasS responds to the A signal and to starvation by autophosphorylating and then phosphorylating SasR, which activates transcription of *4521*. (It appears that both these signals somehow feed into a signaling pathway that results in the autophosphorylation of SasS during starvation, but not during growth, when the set of amino acids comprising the A signal exceeds a threshold value.) The mutation in *sasS* presumably locks it into the autophosphorylating conformation so that it does not need the A signal. Null *sasS* mutants do not express *4521* even in the presence of the A signal, in agreement with the hypothesis that SasS is part of the signal cascade that couples the A signal to the expression of *4521*. Although SasS is clearly required for the transcription of *4521* in response to the A signal, other signaling pathways that respond to the A signal must exist, since *sasS* null mutants are still able to form fruiting bodies and sporulate; that is, they still respond to the A signal, albeit not as well as the wild type.[216]

### The C signal

Mutants defective in making C signal (*csgA* mutants) construct abnormal aggregates and only after a significant delay. Studies with developmentally expressed *lacZ* fusion reporter genes indicate that the C signal acts at a developmental stage after the A signal. (See Note 25 for an explanation of *lacZ* fusions.) The signaling cascade leading to developmental gene expression initiated by the C signal has not yet been characterized. CsgA is a membrane-associated protein that has been localized to the surface of cells. There is evidence that CsgA is a short chain alcohol dehydrogenase.[217, 218]

In contrast to A signaling, C signaling requires close cell–cell contact, although cells will respond to the addition of purified CsgA. Nonmotile cells do not express C-signal-dependent genes or sporulate, despite having the wild-type *csg* gene, because they cannot align themselves properly. However, when the nonmotile cells were placed on an agar surface that had been lightly scratched with emery paper and allowed to settle into the grooves, the cells in the grooves became aligned parallel to each other and were capable of stimulating the expression of C-signal-dependent *lacZ* fusion reporter genes as well as sporulate.[219] Cells not in the grooves did not do so. It is

thought that C signaling requires end-to-end contacts between cells.[220] A hypothesis has been developed to explain how the C signal stimulates aggregation. It has been proposed that the exchange of the C signal between cells that are moving into and entering aggregates decreases the frequency of motility reversals, and that this might be important for cells to continue moving into aggregates.[221]

## 18.19 Virulence Factors: Synthesis in Response to Temperature, pH, Nutrient, Osmolarity

This section discusses some of the regulatory systems that transduce environmental signals to control the transcription of genes that encode virulence factors. Virulence factors are structures or substances that aid pathogenic bacteria in the infection process and/or contribute towards the symptoms of disease (reviewed in Ref. 222 and 223). Virulence factors are generally exported to the bacterial cell surface or beyond. They include pili for adsorption to host tissue, toxins, flagella to aid the bacterium in arriving at the site of colonization, and extracellular enzymes such as proteases. (For example, see Ref. 224.) Some virulence factors produced by *Yersinia* spp. are actually injected into target cells. This is discussed in Section 17.4.4. Virulence factors are not necessarily constitutively produced, and it appears that several are made only during the course of an infection and not when the pathogen is growing outside the host. Certainly this is the case with laboratory cultures, where only under certain conditions of temperature, pH, nutrient composition, iron availability, and osmolarity are certain virulence factors made.[225] How bacteria sense environmental factors such as temperature and osmolarity is not well understood. The student should review the discussion on the response to osmotic pressure and temperature and the regulation of porin synthesis in Section 18.10 for information on a regulatory system influenced by osmotic pressure and temperature. The following discussion focuses on the effect that environmental factors have on the expression of virulence genes. However, as will be made clear from the discussion, the environmental signals that influence the expression of toxin genes inside the host are often not well defined. For a description of the effect of extracellular autoinducers on the transcription of virulence genes, see the discussion in Section 18.20.4.

## 18.19.1 ToxR and cholera

ToxR is a transcription regulatory protein produced by *Vibrio cholerae*, the bacterium that causes cholera. It stimulates the transcription of virulence genes that are in the ToxR regulon. (For a review, see Ref. 226.) To understand the importance of ToxR, it is necessary briefly to describe cholera pathogenesis.

### Cholera

The bacteria grow in the small intestine attached to the intestinal epthelial layer. They swim through the mucus layer and attach to intestinal cells via Type IV pili called TCP pili (toxin-coregulated pilus). The symptoms of cholera include a watery diarrhea, which can result in the loss of as much as 10 to 20 liters of fluid a day, and if untreated, results in dehydration, loss of electrolytes, and death in approximately 60% of infected people. The diarrhea is caused by an enterotoxin called cholera toxin (CT), which is secreted by *V. cholerae*. (Cholera toxin consists of one A subunit and five identical B subunits and is encoded by the *ctxAB* operon that resides in the CTX lysogenic phage. See Note 227 for a description of how cholera toxin works.) *V. cholerae* is spread via the oral–fecal route, and people become ill with cholera when they ingest food or water contaminated with feces.

### Response to temperature, pH, osmolarity, and amino acids

Exactly which environmental stimuli in the intestine influence the activity of ToxR within infected animals is a subject of study.[225, 228] Toxin production is lowered in pure culture when the temperature is increased to 37°C, which is opposite to what one would intuitively believe would happen. Additionaly, low pH in culture stimulates cholera toxin production, but the pH in the intestine where *V. cholerae* grows is believed to be alkaline. This suggests that factors other than temperature and pH increase the activity of ToxR during growth in the intestine. Two of these factors might be osmolarity and the presence of certain amino acids, because toxin production in pure culture is increased under osmolar conditions similar to that in mucus, and in the presence of amino acids likely to be present in mucosal secretions. However, the expression of the ToxR regulon in response to environmental cues can differ with the strain of *V. cholerae*. (See Note 229.)

### The role of ToxR in pathogenesis

The importance of ToxR for pathogenesis is clear, since mutants in which the *toxR* gene has been deleted do not colonize the intestine in human volunteers. ToxR is an *inner membrane bifunctional protein that can sense environmental signals and subsequently activate the transcription of genes in the ToxR regulon* (at least 20 genes). The genes activated by ToxR are called the ToxR-activated genes or *tag* genes, and these are required for efficient colonization of the intestine, toxin production, and survival within the host. The *tag* genes have been identified by screening Tn*phoA* fusions whose transcriptional activity under different growth conditions parallels that of cholera toxin production. (See Note 25 for an explanation of gene fusions and Note 230 for how such fusions were used to detect ToxR-regulated genes.) Genes whose transcription is increased by ToxR include the *ctxAB* operon, which encodes the A and B subunits of cholera toxin, *tcpA*, which encodes the major subunit of the *V. cholerae* pilus, several genes required for pilus assembly and transport, and other genes noted in Fig. 18.27. As mentioned, the pilus is required for *V. cholerae* to bind to the intestinal epithelium and subsequent colonization. Thus it can be considered a virulence factor. As described next, except for the transcription of *ctxAB*, the genes in the ToxR regulon are not activated directly by ToxR.

### How ToxR activates gene expression

In order to study the *toxR* gene, it has been cloned in *E. coli*, where it activates the transcription of a *ctx–lacZ* fusion. (See

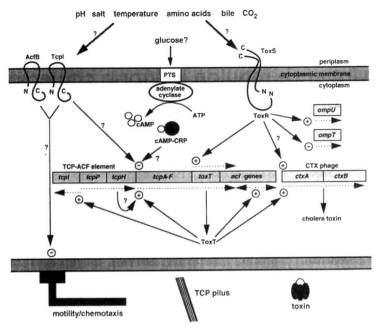

**Fig. 18.27** Model showing regulatory pathways for ToxR regulon. The genes encoding the cholera toxin (*ctxA* and *ctxB*) are encoded on a lysogenic phage (CTX phage). Transcription of the cholera toxin genes can be activated by either ToxR or ToxT. Other genes are: *tcpl* (involved in the negative regulation of pilus production), *tcpP* (involved in the positive regulation of pilus production and ToxR production), *tcpH* (involved in the positive regulation of pilus production), *acf* (accessory colonization factors), *ompU* (outer membrane protein), and *ompT* (outer membrane protein). The *tcp–acf* genes are referred to as a *pathogenicity island*. TcpI is an inner membrane protein that has been compared to the methyl-accepting chemoreceptors that are part of the chemotaxis system. TcpP is thought to be a transcriptional activator of ToxT. TcpH enhances the activity of TcpP. It has been suggested that OmpU might be an adherence factor. The reason for repression of OmpT is not known. Motility and chemotaxis genes are down-regulated by ToxR. It has been speculated that down-regulation of motility genes may keep the bacteria at the intestinal epithelium. *Source:* From Skorupski, K., and R. K. Taylor. Control of the ToxR virulence regulon in *Vibrio cholerae* by environmental stimuli. *Molec. Microbiol.* **25:**1003–1009.

Note 25.) ToxR binds to the promoter region of the *ctxAB* operon, resulting in increased transcription, but its activation of the other genes in the ToxR regulon is indirect. That is to say, it appears that ToxR increases the transcription of a second regulatory gene, *toxT*, whose product directly activates the other genes in the ToxR regulon, including *ctx* and *tcpA*.[231] A model showing the regulatory pathways is shown in Fig. 18.27.

## It is not understood how ToxR transmits environmental signals to the toxR regulon

It is unclear exactly how ToxR receives and transmits environmental signals to the ToxR regulon. It has been concluded, based in part upon amino acid sequence data and the construction of ToxR–PhoA fusions that have alkaline phosphatase activity and also regulate the cholera toxin gene, that ToxR spans the cell membrane with a periplasmic C-terminal region and an N-terminal cytoplasmic region that has a DNA binding domain. (See Ref. 225 and references therein and Note 25 for an explanation as to why alkaline phosphatase activity from a *toxR–phoA* fusion can indicate a C-terminal periplasmic domain.) The C-terminal domain may be sensitive to certain environmental stimuli discussed next. The N-terminal domain is thought to bind to promoter sequences in the DNA of target genes, and it may also respond to intracellular stimuli.

## A signal transducing cascade

How the ToxR protein senses the environmental signals, and whether other components besides ToxR are involved in sensing these signals, is an open question. It has been suggested that another protein, ToxS, which is a membrane-bound protein with a periplasmic domain, may actually receive environmental signals and activate ToxR by interacting with its periplasmic domain. The cytoplasmic domain of ToxR would then activate the *toxT* gene whose product would in turn would activate the genes in the ToxR regulon (Fig. 18.27). However, other signaling proteins are also involved. Two of these are TcpP and TcpH, as described next.

## The toxT *gene is also positively regulated by TcpP*

The *toxT* gene is also positively regulated by another cell membrane-bound transcriptional activator, TcpP, which itself may be activated by a cell membrane-associated protein with a large periplasmic domain, TcpH.[232] Optimal expression of *toxT* requires *both* ToxR and TcpP. It has been speculated that *both* ToxR/ToxS *and* TcpP/TcpH are involved in sensing environmental and cytoplasmic signals and *cooperatively* activate the *toxT* gene.

## 18.19.2 PhoQ/PhoP and Salmonella pathogenesis

### Salmonella pathogenesis

Salmonella spp. cause gastroenteritis, enteric fever, and septicemia (blood infections). During the course of the infections the bacteria are ingested by phagocytic cells such as macrophages and epithelial cells and must survive intracellularly. Indeed, the spread of the bacteria from the intestinal mucosal surface to underlying tissue and the development of gastroenteritis, enteric fever, and speticemia requires the invasion of intestinal epithelial cells.

## The PhoQ/PhoP system signals virulence genes

A two-component system called the PhoQ/PhoP sysem has evolved in *Salmonella* to help them cope with the hostile intracellular macrophage environment (Fig. 18.28). PhoQ is a sensor kinase and PhoP is a response regulator. Together, they are responsible for the induction of 5 genes in *Salmonella* spp. called the *pag* genes (PhoP-activated genes), which are important for the survival of *Salmonella* inside the acid bacteriocidal environment of phagolysosomes of macrophages (they are not induced inside epithelial cells) and for the repression of several other genes called *prg* (PhoP-repressed genes).[233] (The gene *phoP* was originally discovered as a positive regulator for the synthesis of an acid phosphatase, the product of the *phoN* gene. The PhoQ/PhoP system is not related to the two-component PhoB/PhoR system discussed in Section 18.19.1. See Note 234 for a description of how *phoP* was discovered.) Mutants in *phoQ/phoP* are less virulent and have decreased survivability inside macrophages and less resistance to cationic antimicrobial peptides such as defensins, which the bacteria can encounter in the intestinal lumen and within phagolysosomes of neutrophiles and macrophages. (For a review, see Ref 235.) One of the *pag* genes indured by the PhoQ/PhoP system (i.e., *pagC*) is responsible for the synthesis of an envelope protein that supports virulence as well as survival inside macrophages.

## PhoQ/PhoP regulation not restricted to virulence genes; response to Mg$^{2+}$ concentrations

However, the PhoQ/PhoP system is not restricted to the regulation of virulence genes. There are at least 25 genes regulated by the PhoQ/PhoP system, and most of them are not required for virulence. The PhoQ/PhoP system is stimulated by low Mg$^{2+}$ and plays an important role in adapting to low extracellular Mg$^{2+}$ concentrations (micromolar range).[236, 237] Two of the genes activated by PhoQ/PhoP encode proteins for two of the three Mg$^{2+}$ uptake systems. Thus the PhoQ/PhoP system appears to function not only during host invasion but also when the bacteria are living outside their host. Consistent with this conclusion, homologs to PhoP have been found in other bacterial species that are nonpathogenic.

## A second two-component system that interacts with the PhoQ/PhoP system: PmrB/PmrA and the response to pH

Some of the genes regulated by PhoP are also induced when the cells are growing in slightly acidic environments, which they would encounter inside the phagolysosomes of macrophages. But this does not seem to be an effect on PhoQ, which apparently responds only to $Mg^{2+}$. PhoP indirectly regulates the expression of several genes via a second response-regulator protein, PmrA, and it is the PmrA pathway that responds to pH. This story begins with the observation that one of the loci whose transcription is increased by the PhoQ/PhoP system encodes proteins for another two-component system, the PmrB/PmrA system, where PmrB is the histidine kinase and PmrA is the response regulator.[238] Furthermore, some of the PhoP activated genes also require PmrA for induction. Thus it appears that there are two classes of PhoP-activated genes, that is, those that are directly activated by PhoP, and those that are activated by PmrA. According to the model described next, PmrA can be phosphorylated by PhoP or a cognate histidine kinase (i.e., PmrB).

### 1. The model relating PhoQ/PhoP to PmrB/PmrA

The model proposes that PhoQ becomes phosphoryated when the $Mg^{2+}$ concentrations get very low (Fig. 18.28). (Recall that the response-regulator protein, PhoB, that regulates the Pho regulon is phosphorylated when the inorganic phosphate concentrations get very low. See Section 18.9.) Then PhoQ-P transfers the phosphoryl group to PhoP. PhoP-P would then either phosphorylate PmrA (the second response-regulator protein) directly or transfer the phosphoryl group to PmrB (the histidine kinase), which would, in turn, phosphorylate PmrA. PmrA-P would then activate the transcription of target genes. The proposed signaling pathway bears some resemblance to the multicomponent phosphorelay system that operates during the initiation of

sporulation in *B. subtilis*, which also includes two response regulator proteins in series. (See Section 18.15.3.) All genes activated by PhoP, even those that respond to PmrA, are induced by low concentrations of $Mg^{2+}$ because the sensor–histidine kinase, PhoQ, responds to $Mg^{2+}$. Those genes activated by PmrA (including the *pmrCAB* operon that encodes the Pmr A and PmrB proteins) can also be induced by mild acidification As shown by studies with the appropriate mutants, activation by low pH does not require PhoQ or PhoP, indicating that PmrA can receive a signal from a molecule other than PhoP when the pH drops. Perhaps PmrB, the putative cognate sensor for PmrA, is responsive to pH and will phosphorylate PmrA under acidic conditions (around pH 6). See the discussion on the regulation of the formate hydrogen–lyase pathway in Section 18.2.2 for another regulatory system activated by low pH.

## 18.19.3 The bvg genes and pertussis

There are several other regulatory genes that respond to environmental stimuli and promote virulence. (Reviewed in Ref. 225.) These include the *bvgAS* genes (*Bordetella* virulence genes), which are required for the expression of virulence genes (the *vir* regulon) in *Bordetella pertussis*, the causative agent of whooping cough (pertussis).[239] (The *bvgAS* locus was formerly called the *vir* locus.) Growth at 37°C as opposed to 25°C results in increased expression of virulence genes.

### Pertussis

This is primarily a childhood disease in which the bacteria grow attached to the ciliated epithelial cells of the bronchi and trachea. (Virulence is dependent upon bacterial surface adhesins that bind the bacteria to the cilia.) Episodes of severe, spasmodic coughing are followed by a "whoop" during inspiration. The organism is transmitted via respiratory discharges. Children are now vaccinated with the DPT (diptheria, pertussis, tetanus) vaccine. However, prior to the vaccine pertussis was a major childhood killer.

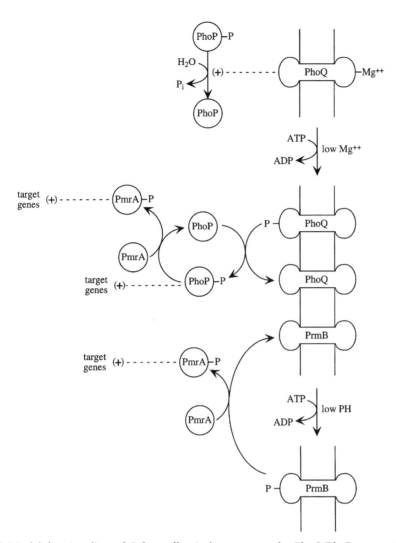

**Fig. 18.28** Model for signaling of *Salmonella* virulence genes: the PhoQ/PhoP system. The model proposes that two transmembrane sensor kinases exist, i.e., PhoQ and PmrB. They each have a periplasmic and cytoplasmic domain. In the presence of millimolar concentrations of $Mg^{2+}$ PhoQ represses transcription of PhoP-dependent virulence genes. It does so either by dephosphorylating its cognate response regulator, PhoP, or by not phosphorylating it. When $Mg^{2+}$ concentrations fall to the micromolar range, then PhoQ autophosphorylates and phosphorylates PhoP. PhoP-P can directly stimulate the transcription of several genes, some of which are important for virulence, and some of which are important for growth in limiting $Mg^{2+}$. PhoP-P can also phosphorylate a second response regulator, PmrA, which stimulates the transcription of a subset of genes activated by the PhoQ/PhoP system. When the cells are grown in moderately acid pH (6.1), a set of genes is induced that does not require PhoQ/PhoM and which is presumably induced in the acidic environment of macrophage phagolysosomes. It has been speculated that PmrB is a sensor kinase that somehow senses pH (either directly or via a second ligand) and phosphorylates PmrA.

### *vir regulon*

Genes induced by the *bvg* genes (i.e. the genes in the *vir* regulon) include the pertussis toxin operon (*ptxA–E*), the adenylate cyclase toxin–hemolysis gene (*cyaA*), the filamentous hemagglutinin locus (*fhaB*), the fimbrial subunit genes (*fim*), and the *bvg* operon.

## The bvg genes encode a two-component signaling system that regulates the vir regulon

According to deduced amino acid sequences, BvgS is a transmembrane sensor kinase with a periplasmic N-terminus region and a cytoplasmic C-terminus region, Similarly, BvgA is proposed to be a response-regulator protein that is phosphorylated by BvgS-P and then activates transcription of the virulence genes, perhaps in some cases by activating other genes that are transcriptional activators. (In *E. coli*, the cloned *bvgAS* locus will activate the *fhaB* and *bvgA* promoters but not the *ptxA–E* or *cyaA* promoters, suggesting that perhaps a signal relay is required for the latter promoters.) Increased expression of virulence genes at 37°C has also been studied in *Shigella*, as discussed next.

### 18.19.4 Virulence genes and bacillary dysentery

#### Bacillary dysentery

*Shigella spp.*, including *S. dysenteriae* and *S. flexneri*, cause bacillary dysentery (shigellosis), which can result in simply a watery diarrhea or be manifested as a severe diarrheal disease of the colon accompanied by passage of bloody, mucoid stools. In the severe cases, the bacteria invade and multiply within the mucosal epithelial cells, causing degeneration of the epithelium and intense inflammation of the colon (colitis). (*Shigella* are facultative intracellular pathogens.) Most *Shigella* species produce a toxin, and the toxin produced by *Shigella dysenteriae* is *shiga toxin*, which is at least partly responsible for damaging the blood vessels, inflammation, and probably the watery diarrhea.

#### Virulence genes

The virulence of *Shigella* is due primarily to a plasmid that encodes the virulence genes. Virulence genes are required for adherence and invasion (via endocytosis) of host cells, spreading (via polymerization of actin filaments), and synthesis of toxin. (For a review of virulence genes in *Shigella* and how the bacteria invade

host cells and spread intercellularly, see Ref. 240. Also, see Note 241.)

#### 1. hns (virR); response to temperature

*Shigella* have a chromosomal regulatory gene called *hns* (*virR*) that encodes a repressor for the virulence genes. The significance of *hns* for this discussion is that the regulation of virulence genes by *hns* is controlled by temperature. *Shigella* strains grown at 30°C are avirulent; that is, the virulence genes are repressed by *hns*, whereas those grown at 37°C are virulent in tissue culture assays and animal assays (i.e., invasion of tissue culture cells and production of keratoconjunctivitis in guinea pigs). Consistent with its role as a repressor gene at 30°C, mutations in *hns* allow expression of the invasive phenotype at both 30 and 37°C. Presumably, turning off virulence genes is an advantage to the bacteria when they are growing outside the host. The product of the *hns* gene is a histone-like protein called H-NS (H1), and binding of H-NS near target genes leads to transcriptional repression, perhaps by interfering with the binding of RNA polymerase. (See the discussion of H-NS and the nucleoid in Section 1.2.6.)

#### 2. virF and virB

The virulence genes are also positively regulated. One of the regulatory genes, *virF*, encodes a positive regulator (VirF) of the virulence gene regulon. It activates some of the *vir* genes directly, and others indirectly by activating the transcription of another positive regulatory gene, *virB*. It is thought that the product of the *hns* gene (H-NS) represses *virF* expression at 30°C, and this accounts for the repression of the other virulence genes. The virulence genes are also repressed when the cells are grown in low-osmolarity media at 37°C, and this was also shown to be due to repression by H-NS in *S. flexneri*.[242]

#### 3. IHF; responses to temperature and osmolarity

When *Shigella* grow outside the host in nature, it is generally in media of low osmolarity as well as at temperatures lower than 37°C. In addition to being repressed by H-NS at 30°C, the expression of virulence genes is also positively stimulated upon a temperature upshift. The positive stimulation appears to be due to another DNA-binding protein, integration host factor (IHF).[243] (See the discussion of IHF in Sections 10.2.5 and 18.8.2 and the discussion of DNA-binding proteins in the nucleoid in Section 1.2.6.) This was determined by examining the expression of a *lacZ* fusion to a virulence gene upon a temperature upshift in wild-type cells and *ihf:Tn10* mutants.

Thus DNA-binding proteins H-NS (a repressor) and IHF (an activator) appear to take part in signaling pathways that detect environmental changes (i.e., temperature and osmolarity) that result in alterations in the expression of virulence genes.

### 18.19.5 Virulence gene expression in Staphylococcus: Stimulation of a two-component system by a peptide pheromone

*Staphylococcus aureus* and other *Staphylococcus* strains secrete several extracellular protein virulence factors. For example, *S. aureus* secretes enterotoxin B and toxic shock syndrome toxin-1, as well as other toxins in the post-exponential phase of growth. The transcription of the toxin genes is regulated by a two-component system consisting of a membrane histidine–kinase called AgrC, a response regulator protein called AgrA, and a small peptide secreted by the cells into the medium.[244] The peptide is a quorum-sensing signal because the amount secreted increases with cell density, and when the concentration of the peptide reaches a threshold concentration, it activates AgrC, which leads to increased transcription of virulence genes.

### 18.19.6 Virulence gene expression in Agrobacterium tumefaciens; Stimulation of a two-component system by phenolic plant exudates

One of the best-characterized systems for the regulation of expression of virulence genes in bacterial–plant interactons is the VirA–VirG two-component system in *Agrobacterium tumefaciens*, the bacterium that causes

crown gall tumors in plants. (For a review, see Ref. 225.)

### Stages in the formation of crown gall tumors

The bacteria gain entrance to the plant tissue via a wound site on the plant and grow extracellularly at the site of infection. Bacteria that are directly attached to plant cells transfer a small piece of DNA, called T-DNA, from a plasmid, called the Ti plasmid (tumor inducing), into the plant cells. The process of transfer of the T-DNA from the bacterium into the plant resembles bacterial conjugation. The T-DNA integrates into the plant genome, and a tumor results. The reason that the tumor forms is that T-DNA encodes plant growth hormones (i.e., auxin and cytokinin) which stimulate growth and division of the plant cells.

### When the T-DNA integrates into the plant genome, the plant makes nutrient that feed the bacteria

The crown gall tumor actually feeds the bacteria. There are genes in the T-DNA that encode enzymes to make organic molecules called opines, which when released by the plant cell can be used as a source of nutrient by the bacteria. Opines are small organic molecules covalently bonded to amino acids. For example, one opine, called octopine, is $N^2$-(1,3-dicarboxyethyl)-L-arginine. The genes that encode enzymes to utilize the opines are in the region of the Ti plasmid that is not transferred. Thus, when the T-DNA becomes incorporated into the plant genome, the plant makes the opines, and the bacteria use these opines for growth. Interestingly, the genes from the Ti plasmid that are expressed in the plant have eukaryotic-type promoters and eukaryotic-type translation initiation regions, whereas the plasmid genes expressed in the bacterium have prokaryotic-type promoters and prokaryotic-type translation initiation regions.

### Plant exudate material stimulates plasmid DNA transfer from bacterium to plant cell

In order for plasmid DNA transfer to occur, the bacterium senses phenolic compounds produced by the wounded plant tissue. The ability to sense the phenolic compounds is due to products of the *vir* genes that are encoded by the plasmid. A signaling system is activated by the phenolic compounds, which induces the transcription of the *tra* genes encoded by the plasmid. The products of the *tra* genes are responsible for the transfer of the T-DNA from the bacterium to the plant cell. Specifically, the phenolic response system is a two-component system (VirA–VirG). VirA is the transmembrane sensor–kinase whose periplasmic domain is thought to interact with the phenolic compounds. VirA is then proposed to autophosphorylate and transfer the phosphoryl group to VirG, a cytoplasmic regulator protein. The phosphorylated form of VirG is believed to be a positive transcription factor for the *tra* genes. The transfer of T-DNA from the Ti plasmid into the plant cell does not result in loss of T-DNA from the plasmid. This is because only one strand of the T-DNA is transferred to the plant cell and a new complementary strand is synthesized in the donor bacterium. The process is very much like the transfer of a single plasmid strand during bacterial conjugation.

## 18.20 Quorum Sensing Using Acylated Homoserine Lactones

Many different gram-negative bacteria are now known to communicate within their own population using acylated homoserine lactones (acylated HSLs) as secreted extracellular signals. When the population of bacteria reach a critical cell-density threshold, the accumulated acylated HSLs generally induce the expression of certain genes that are active only at high cell densities. The signaling systems are called quorom-sensing systems, and may possibly be used to detect other species of cells that also secrete acylated HSLs. (For a review, see Ref. 259.) The first quorum-sensing systems based upon acylated HSLs were discovered in luminescent bacteria, and they serve to activate genes for luminescence. It is now clear that similar systems are widespread among bacteria, although the genes that are regulated differ, according to the bacterium and its ecological niche. This section describes some of these systems, beginning with bacterial luminescence.

## 18.20.1 A brief survey of bioluminescent systems

Bioluminescence is the emission of light by living organisms. The organisms that are bioluminescent reflect a diverse assemblage that includes representatives from the algae (dinoflagellates), fungi, shrimp, insects (fireflies), squid, fish, and bacteria.[245–247] Although some live in the soil or in fresh water, most are marine organisms. In fact, the majority of animal species living at a depth of approximately 200–1000 meters in the ocean are bioluminescent. The specific survival advantage of an organism that is capable of bioluminescence no doubt reflects the organism and its niche. In some cases, bioluminescence is used for intraspecies communication, and to hunt or attract prey. With regards to luminous bacteria, it is reasonable to suggest that the ability to luminesce increases the probablility that the bacterium will be taken up by a host, and thereby be provided a beneficial environment, which includes nutrient. Thus one can imagine that enteric bacteria that normally live in the intestines of fish might increase the probability of being ingested if they emitted light in the dark ocean depths while colonizing particles that the fish might then see and ingest.

### The luminous bacteria

Luminous bacteria live in freshwater and terrestrial environments, but are most common in marine environments. Marine luminescent bacteria can be isolated from the intestines of fish and invertebrates, as saprophytes growing on dead marine animals (e.g., fish, crustacea), as symbionts growing in the light organs of fish (primarily), squids, and pyrosomes (colonial tunicates), and as part of the plankton. The advantages to the bacterial host of harboring luminous bacteria are several, as the following discussion will illustrate. Some species of luminous bacteria live symbiotically in light organs of fish. Light organs in fish consist of tubules or canals, densely packed with extracellular bacteria, that connect to the exterior through pores, from which the bacteria may exit into the intestine or seawater, depending upon the location of the light organ. The majority of bacterial light organs are internal and are derived as outpocketings of the intestinal tract. Probably the purpose that internal light organs serve the fish is one of counterillumination. During counterillumination the fish eliminates its silhouette, which would be produced by down-welling light, by emitting light from its ventral side, thus confounding its predators. The external light organs are at the surface of the fish, for example, special pouches under the eyes. They are used for vision (for seeking prey) and for intraspecies communication. *Photobacterium phosphoreum* and *P. leiognathi*, are symbionts of internal light organs and *Vibrio fischeri* is a symbiont of external light organs. (The genus *Photobacterium* is not easily distinguished from *Vibrio*. See Note 248 for a discussion of this point.) Since bacterial luminescence is continuous, fish have evolved a variety of ways to control the release of the light to the outside. These include the use of pigment (melanophores) screens that block the light organ, shutters, and reflectors. For example, some fish, whose light organ is a pouch under the eye, can block the light by pulling tissue over the pouch (like an eyelid). Other fish can rotate the luminescent pouch downward so that the light does not exit.

All the luminous bacteria are phylogenetically grouped with the enteric bacteria. Luminous bacteria that live in the oceans belong to three genera: *Vibrio*, *Photobacterium*, and *Alteromonas*. It is presumed, or has been demonstrated, that they all can live saprophytically on dead fish or meat, exist free-living as part of the plankton in the oceans, and can also live in the intestines of their respective hosts. Some (*V. fischeri*, *P. phosphoreum*, and *P. leiognathi*) can also live as symbionts in light organs of fish and (except for *P. phosphoreum*) squid. (However, none of the marine luminous bacteria belonging to the *V. harveyi* group, which is discussed later, is known to colonize light organs of fish, although they live in the gut of marine animals as well as in the sea.) The terrestrial and fresh water luminous bacteria are *Xenorhabdus luminescens* and certain species of *Vibrio cholerae*, respectively. *V. cholerae*

inhabits the human intestine and may cause cholera, but the luminescent strains are not known to cause disease. *X. luminescens* grows as a symbiont in the intestines of a nematode, which infects a variety of insects. While the nematode grows and completes its life cycle in the dying insect (e.g., caterpillars), *X. luminescens* is expelled by the nematode into the insect hemolymph, where their growth results in a glowing insect carcass. The developing nematodes feed on the bacteria in the insect hemolymph.

## 18.20.2 Biochemistry of bacterial bioluminescence

### The overall reaction

Luminous bacteria produce light when they simultaneously oxidize reduced riboflavin-5'-phosphate ($FMNH_2$), and a long-chain aldehyde (RCHO), such as $C_{14}$ (tetradecanal), with oxygen. The reaction is catalyzed by a flavin monooxygenase called *luciferase*, as follows:

$$FMNH_2 + RCHO + O_2 \longrightarrow FMN + H_2O$$
$$+ RCOOH + light$$

The reason that light is emitted is that during the reaction the flavin becomes electronically excited and subsequently fluoresces as the electron returns to its ground state. (In certain bacterial strains there may be a second chromophore, bound to a second protein, that emits light.) This is a complicated reaction. Of the two oxygen atoms in $O_2$, one becomes reduced to water, while the other is incorporated into the carboxyl group of the carboxylic acid. A model for how this is coupled to emission of light is discussed in the next section. A schematic illustration of the bioluminescent pathway is shown in Fig. 18.29. The figure points out that in order for luminescence to continue, both $FMNH_2$ and RCHO must be regenerated. The aldehyde (RCHO) is regenerated from the carboxylic acid (RCOOH) via reduction by NADPH, catalyzed by a fatty acid reductase complex consisting of three proteins (a transferase, a reductase, and a synthetase). The $FMNH_2$ is regenerated by the reduction of FMN

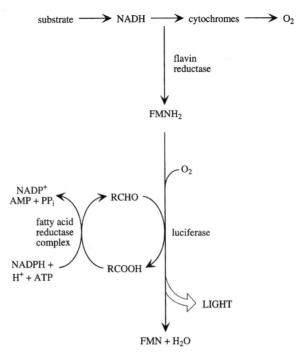

**Fig. 18.29** Bioluminescent pathway in bacteria. The electron transport pathway through luciferase to oxygen is a shunt off the cytochrome pathway. A flavin reductase reduces FMN, which then combines with luciferase and a long-chain aldehyde. The luciferase catalyzes the oxidation of $FMNH_2$ to FMN and the oxidation of the aldehyde to the carboxylic acid with the production of light. The aldehyde is regenerated from the carboxylic acid via a fatty acid reductase complex that uses ATP and NADPH. *Source:* From: Nealson, K. H., and J. W. Hastings. 1992. The luminous bacteria, pp. 625–639. In *The Prokaryotes,* Vol. I, 2nd ed. A. Balows, H. G. Truper, M. Dworkin, W. Harder, and K.-H. Schleifer (Eds.). Springer-Verlag, Berlin.

by NADH, catalyzed by an NADH–FMN oxidoreductase (flavin reductase). Notice that the luciferin pathway can be viewed as a shunt leading from the cytochrome respiratory chain.

*Light emission*

During the luminescent reaction, oxygen reacts with $FMNH_2$ (bound to the luciferase) to form a flavin-4a-hydroperoxide (FMNH-4a-OOH). (See below and Fig. 18.30.) The flavin hydroperoxide then reacts with the aldehyde to form a postulated peroxyhemiacetal intermediate, which decomposes to the carboxylic acid and an enzyme-bound flavin-4a-hydroxide (flavin-4a-OH) with an excited electron. When the electron returns to the ground state, fluorescence occurs.[249] At some point, water is eliminated from the flavin-4a-hydroxide, producing FMN. The reactions shown in Fig. 18.30 can be summarized as follows:

Formation of peroxide: $FMNH_2$ + luciferase + $O_2$ $\longrightarrow$ luciferase–FMNH-4a-OOH

Binding of aldehyde: luciferase–FMNH–OOH + RCHO $\longrightarrow$ luciferase–FMNH–OO–RCHO

Oxidation of aldehyde: luciferase–FMNH–OO–RCHO $\longrightarrow$ luciferase–(FMNH–OH)* + RCOOH

Emission of light: luciferase–(FMNH–OH)* $\longrightarrow$ luciferase–FMNH–OH + light

Elimination of water: luciferase–FMNH–OH $\longrightarrow$ luciferase + FMN + $H_2O$

Sum: $FMNH_2$ + $O_2$ + RCHO $\longrightarrow$ FMN + $H_2O$ + RCHOOH + light

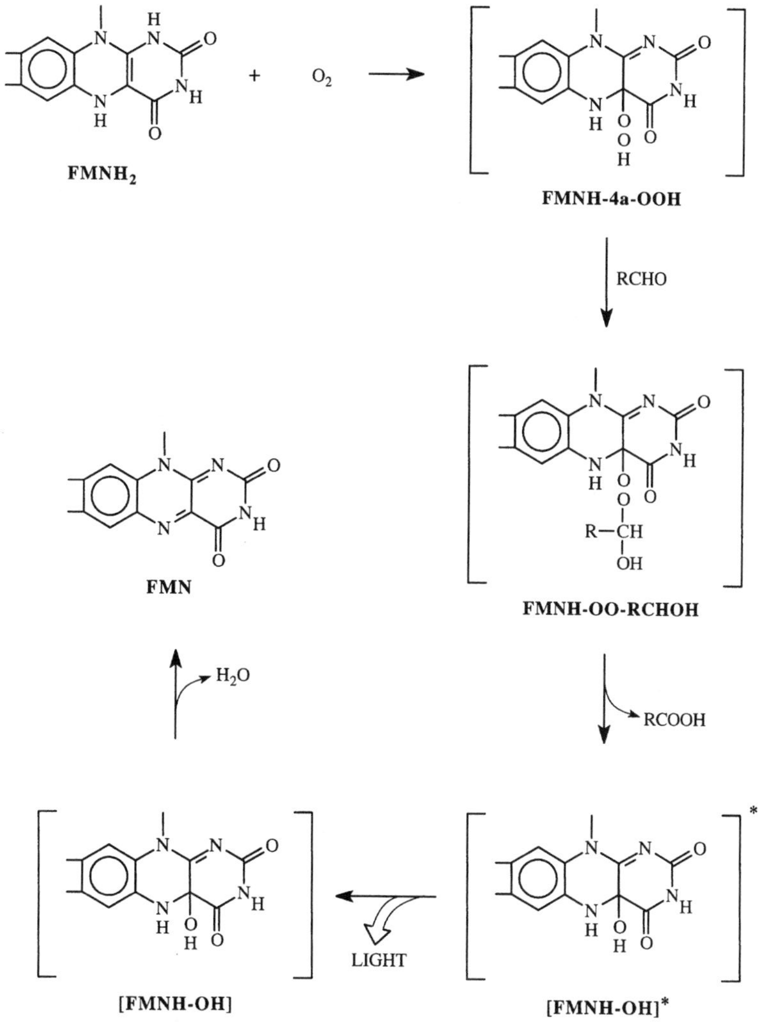

**Fig. 18.30** Model for light emission during luciferase activity. $FMNH_2$ reacts with $O_2$ to form the hydroperoxide derivative, FMNH-4a-OOH. The FMNH-4a-OOH then combines with a long-chain aldehyde, RCHO, to form the peroxyhemiacetal, FMNH-OO-RCHO. The aldehyde becomes oxidized to the carboxylic acid and dissociates, resulting in an electronically excited flavin hydroxide, [FMNH-OH]*. The excited electron in the flavin hydroxide returns to its ground stat, releasing light. The flavin hydroxide loses water to become oxidized flavin, FMN. The overall reaction is that one atom of oxygen becomes reduced to water as it accepts two electrons from $FMNH_2$, and one atom of oxygen becomes incorporated into the carboxyl group as the aldehyde is oxidized to a carboxyl. (For a discussion, see Kurfuerst, M., P. Macheroux, S. Ghisla, and J. W. Hastings. 1987. Isolation and characterization of the transient luciferase-bound flavin-4a-hydroxide in the bacterial luciferase reaction. *Biochem. Biophys. Acta* **924**:104–110.

### 18.20.3 Intercellular signaling, quorum sensing, and the lux genes

It has been known for some time that certain species of luminous bacteria will emit light in cell suspension only when the cell density reaches a certain threshold ("quorum"), such as $> 10^7$/ml.[250] This is because the bacteria will not luminesce until they have accumulated in the extracellular medium a threshold concentration of a species-specific signal for luminescence called *autoinducer*. The autoinducers are N-acyl-L-homoserine lactones (acyl HSLs), that is, homoserine lactone linked to an acyl side chain, the structure of which varies with the specific autodinducer. Before describing the autoinducer and how it works, it is first necessary to introduce the genes that are responsible for bioluminescence. The genes that encode the luciferase proteins and the fatty acid reductase complex components are encoded in the *lux* operon, which consists of a minimum of five genes that are required for luminescence (i.e., *luxCDABE*). *LuxAB* encode the $\alpha$ and $\beta$ subunits, respectively, of the luciferase (a heterodimer), and *luxCDE* encode the three proteins of the fatty acid reductase complex, which is necessary for the synthesis of the long-chain aldehyde. (Depending upon the species of bacterium, other genes may also exist in the operon. These will be mentioned later.) As will be discussed shortly, the expression of these genes is regulated by the autodinducers. The regulation of the *lux* operon in two luminous bacteria, *V. fischeri* and *V. harveyi* has been studied in detail. Interestingly, while both microbes utilize acylHSLs to sense their own density, the mechanism of *lux* gene regulation in *V. fischeri* appears to be very different from that in *V. harveyi*, and a discussion of the *V. harveyi* system is deferred to Section 18.20.5. The *V. fischeri* system, which is mechanistically similar to quorum sensing in many nonluminescent bacteria, will be discussed next. This will be followed by a description of similar quorom-sensing systems in nonluminescent gram-negative bacteria.

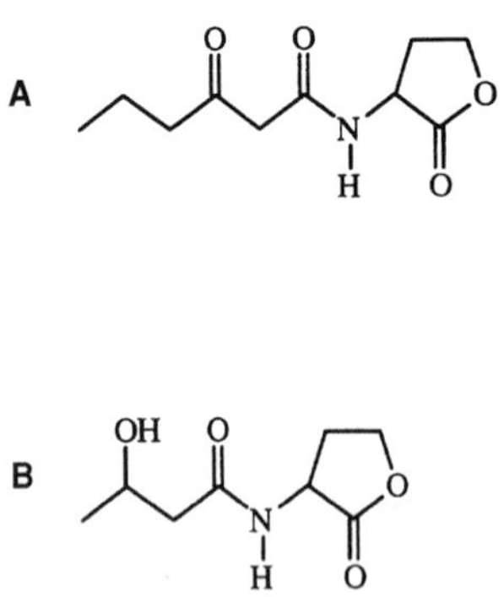

## V. fischeri

*V. fischeri* occurs as a symbiont, where it lives at very high cell densities (i.e., around $10^{11}/ml$) in the light organs of certain marine fish. It also occurs as a free-living bacterium in sea water but at a much lower cell density (i.e., less than $10^2$ cells/ml). The bacteria luminesce only in the light organs of the fish because it is here that the autoinducer produced by the bacteria reaches a sufficient concentration to induce luminescence. The autoinducer acts as a population density signal (quorum-sensing signal) and allows luminescence only when the bacteria are in the light organs of the fish. The autoinducer (AI-I) is $\beta$-ketocaproyl homoserine lactone (N-3-oxohexanoyl-L-homoserine lactone) (Fig. 18.31). It is synthesized by LuxI (autoinducer synthase), which is encoded by the gene *luxI*. The autoinducer is a nonpolar substance that diffuses freely across the bacterial cell membrane and accumulates in the extracellular medium during growth. When the autoinducer reaches a critical concentration inside the cells (the concentrations inside and outside the cells are the same), it binds to a transcription factor, LuxR, encoded by the *luxR* gene, which to the left of the *lux* operon (*luxICDABEG*) and is expressed divergently. The binding of the autoinducer to LuxR enables LuxR to bind to the *lux* regulatory DNA and stimulate transcription of the *lux* operon, perhaps by recruiting $\sigma^{70}$ RNA polymerase.[251, 252] The model is that as the autoinducer re-enters the cell, it binds to LuxR, which may be associated with the cytoplasmic side of the cell membrane.[253] Notice that the operon contains the gene for the synthesis of the autoinducer (i.e., *luxI*). Thus the presence of autoinducer stimulates its own synthesis in a positive feedback loop. (The function of the *luxG* gene is not known, although it is transcribed when the other *lux* genes are activated. Mutations in this gene do not appear to influence luminescence or its regulation.) Recently, *V. fischeri* has been reported to produce two other autoinducers in lesser amounts. These are also acyl derivatives of homoserine lactone, that is, AI-2 (N-hexanoyl-L-homoserine lactone) and AI-3 (N-octanoyl-L-homoserine lactone).[254] The synthesis of AI-3 requires *luxI*, but the synthesis of AI-2 is independent of *luxI*

Fig. 18.31 Autodinducers produced by *Vibrio fischeri* and *V. harveyi*. (A) The *V. fischeri* autoinducer: $\beta$-ketocaproyl homoserine lactone. (B) The *V. harveyi* autodinducer: $\beta$-hydroxybutyryl homoserine lactone.

and requires for its synthesis a newly discovered gene, *ain* (autoinducer), that specifies a protein that is different from AI.[255] All three autoinducers interact with LuxR. It is not obvious why *V. fischeri* should have multiple autoinducers. Although LuxR is the primary regulatory factor for the *lux* genes, several other factors besides LuxR regulate the transcription of the *lux* genes.[256–258]

### 18.20.4 Quorum-sensing using systems similar to LuxR/LuxI by nonluminescent bacteria

Autoinducers similar to the ones produced by luminescent bacteria have been discovered in nonluminescent bacteria, where they regulate the expression of genes unrelated to the *lux* genes. (For reviews, see Refs. 259 and 260.) The components of the quorum-sensing pathways in the systems described next are quite similar to those of the *V. fischeri* system. They employ proteins similar to LuxI and LuxR, and the autoinducers differ from the autoinducers produced by the luminescent bacteria only in the length and nature of the acyl chain attached to the homoserine lactone.

### Agrobacterium tumefaciens

*Agrobacterium tumefaciens* produces crown gall tumors in plants, that is, tumors in the region where the roots and stem meet. (This region is called the crown.) The bacteria cause tumors by transferring a small piece of plasmid DNA, called T-DNA, from a resident plasmid called the Ti plasmid, to the plant. (For more about *Agrobacterium* and crown gall tumors, see Section 18.19.6.) In addition to transferring T-DNA to plants, the Ti plasmid itself can be transferred between *Agrobacterium* cells within the crown gall tumors or in culture with added opines. (Opines are a carbon, energy, and nitrogen source for the bacteria produced by the tumors.) Conjugation between the bacteria requires a high cell density. A quorum-sensing system strikingly similar to the *V. fischeri* system regulates the expression of genes required for conjugal plasmid transfer (the *tra* genes).[261–263] *A. tumefaciens* produces an extracellular conjugation factor, AAI, which is

N-(β-oxo-octan-1-oyl)-L-homoserine lactone (3-oxo-C8-HSL), that is synthesized by the TraI protein. The TraI protein is homologous to LuxI, which is the autoinducer synthase that synthesizes the *V. fischeri* autoinducer. 3-oxo-C8-HSL differs from the *V. fischeri* lux autoinducer (AI) only in the length of the acyl chain. Mutants that do not produce 3-oxo-C8-HSL are unable to conjugate, but will conjugate if 3-oxo-C8-HSL is added to the incubation medium. These results indicate that 3-oxo-C8-HSL serves as a cell-to-cell signal that prepares the population of cells for conjugation. 3-oxo-C8-HSL is a co-inducer for a protein, TraR, a LuxR homolog, which is a positive transcription factor for genes required for conjugation (the *traA* genes). (See Note 264.) Furthermore, the *V. fischeri* autoinducer, AI, will substitute for the *A. tumefaciens* autoinducer, AAI, in activating transcription of the *tra* genes, although not as effectively. Additionally, the presumptive binding sites for TraR (*tra* boxes) in the *A. tumefaciens* genome have sequence similarity to the binding site (*lux* box) for LuxR upstream of the *luxI* promoter in the *V. fischeri* genome.[261] It is clear that there is a great deal of similarity between the LuxI–LuxR system in *V. fischeri* and the 3-oxo-C8-HSL-TrR system in *A. tumefaciens*.

### Pseudomonas aeruginosa

*Pseudomonas aeruginosa* also produces autoinducers. *P. aeruginosa* is one of the more common opportunistic pathogens and can cause infections at different body sites (e.g., urinary tract, lungs, skin), especially in immunosuppressed individuals, burn victims, or individuals who are otherwise weakened by sickness or age. The organism is virulent in part because of extracellular virulence factors (exotoxins, proteases) that it produces. The activation of the virulence genes is due to an autoinducer system similar to the LuxI–LuxR system, which is called LasI–LasR.[265] The autoinducer is called PAI and has been identified as N-(3-oxododecanoyl)-homoserine lactone, which differs from *V. fischeri* AI only in having a longer acyl chain. PAI is synthesized by LasI and interacts with LasR to activate the virulence genes. (*P. aeruginosa* also makes a quorum sensor called RhII-RhIR/C4-HSL that is regulated by LasR–LasI. This is reviewed in Ref. '260.)

## Erwinia carotovora

Autoinducers are also produced by the plant pathogen *Erwinia carotovora*, which causes soft-rot disease in plants.[266] During infection the bacterium produces several plant cell-wall-degrading exoenzymes (including pectin lyase, cellulase, and a protease) that degrade the plant tissue and aid the bacteria in colonizing and propagating in the plant. Mutants that fail to synthesize any of the exoenzymes have decreased virulence. These bacteria have the ExpI–ExpR system, which is homologous to the LuxI–LuxR system. ExpI is required for the synthesis of an autoinducer that appears to be identical to AI produced by *V. fischeri*, and that interacts with ExpR to stimulate transcription of genes that code for the exoenzymes. The evidence that the ExpI–ExpR system is very similar to the LuxI–LuxR system is: (1) *expI* encodes a protein whose amino acid sequence is strikingly similar to LuxI; (2) the wild-type gene of *expI* can complement a mutated *luxI*, and the wild-type *luxI* can complement a mutated *expI*. (Both wild-type genes were introduced into an *expI* mutant of *E. carotovora*, or a *lux* plasmid containing a mutant *luxI* gene was introduced into *E. carotovora* carrying wild-type *expI*. Then *expI*-dependent gene expression or bioluminescence was monitored.) Culture filtrates of *E. carotovora* contained a compound that could substitute for the *V. fischeri* autoinducer in stimulating bioluminescence in *E. coli* carrying a *luxI* mutation; synthetic *V. fischeri* autoinducer restored *expI*-dependent gene expression when added to cultures of *expI* mutants; the product of *expR* shows some sequence similarity to LuxR and LasR. It is also important to note that the expression of the virulence genes (the genes encoding the exoenzymes) increases when the cells enter stationary phase. This suggests that the *expI* gene product is a cell-density signal, as is the *V. fischeri* autoinducer.

## Rhizobium leguminosarum

Rhizobia are gram-negative bacteria that have an interesting symbiotic relationship with leguminous plants. Signals are exchanged between the plant and bacteria, and as a consequence the plant root tissues form nodules that the bacteria enter. This is a multistep process reviewed in Ref. 267. Within the nodules the bacteria differentiate into another cell type called *bacteroids*. A symbiotic relationship exists between the bacteroids and the plant tissue. The bacteroids convert nitrogen gas to ammonium, which the plants use. In return, the plant converts carbon dioxide to organic carbon and feeds the bacteria, mostly dicarboxylic acids. There are three genera of rhizobia. They are *Rhizobium*, *Bradyrhizobium*, and *Azorhizobium*. Recently an autoinducer has been described that is produced by *Rhizobium leguminosarum* biovar viciae, which fixes nitrogen in pea root nodules.[268] The autoinducer is an N-acylated HSL similar to the other autoinducers. The *R. leguminosarum* autoinducer activates RhiR, which is a transcriptional activator that stimulates transcription of the *rhiABC* operon, which is encoded in a symbiotic plasmid carried by *R. leguminosarum*.[269] RhiR belongs to the LuxR family of proteins. The *rhiABC* operon is transcribed by the rhizosphere bacteria but not by the nodule bacteroids. It is not known how the autoinducer is related to the development of the root nodules. Similarly, the role of the RhI proteins has not been established. It has been suggested that perhaps the *rhiABC* gene products allow the rhizobia to transport and/or metabolize nutrients secreted by the pea roots in the rhizosphere.[270]

## E. coli

*E. coli* (and *Salmonella typhimurium*) possesses a LuxR homolog called SdiA.[271, 272] (Reviewed in Ref. 259; the amino acid sequence of SdiA is similar to that of LuxR.) SdiA stimulates the transcription in *E. coli* of the *ftsQAZ* operon, which is required for septation.[271] Presumably, this is part of the complex control of cell division. The transcription mediated by SidA is cell density dependent, and

*E. coli* spent culture supernate stimulates transcription from the SdiA-dependent promoter, indicating the presence of an inducer. Transcription can also be stimulated to a relatively small extent by the addition of *V. fischeri* and *V. harveyi* autoinducers. However, none of the other components of the LuxR/LuxI system appears to be present, and some doubt has been raised as to whether *E. coli* does indeed employ a quorum-sensing system similar to the LuxR/LuxL system.

There is also a report that acylated homoserine lactones in *E. coli* may be involved in the response to starvation.[273] When *E. coli* enters stationary phase or when it is grown in minimal medium, there is increased transcription of the *rpoS* gene (Section 2.2.2). RpoS is a sigma factor ($\sigma^s$) that is required for the transcription of genes that promote long-term survival during stationary phase. Mutants unable to synthesize homoserine (a precursor for threonine biosynthesis) did not synthesize a $\sigma^s$-dependent enzyme (catalase HPII) when grown on minimal media, indicating that $\sigma^s$ synthesis was depressed. However, catalase HPII was synthesized when homoserine or homoserine lactone was included in the growth medium, suggesting that homoserine or one of its metabolic products increased the synthesis of $\sigma^s$. In fact, it was shown that exogenously supplied homoserine lactone did in fact stimulate the synthesis of $\sigma^s$. These are interesting findings whose significance is not yet understood. Perhaps starvation conditions somehow result in the elevation of intracellular homoserine levels, which in turn results in a stimulation of transcription of *rpoS*. This would not be a cell-density signal, since *rpoS* activation is controlled by nutrition (starvation), not cell density.

## 18.20.5 Luminescence by V. harveyi

There are no homologs to *luxI* or *luxR* in the *V. harveyi* chromosome, reflecting the fact that the signaling system is different from that of *V. fischeri*. *V. harveyi* actually has two signaling systems, called signaling system 1 and signaling system 2, neither one of which responds to the autodinducer produced by *V.*

*fischeri.*[274] As demonstrated with null mutants in system 1 and system 2, either system alone is sufficient for cell-density-responsive expression of luminescence.

### Signaling system 1

The system 1 autoinducer is β-hydroxybutyryl HSL (3-OH-C4-HSL), a molecule that is similar to the autoinducer produced by *V. fischeri* but is not active in the *V. fischeri* system (Fig. 18.32). (Nor is the *V. fischeri* autoinducer active with *V. harveyi*.) There is a gene, called *luxR*, which is required *in trans* for transcription of the *lux* operon (*luxCDABEGH*) of *V. harveyi*. LuxR is a different gene from *luxR* in *V. fischeri*, and is not closely linked to *luxCDABEGH*. (The functions of *luxG* and *luxH* are not known. They do not seem to be required for luminescence or the regulation of luminescence in laboratory cultures.) When *luxCDAGBEGH* and *luxR* were introduced into *E. coli*, bioluminescence occurred, but it was not cell density dependent, nor did it require the addition of the autoinducer. Thus LuxR is a regulatory protein necessary for transcription, but it is not responsible for the response to autoinducer. Instead, the genes described next are part of the system that synthesizes and responds to the autoinducer.

One model for the regulation of luminescence in *V. harveyi*, which is based upon the analysis of mutants, is that that there are two signaling systems (system 1 and system 2) that inactivate the putative repressor (LuxO) for transcription of the *lux* operon (*luxCD-ABEGH*) (Fig. 18.32). (Mutations in *luxO* result in constitutive luminsecence, and it has therefore been concluded that LuxO is a negative transcription regulator that is inactivated in the presence of autoinducer.) System 1 includes an autoinducer (β-hydroxybutyryl homoserine lactone) whose synthesis requires the products of the *luxL* and *luxM* genes. (Mutants in *luxL* or *luxM* can be stimulated by adding β-hydroxybutyryl homoserine lactone to the medium.) A gene necessary for response to the autoinducer is *luxN*, which (based upon its nucleotide sequence) codes for a bifunctional sensor kinase/response-regulator pro-

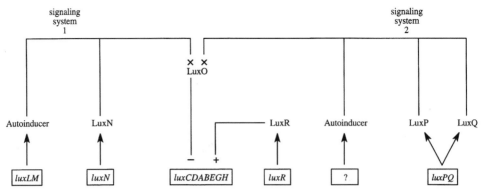

**Fig. 18.32** Model for control of luminescence in *V. harveyi*. Signaling system 1: The *luxLM* genes produce an autoinducer that interacts with LuxN to inactivate a repressor (LuxO) for the *lux* luminescent genes (*luxCDABEGH*). LuxN is thought to be a membrane-bound sensor/kinase that autophosphorylates and then phosphorylates LuxO. LuxO is postulated to be a cytoplasmic regulator protein that is inactivated upon phosphorylation. Signaling system 2: Neither AI-2 nor the the genes responsible have been identified. AI-1 has. Both LuxP and LuxQ are required for the response to autoinducer. LuxP resembles the ribose-binding periplasmic proteins, whereas LuxQ is similar to the sensor-kinase proteins. The autoinducer is postulated to bind to LuxP in the periplasm and be subsequently transferred to LuxQ in the membrane. LuxQ is suggested to autophosphorylate and then to phosphorylate LuxO, thus inactivating its repressor activity. The *luxR* gene codes for a positive transcription factor. For a discussion, see: Bassler, B. L., M. Wright, and M. R. Silverman. 1994. Multiple signaling systems controlling expression of luminescence in *Vibrio harveyi*: Sequence and function of genes encoding a second sensory pathway. *Mol. Microbiol.* 13:273–286.

tein. It is proposed that LuxN responds to the presence of autoinducer by autophosphorylating at a histidine residue in its receiver domain, and then transferring the phosphoryl group to an aspartate residue in its transmitter domain. (This might be an intra- or interphosphoryl transfer.) Because of its hydrophobic regions, it is presumed that LuxN spans the cytoplasmic membrane. LuxN then transfers its phosphoryl group from its aspartate residue to an aspartate residue in LuxO. It is proposed that phosphorylation of LuxO inactivates its repressor activity and stimulates transcription of the *lux* bioluminescence genes.

## Signaling system 2

System 2 involves a second autoinducer, which has not yet been chemically characterized, but has been detected as a stimulatory factor for luminescence (Fig. 18.32). The system 2 sensor is the product of the *luxPQ* genes, which are necessary for the response to exogenously supplied autoinducer. The *luxP* gene codes for a protein whose deduced amino acid sequence resembles that found for the periplasmic ribose-binding proteins in the enteric bacteria, whereas the *luxQ* gene codes for a protein that appears to be a bifunctional sensor kinase/response-regulator protein similar to LuxN. Presumably, it is the LuxQ protein that is the actual sensor for the autoinducer in signaling system 2. The role of LuxP may be to bind the autoinducer in the periplasm and bring it to LuxQ in the cell membrane. Both LuxQ-P and LuxN-P might transfer their phosphoryl group to LuxO, thus inactivating it as a transcription repressor. The student might find it curious that LuxN and LuxQ appear to have both a histidine kinase (receiver) domain and a response-regulator (transmitter) domain. A similar situation occurs in the Arc system, where ArcB has both a receiver and response-regulator domain (Section 18.2.2).[275] (See Note 276 for a discussion of ArcB.)

Many nonluminescent bacteria have been screened to see if they produce an autoinducer that stimulates either system 1 or system 2 in *V. harveyi*.[277] (This was done by using mutants of *V. harveyi* that responded to either AI-1 or AI-2, but not to both.) It was found that culture fluids from some bacteria, including *V. cholerae* and *V. parahaemolyticus*, and *Yersinia enterocolitica*, stimulated luminescence in *V. harveyi*. It is a matter of speculation as to why *V. harveyi* has two autoinducer systems and why it should respond to other species of bacteria. However, a clue is that the nonluminescent bacteria tested were shown to produce a substance that had the activity of AI-2 but not AI-1. Perhaps *V. harveyi* uses system 2 to detect the presence of other species of bacteria, and uses system 1 (which is more sensitive and very specific for *V. harveyi*) as a quorum sensor for its own species.

It is of interest that what began as an investigation in an intriguing but narrow area of research (luminescence in *Vibrio*) has resulted in uncovering what appears to be a general mechanism of signaling and gene regulation (involving acylated homoserine lactone derivatives) that may be widespread among the bacteria.

## 18.21 Summary

Bacteria can sense environmental and cytoplasmic signals and transmit information to the genome or to other parts of the cell to elicit a response. Such signaling systems detect and transmit signals when conditions, such as pH, osmolarity, temperature, starvation, and nutrient source as well as the absence or presence of specific inorganic ions such as phosphate, become altered. The signaling systems also detect signaling molecules secreted by other cells and aid in cooperative cell behavior. Often global transcription regulators are involved. These are transcription regulators that control transcription in noncontiguous operons. A set of noncontiguous operons or genes regulated by the same transcription regulator is called a regulon.

Many of the signaling systems are called "two-component" signaling systems because they have a histidine kinase (component 1) that causes the phosphorylation (transphosphorylation) of a response regulator (component 2). However, as was discussed, other proteins may also be involved in signal transduction. (The histidine kinase is also called a histidine/sensor kinase.)

All the "two-component" systems have four functional components: (1) a sensor that receives the signal; (2) an autophosphorylating histidine protein kinase; (3) a partner response regulator that is the substrate for the histidine kinase; and (4) a phosphatase that removes the phosphate from the regulator. As discussed below, some of the proteins can be bifunctional and carry out two of the activities. During signaling, phosphate (actually the phosphoryl group) travels from ATP to a histidine residue in the histidine kinase to an aspartate residue in the response regulator to water (the phosphatase reaction). The phosphorylated form of the response regulator is the active state and transmits the signal to the genome (stimulation or inhibition of transcription), to the flagellar motor (reversal of turning direction), or to an enzyme. Although many signals cause an increase in phosphorylation of the response regulator, some (e.g., chemoattractants, excess inorganic phosphate, excess ammonia) cause a decrease in phosphorylation (which in some cases is due to dephosphorylation of the response regulator) and therefore an inactivation of the response regulator. All the histidine protein kinases have conserved domains (receiver domains) that bear a startling similarity to each other, and are the basis for classification of these proteins. These conserved domains are in the carboxy terminus of the protein and are the site of the conserved histidine residue that is phosphorylated. The response regulators have conserved amino-terminal domains (transmitter domains) that are believed to interact with the carboxy terminus of the kinase. For several of the response regulators, it has been shown that the phosphoryl group is transferred from the histidine to the conserved aspartate residue in the amino terminus of response regulator.

Although the two-component systems are regulated primarily by the activities of their respective histidine kinases, it has been suggested that acetyl phosphate may be a phosphoryl donor to different response-regulator proteins under certain physiological conditions. This has usually been observed in mutants that lack the cognate histidine kinase. In these mutants acetyl phosphate is able to donate the phosphoryl group to the response regulator. The extent to which this might occur in wild-type cells that have the cogate histidine kinase is generally not yet known. However, analysis of mutants in acetyl phosphate synthesis or utilization in which the levels of acetyl phosphate would be expected to be higher or lower than wild-type cells do suggest that increased levels of acetyl phosphate can inhibit flagellar biosynthesis in *E. coli* wild-type cells by phosphorylating OmpR, which in its phosphorylated form is a negative transcription regulator of regulatory genes required for flagellar biosynthesis. This may account for why *E. coli* grown at higher temperatures have fewer flagella. The reason for this is that the acetyl phosphate levels in cells grown at higher temperatures is elevated because the enzyme that converts acetyl phosphate to acetate (acetate kinase) is less active at the higher temperatures. Presumably this leads to a phosphorylation of OmpR and a consequent repression of the flagellar regulatory genes. All these data suggest the possibility that acetyl phosphate may serve as a global metabolic signal that affects the activity of several different genes by phosphorylating response-regulator proteins.

Responses that use a two-component regulatory system include the Che system in chemotaxis, the ArcA/ArcB system for oxygen regulation of gene expression, the NarL/X/Q system for nitrate regulation, the PHO system for phosphate assimilation, the Ntr system for nitrogen assimilation, the EnvZ system for porin gene expression, the KdpABC system for K⁺ uptake, the RegB/RegA system for the regulation of expression of the genes for the light-harvesting and photoreaction center proteins in *Rhodobacter capsulatus*, the bioluminescence signaling systems in *V. harveyi*, and the dicarboxylic acid permease system in *R. meliloti*. Other two-component signaling systems operate during *Bacillus* sporulation, flagellar synthesis in *E. coli*, and cell-to-cell signaling in myxobacteria. There is also evidence that a two-component signaling system operates to regulate the initiation of DNA replication in *Caulobacter*.

Two-component systems also detect environmental signals and regulate the transcription of virulence genes in certain bacteria, for example, *Agrobacterium*, *Salmonella*, and *Bordetella*. There are several other important regulatory systems in bacterial pathogens that control the expression of virulence genes in response to environmental signals. These may involve a transmembrane sensor, which need not be part of a two-component signaling system, and which responds to environmental signals such as temperature, osmolarity, and pH. Such systems include the ToxR/S system in *Vibrio cholerae*. *Shigella* spp. possess a regulatory gene called the *virR* gene that encodes a protein that is part of a system that regulates the temperature-dependent expression of virulence genes.

The FNR system, which regulates anaerobic gene expression, is not a two-component regulatory system. Neither is there evidence that the formate regulon, which is induced by formate under anaerobic conditions, is controlled by a two-component system.

With respect to the Arc and the FNR systems, it is clear that they represent two systems of gene regulation with which *E. coli* senses anaerobiosis and activates the transcription regulators ArcA and FNR.

The situation is often complex, with multiple regulators controlling the same operon. An example of this is the control of the *cydAB* operon in *E. coli*. This operon encodes cytochrome d oxidase and is regulated by FNR and ArcA, and interestingly by Cra as well. This was demonstrated using *cyd–lacZ* fusions strains that have null mutations in either *cra*, *arcA*, or *fnr* genes.[278] The absence of Cra or ArcA resulted in decreased gene expression by cells under microaerophilic or anaerobic conditions, indicating that Cra and ArcA both stimulate transcription of *cyd*. The results with the *fnr* null mutation indicated that FNR represses *cyd* transcription under anaerobic conditions (but interestingly activates *cyd* transcription under microaerophilic conditions). In mutants that lacked FNR, there appeared to be no effect of a deletion of *cra* on transcription of *cyd*, indicating that the regulatory activity of Cra on *cyd* expression requires FNR. The situation is not unique to *cyd*, as many genes are controlled by multiple transcription factors that may function interdependently.

Catabolite repression refers to the repression by particular carbon sources of expression of genes required for growth on other carbon sources. The result is diauxic growth.

For example, in many bacteria glucose represses genes required for the uptake and/or catabolism of other carbon sources. Some bacteria utilize carboxylic acids in preference to glucose, and it is the carboxylic acids that repress the genes required for glucose catabolism. The mechanisms that underlie catabolite repression vary. These include the cAMP–CRP-dependent system for glucose repression that has been well studied in *E. coli*. This system involves the PTS sugar uptake system. *E. coli* also has a cAMP–CRP-independent system, the Cra system. Cra is a global regulator that influences the direction of carbon flow in either the glycolytic or gluconeogenic directions by controlling the expression of genes encoding enzymes required for the relevant pathways. Cra is an activator for genes required for gluconeogenesis and a repressor for genes required for glucose catabolism. When the cells are growing on glucose or fructose, it is proposed that the levels of fructose-1-phosphate and fructose-1,6-bisphosphate rise and bind to Cra, removing it from the DNA. Thus gluconeogenic genes required for growth on pyruvate, lactate, acetate, and citric acid cycle intermediates would be repressed and glycolytic genes required for growth on carbohydrates would be activated. Catabolite repression in low-GC gram-positive bacteria involves yet a different system, the CcpA system. When these bacteria are growing on glucose, the levels of the glycolytic intermediates fructose-1,6-bisphosphate and 2-phosphoglycerate rise and activate a special kinase that phosphorylates Hpr, forming Hpr(Ser-P). Phosphorylated Hpr may form a ternary complex with the CcpA DNA-binding protein and fructose-1,6-bisphosphate, and this complex binds to the CRE sequence in target genes and represses transcription.

Many gram-negative bacteria have been demonstrated to possess a signaling system based upon acylated homoserine lactones (acyl HSL) that are produced extracellularly and serve as cell-density signals. The acylated homoserine lactone systems were first discovered in luminous bacteria that luminesce only when the population of cells reaches a sufficient cell density in late logarithmic growth. They are thus called quorum-sensing systems. The system in *V. fischeri* includes an acylated homoserine lactone (autoinducer), a positive transcription regulatory factor activated by the autoinducer (LuxR), and an operon containing the gene for autoinducer synthesis (*luxI*). *V. harveyi* has two signaling systems that regulate luminescence, one of which relies on an acylated homoserine lactone similar to that of the *V. fischeri* system. Neither system in *V. harveyi* has components that are homologous to *luxI* or *luxR* in *V. fischeri*. However, systems similar to the LuxR–LuxI system that depend upon acylated homoserine lactone signals have been found in many different gram-negative bacteria. The cell activities that these signaling systems regulate vary with the bacterium and its ecological niche.

Other quorum-sensing systems based upon acylated homoserine lactones include signaling systems that stimulate conjugation in *Agrobacterium*, and systems that stimulate the transcription of virulence genes in *Pseudomonas* and *Erwinia*. Acylated HSL quorum-sensing systems also exist in *Rhizobium*. *E. coli* may respond to a cytoplasmic signal (not a quorum sensor) apparently related to homoserine, which may be involved in stimulating the transcription of some genes in stationary phase or when the bacteria are grown in minimal media. *E. coli* also has a luxR homolog (SdiA), which may be involved in the regulation of cell division.

## Study Questions

1. Describe the components of two-component regulatory systems and how they work. In your answer, explain what is meant by "cross-talk" or "cross-regulation."

2. What criteria must be established in order to characterize a protein as a kinase or regulator protein in a two-component system?

3. What is FNR, and what is its role? Why is it not believed to be part of a two-component regulatory system?

4. It has been suggested that FNR might be activated by some reduced molecule in the cell that rises in concentration in

the absence of oxygen. How might you test whether the redox levels of electron transport carriers are involved?

5. Describe the signaling pathway for chemotaxis in *E. coli*.

6. What causes *E. coli* to swim randomly? Is this the mechanism for all bacteria? Explain the differences. How is random swimming related to chemotaxis?

7. What is the relationship between adaptation and methylation of MCPs?

8. In the Arc and Nar systems, which proteins are thought to be the sensor/kinase proteins, and which ones the regulator proteins? What is the evidence for this?

9. What is the phenotype of an *Fnr⁻* mutant? An *arc⁻* mutant?

10. Describe the role that $P_{II}$ plays in transcriptional regulation of the *glnALG* operon and in the regulation of activity of glutamine synthetase. How are the levels of $P_{II}$ and $P_{II}$-UMP regulated?

11. In what way is catabolite repression by the Cra system and the cAMP system similar? How does it differ from catabolite repression by the CcpA system in *B. subtilis*?

12. How does *B. subtilis* partition Hpr-P between the PTS system and the CcpA catabolite repression system?

13. What is the evidence that acetyl phosphate may be a global signal?

14. Assume that acetyl phosphate phosphorylates OmpR, which then represses the flagellar genes. Explain why the addition of acetate to either wild-type cells or to *pta⁻* cells represses flagellar gene expression but not when added to *ackA⁻* cells.

15. Discuss the role that a two-component phosphorelay system might play in the production of virulence factors by pathogenic bacteria. Give an example of such a two-component system. Describe some experiments to test which environmental factors might provide the necessary signals.

16. Discuss the role that acylated homoserine lactones play in the luminescent bacteria

as well as specific nonluminescent bacteria. Discuss an experimental physiological/biochemical and genetic approach to investigate whether the systems in two different bacteria have interchangeable components.

## NOTES AND REFERENCES

1. Hoch, J. A., and T. J. Silhavy (Eds.) 1995. Two-component signal transduction. ASM Press. Washington, D.C.

2. The transmitter domain of the sensor kinase, which is at the carboxyl terminus of the protein, is generally attached to an *input domain*, e.g., a region that might bind a signaling molecule (ligand-binding site). Thus the sensor kinase would be: N–input domain–transmitter domain–C. The input domain would be expected to be exposed to the signal; e.g., it might be in the periplasm if the sensor kinase is a transmembrane protein in a gram-negative bacterium. Presumably, after binding the signal, the sensor kinase undergoes a conformational change that results in its autophosphorylation in the transmitter domain and subsequent phosphorylation of the response-regulator protein in its receiver domain. As expected, input domains vary, depending upon the sensor kinase. The receiver domain of the response-regulator protein is usually attached to an *output domain* at the amino terminus such as a DNA-binding region. Thus the response regulator would be: N–receiver domain–output domain–C. The sequences of the DNA-binding domains are divided into classes that place the response-regulator proteins into families, e.g., the OmpR family. Not all response regulator proteins have output domains. For example, CheY and SpoOF do not.

3. Hughes, D. A. 1994. Histidine kinases hog the limelight. *Nature* 369:187–188.

4. Wurgler-Murphy, S. M., and H. Saito. 1997. Two-component signal transducers and MAPK cascades. *TIBS* 22:172–176.

5. Saier, M. H., Jr. 1993. Introduction: Protein phosphorylation and signal transduction in bacteria. *J. Cellular Biochem.* 51:1–6.

6. Stock, J. B., A. J. Ninfa, and A. M. Stock. 1989. Protein phosphorylation and regulation of adaptive responses in bacteria. *Microbiol. Rev.* 53:450–490.

7. Parkinson, J. S., and Eric C. Kofoid. 1992. Communication modules in bacterial signaling proteins. *Ann. Rev. Genet.* 26:71–112.

8. Ota, I. M., and A. Varshavsky. 1993. A

yeast protein similar to bacterial two-component regulators. *Science* 262:566–569.

9. Chang, C., S. F. Kwok, A. B. Bleecker, and E. M. Meyerwitz. 1993. *Arabidopsis* ethylene-response gene *etr1*: Similarity of product to two-component regulators. *Science* 262:539–544.

10. Wanner, B. L. 1992. Is cross regulation by phosphorylation of two-component response regulator proteins important in bacteria? *J. Bacteriol.* 174:2053–2058.

11. Reviewed in Antonie van Leeuwenhoek. 1994. Vol. 66, Nos. 1–3.

12. Spiro, S., and J. R. Guest. 1991. Adaptive responses to oxygen limitation in *Escherichia coli*. *TIBS* 16:310–314.

13. Gunsalus, R. P., and S.-J Park. 1994. Aerobic–anaerobic gene regulation in *Escherichia coli*: Control by the ArcAB and FNR regulons. *Res. Microbiol.* 145:437–449.

14. Spangler, W. J., and C. M. Gilmour. 1966. Biochemistry of nitrate respiration in *Pseudomonas stutzeri*. Aerobic and nitrate respiration routes of carbohydrate catabolism. *J. Bacteriol.* 91:245–250.

15. Reviewed in: Berks, B. C., S. J. Ferguson, J. W. B. Moir, and D. J. Richardson. 1995. Enzymes and associated electron transport systems that catalyze the respiratory reduction of nitrogen oxides and oxyanions. *Biochim. et Biophys. Acta* 1232:97–173.

16. Gunsalus, R. P. 1992. Control of electron flow in *Escherichia coli*: Coordinated transcription of respiratory pathway genes. *J. Bacteriol.* 174:7069–7074.

17. Sawers, G., and B. Suppman. 1992. Anaerobic induction of pyruvate formate–lyase gene expression is mediated by the ArcA and FNR proteins. *J. Bacteriol.* 174:3474–3478.

18. Sawers, G. 1993. Specific transcriptional requirements for positive regulation of the anaerobically inducible *pfl* operon by ArcA and FNR. *Molec. Microbiol.* 10:737–747.

19. Anderson, D. I. 1992. Involvement of the Arc system in redox regulation of the cob operon in *Salmonella typhimurium*. *Mol. Microbiol.* 6:1491–1494.

20. Mutations in *arc A* produce a defect in the synthesis of F pili, which results in resistance to male-specific bacteriophages that attach to the F pilus. (See Fig. 18.5.) ArcA also responds to a second sensor/kinase protein (i.e., CpxA), which is necessary for the production of the F pilus in donor strains of *E. coli*. The CpxA gene was originally discovered as a mutation that reduced the efficiency of DNA transfer as a consequence of reduced F-plasmid *tra* gene expression. It is now known that the CpxA protein is an inner membrane protein whose amino acid sequence places it in the class of sensor/kinase proteins. There also exists a cognate response regulator (CpxR) for CpxA shown in Fig. 18.5. The relationship between these two regulatory systems is complex and is discussed in the following papers: Iuchi, S., D. Furlong, and E. C. C. Lin. 1989. Differentiation of *arcA*, *arcB*, and *cpxA* mutant phenotypes of *Escherichia coli* by sex pilus formation and enzyme regulation. *J. Bacteriol.* 171:2889–2893. Dong, J.-M, S. Iuchi, H.-S. Kwan, Z. Lu, and E. C. C. Lin. 1993. The deduced amino acid sequence of the cloned *cpxR* gene suggests the protein is the cognate regulator for the membrane sensor, CpxA, in a two-component signal transductional system of *Escherichia coli*. *Gene* 136:227–230.

21. Iuchi, S., and E. C. C. Lin. 1992. Purification and phosphorylation of the Arc regulatory components of *Escherichia coli*. *J. Bacteriol.* 174:5617–5623.

22. Georgellis, D., A. S. Lynch, and E. C. C. Lin. 1997. In vitro phosphorylation study of the Arc two-component signal transduction system of *Escherichia coli*. *J. Bacteriol.* 179:5429–5435.

23. With regard to the genes for succinic dehydrogenase (*sdhCDAB*), it should be pointed out that the aerobic/anaerobic regulation of transcription of this operon involves more than simply the ArcA/ArcB system. Not only are the levels of succinic dehydrogenase suppressed during anaerobic growth, but they are also lowered in *E. coli* by growth on glucose, even when the cells are grown aerobically, and raised by growth on acetate. Furthermore, mutations in *FNR* increase the levels of transcription of an *sdhC–lacZ* fusion, suggesting that FNR represses the succinate dehydrogenase operon under anaerobic conditions. The mechanism by which glucose lowers the expression of the *sdh* operon is not understood. It appears to be independent of the cAMP–CAP system. See Park, S-J, C.-P. Tseng, and R. P. Gunsalus. 1995. Regulation of succinate dehydrogenase (*sdhCDAB*) operon expression in *Escherichia coli* in response to carbon supply and anaerobiosis: Role of ArcA and FNR. *Molec. Microbiol.* 15:473–482.

24. Iuchi, S., V. Chepuri, H.-A. Fu, R. B. Gennis, and E. C. C. Lin. 1990. Requirement for terminal cytochromes in generation of the aerobic signal for the *arc* regulatory system in *Escherichia coli*: Study utilizing deletions and lac fusions of *cyo* and *cyd*. *J. Bacteriol.* 172:6020–6025.

25. Gene fusions are valuable probes to monitor the expression of genes of interest. Genes are fused to reporter genes whose products are easy to identify. For example, a reporter gene might be *lacZ* (β-galactosidase), *lux* (luciferase),

*cat* (chloramphenicol acetyltransferase), or *phoA* (alkaline phosphatase). Consider a *lacZ* fusion. The fused gene has the promoter region of the target gene but not the promoter for the *lacZ* gene. Expression of the fused gene is therefore under control of the promoter region of the target gene. If the translational initiating region (ribosome binding site) is not present in the reporter gene, then a translational fusion results. The fusions produce a hybrid protein whose amino-terminal end is derived from the target gene and the carboxy-terminal end from $\beta$-galactosidase. The hybrid protein has $\beta$-galactosidase activity. Therefore, an assay for $\beta$-galactosidase is a measure of the expression of the target gene. Thus one can measure the expression of virtually any gene simply by constructing the proper gene fusion and performing an assay for $\beta$-galactosidase. One can construct gene fusions in vitro or in vivo. In vitro construction involves using restriction endonucleases to cut out a portion of the gene with its promotor region from a plasmid containing the cloned DNA, and ligating it to a *lacZ* gene, without its promoter or ribosome binding site, in a second plasmid. The plasmid containing the fused gene is then introduced into the bacterium, and transformants are selected on the basis of being resistant to an antibiotic-resistant marker on the plasmid, and the production of $\beta$-galactosidase. This method was used by Iuchi et al. In vivo construction of gene fusions can also be performed. In this case, one uses Tn5 transposons fused to a promoterless *lacZ* gene. The transposon is introduced into the bacterium where it can recombine with the bacterial chromosome. Cells harboring the transposon are selected using the antibiotic-resistance marker on the transposon. Many of the strains have the transposon inserted into the host bacterial genes in the proper orientation and frame so that $\beta$-galactosidase production is under the control of the promoter of the interrupted gene. Insertion of the transposon into a gene interrupts the gene so that the normal gene product is not made. The gene is identified by mutant analysis.

Cloned genes fused to the *E. coli* periplasmic alkaline phosphatase gene, *phoA*, are also used. These produce a hybrid protein consisting of the N-terminal region of the gene whose transcription is being analyzed and the C-terminal end of alkaline phosphatase. The hybrid protein is missing the N-terminal alkaline phosphatase signal sequence for protein export and relies on the protein export signal sequence of the target gene for export. When the C-terminal region of the hybrid protein containing the alkaline phosphatase region reaches the periplasm, then alkaline phosphatase activity can be measured. Such fusions can be used to examine the transcription of genes that encode secreted proteins such as virulence factors.

26. The control experiments were done to show that, in an $arc^+$ $cyo^+$ $cyd^+$ backround, anaerobiosis repressed *cyo–lacZ* and induced *cyd–lacZ* expression, and that a mutation in the *arc* genes prevented the repression of *cyo–lacZ* and lowered the expression of *cyd–lacZ*. These experiments showed that the fusion genes were regulated by the availability of oxygen, and that the Arc system was responsible for the regulation.

27. Discussed in Iuchi, S., A. Artistarkhov, J. M. Dong, J. S. Taylor, and E. C. C. Lin. 1994. Effects of nitrate respiration on expression of the Arc-controlled operons encoding succinate dehydrogenase and flavin-linked L-lactate dehydrogenase. *J. Bacteriol.* **176**:1695–1701.

28. Lynch, A.S., and E. C. C. Lin. 1996. Responses to molecular oxygen. In: *Escherichia coli and Salmonella: Cellular and Molecular Biology*, Vol. 1, pp. 1526–1538, Neidhardt et al. (Eds.). ASM Press, Washington, D.C.

29. Spiro, S., and J. R. Guest. 1990. FNR and its role in oxygen-regulated gene expression in *Escherichia coli*. *FEMS Microbiol. Rev.* **75**:399–428.

30. Unden, G., S. Becker, J. Bongaerts, G. Holighaus, J. Schirawski, and S. Six. 1995. $O_2$-sensing and $O_2$-dependent gene regulation in facultatively anaerobic bacteria. *Arch. Microbiol.* **164**: 81–90.

31. Jones, H. M., and R. P. Gunsalus. 1987. Regulation of *Escherichia coli* fumarate reductase (*frdABCD*) operon expression by respiratory electron acceptors and the *FNR* gene product. *J. Bacteriol.* **169**:3340–3349.

32. Unden, G., and J. R. Guest. 1985. Isolation and characterization of the FNR protein, the transcriptional regulator of anaerobic electron transport in *Escherichia coli*. *Eur. J. Biochem.* **146**:193–199.

33. It appears that inactivation of FNR is via the reversible oxidation of the $[4Fe–4S]^{2+}$ clusters. The oxidation of the cluster results in the inactivation of the protein and under conditions of prolonged incubation in oxygen by the disassembly and loss of the clusters from the protein. This leads to the release of sulfide ($S^{2-}$) and ferric ion ($Fe^{3+}$). Reassembly of the clusters from sulfide and Fe ions ($Fe^{2+}$ and $Fe^{3+}$) has been demonstrated in vitro under anaerobic conditions and requires a special protein, the NifS protein from *Azotobacter*, which is required for the incorporation of Fe–S clusters into the *Azotobacter* nitrogenase. *E. coli* has a similar protein. See Unden, G., and J. Schirawski. 1997. The oxygen-responsive transcriptional regulator

FNR of *Escherichia coli*: The search for signals and reactions. *Molec. Microbiol.* **25**:205–210.

34. Roualt, T. A., and R. D. Klausner. 1996. Iron–sulfur clusters as biosensors of oxidants and iron. *TIBS* **21**:174–177.

35. Unden, G., and J. Schirawski. 1997. The oxygen-responsive transcriptional regulator FNR of *Escherichia coli*: The search for signals and reactions. *Molec. Microbiol.* **25**:205–210.

36. Jordan, P. A., A. J. Thomson, E. T. Ralph, J. R. Guest, and J. Green. 1997. FNR is a direct oxygen sensor having a biphasic response curve. *FEBS Lett.* **416**:349–352.

37. Nakano, M. M., P. Zuber, P. Glaser, A. Danchin, and F. M. Hulett. 1996. Two-component regulatory proteins ResD–ResE are required for transcriptional activation of *fnr* upon oxygen limitation in *Bacillus subtilis*. *J. Bacteriol.* **178**:3796–3802.

38. These proteins more closely resemble FNR than CRP on the basis of one or more of the following characteristics: DNA-binding specificity, sequence identity, failure to bind cAMP, complementation of an *E. coli* FNR mutant, or being activated under conditions of oxygen limitation. They are involved in the regulation of various physiological reactions such as luminescence (*Vibrio fischeri*), denitrification (*Pseudomonas aeruginosa*), nitrogen fixation (*Rhizobium* species), and haemolysin biosynthesis (*Bordetella pertussis*).

39. Spiro, S. 1994. The FNR family of transcriptional regulators. *Antonie van Leeuwenhoek* **66**:23–36.

40. Compan, I., and D. Touati. 1994. Anaerobic activation of *arcA* transcription in *Escherichia coli*: Roles of Fnr and ArcA. *Mol. Microbiol.* **11**:955–964.

41. *E. coli* actually synthesizes three different formate dehydrogenases (FDH) and three distinct hydrogenases (Hyd). FDH-O is synthesized during aerobic growth and also anaerobically when nitrate is the electron acceptor. FDH-N is made anaerobically when nitrate is present, and FDH-H is made only under fermentation conditions. The FDH-O and FDH-N enzymes oxidize formate and transfer the electrons via quinones to oxygen and nitrate, respectively, during respiration. Accordingly, these are sometimes called respiratory formate dehydrogenases to distinguish them from FDH-H, which functions only during fermentation. (FDH-O is also called formate oxidase.) Since FDH-O is also present in cells grown with nitrate as the electron acceptor, it may also be able to reduce nitrate. FDH-H is part of the formate hydrogen-lyase enzyme complex that oxidizes formate to $CO_2$ and $H_2$ during fermentation. *E. coli* also synthesizes three hydrogenases: Hyd-1, Hyd-2, and Hyd-3. Hyd-1 and Hyd-2 (uptake hydrogenases) oxidize $H_2$ (e.g., during the reduction of fumarate or nitrate). Hyd-3 is part of the formate hydrogen–lyase and is responsible for the reduction of protons, generating hydrogen gas.

42. Reviewed in: Unden, G., S. Becker, J. Bongaerts, J. Schirawski, and S. Six. 1994. Oxygen regulated gene expression in facultatively anaerobic bacteria. *Antonie van Leeuwenhoek* **66**:3–23.

43. The regulon consists of the following genes: *hyc* operon (hydrogenase 3); the *hyp* operon (uptake hydrogenases 1, 2); and *fdhF* (formate dehydrogenase). The *fhlA* gene is encoded in the *hyp* operon.

44. Motteram, P. A. S., J. E. G. McCarthy, S. J. Fergukson, J. B. Jackson, and J. A. Cole. 1981. Energy conservation during the formate-dependent reduction of nitrite by *Escherichia coli*. *FEMS Microbiol. Lett.* **12**:317–320.

45. In *E. coli*, the membrane-bound nitrate reductase is encoded by the *narGHJI* operon. The fumarate reductase is encoded by the *frdABCD* operon. The DMSO/TMAO reductase is encoded by the *dmsABC* operon. And the formate-dependent nitrite reductase, cytochrome $c_{552}$, is encoded by the *nrfABCDEFG* operon.

46. Unden, G., and J. Bongaerts. 1997. Alternative respiratory pathways of *Escherichia coli*: Energetics and transcriptional regulation in response to electron acceptors. *Biochim. Biophys. Acta* **1320**:217–234.

47. Gennis, R. B., and V. Stewart. 1996. Respiration. In: *Escherichia coli and Salmonella: Cellular and Molecular Biology*, Vol. 1, pp. 217–261. Neidhardt et al. (Eds.). ASM Press, Washington, D.C.

48. Rabin, R. S., and V. Stewart. 1993. Dual response regulators (NarL and NarP) interact with dual sensors (NarX and NarQ) to control nitrate- and nitrite-regulated gene expression in *Escherichia coli* K-12. *J. Bacteriol.* **175**:3259–3268.

49. Darwin, A. J., K. L. Tyson, S. J. W. Busby, and V. Stewart. 1997. Differential regulation by the homologous response regulators NarL and NarP of *Escherichia coli* K-12 depends on DNA binding site arrangement. *Molec. Microbiol.* **25**:583–595.

50. Williams, S. B., and V. Stewart. 1997. Discrimination between structurally related ligands nitrate and nitrite controls autokinase activity of the NarX transmemebrane signal transducer of *Escherichia coli* K-12. *Molec. Microbiol.* **26**:911–925.

51. Chiang, R. C., R. Cavicchioli, and R. P. Gunsalus. 1997. "Locked-on" and "locked-off" signal transduction mutations in the periplasmic domain of the *Escherichia coli* NarQ and NarX sensors affect nitrate- and nitrite-dependent regulation by NarL and NarP. *Molec. Microbiol.* 24:1049–1060.

52. Discovery of the *nar* genes: Originally mutants that were defective in nitrate repression of the fumarate reductase gene, *frdA*, were isolated. (The actual screen was to look for mutant colonies in which nitrate did not repress the expression of the fusion gene *frdA–LacZ*.) The mutations were found to be in two genes, called *narL* and *narX*. Whereas *narL* was definitely required for nitrate repression or induction of the nitrate-regulated genes, mutations in *narX* caused only a partial loss of nitrate regulation. It therefore appeared that *narX* was dispensable. Another gene was later discovered, *narQ*, that was also a sensor for nitrate reductase. *E. coli* can use either NarX or NarQ as the sensor when one or the other is inactivated. Cells must be mutated in both *narX* and *narQ* in order to observe the loss of nitrate-dependent repression of fumarate reductase and loss of induction of nitrate reductase.

53. Stewart, V. 1993. Nitrate regulation of anaerobic respiratory gene expression in *Escherichia coli. Mol. Microbiol.* 9:425–434.

54. Schroder, I., C. D. Wolin, R. Cavicchioli, and R. P. Gunsalus. 1994. Phosphorylation and dephosphorylation of the NarQ, NarX, and NarL proteins of the nitrate-dependent two-component regulatory system of *Escherichia coli. J. Bacteriol.* 176:4985–4992.

55. Cavicchioli, R., I. Schroder, M. Constanti, and R. P. Gunsalus. 1995. The NarX and NarQ sensor-transmitter proteins of *Escherichia coli* each require two conserved histidines for nitrate-dependent signal transduction to NarL. *J. Bacteriol.* 177:2416–2424.

56. Cavicchioli, R., I. Schröder, M. Constanti, and R. P. Gunsalus. 1995. The *Escherichia coli* NarQ and NarX regulatory proteins contain two conserved histidines that are required for nitrate dependent signal transduction to NarL. *J. Bacteriol.* 177:2416–2424.

57. Cavicchioli, R., R. C. Chiang, L. V. Kalman, and R. P. Gunsalus. 1996. Role of the periplasmic domain of the *Escherichia coli* NarX sensor-transmitter protein in nitrate-dependent signal transduction and gene regulation. *Molec. Microbiol.* 21:901–911.

58. Reviewed in, Macnab, R. M. 1987. Motility and chemotaxis, pp. 723–759 In: Escherichia coli and Salmonella typhimurium: Cellular and Molecular Biology, Vol. 1. Neidhardt, F. C.,

J. L. Ingraham, K. B. Low, B. Magasanik, M. Schaechter, and H. E. Umbarger (Eds.). ASM Press, Washington, D.C.

59. Reviewed in, Stock, J. B, A. J. Ninfa, and A. M. Stock. 1989. Protein phosphorylation and regulation of adaptive responses in bacteria. *Microbiol. Rev.* 53:450–490.

60. Reviewed in, Manson, M. D. 1992. Bacterial motility and chemotaxis, pp. 277–346. In: *Advances in Microbial Physiology*, Vol. 33. Rose, A. H. (Ed.). Academic Press, New York.

61. Reviewed in Blair, D. F. 1995. How bacteria sense and swim. *Ann. Rev. Microbiol.* 49:489–522.

62. Macnab, R. M., and D. E. Koshland, Jr. 1972. The gradient-sensing mechanism in bacterial chemotaxis. *Proc. Nat. Acad. Sci. USA* 69:2509–2512.

63. Berg, H. C., and D. A. Brown. 1972. Chemotaxis in *Escherichia coli* analysed by three-dimensional tracking. *Nature* 239:500–504.

64. Macnab, R. M., and D. E. Koshland, Jr. 1972. The gradient-sensing mechanism in bacterial chemotaxis. *Proc. Nat. Acad. Sci. USA* 69:2509–2512.

65. Silverman, M., and M. Simon. 1974. Flagellar rotation and the mechanism of bacterial motility. *Nature* 249:73–74.

66. Larsen, S. H., R. W. Reader, E. N. Kort, W-W. Tso, and J. Adler. 1974. Changes in direction of flagellar rotation is the basis of the chemotactic response in *Escherichia coli. Nature* 249:74–77.

67. Iino, T., Y. Komeda, K. Kutsukake, R. M. Macnab, P. Matsumura, J. S. Parkinson, M. I. Simon, and S. Yamaguchi. 1988. New unified nomenclature for the flagellar genes of *Escherichia coli* and *Salmonella typhimurium. Microbiol. Rev.* 52:533–535.

68. Three proteins called FliG, FliM, and FliN, coded for by *fliG*, *fliM*, and *fliN*, are the switch proteins. These proteins are not in the basal body and seem to be peripheral (rather than integral) cell membrane proteins closely associated with the basal body. A description of the flagellar motor is given in Section 1.2.1. Mutants that completely lack these proteins do not make flagella (Fla⁻ phenotype) even though these proteins are not part of the basal body, hook, or filament. Missense mutations in these genes cause a Fla⁻ phenotype, a Che⁻ phenotype (CW or CCW-biased), or paralyzed flagella (Mot phenotype), depending upon the mutation. It therefore appears that these proteins are necessary for basal body synthesis and switching, and perhaps are involved in energy coupling. Two other proteins should be mentioned, even

though they are not chemotaxis proteins. The MotA and MotB proteins are necessary for motor function. Mutations in these genes cause paralyzed flagella. The proteins are integral membrane proteins that are thought to form a ring around the S and P rings of the basal body, and couple the proton potential to flagellar rotation. The MotA protein is a proton channel, and the MotB protein may mediate the interaction of MotA with the basal body.

69. The histidine kinase, CheA, is a dimer. One of the monomers acts as a kinase and phosphorylates a histidine residue on the second monomer to form CheA-P.

70. The MCP proteins are truly remarkable with respect to the different classes of signals that they transduce. Tsr responds to the chemoattractant L-serine. It also responds to the repellants L-leucine, indole, and low external pH. Tsr also responds to weak organic acids such as acetate that act as repellents because they lower the internal pH. In addition, the Tsr protein is a thermoreceptor mediating taxis toward warmer temperatures up to 37°C. The Tar protein in *E. coli* responds to the chemoattractant L-aspartate as well as to maltose bound to the periplasmic maltose binding protein. Tar also detects the repellents $Co^{2+}$ and $Ni^{2+}$. In mutants lacking Tsr, the Tar protein mediates taxis toward higher temperatures. The Trg protein responds to the chemoattractants ribose, glucose, and galactose when they are bound to their respective periplasmic binding proteins. The Tap protein (present in *E. coli* but not *S. typhimurium*) mediates taxis toward a variety of dipepetides when they are bound to DPP, a periplasmic dipeptide binding protein. Trg and Tap from *E. coli* serve as repellent receptors for phenol. Trg is also an attractant receptor, whereas Tap is a repellent receptor for weak organic acids.

71. Lukat, G. S., and J. B. Stock. 1993. Response regulation in bacterial chemotaxis. *J. Cell. Biochemistry* 51:41–46.

72. Welch, M., K. Oosawa, S-I Aizawa, and M. Eisenbach. 1993. Phosphorylation-dependent binding of a signal molecule to the flagellar switch of bacteria. *Proc. Natl. Acad. Sci. USA* 90:8787–8791.

73. Ordal, G. W., L. Màrquez-Magana, and M. J. Chamberlin. 1993. Motility and chemotaxis, pp. 765–784. In: *Bacillus subitilis and Other Gram-Positive Bacteria: Biochemistry, Physiology, and Molecular Genetics*. A. L. Sonenshein, J. A. Hoch, and R. Losick (Eds.). American Society for Microbiology, Washington, D.C.

74. Dunten, P., and D. E. Koshland, Jr. 1991. Tuning the responsiveness of a sensory receptor via covalent modification. *J. Biol. Chem.* 266:1491–1496.

75. Borkovich, K. A., L. A. Alex, and M. I. Simon. 1992. Attenuation of sensory receptor signaling by covalent modification. *Proc. Natl. Acad. Sci. USA* 89:6756–6760.

76. It is possible to construct mutants in the aspartate chemoreceptor protein (Tar) so that glutamine (the amidated form of glutamate) is substituted for the glutamate residues that are normally methylated. The chemotaxis signaling system cannot tell the difference between a chemoreceptor that has glutamine and one that has methylated glutamate. Such mutants show that substituting glutamine for glutamate in a CheR⁻ and CheB⁻ mutant produces a tumbling behavior. The tumbling behavior can be reversed by adding the attractant, aspartate, to the medium, reflecting the fact that the binding of attractant to the chemoreceptor inhibits the kinase activity of CheA. Furthermore, CheR⁻ mutants swim smoothly and CheB⁻ mutants tumble more frequently than wild type, again indicating that methylation stimulates tumbling. The effect of methylation on CheA kinase activity appears to be the dominant reason for adaptation because the affinity of the methylated chemoreceptor for the attractant has been reported to be not much different than for the nonmethylated chemoreceptor. (However, the extent to which methylation changes the affinity of the chemoreceptor to the attractant is a subject of controversy. See Blair, D. F. 1995. How bacteria sense and swim. *Ann. Rev. Microbiol.* 49:489–522) Additional evidence that methylation of Tar stimulates CheA kinase activity is that membranes prepared from cells in which glutamine is substituted for glutamate in Tar, or in which the glutamate residues are methylated because the cells were incubated with aspartate, are more active in stimulating the phosphorylation of CheY than membranes containing unmodified chemoreceptor.

77. Presumably the binding of chemoattractant induces a conformational change in the receptor-transducer protein, exposing additional methylating sites on the receptor–transducer protein, thus increasing the level of methylation.

78. Park, C., D. P. Dutton, and G. L. Hazelbauer. 1990. Effects of glutamines and glutamates at sites of covalent modification of a methyl-accepting transducer. *J. Bacteriol.* 172:7179–7187.

79. Stock, J. B., and M. G. Surette. 1996. Chemotaxis, pp. 1103–1129. In: *Escherichia coli and Salmonella: Cellular and Molecular Biology*. Vol. I. F. C. Neidhardt. (Ed.). ASM Press. Washington, D.C.

80. Stock, J., A. Borczuk, F. Chiou, and J. E. B. Burchenal. 1985. Compensatory mutations in receptor function: A reevaluation of the role of

methylation in bacterila chemotaxis. *Proc. Natl. Acad. Sci. USA* **82**:8364–8368.

81. Stock, J., G. Kersulis, and D. E. Koshland, Jr. 1985. Neither methylating nor demethylating enzymes are required for bacterial chemotaxis. *Cell* **42**:683–690.

82. Stock, J., and A. Stock. 1987. What is the role of receptor methylation in bacterial chemotaxis? *TIBS* **12**:371–375.

83. There are several ways to measure chemotaxis. One way is to use swarm plates. If a chemotactic bacterium is inoculated in the center of semisolid agar fortified by attractants it can metabolize, then as the cells grow, the colony swarms outward from the center. This is because the growing cells create a gradient of attractant as they utilize it. A different assay is a capillary assay. In this assay a capillary is filled with a solution of an attractant and placed in a suspension of cells without attractant. The cells swim into the capillary following the gradient. One then plates out the cells in the capillary to quantify the chemotaxis. A third method is to observe tethered cells with a microscope. The cells can be tethered to a coverslip coated with antiflagellin antibody. The coverslip is then placed over a chamber through which various solutions can be added and the clockwise (CW) or counterclockwise (CCW) rotation of the bacteria can be monitored. For *E. coli*, CCW rotation is equivalent to smooth swimming and CW rotation is equivalent to tumbling. One might measure the percent of cells that rotated CCW over a specific time period. For example, suppose the measurement was for 15 seconds. Upon addition of attractant the percent of cells that rotate CCW without reversing over a 15 second period might be 100% and then fall to the prestimulus level, which might be 80%, in 1 or 2 minutes as adaptation occurred. (As explained in Section 18.4.8, not all bacteria reverse the direction of flagellar rotation.)

84. Eisenbach, M., C. Constantinou, H. Aloni, and M. Shinitzky. 1990. Repellents for *Escherichia coli* operate neither by changing membrane fluidity nor by being sensed by periplasmic receptors during chemotaxis. *J. Bacteriol.* **172**:5218–5224.

85. Reviewed in, Titgemeyer, F. 1993. Signal transduction in chemotaxis mediated by the bacterial phosphotransferase system. *J. Cell. Biochem.* **51**:69–74.

86. Reviewed in Eisenbach, M. 1996. Control of bacterial chemotaxis. *Molec. Microbiol.* **20**:903–910.

87. Armitage, J. P., and R. Schmidtt. 1997. Bacterial chemotaxis: *Rhodobacter sphaeroides* and *Sinorhizobium melloti*—variations on a theme? *Microbiology* **143**:3671–3682.

88. Ward, M. J., A. W. Bell, P. A. Hamblin, H. L. Packer, and J. P. Armitage. 1995. Identification of a chemtoaxis operon with two *cheY* genes in *Rhodobacter sphaeroides*. *Molec. Microbiol.* **17**:357–366.

89. Hamblin, P. A., B. A. Maguire, R. N. Grishanin, and J. P. Armitage. 1997. Evidence for two chemosensory pathways in *Rhodobacter sphaeroides*. *Molec. Microbiol.* **26**:1083–1096.

90. The chemotaxis operon from *R. sphaeroides* was detected using a cloned *cheA* homolog from *Rhizobium meliloti*. The chemotaxis operon was sequenced revealing the presence of the *che* genes.

91. Ward, M. J., D. M. Harrison, M. J. Ebner, and J. P. Armitage. 1995. Identification of a methyl-accepting chemotaxis protein in *Rhodobacter sphaeroides*. *Mol. Microbiol.* **18**:115–121.

92. The *R. sphaeroides* MCP protein was discovered by sequencing a fragment of DNA upstream of the chemotaxis operon. The deduced amino acid sequence of the protein product shows significant similarity to an enteric MCP (MCP Tsr). The protein cross-reacts with antisera to the Mcp A protein of *C. crescentus* and was located primarily in the cytoplasmic fraction as determined by Western blotting. For Western blotting the proteins of the membrane and cytoplasmic cell fractions are first separated by SDS-Page gel electrophoresis. Then the protein bands are electrophoretically transferred to nitrocellulose. The proteins on the nitrocellulose are incubated with antisera, e.g., to anti-McpA antibody, and the bands that bind the antibody are stained using an antirabbit secondary antibody (e.g., goat–antirabbit antibody) conjugated to horseradish peroxidase. An assay for the peroxidase stains the bands.

93. Armitage, J. P. 1997. Behavioral responses of bacteria to light and oxygen. *Arch. Microbiol.* **168**:249–261.

94. Jeziore-Sassoon, Y., P. A. Hamblin, C. A. Bootle-Wilbraham, P. S. Poole, and J. P. Armitage. 1998. Metabolism is required for chemotaxis to sugars in *Rhodobacter sphaeroides*. *Microbiology* **144**:229–239.

95. Part of the evidence that metabolism is required for the chemotactic response to carbohydrate attractants is that mutants of *R. sphaeroides* lacking glucose-6-phosphate dehydrogenase, a key enzyme in the Entner–Doudoroff (and pentose phosphate) pathway, are defective in chemotaxis toward sugars metabolized by this pathway. See Jeziore-Sassoon, Y., P. A. Hamblin, C. A. Bootle-Wilbraham, P. S. Poole, and J. P. Armitage. 1998. Metabolism is

required for chemotaxis to sugars in *Rhodobacter sphaeroides*. *Microbiology* 144:229–239.

96. Packer, H. L., and J. P. Armitage. 1994. The chemokinetic and chemotactic behavior of *Rhodobacter sphaeroides*: Two independent responses. *J. Bacteriol.* 176:206–212.

97. Scnitzer, M. J., S. M. Block, H. C. Berg, and E. M. Purcell. 1990. Strategies for chemotaxis. *Symp. Soc. Gen. Microbiol.* 46:15–34.

98. Gotz, R., and R. Schmitt. 1987. *Rhizobium meliloti* swims by unidirectional, intermittent rotation of right-handed flagellar helices. *J. Bacteriol.* 169:3146–3150.

99. Greck, M., J. Platzer, V. Sourjik, and R. Schmitt. 1955. Analysis of a chemotaxis operon in *Rhizobium meliloti*. *Mol. Microbiol.* 15:989–1000.

100. Sockett, R. E., J. P. Armitage, and M. C. W. Evans. 1987. Methylation-independent and methylation-dependent chemotaxis in *Rhodobacter sphaeroides* and *Rhodospirillum rubrum*. *J. Bacteriol.* 169:5808–5814.

101. Zhulin, I. G., and J. P. Armitage. 1993. Motility, chemokinesis, and methylation-independent chemotaxis in *Azospirillum brasilense*. *J. Bacteriol.* 175:952–958.

102. Hellingwerf, K. J., W. D. Hoff, and W. Crielaard. 1996. Photobiology of microorganisms: How photosensors catch a photon to initialize signalling. *Molec. Microbiol.* 21:683–693.

103. Alam, M., and D. Oesterhelt. 1984. Morphology, function and isolation of halobacterial flagella. *J. Mol. Biol.* 176:459–475.

104. Spudich, J. L. 1993. Color sensing in the Archaea: A eukaryotic-like receptor coupled to a prokaryotic transducer. *J. Bacteriol.* 175:7755–7761.

105. Bickel-Sandkötter, S. 1996. Conversion of energy in halobacteria: ATP synthesis and phototaxis. *Arch. Microbiol.* 166:1–11.

106. Armitage, J. P. 1997. Behavioral responses of bacteria to light and oxygen. *Arch. Microbiol.* 168:249–261.

107. By way of introduction, the student should know that all photosensing involves the absorption of a photon of light by a chromophore bound to a protein. When the photon is absorbed, one or more of the following four events are induced: (1) excitation transfer, (2) electron transfer, (3) $H^+$ transfer, (4) photoisomerization. As we shall see, photoisomerization and $H^+$ transfer occur when the rhodopsins absorb light energy. There are two photosensory rhodopsin pigments called SRI and SRII that are responsible for the photoresponses. SRII is made constitutively at relatively low levels, but SRI is induced to high levels under growth conditions that also induce bacteriorhodopsin, i.e., low oxygen tensions and intense illumination. It is clear that the rhodopsins that function in photoresponses are similar to bacteriorhodopsin and halorhodopsin. (See Sections 3.8.4 and 3.9 for a discussion of bacteriorhodopsin and halorhodopsin, respectively.) These pigment proteins share related sructures and presumably have a common origin. Their primary photochemistry is the same, i.e., a light-induced photoisomerization of the retinal from the all-*trans* isomer to the 13-*cis* form. When the rhodopsins absorb light, a signal is transmitted to the flagellum motor, probably via a membrane-bound signal-transducing protein, HtrI (halobacterial transducer for sensory rhodopsin I) to which SRI is tightly bound. (A second protein, HtrII, is complexed to SRII.) The Htr proteins are similar to the MCP proteins in that they have two membrane-spanning domains and a cytoplasmic domain that is homologous to the MCPs. Presumably a conformational change takes place in Htr when light is absorbed by the chromophore and a signal is thus generated to the Che proteins, which signal the flagellum motor. The model proposes that SRI exists in either one of two conformations, i.e., $SR_{587}$ and $S_{373}$, denoting their maximum absorption peaks. The two conformations are interconvertable. When $SR_{587}$ absorbs light (orange/red), it is converted to $S_{373}$. When $S_{373}$ absorbs light (UV/blue), it is converted back to $SR_{587}$. There are approximately 5000 molecules per cell of SRI, and in the presence of steady background light an equilibrium exists between the two forms. $SR_{587}$ is the receptor for attractant light. The absorption of light by $SR_{587}$ suppresses flagellar reversals, causing the cells to swim toward higher intensities of orange/red light. $S_{373}$ is the receptor for repellent light. The absorption of light by $S_{373}$ increases flagellar reversals, causing the cells to swim away from high intensities of UV/blue light. This is an interesting case of the same photoreceptor serving as either the attractant or repellent light receptor depending upon its conformation. It has been presumed that HtrI undergoes a conformational change upon the absorption of light by SRI and transmits the signal to the flagellar motor. It has been demonstrated that SRI is capable of light-induced proton pumping similar to bacteriorhodopsin proton pumping, but in the presence of tightly bound HtrI the protons are not released into the bulk phase but rather circulate. Either the protons circulate through HtrI residues during SRI photocyle (a direct involvement of Htr1 in proton circulation) or proton circulation is dependent upon conformational interactions between SRI and HtrI (an indirect role for HtrI in proton circulation).

108. Sackett, M. J., J. P. Armitage, E. E. Sherwood, and T. P. Pitta. 1997. Photoresponses

of the purple nonsulfur bacteria *Rhodospirillum centenum* and *Rhodobacter sphaeroides*. *J. Bacteriol.* **179**:6764–6768.

109. Grishanin, R. N., D. E. Gauden, and J. P. Armitage. 1997. Photoresponses in *Rhodobacter sphaeroides*; role of photosynthetic electron transport. *J. Bacteriol.* **179**:24–30.

110. Ragatz, L, Z-Y Jiang, C. Bauer, and H. Gest. 1995. Macroscopic phototactic behavior of the purple photosynthetic bacterium *Rhodospirillum centenum*. *Arch. Microbiol.* **163**:1–6.

111. Jiang, Z-Y, B. G. Rushing, Y. Bai, H. Gest, and C. E. Bauer. 1998. Isolation of *Rhodospirillum centenum* mutants defective in phototactic colony motility by transposon mutagenesis. *J. Bacteriol.* **180**:1248–1255.

112. Sequence analysis of one of the mutants indicates the presence of a photoreceptor that has extensive sequence similarity with MCPs. (Carl Bauer, personal communication.)

113. Shioi, J., R. C. Tribhuwan, S. T. Berg, and B. L. Taylor. 1988. Signal transduction in chemotaxis to oxygen in *Escherichia coli* and *Salmonella typhimurium*. *J. Bacteriol.* **170**:5507–5511.

114. Rowsell, E. H., J. M. Smith, A. Wolfe, and B. L. Taylor. 1995. CheA, CheW, and CheY are required for chemotaxis to oxygen and sugars of the phosphotransferase system in *Escherichia coli*. *J. Bacteriol.* **177**:6011–6014.

115. Bauer, C. E., and T. H. Bird. 1996. Regulatory circuits controlling photosynthesis gene expression. *Cell* **85**:5–8.

116. Mosely, C., J. Y. Suzuki, and C. E. Bauer. 1994. Identification and molecular genetic characterization of a sensor kinase responsible for coordinately regulating light harvesting and reaction center gene expresssion in response to anaerobiosis. *J. Bacteriol.* **176**:7566–7573.

117. Du, S., T. H. Bird, and C. E. Bauer. 1998. DNA binding characteristics of RegA. *J. Biol. Chem.* **273**:18509–18513.

118. Zucconi, A. P., and J. T. Beatty. 1988. Post-transcriptional regulation by light of the steady-state levels of mature B800–850 light-harvesting complexes in *Rhodobacter capsulatus*. *J. Bacteriol.* **170**:877–882.

119. Magasanik, B. 1993. The regulation of nitrogen utilization in enteric bacteria. *J. Cell. Biochem.* **51**:34–40.

120. Magasanik, B. 1988. Reversible phosphorylation of an enhancer binding protein regulates the transcription of bacterial nitrogen utilization genes. *TIBS* **13**:475–479.

121. Merrick, M. J., and R. A. Edwards. 1995. Nitrogen control in bacteria. *Microbiol. Rev.* **59**:604–622.

122. Magasanik, B. 1982. Genetic control of nitrogen assimilation in bacteria. *Ann. Rev. Gen.* **16**:135–168.

123. Reviewed in Magasanik, B. 1996. Regulation of nitrogen utilization, pp. 1344–1356. In: *Escherichia coli and Salmonella, Cellular and Molecular Biology*, Vol. 1. F. C. Neidhardt et al. (Eds.). ASM Press, Washington, D.C.

124. MacNeil, T., G. P. Roberts, D. MacNeil, and B. Tyler. 1982. The products of *glnL* and *glnG* are bifunctional regulatory proteins. *Mol. Gen. Genet.* **188**:325–333.

125. Sigma 54 (encoded by the *rpoN* gene in *E. coli*) polymerases bind to promoters that are unrelated to sigma 70 promoters and transcribe many different genes, i.e., not merely genes in the Ntr regulon. Sigma 54 polymerases can bind to promoter regions of target genes but, unlike sigma 70 polymerases, are unable to produce an open complex without the aid of an activator protein that usually binds 100 to 150 bp upstream of the promoter. The activator proteins are ATP dependent, and in a way that is not understood, the hydrolysis of ATP provides the energy to convert the closed complex to the open complex.

126. The reason for the requirement of NR$_I$-P for transcription is that the sigma 54 polymerase binds to the *glnAp$_2$* promoter but cannot by itself initiate transcription. It is unable to do this because it cannot form an open complex (i.e., it cannot "melt" the DNA double helix around the promoter site to gain access to the transcription start site). However, NR$_I$-P binds to sites on the DNA 100–130 base pairs upstream from the promoter region (called "enhancer sites") and interacts with the RNA polymerase to form the open complex and thus initiate transcription.

127. Dixon, R. 1998. The oxygen-responsive NIFL–NIFA complex: A novel two-component regulatory system controlling nitrogenase synthesis in $\gamma$-*Proteobacteria*. *Arch. Microbiol.* **169**:371–380.

128. Friedman, D. I. 1988. Integration host factor: A protein for all reasons. *Cell* **55**:545–554.

129. References to homologs of PhoR and PhoB can be found in the review by Wanner.[134]

130. van Veen, H. W. 1997. Phosphate transport in prokaryotes: Molecules, mediators and mechanisms. *Antonie van Leeuwenhoek* **72**:299–315.

131. Phosphonates are organophosphates in which there is a direct carbon–phosphorus bond rather than a phosphate ester linkage.

132. The promoters in the PHO regulon are unusual in that they have $-10$ sequences but no consensus $-35$ sequences for $\sigma^{70}$ RNA polymerase. Thus, in the absence of PhoB, RNA polymerase does not bind to the promoter. What the promoters do have, however, are one to three (tandemly repeated) 18-bp sequences called *pho* boxes that are 10 bp upstream of the $-10$ consensus sequence. The phosphorylated PhoB protein binds to the *pho* boxes, and this enables the RNA polymerase to bind to the promoter. It has been postulated that PhoB interacts with the sigma subunit of the RNA polymerase, enabling the polymerase to bind to the promoter. See Makino, K., M. Amemura, S-K Kim, A. Nakata, and H. Shinagawa. 1993. Role of the $\sigma^{70}$ subunit of RNA polymerase in transcriptional activation by activator protein PhoB in *Escherichia coli*. *Genes and Development* 7:149–160.

133. Wanner, B. L. 1993. Gene regulation by phosphate in enteric bacteria. *J. Cell. Biochemistry* 51:47–54.

134. Wanner, B. L. 1996. Phosphorus assimilation and control of the phosphate regulon. pp. 1357–1381. In: *Escherichia coli and Salmonella Cellular and Molecular Biology*. F. C. Neidhardt (Ed.) ASM Press, Washington, D.C.

135. The Pst system is similar to the histidine uptake system described in Section 16.3.3 and belongs to the superfamily of ABC (ATP-binding cassette) transporters. These have four membrane-bound proteins, two of which are transmembrane and probably form a channel, whereas the other two have ATP binding sites. In addition, there is a periplasmic binding protein. Pst A and Pst C are thought to form a transmembrane channel, PstB is the ATPase, which probably exists as a dimer, and PtsS is the periplasmic binding protein.

136. PhoR autophosphorylates in vitro using ATP as the phosphoryl donor and transfers the phosphoryl group to PhoB, but it has not been demonstrated that it dephosphorylates PhoB-P.[134]

137. Nikaido, H., and M. Vaara. 1987. Outer membrane, pp. 7–22. In: *Escherichia coli and Salmonella typhimurium, Cellular and Molecular Biology*. Vol. 1. Neidhardt, F. C., J. L. Ingraham, K. B. Low, B. Magasanik, M. Schaechter, and H. E. Umbarger (Eds.). ASM Press, Washington, D.C.

138. Forst, S. A. and D. L. Roberts. 1994. Signal transduction by the EnvZ–OmpR phosphotransfer system in bacteria. *Res. Microbiol.* 145:363–373.

139. Reviewed in Pratt, L. A. and T. J. Silhavy. 1995. Porin regulon of *Escherichia coli*. In: *Two-Component Signal Transduction*, pp. 105–127.

Hoch, J. A., and T. J. Silhavy (Eds.). ASM Press, Washington, D.C.

140. Waukau, J., and S. Forst. 1992. Molecular analysis of the signaling pathway between EnvZ and OmpR in *Escherichia coli*. *J. Bacteriol.* 174:1522–1527.

141. Aiba, H., T. Mizuno, and S. Mizushima. 1989. Transfer of phosphoryl group between two regulatory proteins involved in osmoregulatory expression of the *ompF* and *ompC* genes in *Escherichia coli*. *J. Biol. Chem.* 264:8563–8567.

142. Aiba, H., F. Nakasai, S. Mizushima, and T. Mizuno. 1989. Evidence for the physiological importance of the phosphotransfer between two regulatory components, EnvZ and OmpR, in osmoregulation in *Escherichia coli*. *J. Biol. Chem.* 264:14090–14094.

143. Ramani, N., L. Huang, and M. Freundlich. 1992. In vitro interactions of integration host factor with the *ompF* promoter–regulatory region of *Escherichia coli*. *Mol. Gen. Genet.* 231:248–255.

144. Mizuno, T., M.-Y. Chou, and M. Inouye. 1984. A unique mechanism regulating gene expression: Translational inhibition by a complementary RNA transcript (micRNA). *Proc. Natl. Acad. Sci. USA* 81:1966–1970.

145. Ramani, N., H. Hedeshian, and M. Freundlich. 1994. *micF* Antisense RNA has a major role in osmoregulation of OmpF in *Escherichia coli*. *J. Bacteriol.* 176:5005–5010.

146. Reviewed in Delihas, N. 1995. Regulation of gene expression by trans-encoded antisense RNAs. *Molec. Microbiol.* 15:411–414.

147. Sugiura, A., K. Nakashima, K. T. Tanaka, and T. Mizuno. 1992. Clarification of the structural and functional features of the osmoregulated kdp operon of *Escherichia coli*. *Molec. Microbiol.* 6:1769–1776.

148. Nakashima, K., H. Sugiura, H. Momoi, and T. Mizuno. 1992. Phosphotransfer signal transduction between two regulatory factors involved in the osmoregulated kdp operon in *Escherichia coli*. *Molec. Microbiol.* 6:1777–1784.

149. Sugiura, A., K. Hirokawa, K. Nakashima, and T. Mizuno. 1994. Signal-sensing mechanisms of the putative osmosensor KdpD in *Escherichia coli*. *Molec. Microbiol.* 14:929–938.

150. McCleary, W. R., J. B. Stock, and A. J. Ninfa. 1993. Is acetyl phosphate a global signal in *Escherichia coli*? *J. Bacteriol.* 175:2793–2798.

151. Feng, J., M. R. Atkinson, W. McCleary, J. B. Stock, B. L. Wanner, and A. J. Ninfa. 1992. Role of phosphorylated metabolic intermediates in the

regulation of glutamine synthetase synthesis in *Escherichia coli. J. Bacteriol.* **174**:6061–6070.

152. Shin, S., and C. Park. 1995. Modulation of flagellar expression in *Escherichia coli* by acetyl phosphate and the osmoregulator OmpR. *J. Bacteriol.* **177**:4696–4702.

153. Dailey, F. E., and H. C. Berg. 1993. Change in direction of flagellar rotation in *Escherichia coli* mediated by acetate kinase. *J. Bacteriol.* **175**:3236–3239.

154. Barak, R., W. N. Abouhamad, and M. Eisenbach. 1998. Both acetate kinase and acetyl coenzyme A synthetase are involved in acetate-stimulated change in the direction of flagellar rotation in *Escherichia coli. J. Bacteriol.* **180**:985–988.

155. For a review of catabolite repression by carboxylic acids, see *Research in Microbiology*, Vol. 147, No. 6–7.

156. Saier, Jr., M. H. 1996. Cyclic AMP-independent catabolite repression in bacteria. *FEMS Microbiol. Lett.* **138**:97–103.

157. Saier, Jr, M. H., T. M. Ramseier, and J. Reizer. 1996. Regulation of carbon utilization. In: *Escherichia coli and Salmonella: Molecular and Cellular Biology.* Vol. 1. pp. 1325–1343. F. C. Neidhardt et al. (Eds.). ASM Press, Washington, D.C.

158. Chin, M. A., B. U. Feucht, and M. H. Saier, Jr. 1987. Evidence for regulation of gluconeogenesis by the fructose phosphotransferase system in *Salmonella typhimurium. J. Bacteriol.* **169**:897–899.

159. Mutants in *cra* show increased levels of enzymes involved in the uptake and catabolism of glucose such as enzyme I and HPr of the PTS system, phosphofructokinase, pyruvate kinase, 6-phosphogluconate dehydratase, and KDPG aldolase. Cra is also a repressor of the *fruBKA* operon that consists of genes required for the fructose-specific PTS system. The *fruBKA* operon codes for three proteins. FruB is an HPr-like protein called Fpr that differs structurally from HPr used in the regular PTS system but can substitute for HPr in mutants unable to synthesize HPr. In fact, *cra* mutants were originally found as suppressor mutants that allowed mutants defective in HPr synthesis (*ptsH* mutants) to grow on PTS carbohydrates. In addition to a C-terminal HPr domain, it has an N-terminal IIA$^{Fru}$ domain. FruK is a fructose-1-phosphate kinase that is necessary to convert the incoming fructose-1-phosphate to fructose-1,6-bisphosphate. FruA is the permease. These genes are required for fructose transport via the fructose-PTS system, and the conversion of the incoming fructose-1-phosphate to fructose-1,6-bisphosphate.

160. Ramseier, T. M. 1996. Cra and the control of carbon flux via metabolic pathways. *Research in Microbiol.* **147**:489–493.

161. Tyson, K., S. Busby, and J. Cole. 1997. Catabolite regulation of two *Escherichia coli* operons encoding nitrite reductases: Role of the Cra protein. *Arch. Microbiol.* **168**:240–244.

162. Saier, M. H., Jr., S. Chauvaux, J. Deutscher, J. Reizer, and J-J. Ye. 1995. Protein phosphorylation and regulation of carbon metabolism in gram-negative versus gram-positive bacteria. *TIBS* **20**:267–271.

163. Hueck, C. J., and W. Hillen. 1995. Catabolite repression in *Bacillus subtilis*: A global regulatory mechanism for the gram-positive bacteria? *Molec. Microbiol.* **15**:395–401.

164. Chauvaux, S., I. T. Paulsen, and M. H. Saier, Jr. 1997. CcpB, a novel transcription factor implicated in catabolite repression in *Bacillus subtilis. J. Bacteriol.* **180**:491–497.

165. Depending upon the gene, CREs have been found to overlap promoters or be located within the reading frame. One suspects that if the CRE overlaps the promoter, then binding of a regulatory protein would prevent initiation of transcription, and if the CRE were in the reading frame, then binding of a regulatory protein would inhibit transcription elongation. The protein that binds to the CREs has not been unambiguously identified.

166. Also present in *B. subtilis* is the PEP-dependent kinase, which can also phosphorylate Hpr and operates during phosphotransferase-mediated glucose uptake. (Although the phosphotransferase system is characteristic of anaerobes and facultative anaerobes, it does occur in some aerobes, e.g., *B. subtilis*.) However, the PEP-dependent kinase is not part of the catabolite repressing system. The reason for this is that the phosphorylation of Hpr by the ATP-dependent kinase is at a serine residue (Ser-46), whereas the phosphorylation of Hpr by the PEP-dependent kinase is at a histidine residue (His-15) and Hpr(His-P) is not active in catabolite repression.

167. One method to isolate mutations in genes required for catabolite repression by glucose is to mutagenize a strain carrying an operon translational fusion of the gene of interest to *lacZ*. One can then screen for blue colonies on media containing glucose plus the alternative carbon source and the indicator X-Gal (5-bromo-4-chloro-3-indolyl-β-D-galactoside). X-Gal is an analog of lactose. It is colorless, but when it is cleaved by β-galactosidase, a blue color is produced. Mutagenesis can be performed using transposons. A transposon is a DNA sequence

that moves from one DNA molecule to another or from one part of a DNA molecule to another part. When the transposon inserts in a gene of interest, it can produce a mutation in that gene because the gene's DNA sequence is disrupted. The transposon can be introduced into the bacterial cell on a plasmid via transformation or conjugation. Transposons can carry antibiotic resistance genes that may be used to select for transformants during the isolation of mutants as well as selection of transformants during cloning of the gene.

168. Ye, J.-J., and M. H. Saier, Jr. 1995. Cooperative binding of lactose and HPr(ser-P) to the lactose:H$^+$ permease of *Lactobacillus brevis*. *Proc. Natl. Acad. Sci. USA* **92**:417–421.

169. Ye, J.-J., and M. H. Saier, Jr. 1995. Purification and characterization of a small membrane-associated sugar–phosphate phosphatase that is allosterically activated by HPr(ser-P) of the phosphotransferase system in *Lactococcus lactis*. *J. Biol. Chem.* **270**:16740–16744.

170. Giblin, L., B. Boesten, S. Turk, P. Hooykaas, and F. O'Gara. 1995. Signal transduction in the *Rhizobium meliloti* dicarboxylic acid transport system. *FEMS Microbiol. Letts.* **126**:25–30.

171. Mandelstam, J., and S. A. Higgs. 1974. Induction of sporulation during synchronized chromosome replication in *Bacillus subtilis*. *J. Bacteriol.* **120**:38–42.

172. Stragier, P. 1994. A few good genes: Developmental loci in *Bacillus subtilis*, pp. 207–245. In: *Regulation of Bacterial Differentiation*. Piggot, P., C. P. Moran, Jr., and P. Youngman (Eds.). Americian Society for Microbiology, Washington, D.C.

173. Stragier, P., and R. Losick. 1996. Molecular genetics of sporulation in *Bacillus subtilis*. *Ann. Rev. Genet.* **30**:297–341.

174. Khvorova, A., L. Zhang, M. L. Higgins, and P. J. Piggot. 1998. The spoIIE locus is involved in the SpoOA-dependent switch in the location of FtsZ rings in *Bacillus subtilis*. *J. Bacteriol.* **180**:1256–1260.

175. Wu, L. J., P. J. Lewis, R. Allmansberger, P. M. Hauser, and J. Ellington. 1995. A conjugation-like mechanism for prespore chromosome partitioning during sporulation in *Bacillus subtilis*. *Genes & Dev.* **9**:1316–1326.

176. Lin, D. C-H, and A. D. Grossman. 1998. Identification and characterization of a bacterial chromosome partitioning site. *Cell* **92**:675–685.

177. Burbulys, D., K. A. Trach, and J. A. Hoch. 1991. Initiation of sporulation in *B. subtilis* is controlled by a multicomponent phosphorelay. *Cell* **64**:545–552.

178. Hoch, J. A. 1993. *SpoO* genes, the phosphorelay, and the initiation of sporulation, pp. 747–755. In: *Bacillus subtilis and Other Gram-Positive Bacteria: Biochemistry, Physiology, and Molecular Genetics.* A. L. Sonenshein, J. A. Hoch, and R. Losick (Eds.). American Society for Microbiology, Washington, D.C.

179. Madhusudan, J. Z., J. A. Hoch, and J. M. Whiteley. 1997. A response regulatory protein with the site of phsophorylation blocked by an arginine interaction: Crystal structure of SpoOF from *Bacillus subtilis*. *Biochem.* **36**:12739–12745.

180. Dartois, V., T. Djavakhishvili, and J. A. Hoch. 1997. KapB is a lipoprotein required for KinB signal transduction and activation of the phosphorelay to sporulation in *Bacillus subtilis*. *Molec. Microbiol.* **26**:1097–1108.

181. KinA has been shown to phosphorylate SpoOF in vitro. KinB has not been purified from the membrane and shown to phosphorylate SpoOF. However, a hybrid protein (encoded by a fused gene) consisting of the N-terminal domain of KinA and the C-terminal domain (the kinase domain) of KinB was isolated and shown to phosphorylate SpoOF in vitro. For further details, see Dartois, V., T. Djavakhishvili, and J. A. Hoch. 1997. KapB is a lipoprotein required for KinB signal transduction and activation of the phosphorelay to sporulation in *Bacillus subtilis*. *Molec. Microbiol.* **26**:1097–1108.

182. Strauch, M. A. 1993. AbrB, a transition state regulator, pp. 757–764, In: *Bacillus subtilis and Other Gram-Positive Bacteria*. Sonenshein, A. L. (Ed.). American Society for Microbiology, Washington, D.C.

183. Perego, M., C. Hanstein, K. M. Welsh, T. Djavakhishvili, P. Glaser, and J. A. Hoch. 1994. Multiple protein–aspartate phosphatases provide a mechanism for the integration of diverse signals in the control of development in *B. subtilis*. *Cell* **79**:1047–1055.

184. Perego, M., and J. A. Hoch. 1996. Protein aspartate phosphatases control the output of two-component signal transduction systems. *TIG* **12**:97–102.

185. Perego, M., and J. A. Hoch. 1996. Cell–cell communication regulates the effects of protein aspartate phosphatases on the phosphorelay controlling development in *Bacillus subtilis*. *Proc. Natl. Acad. Sci. USA* **93**:1549–1553.

186. Perego, M. 1997. A peptide export-import control circuit modulating bacterial development regulates protein phosphatases of the phosphorelay. *Proc. Natl. Acad. Sci. USA* **94**:8612–8617.

187. Solomon, J. M., and A. D. Grossman. 1996. Who's competent and when: Regulation of natural genetic competence in bacteria. *Trends in Gen.* **12**:150–155.

188. Lazazzera, B. A., J. M. Solomon, and A. D. Grossman. 1997. An exported peptide functions intracellularly to contribute to cell density signaling in *B. subtilis. Cell* **89**:917–925.

189. Lazazzera, B. A., and A. D. Grossman. 1997. A regulatory switch involving a Clp ATPase. *BioEssays* **19**:455–458.

190. Dunny, G. M., and B. A. B. Leonard. 1997. Cell–cell communication in gram-positive bacteria. *Ann. Rev. Microbiol.* **51**:527–564.

191. The *srfA* operon encodes the three subunits of the peptide synthetase required to synthesize the cyclic lipopeptide antibiotic surfactin, which is composed of seven amino acids and a β-hydroxy fatty acid. One of the genes in the *srfA* operon is *comS*, which is the only gene in that operon required for competence. Transcription of the *srfA* operon is stimulated by the phosphorylated response regulator protein ComA-P, which binds upstream of the *srfA* promoter. ComA is phosphorylated by the histidine kinase, ComP. The peptide extracellular signaling molecule ComX stimulates the kinase activity of ComP, and the extracellular polypeptide signaling molecule CSF inhibits the dephosphorylation of ComA-P by the RapC phosphatase. Thus both signaling polypeptides increase the amounts of ComA-P, which stimulates the transcription of *comS*. ComS activates ComK, which activates the transcription of all the competence genes required for the uptake and processing of DNA. See text for model of how ComS activates ComK. For reviews, see Solomon, J. M., and A. D. Grossman. 1996. Who's competent and when: Regulation of natural genetic competence in bacteria. *Trends in Gen.* **12**:150–155; and Lazazzera, B. A., and A. D. Grossman. 1997. A regulatory switch involving a Clp ATPase. *BioEssays* **19**:455–458.

192. Amon, A. 1998. Controlling cell cycle and cell fate: Common strategies in prokaryotes and eukaryotes. *Proc. Natl. Acad. Sci. USA* **95**:85–86.

193. Quon, K. C., B. Yang, I. J. Domian, L. Shapiro, and G. T. Marczynski. 1998. Negative control of bacteria DNA replication by a cell cycle regulatory protein that binds at the chromosome origin. *Proc. Natl. Acad. Sci. USA* **95**:120–125.

194. Domian, I. J., K. C. Quon, and L. Shapiro. 1997. Cell type–specific phosphorylation and proteolysis of a transcriptional regulator controls the G1-to-S transition in a bacterial cell cycle. *Cell* **90**:415–424.

195. Kelly, A. J., M. J. Sackett, N. Din, E. Quardokus, and Y. V. Brun. 1998. Cell-cycle-dependent transcriptional and proteolytic regulation of FtsZ in *Caulobacter. Genes and Develop.* **12**:880–893.

196. Quardokus, E. M., N. Din, and Y. V. Brun. 1996. Cell cycle regulation and cell type–specific localization of the FtsZ division initiation protein in *Caulobacter. Proc. Natl. Acad. Sci. USA* **93**:6314–6319.

197. The cellular amounts of any protein can be followed using immunoblot assays. Cell extracts are subjected to SDS-polyacrylamide gel electrophoresis, which separates proteins according to their molecular weights. After electrophoresis, the proteins are transferred as bands to a nitrocellulose membrane ("Western" blotting). Usually electroblotting is used to effect the transfer. A blotting sandwich is constructed where the polyacrylamide gel is covered with a nitrocellulose membrane, and a constant current is applied for several hours. The proteins are electrophoretically transferred to the membrane, where they bind tightly. The protein in question can be detected and its relative amounts measured using specific antibody, which binds to the protein (immunoblotting). Changes in the amount of protein are reflected in changes in the amount of antibody that binds to the protein band. The antibody can be labeled in some way so that the amount that binds to the protein can be measured. For example, the antibody might be radioactive. Sometimes an unlabeled antibody is used that is detected via a "sandwich" reaction with a second antibody. For example, the first antibody might be rabbit IgG raised against the protein to be detected, and the second antibody would be antirabbit IgG. The second antibody might be labeled with a radioactive atom or a fluorescent molecule, or perhaps conjugated to an enzyme such as phosphatase, which can be detected using a colorimetric assay. The sandwich technique is usually more sensitive.

The amounts of a specific protein can also be measured using immunoprecipitation, i.e., precipitating the protein with specific antibody. The cells can be labeled with a radioactive isotope, e.g., [$^{35}$s]methionine, the precipitated protein can be analyzed after SDS polyacrylamide gel electrophoresis, and the radioactivity in the protein band can be measured.

198. Quon, K. C., G. T. Marczynski, and L. Shapiro. 1996. Cell cycle control by an essential bacterial two-component signal transduction protein. *Cell* **84**:83–93.

199. Reviews of different aspects of the biology of myxobacteria, including intercellular signaling, can be found in the following

book: *Myxobacteria II*. 1993. Dworkin, M., and D. Kaiser (Eds.). American Society for Microbiology. Washington, D.C.

200. Rodriguez-Soto, J. P., and D. Kaiser. 1997. Identification and localization of the Tgl protein, required for *Myxococcus xanthus* social motility. *J. Bacteriol.* **179**:4372–4381.

201. Rodriguez-Soto, J. P., and D. Kaiser. 1997. The *tgl* gene: Social motility and stimulation in *Myxococcus xanthus*. *J. Bacteriol.* **179**:4361–4371.

202. Wu, S. S., and D. Kaiser. 1997. Regulation of expression of the *pilA* gene in *Myxococcus xanthus*. *J. Bacteriol.* **179**:7748–7758.

203. Hodgkin, J., and D. Kaiser. 1977. Cell-to-cell stimulation of movement in nonmotile mutants of *Myxococcus*. *Proc. Natl. Acad. Sci. USA* **74**:2938–2942.

204. Rodriguez-Soto, J. P., and D. Kaiser. 1997. The *tgl* gene: Social motility and stimulation in *Myxococcus xanthus*. *J. Bacteriol.* **179**:4361–4371.

205. Wall, D., S. S. Wu, and D. Kaiser. 1998. Contact stimulation of Tgl and type IV pili in *Myxococcus xanthus*. *J. Bacteriol.* **180**:759–761.

206. Rodriguez-Soto, J. P., and D. Kaiser. 1997. Identification and localization of the Tgl protein, which is required for *Myxococcus xanthus* social motility. *J. Bacteriol.* **179**:4372–4381.

207. Hagen, D. C., A. P. Bretscher, and D. Kaiser. 1978. Synergism between morphogenetic mutants of *Myxococcus xanthus*. *J. Bacteriol.* **172**:15–23.

208. Downard, J., and D. Toal. 1995. Branched-chain fatty acids—the case for a novel form of cell–cell signaling during *Myxococcus xanthus* development. *Mol. Microbiol.* **16**:171–175.

209. The tetranucleotide ppGpp increases in intracellular concentration in *M. xanthus* during starvation-induced development. It has been shown that genes expressed during development are stimulated by an increase in the intracellular concentration of ppGpp. This was done by introducing into *M. xanthus* a copy of the *E. coli relA* gene, which encodes the enzyme (p)ppGpp synthetase I required to synthesize ppGpp from ATP and GTP. The *relA* gene was placed in a plasmid under the control of a light-inducible promoter from *M. xanthus* (*carQRS*). The plasmid was introduced into *M. xanthus* by electroporation and it integrated into the *M. xanthus* chromosome at the Mx8 prophage attachment site. Light caused an increase in the intracellular amounts of ppGpp and pppGpp either under starvation or nonstarvation conditions in those strains carrying the *relA* gene under the control of the light-inducible *carQRS* promoter. Importantly, the expression of genes normally expressed during development (monitored by the expression of developmentally specific transcriptional *lacZ* fusions, which are normally expressed early during starvation) was increased by light in the strains carrying the light-inducible *relA* gene. The light-inducible developmentally expressed genes included a *lacZ* fusion gene whose expression normally requires A signal. Mutants unable to synthesize A signal did not show the light-dependent expression of the A-signal-dependent fusion gene. This indicates that A signaling and (p)ppGpp are in the same signaling pathway. One possibility that has been mentioned is that (p)ppGpp increases intracellularly during starvation and that the increased levels of (p)ppGpp signal the cells to make A factor, which serves as an intercellular quorum signal, resulting in the activation of expression of developmental genes. For a further discussion of the role of (p)ppGpp in *M. xanthus* development, see Singer, M., and D. Kaiser. 1995. Ectopic production of guanosine penta- and tetraphosphate can initiate early developmental gene expression in *Myxococcus xanthus*. *Genes & Dev.* **9**:1633–1644.

210. Plamann, L., A. Kuspa, and D. Kaiser. 1992. Proteins that rescue A-signal-defective mutants of *Myxococcus xanthus*. *J. Bacteriol.* **174**:3311–3318.

211. Kuspa, A., L. Plamann, and D. Kaiser. 1992. Identification of heat-stable A-factor from *Myxococcus xanthus*. *J. Bacteriol.* **174**:3319–3326.

212. Kuspa, A., L. Plamann, and D. Kaiser. 1992. A-signaling and the cell density requirement for *Myxococcus xanthus* development. *J. Bacteriol.* **174**:7360–7369.

213. Plamann, L., L. Yonghui, B. Cantwell, and J. Mayor. 1995. The *Myxococcus xanthus asgA* gene encodes a novel signal transduction protein required for multicellular development. *J. Bacteriol.* **177**:2014–2022.

214. Plamann, L., M. Davis, B. Cantwell, and J. Mayor. 1994. Evidence that *asgB* encodes a DNA-binding protein essential for growth and development of *Myxococcus xanthus*. *J. Bacteriol.* **176**:2013–2020.

215. Davis, J., J. Mayor, and L. Plamann. 1995. A missense mutation in *rpoD* results in an A signaling defect in *Myxococcus xanthus*. *Mol. Microbiol.* **18**:943–952.

216. Yang, C., and H. B. Kaplan. 1997. *Myxococcus xanthus sasS* encodes a sensor histidine kinase required for early developmental gene expression. *J. Bacteriol.* **179**:7759–7767.

217. Lee, B-U, K. Lee, J. Mendez, and L. J. Shimkets. 1995. A tactile sensory system of

*Myxococcus xanthus* involves an extracellular NAD(P)$^+$-containing protein. *Genes & Dev.* 9:2964–2973.

218. Lee, K., and L. J. Shimkets. 1996. Suppression of a signaling defect during *Myxococcus xanthus* development. *J. Bacteriol.* 178:977–984.

219. Kim, S. K., and D. Kaiser. 1990. Cell alignment required in differentiation of *Myxococcus xanthus*. *Science* 249:926–928.

220. Sager, B., and K. Kaiser. 1995. Intercellular C-signaling and the traveling waves of *Myxococcus*. *Genes & Dev.* 8:2793–2804.

221. Søgaard-Anderson, L., and D. Kaiser. 1996. C factor, a cell-surface-associated intercellular signaling protein stimulates the cytoplasmic Frz signal transduction system in *Myxococcus xanthus*. *Proc. Natl. Acad. Sci. USA* 93:2675–2679.

222. Finlay, B. B., and S. Falkow. 1997. Common themes in microbial pathogenicity revisited. *Microbiol. and Molec. Biol. Rev.* 61:136–169.

223. Falkow, S. 1997. What is a pathogen? *ASM News* 63:359–365.

224. Beier, D., G. Spohn, R. Rappuoli, and V. Scarlato. 1997. Identification and characterization of an operon of *Helicobacter pylori* that is involved in motility and stress adaptation. *J. Bacteriol.* 179:4676–4683.

225. Miller, J. F., J. J. Mekalanos, and S. Falkow. 1989. Coordinate regulation and sensory transduction in the control of bacterial virulence. *Science* 243:916–922.

226. Skorupski, K., and R. K. Taylor. 1997. Control of the ToxR virulence regulon in *Vibrio cholerae* by environmental stimuli. *Molec. Microbiol.* 25:1003–1009.

227. Cholera toxin consists of one A subunit and five identical B subunits. The pentamer of B subunits binds to GM$_1$ ganglioside (a glycolipid) on the surface of host intestinal epithelial cells. The A subunit is separated from the B pentamer and enters the cytosol, where it is dissociated into two fragments, i.e., A$_1$ and A$_2$, by the reduction of a disulfide bond that links the A$_1$ and A$_2$ subunits together. The A$_1$ subunit activates adenylate cyclase. The activation of adenylate cyclase leads to an increase in cAMP, which leads to secretion of chloride and water. Activation of adenylate cyclase is a multistep process. The A$_1$ subunit catalyzes the transfer of ADP–ribose from NAD$^+$ to a GTP-binding protein (G$_s$) that is associated with adenylate cyclase. ADP–ribosylation of the GTP-binding protein inhibits its GTPase activity, locking the adenylate cyclase into the "on" mode. To understand this, it is necessary to explain how adenylate cyclase is regulated. Adenylate cyclase is a membrane-bound protein that exists in a complex with G$_s$. When G$_s$ binds ATP, it becomes activated and activates adenylate cyclase. But G$_s$ is a GTPase and ordinarily the GTP is hydrolyzed so that G$_s$ has bound GDP and is not active. Hormones such as glucagon or epinephrine bind to a hormone receptor, which then binds to the G$_s$ protein and triggers the exchange of GTP for bound GDP on the G$_s$ protein. This activates the G$_s$ protein (hence adenylate cyclase) until it hydrolyzes the bound GTP. ADP–ribosylation prevents GTP hydrolysis; hence G$_s$ continually activates adenyl cyclase in the presence of the A$_1$ subunit of cholera toxin.

228. Miller, V. L., and J. J. Mekalanos. 1988. A novel suicide vector and its use in construction of insertion mutations: Osmoregulation of outer membrane proteins and virulence determinants in *Vibrio cholerae* requires *toxr*. *J. Bacteriol.* 170:2575–2583.

229. There are two major disease-causing strains of *V. cholerae*. One strain is the classical strain (*V. cholerae* O1), and the other is the the El Tor strain. Higher expression of the ToxR regulon in the classical strain occurs at pH 6.5 and 30°C and lower expression at pH 8.5 and 37°C. The conditions that favor optimal expression in the classical strain are not the same conditions that favor optimal expression in the El Tor strain. This is reviewed in Skorupski, K., and R. K. Taylor. 1997. Control of the ToxR virulence regulon in *Vibrio cholerae* by environmental stimuli. *Molec. Microbiol.* 25:1003–1009.

230. In order to understand how to find and monitor expression of virulence genes using *TnphoA* fusions, it is first necessary to explain what is meant by a transposon. Transposons are DNA elements that move from one part of a DNA molecule to another, or from one DNA molecule to another. Most transposons move almost randomly. The movement is called transposition, and it is catalyzed by enzymes called transposases, which are encoded in the transposon. Bacteria have many different kinds of transposons, some of which carry antibiotic-resistance genes in addition to the transposase genes. The presence of the antibiotic-resistance genes can be used to select for cells carrying the transposon. *TnphoA* is a derivative of the transposon Tn5 carrying a kanamycin-resistance gene that can be used to isolate random gene fusions. The transposon contains an inserted reporter gene, *phoA*. The inserted reporter gene *phoA* encodes the C-terminal end of periplasmic alkaline phosphatase but does not encode the promoter, translational start site, or signal sequence (for export of the protein) of the alkaline phosphatase gene. When *TnphoA* inserts into a gene in the bacterial chromosome

in the right reading frame, then a fusion protein will be synthesized. The fusion protein has the C-terminal end of alkaline phosphatase and the N-terminal end of the protein encoded by the gene into which TnphoA has inserted. The fusion protein will have alkaline phosphatase activity only if TnphoA inserts into a gene that encodes a protein to be exported so that the N-terminal end of the fusion protein has a signal sequence for export. This ensures that the C-terminal end of the fusion protein enters the periplasm, where it can exhibit alkaline phosphatase activity. It is possible to screen the colonies for active alkaline phosphatase. An indicator (XP) is included in the agar. (XP is 5-bromo-4-chloro-3-indolylphosphate. It turns blue when the phosphate is removed.) If active PhoA is present, then it cleaves XP, producing a blue color. Genes detected in this way can be mapped and cloned. To learn how TnphoA is constructed, see Manoil, C., and J. Beckwith. 1985. TnphoA: A transposon probe for protein export signals. Proc. Natl. Acad. Sci. USA 82:8129–8133.

In order to screen for virulence genes that encode exported virulence proteins, Peterson and Mekalanos introduced transposon TnphoA on a plasmid (pRT291) into V. cholerae. The plasmid was introduced into a streptomycin-resistant stain of V. cholerae via conjugation using a donor E. coli strain carrying (pRT291) containing TnphoA. Transconjugates were selected on the basis of resistance to streptomycin and kanamycin. The antibiotic-resistant transconjugates were then mated with E. coli carrying a plasmid (pPH1JI) that was not compatible with pRT291. PPH1JI carries a gentamicin-resistance gene. V. cholerae colonies resistant to gentamicin, kanamycin, and streptomycin, and which produced alkaline phosphatase, were selected on XP agar containing gentamicin, kanamycin, and streptomycin. These carried pPH1JI (conferring resistance to gentamicin) and TnphoA (conferring resistance to kanamycin). Because pRT291 and pPH1JI are incompatible, the kanamycin-resistant colonies carried random insertions of the transposon in the chromosome. In other words, TnphoA transposed (moved) from the plasmid to the bacterial chromosome. The PhoA-positive cells (detected with the XP) were grown in the absence of gentamicin to promote the loss of pPH1JI. In order to find those genes carrying TnphoA insertions that are regulated by ToxR, the production of PhoA in the different isolates was monitored under conditions known to produce high or low levels of cholera toxin and TCPA. Proof that these genes were indeed positively regulated by ToxR was obtained by testing the production of PhoA in toxR⁻ strains. See Peterson, K. M., and J. J. Mekalanos. 1988. Characterization of the Vibrio cholerae ToxR regulon: Identification of novel genes involved in

intestinal colonization. *Infection and Immunity* 56:2822–2829.

231 .DiRita, V. J., C. Parsot, G. Jander, and J. J. Mekalanos. 1991. Regulatory cascade controls virulence in *Vibrio cholerae*. *Proc. Natl. Acad. Sci. USA* 88:5403–5407.

232. Häse, C. C., and J. J. Mekalanos. 1998. TcpP protein is a positive regulator of virulence gene expression in *Vibrio cholerae*. *Proc. Natl. Acad. Sci. USA* 95:730–734.

233. Behlau, I., and S. I. Miller. 1993. A Pho-P repressed gene promotes *Salmonella typhimurium* invasion of epithelial cells. *J. Bacteriol.* 175:4475–4484.

234. An acid phosphatase, the product of the *phoN* gene, is induced under certain growth conditions such as phosphate limitation. A mutant hunt revealed *phoP*, which when expressed constitutively results in high levels of phosphatase synthesis. The reason for this is that PhoP is a positive regulator of the *phoN* gene. It was later discovered that transposon Tn-10 generated mutants that were very sensitive to killing by macrophages mapped to the *phoP* gene earlier discovered as being required for alkaline phosphatase synthesis. See the review by Groisman and Heffron. (Groisman, E. A., and F. Heffron. 1995. Regulation of *Salmonella* virulence by two-component regulatory systems, pp. 319–332. In: *Two-Component Signal Transduction*. J. A. Hoch and T. J. Silhavy (Eds.). ASM Press, Washington, D.C.)

235. Groisman, E. A., and F. Heffron. 1995. Regulation of *Salmonella* virulence by two-component regulatory systems, pp. 319–332. In: *Two-Component Signal Transduction*. J. A. Hoch and T. J. Silhavy (Eds.). ASM Press, Washington, D.C.

236. Véscovi, E. G., F. C. Soncinin, and E. A. Groisman. 1996. Mg⁺⁺ as an extracellular signal: Environmental regulation of *Salmonella* virulence. *Cell* 84:166–174.

237. Soncini, F. C., E. G. Véscovi, F. Solomon, and E. A. Groisman. Molecular basis of the magnesium deprivation response in *Salmonella typhimurium*: Identification of PhoP-regulated genes. *J. Bacteriol.* 178:5092–5099.

238. Soncini, F. C., and E. A. Groisman. 1996. Two-component regulatory systems can interact to process multiple environmental signals. *J. Bacteriol.* 178:6796–6801.

239. Miller, J. F., S. A. Johnson, W. J. Black, D. T. Beattie, J. J. Mekalanos, and S. Falkow. 1992. Constitutive sensory transduction mutations in the *Bordetella pertussis bvgS* gene. *J. Bacteriol.* 174:970–979.

240. Hale, T. L. 1991. Genetic basis of virulence in *Shigella* species. *Microbiol. Rev.* **55**:206–224.

241. *Shigella spp.* as well as *Listeria monocytogenes* invade eukaryotic cells and make a cell surface protein that elicits the formation of a tail of actin at one cell pole when they are growing intracellularly. As a consequence, the bacteria are propelled through the cytoplasm of the host cell. The actin tail also aids in intercellular spread of the bacteria. See Bernardini, M. L., J. Mounier, H. D'Hauteville, M. Coquis-Rondon, and P. J. Sansonetti. 1989. Identification of *icsA*, a plasmid locus of *Shigella flexneri* that governs bacterial intra- and intercellular spread through interaction with F-actin. *Proc. Natl. Acad. Sci. USA* **86**:3867–3871.

242. Porter, M. E., and C. J. Dorman. 1994. A role for H-NS in the thermo-osmotic regulation of virulence gene expression in *Shigella flexneri*. *J. Bacteriol.* **176**:4187–4191.

243. Porter, M. E., and C. J. Dorman. 1997. Positive regulation of *Shigella flexneri* virulence genes by integration host factor. *J. Bacteriol.* **179**:6537–6550.

244. Otto, M., R. Süßmuth, G. Jung, and F. Götz. 1998. Structure of the pheromone peptide of the *Staphylococcus epidermidis agr* system. *FEBS Lett.* **424**:89–94.

245. Haygood, M. G. 1993. Light organ symbioses in fishes. *Crit. Rev. Microbiol.* **19**:191–216.

246. Nealson, K. H., and J. Woodland Hastings. 1991. The Luminous Bacteria, pp. 625–639 In: *The Prokaryotes*, 2nd ed., A. Balows, H. G. Trüper, M. Dworkin, W. Harder, and K.-H. Schleifer (eds.). Springer-Verlag, Berlin.

247. Meighen, E. A. 1991. Molecular biology of bacterial bioluminescence. *Microbiol. Rev.* **55**:123–142.

248. *Photobacterium* and *Vibrio* are not easily distinguished. The principal way in which *Photobacterium* is distinguished from *Vibrio* is the absence of a flagellar sheath, the accumulation of poly-β-hydroxybutyrate (PHB), and the utilization of D-mannitol. However, there is room for confusion, since some *Vibrio* species accumulate PHB and utilize mannitol. A method to distinguish these genera based upon restriction fragment length polymorphism (RFLP) patterns of 16S rDNA has been recently published. See Urakawa, H., K. Kita-Tsukamoto, and K. Ohwada. 1998. A new approach to separate the genus *Photobacterium* from *Vibrio* with RFLP patterns by *HhaI* digestion of PCR-amplified 16S rDNA. *Curr. Microbiol.* **36**:171–174.

249. Kurfuerst, M., P. Macheroux, S. Ghisla, and J. W. Hastings. 1987. Isolation and characterization of the transient, luciferase-bound flavin-4a-hydroxide in the bacterial luciferase reaction. *Bioch. Biophys. Acta* **924**:104–110.

250. Greenberg, E. P. 1997. Quorum sensing in gram-negative bacteria. *ASM News* **63**:371–377.

251. Hanzelaka, B. L., and E. P. Greenberg. 1995. Evidence that the N-terminal region of the *Vibrio fischeri* LuxR protein constitutes an autoinducer-binding domain. *J. Bacteriol.* **177**:815–818.

252. Sitnikov, D. M., J. B. Schineller, J. B., and T. O. Baldwin. 1995. Transcriptional regulation of bioluminescence genes from *Vibrio fischeri*. *Molec. Microbiol.* **17**:801–812.

253. Kolibachuk, D., and E. P. Greenberg. 1993. The *Vibrio fischeri* luminescence gene activator LuxR is a membrane-associated protein. *J. Bacteriol.* **175**:7307–7312

254. Kuo, A., N. V. Blough, and P. V. Dunlap. 1994. Multiple N-acyl-L-homoserine lactone autoinducers of luminescence in the marine symbiotic bacterium *V. fischeri*. *J. Bacteriol.* **176**:7558–7565.

255. Gilson, L., A. Kuo, and P. V. Dunlap. 1995. AinS and a new family of autoinducer synthesis proteins. *J. Bacteriol.* **177**:6946–6951.

256. Sitnikov, D. M., J. B. Schineller, and T. O. Baldwin. 1995. Transcriptional regulation of bioluminescence genes from *Vibrio fischeri*. *Mol. Microbiol.* **17**:801–812.

257. Ulitzur, S., and P. Dunlap. 1995. Regulatory circuitry controlling luminescence autoinduction in *Vibrio fischeri*. *Photobiol. Photochem.* **62**:625–632.

258. Ulitzur, S., A. Matin, C. Fraley, and E. Meighen. 1977. H-NS protein represses transcription of the lux systems of *Vibrio fischeri* and other luminous bacteria cloned into *Escherichia coli*. *Curr. Microbiol.* **35**:336–342.

259. Fuqua, W. C., S. C. Winans., and E. P. Greenberg. 1994. Quorum sensing in bacteria: The LuxR–LuxI family of cell density-responsive transcriptional regulators. *J. Bacteriol.* **176**:269–275.

260. Fuqua, C., and E. P. Greenberg. 1998. Self perception in bacteria: Quorum sensing with acylated homoserine lactones. *Curr. Opinion in Microbiol.* **1**:183–189.

261. Fuqua, C., and S. C. Winans. 1996. Conserved *cis*-acting promoter elements are required for density-dependent transcription of *Agrobacterium tumefaciens* conjugal transfer genes. *J. Bacteriol.* **178**:435–440.

262. Zhang, L., P. J. Murphy, A. Kerr, and M. E. Tate. 1993. *Agrobacterium* conjugation

and gene regulation by N-acyl-L-homoserine lactones. *Nature* 362:446–447.

263. Piper, K. R., S. B. von Bodman, and S. K. Farrand. 1993. Conjugation factor of *Agrobacterium tumefaciens* regulates Ti plasmid transfer by autoinduction. *Nature* 362:448–450.

264. The TraR protein shows 20% amino acid identity and 62% amino acid similarity over a 234 residue overlap with LuxR.

265. Pearson, J. P., K. M. Gray, L. Passador, K. D. Tucker, A. Eberhard, B. H. Iglewski, and E. P. Greenberg. 1994. Structure of the autoinducer required for expression of *Pseudomonas aeruginosa* virulence genes. *Proc. Natl. Acad. Sci. USA* 91:197–201.

266. Pirhonen, M., D. Flego, R. Heikinheimo, and E. T. Palva. 1993. A small diffusible signal molecule is responsible for the global control of virulence and exoenzyme production in the plant pathogen *Erwinia carotovora*. *EMBO J.* 12:2467–2476.

267. Van Rhijn, P., and J. Vanderleyden. The *Rhizobium*–plant symbiosis. *Microbiol. Rev.* 59:124–142.

268. Gray, M. K., J. P. Pearson, J. A. Downie, B. E. A. Boboye, and E. P. Greenberg. 1996. Cell-to-cell signaling in the symbiotic nitrogen-fixing bacterium *Rhizobium leguminosarum*: Autoinduction of a stationary phase and rhizosphere-expressed genes. *J. Bacteriol.* 178:372–376.

269. *Rhizobium* spp. contain large symbiotic plasmids. Some of the plasmid genes are required for nodulation and nitrogen fixation.

270. Cubo, M. t., A. Economou, G. Murphy, A. W. B. Johnston, and J. A. Downie. 1992. Molecular characterization and regulation of the rhizosphere-expressed genes *rhiABCR* that can influence nodulation by *Rhizobium leguminosarum* biovar viciae. *J. Bacteriol.* 174:4026–4035.

271. Ahmer, B. M. M., J. van Reeuwijk, C.CD. Timmers, P. J. Valentine, and F. Heffron. 1998. *Salmonella typhimurium* encodes an SdiA homolog, a putative quorum sensor of the LuxR family, that regulates genes on the virulence plasmid. *J. Bacteriol.* 180:11895–1193.

272. Sitnikov, D. M., J. B. Schineller, and T. O. Baldwin. 1996. Control of cell division

in *Escherichia coli*: Regulation of transcription of *ftsQA* involves both *rpoS* and SdiA-mediated autoinduction. *Proc. Natl. Acad. Sci. USA* 93:336–341.

273. Huisman, G. W., and R. Kolter. 1994. Sensing starvation: A homoserine lactone-dependent signaling pathway in *Escherichia coli*. *Science* 265:537–539.

274. Bassler, B. L., M. Wright, and M. R. Silverman. 1994. Multiple signalling systems controlling expression of luminescence in *Vibrio harveyi*: Sequence and function of genes encoding second sensory pathway. *Mol. Microbiol.* 13:273–286.

275. Iuchi, S., and E. C. C. Lin. 1992. Mutational analysis of signal transduction by ArcB, a membrane sensor protein responsible for anaerobic repression of operons involved in the central aerobic pathways in *Escherichia coli*. *J. Bacteriol.* 174:3972–3980.

276. ArcB autophosphorylates at a histidine residue in its receiver domain and can then transfer the phosphoryl group to an aspartate residue in its response-regulator domain. Mutants that lacked the response-regulator domain in ArcB were still capable of respiratory control, indicating that ArcB can transfer the phosphoryl group directly from its histidine kinase domain to the response-regulator domain in ArcA, but the difference between gene expression under aerobic and anaerobic conditions was not as great as in the wild type. Apparently, phosphorylation of the regulator domain in ArcB modulates the response of ArcB to the level of oxygen. However, most sensor kinases thus far studied do not have response-regulator domains. They simply autophosphorylate and transfer the phosphoryl group from the histidine residue directly to an aspartate residue in a separate response-regulator protein (Section 18.1.2) (Iuchi and Lin, 1992, *J. Bact.* 174:3972).

277. Bassler, B. L., E. P. Greenberg, and A. M. Stevens. Cross-species induction of luminescence in the quorum-sensing bacterium *Vibrio harveyi*. *J. Bacteriol.* 179:4043–4045.

278. Ramseier, T. M., S. Y. Chien, and M. H. Saier, Jr. 1996. Cooperative interaction between Cra and Fnr in the regulation of the *cydAB* operon of *Escherichia coli*. *Curr. Microbiol.* 33:270–274.

# 19

# How Bacteria Respond to Environmental Stress

There are numerous physiological ways in which bacteria respond to environmental stresses such as changes in pH, osmolarity, and temperature, and to starvation. These are of major research interest not only for practical reasons but for what they can teach us about the adaptive responses of bacteria. For example, when E. coli is subjected to starvation or an osmotic upshift in the medium, it induces the synthesis of several proteins whose transcription is dependent upon $\sigma^s$ and that make the cells more resistant to heat, high salt (osmotic shock), near-UV, $H_2O_2$, or prolonged starvation. These and other responses to stress are discussed in Sections 2.2.2, 14.1, and Chapter 15. This chapter discusses responses to elevated temperature (heat-shock response), damage to DNA (the SOS sytem), and toxic forms of oxygen (oxidative stress).

## 19.1 Heat-Shock Response

### 19.1.1 Heat-shock proteins

When bacteria are shifted to a higher temperature, they respond by *transiently* increasing the rate of synthesis of a group of proteins called the heat-shock proteins (hsps) relative to other proteins.[1] (There are also cold-shock proteins induced in bacteria that may function during adaptation to low temperature.[2, 3]) Upon a temperature downshift, the synthesis of heat-shock proteins is decreased. For example, when E. coli is shifted from 30 to 42°C, the cells increase the rate of synthesis of 20 such proteins for 5 to 10 minutes. The increased rate is about 10 to 20 times greater than the majority of the proteins. Actually, this type of response to a temperature shift is not unique to bacteria and occurs in cells of animals (including humans), plants, and eukaryotic microorganisms. In fact, it was first discovered in *Drosophila*, where it is manifested in the synthesis of new chromosomal puffs associated with the production of hsps in salivary gland chromosomes. The heat-shock proteins are classified into families according to their molecular weights. For example, the heat-shock proteins with a molecular weight of about 70 kDa are in the Hsp70 family, those with a molecular weight of around 60 kDa are in the Hsp60 family, and those with a molecular weight of about 40 kDa are in the Hsp40 family. When one examines the proteins that are made during the heat-shock response, they are remarkably similar across phylogenetic lines. For example, the DnaK protein from E. coli is a member of the Hsp70 family and is about 50% homologous with heat-shock proteins in the Hsp70 family found in humans.

What are these proteins, and what do they do? It is clear that in *E. coli* at least some of them are required for growth to occur above 20°C. The evidence for this is that null mutations (loss-of-function mutants) in a gene (*rpoH*) encoding a sigma factor ($\sigma^{32}$) required to transcribe the hsp genes prevent growth above 20°C.[4] (Most of the *E. coli* genes are transcribed by an RNA polymerase that has $\sigma^{70}$ as the sigma factor.) In addition to the absolute requirement for hsp proteins above 20°C, there is an enhanced effect on growth by the hsps at *all* temperatures, as judged by the slower growth of null mutants of *rpoH* compared to wild-type cells.[4]

## 19.1.2 Regulation of sigma factor 32

Following a temperature upshift (e.g, from 30 to 42°C), there is a transient increase in the amount of $\sigma^{32}$ that is responsible for the synthesis of the heat-shock proteins. Several factors contribute towards the accumulation of $\sigma^{32}$ at higher temperatures. During steady-state growth at any temperature $\sigma^{32}$ is an unstable protein, with a half-life of only 1 minute. However, after a shift from 30 to 42°C the protein is stabilized for a few minutes, and as a consequence its amounts increase. Additionally, there is an increase in the rate of translation of the mRNA for $\sigma^{32}$ during the period of increased synthesis of the heat-shock proteins. Interestingly, after a temperature downshift the *activity* of $\sigma^{32}$ decreases, and this, rather than a decrease in concentration of $\sigma^{32}$, results in a lowered rate of synthesis of the heat-shock proteins. Thus the regulation of $\sigma^{32}$ expression is complex and occurs at several levels, including stability of the protein, activity of the protein, and translation of the mRNA. The student is referred to the review by Gross that discusses possible mechanisms of regulation.[5] Also, see the discussion in Section 2.2.2 of sigma factor $\sigma^s$, which directs the transcription of genes induced by starvation and osmotic upshift, and is also regulated at several levels.[6, 7]

Although the heat-shock response is universal, the mechanisms that regulate the synthesis of the heat-shock proteins differ in various prokaryotes. For example, the expression of heat-shock genes in several gram-positive bacteria thus far investigated indicate that an alternative sigma factor is not involved. On the other hand, *Bacillus subtilis* does induce an alternative sigma factor during the heat-shock response.[8]

## 19.1.3 Functions of the E. coli heat-shock proteins

Most of the known heat-shock proteins are involved in (1) the *folding* of newly synthesized proteins or (2) the *proteolysis* of improperly folded or otherwise abnormal proteins.[5] Heat-shock proteins that aid in the folding of newly synthesized proteins belong to a class of proteins called chaperone proteins. A definition of a chaperone protein is a protein that takes part in the folding or assembly of other proteins but is not part of the final protein or protein complex. Chaperone proteins are important for growth at all temperatures, as will be discussed next. (Chaperone proteins often exist as multisubunit complexes of stacked rings sometimes called chaperonins.)

### Chaperone proteins

*1. Role in folding of nascent or newly synthesized proteins*
Newly synthesized proteins may not spontaneously fold into their correct structures without the help of chaperone proteins that transiently bind to the proteins during the folding process. (See the discussion of the role of chaperone proteins in folding proteins in Section 10.3.10.) The chaperone proteins prevent aggregation with other proteins or misfolding and in this way allow the proteins to fold correctly. They are thought to bind to exposed protein surfaces and in so doing prevent these surfaces from improperly interacting with regions from other proteins or regions within the protein. This appears to be especially important at higher temperatures, where improper folding or aggregation may occur more frequently. In addition, some chaperone proteins assist in bringing the target protein to the membrane for protein export. (See later.) Of the 21 hsps of *E. coli* whose functions are known, 8 are chaperone proteins. Three

important chaperone proteins that assist folding are DnaK, DnaJ, and GrpE, which act together, and two other chaperone proteins, GroEL and GroES, which also cooperate to produce correct folding. Exactly how all these chaperone proteins might act and interact in the folding of nascent protein chains is still a matter of speculation. However, it appears that DnaK, DnaJ, and GrpE might cooperate as a team during the early folding process, and GroEL and GroES might participate in later stages of folding. (Reviewed in Ref. 5. See Note 9 for more information on folding.) Given the roles that chaperone proteins are postulated to play, one would expect that they would be required for growth at all temperatures, not merely high temperatures. In fact, this seems to be the case, at least for GroEL, GroES, DnaK, and DnaJ. (Reviewed in Ref. 1.) The fact that mutants that make no $\sigma^{32}$ grow at temperatures below 20°C (albeit much more slowly than the wild type) appears to be due to the recognition of weak promotors in the $\sigma^{32}$ regulon by other sigma factors, resulting in some transcription of the *gro* and *dna* genes.[1, 5]

## 2. Role in export of proteins

Some chaperone proteins seem to play a role in protein export similar to that of SecB, which is discussed in Section 17.1.1. SecB, which is not a heat-shock protein, is the major chaperone protein in *E. coli,* and brings presecretory proteins in a conformation required for export to SecA and the translocation machinery at the membrane. Some of the heat-shock proteins have been reported to function in the export of some of the SecB-independent proteins. The latter are proteins that use the Sec machinery but not SecB. For example, DnaK and DnaJ and GroEL/ES are used for the export of some (not all) SecB-*independent* proteins, and DnaK and DnaJ can substitute for SecB in strains lacking SecB.[10, 11] In support of the conclusion that GroEL plays a role in the export of some secretory proteins, it has been reported that GroEL binds to SecA associated with the membrane.[12] (Recall that SecA delivers the presecretory protein to the transclocase, SecY, in the membrane. See Section 17.1.1.) Other unidentified heat-shock proteins can substitute for SecB in mutants

that do not make SecB or have low amounts of available SecB.[13]

## ATP-dependent proteases

At least 6 of the heat-shock proteins are ATP-dependent proteases.[5, 14] Some of these proteases may function, in part, to degrade improperly folded or denatured proteins at all growth temperatures. Some also play a role in the degradation of specific native proteins during ordinary growth temperatures. The ATP may be required, in part, to unfold the protein prior to proteolysis.

### 1. Lon

Lon is an ATP-dependent protease that is encoded by the *capR* (*lon*) gene.[15] Like the other heat-shock proteins, Lon functions not only during the heat-shock response, but also during ordinary growth temperatures. Lon is important for degrading abnormal proteins, such as proteins that may result from nonsense or missense mutations, or denatured proteins in general. In fact, it appears that Lon is responsible for the degradation of the majority of abnormal proteins in *E. coli*. In addition, several normal cellular proteins are also degraded by Lon, two examples of which follow. Mutations in *capR* (*lon*) give rise to strains that synthesize excess capsular polysaccharide and form mucoid colonies. The reason for this is that Lon recognizes and degrades the RcsA protein, which is a positive regulator of capsular polysaccharide synthesis. Hence, the lack of functional Lon results in increased levels of RcsA and consequently large amounts of capsular polysaccharide accumulate. Lon mutants have an additional phenotype. In the absence of Lon, cells are very sensitive to ultraviolet radiation as a consequence of the uv-induced SOS response discussed in Section 19.2. When the SOS response is induced in *lon* mutants, cell division is inhibited, resulting in cell filamentation and death. The reason why cell division is defective is that an inhibitor of cell septation called SulA accumulates and is stable in uv-irradiated cells in the absence of Lon. (SulA inhibits FtsZ. See Section 2.2.5 for a description of cell division and the role of FtsZ in septum formation and Section 19.3.2 for the role of SulA in the SOS response.) SulA is another substrate for Lon, accounting for why cells that do not make Lon accumulate SulA. In uv-irradiated wild-type cells the inhibition of cell division is transient because Lon degrades the accumulated SulA.

### 2. Other ATP-dependent proteases

Other ATP-dependent proteases in *E. coli* include the ClpP protease (caseinolytic protease) and FtsH, a membrane-bound protein. Unlike Lon and ClpP, which can be dispensed with under standard growth conditions, FtsH is an essential protein. Mutations in *ftsH*, the gene encoding FtsH, are defective in the translocation of certain proteins into the cell membrane as well as the export of some secreted proteins. (See Note 16.) Therefore, it has been proposed that FtsH, in addition to being an ATP-dependent protease that degrades select proteins, is a membrane-bound chaperone protein that in some as yet undefined manner plays a role in the secretion of certain proteins.[14] It has also been suggested that the Clp ATPase family of proteins are chaperone proteins that deliver specific proteins to ClpP for proteolysis. (See Refs. 17 and 18 for a review, and Note 19 for a more complete description of the Clp proteases and their physiological roles.)

## 19.2 Repairing Damaged DNA

There are several ways in which bacteria repair damaged DNA. A more complete discussion of this subject can be found in Refs. 20 and 21.

Some of the repair systems are inducible and are part of the SOS system, which has been extensively studied in *E. coli*.[22, 23]

### 19.2.1 Kinds of DNA damage

A particularly severe type of damage called photodimerization of pyrimidines occurs when DNA bases are excited by ultraviolet radiation (Fig. 19.1). As a consequence of the dimerization of adjacent pyrimidines such as thymine, DNA replication is blocked. Pyrimidine dimers are removed by specific excision processes or photoreactivation, as described next. Other malformities of DNA can include mismatched base pairs in the duplex that arise during DNA replication (discussed in Section 10.1.7), breaks or gaps in the DNA, and bases modified by chemicals, many of which are carcinogens.

### 19.2.2 Repairing uv-damaged DNA by photoreactivation

Photoreactivation of thymine dimers may occur (Fig. 19.2). (For a review of this subject as well as other DNA repair systems and mutagenesis, see Ref. 21.) During photoreactivation a special enzyme called *DNA photolyase*, encoded by the *phr* gene in *E. coli*, reverses the dimerization reaction upon absorbing blue light (300 to 500 nm). The enzyme contains bound $FADH_2$ (reduced flavin adenine dinucleotide) and

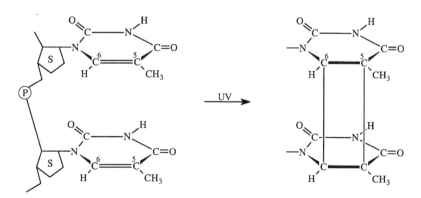

**Fig. 19.1** Formation of thymine dimers by ultraviolet light. Ultraviolet light causes the formation of cyclobutane pyrimidine dimers. Usually this occurs between adjacent thymine dimers in the same strand of DNA and involves C5 and C6 of both molecules.

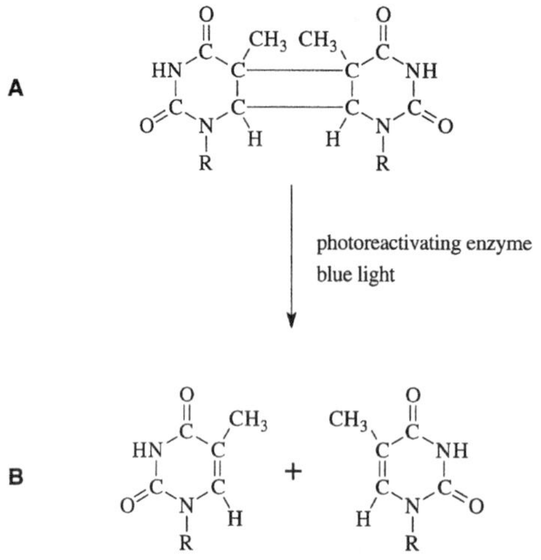

**Fig. 19.2** Photoreactivation of thymine dimers. Thymine dimers form in the presence of ultraviolet light. Most organisms (with the exception of placental mammals) possess a flavoprotein enzyme called photolyase, which will reverse the dimerization upon absorption of blue light (350–500 nm). DNA photolyase in *E. coli* is encoded by the *phr* gene.

in *E. coli*, also a folic acid derivative. 5,10-methenyltetrahydrofolyl)polyglutamate (MTHF), both of which absorb light. The enzyme uses the absorbed light energy to cleave the dimer into two monomers. The primary chromophore is MTHF, which absorbs 60 to 80% of the $A_{385}$ absorbed by the enzyme. The energy is transferred from MTHF to the flavin. The excited flavin molecule cleaves the pyrimidine dimer. (For more details of the model, see Note 24.) DNA photolyase is widespread, ocurring in both prokaryotic and eukaryotic microorganisms, plants, and animals. However, it is not found in placental mammals. Purified DNA photolyases from other organisms have $FADH_2$ but differ with respect to the second chromophore. Some, like *E. coli*, have MTHF, whereas others have 8-hydroxy-5-deazaflavin instead. The folate-containing enzymes have an absorption maximum of about 380 nm, whereas the deazaflavin enzymes have a slightly higher absorption maximum, about 440 nm.

### 19.2.3 Repairing damaged DNA via excision (nucleotide excision repair)

In *E. coli* the products of three genes, *uvrA*, *uvrB*, and *uvrC*, encode an endonuclease (UvrABC endonuclease, sometimes called the ABC excinuclease) that recognizes the distortion in the double helix caused by the uv-induced pyrimidine dimer and cuts the DNA 8 nucleotides away on the 5' side of the dimer and 4 nucleotides away on the 3' side of the dimer. Removal of the segment of DNA, aided by UvrD, which is a helicase (helicase II), leaves a single-stranded gap that is filled with DNA polymerase I and sealed with DNA ligase (Fig. 19.3). The endonuclease has been shown in vitro to recognize other kinds of damage such as chemically modified bases (e.g., alkylated bases) that cause distortion, and similarly remove the damaged DNA. In agreement with this, mutants in the *uvr* genes are very sensitive to killing not only by ultraviolet light but also by different chemicals that damage DNA.

### 19.2.4 Repairing damaged DNA by recombination (postreplication repair)

The excision mechanisms thus far discussed require that one of the strands has no errors and can serve as the template. When both strands are damaged, having, for example, gaps (discontinuities) in the daughter strand made while copying a damaged template, DNA recombination (*recombinational repair*) using RecA is required. For example, this can occur when pyrimidine dimers form as a consequence of UV irradiation and the region of DNA containing the pyrimidine dimer is not replicated. Replication begins again about 1000 bp past the damaged DNA, resulting in a single-stranded gap. The gap is filled, not by replication, but rather by moving a segment of the complementary strand with the correct sequence from the sister duplex into the gap. A model for the repair process is outlined in Fig. 19.4: (A) A gap is present in the daughter strand opposite a lesion in the template strand; (B) a RecA-dependent recombinational event takes place, resulting in the complementary strand from the sister duplex being transferred

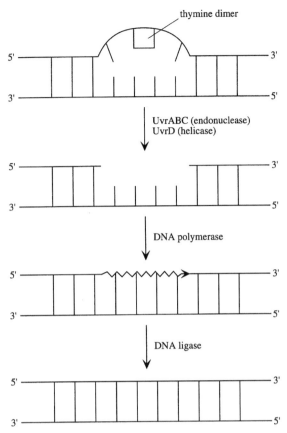

**Fig. 19.3** Excision repair. Certain kinds of DNA damage, e.g., thymine dimers, can be repaired by excision repair. The distortion in the double helix is recognized by a complex of proteins called UvrABC. The figure illustrates that UvrABC is an endonuclease that cuts the damaged DNA 8 nucleotides away on the 5′ side of a thymine dimer and 4 nucleotides away on the 3′ side of the dimer. Removal of the segment of DNA, aided by the helicase UvrD, leaves a single-stranded gap that is filled with DNA polymerase I and sealed with DNA ligase.

to (invades) the gap in the daughter strand; (C) a crossover intermediate forms that after cutting and sealing results in filling the gap with the complementary DNA from the sister strand. The resultant gap in the sister strand is filled with DNA polymerase. The gaps are sealed with DNA ligase. The faulty nucleotide can now be removed using nucleotide excision repair (UvrABC) or simply diluted out during growth. Postreplicative repair is generally error-free, as opposed to SOS mutagenesis, as will be described later.

### 19.2.5 Repair of deaminated bases (base-excision repair)

A common type of DNA damage is the deamination of bases. As a consequence of deamination an amino group is replaced by a keto group. For example, cytosine might be deaminated to uracil, adenine to hypoxanthine, and guanine to xanthine. (See Fig. 9.8 for the structures of the bases.) The deaminated bases pair with the wrong base during replication, resulting in mutations. For example, hypoxanthine pairs with cytosine so that an AT pair is replaced by a GC pair. Also, uracil will pair with adenine resulting in a GC pair being changed to an AT pair. Deamination can occur spontaneously or be caused by various chemical reagents, such as nitrous acid. Deaminated bases (as well as other damaged bases such as alkylated bases) are removed by specific enzymes called *DNA glycosylases*. (See Note 25 for a discussion of alkylation of bases.) The enzyme catalyzes the

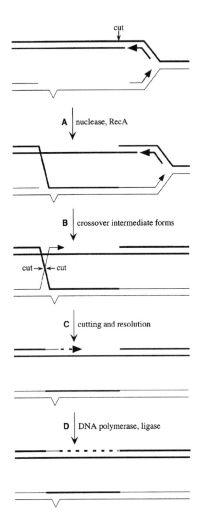

**Fig. 19.4** Postreplication repair (recombinational repair). The presence of a structural deformation in the duplex. For example, one imposed by the presence of a thymine dimer in the template strand causes the DNA polymerase to leave a gap in the copy strand opposite the lesion. The gap is repaired by RecA and other enzymes. (A) The good template sister strand is cleaved and invades the gap in the copy strand. (B and C) Ligation between the 3' end of the invading strand and the 5' end of the copy strand is followed by branch migration, leading to an increase in the length of the exchanged region and the formation of a crossover intermediate. The crossover intermediate is cut, and the breaks are sealed, resulting in the incorporation of the sister strand into the gap in the copy strand, and leaving a gap in the sister strand. (D). DNA polymerase fills in the gap in the sister strand. DNA ligase seals the ends.

breakage of the N-glycosyl bond between the base and the sugar, leaving an apyrimidinic or apurinic site (Fig. 19.5). Such sites are referred to as *AP sites*. Each glycosylase is specific for each type of damaged base that is removed. Thus these repair pathways are sometimes called *specific repair pathways*. After the base has been removed, AP endonucleases cleave the phosphodiester linkage next to the AP site, usually on the 5' side. This generates a free 3' hydroxyl end, which is extended by DNA polymerase I (in *E. coli*) as the 5'-exonuclease activity of DNA polymerase I removes a portion of the damaged strand ahead. After the damaged portion is replaced, DNA ligase seals the nick.

### 19.2.6 The GO system and protection from 8-oxoguanine (7,8-dihydro-8-oxoguanine)

Reactive forms of oxygen such as superoxide radical, hydrogen peroxide, and hydroxyl radicals can oxidatively damage guanine to form 7,8-dihydro-8-oxoguanine (also called 8-oxoguanine, 8-hydroxyguanine, or a GO lesion). (See Ref. 26 for a review and Section 14.1 on oxygen toxicity for a description of how reactive forms of oxygen are made in the cell.) The structure of guanine and 8-oxoguanine are shown in Fig. 19.6A. 8-Oxoguanine mispairs with adenine (Fig. 19.6B) and thus GC pairs are converted to AT pairs, resulting in mutations.[27] The GO system prevents this from happening. The GO system consists of three proteins: MutM, MutT, and MutY. MutM and MutY are base-excision repair systems and rely on N-glycosylase and AP endonuclease activity, as described in Section 19.2.5. MutY is not part of a base-excision repair pathway. (Several other Mut proteins have been previously described. They are MutD, which is the subunit in DNA polymerase III responsible for its 3'-exonuclease editing function, and MutS, MutL, and MutH, which function in mismatch repair. See Section 10.1.7. Mutations in any of the *mut* genes increase the spontaneous rate of mutation because they prevent the repair of damaged DNA or the removal of incorrectly paired nucleotides.)

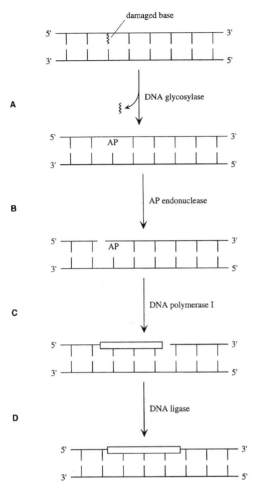

**Fig. 19.5** Base-excision repair. (A) A damaged base is removed by DNA glycosylase leaving an AP site (an apurinic or apyrimidinic site). (B) An AP endonuclease cleaves the phosphodiester bond on the 5′ side of the AP site. (C) DNA polymerase I extends the 3′ end while removing a portion of the damaged strand with its 5′–3′ exonuclease activity. (D) The nick is sealed by DNA ligase. Newly synthesized DNA is shown as a rectangular box.

## MutM

MutM is an N-glycosylase that removes 8-oxoguanine (Fig. 19.7). In addition to N-glycosylase activity, MutM is also an AP endonuclease. After removal of the 8-oxoguanine and cutting of the strand, a portion of the damaged strand is removed by exonuclease activity and repaired by DNA polymerase I and DNA ligase.

## MutY

Occasionally, as a result of replication, an adenylic acid is paired with 8-oxoguanine before the 8-oxoguanine can be removed. (See Fig. 19.6B.) When this happens, MutY, which is an N-glycosylase, removes the adenine to produce an AP site opposite the 8-oxoguanine. The DNA is cut by an AP endonuclease at the AP site, and the DNA is repaired as diagrammed in Fig. 19.7, yielding a C–8-oxoG–C base pair. This avoids the formation of an AT base pair from a GC base pair.

## MutT

Reactive forms of oxygen also convert the guanine in dGTP to 8-oxoguanine to form 8-oxodGTP, which is a substrate for DNA polymerase. MutT is a phosphatase that converts 8-oxodGTP to 8-oxodGMP, thus preventing the incorporation of 8-oxoguanylic acid into DNA.

## 19.3 The SOS Response

When DNA is damaged or DNA replication is inhibited, a signal is produced that results in the induction of over 20 genes, several of which function to repair the damaged DNA. The signal is probably single-stranded DNA that may result from the attempt to replicate damaged DNA, or may arise as a consequence of an interruption of DNA replication. The induced genes include genes for excision repair and recombinational repair, as well as genes that allow DNA polymerase III to copy damaged DNA (Table 19.1). The latter repair system results in mutations, as described next.

### 19.3.1 SOS mutagenesis

When DNA that is damaged as a result of ultraviolet radiation or certain chemicals is repaired as a result of the SOS system in *E. coli*, mutations result. This is because a DNA polymerase (probably a modified DNA polymerase III) reads through damaged DNA and makes errors during the repair of the DNA when nucleotides are inserted opposite

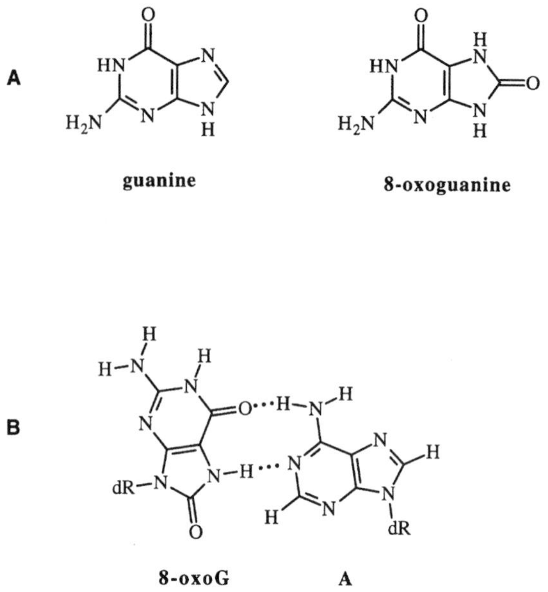

**Fig. 19.6** Structures of guanine and 8-oxoguanine (8-oxoG) and 8-oxoG-A base pairs. Oxygen free radicals can hydroxylate deoxyguanosine at the C8 position. (A) One of the tautomeric forms is 8-oxodeoxyguanosine, which is the major one under physiological conditions. (B) Sometimes the DNA polymerase will insert deoxyAMP rather than deoxyCMP opposite 8-oxoG. Unless this is repaired, it will result in a change of a GC base pair to an AT base pair when the DNA replicates.

**Table 19.1** Some SOS responses and induced genes of *E. coli*

| Response | Induced gene |
|---|---|
| uv mutagenesis of bacterial chromosome | *umuDC, recA* |
| Inhibition of cell division (filamentation) | *sulA* |
| *uvr*-dependent excision repair | *uvrA, uvrB, uvrD* |
| Daughter strand gap repair | *recA, ruvAB*? |
| Double-strand break repair | *recA, recN* |

*Source:* Adapted from Walker, G. C. 1996. The SOS response of *Escherichia coli*. In: *Escherichia coli and Salmonella: Cellular and Molecular Biology*, Vol. 1, pp. 1400–1416. Neidhardt, F. C., et al. (Eds.). ASM Press, Washington, D.C.

lesions in the template strand. Three genes are required for SOS mutagenesis. They are *recA, umuD,* and *umuC*. The latter two are in the *umuDC* operon. Mutagenesis resulting from SOS-mediated DNA repair is sometimes called *error-prone* repair. One role of RecA is to activate the cleavage of UmuD to an active form, UmuD'. A model to explain error-prone repair is that UmuC and UmuD' bind to DNA polymerase, preventing its proofreading and repair function and allowing it to read through the damaged DNA. (See Section 10.1.7 for a discussion of proofreading and error repair during DNA replication.) This could aid in the survival of cells whose DNA is damaged, but would result in increased mutations.

### 19.3.2 Regulation of the SOS response

#### Role of RecA and LexA

When damage occurs to the DNA, RecA becomes activated (RecA*) (Fig. 19.8). As pointed out earlier, evidence suggests that activation of RecA occurs when it binds to regions of single-stranded DNA generated as a result of attempting to replicate the damaged DNA template. (Reviewed in Ref. 23.) LexA, a transcription repressor, then binds to the RecA*–DNA complex, and RecA* promotes the proteolytic cleavage of LexA, thus inactivating the repressor. (RecA* probably stimulates LexA to autodigest rather

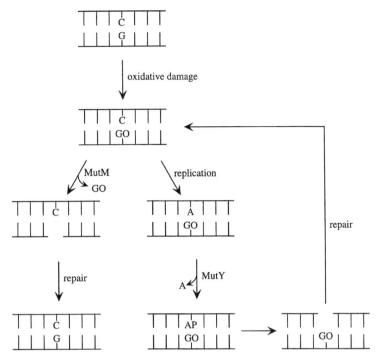

**Fig. 19.7** Repair of 8-oxoguanine (GO) lesions. Oxidative damage results in the oxidation of guanine residues to 8-oxoguanine (GO). The MutM protein removes the 8-oxoguanine, and cuts the damaged DNA with its AP endonuclease activity. The damaged DNA is removed, and the gap is repaired by DNA polymerase and DNA ligase. Sometimes an AMP is inserted opposite the GO. When this happens, the MutY protein can remove the adenine, leaving an apurinic (AP) site. An AP endonuclease activity, which appears to be associated with MutY, cuts the damaged strand near the AP site. A portion of the damaged strand is removed, and the gap is filled by DNA polymerase and sealed with DNA ligase. This generates a GO–C base pair, affording the opportunity for MutM to remove the GO. The MutY protein can correct A/G mismatches as well as A/GO mismatches.

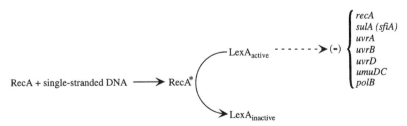

**Fig. 19.8** Regulation of the SOS response. The model proposes that RecA becomes activated when it binds to single-stranded DNA, resulting from replicative bypass of damaged DNA. Activated RecA, i.e., RecA*, then inactivates LexA, which is a repressor for all the genes induced during the SOS response. Inactivation of LexA is due to its autoproteolytic cleavage stimulated by RecA*. There are over 20 genes regulated by LexA. Some of these are shown in the diagram. *recA* encodes the RecA protein. SulA inhibits FtsZ and thus cell division. (See the discussion of Lon in the section on ATP-dependent proteases.) UvrA, UvrB, and UvrD function during excision repair. (See the section on repairing damaged DNA via excision.) PolB is DNA polymerase II. UmuDC allows DNA polymerase to copy damaged DNA. (See the section on SOS mutagenesis.)

than acting as a protease itself.) As a consequence, transcription of the genes in the SOS system is stimulated. The genes discussed in the previous sections that are activated during the SOS response are part of a regulon that is repressed by the negative effector, LexA. These genes are described next.

## Genes whose transcription is stimulated during SOS response

In an undamaged cell LexA represses the transcription of over 20 genes in the SOS regulon, including *recA* and *lexA*. Repression need not be complete, and several of the genes are transcribed at a significant rate even in the presence of LexA. Some of the genes activated during the SOS response as well as the physiological roles for the gene products are listed in Table 19.1. The list does not include *uvrC* or *phr* (the gene that encodes the photoreactivating enzyme) because they are not under LexA control. The product of the *uvrD* gene (helicase II) is not actually required for endonuclease activity but is required for the release of the oligonucleotide fragment and of UvrC from the complex after the cuts in the DNA are made. (These genes are involved in nucleolytic excision repair. See Section 19.2.3.) The list also includes *sulA*, which, as discussed earlier, codes for a protein that inhibits septation and causes the cells to grow as filaments until the DNA is repaired. The levels of SulA increase during the SOS response, but they are quickly reduced by the action of Lon when the DNA is repaired, at which time cell division resumes (Section 19.1.3). Presumably, it is an advantage to the cell not to divide while it is busy repairing its DNA. The SOS response also includes the induction of certain prophages such as λ and P22. This is because RecA* promotes the cleavage of repressors of these bacteriophages in a fashion similar to the induced proteolyisis of LexA. Prophage induction cannot be a benefit to the damaged host, and a rationale for prophage induction is that it benefits the virus, which after induction can find a host whose DNA is not damaged.

## 19.4 Oxidative Stress

### 19.4.1 Toxic forms of oxygen

As discussed in Section 14.1, there are several derivatives of oxygen formed by aerobic organisms that can be toxic to cells. (See Section 14.1 for an explanation of how these toxic forms of oxgyen are formed inside cells.) These include hydroxyl radical (HO·), superoxide radical ($O_2^-$), and hydrogen peroxide ($H_2O_2$), which can damage molecules such as DNA, proteins, and lipids. The superoxide radical is an intermediate in the reduction of oxygen by electron carriers. The hydrogen peroxide is derived from superoxide and can react with transition metal ions to form hydroxyl radicals. (See the discussion of the Fenton reaction in Section 14.1.) There does not exist a mechanism to detoxify hydroxyl radicals, and thus protection from toxic forms of oxygen must rely on eliminating superoxide and hydrogen peroxide. The enzyme *superoxide dismutase* (Sod) converts superoxide to hydrogen peroxide and oxygen, and *catalase* converts hydrogen peroxide to water and oxygen. (See Section 14.1 for the reactions that these enzymes catalyze.) It is of interest to point out that the reactive forms of oxygen as well as nitric oxide (which also induces the superoxide-induced stress response discussed next) are also produced by phagocytic cells in animals as cytotoxic agents to kill bacteria that have been ingested.

### 19.4.2 Proteins made in response to oxidative stress

In order to study the response to oxidative stress, $H_2O_2$, or redox-cycling reagents such as methyl viologen (paraquat) or phenazine methosulfate that cause an increase in production of superoxide by diverting electrons to $O_2$, may be added to the cultures. This results in an increase in the synthesis of many proteins that can be detected by two-dimensional gel electrophoresis. For many of the induced proteins the functions are not known. However, some have an obvious role in protecting the cell from oxidative damage. For example, in *E. coli* the known enzymes include $Mn_2^+$–superoxide dismutase, catalase (HPI catalase), and endonuclease IV. The latter functions in excision repair of damaged DNA, for example, that damaged by superoxide. (Endonuclease IV cuts the DNA at apurinic or apyrimidine sites that are produced by DNA glycosylases that remove damaged bases from DNA.)

### 19.4.3 Transcriptional regulation of genes induced during oxidative stress

In *E. coli* there are two transcriptional regulatory systems that control the production of proteins in response to oxidative stress. These are the SoxRS system and the OxyR system, which control the *soxRS* and *oxyR* regulons, respectively.[28] (A regulon is a set of noncontiguous genes or operons controlled by the same transcription regulator. Other regulons (e.g., the Ntr and PHO regulons) are discussed in Chapter 18.)

### SoxRS

SoxR is an iron–sulfide protein that is activated (probably by undergoing a conformational change) in response to superoxide or to nitric oxide (NO).[29, 30] Activated SoxR, in turn, activates transcription of the *soxS* gene. The SoxS protein then activates the transcription of about 12 target genes whose products confer resistance to a diverse array of toxic elements, including superoxide, nitric oxide produced by activated macrophages, multiple antibiotics, organic solvents, and heavy

metals (Fig. 19.9). The activated genes include *sodA* (codes for $Mn_2^+$–superoxide dismutase) and *nfo* (codes for endonuclease IV).[31] (SoxS autoregulates its own synthesis by being a negative transcription regulator of the *soxS* gene.)

### OxyR

When cells are exposed to $H_2O_2$, OxyR becomes activated, presumably by a signal generated from the $H_2O_2$. There is no increase in the levels of the OxyR protein, simply in its activity. OxyR is also activated in vitro by $O_2$. OxyR is the transcriptional activator of several genes, including *katG* (codes for HPI catalase). In addition, the transcription of the *oxyS* gene is activated. The RNA product of the *oxyS* gene is not translated, and its function is not understood at present.

### 19.5 Summary

Upon a temperature upshift bacteria, as well as other organisms, transiently increase the rates of synthesis of proteins called heat-shock proteins. In *E. coli*, the increased rates of synthesis are due to increased activity of a sigma factor called sigma 32. Although the biological roles for most of these proteins is not known, it is clear that some of them aid bacteria in coping with the increase in temperature. The heat-shock proteins include chaperone proteins that aid in the folding of newly synthesized proteins and also function in the export of certain proteins and ATP-dependent proteases that degrade specific

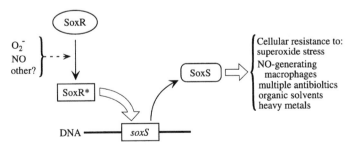

**Fig. 19.9** The SoxRS system. SoxR is activated by oxidative stress and possibly other environmental stress factors. Activated SoxR stimulates the transcription of the *soxS* gene. SoxS stimulates transcription of a set of genes that aid the bacterium in coping with a toxic environment. *Source:* From Hidalgo, E., H. Ding, and B. Demple. 1997. Redox signal transduction via iron–sulfur clusters in the SoxR transcription activator. *TIBS* **22**:207–210.

proteins and abnormal proteins. Several of the heat-shock proteins have been shown to play a role during growth at ordinary temperatures.

DNA can be damaged in several different ways, and various systems exist to repair the damage. There are specific repair systems called base-excision repair systems that remove damaged bases using N-glycosylases and repair the DNA using AP endonucleases, DNA polymerase I, and DNA ligase. Bases can be damaged in several ways, including deamination and alkylation. Guanine can also be oxidatively damaged. The cell protects itself from oxidative guanine damage using the MutM, MutY, and MutT proteins. MutM and MutY are N-glycosylases and function in base-excision systems, but MutT is not. MutT works by preventing the incorporation of oxidized guanylic acid into DNA.

Pyrimidine dimers can be produced by uv irradiation, and this can be corrected by photoreactivation. Pyrimidine dimers can also be removed by N-glycosylases, which are part of a base-excision repair system.

There are also general repair systems. These include nucleotide excision repair, methyl-directed mismatch repair, and postreplication recombination repair.

Bacteria respond to damaged DNA by inducing the genes of the SOS system. A protein, RecA, becomes activated and inactivates a repressor, LexA. As a consequence, the transcription of the genes of the SOS regulon is stimulated. The protein products of these genes repair damaged DNA. The products of the induced genes repair DNA by (1) nucleotide excision or (2) postreplication recombination repair.

Oxidative stress is due to the production of toxic forms of oxygen that damage cell molecules. The transcription of several genes whose protein products protect bacteria from oxidative stress is stimulated. In *E. coli*, these proteins include superoxide dismuatse, catalase, and endonuclease IV. There are two transcriptional regulatory systems that control the expression of genes induced by oxidative stress. These are the SoxRS and the OxyR systems, which consist of transcriptional activators that are activated by superoxide, nitric oxide, and hydrogen peroxide.

## Study Questions

1. What is meant by heat-shock proteins, and what is the evidence that they are required for growth?

2. Which heat-shock proteins are required for growth at all temperatures? Why is that?

3. Discuss the role of Lon in uv resistance and polysaccharide synthesis.

4. Suppose DNA polymerase left a gap in the copy strand. How might this be repaired?

5. Describe two different pathways that can remove or correct thymine dimer damage in the template strand.

6. Mutations in *mutM*, *mutY*, and *mutT* increase the spontaneous rate of mutation. Explain why this is the case. The increase in the mutation rates are additive (i.e., the rates are higher when two or three of the *mut* genes are mutated as compared to when only one of the genes is mutated). Explain why this is the case.

7. What is meant by the SOS response? How are the genes regulated?

8. What is the explanation for SOS mutagenesis?

9. What is meant by oxidative stress, and how does *E. coli* respond to it?

## NOTES AND REFERENCES

1. Georgopoulos, C., D. Ang, A. Maddock, S. Raina, B. Lipinska, and M. Zylicz. 1990. Heat-shock response of *Escherichia coli*. In: *The Bacterial Chromosome*. pp. 405–419. K. Drilica and M. Riley (Eds.). American Society for Microbiology. Washington, D.C.

2. Graumann, P., and M. A. Marahiel. 1996. Some like it cold: Response of microorganisms to cold shock. *Arch. Microbiol.* 166:293–300.

3. Graumann, P., T. M. Wendrich, M. H. W. Weber, K. Schröder, and M. A. Marahiel. 1997. A family of cold shock proteins in *Bacillus subtilis* is essential for cellular growth and for efficient protein synthesis at optimal at optimal and low temperatures. *Molec. Microbiol.* 25:741–756.

4. Zhou, Y-N, K. Noriko, J. W. Erickson, C. A. Gross, and T. Yura. 1988. Isolation and characterization of *Escherichia coli* mutants that lack the heat-shock sigma factor $\sigma^{32}$. *J. Bacteriol.* 170:3640–3649.

5. Gross, C.A. 1996. Function and regulation of the heat-shock proteins. In: *Escherichia coli and Salmonella: Cellular and Molecular Biology*. Vol. 1. pp. 1382–1399. Neidhardt et al. (Eds.). ASM Press, Washington, D.C.

6. Lange, R., and R. Hengge-Aronis. 1994. The cellular concentration of the $\sigma^s$ subunit of RNA polymerase in *Escherichia coli* is controlled at the levels of transcription, translation, and protein stability. *Genes and Dev.* 8:1600–1612.

7. The expression of the *rpoS* gene, which codes for $\sigma^s$ is regulated at the transcriptional and translational levels, as well as by altering the stability of the protein, as shown in Ref. 6. Transcription is stimulated when the cells enter stationary phase in rich media and by an increase in medium osmolarity. However, it is not induced very much when cells enter stationary phase in minimal media. Translation of the mRNA is stimulated when cells enter stationary phase and also during an osmotic upshift. The sigma factor itself is very unstable during exponential growth and becomes stabilized when cells enter stationary phase. The molecular mechanisms responsible for the changes in the rate of translation and for the alterations in protein stability are unknown. However, it is clear that these processes, as well as transcriptional control, are responsible for the increase in $\sigma^s$ in response to starvation and osmolarity signals.

8. Hecker, M., W. Schumann, and U. Völker. 1996. Heat-shock and general stress response in *Bacillus subtilis*. *Molec. Microbiol.* 19:417–428.

9. The crystal structures of GroEL and GroES have been published. This plus electron microscopic studies have suggested that a cavity exists in GroEL within which the target protein is folded. For references, see the review: Rassow, J., O. von Ahsen, U. Bömer, and N. Pfanner. 1997. Molecular chaperones: towards a characterization of the heat-shock protein 70 family. *Trends in Cell Biol.* 7:129–132.

10. Wild, J., E. Altman, T. Yura, and C. A. Gross. 1992. DnaK and DnaJ heat-shock proteins participate in protein export in *Eschericia coli*. *Genes and Development* 6:1165–1172.

11. Kusukawa, N., T. Yura, C. Ueguchi, Y. Akiyama, and K. Ito. 1989. Effects of mutations in heat-shock genes *groES* and *groEL* on protein export in *Escherichia coli*. *EMBO J.* 8:3517–3521.

12. Bochkareva, E. S., M. E. Solovieva, and A. S. Girshovich. 1998. Targeting of GroEL to SecA on the cytoplasmic membrane of *Escherichia coli*. *Proc. Natl. Acad. Sci. USA* 95:478–483.

13. Altman, E., C. A. Kumamoto, and S. D. Emr. 1991. Heat-shock proteins can substitute for SecB function during protein export in *Escherichia coli*. *EMBO J.* 10:239–245.

14. Suzuki, C. K., M. Rep, J. M. van Dijl, K. Suda, L. A. Grivell, and G. Schatz. 1997. ATP-dependent proteases that also chaperone protein biogenesis. *TIBS* 22:118–123.

15. Chung, C. H., and A. L. Goldberg. 1981. The product of the *lon* (*capR*) gene in *Escherichia coli* is the ATP-dependent protease, protease La. *Proc. Natl. Acad. Sci. USA* 78:4931–4935.

16. Apparently FtsH recognizes *stop-transfer* sequences. (See Fig. 16.2.) In wild-type cells, a reporter protein, SecY–PhoA, was used to determine this. SecY is the translocase for protein export and has a *stop-transfer* sequence that keeps it in the cell membrane. PhoA is alkaline phosphatase and is active only when secreted into the periplasm. Wild-type cells place the hybrid protein in the membrane, and PhoA is not active. *ftsH* mutants apparently do not recognize the stop-transfer sequence because they secrete the PhoA domain into the periplasm where it is active. FtsH is also required for the export of $\beta$-lactamase, OmpA, and penicillin-binding protein, PBP3. See the review article by Suzuki et al.[14]

17. Lazazzera, B. A., and A. D. Grossman. 1997. A regulatory switch involving a Clp ATPase. *BioEssays* 19:455–458.

18. Wawrzynow, A., B. Banecki, and M. Zylicz. 1996. The Clp ATPases define a novel class of molecular chaperones. *Mol. Microbiol.* 21:895–899.

19. The Clp ATPase family of chaperone proteins are widespread and have been found in both procaryotes and eucaryotes. They are involved in a variety of different cellular activities associated with the conformation, formation, and disaggregation of protein complexes, as well as delivering specific proteins to ClpP for proteolysis. In *E. coli*, ClpA and ClpX form complexes with ClpP to form the ClpAP and ClpXP proteases. The actual protease is ClpP, which is not homologous to the Clp ATPase family. An example of specific Clp-mediated proteolysis is described in Section 2.2.2, where ClpXP is shown to be responsible for the instability (proteolytic cleavage) of sigma factor $\sigma^S$ in exponential cultures of *E. coli*. It is important to emphasize that the proteolysis by Clp is very specific, and this specificity is due to ClpA or ClpX. ClpA and ClpX bind to the protein substrates, in a reaction that requires ATP, and deliver the polypeptide substrate to ClpP. ATP is also required for the processive cleavage of large polypeptides by ClpP. Perhaps the ATP is required for partial unfolding of the substrate by ClpA or ClpX in preparation for proteolysis by ClpP. Whether or not ClpA

or ClpX dissociate from the complex with ClpP after delivering the protein substrate is not known. If they act as chaperones, then they would be expected to dissociate from the final complex. As discussed in Section 18.14.1, another Clp ATPase, i.e., ClpC of *B. subtilis*, negatively regulates competence.

20. Rupp, W. D. 1996. DNA repair mechanisms. In: *Escherichia coli and Salmonella: Cellular and Molecular Biology*, Vol. 2, pp. 2277–2294. Neidhardt, F. C. et al. (Eds.). ASM Press, Washington, D.C.

21. *DNA Repair and Mutagenesis*. 1995. Friedberg, E. C., G. C. Walker, and W. Siede (Eds.). ASM Press. Washington, D.C.

22. Walker, G. C. 1985. Inducible DNA repair systems. *Ann. Rev. Biochem.* 54:425–457.

23. Walker, G. C. 1996. The SOS response of *Escherichia coli*. In: *Escherichia coli and Salmonella: Cellular and Molecular Biology*, Vol. 1, pp. 1400–1416. Neidhardt, F. C., et al. (Eds.). ASM Press, Washington, D.C.

24. The model for the *E. coli* photolyase proposes that light energy is absorbed by MTHF, raising an electron to an excited state (*MTHF). When the electron returns to the ground state, energy is transferred to deprotonated reduced flavin, i.e., FADH$^-$, exciting an electron to an excited state, i.e., *FADH$^-$. *FADH$^-$ is then proposed to transfer the excited electron to the pyrimidine dimer, resulting in the cleavage of the C–C bonds holding the dimer together. The reaction is completed when the electron is returned to FADH$^0$, regenerating FADH$^-$. For details of the model, see the discussion of photoreactivation of DNA in *DNA Repair and Mutagenesis*, E. C. Friedberg, G. C. Walker, and W. Siede (Eds.). ASM Press, Washington, D.C. (1995).

25. Alkyl groups include methyl (CH$_3$–) and ethyl (CH$_3$CH$_2$–). These can be added to the bases by alkylating agents such as ethylmethanesulfonate (EMS) and nitrosoguanidine (NTG). The alkyl groups are often added to the N3 of adenine and the N7 of guanine to form N-methyl or N-ethyl derivatives of the bases. Nitrosoguanidine will also methylate the O6 of guanine and the O4 of thymine to form the O-methyl derivatives. The alkylated bases mispair, causing mutations.

26. Michaels, M. L., and J. H. Miller. 1992. The GO system protects organisms from the mutagenic effect of the spontaneous lesion 8-hydroxyguanine (7,8-dihydro-8-oxoguanine). *J. Bacteriol.* 174:6321–6325.

27. Kouchakdjian, M., V. Bodepudi, S. Shibutani, M. Eisenberg, F. Johnson, A. P. Grollman, and D. J. Patel. 1991. NMR structural studies of the ionizing radiation adduct 7-hydro-8-oxodeoxyguanosine (8-oxo-7H-dG) opposite deoxyadenosine in a DNA duplex. 8-Oxo-7H-dG(syn)·dA(anti) alignment at lesion site. *Biochem.* 30:1403–1412.

28. Lynch, A. S., and E. C. S. Lin. 1996. Responses to molecular oxygen, pp. 1526–1538. In: *Escherichia coli and Salmonella: Cellular and Molecular Biology*, Vol. 1. F. C. Neidhardt et al. (Eds.). ASM Press, Washington, D.C.

29. Hidalgo, E., H. Ding, and B. Demple. 1997. Redox signal transduction via iron–sulfur clusters in the SoxR transcription activator. *TIBS*, 22: 207–210.

30. As reviewed by Hidalgo et al. (Ref. 24), SoxR is a homodimer and has a pair of [2Fe–2S] clusters, probably in each subunit of the dimer. It is thought that the activity of SoxR is regulated by redox reactions. The midpoint potential for the [2Fe–2S] clusters at pH 7.6 is about −285 mV. When the protein was titrated in vitro, its transcriptional activity was inhibited at −380 mV, i.e., when the FeS centers are more than 95% reduced. This was reversible; i.e., when the FeS centers were reoxidized by titration, activity of the protein was restored. Candidates for the in vivo reductants and oxidants for the FeS centers in SoxR include NADPH, NADH (reductants), O$_2$, O$_2^-$, NO (oxidants). (Nitric oxide, NO, has an unpaired valence electron and is therefore paramagnetic.) The model proposed by Hidalgo et al. is that SoxR binds to the *soxS* promoter regardless of the oxidation state of SoxR, but that the oxidized form of SoxR undergoes an allosteric modification enabling it to activate the *soxS* promoter. Note that this is different from the mechanism proposed for oxygen regulation of FNR activity discussed in Section 18.2.2. It is proposed that oxygen actually results in the loss of the [4Fe–4S] clusters in FNR, thus preventing the protein from specifically binding to DNA.

31. Li, Ziyi, and B. Demple. 1994. SoxS, an activator of superoxide stress genes in *Escherichia coli*. *J. Biol. Chem.* 269:18371–18377.

# INDEX